AF832212

Birkhäuser

Operator Theory: Advances and Applications

Volume 268

Founded in 1979 by Israel Gohberg

Editors:
Joseph A. Ball (Blacksburg, VA, USA)
Albrecht Böttcher (Chemnitz, Germany)
Harry Dym (Rehovot, Israel)
Heinz Langer (Wien, Austria)
Christiane Tretter (Bern, Switzerland)

Associate Editors:
Vadim Adamyan (Odessa, Ukraine)
Wolfgang Arendt (Ulm, Germany)
B. Malcolm Brown (Cardiff, UK)
Raul Curto (Iowa, IA, USA)
Kenneth R. Davidson (Waterloo, ON, Canada)
Fritz Gesztesy (Waco, TX, USA)
Pavel Kurasov (Stockholm, Sweden)
Vern Paulsen (Houston, TX, USA)
Mihai Putinar (Santa Barbara, CA, USA)
Ilya Spitkovsky (Abu Dhabi, UAE)

Honorary and Advisory Editorial Board:
Lewis A. Coburn (Buffalo, NY, USA)
Ciprian Foias (College Station, TX, USA)
J.William Helton (San Diego, CA, USA)
Marinus A. Kaashoek (Amsterdam, NL)
Thomas Kailath (Stanford, CA, USA)
Peter Lancaster (Calgary, Canada)
Peter D. Lax (New York, NY, USA)
Bernd Silbermann (Chemnitz, Germany)
Harold Widom (Santa Cruz, CA, USA)

Subseries
Linear Operators and Linear Systems

Subseries editors:
Daniel Alpay (Orange, CA, USA)
Birgit Jacob (Wuppertal, Germany)
André C.M. Ran (Amsterdam, The Netherlands)

Subseries
Advances in Partial Differential Equations

Subseries editors:
Bert-Wolfgang Schulze (Potsdam, Germany)
Michael Demuth (Clausthal, Germany)
Jerome A. Goldstein (Memphis, TN, USA)
Nobuyuki Tose (Yokohama, Japan)
Ingo Witt (Göttingen, Germany)

More information about this series at http://www.springer.com/series/4850

Albrecht Böttcher • Daniel Potts • Peter Stollmann
David Wenzel
Editors

The Diversity and Beauty of Applied Operator Theory

Birkhäuser

Editors
Albrecht Böttcher
Fakultät für Mathematik
TU Chemnitz
Chemnitz, Germany

Daniel Potts
Fakultät für Mathematik
TU Chemnitz
Chemnitz, Germany

Peter Stollmann
Fakultät für Mathematik
TU Chemnitz
Chemnitz, Germany

David Wenzel
Fakultät für Mathematik
TU Chemnitz
Chemnitz, Germany

ISSN 0255-0156 ISSN 2296-4878 (electronic)
Operator Theory: Advances and Applications
ISBN 978-3-030-09374-7 ISBN 978-3-319-75996-8 (eBook)
https://doi.org/10.1007/978-3-319-75996-8

Mathematics Subject Classification (2010): 47-06, 15B05, 15B52, 42A16, 42C15, 47A10, 47B15, 47B35, 47F05, 47G30, 47L15, 58J40, 81U20, 93C55

© Springer International Publishing AG, part of Springer Nature 2018
Softcover re-print of the Hardcover 1st edition 2018
This work is subject to copyright. All rights are reserved by the Publisher, whether the whole or part of the material is concerned, specifically the rights of translation, reprinting, reuse of illustrations, recitation, broadcasting, reproduction on microfilms or in any other physical way, and transmission or information storage and retrieval, electronic adaptation, computer software, or by similar or dissimilar methodology now known or hereafter developed.
The use of general descriptive names, registered names, trademarks, service marks, etc. in this publication does not imply, even in the absence of a specific statement, that such names are exempt from the relevant protective laws and regulations and therefore free for general use.
The publisher, the authors and the editors are safe to assume that the advice and information in this book are believed to be true and accurate at the date of publication. Neither the publisher nor the authors or the editors give a warranty, express or implied, with respect to the material contained herein or for any errors or omissions that may have been made. The publisher remains neutral with regard to jurisdictional claims in published maps and institutional affiliations.

Printed on acid-free paper

This book is published under the imprint Birkhäuser, www.birkhauser-science.com by the registered company Springer International Publishing AG part of Springer Nature.
The registered company address is: Gewerbestrasse 11, 6330 Cham, Switzerland

Contents

Preface

These are the proceedings of the *International Workshop on Operator Theory and its Applications* (IWOTA) that was held in Chemnitz in 2017. It was the 28th iteration of the event since its initiation in 1981.

The fact that our university was chosen as the venue is a sign of the great appreciation for the longstanding tradition of our work in the field. Operator theory was established in Chemnitz by Siegfried Prößdorf in the 1960s and later advanced by Bernd Silbermann and Georg Heinig. Today it is driven by various research groups, including those of the local organizing committee's members.

Born about a century ago, operator theory now is one of the mathematical keys for the latest progress in science and technology. The methods that operator theory developed and continues to advance are used every day by many people who work in the known application fields of mathematics, engineering, and physics. The aspect of applicability is also reflected in the ensemble of that year's main speakers:

Harm Bart,
Mark Embree,
Fritz Gesztesy,
Frances Kuo,
Christiane Tretter.

They all have made important contributions to both: development of operator theory and concrete practical applications of new conceptual insights.

The list of invited speakers was completed by a healthy mix of young and well-established researchers:

Marcel Hansmann,
Bill Helton,
Rien Kaashoek,
Alexei Karlovich,
Greg Knese,
Marko Lindner,
Alejandra Maestripieri,
Jonathan Partington,
Stefanie Petermichl,
Alexander Pushnitski,
Konrad Schmüdgen,
Carola-Bibiane Schönlieb,
Bernd Silbermann,
Ilya Spitkovsky,
Sanne ter Horst.

We managed to ensure undivided attention for all of them. Moreover, referring to them as "semi plenary speakers" is actually an understatement; since each one had a 45 minutes talk, "semi-sesqui plenaries" would be a more precise, better fitting term.

Traditionally, the IWOTA conferences provide a platform for discussion and exchange of ideas via short talks. This time, they were scheduled in up to only four parallel sessions, and many of these sessions were arranged within a mini symposium. We are happy that especially early career scientists took the opportunity and approached us with well-fitting topics. In summary, seven mini symposia were held:

Functional calculus
(Markus Haase, Christian Le Merdy),
Riemann–Hilbert problems and applications in random matrix theory
(Jani Virtanen),
Structured matrices and operators — in memory of Georg Heinig
(Karla Rost),
New approaches for high-dim. integration in light of physics applications
(Karl Jansen, Frances Kuo),
Semigroups and evolution equations
(András Bátkai, Christian Seifert),
Toeplitz and related operators
(Santeri Miihkinen, Jani Virtanen),
The Rien Kaashoek mini symposium
(Harm Bart, André Ran).

In addition to the thematically more closed symposia, several contributed talks were accepted for presentation. They clearly demonstrate how big operator theory has become, extending over a wide range of topics:

General operator theory,
Differential operators,
Matrix norms and pseudospectra,
Algebras and order relations,
Functional analysis,
Concrete operator theory.

Summing up, 157 participants from almost 40 countries enjoyed a total of 126 talks given at the conference from August 14th to 18th. Many of the talks can be found on the web site

https://www.tu-chemnitz.de/mathematik/iwota2017/

preserved for eternity. We are greatly indebted to the Deutsche Forschungsgemeinschaft (DFG), the president and the chancellor of the TU Chemnitz, and the dean of the Department of Mathematics for their financial support.

We sincerely hope you will also enjoy the 29 articles in this volume of *Operator Theory: Advances and Applications.* We are very thankful that the publisher kindly raised the page number limit. Nevertheless, even more good manuscripts were submitted, and we could not include all of them. So we selected the most beautiful works representing the one or other of the diverse aspects of operator theory.

Chemnitz
January 2018

Albrecht Böttcher, Daniel Potts
Peter Stollmann, David Wenzel

IWOTA

Chemnitz 2017

Participants

Abadias, Luciano
(Zaragoza, Spain)
Adamo, Maria Stella
(Palermo, Italy)
Amenta, Alex
(Delft, Netherlands)
Banert, Michaela
(Chemnitz, Germany)
Bart, Harm
(Rotterdam, Netherlands)
Barta, Tomas
(Praha, Czech Republic)
Bátkai, András
(Feldkirch, Austria)
Batty, Charles
(Oxford, United Kingdom)
Bello-Burguet, Glenier L.
(Madrid, Spain)
Berić, Tomislav
(Zagreb, Croatia)
Blinova, Irina
(St. Petersburg, Russian Fedn.)
Blower, Gordon
(Lancaster, United Kingdom)
Bombach, Clemens
(Chemnitz, Germany)
Böttcher, Albrecht
(Chemnitz, Germany)
Bowkun, Jakob
(Chemnitz, Germany)
Budde, Christian
(Wuppertal, Germany)
Charlier, Christophe
(Bruxelles, Belgium)
Chen, Jinwen
(Beijing, China)
Cho, Muneo
(Hiratsuka, Japan)
Choda, Marie
(Osaka, Japan)
Christensen, Ole
(Lyngby, Denmark)
Corso, Rosario
(Palermo, Italy)
Dhara, Kousik
(Chennai, India)
Didenko, Viktor
(Odessa, Ukraine)
Djikić, Marko
(Niš, Serbia)
Doeraene, Antoine
(Louvain, Belgium)
Dogga, Venku naidu
(Sangareddy, India)
Doktorski, Leo
(Ettlingen, Germany)
Doust, Ian
(Sydney, Australia)
Dragičević, Oliver
(Ljubljana, Slovenia)
Dritschel, Michael
(Newcastle, United Kingdom)
Duduchava, Rolandi
(Tbilisi, Georgia)
Ehrhardt, Torsten
(Santa Cruz, United States)
Embree, Mark
(Blacksburg/VA, United States)
Flemming, Katharina
(Chemnitz, Germany)
Frazho, Arthur
(West Lafayette/IN, Utd. States)
Frymark, Dale
(Waco, United States)
Fulsche, Robert
(Hannover, Germany)
Geher, Gyorgy Pal
(Reading, United Kingdom)
Gesztesy, Fritz
(Waco, United States)
Gohm, Rolf
(Aberystwyth, United Kingdom)
Golla, Ramesh
(Sangareddy, India)

Gonçalves, Helena
(Jena, Germany)
Goncharuk, Anna
(Kharkiv, Ukraine)
Grossmann, Christian
(Dresden, Germany)
Grudsky, Sergei
(Ciudad de México, Mexico)
Guediri, Hocine
(Riyadh, Saudi Arabia)
Gunatillake, Gajath
(Sharjah, Utd. Arab Emirates)
Haase, Markus
(Kiel, Germany)
Hagger, Raffael
(Hannover, Germany)
Hansmann, Marcel
(Chemnitz, Germany)
Hartung, Tobias
(London, United Kingdom)
Hedenmalm, Håkan
(Stockholm, Sweden)
Helton, J. William
(San Diego, United States)
Jaftha, Jacob
(Cape Town, South Africa)
Janse van Rensburg, Dawid
(Potchefstroom, South Africa)
Jansen, Karl
(Zeuthen, Germany)
Jardón Sánchez, Héctor
(Gijón/Xixón, Spain)
Junghanns, Peter
(Chemnitz, Germany)
Kaashoek, Marinus A.
(Amsterdam, Netherlands)
Kaiser, Robert
(Chemnitz, Germany)
Kalmes, Thomas
(Chemnitz, Germany)
Kämmerer, Lutz
(Chemnitz, Germany)
Kapanadze, David
(Tbilisi, Georgia)
Karlovich, Alexei
(Lisboa, Portugal)
Karlovich, Yuri
(Cuernavaca, Mexico)
Kazashi, Yoshihito
(Sydney, Australia)
Kerner, Joachim
(Hagen, Germany)
Kitson, Derek
(Lancaster, United Kingdom)
Kircheis, Melanie
(Chemnitz, Germany)
Klaja, Hubert
(Lille, France)
Knese, Greg
(St. Louis, United States)
Koca, Beyaz Basak
(Istanbul, Turkey)
Kozlowska, Katarzyna
(Reading, United Kingdom)
Kreuter, Marcel
(Ulm, Germany)
Kriegler, Christoph
(Aubiere, France)
Kumar, V. B. Kiran
(Cochin, India)
Kuo, Frances
(Sydney, Australia)
Langenau, Holger
(Chemnitz, Germany)
Łanucha, Bartosz
(Lublin, Poland)
Le Merdy, Christian
(Besançon, France)
Lee, Ji Eun
(Seoul, Korea)
Lee, Mee-Jung
(Seoul, Korea)
Lee, Young Joo
(Gwangju, Korea)
Leiterer, Jürgen
(Berlin, Germany)
Leka, Zoltan
(London, United Kingdom)

Leövey, Hernan
(Villingen-Schwenn., Germany)
Lindner, Marko
(Hamburg, Germany)
Lindström, Mikael
(Turku, Finland)
Mädler, Conrad
(Leipzig, Germany)
Maestripieri, Alejandra
(Buenos Aires, Argentina)
Marchenko, Vitalii
(Kiev, Ukraine)
Mascarenhas, Helena
(Lisboa, Portugal)
Maximenko, Egor
(Ciudad de México, Mexico)
Michael, Isaac
(Waco, United States)
Michalska, Małgorzata
(Lublin, Poland)
Miheisi, Nazar
(London, United Kingdom)
Miihkinen, Santeri
(Joensuu, Finland)
Nakić, Ivica
(Zagreb, Croatia)
Nasdala, Robert
(Chemnitz, Germany)
Nuyens, Dirk
(Leuven, Belgium)
Öztürk, Hasen
(Reading, United Kingdom)
Pannasch, Florian
(Kiel, Germany)
Partington, Jonathan
(Leeds, United Kingdom)
Peruzzetto, Marco
(Kiel, Germany)
Petermichl, Stefanie
(Toulouse, France)
Pietrzycki, Paweł
(Cracow, Poland)
Pik, Derk
(Amsterdam, Netherlands)
Popov, Igor
(St. Petersburg, Russian Fedn.)
Potts, Daniel
(Chemnitz, Germany)
Pushnitski, Alexander
(London, United Kingdom)
Quellmalz, Michael
(Chemnitz, Germany)
Ran, Andre
(Amsterdam, Netherlands)
Rebs, Christian
(Chemnitz, Germany)
Roch, Steffen
(Darmstadt, Germany)
Rocha, Jamilly
(Recife, Brazil)
Rose, Christian
(Chemnitz, Germany)
Rost, Karla
(Chemnitz, Germany)
Sau, Haripada
(Mumbai, India)
Schmüdgen, Konrad
(Leipzig, Germany)
Schönlieb, Carola-Bibiane
(Cambridge, United Kingdom)
Schwenninger, Felix
(Hamburg, Germany)
Seidel, Markus
(Zwickau, Germany)
Seifert, Christian
(Hamburg, Germany)
Seiler, Jörg
(Torino, Italy)
Šemrl, Peter
(Ljubljana, Slovenia)
Shukur, Ali
(Minsk, Belarus)
Silbermann, Bernd
(Chemnitz, Germany)
Singh, Uaday
(Roorkee, India)
Speck, Frank-Olme
(Lisboa, Portugal)

Spitkovsky, Ilya
(Abu Dhabi, Utd. Arab Emirates)
Stahn, Reinhard
(Dresden, Germany)
Stollmann, Peter
(Chemnitz, Germany)
Tanahashi, Kotaro
(Sendai, Japan)
Taskinen, Jari
(Helsinki, Finland)
Tautenhahn, Martin
(Chemnitz, Germany)
ter Horst, Sanne
(Potchefstroom, South Africa)
Thorn, Nicola
(Reading, United Kingdom)
Titkos, Tams
(Budapest, Hungary)
Tomilov, Yuri
(Warsaw, Poland)
Trapani, Camillo
(Palermo, Italy)
Tretter, Christiane
(Bern, Switzerland)
Trostorff, Sascha
(Dresden, Germany)
Trunk, Carsten
(Ilmenau, Germany)
Uhlig, Sven
(Mannheim, Germany)
Undrakh, Batzorig
(Newcastle u. Tyne, Utd. Kingd.)
van Schagen, Frederik
(Amsterdam, Netherlands)
Virosztek, Dniel
(Budapest, Hungary)
Virtanen, Jani
(Reading, United Kingdom)
Volkmer, Toni
(Chemnitz, Germany)
Volmer, Julia
(Zeuthen, Germany)
Wang, Qin
(Shanghai, China)
Waurick, Marcus
(Glasgow, United Kingdom)
Wegert, Elias
(Freiberg, Germany)
Wenzel, David
(Chemnitz, Germany)
Wintermayr, Jens
(Wuppertal, Germany)
Yakubovich, Dmitry
(Madrid, Spain)

Standard versus strict Bounded Real Lemma with infinite-dimensional state space II: The storage function approach

J.A. Ball, G.J. Groenewald and S. ter Horst

Abstract. For discrete-time causal linear input/state/output systems, the Bounded Real Lemma explains (under suitable hypotheses) the contractivity of the values of the transfer function over the unit disk for such a system in terms of the existence of a positive-definite solution of a certain Linear Matrix Inequality (the Kalman–Yakubovich–Popov (KYP) inequality). Recent work has extended this result to the setting of infinite-dimensional state space and associated non-rationality of the transfer function, where at least in some cases unbounded solutions of the generalized KYP-inequality are required. This paper is the second installment in a series of papers on the Bounded Real Lemma and the KYP-inequality. We adapt Willems' storage-function approach to the infinite-dimensional linear setting, and in this way reprove various results presented in the first installment, where they were obtained as applications of infinite-dimensional State-Space-Similarity theorems, rather than via explicit computation of storage functions.

Mathematics Subject Classification (2010). Primary 47A63; Secondary 47A48, 93B20, 93C55, 47A56.

Keywords. KYP-inequality, storage function, bounded real lemma, infinite-dimensional linear system, minimal system.

1. Introduction

This paper is the second installment, following [11], on the infinite-dimensional bounded real lemma for discrete-time systems and the discrete-time Kalman–Yakubovich–Popov (KYP) inequality. In this context, we consider

---

This work is based on the research supported in part by the National Research Foundation of South Africa (Grant Numbers 93039, 90670, and 93406).

© Springer International Publishing AG, part of Springer Nature 2018
A. Böttcher et al. (eds.), *The Diversity and Beauty of Applied Operator Theory*, Operator Theory: Advances and Applications 268, https://doi.org/10.1007/978-3-319-75996-8_1

the discrete-time linear system

$$\Sigma := \begin{cases} \mathbf{x}(n+1) = A\mathbf{x}(n) + B\mathbf{u}(n), \\ \mathbf{y}(n) = C\mathbf{x}(n) + D\mathbf{u}(n), \end{cases} \quad (n \in \mathbb{Z}) \tag{1.1}$$

where $A : \mathcal{X} \to \mathcal{X}$, $B : \mathcal{U} \to \mathcal{X}$, $C : \mathcal{X} \to \mathcal{Y}$ and $D : \mathcal{U} \to \mathcal{Y}$ are bounded linear Hilbert space operators, i.e., $\mathcal{X}$, $\mathcal{U}$ and $\mathcal{Y}$ are Hilbert spaces and the *system matrix* associated with Σ takes the form

$$M = \begin{bmatrix} A & B \\ C & D \end{bmatrix} : \begin{bmatrix} \mathcal{X} \\ \mathcal{U} \end{bmatrix} \to \begin{bmatrix} \mathcal{X} \\ \mathcal{Y} \end{bmatrix}. \tag{1.2}$$

We refer to the pair (C, A) as the *output pair* and to the pair (A, B) as the *input pair*. In this case input sequences $\mathbf{u} = (\mathbf{u}(n))_{n\in\mathbb{Z}}$, with $\mathbf{u}(n) \in \mathcal{U}$, are mapped to output sequences $\mathbf{y} = (\mathbf{y}(n))_{n\in\mathbb{Z}}$, with $\mathbf{y}(n) \in \mathcal{Y}$, through the state sequence $\mathbf{x} = (\mathbf{x}(n))_{n\in\mathbb{Z}}$, with $\mathbf{x}(n) \in \mathcal{X}$. A system trajectory of the system Σ is then any triple $(\mathbf{u}(n), \mathbf{x}(n), \mathbf{y}(n))_{n\in\mathbb{Z}}$ of input, state and output sequences that satisfy the system equations (1.1).

With the system Σ we associate the *transfer function* given by

$$F_\Sigma(\lambda) = D + \lambda C(I - \lambda A)^{-1}B. \tag{1.3}$$

Since A is bounded, F_Σ is defined and analytic on a neighborhood of 0 in $\mathbb{C}$. We are interested in the case where F_Σ admits an analytic continuation to the open unit disk $\mathbb{D}$ such that the supremum norm $\|F_\Sigma\|_\infty$ of F_Σ over $\mathbb{D}$ is at most one, i.e., F_Σ has analytic continuation to a function in the Schur class

$$\mathcal{S}(\mathcal{U}, \mathcal{Y}) = \left\{ F\colon \mathbb{D} \underset{\text{holo}}{\mapsto} \mathcal{L}(\mathcal{U}, \mathcal{Y})\colon \|F(\lambda)\| \leq 1 \text{ for all } z \in \mathbb{D} \right\}.$$

Sometimes we also consider system trajectories $(\mathbf{u}(n), \mathbf{x}(n), \mathbf{y}(n))_{n\geq n_0}$ of the system Σ that are initiated at a certain time $n_0 \in \mathbb{Z}$, in which case the input, state and output at time $n < n_0$ are set equal to zero, and we only require that the system equations (1.1) are satisfied for $n \geq n_0$. Although technically such trajectories are not system trajectories for Σ, but rather correspond to trajectories of the corresponding singly-infinite forward-time system rather than the bi-infinite system Σ, the transfer function of this singly-infinite forward-time system coincides with the transfer function F_Σ of Σ. Hence for the sake of the objective, determining whether $F_\Sigma \in \mathcal{S}(\mathcal{U}, \mathcal{Y})$, there is no problem with considering such singly-infinite system trajectories.

Before turning to the infinite-dimensional setting, we first discuss the case where $\mathcal{U}$, $\mathcal{X}$, $\mathcal{Y}$ are all finite-dimensional. If in this case one considers the parallel situation in continuous time rather than in discrete time, these ideas have origins in circuit theory, specifically conservative or passive circuits. An important question in this context is to identify which rational matrix functions, analytic on the left half-plane (rather than the unit disk $\mathbb{D}$), arise from a lossless or dissipative circuit in this way (see, e.g., Belevitch [12]).

According to Willems [28, 29], a linear system Σ as in (1.1) is *dissipative* (with respect to *supply rate* $s(u, y) = \|u\|^2 - \|y\|^2$) if it has a *storage function* $S\colon \mathcal{X} \to \mathbb{R}_+$, where $S(x)$ is to be interpreted as a measure of the *energy* stored by the system when it is in state x. Such a storage function S is assumed to satisfy the dissipation inequality

$$S(\mathbf{x}(n+1)) - S(\mathbf{x}(n)) \leq \|\mathbf{u}(n)\|^2 - \|\mathbf{y}(n)\|^2 \tag{1.4}$$

over all trajectories $(\mathbf{u}(n), \mathbf{x}(n), \mathbf{y}(n))_{n\in\mathbb{Z}}$ of the system Σ as well as the additional normalization condition that $S(0) = 0$. The dissipation inequality can be interpreted as saying that for the given system trajectory, the energy stored in the system $(S(\mathbf{x}(n+1)) - S(\mathbf{x}(n)))$ when going from state $x(n)$ to $x(n+1)$ can be no more than the difference between the energy that enters the system $(\|\mathbf{u}(n)\|^2)$ and the energy that leaves the system $(\|\mathbf{y}(n)\|^2)$ at time n.

For our discussion here we shall only be concerned with the so-called *scattering supply rate* $s(u, y) = \|u\|^2 - \|y\|^2$. It is not hard to see that a consequence of the dissipation inequality (1.4) on system trajectories is that the transfer function F_Σ is in the Schur class $\mathcal{S}(\mathcal{U}, \mathcal{Y})$. The results extend to nonlinear systems as well (see [28]), where one talks about the system having L^2-gain at most 1 rather the system having transfer function in the Schur class.

In case the system Σ is finite-dimensional and minimal (as defined in the statement of Theorem 1.1 below), one can show that the smallest storage function, the *available storage* S_a, and the largest storage function, the *required supply* S_r, are *quadratic*, provided storage functions for Σ exist. That S_a and S_r are quadratic means that there are positive-definite matrices H_a and H_r so that S_a and S_r have the quadratic form

$$S_a(x) = \langle H_a x, x\rangle, \quad S_r(x) = \langle H_r x, x\rangle$$

with H_a and H_r actually being positive-definite. For a general quadratic storage function $S_H(x) = \langle Hx, x\rangle$ for a positive-definite matrix H, it is not hard to see that the dissipation inequality (1.4) assumes the form of a linear matrix inequality (LMI):

$$\begin{bmatrix} A & B \\ C & D \end{bmatrix}^* \begin{bmatrix} H & 0 \\ 0 & I_{\mathcal{Y}} \end{bmatrix} \begin{bmatrix} A & B \\ C & D \end{bmatrix} \preceq \begin{bmatrix} H & 0 \\ 0 & I_{\mathcal{U}} \end{bmatrix}. \tag{1.5}$$

This is what we shall call the *Kalman–Yakubovich–Popov* or KYP inequality (with solution H for given system matrix $M = [\begin{smallmatrix} A & B \\ C & D \end{smallmatrix}]$).

Conversely, if one starts with a finite-dimensional, minimal, linear system Σ as in (1.1) for which the transfer function F_Σ is in the Schur class, it is possible to show that there exist quadratic storage functions S_H for the system satisfying the coercivity condition $S_H(x) \geq \delta\|x\|^2$ for some $\delta > 0$ (i.e., with H strictly positive-definite). This is the storage-function interpretation behind the following result, known as the *Kalman–Yakubovich–Popov lemma.*

Theorem 1.1 (Standard Bounded Real Lemma (see [1])). *Let Σ be a discrete-time linear system as in* (1.1) *with $\mathcal{X}$, $\mathcal{U}$ and $\mathcal{Y}$ finite-dimensional, say $\mathcal{U} = \mathbb{C}^r$, $\mathcal{Y} = \mathbb{C}^s$, $\mathcal{X} = \mathbb{C}^n$, so that the system matrix M has the form*

$$M = \begin{bmatrix} A & B \\ C & D \end{bmatrix} : \begin{bmatrix} \mathbb{C}^n \\ \mathbb{C}^r \end{bmatrix} \to \begin{bmatrix} \mathbb{C}^n \\ \mathbb{C}^s \end{bmatrix} \tag{1.6}$$

and the transfer function F_Σ *is equal to a rational matrix function of size* $s \times r$. *Assume that the realization* (A, B, C, D) *is* minimal, *i.e., the output pair* (C, A) *is* observable *and the input pair* (A, B) *is* controllable:

$$\bigcap_{k=0}^{n} \operatorname{Ker} CA^k = \{0\} \quad \text{and} \quad \operatorname*{span}_{k=0,1,\dots,n-1} \operatorname{Im} A^k B = \mathcal{X} = \mathbb{C}^n. \tag{1.7}$$

Then F_Σ *is in the Schur class* $\mathcal{S}(\mathbb{C}^r, \mathbb{C}^s)$ *if and only if there is an* $n \times n$ *positive-definite matrix* H *satisfying the KYP-inequality* (1.5).

There is also a *strict* version of the Bounded Real Lemma. The associated storage function required is a *strict storage function*, i.e., a function $S\colon \mathcal{X} \to \mathbb{R}_+$ for which there is a number $\delta > 0$ so that

$$S(\mathbf{x}(n+1)) - S(\mathbf{x}(n)) + \delta\|x(n)\|^2 \leq (1-\delta)\|\mathbf{u}(n)\|^2 - \|\mathbf{y}(n)\|^2 \tag{1.8}$$

holds over all system trajectories $(\mathbf{u}(n), \mathbf{x}(n), \mathbf{y}(n))_{n\in\mathbb{Z}}$, in addition to the normalization condition $S(0) = 0$. If $S_H(x) = \langle Hx, x\rangle$ is a quadratic strict storage function, then the associated linear matrix inequality is the *strict KYP-inequality*

$$\begin{bmatrix} A & B \\ C & D \end{bmatrix}^* \begin{bmatrix} H & 0 \\ 0 & I_{\mathcal{Y}} \end{bmatrix} \begin{bmatrix} A & B \\ C & D \end{bmatrix} \prec \begin{bmatrix} H & 0 \\ 0 & I_{\mathcal{U}} \end{bmatrix}. \tag{1.9}$$

In this case, one also arrives at a stronger condition on the transfer function F_Σ, namely that it has an analytic continuation to a function in the *strict Schur class*

$$\mathcal{S}^o(\mathcal{U}, \mathcal{Y}) = \left\{ F\colon \mathbb{D} \underset{\text{holo}}{\mapsto} \mathcal{L}(\mathcal{U}, \mathcal{Y}) \colon \sup_{z\in\mathbb{D}} \|F(z)\| \leq \rho \text{ for some } \rho < 1 \right\}.$$

Note, however, that the strict KYP-inequality implies that A is stable, so that in case (1.9) holds, F_Σ is in fact analytic on $\mathbb{D}$. This is the storage-function interpretation of the following strict Bounded Real Lemma, in which one replaces the minimality condition with a stability condition.

Theorem 1.2 (Strict Bounded Real Lemma (see [24])). *Suppose that the discrete-time linear system* Σ *is as in* (1.1) *with* $\mathcal{X}$, $\mathcal{U}$ *and* $\mathcal{Y}$ *finite-dimensional, say* $\mathcal{U} = \mathbb{C}^r$, $\mathcal{Y} = \mathbb{C}^s$, $\mathcal{X} = \mathbb{C}^n$, *i.e., the system matrix* M *is as in* (1.6). *Assume that* A *is* stable, *i.e., all eigenvalues of* A *are inside the open unit disk* $\mathbb{D}$, *so that* $r_{\text{spec}}(A) < 1$ *and the transfer function* $F_\Sigma(z)$ *is analytic on a neighborhood of* $\overline{\mathbb{D}}$. *Then* $F_\Sigma(z)$ *is in the strict Schur class* $\mathcal{S}^o(\mathbb{C}^r, \mathbb{C}^s)$ *if and only if there is a positive-definite matrix* $H \in \mathbb{C}^{n\times n}$ *so that the strict KYP-inequality* (1.9) *holds.*

We now turn to the general case, where the state space $\mathcal{X}$ and the input space $\mathcal{U}$ and the output space $\mathcal{Y}$ are allowed to be infinite-dimensional. In this case, the results are more recent, depending on the precise hypotheses.

For generalizations of Theorem 1.1, much depends on what is meant by minimality of Σ, and hence by the corresponding notions of controllable and observable. Here are the three possibilities for controllability of an input pair (A, B) which we shall consider. The third notion involves the controllability operator $\mathbf{W}_c$ associated with the pair (A, B) tailored to the Hilbert space setup which in general is a closed, possibly unbounded operator with domain $\mathcal{D}(\mathbf{W}_c)$ dense in $\mathcal{X}$ mapping into the Hilbert space $\ell^2_{\mathcal{U}}(\mathbb{Z}_-)$ of $\mathcal{Y}$-valued sequences supported on the negative integers $\mathbb{Z}_- = \{-1, -2, -3, \dots\}$, as well as the observability operator $\mathbf{W}_o$ associated with the pair (C, A), which has similar properties. We postpone precise definitions and properties of these operators to Section 2.

For an input pair (A, B) we define the following notions of controllability:

- (A, B) is *(approximately) controllable* if the reachability space
$$\operatorname{Rea}(A|B) = \operatorname{span}\{\operatorname{Im} A^k B \colon k = 0, 1, 2, \dots\} \tag{1.10}$$
is dense in $\mathcal{X}$.
- (A, B) is *exactly controllable* if the reachability space $\operatorname{Rea}(A|B)$ is equal to $\mathcal{X}$, i.e., each state vector $x \in \mathcal{X}$ has a representation as a finite linear combination $x = \sum_{k=0}^{K} A^k B u_k$ for a choice of finitely many input vectors $u_0, u_1, \dots, u_K$ (also known as every x is a *finite-time reachable state* (see [22, Definition 3.3]).
- (A, B) is *ℓ^2-exactly controllable* if the ℓ^2-adapted controllability operator $\mathbf{W}_c$ has range equal to all of $\mathcal{X}$: $\mathbf{W}_c \mathcal{D}(\mathbf{W}_c) = \mathcal{X}$.

If (C, A) is an output pair, we have the dual notions of observability:

- (C, A) is *(approximately) observable* if the input pair (A^*, C^*) is (approximately) controllable, i.e., if the observability space
$$\operatorname{Obs}(C|A) = \operatorname{span}\{\operatorname{Im} A^{*k} C^* \colon k = 0, 1, 2, \dots\} \tag{1.11}$$
is dense in $\mathcal{X}$, or equivalently, if $\cap_{k=0}^{\infty} \ker CA^k = \{0\}$.
- (C, A) is *exactly observable* if the observability subspace $\operatorname{Obs}(C|A)$ is the whole space $\mathcal{X}$.
- (C, A) is *ℓ^2-exactly observable* if the adjoint input pair (A^*, C^*) is ℓ^2-exactly controllable, i.e., if the adjoint $\mathbf{W}_o^*$ of the ℓ^2-adapted observability operator $\mathbf{W}_o$ has full range: $\mathbf{W}_o^* \mathcal{D}(\mathbf{W}_o^*) = \mathcal{X}$.

Then we say that the system $\Sigma \sim (A, B, C, D)$ is

- *minimal* if (A, B) is controllable and (C, A) is observable,
- *exactly minimal* if both (A, B) is exactly controllable and (C, A) is exactly observable, and
- *ℓ^2-exactly minimal* if both (A, B) is ℓ^2-exactly controllable and (C, A) is ℓ^2-exactly observable.

Despite the fact that the operators A, B, C and D associated with the system Σ are all bounded, in the infinite-dimensional analogue of the KYP-inequality (1.5) unbounded solutions H may appear. We therefore have to be more precise concerning the notion of positive-definiteness we employ. Suppose that H is a (possibly unbounded) selfadjoint operator H on a Hilbert space $\mathcal{X}$ with domain $\mathcal{D}(H)$ dense in $\mathcal{X}$; we refer to [26] for background and details on this class and other classes of unbounded operators. Then we shall say:

- H is *strictly positive-definite* (written $H \succ 0$) if there is a $\delta > 0$ so that $\langle Hx, x\rangle \geq \delta\|x\|^2$ for all $x \in \mathcal{D}(H)$;
- H is *positive-definite* if $\langle Hx, x\rangle > 0$ for all nonzero $x \in \mathcal{D}(H)$;
- H is *positive-semidefinite* (written $H \succeq 0$) if $\langle Hx, x\rangle \geq 0$ for $x \in \mathcal{D}(H)$.

We also note that any (possibly unbounded) positive-semidefinite operator H has a positive-semidefinite square root $H^{\frac{1}{2}}$; as $H = H^{\frac{1}{2}} \cdot H^{\frac{1}{2}}$, we have

$$\mathcal{D}(H) = \{x \in \mathcal{D}(H^{\frac{1}{2}})\colon H^{\frac{1}{2}}x \in \mathcal{D}(H^{\frac{1}{2}})\} \subset \mathcal{D}(H^{\frac{1}{2}}).$$

See, e.g., [26] for details.

Since solutions H to the corresponding KYP-inequality may be unbounded, the KYP-inequality cannot necessarily be written in the LMI form (1.5), but rather, we require a spatial form of (1.5) on the appropriate domain: For a (possibly unbounded) positive-definite operator H on $\mathcal{X}$ satisfying

$$A\mathcal{D}(H^{\frac{1}{2}}) \subset \mathcal{D}(H^{\frac{1}{2}}), \quad B\mathcal{U} \subset \mathcal{D}(H^{\frac{1}{2}}), \tag{1.12}$$

the spatial form of the KYP-inequality takes the form

$$\left\| \begin{bmatrix} H^{\frac{1}{2}} & 0 \\ 0 & I_{\mathcal{U}} \end{bmatrix} \begin{bmatrix} x \\ u \end{bmatrix} \right\|^2 - \left\| \begin{bmatrix} H^{\frac{1}{2}} & 0 \\ 0 & I_{\mathcal{Y}} \end{bmatrix} \begin{bmatrix} A & B \\ C & D \end{bmatrix} \begin{bmatrix} x \\ u \end{bmatrix} \right\|^2 \geq 0 \tag{1.13}$$

$$(x \in \mathcal{D}(H^{\frac{1}{2}}), u \in \mathcal{U}).$$

The corresponding notion of a storage function will then be allowed to assume $+\infty$ as a value; this will be made precise in Section 3.

With all these definitions out of the way, we can state the following three distinct generalizations of Theorem 1.1 to the infinite-dimensional situation.

Theorem 1.3 (Infinite-dimensional standard Bounded Real Lemma). *Let Σ be a discrete-time linear system as in* (1.1) *with system matrix M as in* (1.2) *and transfer function F_Σ defined by* (1.3).

(1) *Suppose that the system Σ is minimal, i.e., the input pair (A, B) is controllable and the output pair (C, A) is observable. Then the transfer function F_Σ has an analytic continuation to a function in the Schur class $\mathcal{S}(\mathcal{U}, \mathcal{Y})$ if and only if there exists a positive-definite solution H of the KYP-inequality in the following generalized sense: H is a closed, possibly unbounded, densely defined, positive-definite* (*and hence injective*) *operator on $\mathcal{X}$ such that $\mathcal{D}(H^{\frac{1}{2}})$ satisfies* (1.12) *and H solves the spatial KYP-inequality* (1.13).

(2) *Suppose that Σ is exactly minimal. Then the transfer function F_Σ has an analytic continuation to a function in the Schur class $\mathcal{S}(\mathcal{U}, \mathcal{Y})$ if and only if there exists a bounded, strictly positive-definite solution H of the KYP-inequality* (1.5). *In this case A has a spectral radius of at most one, and hence F_Σ is in fact analytic on $\mathbb{D}$.*

(3) *Statement* (2) *above continues to hold if the "exactly minimal" hypothesis is replaced by the hypothesis that Σ be "ℓ^2-exactly minimal."*

We shall refer to a closed, densely defined, positive-definite solution H of (1.12)–(1.13) as a positive-definite solution of the *generalized KYP-inequality*.

The paper of Arov–Kaashoek–Pik [6] gives a penetrating treatment of item (1) in Theorem 1.3, including examples to illustrate various subtleties surrounding this result—e.g., the fact that the result can fail if one insists on classical bounded and boundedly invertible selfadjoint solutions of the KYP-inequality. We believe that items (2) and (3) appeared for the first time in [11], where also a sketch of the proof of item (1) is given. The idea behind the proofs of items (1)–(3) in [11] is to combine the result that a Schur-class function S always has a contractive realization (i.e., such an S can be realized as $S = F_\Sigma$ for a system Σ as in (1.1) with system matrix M in (1.2) a contraction operator) with variations of the State-Space-Similarity Theorem (see [11, Theorem 1.5]) for the infinite-dimensional situation under the conditions that hold in items (1)–(3); roughly speaking, under appropriate hypothesis, a State-Space-Similarity Theorem says that two systems Σ and Σ' whose transfer functions coincide on a neighborhood of zero, necessarily can be transformed (in an appropriate sense) from one to other via a change of state-space coordinates.

In the present paper we revisit these three results from a different point of view: we adapt Willems' variational formulas to the infinite-dimensional setting, and in this context present the available storage S_a and required supply S_r, as well as an ℓ_2-regularized version $\underline{S}_r$ of the required supply. It is shown, under appropriate hypothesis, that these are storage functions, with S_a and $\underline{S}_r$ being quadratic storage functions, i.e., S_a agrees with $S_{H_a}(x) = \|H_a^{\frac{1}{2}} x\|^2$ and $\underline{S}_r(x) = S_{H_r}(x) = \|H_r^{\frac{1}{2}} x\|^2$ for x in a suitably large subspace of $\mathcal{X}$, where H_a and H_r are possibly unbounded, positive-definite density operators, which turn out to be positive-definite solutions to the generalized KYP-inequality. In this way we will arrive at a proof of item (1). Further analysis of the behavior of H_a and H_r, under additional restrictions on Σ, lead to proofs of items (2) and (3), as well as the following version of the strict Bounded Real Lemma for infinite-dimensional systems, which is a much more straightforward generalization of the result in the finite-dimensional case (Theorem 1.2).

Theorem 1.4 (Infinite-dimensional strict Bounded Real Lemma). *Let Σ be a discrete-time linear system as in* (1.1) *with system matrix M as in* (1.2) *and transfer function F_Σ defined by* (1.3). *Assume that A is exponentially stable, i.e., $r_{\rm spec}(A) < 1$. Then the transfer function F_Σ is in the strict Schur class $\mathcal{S}^o(\mathcal{U}, \mathcal{Y})$ if and only if there exists a bounded strictly positive-definite solution H of the strict KYP-inequality* (1.9).

Theorem 1.2 was proved by Petersen–Anderson–Jonkheere [24] for the continuous-time finite-dimensional setting by using what we shall call an ϵ-regularization procedure to reduce the result to the standard case Theorem 1.1. In [11] we show how this same idea can be used in the infinite-dimensional setting to reduce the hard direction of Theorem 1.4 to the result of either of item (2) or item (3) in Theorem 1.3. For the more general nonlinear setting, Willems [28] was primarily interested in what storage functions look like assuming that they exist, while in [29] for the finite-dimensional linear setting he reduced the existence problem to the existence theory for Riccati matrix equations. Here we solve the existence problem for the more general infinite-dimensional linear setting by converting Willems' variational formulation of the available storage S_a and an ℓ^2-regularized version $\underline{S}_r$ of his required supply S_r to an operator-theoretic formulation amenable to explicit analysis.

This paper presents a more unified approach to the different variations of the Bounded Real Lemma, in the sense that we present a pair of concretely defined, unbounded, positive-definite operators H_a and H_r that, under the appropriate conditions, form positive-definite solutions to the generalized KYP-inequality, and that have the required additional features under the additional conditions in items (2) and (3) of Theorem 1.3 as well as Theorem 1.4. We also make substantial use of connections with corresponding objects for the adjoint system Σ^* (see (5.1)) to complete the analysis and arrive at some order properties for the set of all solutions of the generalized KYP-inequality which are complementary to those in [6].

The paper is organized as follows. Besides the current introduction, the paper consists of seven sections. In Section 2 we give the definitions of the observability operator $\mathbf{W}_o$ and controllability operator $\mathbf{W}_c$ associated with the system Σ in (1.1) and recall some of their basic properties. In Section 3 we define what is meant by a storage function in the context of infinite-dimensional discrete-time linear systems Σ of the form (1.1) as well as strict and quadratic storage functions, and we clarify the relations between quadratic (strict) storage functions and solutions to the (generalized) KYP-inequality. Section 4 is devoted to the available storage S_a and required supply S_r, two examples of storage functions, in case the transfer function of Σ has an analytic continuation to a Schur-class function. It is shown that S_a and an ℓ^2-regularized version $\underline{S}_r$ of S_r in fact agree with quadratic storage functions on suitably large domain via explicit constructions of two closed, densely defined, positive-definite operators H_a and H_r that exhibit S_a and $\underline{S}_r$ as quadratic storage functions S_{H_a} and S_{H_r}. In Section 5 we make explicit the theory for the adjoint system Σ^* and the duality connections between Σ and Σ^*. In Section 6 we study the order properties of a class of solutions of the generalized KYP-inequality, and obtain the conditions under which H_a and H_r are bounded and/or boundedly invertible and thereby solutions of the classical KYP-inequality. These results are then used in Section 7 to give proofs of Theorems 1.3 and 1.4 via the storage function approach.

2. Review: minimality, controllability, observability

In this section we recall the definitions of the observability operator $\mathbf{W}_o$ and controllability operator $\mathbf{W}_c$ associated with the discrete-time linear system Σ given by (1.1) and various of their basic properties which will be needed in the sequel. Detailed proofs of most of these results as well as additional properties can be found in [11, Section 2].

For the case of a general system Σ, following [11, Section 2], we define the *observability operator* $\mathbf{W}_o$ associated with Σ to be the possibly unbounded operator with domain $\mathcal{D}(\mathbf{W}_o)$ in $\mathcal{X}$ given by

$$\mathcal{D}(\mathbf{W}_o) = \{x \in \mathcal{X} \colon \{CA^n x\}_{n\geq 0} \in \ell^2_{\mathcal{Y}}(\mathbb{Z}_+)\} \tag{2.1}$$

with action given by

$$\mathbf{W}_o x = \{CA^n x\}_{n\geq 0} \text{ for } x \in \mathcal{D}(\mathbf{W}_o). \tag{2.2}$$

Dually, we define the *adjoint controllability operator* $\mathbf{W}_c^*$ associated with Σ to have domain

$$\mathcal{D}(\mathbf{W}_c^*) = \{x \in \mathcal{X} \colon \{B^* A^{*(-n-1)} x\}_{n\leq -1} \in \ell^2_{\mathcal{U}}(\mathbb{Z}_-)\} \tag{2.3}$$

with action given by

$$\mathbf{W}_c^* x = \{B^* A^{*(-n-1)} x\}_{n\leq -1} \text{ for } x \in \mathcal{D}(\mathbf{W}_c^*). \tag{2.4}$$

It is directly clear from the definitions of $\mathbf{W}_o$ and $\mathbf{W}_c^*$ that

$$\ker \mathbf{W}_o = \operatorname{Obs}(C|A)^\perp \quad \text{and} \quad \ker \mathbf{W}_c^* = \operatorname{Rea}(A|B)^\perp. \tag{2.5}$$

We next summarize the basic properties of $\mathbf{W}_c$ and $\mathbf{W}_o$.

Proposition 2.1 (Proposition 2.1 in [11]). *Let Σ be a system as in* (1.1) *with observability operator $\mathbf{W}_o$ and adjoint controllability operator $\mathbf{W}_c^*$ as in* (2.1)–(2.4). *Basic properties of the controllability operator $\mathbf{W}_c$ are:*

(1) *It is always the case that $\mathbf{W}_o$ is a closed operator on its domain* (2.1).

(2) *If $\mathcal{D}(\mathbf{W}_o)$ is dense in $\mathcal{X}$, then the adjoint $\mathbf{W}_o^*$ of $\mathbf{W}_o$ is a closed and densely defined operator, by a general property of adjoints of closed operators with dense domain. Concretely for the case here, $\mathcal{D}(\mathbf{W}_o^*)$ contains the dense linear manifold $\ell_{\text{fin},\mathcal{Y}}(\mathbb{Z}_+)$ consisting of finitely supported sequences in $\ell^2_{\mathcal{Y}}(\mathbb{Z}_+)$. In general, one can characterize $\mathcal{D}(\mathbf{W}_o^*)$ explicitly as the set of all $\mathbf{y} \in \ell^2_{\mathcal{Y}}(\mathbb{Z}_+)$ such that there exists a vector $x_o \in \mathcal{X}$ such that the limit*

$$\lim_{K\to\infty} \left\langle x, \sum_{k=0}^{K} A^{*k} C^* \mathbf{y}(k) \right\rangle_{\mathcal{X}}$$

exists for each $x \in \mathcal{D}(\mathbf{W}_o)$ and is given by

$$\lim_{K\to\infty} \left\langle x, \sum_{k=0}^{K} A^{*k} C^* \mathbf{y}(k) \right\rangle_{\mathcal{X}} = \langle x, x_o \rangle_{\mathcal{X}}, \tag{2.6}$$

with action of $\mathbf{W}_c$ then given by

$$\mathbf{W}_o^* \mathbf{y} = x_o \tag{2.7}$$

where x_o is as in (2.6). *In particular, $\ell_{\text{fin},\mathcal{Y}}(\mathbb{Z}_+)$ is contained in $\mathcal{D}(\mathbf{W}_o^*)$ and the observability space defined in* (1.11) *is given by*

$$\operatorname{Obs}(C|A) = \mathbf{W}_o^* \ell_{\text{fin},\mathcal{Y}}(\mathbb{Z}_+).$$

Thus, if in addition (C, A) is observable, then $\mathbf{W}_o^*$ has dense range.

Dual properties of the controllability operator $\mathbf{W}_c^*$ are:

(3) *It is always the case that the adjoint controllability operator $\mathbf{W}_c^*$ is closed on its domain* (2.3).

(4) *If $\mathcal{D}(\mathbf{W}_c^*)$ is dense in $\mathcal{X}$, then the controllability operator $\mathbf{W}_c = (\mathbf{W}_c^*)^*$ is closed and densely defined by a general property of the adjoint of a closed and densely defined operator. Concretely for the case here, $\mathcal{D}(\mathbf{W}_c)$ contains the dense linear manifold $\ell_{\mathrm{fin},\mathcal{U}}(\mathbb{Z}_-)$ of finitely supported sequences in $\ell^2_{\mathcal{U}}(\mathbb{Z}_-)$. In general, one can characterize $\mathcal{D}(\mathbf{W}_c)$ explicitly as the set of all $\mathbf{u} \in \ell^2_{\mathcal{U}}(\mathbb{Z}_-)$ such that there exists a vector $x_c \in \mathcal{X}$ so that*

$$\lim_{K\to\infty} \left\langle x, \sum_{k=-K}^{-1} A^{-k-1}B\mathbf{u}(k)\right\rangle_{\mathcal{X}}$$

exists for each $x \in \mathcal{D}(\mathbf{W}_c^*)$ and is given by

$$\lim_{K\to\infty} \left\langle x, \sum_{k=-K}^{-1} A^{-k-1}B\mathbf{u}(k)\right\rangle_{\mathcal{X}} = \langle x, x_c\rangle_{\mathcal{X}}, \tag{2.8}$$

and action of $\mathbf{W}_c$ then given by

$$\mathbf{W}_c\mathbf{u} = x_c \tag{2.9}$$

where x_c is as in (2.8). *In particular, the reachability space* $\mathrm{Rea}(A|B)$ *is equal to $\mathbf{W}_c\ell_{\mathrm{fin},\mathcal{U}}(\mathbb{Z}_-)$. Thus, if in addition (A, B) is controllable, then $\mathbf{W}_c$ has dense range.*

For systems Σ as in (1.1), without additional conditions, it can happen that $\mathbf{W}_o$ and/or $\mathbf{W}_c^*$ are not densely defined, and therefore the adjoints $\mathbf{W}_o^*$ and $\mathbf{W}_c$ are at best linear relations and difficult to work with. However, our interest here is the case where the transfer function F_Σ has analytic continuation to a bounded function on the unit disk (or even in the Schur class, i.e., norm-bounded by 1 on the unit disk). In this case the multiplication operator

$$M_{F_\Sigma} : f(\lambda) \mapsto F_\Sigma(\lambda)f(\lambda) \tag{2.10}$$

is a bounded operator from $L^2_{\mathcal{U}}(\mathbb{T})$ to $L^2_{\mathcal{Y}}(\mathbb{T})$ and hence also its compression to a map “from past to future”

$$\mathbb{H}_{F_\Sigma} = P_{H^2_{\mathcal{Y}}(\mathbb{D})} M_{F_\Sigma}|_{H^2_{\mathcal{U}}(\mathbb{D})^\perp}, \tag{2.11}$$

often called the *Hankel operator* with symbol F_Σ, is also bounded (by $\|M_{F_\Sigma}\|$). If we take inverse Z-transform to represent $L^2(\mathbb{T})$ as $\ell^2(\mathbb{Z})$, $H^2(\mathbb{D})$ as $\ell^2(\mathbb{Z}_+)$ and $H^2(\mathbb{D})^\perp$ as $\ell^2(\mathbb{Z}_-)$, then the frequency-domain Hankel operator

$$\mathbb{H}_{F_\Sigma} : H^2_{\mathcal{U}}(\mathbb{D})^\perp \to H^2_{\mathcal{Y}}(\mathbb{D})$$

given by (2.11) transforms via inverse Z-transform to the time-domain Hankel operator $\mathfrak{H}_{F_\Sigma}$ with matrix representation

$$\mathfrak{H}_{F_\Sigma} = [CA^{i-j-1}B]_{i\geq 0, j<0} : \ell^2_{\mathcal{U}}(\mathbb{Z}_-) \to \ell^2_{\mathcal{Y}}(\mathbb{Z}_+). \tag{2.12}$$

We conclude that the Hankel matrix $\mathfrak{H}_{F_\Sigma}$ is bounded as an operator from $\ell^2_{\mathcal{U}}(\mathbb{Z}_-)$ to $\ell^2_{\mathcal{Y}}(\mathbb{Z}_+)$ whenever F_Σ has analytic continuation to an H^∞ function. From the matrix representation (2.12) we see that the Hankel matrix formally has a factorization

$$\mathfrak{H}_{F_\Sigma} = \mathrm{col}[CA^i]_{i\geq 0} \cdot \mathrm{row}[A^{-j-1}B]_{j<0} = \mathbf{W}_o \cdot \mathbf{W}_c. \tag{2.13}$$

It can happen that $\mathfrak{H}_{F_\Sigma}$ is bounded while $\mathbf{W}_o$ and $\mathbf{W}_c$ are unbounded. Nevertheless, from the fact that $\mathfrak{H}_{F_\Sigma}$ is bounded one can see that $\mathrm{Rea}(A|B)$ is in $\mathcal{D}(\mathbf{W}_o)$ and

$$\mathfrak{H}_{F_\Sigma}\mathbf{u} = \mathbf{W}_o\left(\sum_{k=K}^{-1} A^{-1-k}B\mathbf{u}(k)\right) \in \ell^2_{\mathcal{Y}}(\mathbb{Z}_+).$$

for each finitely supported input string $\mathbf{u}(K), \dots, \mathbf{u}(-1)$. If we assume that (A, B) is controllable, we conclude that $\mathbf{W}_o$ is densely defined. Similarly, by working with boundedness of $\mathfrak{H}^*_{F_\Sigma}$ one can show that boundedness of F_Σ on $\mathbb{D}$ leads to $\mathcal{D}(\mathbf{W}_c^*)$ containing the observability space $\mathrm{Obs}(C|A)$; hence if we assume that (C, A) is observable, we get that $\mathbf{W}_c^*$ is densely defined. With these observations in hand, the following precise version of the formal factorization (2.13) for the case where $\mathbf{W}_o$ and $\mathbf{W}_c$ may be unbounded becomes plausible.

Proposition 2.2 (Corollary 2.4 and Proposition 2.6 in [11]). *Suppose that the system Σ given by* (1.1) *has transfer function F_Σ with analytic continuation to an H^∞-function on the unit disk $\mathbb{D}$.*

(1) *Assume that $\mathcal{D}(\mathbf{W}_c^*)$ is dense in $\mathcal{X}$ (as is the case if (C, A) is observable). Then $\mathcal{D}(\mathbf{W}_o)$ contains* $\mathrm{Im}\,\mathbf{W}_c = \mathbf{W}_c\mathcal{D}(\mathbf{W}_c)$ *and*

$$\mathfrak{H}_{F_\Sigma}|_{\mathcal{D}(\mathbf{W}_c)} = \mathbf{W}_o\mathbf{W}_c. \tag{2.14}$$

In particular, as $\ell_{\mathrm{fin},\mathcal{U}}(\mathbb{Z}_-) \subset \mathcal{D}(\mathbf{W}_c)$ and $\mathbf{W}_c\ell_{\mathrm{fin},\mathcal{U}}(\mathbb{Z}_-) = \mathrm{Rea}(A|B)$ (from Proposition 2.1 (4)*), it follows that* $\mathrm{Rea}(A|B) \subset \mathcal{D}(\mathbf{W}_o)$.

(2) *Assume that $\mathcal{D}(\mathbf{W}_o)$ is dense in $\mathcal{X}$ (as is the case if (A, B) is controllable). Then $\mathcal{D}(\mathbf{W}_c^*)$ contains* $\mathrm{Im}\,\mathbf{W}_o^* = \mathbf{W}_o^*\mathcal{D}(\mathbf{W}_o^*)$ *and*

$$\mathfrak{H}^*_{F_\Sigma}|_{\mathcal{D}(\mathbf{W}_o^*)} = \mathbf{W}_c^*\mathbf{W}_o^*. \tag{2.15}$$

In particular, as $\ell_{\mathrm{fin},\mathcal{Y}}(\mathbb{Z}) \subset \mathcal{D}(\mathbf{W}_o^*)$ and $\mathbf{W}_o^*\ell_{\mathrm{fin},\mathcal{Y}}(\mathbb{Z}_+) = \mathrm{Obs}(C|A)$ (from Proposition 2.1 (2)*), it follows that* $\mathrm{Obs}(C|A) \subset \mathcal{D}(\mathbf{W}_c^*)$.

(3) *In case the system matrix $M = \left[\begin{smallmatrix} A & B \\ C & D\end{smallmatrix}\right]$ is contractive, then $\mathbf{W}_o$ and $\mathbf{W}_c$ also are bounded contraction operators and we have the bounded-operator factorizations*

$$\mathfrak{H}_{F_\Sigma} = \mathbf{W}_o\mathbf{W}_c, \quad (\mathfrak{H}_{F_\Sigma})^* = \mathbf{W}_c^*\mathbf{W}_o^*. \tag{2.16}$$

The following result from [11] describes the implications of ℓ^2-exact controllability and ℓ^2-exact observability on the operators $\mathbf{W}_o$ and $\mathbf{W}_c$

Proposition 2.3 (Corollary 2.5 in [11]). *Let Σ be a discrete-time linear system as in* (1.1) *with system matrix M as in* (1.2)*. Assume that the transfer function F_Σ defined by* (1.3) *has an analytic continuation to an H^∞-function on $\mathbb{D}$.*

(1) *If Σ is ℓ^2-exactly controllable, then $\mathbf{W}_o$ is bounded.*
(2) *If Σ is ℓ^2-exactly observable, then $\mathbf{W}_c$ is bounded.*
(3) *Σ is ℓ^2-exactly minimal, i.e., both ℓ^2-exactly controllable and ℓ^2-exactly observable, then $\mathbf{W}_o$ and $\mathbf{W}_c^*$ are both bounded and bounded below.*

The following result will be useful in the sequel.

Proposition 2.4. *Suppose that the discrete-time linear system* Σ *given by* (1.1) *is minimal and that its transfer function* F_Σ *has analytic continuation to an* H^∞*-function on* $\mathbb{D}$, *so (by Propositions* 2.1 *and* 2.2*)* $\mathcal{D}(\mathbf{W}_c^*) \supset \mathrm{Obs}(C|A)$ *is dense in* $\mathcal{X}$ *and* $\mathbf{W}_c = (\mathbf{W}_c^*)^*$ *is densely defined with dense range* $\mathrm{Im}(\mathbf{W}_c) \supset \mathrm{Rea}(A|B)$.

(1) *Suppose that* $(\mathbf{u}(n), \mathbf{x}(n), \mathbf{y}(n))_{n \ge n_{-1}}$ *is a system trajectory of* Σ *with initialization* $\mathbf{x}(n_{-1}) = 0$. *Define an input string* $\mathbf{u}' \in \ell_{\mathrm{fin},\mathcal{U}}(\mathbb{Z}_-)$ *by*

$$\mathbf{u}'(n) = \begin{cases} 0 & \text{if } n < n_{-1}, \\ \mathbf{u}(n) & \text{if } n_{-1} \le n < 0. \end{cases}$$

Then $\mathbf{x}(0) = \mathbf{W}_c \mathbf{u}'$.

(2) *Suppose that* $\mathbf{u} \in \ell^2_{\mathcal{U}}(\mathbb{Z}_-)$ *is in* $\mathcal{D}(\mathbf{W}_c)$ *and* $\widetilde{u} \in \mathcal{U}$. *Define a new input string* $\mathbf{u}' \in \ell^2_{\mathcal{U}}(\mathbb{Z}_-)$ *by*

$$\mathbf{u}'(n) = \begin{cases} \mathbf{u}(n+1) & \text{if } n < -1, \\ \widetilde{u} & \text{if } n = -1. \end{cases}$$

Then $\mathbf{u}' \in \mathcal{D}(\mathbf{W}_c)$ *and*

$$\mathbf{W}_c \mathbf{u}' = A \mathbf{W}_c \mathbf{u} + B \widetilde{u}.$$

Proof. We start with item (1). From item (4) of Proposition 2.1 see that $\ell_{\mathrm{fin},\mathcal{U}}(\mathbb{Z}_+)$ is contained in $\mathcal{D}(\mathbf{W}_c)$, and thus $\mathbf{u}' \in \mathcal{D}(\mathbf{W}_c)$. From formula (2.8) for the action of $\mathbf{W}_c$ on its domain we obtain that

$$\mathbf{W}_c \mathbf{u}' = \sum_{k \in \mathbb{Z}_-} A^{-k-1} B \mathbf{u}'(k) = \sum_{k=n_{-1}}^{-1} A^{-k-1} B \mathbf{u}(k) \tag{2.17}$$

where the sum is well defined since there are only finitely many nonzero terms. By a standard induction argument, using the input-state equation in (1.1), one verifies that this is the formula for $\mathbf{x}(0)$ for a system trajectory $(\mathbf{u}(n), \mathbf{x}(n), \mathbf{y}(n))_{n \ge n_{-1}}$ with initialization $\mathbf{x}(n_{-1}) = 0$. This verifies (1).

As for item (2), it is easily verified that $\mathcal{D}(\mathbf{W}_c^*)$ is invariant under A^* and that the following intertwining condition holds:

$$\mathbf{W}_c^* A^*|_{\mathcal{D}(\mathbf{W}_c^*)} = \mathcal{S}_- \mathbf{W}_c^*,$$

with $\mathcal{S}_-$ the truncated right shift operator on $\ell^2_{\mathcal{U}}(\mathbb{Z}_-)$ given by

$$(\mathcal{S}_- \mathbf{u})(n) = \mathbf{u}(n-1) \text{ for } n \in \mathbb{Z}_-.$$

The adjoint version of this is that $\mathcal{D}(\mathbf{W}_c)$ is invariant under the untruncated left shift operator $\mathcal{S}_-^*$ on $\ell^2_{\mathcal{U}}(\mathbb{Z}_-)$

$$(\mathcal{S}_-^* \mathbf{u})(n) = \begin{cases} \mathbf{u}(n+1) & \text{if } n < -1, \\ 0 & \text{if } n = -1 \end{cases}$$

and we have the intertwining condition

$$\mathbf{W}_c \mathcal{S}_-^*|_{\mathcal{D}(\mathbf{W}_c)} = A \mathbf{W}_c.$$

Next note that $\mathcal{S}_-^*\mathbf{u} = \mathbf{u}' - \Pi_{-1}\widetilde{u}$, with $\Pi_{-1} : \mathcal{U} \to \ell^2_{\mathcal{U}}(\mathbb{Z}_-)$ the embedding of $\mathcal{U}$ into the -1-th entry of $\ell^2_{\mathcal{U}}(\mathbb{Z}_-)$. This implies that

$$\mathbf{u}' = \mathcal{S}_-^*\mathbf{u} + \Pi_{-1}\widetilde{u} \in \mathcal{S}_-^*\mathcal{D}(\mathbf{W}_c) + \ell_{\mathrm{fin},\mathcal{U}}(\mathbb{Z}_-) \subset \mathcal{D}(\mathbf{W}_c),$$

and

$$A\mathbf{W}_c\mathbf{u} = \mathbf{W}_c\mathcal{S}_-^*|_{\mathcal{D}(\mathbf{W}_c)}\mathbf{u} = \mathbf{W}_c(\mathbf{u}' - \Pi_{-1}\widetilde{u}) = \mathbf{W}_c\mathbf{u}' - B\widetilde{u}, \tag{2.18}$$

which provides the desired identity. □

Remark 2.5. It is of interest to consider the shift $\mathbf{W}_c^{(1)}$ of the controllability operator $\mathbf{W}_c$ to the interval $(-\infty, 0]$ in place of $\mathbb{Z}_- = (-\infty, 0)$, i.e.,

$$\mathbf{W}_c^{(1)} = \mathbf{W}_c\tau^{-1}$$

where the map τ transforms sequences $\mathbf{u}$ supported on $\mathbb{Z}_- = (-\infty, 0)$ to sequences $\mathbf{u}'$ supported on $(-\infty, 0]$ according to the action

$$(\tau\mathbf{u})(n) = \mathbf{u}(n+1) \quad \text{for } n < 0$$

with inverse given by

$$(\tau^{-1}\mathbf{v})(n) = \mathbf{v}(n-1) \quad \text{for } n \leq 0.$$

For all $\mathbf{u} \in \ell^2_{\mathcal{U}}(\mathbb{Z}_-)$ and $\widetilde{u} \in \mathcal{U}$, define a sequence $(\mathbf{u}, \widetilde{u}) \in \ell^2_{\mathcal{U}}((-\infty, 0])$ by

$$(\mathbf{u}, \widetilde{u})(n) = \begin{cases} \mathbf{u}(n) & \text{if } n \in \mathbb{Z}_-, \\ \widetilde{u} & \text{if } n = 0. \end{cases}$$

The result of item (2) in Proposition 2.4 can be interpreted as saying: given $\mathbf{u} \in \ell^2_{\mathcal{U}}(\mathbb{Z}_-)$ and $\widetilde{u} \in \mathcal{U}$ we have

$$(\mathbf{u}, \widetilde{u}) \in \mathcal{D}(\mathbf{W}_c^{(1)}) \iff \mathbf{u} \in \mathcal{D}(\mathbf{W}_c)$$

and in that case $\mathbf{W}_c^{(1)}(\mathbf{u}, \widetilde{u}) = A\mathbf{W}_c\mathbf{u} + B\widetilde{u}$.

3. Storage functions

In the case of systems with an infinite-dimensional state space we allow storage functions to also attain $+\infty$ as a value. Set $[0, \infty] := \mathbb{R}_+ \cup \{+\infty\}$. Then, given a discrete-time linear system Σ as in (1.1), we say that a function $S \colon \mathcal{X} \to [0, \infty]$ is a *storage function* for the system Σ if the dissipation inequality

$$S(\mathbf{x}(n+1)) \leq S(\mathbf{x}(n)) + \|\mathbf{u}(n)\|^2_{\mathcal{U}} - \|\mathbf{y}(n)\|^2_{\mathcal{Y}} \text{ for } n \geq N_0 \tag{3.1}$$

holds along all system trajectories $(\mathbf{u}(n), \mathbf{x}(n), \mathbf{y}(n))_{n \geq N_0}$ with state initialization $x(N_0) = x_0$ for some $x_0 \in \mathcal{X}$ at some $N_0 \in \mathbb{Z}$, and S is normalized to satisfy

$$S(0) = 0. \tag{3.2}$$

As a first result we show that existence of a storage function for Σ is a sufficient condition for the transfer function to have an analytic continuation to a Schur-class function.

Proposition 3.1. *Suppose that the system* Σ *in* (1.1) *has a storage function* S*. Then the transfer function* F_Σ *of* Σ *defined in* (1.3) *has an analytic continuation to a function in the Schur class* $\mathcal{S}(\mathcal{U}, \mathcal{Y})$*.*

The proof of Proposition 3.1 relies on the following observation, which will also be of use in the sequel.

Lemma 3.2. *Suppose that the system* Σ *in* (1.1) *has a storage function* S*. For each system trajectory* $(\mathbf{u}(n), \mathbf{x}(n), \mathbf{y}(n))_{n\in\mathbb{Z}}$ *and* $N_0 \in \mathbb{Z}$ *so that* $\mathbf{x}(N_0) = 0$*, the following inequalities hold for all* $N \in \mathbb{Z}_+$*:*

$$S(\mathbf{x}(N_0 + N + 1)) \le \sum_{n=N_0}^{N_0+N} \|\mathbf{u}(n)\|_{\mathcal{U}}^2 - \sum_{n=N_0}^{N_0+N} \|\mathbf{y}(n)\|_{\mathcal{Y}}^2; \tag{3.3}$$

$$\sum_{n=N_0}^{N_0+N} \|\mathbf{y}(n)\|_{\mathcal{Y}}^2 \le \sum_{n=N_0}^{N_0+N} \|\mathbf{u}(n)\|_{\mathcal{U}}^2. \tag{3.4}$$

Proof. By the translation invariance of the system Σ we may assume without loss of generality that $N_0 = 0$, i.e., $\mathbf{x}(0) = 0$. From (3.1) and (3.2) we get

$$S(\mathbf{x}(1)) \le \|\mathbf{u}(0)\|^2 - \|\mathbf{y}(0)\|^2 + S(0) = \|\mathbf{u}(0)\|^2 - \|\mathbf{y}(0)\|^2 < \infty.$$

Inductively, suppose that $S(\mathbf{x}(n)) < \infty$. Then (3.1) gives us

$$S(\mathbf{x}(n+1)) \le \|\mathbf{u}(n)\|_{\mathcal{U}}^2 - \|\mathbf{y}(n)\|_{\mathcal{Y}}^2 + S(\mathbf{x}(n)) < \infty.$$

We may now rearrange the dissipation inequality for $n \in \mathbb{Z}_+$ in the form

$$S(\mathbf{x}(n+1)) - S(\mathbf{x}(n)) \le \|\mathbf{u}(n)\|^2 - \|\mathbf{y}(n)\|^2 \quad (n \in \mathbb{Z}_+). \tag{3.5}$$

Summing from $n = 0$ to $n = N$ gives

$$0 \le S(\mathbf{x}(N+1)) \le \sum_{n=0}^{N} \|\mathbf{u}(n)\|_{\mathcal{U}}^2 - \sum_{n=0}^{N} \|\mathbf{y}(n)\|_{\mathcal{Y}}^2,$$

which leads to

$$\sum_{n=0}^{N} \|\mathbf{y}(n)\|_{\mathcal{Y}}^2 \le \sum_{n=0}^{N} \|\mathbf{u}(n)\|_{\mathcal{U}}^2 \text{ for all } N \in \mathbb{Z}_+.$$

These inequalities prove (3.3) and (3.4) for $N_0 = 0$. As observed above, the case of $N_0 \neq 0$ is then obtained by translation of the system trajectory. $\square$

Proof of Proposition 3.1. Let $\mathbf{u} \in \ell_{\mathcal{U}}^2(\mathbb{Z}_+)$ and run the system Σ with input sequence $\mathbf{u}$ and initial condition $\mathbf{x}(0) = 0$. From Lemma 3.2, with $N_0 = 0$, we obtain that for each $N \in \mathbb{Z}_+$ we have

$$\sum_{n=0}^{N} \|\mathbf{y}(n)\|_{\mathcal{Y}}^2 \le \sum_{n=0}^{N} \|\mathbf{u}(n)\|_{\mathcal{U}}^2 \quad \text{for all} \quad N \in \mathbb{Z}_+.$$

Letting $N \to \infty$, we conclude that $\mathbf{u} \in \ell_{\mathcal{U}}^2(\mathbb{Z}_+)$ implies that the output sequence $\mathbf{y}$ is in $\ell_{\mathcal{Y}}^2(\mathbb{Z}_+)$ with $\|\mathbf{y}\|_{\ell_{\mathcal{Y}}^2(\mathbb{Z}_+)}^2 \le \|\mathbf{u}\|_{\ell_{\mathcal{U}}^2(\mathbb{Z}_+)}^2$.

Write $\widehat{u}$ and $\widehat{y}$ for the Z-transforms of $\mathbf{u}$ and $\mathbf{y}$, respectively, i.e., denote $\widehat{u}(z) = \sum_{n=0}^{\infty} \mathbf{u}(n) z^n$ and $\widehat{y}(z) = \sum_{n=0}^{\infty} \mathbf{y}(n) z^n$. Since we have imposed zero-initial condition on the state, it now follows that $\widehat{y}(z) = F_\Sigma(z)\widehat{u}(z)$ in a neighborhood of 0. Since $\mathbf{u}$ was chosen arbitrarily in $\ell^2_{\mathcal{U}}(\mathbb{Z}_+)$, we see that $\widehat{u}$ is an arbitrary element of $H^2_{\mathcal{U}}(\mathbb{D})$. Thus, the multiplication operator $M_{F_\Sigma}\colon \widehat{u} \mapsto F_\Sigma \cdot \widehat{u}$ maps $H^2_{\mathcal{U}}(\mathbb{D})$ into $H^2_{\mathcal{Y}}(\mathbb{D})$. In particular, taking $\widehat{u} \in H^2_{\mathcal{U}}(\mathbb{D})$ constant, it follows that F_Σ has an analytic continuation to $\mathbb{D}$. Furthermore, the inequality

$$\|F_\Sigma \widehat{u}\|_{H^2_{\mathcal{Y}}(\mathbb{D})} = \|\widehat{y}\|_{H^2_{\mathcal{Y}}(\mathbb{D})} = \|\mathbf{y}\|^2_{\ell^2_{\mathbb{Z}_+}(\mathcal{Y})} \leq \|\mathbf{u}\|^2_{\ell^2_{\mathbb{Z}_+}(\mathcal{U})} = \|\widehat{u}\|^2_{H^2_{\mathcal{U}}(\mathbb{D})},$$

implies that the operator norm of the multiplication operator M_{F_Σ} from $H^2_{\mathcal{U}}(\mathbb{D})$ to $H^2_{\mathcal{Y}}(\mathbb{D})$ is at most 1. It is well known that the operator norm of M_{F_Σ} is the same as the supremum norm $\|F_\Sigma\|_\infty = \sup_{z\in\mathbb{D}} \|F_\Sigma(z)\|$. Hence we obtain that the analytic continuation of F_Σ is in the Schur class $\mathcal{S}(\mathcal{U}, \mathcal{Y})$. □

We shall see below (see Proposition 4.2) that conversely, if the transfer function F_Σ admits an analytic continuation to a Schur-class function, then a storage function for Σ exists.

Quadratic storage functions. The class of storage functions associated with solutions to the generalized KYP-inequality (1.12)–(1.13) are the so-called *quadratic storage functions* described next. We shall say that a storage function S is *quadratic* in case there is a positive-semidefinite operator H on the state space $\mathcal{X}$ so that S has the form

$$S(x) = S_H(x) = \begin{cases} \|H^{\frac{1}{2}}x\|^2 & \text{for } x \in \mathcal{D}(H^{\frac{1}{2}}), \\ +\infty & \text{otherwise.} \end{cases} \tag{3.6}$$

If in addition to F_Σ having an analytic continuation to a Schur-class function it is assumed that Σ is minimal, it can in fact be shown (see Theorem 4.9 below) that quadratic storage functions for Σ exist; for the finite-dimensional case see [29].

Proposition 3.3. *Suppose that the function $S\colon \mathcal{X} \to [0,\infty]$ has the form* (3.6) *for a (possibly) unbounded positive-semidefinite operator H on $\mathcal{X}$. Then S_H is a storage function for Σ if and only if H is a positive-semidefinite solution of the generalized KYP-inequality* (1.12)–(1.13). *Moreover, S is* nondegenerate *in the sense that $S_H(x) > 0$ for all nonzero x in $\mathcal{X}$ if and only if H is positive-definite.*

Proof. Suppose that H solves (1.12)–(1.13). It is clear that $S(0) = \|H^{\frac{1}{2}}0\|^2 = 0$, so in order to conclude that S is a storage function it remains to verify the dissipation inequality (3.1). Let $(\mathbf{u}(n), \mathbf{x}(n), \mathbf{y}(n))_{n\geq N_0}$ be a system trajectory with state initialization $\mathbf{x}(n_0) = x_0$ for some $x_0 \in \mathcal{X}$ and $N_0 \in \mathbb{Z}$. Fix $n \geq N_0$. If $\mathbf{x}(n) \notin \mathcal{D}(H^{\frac{1}{2}})$, then $S_H(\mathbf{x}(n)) = \infty$ and the dissipation inequality (3.1) is automatically satisfied. If $\mathbf{x}(n) \in \mathcal{D}(H^{\frac{1}{2}})$, then (1.12) implies that $\mathbf{x}(n+1) = A\mathbf{x}(n) + B\mathbf{u}(n) \in \mathcal{D}(H^{\frac{1}{2}})$. Thus $S_H(\mathbf{x}(n+1)) < \infty$. Replacing x by $\mathbf{x}(n)$ and u by $\mathbf{u}(n)$ in (1.13) and applying (1.1) we obtain that

$$\left\| \begin{bmatrix} H^{\frac{1}{2}} & 0 \\ 0 & I_{\mathcal{U}} \end{bmatrix} \begin{bmatrix} \mathbf{x}(n) \\ \mathbf{u}(n) \end{bmatrix} \right\|^2 - \left\| \begin{bmatrix} H^{\frac{1}{2}} & 0 \\ 0 & I_{\mathcal{Y}} \end{bmatrix} \begin{bmatrix} \mathbf{x}(n+1) \\ \mathbf{y}(n) \end{bmatrix} \right\|^2 \geq 0.$$

This can be rephrased in terms of S_H as

$$S_H(\mathbf{x}(n)) + \|\mathbf{u}(n)\|^2 - S_H(\mathbf{x}(n+1)) - \|\mathbf{y}(n)\|^2 \geq 0,$$

so that (3.1) appears after adding $S_H(\mathbf{x}(n+1))$ on both sides.

Conversely, suppose that S_H is a storage function. Take $x \in \mathcal{X}$ and $u \in \mathcal{U}$ arbitrarily. Let $(\mathbf{u}(n), \mathbf{x}(n), \mathbf{y}(n))_{n\geq 0}$ be any system trajectory with initialization $\mathbf{x}(0) = x$ and with $\mathbf{u}(0) = u$. Then the dissipation inequality (3.1) with $n = 0$ gives us

$$S_H(Ax + Bu) \leq S_H(x) + \|u\|^2 - \|y\|^2, \quad \text{with} \quad y = Cx + Du. \tag{3.7}$$

In particular, $S_H(x) < \infty$ implies that $S_H(Ax + Bu) < \infty$ (equivalently, $x \in \mathcal{D}(H^{\frac{1}{2}})$ implies $Ax + Bu \in \mathcal{D}(H^{\frac{1}{2}})$). Specifying $u = 0$ shows that $A\mathcal{D}(H^{\frac{1}{2}}) \subset \mathcal{D}(H^{\frac{1}{2}})$ and specifying $x = 0$ shows $B\mathcal{U} \subset \mathcal{D}(H^{\frac{1}{2}})$. Thus (1.12) holds. Bringing $\|y\|^2$ in (3.7) to the other side and writing out S_H gives

$$\|H^{\frac{1}{2}}(Ax + Bu)\|^2 + \|Cx + Du\|^2 \leq \|H^{\frac{1}{2}}x\|^2 + \|u\|^2,$$

which provides (1.13). □

We say that a function $S \colon \mathcal{X} \to \mathbb{R}_+ = [0, \infty)$ is a *strict storage function* for the system Σ in (1.1) if the strict dissipation inequality (1.8) holds, i.e., if there exists a $\delta > 0$ so that

$$S(\mathbf{x}(n+1)) - S(\mathbf{x}(n)) + \delta\|x(n)\|^2 \leq (1-\delta)\|\mathbf{u}(n)\|^2 - \|\mathbf{y}(n)\|^2 \quad (n \geq N_0) \tag{3.8}$$

holds for all system trajectories $\{\mathbf{u}(n), \mathbf{x}(n), \mathbf{y}(n)\}_{n\geq N_0}$, initiated at some $N_0 \in \mathbb{Z}$. Note that strict storage functions are not allowed to attain $+\infty$ as a value. The significance of the existence of a strict storage function for a system Σ is that it guarantees that the transfer function F_Σ has analytic continuation to a H^∞-function with H^∞-norm strictly less than 1 as well as a coercivity condition on S, i.e., we have the following strict version of Proposition 3.1.

Proposition 3.4. *Suppose that the system Σ in* (1.1) *has a strict storage function S. Then*

(1) *the transfer function F_Σ has analytic continuation to a function in H^∞ on the unit disk $\mathbb{D}$ with H^∞-norm strictly less than 1, and*
(2) *S satisfies a coercivity condition, i.e., there is a $\delta > 0$ so that*

$$S(x) \geq \delta\|x\|^2 \quad (x \in \mathcal{X}). \tag{3.9}$$

Proof. Assume that $S \colon \mathcal{X} \to [0, \infty)$ is a strict storage function for Σ. Then for each system trajectory $(\mathbf{u}(n), \mathbf{x}(n), \mathbf{y}(n)))_{n\geq 0}$ with initialization $\mathbf{x}(0) = 0$, the strict dissipation inequality (3.8) gives that there is a $\delta > 0$ so that for $n \geq 0$ we have

$$\begin{aligned} S(\mathbf{x}(n+1)) - S(\mathbf{x}(n)) &\leq -\delta\|x\|^2 + (1-\delta)\|\mathbf{u}(n)\|^2 - \|\mathbf{y}(n)\|^2 \\ &\leq (1-\delta)\|\mathbf{u}(n)\|^2 - \|\mathbf{y}(n)\|^2. \end{aligned}$$

Summing up over $n = 0, 1, 2, \ldots, N$ for some $N \in \mathbb{N}$ for a system trajectory $(\mathbf{u}(n), \mathbf{x}(n), \mathbf{y}(n))_{n\geq 0}$ subject to initialization $\mathbf{x}(0) = 0$ then gives

$$0 \leq S(\mathbf{x}(N+1)) = S(\mathbf{x}(N+1)) - S(\mathbf{x}(0)) \leq (1-\delta)\sum_{n=0}^{N} \|\mathbf{u}(n)\|^2 - \sum_{n=0}^{N} \|\mathbf{y}(n)\|^2.$$

By restricting to input sequences $\mathbf{u} \in \ell^2_{\mathcal{U}}(\mathbb{Z}_+)$, it follows that the corresponding output sequences satisfy $\mathbf{y} \in \ell^2_{\mathcal{Y}}(\mathbb{Z}_+)$ and $\|\mathbf{y}\|^2_{\ell^2_{\mathcal{U}}(\mathbb{Z}_+)} \leq (1-\delta)\|\mathbf{u}\|^2_{\ell^2_{\mathcal{Y}}(\mathbb{Z}_+)}$. Taking Z-transform and using the Plancherel theorem then gives

$$\|M_{F_\Sigma}\widehat{\mathbf{u}}\|^2_{H^2_{\mathcal{Y}}(\mathbb{D})} = \|\widehat{\mathbf{y}}\|^2_{H^2_{\mathcal{Y}}(\mathbb{D})} \leq (1-\delta)\|\widehat{\mathbf{u}}\|^2_{H^2_{\mathcal{U}}(\mathbb{D})}.$$

Thus $\|M_{F_\Sigma}\| \leq \sqrt{1-\delta} < 1$. This implies F_Σ has analytic continuation to an $\mathcal{L}(\mathcal{U}, \mathcal{Y})$-valued H^∞ function with H^∞-norm at most $\|M_{F_\sigma}\| \leq \sqrt{1-\delta} < 1$.

To this point we have not made use of the presence of the term $\delta\|x(n)\|^2$ in the strict dissipation inequality (3.8). We now show how the presence of this term leads to the validity of the coercivity condition (3.9) on S. Let x_0 be any state in $\mathcal{X}$ and let $(\mathbf{u}(n), \mathbf{x}(n), \mathbf{y}(n))_{n\geq 0}$ be any system trajectory with initialization $\mathbf{x}(0) = x_0$ and $\mathbf{u}(0) = 0$. Then the strict dissipation inequality (3.8) with $n = 0$ gives us

$$\delta\|x_0\|^2 = \delta\|\mathbf{x}(0)\|^2 \leq S(\mathbf{x}(1)) + \delta\|\mathbf{x}(0)\|^2 + \|\mathbf{y}(0)\|^2 \leq S(\mathbf{x}(0)) = S(x_0),$$

i.e., $S(x_0) \geq \delta\|x_0\|^2$ for each $x_0 \in \mathcal{X}$, verifying the validity of (3.9). □

The following result classifies which quadratic storage functions S_H are strict storage functions.

Proposition 3.5. *Suppose that $S = S_H$ is a quadratic storage function for the system Σ in (1.1). Then S_H is a strict storage function for Σ if and only if H is a bounded positive-semidefinite solution of the strict KYP-inequality (1.9). Any such solution is in fact strictly positive-definite.*

Proof. Suppose that S_H is a strict storage function for Σ. Then by definition $S_H(x) < \infty$ for all $x \in \mathcal{X}$. Hence $\mathcal{D}(H) = \mathcal{X}$. By the Closed Graph Theorem, it follows that H is bounded. As a consequence of Proposition 3.4, S_H is coercive and hence H is strictly positive-definite. The strict dissipation inequality (3.8) expressed in terms of H and the system matrix $\left[\begin{smallmatrix} A & B \\ C & D \end{smallmatrix}\right]$ becomes

$$\|H^{\frac{1}{2}}(Ax + Bu)\|^2 - \|H^{\frac{1}{2}}x\|^2 + \delta\|x\|^2 \leq (1-\delta)\|u\|^2 - \|Cx + Du\|^2$$

for all $x \in \mathcal{X}$ and $u \in \mathcal{U}$. This can be expressed more succinctly as

$$\left\langle \begin{bmatrix} H & 0 \\ 0 & I \end{bmatrix} \begin{bmatrix} A & B \\ C & D \end{bmatrix} \begin{bmatrix} x \\ u \end{bmatrix}, \begin{bmatrix} A & B \\ C & D \end{bmatrix} \begin{bmatrix} x \\ u \end{bmatrix} \right\rangle - \left\langle \begin{bmatrix} H & 0 \\ 0 & I \end{bmatrix} \begin{bmatrix} x \\ u \end{bmatrix}, \begin{bmatrix} x \\ u \end{bmatrix} \right\rangle$$
$$\leq -\delta \left\langle \begin{bmatrix} x \\ u \end{bmatrix}, \begin{bmatrix} x \\ u \end{bmatrix} \right\rangle$$

for all $x \in \mathcal{X}$ and $u \in \mathcal{U}$, for some $\delta > 0$. This is just the spatial version of (1.9), so H is a strictly positive-definite solution of the strict KYP-inequality (1.9). By reversing the steps one sees that $H \succeq 0$ being a solution of the strict KYP-inequality (1.9) implies that S_H is a strict storage function.

As a consequence of Proposition 3.4 we see that then S_H satisfies a coercivity condition (3.9), so necessarily H is strictly positive-definite. □

4. The available storage and required supply

In Proposition 3.1 we showed that the existence of a storage function (which is allowed to attain the value $+\infty$) for a discrete-time linear system Σ implies that the transfer function F_Σ associated with Σ is equal to a Schur-class function on a neighborhood of 0. In this section we investigate the converse direction. Specifically, we give explicit variational formulas for three storage functions, referred to as the available storage function S_a (defined in (4.1)) the required supply function S_r (defined in (4.2)) and the "regularized" version $\underline{S}_r$ of the required supply (defined in (4.18)). Let $\boldsymbol{\mathcal{U}}$ denote the space of all functions $n \mapsto u(n)$ from the integers $\mathbb{Z}$ into the input space $\mathcal{U}$. Then S_a is given by

$$S_a(x_0) = \sup_{\mathbf{u}\in\boldsymbol{\mathcal{U}},\, n_1\geq 0} \sum_{n=0}^{n_1} (\|\mathbf{y}(n)\|^2 - \|\mathbf{u}(n)\|^2) \tag{4.1}$$

with the supremum taken over all system trajectories $(\mathbf{u}(n), \mathbf{x}(n), \mathbf{y}(n))_{n\geq 0}$ with initialization $\mathbf{x}(0) = x_0$, while S_r is given by

$$S_r(x_0) = \inf_{\mathbf{u}\in\boldsymbol{\mathcal{U}},\, n_{-1}<0} \sum_{n=n_{-1}}^{-1} (\|\mathbf{u}(n)\|^2 - \|\mathbf{y}(n)\|^2) \tag{4.2}$$

with the infimum taken over all system trajectories $(\mathbf{u}(n), \mathbf{x}(n), \mathbf{y}(n))_{n\geq n_{-1}}$ subject to the initialization condition $\mathbf{x}(n_{-1}) = 0$ and the condition $\mathbf{x}(0) = x_0$.

The proof that S_a and S_r are storage functions whenever F_Σ is in the Schur class requires the following preparatory lemma. We shall use the following notation. For an arbitrary Hilbert space $\mathcal{Z}$, write P_+ and P_- for the orthogonal projections onto $\ell^2_{\mathcal{Z}}(\mathbb{Z}_+)$ and $\ell^2_{\mathcal{Z}}(\mathbb{Z}_-)$, respectively, acting on $\ell^2_{\mathcal{Z}}(\mathbb{Z})$. For integers $m \leq n$, we write $P_{[m,n]}$ for the orthogonal projection on the subspace of sequences in $\ell^2_{\mathcal{Z}}(\mathbb{Z})$ with support on the coordinate positions $m, m+1, \dots, n$.

Lemma 4.1. *Let Σ be as in* (1.1) *and suppose that its transfer function F_Σ is in the Schur class. Then, for each system trajectory $(\mathbf{u}(n), \mathbf{x}(n), \mathbf{y}(n))_{n\geq 0}$ with initialization $\mathbf{x}(0) = 0$, the inequality*

$$\sum_{n=0}^{N} \|\mathbf{y}(n)\|^2 \leq \sum_{n=0}^{N} \|\mathbf{u}(n)\|^2 \tag{4.3}$$

holds for all $N \in \mathbb{Z}_+$.

Proof. As we have already observed, the fact that F_Σ is in the Schur class $\mathcal{S}(\mathcal{U}, \mathcal{Y})$ implies that the multiplication operator M_{F_Σ} (2.10) has norm at most 1 as an operator from $L^2_{\mathcal{U}}(\mathbb{T})$ to $L^2_{\mathcal{Y}}(\mathbb{T})$. If we apply the inverse Z-transform to the full operator M_{F_Σ}, not just to the compression $\mathbb{H}_{F_\Sigma}$ as was done to arrive at the Hankel operator $\mathfrak{H}_{F_\Sigma}$ in (2.12), we get the *Laurent operator*

$$\mathfrak{L}_{F_\Sigma} = \left[\begin{array}{ccc|ccc} \ddots & \ddots & \ddots & \ddots & \ddots & \ddots \\ \ddots & F_0 & 0 & 0 & 0 & \ddots \\ \ddots & F_1 & F_0 & 0 & 0 & \ddots \\ \hline \ddots & F_2 & F_1 & F_0 & 0 & \ddots \\ \ddots & F_3 & F_2 & F_1 & F_0 & \ddots \\ \ddots & \ddots & \ddots & \ddots & \ddots & \ddots \end{array}\right] : \ell^2_{\mathcal{U}}(\mathbb{Z}) \to \ell^2_{\mathcal{Y}}(\mathbb{Z}), \tag{4.4}$$

where $F_0, F_1, F_2, \dots$ are the Taylor coefficients of F_Σ:

$$F_n = \begin{cases} D & \text{if } n = 0, \\ CA^{n-1}B & \text{if } n \geq 1. \end{cases} \tag{4.5}$$

It is convenient to write $\mathfrak{L}_{F_\Sigma}$ as a 2×2-block matrix with respect to the decomposition $\ell^2_{\mathcal{U}}(\mathbb{Z}) = \left[\begin{smallmatrix} \ell^2_{\mathcal{U}}(\mathbb{Z}_-) \\ \ell^2_{\mathcal{U}}(\mathbb{Z}_+) \end{smallmatrix}\right]$ of the domain and the analogous decomposition $\ell^2_{\mathcal{Y}}(\mathbb{Z}) = \left[\begin{smallmatrix} \ell^2_{\mathcal{Y}}(\mathbb{Z}_-) \\ \ell^2_{\mathcal{Y}}(\mathbb{Z}_+) \end{smallmatrix}\right]$ of the range; the result is

$$\mathfrak{L}_{F_\Sigma} = \left[\begin{array}{c|c} \widetilde{\mathfrak{T}}_{F_\Sigma} & 0 \\ \hline \mathfrak{H}_{F_\Sigma} & \mathfrak{T}_{F_\Sigma} \end{array}\right] : \begin{bmatrix} \ell^2_{\mathcal{U}}(\mathbb{Z}_-) \\ \ell^2_{\mathcal{U}}(\mathbb{Z}_+) \end{bmatrix} \to \begin{bmatrix} \ell^2_{\mathcal{Y}}(\mathbb{Z}_-) \\ \ell^2_{\mathcal{Y}}(\mathbb{Z}_+) \end{bmatrix}. \tag{4.6}$$

Here $\mathfrak{H}_{F_\Sigma} : \ell^2_-(\mathcal{U}) \to \ell^2_+(\mathcal{Y})$ denotes the Hankel operator associated with F_Σ already introduced in (2.12), $\mathfrak{T}_{F_\Sigma} : \ell^2_+(\mathcal{U}) \to \ell^2_+(\mathcal{Y})$ the Toeplitz operator associated with F_Σ, and $\widetilde{\mathfrak{T}}_{F_\Sigma}$ the Toeplitz operator acting from $\ell^2_{\mathcal{U}}(\mathbb{Z}_-)$ to $\ell^2_{\mathcal{Y}}(\mathbb{Z}_-)$ associated with F_Σ. From the assumption that F_Σ is in the Schur class $\mathcal{S}(\mathcal{U}, \mathcal{Y})$, it follows that M_{F_Σ} is contractive, and hence also each of the operators $\widetilde{\mathfrak{T}}_{F_\Sigma}$, $\mathfrak{H}_{F_\Sigma}$, and $\mathfrak{T}_{F_\Sigma}$ is contractive. From the lower triangular form of $\mathfrak{T}_{F_\Sigma}$ we see in addition that $\mathfrak{T}_{F_\Sigma}$ has the *causality property*:

$$P_{[0,N]}\mathfrak{T}_{F_\Sigma} = P_{[0,N]}\mathfrak{T}_{F_\Sigma}P_{[0,N]} \quad (N \geq 0). \tag{4.7}$$

Now suppose that $(\mathbf{u}(n), \mathbf{x}(n), \mathbf{y}(n))_{n\geq 0}$ is a system trajectory on $\mathbb{Z}_+$ with initialization $\mathbf{x}(0) = 0$. In this case the infinite matrix identity $\mathbf{y} = \mathfrak{T}_{F_\Sigma}\mathbf{u}$ holds formally. For $N \in \mathbb{Z}_+$ we have $P_{[0,N]}\mathbf{u} \in \ell^2_{\mathcal{U}}(\mathbb{Z}_+)$, and by the causality property,

$$P_{[0,N]}\mathfrak{T}_{F_\Sigma}P_{[0,N]}\mathbf{u} = P_{[0,N]}\mathfrak{T}_{F_\Sigma}\mathbf{u} = P_{[0,N]}\mathbf{y}.$$

Since $\mathfrak{T}_{F_\Sigma}$ is contractive, so is $P_{[0,N]}\mathfrak{T}_{F_\Sigma}P_{[0,N]}$ and thus the above identity shows that $\|P_{[0,N]}\mathbf{y}\| \leq \|P_{[0,N]}\mathbf{u}\|$, or, equivalently,

$$\sum_{n=0}^{N} \|\mathbf{y}(n)\|^2 \leq \sum_{n=0}^{N} \|\mathbf{u}(n)\|^2 \tag{4.8}$$

holds for each system trajectory $(\mathbf{u}(n), \mathbf{x}(n), \mathbf{y}(n))_{n\geq 0}$ with $\mathbf{x}(0) = 0$. □

The proof of the following result is an adaptation of the proofs of Theorems 1 and 2 for the continuous-time setting in [28].

Proposition 4.2. *Assume that the discrete-time linear system Σ has a transfer function F_Σ which has an analytic continuation to a function in the Schur class $\mathcal{S}(\mathcal{U},\mathcal{Y})$. Define S_a and S_r by (4.1) and (4.2). Then*

(1) *S_a is a storage function,*
(2) *S_r is a storage function, and*
(3) *for each storage function S for Σ we have*

$$S_a(x_0) \leq S(x_0) \leq S_r(x_0) \text{ for all } x_0 \in \mathcal{X}.$$

Proof. The proof consists of three parts, corresponding to the three assertions of the proposition.

(1) To see that $S_a(x_0) \geq 0$ for all $x_0 \in \mathcal{X}$, choose $\mathbf{x}(0) = x_0$ and $\mathbf{u}(n) = 0$ for $n \geq 0$ to generate a system trajectory $(\mathbf{u}(n), \mathbf{x}(n), \mathbf{y}(n))_{n\geq 0}$ such that $\sum_{n=0}^{n_1}(\|\mathbf{y}(n)\|^2 - \|\mathbf{u}(n)\|^2) = \sum_{n=0}^{n_1}\|\mathbf{y}(n)\|^2 \geq 0$ for all $n_1 \geq 0$. From the definition (4.1), we see that $S_a(x_0) \geq 0$.

By Lemma 4.1, each system trajectory $(\mathbf{u}(n), \mathbf{x}(n), \mathbf{y}(n))_{n\geq 0}$ with initialization $\mathbf{x}(0) = 0$ satisfies the inequality

$$\sum_{n=0}^{n_1} \|\mathbf{y}(n)\|^2_{\mathcal{Y}} \leq \sum_{n=0}^{n_1} \|\mathbf{u}(n)\|^2_{\mathcal{U}} \quad (n_1 \in \mathbb{Z}_+).$$

This observation leads to the conclusion that $S_a(0) \leq 0$. Hence $S_a(0) = 0$ and thus S_a satisfies the normalization (3.2).

Now let $\{\widetilde{\mathbf{u}}(n), \widetilde{\mathbf{x}}(n), \widetilde{\mathbf{y}}(n)\}_{n\geq N_0}$ be any system trajectory initiated at some $N_0 \in \mathbb{Z}$. We wish to show that this trajectory satisfies the dissipation inequality (3.1). It is convenient to rewrite this condition in the form

$$\|\widetilde{\mathbf{y}}(n)\|^2_{\mathcal{Y}} - \|\widetilde{\mathbf{u}}(n)\|^2_{\mathcal{U}} + S_a(\widetilde{\mathbf{x}}(n+1)) \leq S_a(\widetilde{\mathbf{x}}(n)) \quad (n \in \mathbb{Z}).$$

By translation invariance of the system equations (1.1), without loss of generality we may take $n = 0$, so we need to show

$$\|\widetilde{\mathbf{y}}(0)\|^2_{\mathcal{Y}} - \|\widetilde{\mathbf{u}}(0)\|^2_{\mathcal{U}} + S_a(\widetilde{\mathbf{x}}(1)) \leq S_a(\widetilde{\mathbf{x}}(0)). \tag{4.9}$$

We rewrite the definition (4.1) for $S_a(\widetilde{\mathbf{x}}(1))$ in the form

$$S_a(\widetilde{\mathbf{x}}(1)) = \sup_{\mathbf{u}\in\boldsymbol{\mathcal{U}}, n_1\geq 0} \sum_{n=0}^{n_1} \left(\|\mathbf{y}(n)\|^2_{\mathcal{Y}} - \|\mathbf{u}(n)\|^2_{\mathcal{U}}\right),$$

where the system trajectory $(\mathbf{u}(n), \mathbf{x}(n), \mathbf{y}(n))_{n\geq 0}$ is subject to the initialization $\mathbf{x}(0) = \widetilde{\mathbf{x}}(1)$. Again making use of the translation invariance of the system equations, we may rewrite this in the form

$$S_a(\widetilde{\mathbf{x}}(1)) = \sup_{\mathbf{u}\in\boldsymbol{\mathcal{U}}, n_1\geq 1} \sum_{n=1}^{n_1} \left(\|\mathbf{y}(n)\|^2_{\mathcal{Y}} - \|\mathbf{u}(n)\|^2_{\mathcal{U}}\right),$$

where $(\mathbf{u}(n), \mathbf{x}(n), \mathbf{y}(n))_{n\geq 0}$ is a system trajectory with initialization now given by $\mathbf{x}(1) = \widetilde{\mathbf{x}}(1)$. Substituting this expression for $S(\widetilde{\mathbf{x}}(1))$, the left-hand side of (4.9) reads

$$\|\widetilde{\mathbf{y}}(0)\|_{\mathcal{Y}}^2 - \|\widetilde{\mathbf{u}}(0)\|_{\mathcal{U}}^2 + \sup_{\mathbf{u}\in\boldsymbol{\mathcal{U}}, n_1\geq 1} \sum_{n=1}^{n_1} \left(\|\mathbf{y}(n)\|_{\mathcal{Y}}^2 - \|\mathbf{u}(n)\|_{\mathcal{U}}^2\right).$$

This quantity indeed is bounded above by

$$S_a(\widetilde{\mathbf{x}}(0)) = \sup_{\mathbf{u}\in\boldsymbol{\mathcal{U}}, n_1\geq 0} \sum_{n=0}^{n_1} \left(\|\mathbf{y}(n)\|_{\mathcal{Y}}^2 - \|\mathbf{u}(n)\|_{\mathcal{U}}^2\right),$$

with $(\mathbf{u}(n), \mathbf{x}(n), \mathbf{y}(n))_{n\geq 0}$ a system trajectory subject to the initialization $\widetilde{\mathbf{x}}(0) = \mathbf{x}(0)$. Hence the inequality (4.9) follows as required, and S_a is a storage function for Σ.

(2) Let $(\mathbf{u}(n), \mathbf{x}(n), \mathbf{y}(n))_{n\geq n_{-1}}$ be a system trajectory with zero-initialization of the state at $n_{-1} < 0$, subject also to $\mathbf{x}(0) = x_0$. Applying the result of Lemma 4.1 to this system trajectory, using the translation invariance property of Σ to get a sum in (4.3) starting at n_{-1} and ending at 0, it follows that $S_r(x_0) \geq 0$ for all x_0 in $\operatorname{Rea}(A|B)$. In case $x_0 \notin \operatorname{Rea}(A|B)$, i.e., x_0 is not reachable in finitely many steps via some input signal $\mathbf{u}(n)$ $(n_{-1} \leq n < 0)$ with $\mathbf{x}(n_{-1}) = 0$, then the definition of S_r in (4.2) gives us $S_r(x) = +\infty \geq 0$. By choosing $n_{-1} = -1$ with $\mathbf{u}(-1) = 0$, we see that $S_r(0) \leq 0$. Since $S_r(x_0) \geq 0$ for each $x_0 \in \mathcal{X}$, it follows that S_r also satisfies the normalization (3.2).

An argument similar to that used in part (1) of the proof shows that S_r satisfies (3.1). Indeed, note that it suffices to show that for each system trajectory $\{\widetilde{\mathbf{u}}(n), \widetilde{\mathbf{x}}(n), \widetilde{\mathbf{y}}(n)\}_{n\geq 0}$ we have

$$\begin{aligned} S_r(\widetilde{\mathbf{x}}(1)) &\leq \|\widetilde{\mathbf{u}}(0)\|_{\mathcal{U}}^2 - \|\widetilde{\mathbf{y}}(0)\|_{\mathcal{Y}}^2 + S_r(\widetilde{\mathbf{x}}(0)) \qquad (4.10)\\ &= \inf_{\mathbf{u}\in\boldsymbol{\mathcal{U}},\, n_{-1}<0} \left\{ \|\widetilde{\mathbf{u}}(0)\|_{\mathcal{U}}^2 - \|\widetilde{\mathbf{y}}(0)\|_{\mathcal{Y}}^2 + \sum_{n=n_{-1}}^{-1} \left(\|\mathbf{u}(n)\|_{\mathcal{U}}^2 - \|\mathbf{y}(n)\|_{\mathcal{Y}}^2\right) \right\} \end{aligned}$$

where $(\mathbf{u}(n), \mathbf{x}(n), \mathbf{y}(n))_{n\geq n_{-1}}$ is a system trajectory subject to the initial condition $\mathbf{x}(n_{-1}) = 0$ and the terminal condition $\mathbf{x}(0) = \widetilde{x}(0)$. Rewrite the definition of $S_r(\widetilde{\mathbf{x}}(1))$ as

$$S_r(\widetilde{\mathbf{x}}(1)) = \inf_{\mathbf{u}\in\boldsymbol{\mathcal{U}},\, n_{-1}<1} \sum_{n=n_{-1}}^{0} \left(\|\mathbf{u}(n)\|_{\mathcal{U}}^2 - \|\mathbf{y}(n)\|_{\mathcal{Y}}^2\right),$$

with the system trajectory $(\mathbf{u}(n), \mathbf{x}(n), \mathbf{y}(n))_{n\in\mathbb{Z}}$ subject to the initial and terminal conditions $\mathbf{x}(n_{-1}) = 0$ and $\mathbf{x}(1) = \widetilde{\mathbf{x}}(1)$. Now recognize the argument of the inf in the right-hand side of (4.10) as part of the competition in the infimum defining $S_r(\widetilde{\mathbf{x}}(1))$ to deduce the inequality (4.10).

(3) Let S be any storage function for Σ and $(\mathbf{u}(n), \mathbf{x}(n), \mathbf{y}(n))_{n\geq 0}$ any system trajectory with initialization $\mathbf{x}(0) = x_0$. Iteration of the dissipation inequality (3.1) for S along the system trajectory $(\mathbf{u}(n), \mathbf{x}(n), \mathbf{y}(n))_{n\geq 0}$ as in

the proof of Lemma 3.2 yields

$$0 \le S(n_1+1) \le S(x_0) + \sum_{n=0}^{n_1} (\|\mathbf{u}(n)\|^2 - \|\mathbf{y}(n)\|^2)$$

or

$$\sum_{n=0}^{n_1} (\|\mathbf{y}(n)\|^2 - \|\mathbf{u}(n)\|^2) \le S(x_0).$$

Taking the supremum in the left-hand side of the above inequality over all such system trajectories $(\mathbf{u}(n), \mathbf{x}(n), \mathbf{y}(n))_{n\ge 0}$ and over all $n_1 \ge 0$ yields $S_a(x_0) \le S(x_0)$, and the first part of (3) is verified.

Next let $x_0 \in \mathcal{X}$ be arbitrary. If $(\mathbf{u}(n), \mathbf{x}(n), \mathbf{y}(n))_{n\ge n_{-1}}$ is any system trajectory with state-initialization $x(n_{-1}) = 0$ and $x(0) = x_0$, applying Lemma 3.2 with $N_0 = n_{-1}$ and $N = -1 - n_{-1}$ gives us that

$$S(x_0) \le \sum_{n=n_{-1}}^{-1} \left(\|\mathbf{u}(n)\|^2_{\mathcal{U}} - \|\mathbf{y}(n)\|^2_{\mathcal{Y}}\right). \tag{4.11}$$

Taking the infimum of the right-hand side over all such system trajectories gives us $S(x_0) \le S_r(x_0)$. Here we implicitly assumed that the state $x_0 \in \mathcal{X}$ is reachable. If x_0 is not reachable, there are no such system trajectories, and taking the infimum over an empty set leads to $S_r(x_0) = \infty$, in which case $S(x_0) \le S_r(x_0)$ is also valid. Hence $S(x_0) \le S_r(x_0)$ holds for all possible $x_0 \in \mathcal{X}$. This completes the verification of the second part of (3). □

Combining Proposition 4.2 with Proposition 3.1 leads to the following.

Corollary 4.3. *A discrete-time linear system Σ in (1.1) has a transfer function F_Σ with an analytic continuation in the Schur class if and only if Σ has a storage function S.*

Proof. The sufficiency is Proposition 3.1. For the necessity direction, by Proposition 4.2 we may choose S equal to either S_a or S_r. □

We next impose a minimality assumption on Σ and in addition assume that F_Σ has an analytic continuation in the Schur class $\mathcal{S}(\mathcal{U}, \mathcal{Y})$, i.e., we make the following assumptions:

$$\begin{cases} \Sigma \textit{ is minimal, i.e., } (C, A) \textit{ is observable and } (A, B) \textit{ is controllable,} \\ \textit{and } F_\Sigma \textit{ has an analytic continuation to a function in } \mathcal{S}(\mathcal{U}, \mathcal{Y}). \end{cases} \tag{4.12}$$

Our next goal is to understand storage functions from a more operator-theoretic point of view. We first need some preliminaries.

Recall the Laurent operator $\mathfrak{L}_{F_\Sigma}$ in (4.4). From the 2×2-block form for $\mathfrak{L}_{F_\Sigma}$ in (4.6), we see that

$$\begin{aligned} I - \mathfrak{L}_{F_\Sigma}\mathfrak{L}_{F_\Sigma}^* &= \begin{bmatrix} D^2_{\widetilde{\mathfrak{T}}^*_{F_\Sigma}} & -\widetilde{\mathfrak{T}}_{F_\Sigma}\mathfrak{H}^*_{F_\Sigma} \\ -\mathfrak{H}_{F_\Sigma}\widetilde{\mathfrak{T}}^*_{F_\Sigma} & D^2_{\mathfrak{T}^*_{F_\Sigma}} - \mathfrak{H}_{F_\Sigma}\mathfrak{H}^*_{F_\Sigma} \end{bmatrix}; \\ I - \mathfrak{L}^*_{F_\Sigma}\mathfrak{L}_{F_\Sigma} &= \begin{bmatrix} D^2_{\widetilde{\mathfrak{T}}_{F_\Sigma}} - \mathfrak{H}^*_{F_\Sigma}\mathfrak{H}_{F_\Sigma} & -\mathfrak{H}^*_{F_\Sigma}\mathfrak{T}_{F_\Sigma} \\ -\mathfrak{T}^*_{F_\Sigma}\mathfrak{H}_{F_\Sigma} & D^2_{\mathfrak{T}_{F_\Sigma}} \end{bmatrix}. \end{aligned} \tag{4.13}$$

where in general the notation D_X for the defect operator $D_X = (I - X^*X)^{\frac{1}{2}}$ of a contraction operator X is used. Since F_Σ is assumed to be a Schur-class function, $\mathfrak{T}_{F_\Sigma}$ and $\widetilde{\mathfrak{T}}_{F_\Sigma}$ are contractions, and hence $D_{\mathfrak{T}_{F_\Sigma}}$, $D_{\mathfrak{T}^*_{F_\Sigma}}$, $D_{\widetilde{\mathfrak{T}}_{F_\Sigma}}$ and $D_{\widetilde{\mathfrak{T}}^*_{F_\Sigma}}$ are well defined.

Lemma 4.4. *Let the discrete-time linear system Σ in* (1.1) *satisfy the assumptions* (4.12). *The available storage function S_a and required supply function S_r can then be written in operator form as*

$$S_a(x_0) = \sup_{\mathbf{u}\in\ell^2_{\mathcal{U}}(\mathbb{Z}_+)} \|\mathbf{W}_o x_0 + \mathfrak{T}_{F_\Sigma}\mathbf{u}\|^2_{\ell^2_{\mathcal{Y}}(\mathbb{Z}_+)} - \|\mathbf{u}\|^2_{\ell^2_{\mathcal{U}}(\mathbb{Z}_+)} \quad (x_0 \in \mathcal{D}(\mathbf{W}_o)), \tag{4.14}$$

$$S_r(x_0) = \inf_{\mathbf{u}\in\ell_{\text{fin},\mathcal{U}}(\mathbb{Z}_-):\, x_0=\mathbf{W}_c\mathbf{u}} \|D_{\widetilde{\mathfrak{T}}_{F_\Sigma}}\mathbf{u}\|^2 \quad (x_0 \in \mathcal{X}), \tag{4.15}$$

and $S_a(x_0) = +\infty$ for $x_0 \notin \mathcal{D}(\mathbf{W}_o)$. Here $\mathbf{W}_o$ and $\mathbf{W}_c$ are the observability and controllability operators defined via (2.1)–(2.4) *and $\ell_{\text{fin},\mathcal{U}}(\mathbb{Z}_-)$ is the linear manifold of finitely supported sequences in $\ell^2_{\mathcal{U}}(\mathbb{Z}_-)$. In particular, $S_r(x_0) < \infty$ if and only if $x_0 \in \operatorname{Rea}(A|B)$.*

Proof. We shall use the notation $P_\pm$ and $P_{[m,n]}$ as introduced in the discussion immediately preceding the statement of Lemma 4.1.

We start with S_a. For each system trajectory $(\mathbf{u}(n), \mathbf{x}(n), \mathbf{y}(n))_{n\geq 0}$ with initialization $\mathbf{x}(0) = x_0$ and with $\mathbf{u} \in \ell^2_{\mathcal{U}}(\mathbb{Z}_+)$ by linearity we have

$$\mathbf{y} = \mathbf{W}_o x_0 + \mathfrak{T}_{F_\Sigma}\mathbf{u}.$$

Now note that, for each system trajectory $(\mathbf{u}(n), \mathbf{x}(n), \mathbf{y}(n))_{n\geq 0}$ with initialization $\mathbf{x}(0) = x_0$ but with $\mathbf{u}$ not necessarily in $\ell^2_{\mathcal{U}}(\mathbb{Z}_+)$ and with $n_1 \geq 0$, by the causality property (4.7), as in the proof of Lemma 4.1 we see that we can replace $\mathbf{u}$ with $P_{[0,n_1]}\mathbf{u} \in \ell_{\text{fin},\mathcal{U}}(\mathbb{Z}_+) \subset \ell^2_{\mathcal{U}}(\mathbb{Z}_+)$ within the supremum in (4.1) without changing the value. Therefore, the value of S_a at x_0 can be rewritten in operator form as

$$S_a(x_0) = \sup_{\substack{\mathbf{u}\in\ell_{\text{fin},\mathcal{U}}(\mathbb{Z}_+) \\ n_1\geq 0}} \|P_{[0,n_1]}(\mathbf{W}_o x_0 + \mathfrak{T}_{F_\Sigma}\mathbf{u})\|^2_{\ell^2_{\mathcal{Y}}(\mathbb{Z}_+)} - \|P_{[0,n_1]}\mathbf{u}\|^2_{\ell^2_{\mathcal{U}}(\mathbb{Z}_+)} \tag{4.16}$$

where we use the notation $\ell_{\text{fin},\mathcal{U}}(\mathbb{Z}_+)$ for $\mathcal{U}$-valued sequences on $\mathbb{Z}_+$ of finite support.

If $x_0 \notin \mathcal{D}(\mathbf{W}_o)$ so that $\mathbf{W}_o x_0 \notin \ell^2_{\mathcal{Y}}(\mathbb{Z}_+)$, the above formulas are to be interpreted algebraically, and we may choose $\mathbf{u} = 0$ and take the limit as $n_1 \to \infty$ to see that $\mathbf{W}_o(x_0) = +\infty$.

Now assume $x_0 \in \mathcal{D}(\mathbf{W}_o)$. Fix $\mathbf{u} \in \ell_{\text{fin},\mathcal{U}}(\mathbb{Z}_+)$ and take the limit as $n_1 \to +\infty$ in the right-hand side of (4.16) to see that an equivalent expression for $S_a(x_0)$ is

$$S_a(x_0) = \sup_{\mathbf{u}\in\ell_{\text{fin},\mathcal{U}}(\mathbb{Z}_+)} \|\mathbf{W}_o x_0 + \mathfrak{T}_{F_\Sigma}\mathbf{u}\|^2 - \|\mathbf{u}\|^2.$$

Since $\ell_{\text{fin},\mathcal{U}}(\mathbb{Z}_+)$ is dense in $\ell^2_{\mathcal{U}}(\mathbb{Z}_+)$ and $\mathfrak{T}_{F_\Sigma}$ is a bounded operator, we see that another equivalent expression for $S_a(x_0)$ is (4.14). This completes the verification of (4.14).

We next look at S_r. Let $(\mathbf{u}(n), \mathbf{x}(n), \mathbf{y}(n))_{n\geq n_{-1}}$ be any system trajectory with initialization $x(n_{-1}) = 0$ for some $n_{-1} < 0$. Let us identify $\mathbf{u}$ with an element $\mathbf{u} \in \ell_{\text{fin},\mathcal{U}}(\mathbb{Z}_-)$ by ignoring the values of $\mathbf{u}$ on $\mathbb{Z}_+$ and defining $\mathbf{u}(n) = 0$ for $n < n_{-1}$. Then, as a consequence of item (1) in Proposition 2.4, the constraint $\mathbf{x}(0) = x_0$ in (4.2) can be written in operator form as $\mathbf{W}_c\mathbf{u} = x_0$. Furthermore, since $(\mathbf{u}(n), \mathbf{x}(n), \mathbf{y}(n))_{n\geq n_{-1}}$ is a system trajectory with zero state initialization at n_{-1}, it follows that

$$\mathbf{y}|_{\mathbb{Z}_-} = \widetilde{\mathfrak{T}}_{F_\Sigma}\mathbf{u}.$$

We conclude that a formula for S_r equivalent to (4.2) is

$$S_r(x_0) = \inf_{\mathbf{u}\in\ell^2_{\text{fin},\mathcal{U}}(\mathbb{Z}_-):\, \mathbf{W}_c\mathbf{u}=x_0} \|\mathbf{u}\|^2 - \|\widetilde{\mathfrak{T}}_{F_\Sigma}\mathbf{u}\|^2$$

which in turn has the more succinct formulation (4.15). If $x_0 \in \text{Rea}(A|B)$, then the infimum in (4.15) is taken over a nonempty set, so that $S_r(x_0) < \infty$. On the other hand, if $x_0 \notin \text{Rea}(A|B)$, then the infimum is taken over an empty set, so that $S_r(x_0) = \infty$. □

To compute storage functions more explicitly for the case where assumptions (4.12) are in place, it will be convenient to restrict to what we shall call *ℓ^2-regular storage functions* S, namely, storage functions S which assume finite values on $\operatorname{Im}\mathbf{W}_c$:

$$x_0 = \mathbf{W}_c\mathbf{u} \text{ where } \mathbf{u} \in \mathcal{D}(\mathbf{W}_c) \Rightarrow S(x_0) < \infty. \tag{4.17}$$

We shall see in the next result that S_a is ℓ^2-regular. However, unless if $\text{Rea}(A|B)$ is equal to the range of $\mathbf{W}_c$, the required supply S_r will not be ℓ^2-regular (by the last assertion of Lemma 4.4).

To remedy this situation, we introduce the following modification $\underline{S}_r$ of the required supply S_r, which we shall call the *ℓ^2-regularized required supply*:

$$\underline{S}_r(x_0) = \inf_{\mathbf{u}\in\mathcal{D}(\mathbf{W}_c):\, \mathbf{W}_c\mathbf{u}=x_0} \sum_{n=-\infty}^{-1} (\|\mathbf{u}(n)\|^2 - \|\mathbf{y}(n)\|^2) \tag{4.18}$$

where $\mathbf{u} \in \ell^2_{\mathcal{U}}(\mathbb{Z}_-)$ determines $\mathbf{y} \in \ell^2_{\mathcal{Y}}(\mathbb{Z}_-)$ via the system input/output map: $\mathbf{y} = \widetilde{\mathfrak{T}}_{F_\Sigma}\mathbf{u}$. Thus formula (4.18) can be written more succinctly in operator form as

$$\begin{aligned}\underline{S}_r(x_0) &= \inf_{\mathbf{u}\in\mathcal{D}(\mathbf{W}_c):\ \mathbf{W}_c\mathbf{u}=x_0} \|\mathbf{u}\|^2_{\ell^2_{\mathcal{U}}(\mathbb{Z}_-)} - \|\widetilde{\mathfrak{T}}_{F_\Sigma}\mathbf{u}\|^2_{\ell^2_{\mathcal{Y}}(\mathbb{Z}_-)} \\ &= \inf_{\mathbf{u}\in\mathcal{D}(\mathbf{W}_c):\ \mathbf{W}_c\mathbf{u}=x_0} \|D_{\widetilde{\mathfrak{T}}_{F_\Sigma}}\mathbf{u}\|^2_{\ell^2_{\mathcal{Y}}(\mathbb{Z}_-)} \text{ for } x_0\in\operatorname{Im}\mathbf{W}_c. \end{aligned} \tag{4.19}$$

It is clear that $\underline{S}_r(x_0) < \infty$ if and only if $x_0 \in \operatorname{Im}\mathbf{W}_c$. Since the objective in the infimum defining $\underline{S}_r$ in (4.19) is the same as the objective in the infimum defining S_r in (4.15) but the former infimum is taken over an a priori larger set, it follows directly that $S_r(x_0) \geq \underline{S}_r(x_0)$ for all $x_0 \in \mathcal{X}$, as can also be seen as a consequence of Proposition 4.2 once we show that $\underline{S}_r$ is a storage function for Σ. From either of the formulas we see that $0 \leq \underline{S}_r(x_0)$ and that $\underline{S}_r(x_0) < \infty$ exactly when x_0 is in the range of $\mathbf{W}_c$. Hence once we show that $\underline{S}_r$ is a storage function, it follows that $\underline{S}_r$ is an ℓ^2-regular storage function and is a candidate to be the largest such. However at this stage we have only partial results in this direction, as laid out in the next result.

Proposition 4.5. *Assume that Σ is a system satisfying the assumptions* (4.12) *and let the function $\underline{S}_r$:* $\operatorname{Im}\mathbf{W}_c \to \mathbb{R}_+$ *be given by* (4.19). *Then:*

(1) *S_a and $\underline{S}_r$ are ℓ^2-regular storage functions.*

(2) *$\underline{S}_r$ is "almost" the largest ℓ^2-regular storage function in the following sense: if S is another ℓ^2-regular storage function such that either*

(a) *S is $\mathcal{D}(\mathbf{W}_c^*)$-weakly continuous in the sense that: given a sequence $\{x_n\} \subset \operatorname{Im}\mathbf{W}_c$ and $x_c \in \operatorname{Im}\mathbf{W}_c$ such that*

$$\lim_{n\to\infty}\langle x, x_n\rangle_{\mathcal{X}} = \langle x, x_c\rangle_{\mathcal{X}} \text{ for all } x\in\mathcal{D}(\mathbf{W}_c^*),$$

then $\lim_{n\to\infty} S(x_n) = S(x_0)$, or

(b) *$\mathbf{W}_c$ is bounded and S is continuous on $\mathcal{X}$ (with respect to the norm topology on $\mathcal{X}$),*

then $S(x_0) \leq \underline{S}_r(x_0)$ for all $x_0 \in \mathcal{X}$.

Proof. We first prove item (1), starting with the claim for S_a. Since by assumption Σ is minimal and F_Σ has an analytic continuation to a Schur-class function, by item (1) of Proposition 2.2, $\operatorname{Im}\mathbf{W}_c \subset \mathcal{D}(\mathbf{W}_o)$. So on $\operatorname{Im}\mathbf{W}_c$, the available storage S_a is given by (4.14). It remains to show that for $x_0 \in \operatorname{Im}\mathbf{W}_c$ the formula for $S_a(x_0)$ in (4.14) gives a finite value. So assume $x_0 \in \operatorname{Im}\mathbf{W}_c$, say $x_0 = \mathbf{W}_c\mathbf{u}_-$ for a $\mathbf{u}_- \in \ell^2_{\mathcal{U}}(\mathbb{Z}_-)$. Choose $\mathbf{u}_+ \in \ell^2_{\mathcal{U}}(\mathbb{Z}_+)$ arbitrarily and define $\mathbf{u} \in \ell^2_{\mathcal{U}}(\mathbb{Z})$ by setting $P_-\mathbf{u} = \mathbf{u}_-$ and $P_+\mathbf{u} = \mathbf{u}_+$. Then $\mathbf{W}_o x_0 = \mathbf{W}_o\mathbf{W}_c\mathbf{u}_- = \mathfrak{H}_{F_\Sigma}\mathbf{u}_-$. Thus, using the decomposition of $\mathfrak{L}_{F_\Sigma}$ in (4.6) and the fact that $\|\mathfrak{L}_{F_\Sigma}\| \leq 1$, we find that

$$\begin{aligned}\|\mathbf{W}_o x_0 + \mathfrak{T}_{F_\Sigma}\mathbf{u}_+\|^2 - \|\mathbf{u}_+\|^2 &= \|\mathfrak{H}_{F_\Sigma}\mathbf{u}_- + \mathfrak{T}_{F_\Sigma}\mathbf{u}_+\|^2 - \|\mathbf{u}_+\|^2 \\ &= \|P_+\mathfrak{L}_{F_\Sigma}\mathbf{u}\|^2 - \|P_+\mathbf{u}\|^2 = \|P_-\mathbf{u}\|^2 + \|P_+\mathfrak{L}_{F_\Sigma}\mathbf{u}\|^2 - \|\mathbf{u}\|^2 \\ &\leq \|P_-\mathbf{u}\|^2 = \|\mathbf{u}_-\|^2. \end{aligned}$$

Since the upper bound $\|\mathbf{u}_-\|^2$ is independent of the choice of $\mathbf{u}_+ \in \ell^2_{\mathcal{U}}(\mathbb{Z}_+)$, we can take the supremum over all $\mathbf{u}_+ \in \ell^2_{\mathcal{U}}(\mathbb{Z}_+)$ to arrive at the inequality $S_a(x_0) \leq \|\mathbf{u}_-\|^2 < \infty$.

Next we prove the statement of item (1) concerning $\underline{S}_r$. By the discussion immediately preceding the statement of the proposition, it follows that $\underline{S}_r$ is an ℓ^2-regular storage function once we show that $\underline{S}_r$ is a storage function, that is, $\underline{S}_r(0) = 0$ and that $\underline{S}_r$ satisfies the dissipation inequality (3.1).

If $x_0 = 0$, we can choose $\mathbf{u} = 0$ as the argument in the right-hand side of (4.19) to conclude that $\underline{S}_r(0) \leq 0$. As we have already seen that $\underline{S}_r(x_0) \geq 0$ for all x_0, we conclude that $\underline{S}_r(0) = 0$.

To complete the proof of item (1), it remains to show that $\underline{S}_r$ satisfies the dissipation inequality (3.1). By shift invariance we may take $n = N_0 = 0$ in (3.1). If $\mathbf{x}(0) \notin \operatorname{Im} \mathbf{W}_c$, then $\underline{S}_r(x_0) = \infty$ and (3.1) holds trivially. We therefore assume that $(\widetilde{\mathbf{u}}(n), \widetilde{\mathbf{x}}(n), \widetilde{\mathbf{y}}(n))_{n \geq 0}$ is a system trajectory with initialization $\widetilde{\mathbf{x}}(0) = x_0 = \mathbf{W}_c \mathbf{u}_-$ for some $\mathbf{u}_- \in \mathcal{D}(\mathbf{W}_o)$ and the problem is to show

$$\begin{aligned} \underline{S}_r(\widetilde{\mathbf{x}}(1)) &\leq \|\widetilde{\mathbf{u}}(0)\|^2 - \|\widetilde{\mathbf{y}}(0)\|^2 + \underline{S}_r(\widetilde{\mathbf{x}}(0)) \qquad (4.20) \\ &= \inf_{\substack{\mathbf{u} \in \mathcal{D}(\mathbf{W}_c) \\ \mathbf{W}_c \mathbf{u} = \widetilde{\mathbf{x}}(0)}} \left[\|\widetilde{\mathbf{u}}(0)\|^2 - \|\widetilde{\mathbf{y}}(0)\|^2 + \sum_{n=-\infty}^{-1} \left(\|\mathbf{u}(n)\|^2 - \|\mathbf{y}(n)\|^2 \right) \right], \end{aligned}$$

where $\mathbf{y} = \widetilde{\mathfrak{T}}_{F_\Sigma} \mathbf{u}$. As $(\widetilde{\mathbf{u}}(n), \widetilde{\mathbf{x}}(n), \widetilde{\mathbf{y}}(n))_{n \geq 0}$ is a system trajectory initiated at 0, we know that $\widetilde{\mathbf{x}}(1) = A\widetilde{\mathbf{x}}(0) + B\widetilde{\mathbf{u}}(0)$ and $\widetilde{\mathbf{y}}(0) = C\widetilde{\mathbf{x}}(0) + D\widetilde{\mathbf{u}}(0)$. On the other hand, by translation-invariance of the system equations (1.1) we may rewrite the formula (4.18) for $\underline{S}_r(\widetilde{\mathbf{x}}(1))$ as

$$\underline{S}_r(\widetilde{\mathbf{x}}(1)) = \inf_{\mathbf{u}' \in \mathcal{D}(\mathbf{W}_c^{(1)}) \colon \mathbf{W}_c^{(1)} \mathbf{u} = \widetilde{\mathbf{x}}(1)} \sum_{n=-\infty}^{0} \left(\|\mathbf{u}'(n)\|^2 - \|\mathbf{y}'(n)\|^2 \right), \qquad (4.21)$$

where $\mathbf{W}_c^{(1)}$ is the shifted observability operator discussed in Remark 2.5 and where $\mathbf{y}' = \widetilde{\mathfrak{T}}_{F_\Sigma}^{(1)} \mathbf{u}'$; here now $\mathbf{u}'$ is supported on $(-\infty, 0]$ rather than on $\mathbb{Z}_- = (-\infty, 0)$ and $\widetilde{\mathfrak{T}}_{F_\Sigma}^{(1)}$ is the shift of $\widetilde{\mathfrak{T}}_{F_\Sigma}$ from the interval $\mathbb{Z}_-$ to the interval $(-\infty, 0]$. Let us write sequences $\mathbf{u}' \in \ell^2_{\mathcal{U}}((-\infty, 0])$ in the form $\mathbf{u}' = (\mathbf{v}', v')$ as in Remark 2.5 where $\mathbf{v}' \in \ell^2_{\mathcal{U}}(\mathbb{Z}_-)$ and $v' \in \mathcal{U}$. As observed in Remark 2.5,

$$\mathbf{W}_c^{(1)}(\mathbf{v}', v) = A\mathbf{W}_c \mathbf{v}' + Bv'.$$

Furthermore, from the structure of the Laurent operator $\mathfrak{L}_{F_\Sigma}$ (4.6) we read off that

$$\widetilde{\mathfrak{T}}_{F_\Sigma}^{(1)}(\mathbf{v}', v') = \left(\widetilde{\mathfrak{T}}_{F_\Sigma} \mathbf{v}', \sum_{k=-\infty}^{-1} CA^{-k-1} B\mathbf{v}'(k) + Dv' \right) \qquad (4.22)$$

where the series converges at least in the weak topology of $\mathcal{Y}$. For $\mathbf{v}' \in \mathcal{D}(\mathbf{W}_c)$, we know from Proposition 2.1 that $\mathbf{W}_c \mathbf{v}'$ is given by

$$\mathbf{W}_c \mathbf{v}' = \sum_{k=-\infty}^{-1} A^{-k-1} B\mathbf{v}'(k) \qquad (4.23)$$

where the series converges $\mathcal{D}(\mathbf{W}_c^*)$-weakly. We also know under our standing assumption (4.12) that $\operatorname{Obs}(C|A) \subset \mathcal{D}(\mathbf{W}_c^*)$ (see Proposition 2.2 (2)), and hence in particular $C^*y \in \mathcal{D}(\mathbf{W}_c^*)$ for all $y \in \mathcal{Y}$. This observation combined with the formula (4.23) implies that

$$C\mathbf{W}_c\mathbf{v}' = \sum_{k=-\infty}^{-1} CA^{-k-1}B\mathbf{v}'(k)$$

where the series converges weakly in $\mathcal{Y}$. This combined with (4.22) gives us

$$\widetilde{\mathfrak{T}}_{F_\Sigma}^{(1)}(\mathbf{v}', u) = \left(\widetilde{\mathfrak{T}}_{F_\Sigma}\mathbf{v}', C\mathbf{W}_c\mathbf{v}' + Dv'\right).$$

Thus the formula (4.21) for $\underline{S}_r(\widetilde{\mathbf{x}}(1))$ can be written out in more detail as

$$\underline{S}_r(\widetilde{\mathbf{x}}(1)) = \inf_{(\mathbf{v}', v')\in\mathcal{T}'}\left\{(\|\mathbf{v}'\|^2 - \|\widetilde{\mathfrak{T}}_{F_\Sigma}\mathbf{v}'\|^2) + \|u\|^2 - \|C\mathbf{W}_c\mathbf{v}' + Du\|^2\right\} \tag{4.24}$$

where

$$\mathcal{T}' := \{\mathbf{v}' \in \mathcal{D}(\mathbf{W}_c),\ v' \in \mathcal{U} \colon A\mathbf{W}_c\mathbf{v}' + Bv' = \widetilde{\mathbf{x}}(1)\}. \tag{4.25}$$

Note that the infimum (4.20) can be identified with the infimum (4.24) if we restrict the free parameter $(\mathbf{v}', v')$ to lie in the subset

$$\mathcal{T} = \{(\mathbf{v}', v') \in \mathcal{T}' \colon \mathbf{W}_c\mathbf{v}' = \widetilde{\mathbf{x}}(0), \quad v' = \widetilde{\mathbf{u}}(0)\}.$$

As the infimum of an objective function over a given set $\mathcal{T}$ is always bounded above by the infimum of the same objective function over a smaller set $\mathcal{T}' \subset \mathcal{T}$, the inequality (4.20) now follows as wanted.

It remains to address item (2), i.e., to show that $S(x_0) \le \underline{S}_r(x_0)$ for any other storage function S satisfying appropriate hypotheses. If $x_0 \notin \operatorname{Im}\mathbf{W}_c$, $\underline{S}_r(x_0) = \infty$ and the desired inequality holds trivially, so we assume that $x_0 = \mathbf{W}_c\mathbf{u}$ for some $\mathbf{u} \in \mathcal{D}(\mathbf{W}_c)$. Let us approximate $\mathbf{u}$ by elements of $\ell_{\text{fin},\mathcal{U}}(\mathbb{Z}_-)$ in the natural way:

$$\mathbf{u}_K(n) = \begin{cases} \mathbf{u}(n) & \text{for } -K \le n \le -1, \\ 0 & \text{for } n < -K \end{cases}$$

for $K = 1, 2, \dots$, and set $x_K = \mathbf{W}_c\mathbf{u}_K$. We let $(\mathbf{u}(n), \mathbf{x}(n), \mathbf{y}(n))_{n\ge -K}$ be a system trajectory with $\mathbf{u}(n) = \mathbf{u}_K(n)$ and with the state initialization $\mathbf{x}(-K) = 0$. Then, as $\mathbf{x}(0)$ will then be equal to x_K, iteration of the dissipation inequality (3.1) gives us

$$S(x_K) \le \sum_{n=-K}^{-1}\left(\|\mathbf{u}_K(n)\|^2 - \|\widetilde{\mathfrak{T}}_{F_\Sigma}\mathbf{u}_K(n)\|^2\right). \tag{4.26}$$

We seek to let $K \to \infty$ in this inequality. As $\mathbf{u}_K \to \mathbf{u}$ in the norm topology of $\ell^2_{\mathcal{U}}(\mathbb{Z}_-)$ and $\|\widetilde{\mathfrak{T}}_{F_\Sigma}\| \le 1$ since F is in the Schur class by assumption, it is clear that the right-hand side of (4.26) converges to

$$\|\mathbf{u}\|^2_{\ell^2_{\mathcal{U}}(\mathbb{Z}_-)} - \|\widetilde{\mathfrak{T}}_{F_\Sigma}\mathbf{u}\|^2_{\ell^2_{\mathcal{U}}(\mathbb{Z}_-)} = \|D_{\widetilde{\mathfrak{T}}_{F_\Sigma}}\mathbf{u}\|^2_{\ell^2_{\mathcal{U}}(\mathbb{Z}_-)}$$

as $K \to \infty$. On the other hand, as a consequence of the characterization (2.8) of the action of $\mathbf{W}_c$, it follows that $x_K = \mathbf{W}_c\mathbf{u}_K$ converges to $x_0 = \mathbf{W}_c\mathbf{u}$

in the $\mathcal{D}(\mathbf{W}_c^*)$-weak sense. Hence, if S is continuous with respect to the $\mathcal{D}(\mathbf{W}_c^*)$-weak topology as described in the statement of item (a), we see that $S(x_K) \to S(x_0)$ as $K \to \infty$ and we arrive at the limiting version of inequality (4.26):

$$S(x_0) \leq \|\mathbf{u}\|^2 - \|\widetilde{\mathfrak{T}}_{F_\Sigma}\mathbf{u}\|^2 = \|D_{\widetilde{\mathfrak{T}}_{F_\Sigma}}\mathbf{u}\|^2_{\ell^2_{\mathcal{U}}(\mathbb{Z}_-)}. \tag{4.27}$$

We may now take the infimum over all $\mathbf{u} \in \mathcal{D}(\mathbf{W}_c)$ with $\mathbf{W}_c\mathbf{u} = x_0$ to arrive at the desired inequality $S(x_0) \leq \underline{S}_r(x_0)$. This proves item (a) of (2). If $\mathbf{W}_c$ is bounded, then $x_K = \mathbf{W}_c\mathbf{u}_K$ converges in norm to $\mathbf{W}_c\mathbf{u} = x_0$. If S is continuous with respect to the norm topology on $\mathcal{X}$, then $S(x_K) \to S(x_0)$ and we again arrive at the limit inequality (4.27), from which the desired inequality $S(x_0) \leq \underline{S}_r(x_0)$ again follows. This completes the verification of item (2) in Proposition 4.5. □

Remark 4.6. Note that the fact that S_a is ℓ^2-regular can alternatively be seen from the fact that $\underline{S}_r$ is an ℓ^2-regular storage function combined with the first inequality in item (3) of Proposition 4.2.

Collecting some of the observations on the boundedness of S_a and $\underline{S}_r$ from the above results we obtain the following corollary. The inequalities in (4.28) follow directly from (4.14) and (4.19).

Corollary 4.7. *Assume Σ as in (1.1) is a system satisfying the assumptions (4.12). Define S_a by (4.1) and $\underline{S}_r$ by (4.18). For $x_0 \in \operatorname{Im} \mathbf{W}_o$ we have*

$$\|\mathbf{W}_o x_0\|^2 \leq S_a(x_0) \leq \underline{S}_r(x_0) \leq \|\mathbf{u}_-\|^2 \tag{4.28}$$

for all $\mathbf{u}_- \in \mathcal{D}(\mathbf{W}_c)$ with $x_0 = \mathbf{W}_c\mathbf{u}_-$, with the last inequality being vacuous if $x_0 \notin \operatorname{Im} \mathbf{W}_c$, in which case $\underline{S}_r(x_0) = \infty$. Hence

$$\underline{S}_r(x_0) < \infty \iff x_0 \in \operatorname{Im} \mathbf{W}_c,$$
$$x_0 \in \operatorname{Im} \mathbf{W}_c \implies S_a(x_0) < \infty \implies x_0 \in \mathcal{D}(\mathbf{W}_o).$$

In particular, $\underline{S}_r$ is finite-valued if and only if $\operatorname{Im} \mathbf{W}_c = \mathcal{X}$, that is, Σ is ℓ^2-exactly controllable, and S_a is finite-valued in case Σ is ℓ^2-exactly controllable.

Since F_Σ is assumed to be a Schur-class function, $\mathfrak{L}_{F_\Sigma}$ is a contraction, so that $I - \mathfrak{L}_{F_\Sigma}\mathfrak{L}_{F_\Sigma}^*$ and $I - \mathfrak{L}_{F_\Sigma}^*\mathfrak{L}_{F_\Sigma}$ are positive-semidefinite operators. We can thus read off from the $(2,2)$-entry in the right-hand side of the first identity and the $(1,1)$-entry in the right-hand side of the second identity of (4.13) that

$$D^2_{\mathfrak{T}^*_{F_\Sigma}} \succeq \mathfrak{H}_{F_\Sigma}\mathfrak{H}_{F_\Sigma}^* \quad \text{and} \quad D^2_{\widetilde{\mathfrak{T}}_{F_\Sigma}} \succeq \mathfrak{H}_{F_\Sigma}^*\mathfrak{H}_{F_\Sigma}. \tag{4.29}$$

The observability and controllability assumptions of (4.12) imply that the observability operator $\mathbf{W}_o : \mathcal{D}(\mathbf{W}_o) \to \ell^2_{\mathcal{Y}}(\mathbb{Z}_+)$ and the controllability operator $\mathbf{W}_c : \mathcal{D}(\mathbf{W}_c) \to \mathcal{X}$ are closed densely defined operators that satisfy the properties listed in Propositions 2.1 and 2.2. As spelled out in Proposition 2.2, the Hankel operator $\mathfrak{H}_{F_\Sigma}$ admits the factorizations

$$\mathfrak{H}_{F_\Sigma}|_{\mathcal{D}(\mathbf{W}_c)} = \mathbf{W}_o\mathbf{W}_c \quad \text{and} \quad \mathfrak{H}_{F_\Sigma}^*|_{\mathcal{D}(\mathbf{W}_o^*)} = \mathbf{W}_c^*\mathbf{W}_o^*. \tag{4.30}$$

Using the Douglas factorization lemma [16] together with the factorizations (4.30), we arrive at the following result. The proof also requires use of the Moore–Penrose generalized inverse $X^\dagger$ of a densely defined closed linear Hilbert-space operator $X : \mathcal{D}(X) \to \mathcal{H}_2$, with $\mathcal{D}(X) \subset \mathcal{H}_1$: we define $X^\dagger \colon \mathcal{D}(X^\dagger) = (\operatorname{Im} X \oplus (\operatorname{Im} X)^\perp) \to \mathcal{H}_1$ by

$$\begin{cases} X^\dagger(Xh_1) = P_{(\operatorname{Ker} X)^\perp} h_1, \\ X^\dagger|_{(\operatorname{Im} X)^\perp} = 0. \end{cases} \tag{4.31}$$

Then $X^\dagger$ is also closed and has the properties

$$X^\dagger X = P_{(\operatorname{Ker} X)^\perp}|_{\mathcal{D}(X)}, \quad XX^\dagger = P_{\overline{\operatorname{Im} X}}|_{\operatorname{Im} X \oplus (\operatorname{Im} X)^\perp}.$$

In particular, if X is bounded and surjective, then $X^\dagger$ is a bounded right inverse of X, and, if X is bounded, bounded below and injective, then $X^\dagger$ is a bounded left inverse of X.

Lemma 4.8. *Let the discrete-time linear system* Σ *in* (1.1) *satisfy the assumptions in* (4.12). *Then:*

(1) *There exists a unique closable operator* X_a *with domain* $\operatorname{Im} \mathbf{W}_c$ *mapping into* $(\operatorname{Ker} D_{\mathfrak{T}_{F_\Sigma}^*})^\perp \subset \ell^2_{\mathcal{Y}}(\mathbb{Z}_+)$ *so that we have the factorization*

$$\mathbf{W}_o|_{\operatorname{Im} \mathbf{W}_c} = D_{\mathfrak{T}_{F_\Sigma}^*} X_a. \tag{4.32}$$

Moreover, if we let $\overline{X}_a$ *denote the closure of* X_a, *then* $\overline{X}_a$ *is injective.*

(2) *There exists a unique closable operator* X_r *with domain* $\operatorname{Im} \mathbf{W}_o^*$ *mapping into* $(\operatorname{Ker} D_{\widetilde{\mathfrak{T}}_{F_\Sigma}})^\perp \subset \ell^2_{\mathcal{U}}(\mathbb{Z}_-)$ *so that we have the factorization*

$$\mathbf{W}_c^*|_{\operatorname{Im} \mathbf{W}_o^*} = D_{\widetilde{\mathfrak{T}}_{F_\Sigma}} X_r. \tag{4.33}$$

Moreover, if we let $\overline{X}_r$ *denote the closure of* X_r, *then* $\overline{X}_r$ *is injective.*

Proof. As statement (2) is just a dual version of statement (1), we only discuss the proof of (1) in detail.

Apply the Douglas factorization lemma to the first of the inequalities in (4.29) to get the existence of a unique contraction operator

$$Y_a : \ell^2_{\mathcal{U}}(\mathbb{Z}_-) \to (\operatorname{Ker} D_{\mathfrak{T}_{F_\Sigma}^*})^\perp \subset \ell^2_{\mathcal{Y}}(\mathbb{Z}_+)$$

such that

$$D_{\mathfrak{T}_{F_\Sigma}^*} Y_a = \mathfrak{H}_{F_\Sigma}, \quad \text{so that, by (4.30),} \quad D_{\mathfrak{T}_{F_\Sigma}^*} Y_a|_{\mathcal{D}(\mathbf{W}_c)} = \mathbf{W}_o \mathbf{W}_c.$$

If we let $\mathbf{W}_c^\dagger$ be the Moore–Penrose generalized inverse (4.31) of $\mathbf{W}_c$, then

$$\mathbf{W}_c^\dagger(x) = \arg\min \{\|\mathbf{u}\|^2_{\ell^2_{\mathcal{U}}(\mathbb{Z}_-)} : \mathbf{u} \in \mathcal{D}(\mathbf{W}_c),\ x = \mathbf{W}_c \mathbf{u}\} \quad (x \in \operatorname{Im} \mathbf{W}_c).$$

Since $\mathbf{W}_c$ is closed, $\operatorname{Ker} \mathbf{W}_c$ is a closed subspace of $\ell^2_{\mathcal{U}}(\mathbb{Z}_-)$ and for all $\mathbf{u} \in \mathcal{D}(\mathbf{W}_c)$ with $x = \mathbf{W}_c \mathbf{u}$ we have $\mathbf{W}_c^\dagger(x) = \mathbf{u} - P_{\operatorname{Ker} \mathbf{W}_c} \mathbf{u}$. We next define $X_a \colon \operatorname{Im} \mathbf{W}_c \to \ell^2_{\mathcal{Y}}(\mathbb{Z}_+)$ by

$$X_a = Y_a \mathbf{W}_c^\dagger.$$

Then X_a is a well-defined, possibly unbounded, operator on the dense domain $\mathcal{D}(X_a) = \operatorname{Im} \mathbf{W}_c$. Moreover we have

$$D_{\mathfrak{T}^*_{F_\Sigma}} X_a = D_{\mathfrak{T}^*_{F_\Sigma}} Y_a \mathbf{W}_c^\dagger = \mathfrak{H}_{F_\Sigma} \mathbf{W}_c^\dagger = \mathbf{W}_o \mathbf{W}_c \mathbf{W}_c^\dagger = \mathbf{W}_o|_{\operatorname{Im} \mathbf{W}_c}.$$

Hence X_a provides the factorization (4.32). Furthermore, $X_a = Y_a \mathbf{W}_c^\dagger$ implies that $\operatorname{Im} X_a \subset \operatorname{Im} Y_a$, so that $\operatorname{Im} X_a \subset (\operatorname{Ker} D_{\mathfrak{T}^*_{F_\Sigma}})^\perp$. Moreover, from the factorization (4.32) we see that this property makes the choice of X_a unique.

We now check that X_a so constructed is closable. Suppose that $\{x_0^{(k)}\}_{k\geq 0}$ is a sequence of vectors in $\operatorname{Im} \mathbf{W}_c$ such that $\lim_{k\to\infty} x_0^{(k)} = 0$ in $\mathcal{X}$-norm, while $\lim_{k\to\infty} X_a x_0^{(k)} = \mathbf{y}$ in $\ell^2_{\mathcal{Y}}(\mathbb{Z}_+)$-norm. As $D_{\mathfrak{T}^*_{F_\Sigma}}$ is bounded, it follows that

$$\lim_{k\to\infty} \mathbf{W}_o x_0^{(k)} = \lim_{k\to\infty} D_{\mathfrak{T}^*_{F_\Sigma}} X_a x_0^{(k)} = D_{\mathfrak{T}^*_{F_\Sigma}} \mathbf{y} \text{ in } \ell^2_{\mathcal{Y}}(\mathbb{Z}_+)\text{-norm}.$$

Since $\mathbf{W}_o$ is a closed operator and we have $x_0^{(k)} \to 0$ in $\mathcal{X}$-norm, it follows that $D_{\mathfrak{T}^*_{F_\Sigma}} \mathbf{y} = 0$. As $\operatorname{Im} X_a \subset (\operatorname{Ker} D_{\mathfrak{T}^*_{F_\Sigma}})^\perp$ and $X_a x_0^{(k)} \to \mathbf{y}$, we also have that $\mathbf{y} \in (\operatorname{Ker} D_{\mathfrak{T}^*_{F_\Sigma}})^\perp$. It follows that $\mathbf{y} = 0$, and hence X_a is closable.

Let $\overline{X}_a$ be the closure of X_a. We check that $\overline{X}_a$ is injective as follows. The vector x_0 being in $\mathcal{D}(\overline{X}_a)$ means that there is a sequence of vectors $\{x_0^{(k)}\}_{k\geq 1}$ contained in $\mathcal{D}(X_a)$ with $\lim_{k\to\infty} x_0^{(k)} = x_0$ in $\mathcal{X}$ and $\lim_{k\to\infty} X_a x_0^{(k)} = \mathbf{y}$ for some $\mathbf{y} \in \ell^2_{\mathcal{Y}}(\mathbb{Z}_+)$. The condition that $\overline{X}_a x = 0$ means that in addition $\mathbf{y} = 0$. Since $D_{\mathfrak{T}^*_{F_\Sigma}}$ is bounded, it then follows that $\lim_{k\to\infty} D_{\mathfrak{T}^*_{F_\Sigma}} X_a x_0^{(k)} = 0$, or, by (4.32)

$$\lim_{k\to\infty} \mathbf{W}_o x_0^{(k)} = 0.$$

As we also have $\lim_{k\to\infty} x_0^{(k)} = x_0$ in $\mathcal{X}$ and $\mathbf{W}_o$ is a closed operator, it follows that $x_0 \in \mathcal{D}(\mathbf{W}_o)$ and $\mathbf{W}_o x_0 = 0$. As $\mathbf{W}_o$ is injective, it follows that $x_0 = 0$. We conclude that $\overline{X}_a$ is injective as claimed. □

Using the closed operators $\overline{X}_a$ and $\overline{X}_r$ defined in Lemma 4.8 we now define (possibly unbounded) positive-definite operators H_a and H_r so that the storage functions S_a and $\underline{S}_r$ have the quadratic forms $S_a = S_{H_a}$ and $\underline{S}_r = S_{H_r}$ as in (3.6).

We start with H_a. Since $\overline{X}_a$ is closed, there is a good polar factorization

$$\overline{X}_a = U_a |\overline{X}_a|$$

(see [26, Theorem VIII.32]); in detail, $\overline{X}_a^* \overline{X}_a$ is selfadjoint with positive selfadjoint square-root $|\overline{X}_a| = (\overline{X}_a^* \overline{X}_a)^{\frac{1}{2}}$ satisfying $\mathcal{D}(|\overline{X}_a|) = \mathcal{D}(\overline{X}_a)$, U_a is a partial isometry with initial space equal to $(\operatorname{Ker} \overline{X}_a)^\perp$ and final space equal to $\overline{\operatorname{Im} X_a}$ so that we have the factorization $\overline{X}_a = U_a |\overline{X}_a|$.

Now set

$$H_a = \overline{X}_a^* \overline{X}_a, \quad H_a^{\frac{1}{2}} = |\overline{X}_a|. \tag{4.34}$$

As noted in Lemma 4.8, $\overline{X}_a$ is injective, and thus H_a and $H_a^{\frac{1}{2}}$ are injective as well, and as a result U_a is an isometry.

We proceed with the definition of H_r. As the properties of $\overline{X}_r$ parallel those of $\overline{X}_a$, also $\overline{X}_r$ has a good polar decomposition $\overline{X}_r = U_r|\overline{X}_r|$ with $|\overline{X}_r|$ and U_r having similar properties as $|\overline{X}_a|$ and U_a, in particular, $\overline{X}_r^*\overline{X}_r$ and $|\overline{X}_r|$ are injective and U_r is an isometry. We then define

$$H_r = \left(\overline{X}_r^*\overline{X}_r\right)^{-1}, \quad H_r^{\frac{1}{2}} = |\overline{X}_r|^{-1}. \tag{4.35}$$

We shall also need a modification of the factorization (4.33). For $\mathbf{u} \in \mathcal{D}(\mathbf{W}_c)$ and $x \in \operatorname{Im} \mathbf{W}_o^*$, let us note that

$$\begin{aligned} \langle \mathbf{W}_c\mathbf{u}, x\rangle_{\mathcal{X}} &= \langle \mathbf{u}, \mathbf{W}_c^* x\rangle_{\ell^2_{\mathcal{U}}(\mathbb{Z}_-)} = \langle \mathbf{u}, D_{\widetilde{\mathfrak{T}}_{F_\Sigma}} X_r x\rangle_{\ell^2_{\mathcal{U}}(\mathbb{Z}_-)} \quad \text{(by (4.33))} \\ &= \langle D_{\widetilde{\mathfrak{T}}_{F_\Sigma}} \mathbf{u}, X_r x\rangle_{\ell^2_{\mathcal{U}}(\mathbb{Z}_-)}. \end{aligned}$$

The end result is that then $D_{\widetilde{\mathfrak{T}}_{F_\Sigma}} \mathbf{u}$ is in $\mathcal{D}(X_r^*)$ and $X_r^* D_{\widetilde{\mathfrak{T}}_{F_\Sigma}} \mathbf{u} = \mathbf{W}_c\mathbf{u}$. In summary we have the following adjoint version of the factorization (4.33):

$$\mathbf{W}_c = X_r^* D_{\widetilde{\mathfrak{T}}_{F_\Sigma}}|_{\mathcal{D}(\mathbf{W}_c)}. \tag{4.36}$$

In the following statement we use the notion of a *core* of a closed, densely defined operator Γ between two Hilbert spaces $\mathcal{H}$ and $\mathcal{K}$ (see [26] or [20]), namely: a dense linear submanifold $\mathcal{D}$ is said to be a *core* for the closed, densely defined operator X with domain $\mathcal{D}(X)$ in $\mathcal{H}$ mapping into $\mathcal{K}$ if, given any $x \in \mathcal{D}(X)$, there is a sequence $\{x_n\}_{n\geq 1}$ of points in $\mathcal{D}$ such that $\lim_{n\to\infty} x_n = x$ and also $\lim_{n\to\infty} Xx_n = Xx$.

Theorem 4.9. *Let the discrete-time linear system* Σ *in* (1.1) *satisfy the assumptions in* (4.12). *Define* X_a, $\overline{X}_a$, X_r, $\overline{X}_r$ *as in Lemma* 4.8 *and the closed operators* H_a *and* H_r *as in the preceding discussion. Then the available storage function* S_a *and required supply function* S_r *are given by*

$$S_a(x_0) = \|\overline{X}_a x_0\|^2 = \|H_a^{\frac{1}{2}} x_0\|^2 \qquad (x_0 \in \operatorname{Im} \mathbf{W}_c), \tag{4.37}$$

$$\underline{S}_r(x_0) = \||\overline{X}_r|^{-1} x_0\|^2 = \|H_r^{\frac{1}{2}} x_0\|^2 \qquad (x_0 \in \operatorname{Im} \mathbf{W}_c). \tag{4.38}$$

In particular, the available storage S_a *and* ℓ^2*-regularized required supply* $\underline{S}_r$ *agree with quadratic storage functions on* $\operatorname{Im} \mathbf{W}_c$.

Moreover, $\operatorname{Im} \mathbf{W}_c$ *is a core for* $H_a^{\frac{1}{2}}$ *and* $\operatorname{Im} \mathbf{W}_o^*$ *is a core for* $H_r^{-\frac{1}{2}}$.

Proof. By Lemma 4.8, in the operator form of S_a derived in Lemma 4.4 we can replace $\mathbf{W}_o x_0$ by $D_{\mathfrak{T}^*_{F_\Sigma}} X_a x_0$, leading to

$$S_a(x_0) = \sup_{\mathbf{u}\in\ell^2_{\mathcal{U}}(\mathbb{Z}_+)} \|D_{\mathfrak{T}^*_{F_\Sigma}} X_a x_0 + \mathfrak{T}_{F_\Sigma}\mathbf{u}\|^2_{\ell^2_{\mathcal{Y}}(\mathbb{Z}_+)} - \|\mathbf{u}\|^2_{\ell^2_{\mathcal{U}}(\mathbb{Z}_+)}. \tag{4.39}$$

For $x_0 \in \operatorname{Im} \mathbf{W}_c$ and each $\mathbf{u} \in \ell^2_{\mathcal{U}}(\mathbb{Z}_+)$ we have

$$\begin{aligned}
&\|D_{\mathfrak{T}^*_{F_\Sigma}} X_a x_0 + \mathfrak{T}_{F_\Sigma} \mathbf{u}\|^2 - \|\mathbf{u}\|^2 = \\
&\quad = \|D_{\mathfrak{T}^*_{F_\Sigma}} X_a x_0\|^2 + 2\operatorname{Re} \langle D_{\mathfrak{T}^*_{F_\Sigma}} X_a x_0, \mathfrak{T}_{F_\Sigma} \mathbf{u}\rangle + \|\mathfrak{T}_{F_\Sigma} \mathbf{u}\|^2 - \|\mathbf{u}\|^2 \\
&\quad = \|D_{\mathfrak{T}^*_{F_\Sigma}} X_a x_0\|^2 + 2\operatorname{Re} \langle D_{\mathfrak{T}^*_{F_\Sigma}} X_a x_0, \mathfrak{T}_{F_\Sigma} \mathbf{u}\rangle - \|D_{\mathfrak{T}_{F_\Sigma}} \mathbf{u}\|^2 \\
&\quad = \|D_{\mathfrak{T}^*_{F_\Sigma}} X_a x_0\|^2 + 2\operatorname{Re} \langle X_a x_0, D_{\mathfrak{T}^*_{F_\Sigma}} \mathfrak{T}_{F_\Sigma} \mathbf{u}\rangle - \|D_{\mathfrak{T}_{F_\Sigma}} \mathbf{u}\|^2 \\
&\quad = \|D_{\mathfrak{T}^*_{F_\Sigma}} X_a x_0\|^2 + 2\operatorname{Re} \langle X_a x_0, \mathfrak{T}_{F_\Sigma} D_{\mathfrak{T}_{F_\Sigma}} \mathbf{u}\rangle - \|D_{\mathfrak{T}_{F_\Sigma}} \mathbf{u}\|^2 \\
&\quad = \|D_{\mathfrak{T}^*_{F_\Sigma}} X_a x_0\|^2 + 2\operatorname{Re} \langle \mathfrak{T}^*_{F_\Sigma} X_a x_0, D_{\mathfrak{T}_{F_\Sigma}} \mathbf{u}\rangle - \|D_{\mathfrak{T}_{F_\Sigma}} \mathbf{u}\|^2 \\
&\quad = \|D_{\mathfrak{T}^*_{F_\Sigma}} X_a x_0\|^2 + \|\mathfrak{T}^*_{F_\Sigma} X_a x_0\|^2 - \|\mathfrak{T}^*_{F_\Sigma} X_a x_0 - D_{\mathfrak{T}_{F_\Sigma}} \mathbf{u}\|^2 \\
&\quad = \|X_a x_0\|^2 - \|\mathfrak{T}^*_{F_\Sigma} X_a x_0 - D_{\mathfrak{T}_{F_\Sigma}} \mathbf{u}\|^2.
\end{aligned}$$

By construction, we have $\operatorname{Im} X_a \subset (\operatorname{Ker} D_{\mathfrak{T}^*_{F_\Sigma}})^\perp = \overline{\operatorname{Im}} D_{\mathfrak{T}^*_{F_\Sigma}}$. Using that $\mathfrak{T}^*_{F_\Sigma} D_{\mathfrak{T}^*_{F_\Sigma}} = D_{\mathfrak{T}_{F_\Sigma}} \mathfrak{T}^*_{F_\Sigma}$, we obtain

$$\mathfrak{T}^*_{F_\Sigma} \overline{\operatorname{Im}} D_{\mathfrak{T}^*_{F_\Sigma}} \subset \overline{\operatorname{Im}} D_{\mathfrak{T}_{F_\Sigma}}.$$

Thus $\operatorname{Im} \mathfrak{T}^*_{F_\Sigma} X_a \subset \overline{\operatorname{Im}} D_{\mathfrak{T}_{F_\Sigma}}$. Hence there is a sequence $\mathbf{u}_k$ of input signals in $\ell^2_{\mathcal{U}}(\mathbb{Z}_+)$ so that $\|\mathfrak{T}^*_{F_\Sigma} X_a x_0 - D_{\mathfrak{T}_{F_\Sigma}} \mathbf{u}_k\| \to 0$ as $k \to \infty$. We conclude that for $x_0 \in \operatorname{Im} \mathbf{W}_c$ the supremum in (4.39) is given by

$$S_a(x_0) = \|X_a x_0\|^2 = \|\overline{X}_a x_0\|^2 = \|H_a^{\frac{1}{2}} x_0\|^2.$$

Let $x_0 \in \operatorname{Im} \mathbf{W}_c$. Given a $\mathbf{u} \in \mathcal{D}(\mathbf{W}_c)$, by the factorization (4.36) we see that $\mathbf{W}_c \mathbf{u} = x_0$ if and only if $X_r^* D_{\widetilde{\mathfrak{T}}_{F_\Sigma}} \mathbf{u} = x_0$. Therefore, we have

$$\underline{S}_r(x_0) = \inf_{\mathbf{u} \in \mathcal{D}(\mathbf{W}_c),\, X_r^* D_{\widetilde{\mathfrak{T}}_{F_\Sigma}} \mathbf{u} = x_0} \|D_{\widetilde{\mathfrak{T}}_{F_\Sigma}} \mathbf{u}\|^2 = \inf_{\mathbf{v} \in D_{\widetilde{\mathfrak{T}}_{F_\Sigma}} \mathcal{D}(\mathbf{W}_c),\, X_r^* \mathbf{v} = x_0} \|\mathbf{v}\|^2.$$

A general property of operator closures is $\overline{X}_r^* = X_r^*$. Hence

$$\underline{S}_r(x_0) = \inf_{\mathbf{v} \in D_{\widetilde{\mathfrak{T}}_{F_\Sigma}} \mathcal{D}(\mathbf{W}_c),\, \overline{X}_r^* \mathbf{v} = x_0} \|\mathbf{v}\|^2. \tag{4.40}$$

As $x_0 \in \operatorname{Im} \mathbf{W}_c$ by assumption, the factorization (4.36) gives us a $\mathbf{u}_0 \in \mathcal{D}(\mathbf{W}_c)$ so that

$$x_0 = \overline{X}_r^* D_{\widetilde{\mathfrak{T}}_{F_\Sigma}} \mathbf{u}_0. \tag{4.41}$$

In particular, x_0 has the form $x_0 = \overline{X}_r^* \mathbf{v}_0$ with $\mathbf{v}_0 \in D_{\widetilde{\mathfrak{T}}_{F_\Sigma}} \mathcal{D}(\mathbf{W}_c)$. From (4.41) we see that the general solution $\mathbf{v} \in \mathcal{D}(\overline{X}_r^*)$ of $x_0 = \overline{X}_r^* \mathbf{v}$ is

$$\mathbf{v} = D_{\widetilde{\mathfrak{T}}_{F_\Sigma}} \mathbf{u}_0 + k \text{ where } k \in \operatorname{Ker} \overline{X}_r^*. \tag{4.42}$$

By construction the target space for X_r (and $\overline{X}_r$) is $(\operatorname{Ker} D_{\widetilde{\mathfrak{T}}_{F_\Sigma}})^\perp$ so the domain space for $\overline{X}_r^*$ is $(\operatorname{Ker} D_{\widetilde{\mathfrak{T}}_{F_\Sigma}})^\perp$ and $\operatorname{Ker} \overline{X}_r^* \subset \overline{\operatorname{Im}} D_{\widetilde{\mathfrak{T}}_{F_\Sigma}}$. Hence the infimum in (4.40) remains unchanged if we relax the constraint $\mathbf{v} \in D_{\widetilde{\mathfrak{T}}_{F_\Sigma}} \mathcal{D}(\mathbf{W}_c)$

to just $\mathbf{v} \in \mathcal{D}(\overline{X}_r^*)$, i.e.,

$$\underline{S}_r(x_0) = \inf_{\mathbf{v} \in \mathcal{D}(\overline{X}_r^*),\, \overline{X}_r^* \mathbf{v} = x_0} \|\mathbf{v}\|^2. \tag{4.43}$$

In terms of the polar decomposition $\overline{X}_r = U_r |\overline{X}_r|$ for $\overline{X}_r$, we have

$$\overline{X}_r^* = |\overline{X}_r| U_r^*$$

with

$$\mathcal{D}(\overline{X}_r^*) = \{\mathbf{u} \in \overline{\operatorname{Im}} D_{\widetilde{\mathfrak{T}}_{F_\Sigma}} : U_r^* \mathbf{u} \in \mathcal{D}(|\overline{X}_r|) = \mathcal{D}(\overline{X}_r)\}.$$

Since $|\overline{X}_r|$ is injective and U_r is an isometry with range equal to $(\operatorname{Ker} \overline{X}_r^*)^\perp$, the constraint $|\overline{X}_r| U_r^* \mathbf{v} = \overline{X}_r^* \mathbf{v} = x_0$ is equivalent to

$$P_{(\operatorname{Ker} \overline{X}_r^*)^\perp} \mathbf{v} = U_r U_r^* \mathbf{v} = U_r |\overline{X}_r|^{-1} x_0.$$

As we want to minimize $\|\mathbf{v}\|^2$ with $P_{(\operatorname{Ker} \overline{X}_r^*)^\perp} \mathbf{v}$ equal to $U_r |\overline{X}_r^*|^{-1} x_0 \in \mathcal{D}(\overline{X}_r)$, it is clear that this is achieved at $\mathbf{v}_{\rm opt} = U_r |\overline{X}_r^*|^{-1} x_0$, so that

$$\underline{S}_r(x_0) = \|\mathbf{v}_{\rm opt}\|^2 = \|U_r |\overline{X}_r|^{-1} x_0\|^2 = \||\overline{X}_r|^{-1} x_0\|^2 = \|H_r^{\frac{1}{2}} x_0\|^2,$$

as claimed.

It remains to verify the last assertion concerning the core properties of $\operatorname{Im} \mathbf{W}_c$ and $\operatorname{Im} \mathbf{W}_o^*$. By definition $H_a^{\frac{1}{2}} = |\overline{X}_a|$ where $\overline{X}_a$ is defined to be the closure of the $X_a = \overline{X}_a|_{\operatorname{Im} \mathbf{W}_c}$. Hence $\operatorname{Im} \mathbf{W}_c$ by definition is a core for $\overline{X}_a$ from which it immediately follows that $\operatorname{Im} \mathbf{W}_c$ is a core for $H_a^{\frac{1}{2}} = |\overline{X}_a|$. That $\operatorname{Im} \mathbf{W}_o^*$ is a core for $H_r^{-\frac{1}{2}} = |\overline{X}_r|$ follows in the same way via a dual analysis. □

5. The dual system Σ^*

In this section we develop a parallel theory for the dual system Σ^* of Σ, which is the system with system matrix equal to the adjoint of (1.2) evolving in backwards time.

5.1. Controllability, observability, minimality and transfer function for the dual system

With the discrete-time linear system Σ given by (1.1) with system matrix $M = \left[\begin{smallmatrix} A & B \\ C & D \end{smallmatrix}\right]$ we will associate the dual system Σ^* with system matrix $M^* = \left[\begin{smallmatrix} A^* & C^* \\ B^* & D^* \end{smallmatrix}\right] : \left[\begin{smallmatrix} \mathcal{X} \\ \mathcal{Y} \end{smallmatrix}\right] \to \left[\begin{smallmatrix} \mathcal{X} \\ \mathcal{U} \end{smallmatrix}\right]$. It will be convenient for our formalism here to let the dual system evolve in backwards time; we therefore define the system Σ^* to be given by the system input/state/output equations

$$\Sigma^* : = \left\{ \begin{aligned} \mathbf{x}_*(n-1) &= A^* \mathbf{x}_*(n) + C^* \mathbf{u}_*(n), \\ \mathbf{y}_*(n) &= B^* \mathbf{x}_*(n) + D^* \mathbf{u}_*(n). \end{aligned} \right. \tag{5.1}$$

If we impose a final condition $\mathbf{x}_*(-1) = x_0$ and feed in an input-sequence $\{\mathbf{u}(n)\}_{n\in\mathbb{Z}_-}$, one can solve recursively to get, for $n \le -1$,

$$\begin{cases} \mathbf{x}_*(n) = A^{*-n-1}x_0 + \sum_{j=n+1}^{-1} A^{*-n+j}C^*\mathbf{u}_*(j), \\ \mathbf{y}_*(n) = B^*A^{*-n-1}x_0 + \sum_{j=n+1}^{-1} B^*A^{*-n+j-1}C^*\mathbf{u}_*(j) + D^*\mathbf{u}_*(n). \end{cases}$$

Alternatively, the Z-transform $\{\mathbf{x}_*(n)\}_{n\in\mathbb{Z}_-} \mapsto \widehat{\mathbf{x}}_*(\lambda) = \sum_{n=-\infty}^{-1} \mathbf{x}_*(n)\lambda^n$ may be applied directly to the system equations (5.1). Combining this with the observation that

$$\sum_{n=-\infty}^{-1} \mathbf{x}_*(n-1)\lambda^n = \lambda\left(\sum_{n=-\infty}^{-1} \mathbf{x}_*(n-1)\lambda^{n-1}\right) = \lambda\left(\sum_{n=-\infty}^{-2} \mathbf{x}_*(n)\lambda^n\right)$$
$$= \lambda\left(\widehat{\mathbf{x}}_*(\lambda) - x_0\lambda^{-1}\right) = \lambda\widehat{\mathbf{x}}_*(\lambda) - x_0,$$

converts the first system equation in (5.1) to

$$\lambda\widehat{\mathbf{x}}_*(\lambda) - x_0 = A^*\widehat{\mathbf{x}}_*(\lambda) + C^*\widehat{\mathbf{u}}_*(\lambda),$$

leading to the Z-transformed version of the whole system:

$$\begin{cases} \widehat{\mathbf{x}}_*(\lambda) = (\lambda I - A^*)^{-1}x_0 + (\lambda I - A^*)^{-1}C^*\widehat{\mathbf{u}}_*(\lambda), \\ \widehat{\mathbf{y}}_*(\lambda) = B^*(\lambda I - A^*)^{-1}x_0 + F_{\Sigma^*}(\lambda)\widehat{\mathbf{u}}_*(\lambda), \end{cases}$$

where the *transfer function* $F_{\Sigma^*}(\lambda)$ for the system Σ^* is then given by

$$\begin{aligned} F_{\Sigma^*}(\lambda) &= D^* + B^*(\lambda I - A^*)^{-1}C^* \\ &= D^* + \lambda^{-1}(I - \lambda^{-1}A^*)^{-1}C^* = F_\Sigma(1/\overline{\lambda})^* \end{aligned} \tag{5.2}$$

which is an analytic function on a neighborhood of the point at ∞ in the complex plane. Moreover, F_{Σ^*} has analytic continuation to a function analytic on the exterior of the unit disk $\mathbb{D}_e := \{\lambda \in \mathbb{C} \colon |\lambda| > 1\} \cup \{\infty\}$ exactly when F_Σ has analytic continuation to a function analytic on the unit disk $\mathbb{D}$ with equality of corresponding ∞-norms:

$$\|F_{\Sigma_*}\|_{\infty,\mathbb{D}_e} := \sup_{\lambda\in\mathbb{D}_e} \|F_{\Sigma_*}(\lambda)\| = \sup_{\lambda\in\mathbb{D}} \|F_\Sigma(\lambda)\| =: \|F_\Sigma\|_{\infty,\mathbb{D}}.$$

All the analysis done up to this point for the system Σ has a dual analogue for the system Σ^*. In particular, the observability operator $\mathbf{W}_{*o}$ for the dual system is obtained by running the system (5.1) with final condition $\mathbf{x}_*(-1) = x_0$ and input string $\mathbf{u}_*(n) = 0$ for $n \le -1$, resulting in the output string $\{B^*A^{*(-n-1)}x_0\}_{n\in\mathbb{Z}_-}$. Since we are interested in a setting with operators on ℓ^2, we define the *observability operator* $\mathbf{W}_{*o}$ for Σ^* to have domain

$$\mathcal{D}(\mathbf{W})_{*o} = \{x_0 \in \mathcal{X} \colon \{B^*A^{*(-n-1)}x_0\}_{n\in\mathbb{Z}_-} \in \ell^2_{\mathcal{U}}(\mathbb{Z}_-)\}$$

with action given by

$$\mathbf{W}_{*o}x_0 = \{B^*A^{*(-n-1)}x_0\}_{n\in\mathbb{Z}_-} \text{ for } x_0 \in \mathcal{D}(\mathbf{W}_{*o}).$$

Note that $\mathbf{W}_{*o}$ so defined is exactly the same as the adjoint controllability operator $\mathbf{W}_c^*$ for the original system (2.3)–(2.4), and in fact viewing this operator as $\mathbf{W}_{*o}$ gives a better control-theoretic interpretation for this operator. Similarly it is natural to define the adjoint controllability operator for the adjoint system $(\mathbf{W}_{*c})^*$ by

$$\mathcal{D}((\mathbf{W}_{*c})^*) = \{x_0 \in \mathcal{X} \colon \{CA^n x_0\}_{n\in\mathbb{Z}_+} \in \ell^2_{\mathcal{Y}}(\mathbb{Z}_+)\} = \mathcal{D}(\mathbf{W}_o)$$

with action given by

$$\mathbf{W}_{*c}^* x_0 = \{CA^n x_0\}_{n\in\mathbb{Z}_+} = \mathbf{W}_o x_0.$$

In view of the equalities

$$\mathbf{W}_{*o} = \mathbf{W}_c^*, \quad (\mathbf{W}_{*c})^* = \mathbf{W}_o, \quad (\mathbf{W}_{*o})^* = \mathbf{W}_c, \quad \mathbf{W}_{*c} = \mathbf{W}_o^*, \tag{5.3}$$

one can work out the dual analogue of Proposition 2.1, either by redoing the original proof with the backward-time system Σ^* in place of the forward-time system Σ, or simply by making the substitutions (5.3) in the statement of the results.

Let us now assume that F_Σ has analytic continuation to a bounded analytic $\mathcal{L}(\mathcal{U},\mathcal{Y})$-valued function on the unit disk, or equivalently, F_{Σ^*} has analytic continuation to a bounded analytic $\mathcal{L}(\mathcal{Y},\mathcal{U})$-valued function on the exterior of the unit disk $\mathbb{D}_e$. Then F_Σ and F_{Σ^*} can be identified via strong nontangential boundary-value limits with L^∞-functions on the unit circle $\mathbb{T}$; the relation between these boundary-value functions is simply

$$F_{\Sigma^*}(\lambda) = F_\Sigma(\lambda)^* \quad (\text{a.e. } \lambda \in \mathbb{T})$$

with the consequence that the associated multiplication operators

$$M_{F_\Sigma} \colon L^2_{\mathcal{U}}(\mathbb{T}) \to L^2_{\mathcal{Y}}(\mathbb{T}), \quad M_{F_{\Sigma^*}} \colon L^2_{\mathcal{Y}}(\mathbb{T}) \to L^2_{\mathcal{U}}(\mathbb{T})$$

given by

$$M_{F_\Sigma} \colon \widehat{\mathbf{u}}(\lambda) \mapsto \widehat{F}_\Sigma(\lambda)\cdot\widehat{\mathbf{u}}(\lambda), \quad M_{F_{\Sigma^*}} \colon \widehat{\mathbf{u}}_*(\lambda) \mapsto \widehat{F}_{\Sigma_*}(\lambda)\cdot\widehat{\mathbf{u}}_*(\lambda)$$

are adjoints of each other:

$$(M_{F_\Sigma})^* = M_{F_{\Sigma^*}}.$$

Notice also that M_{F_Σ} maps $H^2_{\mathcal{U}}(\mathbb{D})$ into $H^2_{\mathcal{Y}}(\mathbb{D})$, while $M_{F_{\Sigma^*}} = M_{F_\Sigma}^*$ maps $(H^2_{\mathcal{Y}})^\perp := L^2_{\mathcal{Y}}(\mathbb{T}) \ominus H^2_{\mathcal{Y}}(\mathbb{D}) \cong H^2_{\mathcal{Y}}(\mathbb{D}_e)$ into $(H^2_{\mathcal{U}})^\perp := L^2_{\mathcal{U}}(\mathbb{T}) \ominus H^2_{\mathcal{U}}(\mathbb{D}) \cong H^2_{\mathcal{U}}(\mathbb{D}_e)$.

It is natural to define the frequency-domain Hankel operator $\mathbb{H}_{F_{\Sigma^*}}$ for the adjoint system as the operator from $H^2_{\mathcal{Y}}(\mathbb{D}_e)^\perp = H^2_{\mathcal{Y}}(\mathbb{D})$ (the past from the point of view of the backward-time system Σ^*) to $H^2_{\mathcal{U}}(\mathbb{D}_e) = H^2_{\mathcal{U}}(\mathbb{D})^\perp$ (the future from the point of view of Σ^*) by

$$\mathbb{H}_{F_{\Sigma^*}} = P_{H^2_{\mathcal{U}}(\mathbb{D})^\perp} M_{F_{\Sigma^*}}|_{H^2_{\mathcal{Y}}(\mathbb{D})} = (\mathbb{H}_{F_\Sigma})^*. \tag{5.4}$$

After application of the inverse Z-transform, we see that the time-domain version $\mathfrak{H}_{F_{\Sigma^*}}$ of the Hankel operator for Σ^* is just the adjoint $(\mathfrak{H}_{F_\Sigma})^*$ of the time-domain version of the Hankel operator for Σ, namely

$$\mathfrak{H}_{F_{\Sigma^*}} = [B^* A^{*(-i+j-1)} C^*]_{i<0,j\geq 0} \colon \ell^2_{\mathcal{Y}}(\mathbb{Z}_+) \to \ell^2_{\mathcal{U}}(\mathbb{Z}_-),$$

from which we see immediately the formal factorization

$$\mathfrak{H}_{F_{\Sigma^*}} = \mathrm{col}_{i<0}[B^* A^{*(-i-1)}] \cdot \mathrm{row}_{j\geq 0}[A^{*j} C^*] = \mathbf{W}_{*o} \mathbf{W}_{*c} = \mathbf{W}_c^* \mathbf{W}_o^*. \quad (5.5)$$

With all these observations in place, it is straightforward to formulate the dual version of Proposition 2.2, again, either by redoing the proof of Proposition 2.2 with the backward-time system Σ^* in place of the forward-time system Σ, or by simply substituting the identifications (5.3) and (5.4).

Note next that an immediate consequence of the identifications (5.3) is that ℓ^2-exact controllability for Σ is the same as ℓ^2-exact observability for Σ^* and ℓ^2-exact observability for Σ is the same as ℓ^2-exact controllability for Σ^*. With this observation in hand, the dual version of Proposition 2.3 is immediate.

5.2. Storage functions for the adjoint system

Let S_* be a function from $\mathcal{X}$ to $[0, \infty]$. In parallel with what is done in Section 3, we define S_* to be a *storage function for the system* Σ^* if

$$S_*(\mathbf{x}_*(n-1))) \leq S_*(\mathbf{x}_*(n)) + \|\mathbf{u}_*(n)\|^2 - \|\mathbf{y}_*(n)\|^2_{\mathcal{Y}} \text{ for } n \leq N_0 \quad (5.6)$$

holds over all system trajectories $(\mathbf{u}_*(n), \mathbf{x}_*(n), \mathbf{y}_*(n))_{n \leq N_0}$ of the system Σ^* in (5.1) with state initialization $\mathbf{x}_*(N_0) = x_0$ for some $x_0 \in \mathcal{X}$ at some $N_0 \in \mathbb{Z}$, and S_* is normalized to satisfy

$$S_*(0) = 0. \quad (5.7)$$

Then by redoing the proof of Proposition 3.1 with the backward-time system Σ_* in place of the forward-time system Σ, we arrive at the following dual version of Proposition 3.1.

Proposition 5.1. *Suppose that the system* Σ^* *in* (5.1) *has a storage function* S_* *as in* (5.6) *and* (5.7). *Then the transfer function* F_{Σ^*} *of* Σ^* *defined by* (5.2) *has an analytic continuation to the exterior unit disk* $\mathbb{D}_e$ *in the Schur class* $\mathcal{S}_{\mathbb{D}_e}(\mathcal{Y}, \mathcal{U})$.

Note that by the duality considerations already discussed above, an equivalent conclusion is that F_Σ has analytic continuation to the unit disk in the Schur class $\mathcal{S}(\mathcal{U}, \mathcal{Y})$ over the unit disk.

We say that S_* is a *quadratic storage function* for Σ^* if S_* is a storage function of the form

$$S_*(x) = S_{H_*}(x) = \begin{cases} \|H_*^{\frac{1}{2}} x\|^2 & \text{for } x \in \mathcal{D}(H_*^{\frac{1}{2}}), \\ +\infty & \text{otherwise,} \end{cases} \quad (5.8)$$

where H_* is a (possibly) unbounded positive-semidefinite operator on $\mathcal{X}$. To analyze quadratic storage functions for Σ^*, we introduce the adjoint KYP-inequality: we say that the bounded selfadjoint operator H on $\mathcal{X}$ satisfies the *adjoint KYP-inequality* if

$$\begin{bmatrix} A & B \\ C & D \end{bmatrix} \begin{bmatrix} H_* & 0 \\ 0 & I_{\mathcal{U}} \end{bmatrix} \begin{bmatrix} A & B \\ C & D \end{bmatrix}^* \preceq \begin{bmatrix} H_* & 0 \\ 0 & I_{\mathcal{U}} \end{bmatrix}. \quad (5.9)$$

More generally, for a (possibly) unbounded positive-semidefinite operator H_* on $\mathcal{X}$, we say that H_* satisfies the *generalized KYP-inequality* if, for all $x \in \mathcal{D}(H_*^{\frac{1}{2}})$ we have

$$A^*\mathcal{D}(H_*^{\frac{1}{2}}) \subset \mathcal{D}(H_*^{\frac{1}{2}}), \quad C^*\mathcal{Y} \subset \mathcal{D}(H_*^{\frac{1}{2}}), \tag{5.10}$$

and for all $x_* \in \mathcal{D}(H_*^{\frac{1}{2}})$ and $u_* \in \mathcal{Y}$ we have

$$\left\| \begin{bmatrix} H_*^{\frac{1}{2}} & 0 \\ 0 & I_{\mathcal{Y}} \end{bmatrix} \begin{bmatrix} x_* \\ u_* \end{bmatrix} \right\|^2 - \left\| \begin{bmatrix} H_*^{\frac{1}{2}} & 0 \\ 0 & I_{\mathcal{U}} \end{bmatrix} \begin{bmatrix} A^* & C^* \\ B^* & D^* \end{bmatrix} \begin{bmatrix} x_* \\ u_* \end{bmatrix} \right\|^2 \geq 0. \tag{5.11}$$

Then the dual version of Proposition 3.3 is straightforward.

Proposition 5.2. *Suppose that the function S_* has the form* (5.8) *for a (possibly) unbounded positive-semidefinite operator H_* on $\mathcal{X}$. Then S_* is a storage function for Σ^* if and only if H_* is a solution of the generalized adjoint-KYP-inequality* (5.10)–(5.11). *In particular, S_* is a finite-valued storage function for Σ^* if and only if H is a bounded positive-semidefinite operator satisfying the adjoint KYP-inequality* (5.9).

We next discuss a direct connection between positive-definite solutions H of the KYP-inequality (1.5) and positive-definite solutions H_* of the adjoint KYP-inequality (5.9). First let us suppose that H is a bounded strictly positive-definite solution of the KYP-inequality (1.5). Set

$$Q = \begin{bmatrix} H^{\frac{1}{2}} & 0 \\ 0 & I_{\mathcal{Y}} \end{bmatrix} \begin{bmatrix} A & B \\ C & D \end{bmatrix} \begin{bmatrix} H^{-\frac{1}{2}} & 0 \\ 0 & I_{\mathcal{U}} \end{bmatrix}.$$

Then the KYP-inequality (1.5) is equivalent to $Q^*Q \preceq I$, i.e., the fact that the operator $Q\colon \left[\begin{smallmatrix} \mathcal{X} \\ \mathcal{U} \end{smallmatrix}\right] \to \left[\begin{smallmatrix} \mathcal{X} \\ \mathcal{Y} \end{smallmatrix}\right]$ is a contraction operator. But then the adjoint Q^* of Q is also a contraction operator, i.e., $QQ^* \preceq I$. Writing out

$$Q^* = \begin{bmatrix} H^{-\frac{1}{2}} & 0 \\ 0 & I_{\mathcal{U}} \end{bmatrix} \begin{bmatrix} A^* & C^* \\ B^* & D^* \end{bmatrix} \begin{bmatrix} H^{\frac{1}{2}} & 0 \\ 0 & I_{\mathcal{Y}} \end{bmatrix}$$

and rearranging gives

$$\begin{bmatrix} A & B \\ C & D \end{bmatrix} \begin{bmatrix} H^{-1} & 0 \\ 0 & I_{\mathcal{U}} \end{bmatrix} \begin{bmatrix} A^* & C^* \\ B^* & D^* \end{bmatrix} \preceq \begin{bmatrix} H^{-1} & 0 \\ 0 & I_{\mathcal{Y}} \end{bmatrix},$$

i.e., $H_* := H^{-1}$ is a solution of the adjoint KYP-inequality (5.9) for the adjoint system Σ_*. Conversely, by flipping the roles of Σ and Σ^* and using that $\Sigma^{**} = \Sigma$, we see that if H_* is a bounded, strictly positive-definite solution of the adjoint KYP-inequality (5.9), then $H := H_*^{-1}$ is a bounded, strictly positive-definite solution of the KYP-inequality (1.5).

The same correspondence between solutions of the generalized KYP-inequality (1.12)–(1.13) for Σ and solutions of the generalized KYP-inequality for the adjoint system (5.10)–(5.11) continues to hold, but the details are more delicate, as explained in the following proposition. For an alternative proof see Proposition 4.6 in [6].

Proposition 5.3. *Suppose Σ in* (1.1) *is a linear system with system matrix $M = \left[\begin{smallmatrix} A & B \\ C & D \end{smallmatrix}\right]$ while Σ^* is the adjoint system* (5.1) *with system matrix $M^* = \left[\begin{smallmatrix} A^* & C^* \\ B^* & D^* \end{smallmatrix}\right]$. Then the (possibly unbounded) positive-definite operator H is a solution of the generalized KYP-inequality* (1.12)–(1.13) *for Σ if and only if H^{-1} is a positive-definite solution of the generalized KYP-inequality* (5.10)–(1.13) *for Σ^*.*

Proof. Suppose that the positive-definite operator H with dense domain $\mathcal{D}(H)$ in $\mathcal{X}$ solves the generalized KYP-inequality (1.12)–(1.13). Define an operator $Q\colon \left[\begin{smallmatrix} \operatorname{Im} H^{\frac{1}{2}} \\ \mathcal{U} \end{smallmatrix}\right] \to \left[\begin{smallmatrix} \operatorname{Im} H^{\frac{1}{2}} \\ \mathcal{Y} \end{smallmatrix}\right]$ by

$$Q\colon \begin{bmatrix} H^{\frac{1}{2}} & 0 \\ 0 & I_{\mathcal{U}} \end{bmatrix} \begin{bmatrix} x \\ u \end{bmatrix} \mapsto \begin{bmatrix} H^{\frac{1}{2}} & 0 \\ 0 & I_{\mathcal{Y}} \end{bmatrix} \begin{bmatrix} A & B \\ C & D \end{bmatrix} \begin{bmatrix} x \\ u \end{bmatrix}$$

for $x \in \mathcal{D}(H^{\frac{1}{2}})$ and $u \in \mathcal{U}$. We can write the formula for Q more explicitly in terms of $x' = H^{\frac{1}{2}} x \in \operatorname{Im} H^{\frac{1}{2}}$ as

$$Q\colon \begin{bmatrix} x' \\ u \end{bmatrix} \mapsto \begin{bmatrix} H^{\frac{1}{2}} & 0 \\ 0 & I_{\mathcal{Y}} \end{bmatrix} \begin{bmatrix} A & B \\ C & D \end{bmatrix} \begin{bmatrix} H^{-\frac{1}{2}} & 0 \\ 0 & I_{\mathcal{U}} \end{bmatrix} \begin{bmatrix} x' \\ u \end{bmatrix}$$

for $x' \in \operatorname{Im} H^{\frac{1}{2}}$ and $u \in \mathcal{U}$. The content of the generalized KYP-inequality (1.12)–(1.13) is that Q is a well-defined contraction operator from $\left[\begin{smallmatrix} \operatorname{Im} H^{\frac{1}{2}} \\ \mathcal{U} \end{smallmatrix}\right]$ into $\left[\begin{smallmatrix} \mathcal{X} \\ \mathcal{Y} \end{smallmatrix}\right]$ and hence has a uniquely determined contractive extension to a contraction operator from $\left[\begin{smallmatrix} \mathcal{X} \\ \mathcal{U} \end{smallmatrix}\right]$ to $\left[\begin{smallmatrix} \mathcal{X} \\ \mathcal{Y} \end{smallmatrix}\right]$. Let us now choose arbitrary vectors $x \in \mathcal{D}(H^{\frac{1}{2}})$, $x_* \in \mathcal{D}(H^{-\frac{1}{2}}) = \operatorname{Im} H^{\frac{1}{2}}$, $u \in \mathcal{U}$, $u_* \in \mathcal{Y}$ and set $x' = H^{\frac{1}{2}} x$, $x'_* = H^{-\frac{1}{2}} x_*$. Then we compute on the one hand

$$\begin{aligned}
\left\langle \begin{bmatrix} A & B \\ C & D \end{bmatrix} \begin{bmatrix} x \\ u \end{bmatrix}, \begin{bmatrix} x_* \\ u_* \end{bmatrix} \right\rangle &= \left\langle \begin{bmatrix} A & B \\ C & D \end{bmatrix} \begin{bmatrix} H^{-\frac{1}{2}} & 0 \\ 0 & I \end{bmatrix} \begin{bmatrix} x' \\ u \end{bmatrix}, \begin{bmatrix} H^{\frac{1}{2}} x'_* \\ u_* \end{bmatrix} \right\rangle \\
&= \left\langle \begin{bmatrix} H^{\frac{1}{2}} & 0 \\ 0 & I \end{bmatrix} \begin{bmatrix} A & B \\ C & D \end{bmatrix} \begin{bmatrix} H^{-\frac{1}{2}} & 0 \\ 0 & I \end{bmatrix} \begin{bmatrix} x' \\ u \end{bmatrix}, \begin{bmatrix} x'_* \\ u_* \end{bmatrix} \right\rangle \\
&= \left\langle Q \begin{bmatrix} x' \\ u \end{bmatrix}, \begin{bmatrix} x'_* \\ u_* \end{bmatrix} \right\rangle = \left\langle \begin{bmatrix} x' \\ u \end{bmatrix}, Q^* \begin{bmatrix} x'_* \\ u_* \end{bmatrix} \right\rangle \\
&= \left\langle \begin{bmatrix} H^{\frac{1}{2}} x \\ u \end{bmatrix}, Q^* \begin{bmatrix} H^{-\frac{1}{2}} x_* \\ u_* \end{bmatrix} \right\rangle
\end{aligned}$$

while on the other hand

$$\left\langle \begin{bmatrix} A & B \\ C & D \end{bmatrix} \begin{bmatrix} x \\ u \end{bmatrix}, \begin{bmatrix} x_* \\ u_* \end{bmatrix} \right\rangle = \left\langle \begin{bmatrix} x \\ u \end{bmatrix}, \begin{bmatrix} A^* & C^* \\ B^* & D^* \end{bmatrix} \begin{bmatrix} x_* \\ u_* \end{bmatrix} \right\rangle.$$

We thus conclude that

$$\left\langle \begin{bmatrix} H^{\frac{1}{2}} & 0 \\ 0 & I \end{bmatrix} \begin{bmatrix} x \\ u \end{bmatrix}, Q^* \begin{bmatrix} H^{-\frac{1}{2}} x_* \\ u_* \end{bmatrix} \right\rangle = \left\langle \begin{bmatrix} x \\ u \end{bmatrix}, \begin{bmatrix} A^* & C^* \\ B^* & D^* \end{bmatrix} \begin{bmatrix} x_* \\ u_* \end{bmatrix} \right\rangle$$

for all $\begin{bmatrix} x \\ u \end{bmatrix}$ in $\mathcal{D}\left(\begin{bmatrix} H^{\frac{1}{2}} & 0 \\ 0 & I \end{bmatrix}\right)$. Hence

$$Q^* \begin{bmatrix} H^{-\frac{1}{2}} x_* \\ u_* \end{bmatrix} \in \mathcal{D}\left(\begin{bmatrix} H^{\frac{1}{2}} & 0 \\ 0 & I \end{bmatrix}^*\right) = \mathcal{D}\left(\begin{bmatrix} H^{\frac{1}{2}} & 0 \\ 0 & I \end{bmatrix}\right) \tag{5.12}$$

and

$$\begin{bmatrix} H^{\frac{1}{2}} & 0 \\ 0 & I \end{bmatrix} Q^* \begin{bmatrix} H^{-\frac{1}{2}} x_* \\ u_* \end{bmatrix} = \begin{bmatrix} A^* & C^* \\ B^* & D^* \end{bmatrix} \begin{bmatrix} x_* \\ u_* \end{bmatrix} \tag{5.13}$$

where $x_* \in \mathcal{D}(H^{-\frac{1}{2}})$ and $u_* \in \mathcal{Y}$ are arbitrary. From the formula (5.13) we see that $A^* \colon \mathcal{D}(H^{-\frac{1}{2}}) \to \operatorname{Im} H^{\frac{1}{2}} = \mathcal{D}(H^{-\frac{1}{2}})$ and $C^* \colon \mathcal{Y} \to \operatorname{Im} H^{\frac{1}{2}} = \mathcal{D}(H^{-\frac{1}{2}})$, i.e., condition (5.10) holds with $H_* = H^{-1}$. Let us now rewrite equation (5.13) in the form

$$Q^* \begin{bmatrix} H^{-\frac{1}{2}} x_* \\ u_* \end{bmatrix} = \begin{bmatrix} H^{-\frac{1}{2}} & 0 \\ 0 & I \end{bmatrix} \begin{bmatrix} A^* & C^* \\ B^* & D^* \end{bmatrix} \begin{bmatrix} x_* \\ y_* \end{bmatrix}.$$

Using that Q^* is a contraction operator now gives us the spatial KYP-inequality (5.11) with $H_* = H^{-1}$. This completes the proof of Proposition 5.3. □

We next pursue the dual versions of the results of Section 4 concerning the *available storage* and *required supply* as well as the ℓ^2-regularized required supply.

First of all let us note that the Laurent operator $\mathfrak{L}_{F_{\Sigma^*}}$ of F_{Σ^*}, i.e., the inverse Z-transform version of the multiplication operator $M_{F_{\Sigma^*}} = (M_{F_\Sigma})^*$, is just the adjoint of the Laurent operator $\mathfrak{L}_{F_\Sigma}$ given by (4.4). We can rewrite $\mathfrak{L}_{F_{\Sigma^*}}$ in the convenient block form

$$\mathfrak{L}_{F_{\Sigma^*}} = \left[\begin{array}{c|c} \mathfrak{T}_{F_{\Sigma^*}} & \mathfrak{H}_{F_{\Sigma^*}} \\ \hline 0 & \widetilde{\mathfrak{T}}_{F_{\Sigma^*}} \end{array}\right] = \left[\begin{array}{c|c} (\widetilde{\mathfrak{T}}_{F_\Sigma})^* & (\mathfrak{H}_{F_\Sigma})^* \\ \hline 0 & (\mathfrak{T}_{F_\Sigma})^* \end{array}\right] \tag{5.14}$$

where the Toeplitz operators associated with the adjoint system Σ^* are given by

$$\mathfrak{T}_{F_{\Sigma^*}} = (\mathfrak{L}_{F_\Sigma})^*|_{\ell^2_{\mathcal{Y}}(\mathbb{Z}_-)} = (\widetilde{\mathfrak{T}}_{F_\Sigma})^*,$$
$$\widetilde{\mathfrak{T}}_{F_{\Sigma^*}} = P_{\ell^2_{\mathcal{U}}(\mathbb{Z}_+)}(\mathfrak{L}_{F_\Sigma})^*|_{\ell^2_{\mathcal{Y}}(\mathbb{Z}_+)} = (\mathfrak{T}_{F_\Sigma})^*$$

and where the Hankel operator for the adjoint system (already introduced as the inverse Z-transform version of the frequency-domain Hankel operator $\mathbb{H}_{F_{\Sigma^*}}$ given by (5.4)) has the explicit representation in terms of the Laurent operator $\mathfrak{L}_{F_{\Sigma^*}} = (\mathfrak{L}_{F_\Sigma})^*$:

$$\mathfrak{H}_{F_{\Sigma^*}} = P_{\ell^2_{\mathcal{U}}(\mathbb{Z}_-)}(\mathfrak{L}_{F_\Sigma})^*|_{\ell^2_{\mathcal{U}}(\mathbb{Z}_+)}.$$

Let $\boldsymbol{\mathcal{U}}_*$ be the space of all functions $n \mapsto \mathbf{u}_*(n)$ from the integers $\mathbb{Z}$ into the input space $\mathcal{Y}$ for the adjoint system Σ^*. We define the available storage for the adjoint system S_{*a} by

$$S_{*a}(x_0) = \sup_{\mathbf{u} \in \boldsymbol{\mathcal{U}}_*, n_{-1} < 0} \sum_{n=n_{-1}}^{n=-1} (\|\mathbf{y}_*(n)\|^2 - \|\mathbf{u}_*(n)\|^2) \tag{5.15}$$

where the supremum is taken over all adjoint-system trajectories

$$(\mathbf{u}_*(n), \mathbf{x}_*(n), \mathbf{y}_*(n))_{n \le -1}$$

(specified by the adjoint-system equations (5.1) running in backwards time) with final condition $\mathbf{x}_*(-1) = x_0$. Similarly, the dual required supply S_{*r} is given by

$$S_{*r}(x_0) = \inf_{\mathbf{u} \in \boldsymbol{\mathcal{U}},\, n_1 \ge 0} \sum_{n=0}^{n_1} (\|\mathbf{u}_*(n)\|^2 - \|\mathbf{y}_*(n)\|^2) \tag{5.16}$$

where the infimum is taken over system trajectories $(\mathbf{u}_*(n), \mathbf{x}_*(n), \mathbf{y}_*(n))_{n \le n_1}$ subject to the boundary conditions $\mathbf{x}_*(n_1) = 0$ and $\mathbf{x}(-1) = x_0$. Then one applies the analysis behind the proof of Proposition 4.2 to the backward-time system Σ^* in place of the forward-time system Σ to see that S_{*a} and S_{*r} are both storage functions for Σ^* and furthermore $S_{*a}(x_0) \le S_*(x_0) \le S_{*r}(x_0)$, $x_0 \in \mathcal{X}$, for any other Σ^*-storage function S_*. We shall however be primarily interested in the ℓ^2-regularized dual required supply $\underline{S}_{*r}$, rather than in S_{*r}, defined by

$$\underline{S}_{*r}(x_0) = \inf_{\mathbf{u} \in \mathcal{D}(\mathbf{W}_{*c}) :\, \mathbf{W}_{*c}\mathbf{u} = x_0} \sum_{n=0}^{\infty} (\|\mathbf{u}_*(n)\|^2 - \|\mathbf{y}_*(n)\|^2). \tag{5.17}$$

Furthermore, by working out the backward-time analogues of the analysis in Section 4, one can see that $\underline{S}_{*r}$ is also a storage function for Σ^*, and that the definitions of S_{*a} and S_{*r} can be reformulated in a more convenient operator-theoretic form:

$$\begin{aligned} S_{*a}(x_0) &= \sup_{\mathbf{u}_* \in \ell^2_{\mathcal{Y}}(\mathbb{Z}_-)} \|\mathbf{W}_{*o} x_0 + \mathfrak{T}_{F_{\Sigma^*}} \mathbf{u}_*\|^2_{\ell^2_{\mathcal{U}}(\mathbb{Z}_-)} - \|\mathbf{u}_*\|^2_{\ell^2_{\mathcal{Y}}(\mathbb{Z}_-)} \\ &= \sup_{\mathbf{u}_* \in \ell^2_{\mathcal{Y}}(\mathbb{Z}_-)} \|\mathbf{W}_c^* x_0 + \widetilde{\mathfrak{T}}^*_{F_\Sigma} \mathbf{u}_*\|^2_{\ell^2_{\mathcal{U}}(\mathbb{Z}_-)} - \|\mathbf{u}_*\|^2_{\ell^2_{\mathcal{Y}}(\mathbb{Z}_-)} \end{aligned} \tag{5.18}$$

for $x_0 \in \mathcal{D}(\mathbf{W}_c^*)$, with $S_{*a}(x_0) = +\infty$ if $x_0 \notin \mathcal{D}(\mathbf{W}_c^*)$, while

$$\begin{aligned} \underline{S}_{*r}(x_0) &= \inf_{\mathbf{u}_* \in \mathcal{D}(\mathbf{W}_{*c}) :\, \mathbf{W}_{*c}\mathbf{u}_* = x_0} \|\mathbf{u}_*\|^2_{\ell^2_{\mathcal{Y}}(\mathbb{Z}_+)} - \|\widetilde{\mathfrak{T}}_{F_{\Sigma^*}} \mathbf{u}_*\|^2_{\ell^2_{\mathcal{U}}(\mathbb{Z}_+)} \\ &= \inf_{\mathbf{u}_* \in \mathcal{D}(\mathbf{W}_o^*) :\, \mathbf{W}_o^*\mathbf{u}_* = x_0} \|\mathbf{u}_*\|^2_{\ell^2_{\mathcal{Y}}(\mathbb{Z}_+)} - \|\mathfrak{T}^*_{F_\Sigma} \mathbf{u}_*\|^2_{\ell^2_{\mathcal{U}}(\mathbb{Z}_+)} \\ &= \inf_{\mathbf{u}_* \in \mathcal{D}(\mathbf{W}_o^*),\, \mathbf{W}_o^*\mathbf{u}_* = x_0} \|D_{\mathfrak{T}^*_{F_\Sigma}} \mathbf{u}_*\|^2. \end{aligned} \tag{5.19}$$

By notational adjustments to the arguments in the proof of Theorem 4.9, we arrive at the following formulas for S_{*a} and $\underline{S}_{*r}$ on $\operatorname{Im} \mathbf{W}_o^*$.

Theorem 5.4. *Let the operators $\overline{X}_a$, $\overline{X}_r$ be as in Lemma 4.8 and define operators H_a and H_r as in* (4.34) *and* (4.35)*. Then the dual available storage S_{*a} and the dual ℓ^2-regularized required supply are given (on a suitably restricted domain) by*

$$S_{*a}(x_0) = \|\overline{X}_r x_0\|^2 = \|H_r^{-\frac{1}{2}} x_0\|^2 \text{ for } x_0 \in \operatorname{Im} \mathbf{W}_o^*, \tag{5.20}$$

$$\underline{S}_{*r}(x_0) = \||\overline{X}_a|^{-1} x_0\|^2 = \|H_a^{-\frac{1}{2}} x_0\|^2 \text{ for } x_0 \in \operatorname{Im} \mathbf{W}_o^*. \tag{5.21}$$

Let us associate extended-real-valued functions S_{H_a}, S_{H_r}, $S_{H_r^{-1}}$, $S_{H_a^{-1}}$ with the positive-definite operators H_a, H_r, H_r^{-1}, H_a^{-1} as in (3.6). Theorems 4.9 and 5.4 give us the close relationship between these functions and the functions S_a, $\underline{S}_r$ (storage functions for Σ) and S_{*a}, $\underline{S}_{*r}$ (storage functions for Σ^*), namely:

$$\begin{aligned} S_a(x) &= S_{H_a}(x), \quad \underline{S}_r(x) = S_{H_r}(x) \quad \text{for } x \in \operatorname{Im} \mathbf{W}_c, \\ S_{*a}(x) &= S_{H_r^{-1}}(x), \quad \underline{S}_{*r}(x) = S_{H_a^{-1}}(x) \quad \text{for } x \in \operatorname{Im} \mathbf{W}_o^*. \end{aligned} \tag{5.22}$$

In general we do not assert that equality holds in any of the four equalities in (5.22) for all $x \in \mathcal{X}$. Nevertheless it is the case that S_{H_a} and S_{H_r} are storage functions for Σ and $S_{H_r^{-1}}$ and $S_{H_a^{-1}}$ are storage functions for Σ^*, as we now explain.

Proposition 5.5. *Let H_a, H_r, H_r^{-1}, H_a^{-1} be the positive-definite operators as in Theorems* 4.9 *and* 5.4. *Then the following hold:*

(1) *S_{H_a} and S_{H_r} are nondegenerate storage functions for Σ, or equivalently, H_a and H_r are positive-definite solutions of the generalized KYP-inequality* (1.12)–(1.13) *for Σ.*
(2) *$S_{H_r^{-1}}$ and $S_{H_a^{-1}}$ are storage functions for Σ^*, or equivalently, H_r^{-1} and H_a^{-1} are positive-definite solutions of the generalized KYP-inequality* (5.10)–(5.11) *for Σ^*.*

Proof. The fact that S_H is a nondegenerate storage function for Σ (respectively Σ^*) if and only if H is a positive-definite solution of the generalized KYP-inequality for Σ (respectively Σ^*) is a consequence of Proposition 3.3 and its dual Proposition 5.2. We shall use these formulations interchangeably.

We know that $S_{H_a}(x) = S_a(x)$ for $x \in \operatorname{Im} \mathbf{W}_c$. Furthermore as a consequence of (2.18) with $\widetilde{u} = 0$ and of (2.17) with $n_{-1} = -1$, we see that $\operatorname{Im} \mathbf{W}_c$ is invariant under A and contains $\operatorname{Im} B$. Thus condition (1.12) holds with $\operatorname{Im} \mathbf{W}_c$ in place of $\mathcal{D}(H^{\frac{1}{2}})$. The facts that S_{H_a} agrees with S_a on $\operatorname{Im} \mathbf{W}_c$ and that S_a is a storage function for Σ imply that the inequality (1.13) holds for $x \in \operatorname{Im} \mathbf{W}_c$ and $u \in \mathcal{U}$:

$$\left\| \begin{bmatrix} H_a^{\frac{1}{2}} & 0 \\ 0 & I_{\mathcal{U}} \end{bmatrix} \begin{bmatrix} x \\ u \end{bmatrix} \right\|^2 - \left\| \begin{bmatrix} H_a^{\frac{1}{2}} & 0 \\ 0 & I_{\mathcal{Y}} \end{bmatrix} \begin{bmatrix} A & B \\ C & D \end{bmatrix} \begin{bmatrix} x \\ u \end{bmatrix} \right\|^2 \geq 0. \tag{5.23}$$

As noted at the end of Theorem 4.9, $\operatorname{Im} \mathbf{W}_c$ is a core for $H_a^{\frac{1}{2}}$; hence, given $x \in \mathcal{D}(H_a)$, there is a sequence of points $\{x_n\}_{n \geq 1}$ contained in $\operatorname{Im} \mathbf{W}_c$ such that $\lim_{n \to \infty} x_n = x$ and $\lim_{n \to \infty} H^{\frac{1}{2}} x_n = H^{\frac{1}{2}} x$. As each $x_n \in \operatorname{Im} \mathbf{W}_c$, we know that the inequality (5.23) holds with x_n in place of x for all $n = 1, 2, \dots$. We may now take limits in this inequality to see that the inequality continues to hold with $x = \lim_{n \to \infty} x_n \in \mathcal{D}(H_a^{\frac{1}{2}})$, i.e., condition (1.13) holds with H_a in place of H. Thus H is a solution of the generalized KYP-inequality for Σ. That H_r^{-1} is a solution of the generalized KYP-inequality for Σ^* now follows by applying the same analysis to Σ^* rather than to Σ. Finally, the fact that H_a (respectively, H_r^{-1}) is a positive-definite solution of the generalized

KYP-inequality for Σ (respectively for Σ^*) implies that H_a^{-1} (respectively, H_r) is a positive-definite solution of the generalized KYP-inequality for Σ^* (respectively, Σ) as a consequence of Proposition 5.3. □

6. Order properties of solutions of the generalized KYP-inequality and finer results for special cases

We have implicitly been using an order relation on storage functions, namely: we say that $S_1 \leq S_2$ if $S_1(x_0) \leq S_2(x_0)$ for all $x_0 \in \mathcal{X}$. For the case of quadratic storage functions S_{H_1} and S_{H_2} where H_1 and H_2 are two positive-semidefinite solutions of the generalized KYP-inequality (1.12)–(1.13), the induced ordering $\leq$ on positive-semidefinite (possibly unbounded) operators can be defined as follows: given two positive-semidefinite operators H_1 with dense domain $\mathcal{D}(H_1)$ and H_2 with dense domain $\mathcal{D}(H_2)$ in $\mathcal{X}$, we say that $H_1 \leq H_2$ if $\mathcal{D}(H_2^{\frac{1}{2}}) \subset \mathcal{D}(H_1^{\frac{1}{2}})$ and

$$\|H_1^{\frac{1}{2}}x\|^2 \leq \|H_2^{\frac{1}{2}}x\|^2 \text{ for all } x \in \mathcal{D}(H_2^{\frac{1}{2}}). \tag{6.1}$$

In case H_1 and H_2 are bounded positive-semidefinite operators, one can see that $H_1 \leq H_2$ is equivalent to $H_1 \preceq H_2$ in the sense of the inequality between quadratic forms: $\langle H_1x, x\rangle \leq \langle H_2x, x\rangle$, i.e., in the Loewner partial order: $H_2 - H_1 \succeq 0$. This ordering $\leq$ on (possibly unbounded) positive-semidefinite operators has appeared in the more general context of closed quadratic forms S_H (not necessarily storage functions for some dissipative system Σ) and associated semibounded selfadjoint operators H (not necessarily solving some generalized KYP-inequality); see formula (2.17) and the subsequent remark in the book of Kato [20]. This order has been studied in the setting of solutions of a generalized KYP-inequality in the paper of Arov–Kaashoek–Pik [6]. Here we offer a few additional such order properties which follow from the results developed here. Recall that the notion of a *core* of a closed, densely defined linear operator was introduced in the paragraph preceding Theorem 4.9.

Theorem 6.1. *Assume that the system* Σ *in* (1.1) *satisfies the standing assumption* (4.12) *and* H_a *and* H_r *are defined by* (4.34) *and* (4.35). *Let* H *be any positive-definite solution of the generalized KYP-inequality* (1.12)–(1.13).

(1) *Assume that* $\operatorname{Im} \mathbf{W}_c$ *is a core for* $H^{\frac{1}{2}}$. *Then we have the operator inequality*

$$H_a \leq H \tag{6.2}$$

and furthermore $\operatorname{Im} \mathbf{W}_o^* \subset \mathcal{D}(H^{-\frac{1}{2}})$.

(2) *Assume that* $\operatorname{Im} \mathbf{W}_o^*$ *is a core for* $H^{-\frac{1}{2}}$. *Then we have the operator inequality*

$$H \leq H_r \tag{6.3}$$

and furthermore $\operatorname{Im} \mathbf{W}_c \subset \mathcal{D}(H^{\frac{1}{2}})$.

Proof. We deal with (1) and (2) in turn.

(1) Suppose that H is a positive-definite solution of the generalized KYP-inequality such that $\operatorname{Im}\mathbf{W}_c$ is a core for $H^{\frac{1}{2}}$. From Theorem 4.9, we know that $S_a(x) = \|H_a^{\frac{1}{2}}x\|$ for $x \in \operatorname{Im}\mathbf{W}_c$. Since S_a is the smallest storage function (see Proposition 4.2) and S_H is a storage function, it follows that

$$\|H_a^{\frac{1}{2}}x\|^2 = S_a(x) \leq S_H(x) = \|H^{\frac{1}{2}}x\|^2 \text{ for } x \in \operatorname{Im}\mathbf{W}_c. \tag{6.4}$$

Let now x be an arbitrary point of $\mathcal{D}(H^{\frac{1}{2}})$. Since $\operatorname{Im}\mathbf{W}_c$ is a core for $H^{\frac{1}{2}}$, we can find a sequence $\{x_n\}_{n\geq 1}$ of points in $\operatorname{Im}\mathbf{W}_c$ such that $x_n \to x$ and $H^{\frac{1}{2}}x_n \to H^{\frac{1}{2}}x$. In particular $H^{\frac{1}{2}}x_n$ is a Cauchy sequence and the inequality

$$\begin{aligned}\|H_a^{\frac{1}{2}}x_n - H_a^{\frac{1}{2}}x_m\|^2 &= \|H_a^{\frac{1}{2}}(x_n - x_m)\|^2\\ &\leq \|H^{\frac{1}{2}}(x_n - x_m)\|^2 = \|H^{\frac{1}{2}}x_n - H^{\frac{1}{2}}x_m\|^2\end{aligned}$$

implies that $\{H_a^{\frac{1}{2}}x_n\}_{n\geq 1}$ is Cauchy as well, so converges to some $y \in \mathcal{X}$. As H_a is closed, we get that $x \in \mathcal{D}(H_a)$ and $y = H_a^{\frac{1}{2}}x$. We may then take limits in the inequality $\|H_a^{\frac{1}{2}}x_n\|^2 \leq \|H^{\frac{1}{2}}x_n\|^2$ holding for all n (a consequence of (6.4)) to conclude that $\|H_a^{\frac{1}{2}}x\|^2 \leq \|H^{\frac{1}{2}}x\|^2$, i.e., $H_a \leq H$, i.e., (6.2) holds.

Recall next from Corollary 4.7 that $\|\mathbf{W}_o x_0\|^2 \leq S_a(x_0)$, where we now also know from Theorem 4.9 that $S_a(x_0) = \|H^{\frac{1}{2}}x_0\|^2$ for $x_0 \in \operatorname{Im}\mathbf{W}_c$. We thus have the chain of operator inequalities

$$\mathbf{W}_o^*\mathbf{W}_o \leq H_a \leq H.$$

By Proposition 3.4 in [5], we may equivalently write

$$H^{-1} \leq H_a^{-1} \leq (\mathbf{W}_o^*\mathbf{W}_o)^{-1}.$$

In particular $\mathcal{D}(|\mathbf{W}_o|^{-1}) \subset \mathcal{D}(H^{-\frac{1}{2}})$. If we introduce the polar decomposition $\mathbf{W}_o = U_o|\mathbf{W}_o|$ for $\mathbf{W}_o$, we see that $\mathbf{W}_o^* = |\mathbf{W}_o|U_o^*$ and $\operatorname{Im}\mathbf{W}_o^* = \operatorname{Im}|\mathbf{W}_o|$. Thus

$$\mathcal{D}(|\mathbf{W}_o|^{-1}) = \operatorname{Im}|\mathbf{W}_o| = \operatorname{Im}\mathbf{W}_o^*$$

and it follows that $\operatorname{Im}\mathbf{W}_o^* \subset \mathcal{D}(H^{-\frac{1}{2}})$ and the verification of (1) is complete.

(2) We now suppose that H is a positive-definite solution of the generalized KYP-inequality such that $\operatorname{Im}\mathbf{W}_o^*$ is a core for $H^{-\frac{1}{2}}$. By applying the result of part (1) to the adjoint system Σ^*, we see that $H_r^{-1} \leq H^{-1}$ and that $\operatorname{Im}\mathbf{W}_c \subset H^{\frac{1}{2}}$. If we apply the result of Proposition 3.4 in [5], we see that $H_r^{-1} \leq H^{-1}$ implies that (or is actually equivalent to) $H \leq H_r$, completing the verification of (2). □

Remark 6.2. By the last assertion in Theorem 4.9, we know that $\operatorname{Im}\mathbf{W}_c$ is a core for $H_a^{\frac{1}{2}}$ and that $\operatorname{Im}\mathbf{W}_o^*$ is a core for $H_r^{-\frac{1}{2}}$. Also by Proposition 5.5 we know that H_a and H_r are positive-definite solutions of the generalized KYP-inequality for Σ. Thus item (1) in Theorem 6.1 may be rephrased as follows:

- *The set $\mathcal{GS}_c$ consisting of all positive-definite solutions H of the generalized KYP-inequality* (1.12)–(1.13) *for* Σ *such that* $\operatorname{Im} \mathbf{W}_c$ *is a core for* $H^{\frac{1}{2}}$ *has the solution* H_a *as a minimal element with respect to the ordering* $\leq$.

Similarly item (2) in Theorem 6.1 may be rephrased as:

- *The set $\mathcal{GS}_o$ consisting of all positive-definite solutions H of the generalized KYP-inequality* (1.12)–(1.13) *such that* $\operatorname{Im} \mathbf{W}_o^*$ *is a core for* $H^{-\frac{1}{2}}$ *has the solution* H_r *as a maximal element with respect to the ordering* $\leq$.

It would be tempting to say:

- *The set $\mathcal{GS}$ consisting of all positive-definite solutions H of the generalized KYP-inequality* (1.12)–(1.13) *such that* $\operatorname{Im} \mathbf{W}_c$ *is a core for* $H^{\frac{1}{2}}$ *and* $\operatorname{Im} \mathbf{W}_o^*$ *is a core for* $H^{-\frac{1}{2}}$ *has* H_a *as a minimal element and* H_r *as a maximal element with respect to the ordering* $\leq$.

However, while the above results imply that $\operatorname{Im} \mathbf{W}_c \subset \mathcal{D}(H_r^{\frac{1}{2}})$ and that $\operatorname{Im} \mathbf{W}_o^* \subset \mathcal{D}(H_a^{-\frac{1}{2}})$, we have not been able to show in general that $\operatorname{Im} \mathbf{W}_c$ is a core for $H_r^{\frac{1}{2}}$ or that $\operatorname{Im} \mathbf{W}_o^*$ is a core for $H_a^{-\frac{1}{2}}$. Such a more satisfying symmetric statement does hold in the pseudo-similarity framework for the analysis of solutions of generalized KYP-inequalities (see Proposition 5.8 in [6]).

We now consider the case that Σ is not only controllable and/or observable, but has the stronger ℓ^2-exact controllability or ℓ^2-exact observability condition, or both, i.e., ℓ^2-exact minimality. We first consider the implications on H_a and H_r.

Proposition 6.3. *Let* Σ *be a system as in* (1.1) *such that assumption* (4.12) *holds.*

(1) *If* Σ *is* ℓ^2*-exactly controllable, then* H_a *and* H_r *are bounded.*
(2) *If* Σ *is* ℓ^2*-exactly observable, then* H_a *and* H_r *are boundedly invertible.*
(3) Σ *is* ℓ^2*-exactly minimal, i.e., both* ℓ^2*-exactly controllable and* ℓ^2*-exactly observable, then* H_a *and* H_r *are both bounded and boundedly invertible.*

Proof. We discuss each of (1), (2), (3) in turn.

(1) Item (1) follows directly from the fact that $\operatorname{Im} \mathbf{W}_c$ is contained in both $\mathcal{D}(H_a)$ and $\mathcal{D}(H_r)$ together with the Closed Graph Theorem.

(2) From the last assertion in Theorem 4.9, we know that $\operatorname{Im} \mathbf{W}_c$ is a core for $H_a^{\frac{1}{2}}$. Then item (1) in Theorem 6.1 implies that $\operatorname{Im} \mathbf{W}_o^* \subset \mathcal{D}(H_a^{-\frac{1}{2}})$. If $\operatorname{Im} \mathbf{W}_o^* = \mathcal{X}$, the Closed Graph Theorem then gives us that $H_a^{-\frac{1}{2}}$ is bounded.

Also part of the last assertion of Theorem 4.9 is the statement that $\mathbf{W}_o^*$ is a core for $H_r^{-\frac{1}{2}}$, so in particular $\operatorname{Im} \mathbf{W}_o^* \subset \mathcal{D}(H_r^{-\frac{1}{2}})$. Then again the Closed Graph Theorem implies that $H_r^{-\frac{1}{2}}$ is bounded.

(3) Simply combine the results of items (1) and (2). □

Next we consider general positive-definite solutions to the generalized KYP-inequality.

Proposition 6.4. *Suppose that* Σ *is a system as in* (1.1) *such that assumption* (4.12) *holds and that* H *is any positive-definite solution of the generalized KYP-inequality.*

(1) *Suppose that* Σ *is* ℓ^2*-exactly controllable and that* $\operatorname{Im}\mathbf{W}_c \subset \mathcal{D}(H^{\frac{1}{2}})$ (*as is the case, e.g., if* $\operatorname{Im}\mathbf{W}_o^*$ *is a core for* $H^{-\frac{1}{2}}$). *Then* H *is bounded and furthermore*

$$H_a \le H.$$

(2) *Suppose that* Σ *is* ℓ^2*-exactly observable and that* $\operatorname{Im}\mathbf{W}_o^* \subset \mathcal{D}(H^{-\frac{1}{2}})$ (*as is the case, e.g., if* $\operatorname{Im}\mathbf{W}_c$ *is a core for* $H^{\frac{1}{2}}$). *Then* H^{-1} *is bounded and furthermore*

$$H \le H_r.$$

(3) *Suppose that* Σ *is both* ℓ^2*-exactly controllable and* ℓ^2*-exactly observable and that either* (a) $\operatorname{Im}\mathbf{W}_c \subset \mathcal{D}(H^{\frac{1}{2}})$ *or* (b) $\operatorname{Im}\mathbf{W}_o^* \subset \mathcal{D}(H^{-\frac{1}{2}})$. *Then* H *is bounded and boundedly invertible and we have the inequality chain*

$$H_a \le H \le H_r. \tag{6.5}$$

Proof. First note that the fact that the parenthetical hypotheses in items (1) and (2) are stronger than the given hypotheses is a consequence of the final assertions in parts (1) and (2) of Theorem 6.1. We now deal with the rest of (1), (2), (3).

(1) If we assume that $\mathcal{X} = \operatorname{Im}\mathbf{W}_c \subset \mathcal{D}(H^{\frac{1}{2}})$, then $H^{\frac{1}{2}}$ (and hence also H) is bounded by the Closed Graph Theorem. Moreover, as $\operatorname{Im}\mathbf{W}_c = \mathcal{D}(H^{\frac{1}{2}})$, in particular $\operatorname{Im}\mathbf{W}_c$ is a core for $H^{\frac{1}{2}}$ and the inequality $H_a \le H$ follows from Theorem 6.1 (1).

(2) Similarly, if we assume $\mathcal{X} = \operatorname{Im}\mathbf{W}_o^* \subset \mathcal{D}(H^{-\frac{1}{2}})$, then $H^{-\frac{1}{2}}$ is bounded by the Closed Graph Theorem. As $\operatorname{Im}\mathbf{W}_o^* = \mathcal{D}(H^{-\frac{1}{2}})$, in particular $\operatorname{Im}\mathbf{W}_o^*$ is a core for $H^{-\frac{1}{2}}$ and $H \le H_r$ follows as a consequence of Theorem 6.1 (2).

(3) If $\mathcal{X} = \operatorname{Im}\mathbf{W}_c \subset \mathcal{D}(H^{\frac{1}{2}})$, then in fact $\operatorname{Im}\mathbf{W}_c = \mathcal{D}(H^{\frac{1}{2}})$ so $\operatorname{Im}\mathbf{W}_c$ is a core for $H^{\frac{1}{2}}$. By Theorem 6.1, it follows that $\operatorname{Im}\mathbf{W}_o^* \subset \mathcal{D}(H^{-\frac{1}{2}})$ and hence hypothesis (b) is a consequence of hypothesis (a) when combined with all the other hypotheses in (3). Similarly hypothesis (a) is a consequence of hypothesis (b). Hence there is no loss of generality in assuming that both (a) and (b) hold. Then the verification of (3) is completed by simply combining the results of (1) and (2). □

7. Proofs of Bounded Real Lemmas

We now put all the pieces together to give a storage-function proof of Theorem 1.3.

Proof of Theorem 1.3. We are given a minimal system Σ as in (1.1) with transfer function F_Σ in the Schur class $\mathcal{S}(\mathcal{U}, \mathcal{Y})$.

Proof of sufficiency. For the sufficiency direction, we assume either that there exists a positive-definite solution H of the generalized KYP-inequality (1.12)–(1.13) (statement (1)) or a bounded and boundedly invertible solution H of the KYP-inequality (1.5) (statements (2) and (3)). As the latter case is a particular version of the former case, it suffices to assume that we have a positive-definite solution of the generalized KYP-inequality (1.12)–(1.13). We are to show that then F_Σ is in the Schur class $\mathcal{S}(\mathcal{U}, \mathcal{Y})$.

Given such a generalized solution of the KYP-inequality, Proposition 3.3 guarantees us that S_H is an (even quadratic) storage function for Σ. Then F_Σ has analytic continuation to a Schur-class function by Proposition 3.1.

Proof of necessity in statement (1). We assume that Σ is minimal and that F_Σ has analytic continuation to a Schur-class function, i.e., assumption (4.12) holds. Then Proposition 5.5 gives us two choices H_a and H_r of positive-definite solutions of the generalized KYP-inequality (1.12)–(1.13).

Proof of necessity in statement (2). We assume that Σ is exactly controllable and exactly observable with transfer function F_Σ having analytic continuation to the Schur class. From Proposition 2.2 (1) we see that $\operatorname{Im} \mathbf{W}_c \supset \operatorname{Rea}(A|B) = \mathcal{X}$ and that $\mathcal{D}(\mathbf{W}_o) \supset \operatorname{Rea}(A|B) = \mathcal{X}$ while from item (2) in the same proposition we see that $\operatorname{Im} \mathbf{W}_o^* \supset \operatorname{Obs}(C|A) = \mathcal{X}$ and that $\mathcal{D}(\mathbf{W}_c^*) \supset \operatorname{Obs}(C|A) = \mathcal{X}$. Hence by the Closed Graph Theorem, in fact $\mathbf{W}_c$ and $\mathbf{W}_o^*$ are bounded in addition to being surjective. In particular Σ is ℓ^2-exactly controllable and ℓ^2-exactly observable, so this case actually falls under item (3) of Theorem 1.3, which we will prove next.

Proof of necessity in statement (3). We now assume that Σ is ℓ^2-exactly controllable and ℓ^2-exactly observable with F_Σ having analytic continuation to a function in the Schur class $\mathcal{S}(\mathcal{U}, \mathcal{Y})$ and we want to produce a bounded and boundedly invertible solution H of the KYP-inequality (1.5). In particular, Σ is minimal (controllable and observable), so Proposition 5.5 gives us two solutions H_a and H_r of the generalized KYP-inequality. But any solution H of the generalized KYP-inequality becomes a solution of the standard KYP-inequality (1.5) if it happens to be the case that H is bounded. By the result of item (3) in Proposition 6.3, both H_a and H_r are bounded and boundedly invertible under our ℓ^2-minimality assumptions. Thus in this case H_a and H_r serve as two choices for bounded, strictly positive-definite solutions of the KYP-inequality, as needed. □

We are now ready also for a storage-function proof of Theorem 1.4.

Proof of Theorem 1.4. The standing assumption for both directions is that Σ is a linear system as in (1.1) with exponentially stable state operator A.

Proof of necessity. Assume that there exists a bounded strictly positive-definite solution H of the strict KYP-inequality. By Proposition 3.5, S_H is a strict storage function for Σ. Then by Proposition 3.4, F_Σ has analytic continuation to an $\mathcal{L}(\mathcal{U}, \mathcal{Y})$-valued H^∞-function with H^∞-norm strictly less than 1 as wanted. The fact that A is exponentially stable implies that F_Σ has analytic continuation to a slightly larger disk beyond $\mathbb{D}$, and the fact that

H is strictly positive-definite implies that S_H has the additional coercivity property $S_H(x) \geq \epsilon_0 \|x\|^2$ for some $\epsilon_0 > 0$.

Proof of sufficiency. We are assuming that Σ has state operator A exponentially stable and with transfer function F_Σ in the strict Schur class. The exponential stability of A (i.e., A has spectral radius $r_{\rm spec}(A) < 1$) means that the series

$$\mathbf{W}_o^* \mathbf{y} = \sum_{k=0}^{\infty} A^{*k} C^* \mathbf{y}(k) \ \ (\mathbf{y} \in \ell^2_{\mathcal{Y}}(\mathbb{Z}_+)), \quad \mathbf{W}_c \mathbf{u} = \sum_{k=0}^{\infty} A^k B \mathbf{u}(k) \ \ (\mathbf{u} \in \ell^2_{\mathcal{U}}(\mathbb{Z}_-))$$

are norm-convergent (not just in the weak sense as in Proposition 2.1), and hence $\mathbf{W}_c$ and $\mathbf{W}_o$ are bounded. However it need not be the case that $\mathbf{W}_c$ or $\mathbf{W}_o^*$ be surjective, so we are not in a position to apply part (3) of Theorem 1.3 to the system Σ. The adjustment for handling this difficulty which also ultimately produces bounded and boundedly invertible solutions of the strict KYP-inequality (1.9) is what we shall call *ϵ-regularization reduction.* It goes back at least to Petersen–Anderson–Jonkheere [24] for the finite-dimensional case, and was extended to the infinite-dimensional case in our previous paper [11]. We recall the procedure here for completeness and because we refer to it in a subsequent remark.

Since $r_{\rm spec}(A) < 1$, the resolvent expression $(I - \lambda A)^{-1}$ is uniformly bounded for all λ in the unit disk $\mathbb{D}$. Since we are now assuming that F_Σ is in the strict Schur class, it follows that we can choose $\epsilon > 0$ sufficiently small so that the augmented matrix function

$$F_\epsilon(\lambda) := \begin{bmatrix} F(\lambda) & \epsilon\lambda C(I - \lambda A)^{-1} \\ \epsilon\lambda(I - \lambda A)^{-1} B & \epsilon^2 \lambda (I - \lambda A)^{-1} \\ \epsilon I_{\mathcal{U}} & 0 \end{bmatrix} \tag{7.1}$$

is in the strict Schur class $\mathcal{S}^o(\mathcal{U} \oplus \mathcal{X}, \mathcal{Y} \oplus \mathcal{X} \oplus \mathcal{U})$. Note that

$$F_\epsilon(\lambda) = \begin{bmatrix} D & 0 \\ 0 & 0 \\ \epsilon I_{\mathcal{U}} & 0 \end{bmatrix} + \lambda \begin{bmatrix} C \\ \epsilon I_{\mathcal{X}} \\ 0 \end{bmatrix} (I - \lambda A)^{-1} \begin{bmatrix} B & \epsilon I_{\mathcal{X}} \end{bmatrix}$$

and hence

$$M_\epsilon = \begin{bmatrix} \mathbf{A} & \mathbf{B} \\ \mathbf{C} & \mathbf{D} \end{bmatrix} := \left[\begin{array}{c|cc} A & B & \epsilon I_{\mathcal{X}} \\ \hline C & D & 0 \\ \epsilon I_{\mathcal{X}} & 0 & 0 \\ 0 & \epsilon I_{\mathcal{U}} & 0 \end{array}\right] \tag{7.2}$$

is a realization for $F_\epsilon(\lambda)$. Suppose that we can find a bounded and boundedly invertible positive-definite operator H satisfying the KYP-inequality (1.5) associated with the system Σ_ϵ:

$$\begin{bmatrix} \mathbf{A}^* & \mathbf{C}^* \\ \mathbf{B}^* & \mathbf{D}^* \end{bmatrix} \begin{bmatrix} H & 0 \\ 0 & I_{\mathcal{Y} \oplus \mathcal{X} \oplus \mathcal{U}} \end{bmatrix} \begin{bmatrix} \mathbf{A} & \mathbf{B} \\ \mathbf{C} & \mathbf{D} \end{bmatrix} \preceq \begin{bmatrix} H & 0 \\ 0 & I_{\mathcal{U} \oplus \mathcal{X}} \end{bmatrix}. \tag{7.3}$$

Spelling this out gives

$$\begin{bmatrix} A^*HA + C^*C + \epsilon^2 I_{\mathcal{X}} & A^*HB + C^*D & \epsilon A^*H \\ B^*HA + D^*C & B^*HB + D^*D + \epsilon^2 I_{\mathcal{U}} & \epsilon B^*H \\ \epsilon HA & \epsilon HB & \epsilon^2 H \end{bmatrix} \preceq \begin{bmatrix} H & 0 & 0 \\ 0 & I_{\mathcal{U}} & 0 \\ 0 & 0 & I_{\mathcal{X}} \end{bmatrix}.$$

By crossing off the third row and third column, we get the inequality

$$\begin{bmatrix} A^*HA + C^*C + \epsilon^2 I_{\mathcal{X}} & A^*HB + C^*D \\ B^*HA + D^*C & B^*HB + D^*D + \epsilon^2 I_{\mathcal{U}} \end{bmatrix} \preceq \begin{bmatrix} H & 0 \\ 0 & I_{\mathcal{U}} \end{bmatrix}$$

or

$$\begin{bmatrix} A^* & C^* \\ B^* & D^* \end{bmatrix} \begin{bmatrix} H & 0 \\ 0 & I_{\mathcal{Y}} \end{bmatrix} \begin{bmatrix} A & B \\ C & D \end{bmatrix} + \epsilon^2 \begin{bmatrix} I_{\mathcal{X}} & 0 \\ 0 & I_{\mathcal{U}} \end{bmatrix} \preceq \begin{bmatrix} H & 0 \\ 0 & I_{\mathcal{U}} \end{bmatrix}$$

leading us to the strict KYP-inequality (1.9) for the original system Σ as wanted.

It remains only to see why there is a bounded and boundedly invertible solution H of (7.3). It is easily checked that the system Σ_ϵ is exactly controllable and exactly minimal, since $\mathbf{B}$ and $\mathbf{C}^*$ are both already surjective; as observed in the proof of necessity in item (2) of Theorem 1.3, since F_{Σ_ϵ} is in the Schur class it then follows that Σ_ϵ is ℓ^2-exactly controllable and ℓ^2-exactly observable as well. Hence we can appeal to either items (2) or (3) of Theorem 1.3 to conclude that indeed the KYP-inequality (7.3) has a bounded and boundedly invertible positive-definite solution. This is what is done in [11], where the State-Space-Similarity approach is used to prove items (2) and (3) in Theorem 1.3 rather the storage-function approach as is done here. □

Remark 7.1. Let Σ and F_Σ satisfy the conditions of the strict Bounded Real Lemma (Theorem 1.4). Define the ϵ-augmented system Σ_ϵ as in (7.2). We then obtain bounded, strictly positive-definite solutions $H_{a,\epsilon}$ and $H_{r,\epsilon}$ of the strict KYP-inequality (1.9), and consequently, by Proposition 6.4 (3) all bounded or bounded below solutions H to the generalized KYP-inequality (1.12)–(1.13) for Σ_ϵ satisfy $H_{a,\epsilon} \leq H \leq H_{r,\epsilon}$ and hence are in fact bounded, strictly positive-definite solutions to the KYP-inequality (1.5) for the original system Σ. An application of Theorem 6.1 together with the observation that $\mathcal{D}$ being a core for the bounded operator X on $\mathcal{X}$ is the same as $\mathcal{D}$ being dense in $\mathcal{X}$ leads to the conclusion that the operators H_a and H_r associated with the original system satisfy $H_a \leq H_{a,\epsilon}$ and $H_r^{-1} \leq H_{r,\epsilon}^{-1}$ and hence are bounded. However, this by itself is not enough to conclude that H_a and H_r^{-1} are also bounded below.

Acknowledgements. This work is based on the research supported in part by the National Research Foundation of South Africa. Any opinion, finding and conclusion or recommendation expressed in this material is that of the authors, and the NRF does not accept any liability in this regard.

It is a pleasure to thank Olof Staffans for enlightening discussion (both verbal and written) and Mikael Kurula for his useful comments while visiting the first author in July 2017.

References

[1] B.D.O. Anderson and S. Vongpanitlerd, *Network Analysis and Synthesis: A Modern Systems Theory Approach*, Prentice-Hall, Englewood Cliffs, 1973.

[2] D.Z. Arov, Scattering theory with dissipation of energy, *Dokl. Akad. Nauk SSSR*, **216**(1974), 713–716 [Russian]; English (with addenda): *Sov. Math. Dokl.*, **15** (1974), 149–162.

[3] D.Z. Arov, Stable dissipative linear stationary dynamical scattering systems, *J. Operator Theory*, **1** (1979), 95–126 [Russian]; English translation in: *Interpolation Theory, Systems Theory and Related Topics: The Harry Dym Anniversary Volume*, pp. 99–136, Oper. Theory Adv. Appl. **134**, Birkhäuser, Basel, 2002.

[4] D.Z. Arov, M.A. Kaashoek, and D.R. Pik, Minimal and optimal linear discrete time-invariant dissipative scattering systems, *Integr. Equ. Oper. Theory*, **29** (1997), 127–154.

[5] D.Z. Arov, M.A. Kaashoek, and D.R. Pik, Minimal representations of a contractive operator as a product of two bounded operators, *Acta Sci. Math. (Szeged)*, **71** (2005), 313–336.

[6] D.Z. Arov, M.A. Kaashoek, and D.R. Pik, The Kalman-Yakubovich-Popov inequality for discrete time systems of infinite dimension, *J. Operator Theory*, **55** (2006), 393–438.

[7] D.Z. Arov, M.A. Kaashoek, and D.R. Pik, Generalized solutions of Riccati equalities and inequalities, *Methods Funct. Anal. Topology*, **22** (2016), 95–116.

[8] D.Z. Arov and O.J. Staffans, The infinite-dimensional continuous time Kalman-Yakubovich-Popov inequality, in: *The Extended Field of Operator Theory*, pp. 37–72, Oper. Theory Adv. Appl. **171**, Birkhäuser, Basel, 2007.

[9] J.A. Ball and N. Cohen, De Branges-Rovnyak operator models and systems theory: a survey, in: *Topics in Matrix and Operator Theory*, Oper. Theory Adv. Appl. **50**, Birkhäuser, Basel, 1991, 93–136.

[10] J.A. Ball, G. Groenewald, and T. Malakorn, Bounded real lemma for structured noncommutative multidimensional linear systems and robust control, *Multidimens. Syst. Signal Process.*, **17** (2006), 119–150.

[11] J.A. Ball, G. Groenewald, and S. ter Horst, Standard versus strict bounded real lemma with infinite-dimensional state space I: The state-space-similarity approach, *J. Operator Theory*, to appear.

[12] V. Belevitch, *Classical Network Theory*, Holden-Day, San Francisco, Calif.-Cambridge-Amsterdam, 1968.

[13] A. Ben-Artzi, I. Gohberg, and M.A. Kaashoek, Discrete nonstationary bounded real lemma in indefinite metrics, the strict contractive case, in: *Operator Theory and Boundary Eigenvalue Problems (Vienna, 1993)*, Oper. Theory Adv. Appl. **80**, Birkhäuser, Basel, 1995, 49–78.

[14] A.N. Chakhchoukh and M.R. Opmeer, The state space isomorphism theorem for discrete-time infinite-dimensional systems, *Integr. Equ. Oper. Theory*, **84** (2016), 105–120.

[15] G.E. Dullerud and S. Lall, A new approach for analysis and synthesis of time-varying systems, *IEEE Trans. Automat. Control*, **44** (1999), 1486–1497.

[16] R.G. Douglas, On majorization, factorization, and range inclusion of operators in Hilbert space, *Proc. Amer. Math. Soc.*, **17** (1966), 413–415.

[17] C. Foias, A.E. Frazho, I. Gohberg, and M.A. Kaashoek, *Metric Constrained Interpolation, Commutant Lifting and Systems*, Oper. Theory Adv. Appl. **100**, Birkhäuser, Basel. 1998.

[18] A. Halanay and V. Ionescu, *Time-Varying Discrete Linear Systems*, Birkhäuser Verlag, Basel, 1994.

[19] J.W. Helton, Discrete time systems, operator models, and scattering theory, *J. Functional Analysis*, **16** (1974), 15–38.

[20] T. Kato, *Perturbation Theory for Linear Operators*, Grundlehren Math. Wiss. **132**, Springer Verlag, Berlin-Heidelberg, 1980.

[21] B. Sz.-Nagy, C. Foias, H. Bercovici, and L. Kérchy, *Harmonic Analysis of Operators on Hilbert Space. Revised and enlarged edition*, Universitext, Springer, New York, 2010.

[22] M.R. Opmeer and O.J. Staffans, Optimal state feedback input-output stabilization of infinite-dimensional discrete time-invariant linear systems, *Compl. Anal. Oper. Theory*, **2** (2008), 479–510.

[23] M.R. Opmeer and O.J. Staffans, Optimal input-output stabilization of infinite-dimensional discrete time-invariant linear systems by output injection, *SIAM J. Control Optim.*, **48** (2010), 5084–5107.

[24] I.R. Petersen, B.D.O. Anderson, and E.A. Jonckheere, A first principles solution to the non-singular H^∞ control problem, *Internat. J. Robust Nonlinear Control*, **1** (1991), 171–185.

[25] A. Rantzer, On the Kalman-Yakubovich-Popov lemma, *Systems & Control Letters*, **28** (1996), 7–10.

[26] M. Reed and B. Simon, *Methods of Mathematical Physics I: Functional Analysis* Academic Press, San Diego, 1980.

[27] O.J. Staffans, Quadratic optimal control of well-posed linear systems, *SIAM J. Control Optim.*, **37** (1998), 131–164.

[28] J.C. Willems, Dissipative dynamical systems Part I: General theory, *Arch. Rational Mech. Anal.*, **45** (1972), 321–351.

[29] J.C. Willems, Dissipative dynamical systems Part II: Linear systems with quadratic supply rates, *Arch. Rational Mech. Anal.*, **45** (1972), 352–393.

[30] V.A. Yakubovich, The frequency theorem for the case in which the state space and the control space are Hilbert spaces, and its application in certain problems in the synthesis of optimal control. I, *Sibirsk. Mat. Ž.*, **15** (1974), 639–668 [Russian]; translation in: *Sib. Math. J.*, **15** (1974), 457–476 (1975).

[31] V.A. Yakubovich, The frequency theorem for the case in which the state space and the control space are Hilbert spaces, and its application in certain problems in the synthesis of optimal control. II, *Sibirsk. Mat. Ž.*, **16** (1975), 1081–1102 [Russian]; translation in: *Sib. Math. J.* **16** (1974), 828–845 (1976).

J.A. Ball
Department of Mathematics, Virginia Tech, Blacksburg, VA 24061-0123, USA
e-mail: `joball@math.vt.edu`

G.J. Groenewald and S. ter Horst
Department of Mathematics, Unit for BMI, North-West University
Potchefstroom 2531, South Africa
e-mail: `Gilbert.Groenewald@nwu.ac.za`
`Sanne.TerHorst@nwu.ac.za`

Eigenvalues of even very nice Toeplitz matrices can be unexpectedly erratic

Mauricio Barrera, Albrecht Böttcher,
Sergei M. Grudsky and Egor A. Maximenko

Abstract. It was shown in a series of recent publications that the eigenvalues of $n \times n$ Toeplitz matrices generated by so-called simple-loop symbols admit certain regular asymptotic expansions into negative powers of $n+1$. On the other hand, recently two of the authors considered the pentadiagonal Toeplitz matrices generated by the symbol $g(x) = (2\sin(x/2))^4$, which does not satisfy the simple-loop conditions, and derived asymptotic expansions of a more complicated form. Here we use these results to show that the eigenvalues of the pentadiagonal Toeplitz matrices do not admit the expected regular asymptotic expansion. This also delivers a counter-example to a conjecture by Ekström, Garoni, and Serra-Capizzano and reveals that the simple-loop condition is essential for the existence of the regular asymptotic expansion.

Mathematics Subject Classification (2010). Primary 15B05; Secondary 15A18, 41A60, 47B35, 65F15.

Keywords. Toeplitz matrix, eigenvalue, spectral asymptotics.

1. Main results

This paper is on the eigenvalues of the $n \times n$ analog $T_n(g)$ of the symmetric pentadiagonal Toeplitz matrix

$$T_6(g) = \begin{pmatrix} 6 & -4 & 1 & & & \\ -4 & 6 & -4 & 1 & & \\ 1 & -4 & 6 & -4 & 1 & \\ & 1 & -4 & 6 & -4 & 1 \\ & & 1 & -4 & 6 & -4 \\ & & & 1 & -4 & 6 \end{pmatrix}.$$

Research of the first author is partially supported by CONACYT scholarship.
Research of the third author is partially supported by CONACYT grant 238630.
Research of the fourth author is partially supported by IPN-SIP projects.

© Springer International Publishing AG, part of Springer Nature 2018
A. Böttcher et al. (eds.), *The Diversity and Beauty of Applied Operator Theory*, Operator Theory: Advances and Applications 268, https://doi.org/10.1007/978-3-319-75996-8_2

These matrices are generated by the Fourier coefficients of the function

$$g(x) = e^{-2ix} - 4e^{-ix} + 6 - 4e^{ix} + e^{2ix}$$
$$= (2 - e^{-ix} - e^{ix})^2 = (2 - 2\cos x)^2 = \left(2\sin\frac{x}{2}\right)^4. \tag{1.1}$$

Previous results, and we will say more about them below, raise the expectation that, given any natural number p, the eigenvalues $\lambda_{n,1} < \cdots < \lambda_{n,n}$ of $T_n(g)$ admit an asymptotic expansion

$$\lambda_{n,j} = \sum_{k=0}^{p} \frac{f_k\left(\frac{j\pi}{n+1}\right)}{(n+1)^k} + O\left(\frac{1}{(n+1)^{p+1}}\right) \quad \text{as } n \to \infty \tag{1.2}$$

with the error term being uniform in $1 \le j \le n$ and with continuous functions $f_0, \ldots, f_p : [0, \pi] \to \mathbb{R}$. The following theorem, which is the main result of the present paper, shows that this is surprisingly false for $p = 4$.

Theorem 1.1. *Let g and $T_n(g)$ be as above. There do not exist continuous functions $f_0, \ldots, f_4 \colon [0, \pi] \to \mathbb{R}$ and numbers $C > 0$, $N \in \mathbb{N}$ such that*

$$\left| \lambda_{n,j} - \sum_{k=0}^{4} \frac{f_k\left(\frac{j\pi}{n+1}\right)}{(n+1)^k} \right| \le \frac{C}{(n+1)^5} \tag{1.3}$$

for every $n \ge N$ and every $j \in \{1, \ldots, n\}$.

Unfortunately, we are confronted with an unpleasant complication. We call it the $n, n+1, n+2$ problem. In (1.2) and (1.3) we used the denominator $n + 1$. This denominator is very convenient when tackling simple-loop symbols. However, when dealing with the symbol (1.1), the denominator $n + 2$ is naturally emerging. See Remark 6.6. Therefore we decided to work mostly with $n+2$ in this paper. We will denote the coefficient functions by f_k if the denominator is $n + 1$ and by d_k in case it is $n + 2$. To avoid any confusion, let us state the $n + 2$ result we will prove.

Theorem 1.2. *Let g and $T_n(g)$ be as above and let $p \ge 0$ be an integer.*

(a) *There exist continuous functions $d_0, \ldots, d_p \colon [0, \pi] \to \mathbb{R}$ and a number $D_p > 0$ such that*

$$\left| \lambda_{n,j} - \sum_{k=0}^{p} \frac{d_k\left(\frac{j\pi}{n+2}\right)}{(n+2)^k} \right| \le \frac{D_p}{(n+2)^{p+1}} \tag{1.4}$$

whenever $n \ge 1$ and $\frac{p}{2}\log(n + 2) \le j \le n$. These functions $d_0, \ldots, d_p$ are uniquely determined.

(b) *There is a constant $C > 0$ such that*

$$\left| \lambda_{n,j} - \sum_{k=0}^{3} \frac{d_k\left(\frac{j\pi}{n+2}\right)}{(n+2)^k} \right| \le \frac{C}{(n+2)^4} \tag{1.5}$$

for all $n \ge 1$ and all $j \in \{1, \ldots, n\}$.

(c) *However, there do not exist numbers $C > 0$ and $N \in \mathbb{N}$ such that*

$$\left| \lambda_{n,j} - \sum_{k=0}^{4} \frac{d_k\left(\frac{j\pi}{n+2}\right)}{(n+2)^k} \right| \leq \frac{C}{(n+2)^5} \tag{1.6}$$

for all $n \geq N$ and all $j \in \{1, \ldots, n\}$.

In the final section of the paper we will pass from $n+2$ to $n+1$ and prove Theorem 1.1.

Part (b) of Theorem 1.2 might suggest that all eigenvalues $\lambda_{n,j}$ are *moderately well* approximated by the sums $\sum_{k=0}^{3} d_k\left(\frac{j\pi}{n+2}\right)/(n+2)^k$. In fact, as we will show in Remark 7.4, this approximation is *extremely bad* for the first eigenvalues, in the sense that the corresponding relative errors do not converge to zero. However, as Theorem 1.2 (a) shows, asymptotic expansions of the form (1.2) for $p = 2, 3, 4, \ldots$ can be used outside a small neighborhood of the point at which the symbol has a zero of order greater than 2.

It is well known that $\lambda_{n,j} = g(j\pi/n) + O(1/n)$, uniformly in j, implying that (1.2) and (1.4) hold for $p = 0$ with $f_0 = d_0 = g$. Figure 1 shows the plot of the symbol g (from 0 to π) and the eigenvalues of $T_{64}(g)$ as the points $(j\pi/65, \lambda_{64,j})$ and $(j\pi/66, \lambda_{64,j})$ with $n+1 = 65$ and $n+2 = 66$, respectively. Notice that the approximation of $\lambda_{n,j}$ by $g(j\pi/(n+2))$ is not very good for large values of j. It is seen that the approximation of $\lambda_{n,j}$ by $g(j\pi/(n+1))$ is better.

We will compute the functions $d_1, \ldots, d_4$ of Theorem 1.2. Knowledge of these functions allows us to illustrate the higher-order asymptotics of the eigenvalues and to depict the expected behavior for $p = 0, 1, 2, 3$ and the erratic behavior for $p = 4$. Put

$$\Omega_{p+1,n,j} := (n+2)^{p+1} \left(\lambda_{n,j} - \sum_{k=0}^{p} \frac{d_k\left(\frac{j\pi}{n+2}\right)}{(n+2)^k} \right).$$

In Figure 2, we see a perfect matching between $\Omega_{p,64,j}$ and $d_p(j\pi/66)$ for $p = 1, 2, 3, 4$, except for $p = 4$ and $j = 1, 2$. The gap between $d_4(\pi/66)$ and $\Omega_{4,64,1}$ shows that the asymptotics of $\lambda_{n,1}$ does not obey the regular rule with the functions d_0, d_1, d_2, d_3, d_4.

Of course, the erratic behavior of the first two eigenvalues in subplot (d) of Figure 2 might be caused by the circumstance that $n = 64$ is not yet large enough. Figure 3 reveals that this behavior persists when passing to larger n. In that figure we see the first piece of the graph of d_4 and the points $(j\pi/(n+2), \Omega_{4,n,j})$ for $1 \leq j \leq 64$ and $n = 1024$. Now the first three eigenvalues show distinct irregularity.

Figures 4 and 5 show what happens for $p = 5$ and for the matrix dimensions $n = 64$ and $n = 1024$.

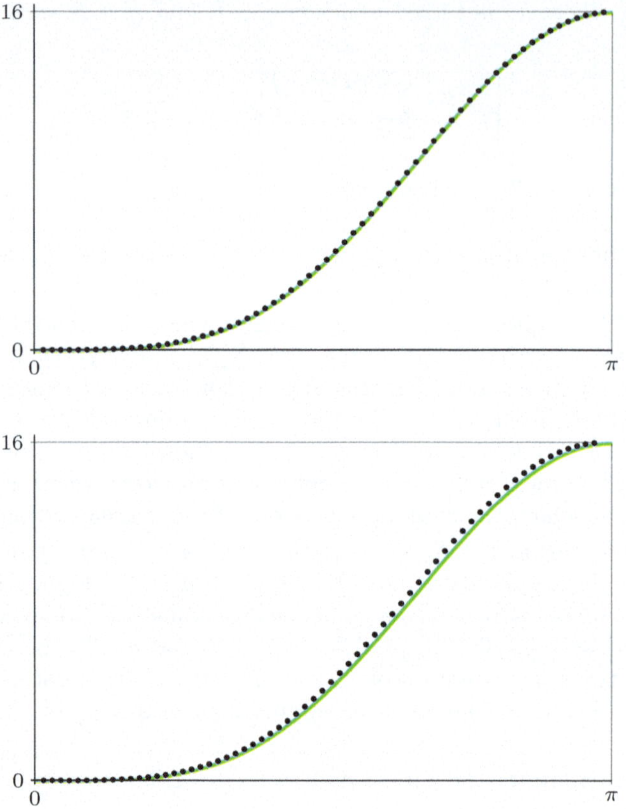

FIGURE 1. Above is the plot of the function g and the points $(j\pi/65, \lambda_{64,j})$ for $1 \leq j \leq 64$. Below we see the plot of g and the points $(j\pi/66, \lambda_{64,j})$ for $1 \leq j \leq 64$.

2. Prehistory

It was the previous papers [6, 9, 2, 5] that were devoted to regular asymptotic expansions for the eigenvalues of Toeplitz matrices with so-called simple-loop symbols. We recall that, in a more general context, the starting point is a 2π-periodic bounded function $g : \mathbb{R} \to \mathbb{R}$ with Fourier series $g(x) \sim \sum_{k=-\infty}^{\infty} \hat{g}_k e^{ikx}$. The $n \times n$ Toeplitz matrix generated by g is the matrix $T_n(g) = (\hat{g}_{j-k})_{j,k=1}^{n}$. The function g is referred to as the symbol of the matrix sequence $\{T_n(g)\}_{n=1}^{\infty}$. Examples of simple-loop symbols are even 2π-periodic C^∞ functions $g : \mathbb{R} \to \mathbb{R}$ satisfying $g'(x) > 0$ for every x in $(0, \pi)$, $g'(0) = 0$, $g''(0) > 0$, $g'(\pi) = 0$, $g''(\pi) < 0$. The requirement that g be a real-valued and even function implies that the matrices $T_n(g)$ are real and symmetric.

In the beginning of Section 7 of [2], we also noted that the mere existence of such regular asymptotic expansions already helps to approximate

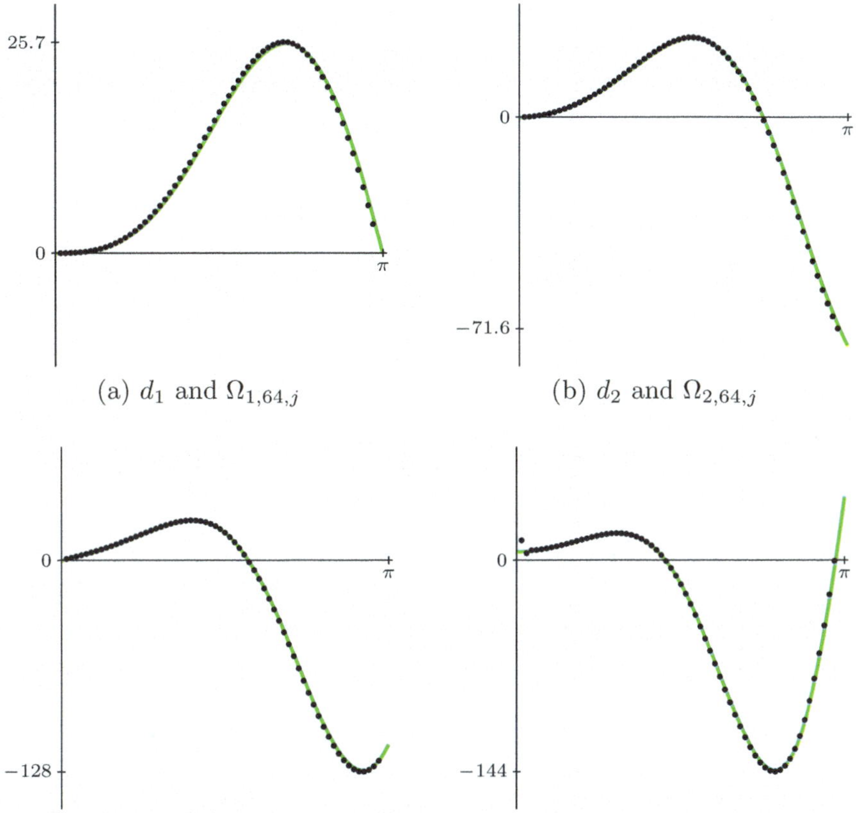

(a) d_1 and $\Omega_{1,64,j}$ (b) d_2 and $\Omega_{2,64,j}$ (c) d_3 and $\Omega_{3,64,j}$ (d) d_4 and $\Omega_{4,64,j}$

FIGURE 2. In subplot (a), we see the graph of d_1 and the values of $\Omega_{1,64,j}$, shown as the points $(j\pi/66, \Omega_{1,64,j})$. On subplot (b), we see d_2 and $\Omega_{2,64,j}$, etc.

the eigenvalues of large matrices by using the eigenvalues of small matrices and some sort of extrapolation.

Ekström, Garoni, and Serra-Capizzano [10] worked out the idea of such extrapolation in detail. They also emphasized that the symbols of interest in connection with the discretization of differential equations are of the form

$$g_m(x) = (2 - 2\cos x)^m = \left(2\sin\frac{x}{2}\right)^{2m}. \tag{2.1}$$

In the simplest case $m = 1$, the matrices $T_n(g_1)$ are the $n \times n$ analogs of the tridiagonal Toeplitz matrix

$$T_4(g_1) = \begin{pmatrix} 2 & -1 & & \\ -1 & 2 & -1 & \\ & -1 & 2 & -1 \\ & & -1 & 2 \end{pmatrix}.$$

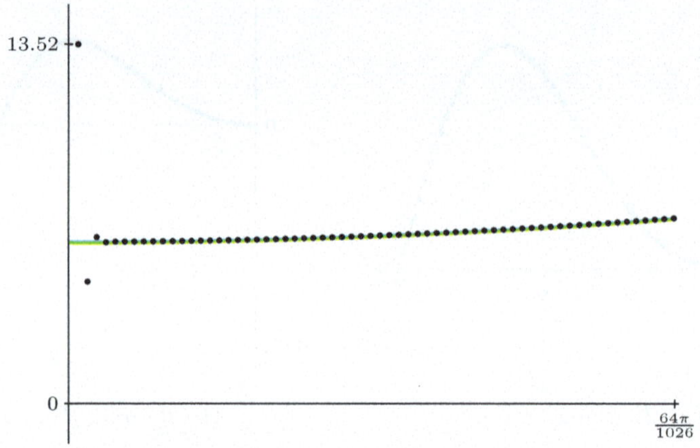

FIGURE 3. The graph of the function d_4 and the points $(j\pi/1026, \Omega_{4,1024,j})$ for $j = 1, \ldots, 64$.

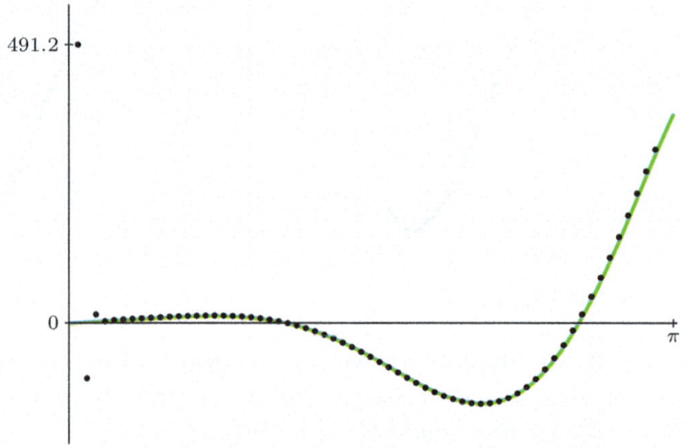

FIGURE 4. The graph of d_5 and the points $(j\pi/66, \Omega_{5,64,j})$.

The eigenvalues of these matrices are known exactly,

$$\lambda_{n,j} = 2 - 2\cos\frac{j\pi}{n+1} = \left(2\sin\frac{j\pi}{2n+2}\right)^2,$$

and hence they obey the regular asymptotics (1.2) with $f_0 = g$ and $f_k = 0$ for $k \geq 1$. A crucial observation of [10] is that the symbols g_m are no longer simple-loop symbols for $m \geq 2$, because then the second derivative at 0 vanishes. Our concrete symbol (1.1) is just g_2 and hence not a simple-loop symbol. Ekström, Garoni, and Serra-Capizzano nevertheless conjectured that the regular asymptotic expansions stay true for smooth even real-valued symbols that are monotone on $[0, \pi]$ and that may have a minimum or a maximum

FIGURE 5. The first piece of the graph of d_5 (green) and the points $(j\pi/1026, \Omega_{5,1024,j})$ for $j = 1, \ldots, 64$. The plot of d_5 cannot be distinguished from the abscissa axis.

of higher order. They verified this conjecture numerically for some examples and for small values of p. This conjecture has attracted a lot of attention.

Independently and at the same time, two of us [1] considered just the symbol (1.1) and derived exact equations and asymptotic expansions for the eigenvalues of $T_n(g)$. Later, when paper [10] came to our attention, we realized to our surprise that the results of [1] imply that for the symbol (1.1) the eigenvalues do not admit a regular asymptotic expansion of the form (1.2) with $p = 4$. This is what Theorem 1.1 says and this is a counter-example to the conjecture by Ekström, Garoni, and Serra-Capizzano.

The rest of the paper is organized as follows. In Sections 3 and 4 we provide some general facts about regular asymptotic expansions. In Section 5, using formulas and ideas from [1], we show that an analog of (1.3) is true for the eigenvalues that are not too close to the minimum of the symbol, namely, for $2\log(n+2) \le j \le n$, and provide recipes to compute the corresponding coefficients. On the other hand, in Section 6 we deduce an asymptotic formula for the first eigenvalue. In Section 8 we prove that the asymptotics from Sections 5 and 6 cannot be joined.

3. Regular expansions of the eigenvalues

In this and the following sections, we work in abstract settings and use the denominator $n + s$, where s is an arbitrary positive constant ("shift"). This allows us to unify the situations with $n + 1$ and $n + 2$ and to simplify the subsequent references in the last sections of the paper.

We first introduce some notation and recall some facts. Given a 2π-periodic bounded real-valued function g on the real line, we denote by

$$\lambda_{n,1}, \ldots, \lambda_{n,n}$$

the eigenvalues of the corresponding Toeplitz matrices $T_n(g)$, ordered in the ascending order: $\lambda_{n,1} \leq \cdots \leq \lambda_{n,n}$. Using the first Szegő limit theorem and criteria for weak convergence of probability measures, we proved in [4, 3] that if the essential range of g is a segment of the real line, then $\lambda_{n,j}$ can be uniformly approximated by the values of the quantile function Q (associated to g) at the points $j/(n+s)$:

$$\max_{1\leq j\leq n} \left| \lambda_{n,j} - Q\left(\frac{j}{n+s}\right) \right| = o(1) \quad \text{as} \quad n \to \infty. \tag{3.1}$$

If g is continuous, even, and strictly increasing on $[0,\pi]$, then $Q(x)$ is just $g(\pi x)$. Denote by $u_{n,j}$ the points of the uniform mesh $j\pi/(n+s)$, where $j \in \{1,\ldots,n\}$. Then (3.1) can be rewritten in the form

$$\max_{1\leq j\leq n} |\lambda_{n,j} - g(u_{n,j})| = o(1) \quad \text{as} \quad n \to \infty. \tag{3.2}$$

Trench proved [14] that for this class of symbols the eigenvalues are all distinct:

$$g(0) < \lambda_{n,1} < \cdots < \lambda_{n,n} < g(\pi).$$

Thus, there exist real numbers $\varphi_{n,1},\ldots,\varphi_{n,n}$ such that

$$0 < \varphi_{n,1} < \ldots < \varphi_{n,n} < \pi$$

and $\lambda_{n,j} = g(\varphi_{n,j})$. Taking into account (3.2), we can try to use $u_{n,j}$ as an initial approximation for $\varphi_{n,j}$. This approximation can be very inaccurate, but it is better than nothing.

Now let J be an arbitrary set of integer pairs (n,j) such that $1 \leq j \leq n$ for every (n,j) in J. Suppose that for each (n,j) in J the number $\varphi_{n,j}$ is the unique solution of the equation

$$x = u_{n,j} + \frac{\eta(x)}{n+s} + \rho_{n,j}(x), \tag{3.3}$$

where η is an infinitely smooth real-valued function on $[0,\pi]$ and $\{\rho_{n,j}\}_{(n,j)\in J}$ is a family of infinitely smooth real-valued functions on $[0,\pi]$ such that

$$\sup_{0\leq x\leq \pi} \sup_{j:(n,j)\in J} |\rho_{n,j}(x)| = O\left(\frac{1}{(n+s)^p}\right) \tag{3.4}$$

for some p in $\mathbb{N}$.

In the simple-loop case, the function ρ_n did not depend on j, and J was of the form $\{(n,j)\colon n \geq N,\ 1 \leq j \leq n\}$ for some N.

Let us show how to derive asymptotic expansions of $\varphi_{n,j}$ and $\lambda_{n,j}$ from equation (3.3).

Proposition 3.1. *Let η be an infinitely smooth real-valued function on $[0,\pi]$, and $\{\rho_{n,j}\}_{(n,j)\in J}$ be a family of real-valued functions on $[0,\pi]$ satisfying (3.4) for some natural number p. Suppose that for all (n,j) in J equation (3.3) has a unique solution $\varphi_{n,j}$. Then there exists a sequence of real-valued infinitely* *smooth functions* $c_0, c_1, c_2, \dots$ *defined on* $[0, \pi]$ *such that there is a number* $r_p > 0$ *ensuring that, for all* (n, j) *in* J,

$$\left| \varphi_{n,j} - \sum_{k=0}^{p} \frac{c_k(u_{n,j})}{(n+s)^k} \right| \le \frac{r_p}{(n+s)^{p+1}}. \tag{3.5}$$

Furthermore, if g *is an infinitely smooth* 2π*-periodic real-valued even function on* $\mathbb{R}$, *strictly increasing on* $[0, \pi]$, *then there exists a sequence of real-valued infinitely smooth functions* $d_0, d_1, d_2, \dots$ *defined on* $[0, \pi]$ *such that the numbers* $\lambda_{n,j} := g(\varphi_{n,j})$ *can be approximated as follows: there exists an* R_p *such that, for all* (n, j) *in* J,

$$\left| \lambda_{n,j} - \sum_{k=0}^{p} \frac{d_k(u_{n,j})}{(n+s)^k} \right| \le \frac{R_p}{(n+s)^{p+1}}. \tag{3.6}$$

Proof. This proposition was essentially proved in [2, 5], with a slightly different notation and reasoning, including a justification of the fixed-point method. Here we propose a simpler proof. Our goal is to show that (3.5) and (3.6) are direct and trivial consequences of the main equation (3.3).

In order to simplify notation, we denote by $O(1/(n+s)^p)$ any expression that may depend on n and j but can be estimated from above by $C/(n+s)^p$ with C independent of n or j. Then (3.3) implies that

$$\varphi_{n,j} = u_{n,j} + O\left(\frac{1}{n+s}\right).$$

Substitute this expression into (3.3) and expand η by Taylor's formula around the point $u_{n,j}$:

$$\begin{aligned} \varphi_{n,j} &= u_{n,j} + \frac{\eta\left(u_{n,j} + O(\frac{1}{n+s})\right)}{n+s} + O\left(\frac{1}{(n+s)^2}\right) \\ &= u_{n,j} + \frac{\eta(u_{n,j})}{n+s} + O\left(\frac{1}{(n+s)^2}\right). \end{aligned}$$

Substituting the last expression into (3.3) and expanding η by Taylor's formula around $u_{n,j}$ we get

$$\begin{aligned} \varphi_{n,j} &= u_{n,j} + \frac{\eta\left(u_{n,j} + \frac{\eta(u_{n,j})}{n+s} + O(\frac{1}{(n+s)^2})\right)}{n+s} + O\left(\frac{1}{(n+s)^3}\right) \\ &= u_{n,j} + \frac{\eta(u_{n,j})}{n+s} + \frac{\eta(u_{n,j})\eta'(u_{n,j})}{(n+s)^2} + O\left(\frac{1}{(n+s)^3}\right). \end{aligned}$$

This "Münchhausen trick" can be applied again and again (we refer to the story when Baron von Münchhausen saved himself from being drowned in a swamp by pulling on his own hair), yielding an asymptotic expansion of the form (3.5) of any desired order p.

The first of the functions c_k are

$$
\begin{aligned}
&c_0(x) = x, \quad c_1(x) = \eta(x), \quad c_2(x) = \eta(x)\eta'(x),\\
&c_3 = \eta(\eta')^2 + \frac{1}{2}\eta^2\eta'', \quad c_4 = \eta(\eta')^3 + \frac{3}{2}\eta^2\eta'\eta'' + \frac{1}{6}\eta^3\eta''', \\
&c_5 = \eta(\eta')^4 + 3\eta^2(\eta'^2)\eta'' + \frac{1}{2}\eta^3(\eta'')^2 + \frac{2}{3}\eta^3\eta'\eta''' + \frac{1}{24}\eta^4\eta^{(4)}.
\end{aligned}
\tag{3.7}
$$

By induction on p it is straightforward to show that c_k is a uniquely determined polynomial in $\eta, \eta', \dots, \eta^{(k-1)}$ also for $k \geq 6$.

Once we have the asymptotic formulas for $\varphi_{n,j}$, we can use the formula $\lambda_{n,j} = g(\varphi_{n,j})$ and expand the function g by Taylor's formula around the point $u_{n,j}$ to get

$$
\begin{aligned}
\lambda_{n,j} &= g\left(u_{n,j} + \sum_{k=1}^{p} \frac{c_k(u_{n,j})}{(n+s)^k} + O\left(\frac{1}{(n+s)^{p+1}}\right)\right)\\
&= \sum_{m=0}^{p} \frac{g^{(m)}(u_{n,j})}{m!} \left(\sum_{k=1}^{p} \frac{c_k(u_{n,j})}{(n+s)^k} + O\left(\frac{1}{(n+s)^{p+1}}\right)\right)^m\\
&\quad + O((\varphi_{n,j} - u_{n,j})^{p+1}).
\end{aligned}
$$

Expanding the powers, regrouping the summands, and writing $\varphi_{n,j} - u_{n,j}$ as $O(1/(n+s))$, we obtain a regular asymptotic formula for $\lambda_{n,j}$:

$$
\lambda_{n,j} = \sum_{k=0}^{p} \frac{d_k(u_{n,j})}{(n+s)^k} + O\left(\frac{1}{(n+s)^{p+1}}\right). \tag{3.8}
$$

The first of the functions $d_0, d_1, d_2, \dots$ can be computed by the formulas

$$
\begin{aligned}
&d_0 = g, \quad d_1 = g'c_1, \quad d_2 = g'c_2 + \frac{1}{2}g''c_1^2,\\
&d_3 = g'c_3 + g''c_1c_2 + \frac{1}{6}g'''c_1^3,\\
&d_4 = g'c_4 + g''\left(c_1c_3 + \frac{1}{2}c_2^2\right) + \frac{1}{2}g'''c_1^2c_2 + \frac{1}{24}g^{(4)}c_1^4,\\
&d_5 = g'c_5 + g''(c_2c_3 + c_1c_4) + \frac{1}{2}g'''(c_1^2c_3 + c_1c_2^2) + \frac{1}{6}g^{(4)}c_1^3c_2 + \frac{1}{120}g^{(5)}c_1^5.
\end{aligned}
\tag{3.9}
$$

It can again be proved by induction on p that the functions $c_0, c_1, c_2, \dots$ are polynomials in $\eta, \eta', \eta'', \dots$ and that the functions $d_0, d_1, d_2, \dots$ are polynomials in $c_0, c_1, c_2, \dots$ and $g, g', g'', \dots$. As a consequence, all the functions c_k and d_k are infinitely smooth. □

Remark 3.2. The expressions (3.7) and (3.9) can be easily derived with various computer algebra systems. For example, in SageMath we used the following commands (the expression $1/n$ is denoted by h):

```
var('u, h, c1, c2, c3, c4, c5'); (eta, g) = function('eta, g')
phiexpansion1 = u + h * eta(u)
phiexpansion2 = u + h * taylor(eta(phiexpansion1), h, 0, 2)
phiexpansion3 = u + h * taylor(eta(phiexpansion2), h, 0, 3)
```

```
phiexpansion4 = u + h * taylor(eta(phiexpansion3), h, 0, 4)
phiexpansion5 = u + h * taylor(eta(phiexpansion4), h, 0, 5)
print(phiexpansion5.coefficients(h))
phiformal5 = u + c1*h + c2*h^2 + c3*h^3 + c4*h^4 + c5*h^5
lambdaexpansion5 = taylor(g(phiformal5), h, 0, 5)
print(lambdaexpansion5.coefficients(h))
```

We also performed similar computations in Wolfram Mathematica, starting with

```
phiexpansion0 = u + O[h]
phiexpansion1 = Series[u + h * eta[phiexpansion0], {h, 0, 1}]
```

Remark 3.3. If the functions $d_0, d_1, \dots$ are infinitely smooth, then one can transform an asymptotic expansion into negative powers of $n + s_1$ into an asymptotic expansion in negative powers of $n + s_2$. For example, suppose we have

$$\lambda_{n,j} = \sum_{k=0}^{p} \frac{d_k(u_{n,j})}{(n+2)^k} + O\left(\frac{1}{(n+2)^{p+1}}\right),$$

and we want

$$\lambda_{n,j} = \sum_{k=0}^{p} \frac{f_k(u_{n,j})}{(n+1)^k} + O\left(\frac{1}{(n+1)^{p+1}}\right).$$

For $k = 0, 1$, we have

$$\begin{aligned} d_k\left(\frac{j\pi}{n+2}\right) &= d_k\left(\frac{j\pi}{(n+1)\left(1+\frac{1}{n+1}\right)}\right) \\ &= d_k\left(\frac{j\pi}{n+1} - \frac{j\pi}{(n+1)^2} + O\left(\frac{1}{(n+1)^2}\right)\right) \\ &= d_k\left(\frac{j\pi}{n+1}\right) - d_k'\left(\frac{j\pi}{n+1}\right)\frac{j\pi}{n+1}\frac{1}{n+1} + O\left(\frac{1}{(n+1)^2}\right), \end{aligned}$$

and thus

$$\begin{aligned} d_0\left(\frac{j\pi}{n+2}\right) &+ d_1\left(\frac{j\pi}{n+2}\right)\frac{1}{n+2} + O\left(\frac{1}{(n+2)^2}\right) \\ &= d_0\left(\frac{j\pi}{n+1}\right) - d_0'\left(\frac{j\pi}{n+1}\right)\frac{j\pi}{n+1}\frac{1}{n+1} \\ &+ d_1\left(\frac{j\pi}{n+1}\right)\frac{1}{n+1} + O\left(\frac{1}{(n+1)^2}\right), \end{aligned}$$

resulting in the equalities $f_0(x) = d_0(x)$ and $f_1(x) = d_1(x) - x d_0'(x)$.

Remark 3.4. The hard part of the work in [2, 5] was to derive equation (3.3) and an explicit formula for η, to verify that η is sufficiently smooth, to establish upper bounds for the functions ρ_n, and to prove that (3.3) has a unique solution for every n large enough and for every j. Moreover, all this work was done under the assumption that g has some sort of smoothness of a finite order. In Proposition 3.1 we just require all these properties.

4. Uniqueness of the regular asymptotic expansion

As in the previous section, we fix some $s > 0$.

If there exists an asymptotic expansion of the form (3.8), then the functions $d_0, d_1, d_2, \ldots$ are uniquely determined. Let us state and prove this fact formally. Instead of requiring (3.8) for all n and j, we assume it holds for a set of pairs (n, j) such that the quotients $u_{n,j} := j\pi/(n + s)$ "asymptotically fill" $[0, \pi]$. Here is the corresponding technical definition.

Definition 4.1. Let J be a subset of $\mathbb{N}^2$. We say that J *asymptotically fills* $[0, \pi]$ *by quotients* if for every x in $[0, \pi]$, every N in $\mathbb{N}$, and every $\delta > 0$ there is a pair of numbers (n, j) in J such that $n \geq N$, $1 \leq j \leq n$, and $|u_{n,j} - x| \leq \delta$.

It is easy to see that J asymptotically fills $[0, \pi]$ by quotients if and only if the set $\{u_{n,j} : (n, j) \in J\}$ is dense in $[0, \pi]$.

Proposition 4.2. *Let $p \geq 0$ be an integer, let $d_0, d_1, \ldots, d_p$ and $\widetilde{d}_0, \widetilde{d}_1, \ldots, \widetilde{d}_p$ be continuous functions on $[0, \pi]$, let $C > 0$, and let J be a subset of $\mathbb{N}^2$ such that J asymptotically fills $[0, \pi]$ by quotients. Suppose that for every pair (n, j) in J the inequalities*

$$\left| \lambda_{n,j} - \sum_{k=0}^{p} \frac{d_k(u_{n,j})}{(n+s)^k} \right| \leq \frac{C}{(n+s)^{p+1}}, \quad \left| \lambda_{n,j} - \sum_{k=0}^{p} \frac{\widetilde{d}_k(u_{n,j})}{(n+s)^k} \right| \leq \frac{C}{(n+s)^{p+1}}$$

hold. Then $d_k(x) = \widetilde{d}_k(x)$ for every $k \in \{0, \ldots, p\}$ and every $x \in [0, \pi]$.

Proof. Denote the function $d_p - \widetilde{d}_p$ by h_p. It is clear that $h_0 = 0$. Proceeding by mathematical induction on p, we assume that h_k is the zero constant for every k with $k < p$, and we have to show that h_p is the zero constant.

Let $x \in [0, \pi]$ and $\varepsilon > 0$. Using the continuity of h_p at the point x, choose $\delta > 0$ such that $|y - x| \leq \delta$ implies

$$|h_p(y) - h_p(x)| \leq \frac{\varepsilon}{2}.$$

Take N such that

$$\frac{2C}{N+s} \leq \frac{\varepsilon}{2}.$$

After that, pick n and j such that $(n, j) \in J$, $n \geq N$, and $|u_{n,j} - x| \leq \delta$. Then

$$\left| \frac{d_p(u_{n,j})}{(n+s)^p} - \frac{\widetilde{d}_p(u_{n,j})}{(n+s)^p} \right| \leq \frac{2C}{(n+s)^{p+1}},$$

which implies

$$|h_p(u_{n,j})| \leq \frac{2C}{n+s} \leq \frac{2C}{N+s} \leq \frac{\varepsilon}{2}.$$

Finally,

$$|h_p(x)| \leq |h_p(x) - h_p(u_{n,j})| + |h_p(u_{n,j})| \leq \frac{\varepsilon}{2} + \frac{\varepsilon}{2} = \varepsilon.$$

As $\varepsilon > 0$ can be chosen arbitrarily, it follows that h_p is identically zero. □

5. An example with a minimum of the fourth order

We now consider the pentadiagonal Toeplitz matrices generated by the trigonometric polynomial

$$g(x) = \left(2\sin\frac{x}{2}\right)^4. \tag{5.1}$$

The function g takes real values, is even, and strictly increases on $[0, \pi]$. Nevertheless, g does not belong to the simple-loop class, because g has a minimum of the fourth order: $g(0) = g'(0) = g''(0) = g'''(0) = 0$, $g^{(4)}(0) > 0$.

The purpose of this section is to recall some results of [1] and to derive some new corollaries. We begin by introducing some auxiliary functions:

$$\beta(x) := 2\operatorname{arcsinh}\left(\sin\frac{x}{2}\right) = 2\ln\left(\sin\frac{x}{2} + \sqrt{1 + \left(\sin\frac{x}{2}\right)^2}\right),$$

$$f(x) := \beta'(x) = \frac{\cos\frac{x}{2}}{\sqrt{1 + \left(\sin\frac{x}{2}\right)^2}},$$

$$\eta_n^{\text{odd}}(x) := 2\arctan\left(\frac{1}{f(x)}\coth\frac{(n+2)\beta(x)}{2}\right),$$

$$\eta_n^{\text{even}}(x) := 2\arctan\left(\frac{1}{f(x)}\tanh\frac{(n+2)\beta(x)}{2}\right),$$

$$\eta_{n,j}(x) := \begin{cases} \eta_n^{\text{odd}}(x), & \text{if } j \text{ is odd,} \\ \eta_n^{\text{even}}(x), & \text{if } j \text{ is even.} \end{cases}$$

As previously, we denote by $\varphi_{n,j}$ the points in $(0, \pi)$ such that $\lambda_{n,j} = g(\varphi_{n,j})$. In this example, we let $u_{n,j}$ stand for $j\pi/(n+2)$.

In [1, Theorems 2.1 and 2.3], two of us used Elouafi's formulas [11] for the determinants of Toeplitz matrices and derived exact equations for the eigenvalues of $T_n(g)$. Namely, it was proved that there exists an N_0 such that if $n \geq N_0$ and $j \in \{1, \ldots, n\}$, then $\varphi_{n,j}$ is the unique solution in the interval $(u_{n,j}, u_{n,j+1})$ of the equation

$$x = u_{n,j} + \frac{\eta_{n,j}(x)}{n+2}. \tag{5.2}$$

The corresponding equation in [1] is written in a slightly different (but equivalent) form, without joining the cases of odd and even values of j.

Equation (5.2) is hard to derive but easy to verify numerically. We computed the eigenvalues by general numerical methods in Wolfram Mathematica, using high-precision arithmetic with 100 decimal digits after the floating point, and obtained coincidence in (5.2) up to 99 decimal digits for each n from 10 to 100 and for each j from 1 to n.

Equation (5.2) is more complicated than (3.3), in the sense that now instead of one function η we have a family of functions, depending on n and on the parity of j.

Note that if x is not too close to zero, then $\beta(x)$ is away from zero. Thus, when n is large enough, the product $\frac{n+2}{2}\beta(x)$ is large and the expressions

$\tanh \frac{(n+2)\beta(x)}{2}$ and $\coth \frac{(n+2)\beta(x)}{2}$ are very close to 1. Denote by η the function obtained from η_n^{odd} and η_n^{even} by neglecting these expressions, that is,

$$\eta(x) := 2\arctan\left(\frac{1}{f(x)}\right), \tag{5.3}$$

and put

$$\rho_{n,j}(x) := \frac{\eta_{n,j}(x) - \eta(x)}{n+2}.$$

Then the main equation (5.2) takes the form (3.3) with $s = 2$:

$$x = u_{n,j} + \frac{\eta(x)}{n+2} + \rho_{n,j}(x). \tag{5.4}$$

So, for each (n, j) with $n \geq N_0$ and $1 \leq j \leq n$ the number $\varphi_{n,j}$ is the unique solution of (5.4) in the interval $(u_{n,j}, u_{n,j+1})$.

Figure 6 shows that the functions η_{64}^{odd}, $\eta_{64}^{\mathrm{even}}$, and η almost coincide outside a small neighborhood of zero.

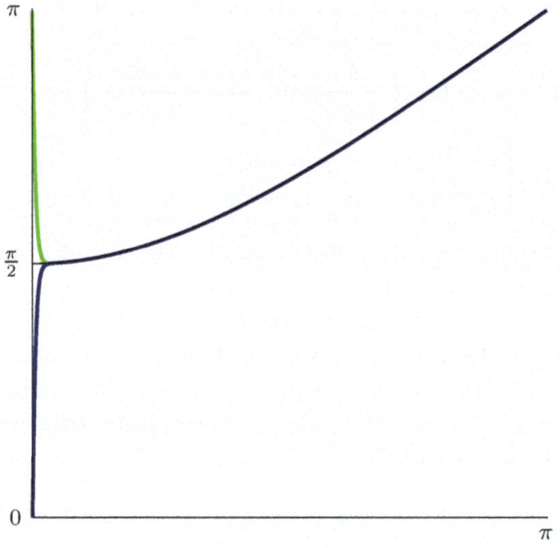

FIGURE 6. Plots of η_{64}^{odd} (green), $\eta_{64}^{\mathrm{even}}$ (blue), and η (black)

The following lemma provides us with upper estimates for $\rho_{n,j}(x)$.

Lemma 5.1. *Let $n, j \in \mathbb{N}$. If $1 \leq j \leq n/2$, then*

$$\sup_{u_{n,j} \leq x \leq u_{n,j+1}} |\rho_{n,j}(x)| \leq \frac{6e^{-2j}}{n+2}. \tag{5.5}$$

If $n/2 \leq j \leq n$, then

$$\sup_{u_{n,j} \leq x \leq u_{n,j+1}} |\rho_{n,j}(x)| \leq 6e^{-(n+2)\pi/2}. \tag{5.6}$$

Proof. First suppose that $1 \le j \le n/2$ and $u_{n,j} \le x \le u_{n,j+1}$. Then

$$\frac{j\pi}{n+2} \le x \le \frac{(j+1)\pi}{n+2} \le \frac{\pi}{2}.$$

It is readily verified that $\beta(x) \ge 2x/\pi$ for every x in $[0, \pi/2]$. Consequently,

$$\frac{(n+2)\beta(x)}{2} \ge j.$$

It is also easy to see that $0 \le 1 - \tanh(y) \le 2e^{-2y}$ and $0 \le \coth(y) - 1 \le 3e^{-2y}$ for $y \ge 1$, $f(x) > 1/2$ for x in $[0, \pi/2]$, and that arctan is Lipschitz continuous with coefficient 1. Thus

$$|\eta_{n,j}(x) - \eta(x)| \le 6e^{-2j},$$

which yields (5.5).

Now consider the case $n/2 \le j \le n$. Here we use the two estimates $\beta(x) \ge x/2$ and $f(x) > 1/(n+2)$ to obtain

$$\frac{(n+2)\beta(x)}{2} \ge \frac{(n+2)\pi}{4},$$

$$|\eta_{n,j}(x) - \eta(x)| \le 6(n+2)e^{-(n+2)\pi/2},$$

which results in (5.6). □

The next proposition is similar to Theorem 2.3 from [1], but here we join the cases of odd and even values of j and get rid of the additional requirement that $n \ge N_0$. We rely on [1] for the existence of the solution but use a simpler argument to prove the uniqueness.

Proposition 5.2. *For all $n \ge 1$ and all $j \in \{1, \dots, n\}$, the number $\varphi_{n,j}$ is the unique solution of the equation (5.2) on the interval $(u_{n,j}, u_{n,j+1})$.*

Proof. Let $n \ge 1$. For each $j \in \{1, \dots, n\}$, the main equation (5.2) can be written in the form

$$(n+2)x - \eta_{n,j}(x) = j\pi. \tag{5.7}$$

By Theorem 2.1 from [1], if x belongs to $(0, \pi)$ and satisfies (5.7) for some integer j, then the number $g(x)$ is an eigenvalue of $T_n(g)$.

Notice that $f(x) > 0$ and $\beta(x) > 0$ for every $x \in (0, \pi)$. Using the definitions of tanh, coth, and arctan, we conclude that $0 < \eta_{n,j}(x) < \pi$ for each $x \in (0, \pi)$; see also Figure 6. Denote the left-hand side of (5.7) by $F_{n,j}(x)$. Then

$$F_{n,j}(u_{n,j}) = j\pi - \eta_{n,j}(u_{n,j}) < j\pi,$$
$$F_{n,j}(u_{n,j+1}) = (j+1)\pi - \eta_{n,j}(u_{n,j+1}) > j\pi.$$

Hence, by the intermediate value theorem, equation (5.7) has at least one solution in the interval $(u_{n,j}, u_{n,j+1})$. At this moment we do not know whether this solution is unique. So let us, for each j, denote by $\psi_{n,j}$ one of the solutions of (5.7) on $(u_{n,j}, u_{n,j+1})$.

Contrary to what we want, assume that for some $j \in \{1, \dots, n\}$ equation (5.7) has another solution x belonging to $(u_{n,j}, u_{n,j+1})$. The $n+1$ numbers $\psi_{n,1}, \dots, \psi_{n,n}, x$ are different. Since g is strictly increasing on $[0, \pi]$, the corresponding eigenvalues $g(\psi_{n,1}), \dots, g(\psi_{n,n}), g(x)$ are different, too. This contradicts the fact that the matrix $T_n(g)$ has only n eigenvalues.

We conclude that for each j equation (5.7) has only one solution $\psi_{n,j}$ in $(u_{n,j}, u_{n,j+1})$. The numbers $\psi_{n,j}$ satisfy $\psi_{n,1} < \ldots < \psi_{n,n}$, and their images under g are eigenvalues of $T_n(g)$, so $g(\psi_{n,j}) = \lambda_{n,j}$ and $\psi_{n,j} = \varphi_{n,j}$ for all j in $\{1, \dots, n\}$. □

The next proposition gives asymptotic formulas for the eigenvalues $\lambda_{n,j}$ provided j is "not too small". It mimics Theorem 2.6 from [1], the novelty being that we here join the cases of odd and even values of j and state the result for an arbitrary order p.

Proposition 5.3. *For every $p \in \mathbb{N}$, the functions $\rho_{n,j}$ admit the asymptotic upper estimate*

$$\max_{(p/2)\log(n+2) \le j \le n} \ \sup_{x \in [u_{n,j}, u_{n,j+1}]} |\rho_{n,j}(x)| = O\left(\frac{1}{n^{p+1}}\right). \tag{5.8}$$

Moreover, for every $p \in \mathbb{N}$, every $n \in \mathbb{N}$, and every j satisfying

$$\frac{p}{2}\log(n+2) \le j \le n, \tag{5.9}$$

the numbers $\varphi_{n,j}$ and $\lambda_{n,j}$ have asymptotic expansions of the form

$$\varphi_{n,j} = \sum_{k=0}^{p} \frac{c_k(u_{n,j})}{(n+2)^k} + O\left(\frac{1}{(n+2)^{p+1}}\right), \tag{5.10}$$

$$\lambda_{n,j} = \sum_{k=0}^{p} \frac{d_k(u_{n,j})}{(n+2)^k} + O\left(\frac{1}{(n+2)^{p+1}}\right), \tag{5.11}$$

where the upper estimates of the residue terms are uniform in j, the functions c_k and d_k are infinitely smooth and can be expressed in terms of η and g by the formulas shown in the proof of Proposition 3.1.

Proof. We have to verify the upper bound (5.8). The other statements then follow from Proposition 3.1. Let $p, n \in \mathbb{N}$ and j satisfy (5.9). If $j \le n/2$, then (5.5) gives

$$\frac{6e^{-2j}}{n+2} \le \frac{6e^{-p\log(n+2)}}{n+2} = \frac{6}{(n+2)^{p+1}},$$

while if $j > n/2$, we obtain from (5.6) that

$$e^{-(n+2)\pi/2} = O\left(\frac{1}{n^{p+1}}\right).$$

Joining these two cases we arrive at (5.8). □

In Proposition 3.1 we expressed the first of the coefficients c_k and d_k in terms of the first derivatives of g and η. Here are explicit formulas for $g', \ldots, g^{(5)}$:

$$\begin{aligned} g'(x) &= 23\cos\frac{x}{2}\left(\sin\frac{x}{2}\right)^3, \qquad g''(x) = 16(1+2\cos(x))\left(\sin\frac{x}{2}\right)^2, \\ g'''(x) &= -8\sin(x)+16\sin(2x), \quad g^{(4)}(x) = -8\cos(x)+32\cos(2x), \\ g^{(5)}(x) &= 8\sin(x) - 64\sin(2x). \end{aligned} \tag{5.12}$$

For $\eta', \ldots, \eta^{(4)}$ we have

$$\begin{aligned} \eta'(x) &= \frac{\sin\frac{x}{2}}{\left(1+\left(\sin\frac{x}{2}\right)^2\right)^{1/2}}, \qquad \eta''(x) = \frac{\sqrt{2}\cos\frac{x}{2}}{(3-\cos(x))^{3/2}}, \\ \eta'''(x) &= -\frac{5\sin\frac{x}{2}+\sin\frac{3x}{2}}{\sqrt{2}(3-\cos(x))^{5/2}}, \quad \eta^{(4)}(x) = \frac{-4\cos\frac{x}{2}+19\cos\frac{3x}{2}+\cos\frac{5x}{2}}{2\sqrt{2}(3-\cos(x))^{7/2}}. \end{aligned} \tag{5.13}$$

Numerical test 5.4. In order to test (5.11) numerically for $p = 4$, we computed $g', \ldots, g^{(4)}$ by (5.12), $\eta, \eta', \ldots, \eta^{(3)}$ by (5.3) and (5.13), $c_0, c_1, \ldots, c_4$ by (3.7) and $d_0, d_1, \ldots, d_4$ by (3.9). The exact eigenvalues were computed by simple iteration in equation (5.4) and independently by general eigenvalue algorithms (for $n \le 1024$). All computations were made in high-precision arithmetic with 100 decimal digits after the floating point, in SageMath and independently in Wolfram Mathematica. Denote by $E_{n,4}$ the maximal error in (5.11), with $p = 4$:

$$E_{n,4} := \max_{2\log(n+2)\le j\le n} \left| \lambda_{n,j} - \sum_{k=0}^{4} \frac{d_k(u_{n,j})}{(n+2)^k} \right|.$$

The following table shows $E_{n,4}$ and $(n+2)^5 E_{n,4}$ for various values of n.

	$n = 64$	$n = 256$	$n = 1024$	$n = 4096$	$n = 16384$
$E_{n,4}$	$2.4 \cdot 10^{-7}$	$3.1 \cdot 10^{-10}$	$3.2 \cdot 10^{-13}$	$3.2 \cdot 10^{-16}$	$3.1 \cdot 10^{-19}$
$(n+2)^5 E_{n,4}$	306.72	354.87	366.61	369.52	370.25

We see that the numbers $E_{n,4}$ really behave like $O(1/(n+2)^5)$.

6. An asymptotic formula for the first eigenvalues in the example

In this section we study the asymptotic behavior of $\lambda_{n,j}$ as n tends to ∞, considering j as a fixed parameter.

Using the definition of arctan and the formula for $\tan(x + j\pi/2)$, we can rewrite equation (5.2) in the equivalent form

$$f(x)^{(-1)^{j+1}} \tanh\frac{(n+2)\beta(x)}{2} = (-1)^j \tan\frac{(n+2)x}{2}. \tag{6.1}$$

The first factor on the left-hand side of (6.1) is just $f(x)$ for odd values of j and $1/f(x)$ for even values of j. We know that

$$\frac{j\pi}{n+2} \le \varphi_{n,j} \le \frac{(j+1)\pi}{n+2},$$

and it is natural to expect that the product $(n+2)\varphi_{n,j}$ has a finite limit α_j as n tends to infinity and j is fixed. Assuming this and taking into account that

$$f(x) \to 1, \quad \beta(x) \sim x, \quad \text{as} \quad x \to 0,$$

we can pass to the limit in (6.1) to obtain a simple transcendental equation for α_j. This is an informal motivation of the following formal reasoning.

For each j in $\mathbb{N}$, denote by α_j the unique real number that belongs to the interval $(j\pi, (j+1)\pi)$ and satisfies

$$\tanh \frac{\alpha_j}{2} = (-1)^j \tan \frac{\alpha_j}{2}. \tag{6.2}$$

Figure 7 shows both sides of (6.2) for $j = 1, 2, 3$.

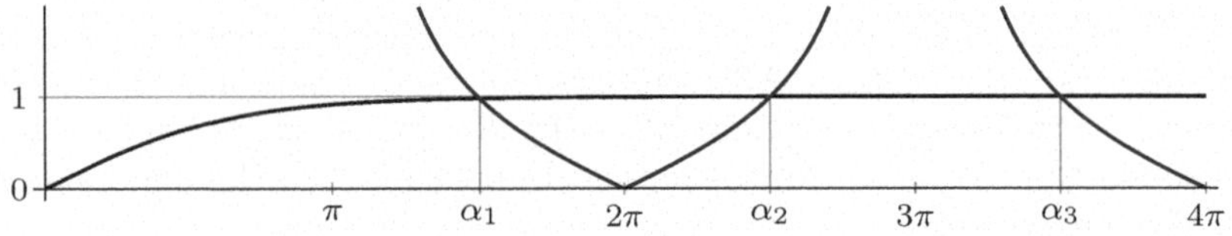

FIGURE 7. The left-hand side (black) and the right-hand side (blue) of (6.2), for $j = 1$ on $(\pi, 2\pi)$, for $j = 2$ on $(2\pi, 3\pi)$ and for $j = 3$ on $(3\pi, 4\pi)$.

For each j, the transcendental equation (6.2) is easy to solve by numerical methods. Approximately,

$$\alpha_1 \approx 4.73004, \qquad \alpha_2 \approx 7.85320, \qquad \alpha_3 \approx 10.99561.$$

It follows from (6.2) that $\alpha_j > \frac{(2j+1)\pi}{2}$ if j is odd and $\alpha_j < \frac{(2j+1)\pi}{2}$ if j is even. In particular,

$$\alpha_1 > \frac{3\pi}{2}, \qquad \alpha_2 < \frac{5\pi}{2}, \qquad \alpha_3 > \frac{7\pi}{2}.$$

We remark that differences between α_j and $(2j+1)\pi/2$ are extremely small:

$$\alpha_1 - \frac{3\pi}{2} \approx 1.8 \cdot 10^{-2}, \quad \alpha_2 - \frac{5\pi}{2} \approx -7.8 \cdot 10^{-4}, \quad \alpha_3 - \frac{7\pi}{2} \approx 3.3 \cdot 10^{-5}.$$

Contrary to the general agreement of this paper, the upper estimates of the residual terms in the following proposition are not uniform in j. Thus we use the notation O_j instead of O.

Proposition 6.1. *Let g be the function defined by* (5.1) *and define $\varphi_{n,j} \in (0,\pi)$ by $\lambda_{n,j} = g(\varphi_{n,j})$. Then for each fixed j in $\mathbb{N}$, $\varphi_{n,j}$ and $\lambda_{n,j}$ satisfy the asymptotic formulas*

$$\varphi_{n,j} = \frac{\alpha_j}{n+2} + O_j\left(\frac{1}{(n+2)^3}\right), \tag{6.3}$$

$$\lambda_{n,j} = \left(\frac{\alpha_j}{n+2}\right)^4 + O_j\left(\frac{1}{(n+2)^6}\right). \tag{6.4}$$

Proof. Fix j in $\mathbb{N}$. We are going to treat (6.1) by asymptotic methods, as n tends to infinity. Put

$$\delta_{n,j} := (n+2)\varphi_{n,j} - \alpha_j,$$

i.e., represent the product $(n+2)\varphi_{n,j}$ in the form

$$(n+2)\varphi_{n,j} = \alpha_j + \delta_{n,j}.$$

It is easy to verify that, as $x \to 0$,

$$f(x) = 1 + O(x^2), \qquad \beta(x) = x + O(x^3).$$

Moreover, we know that $\frac{j\pi}{n+2} \le \varphi_{n,j} \le \frac{(j+1)\pi}{n+2}$ and thus $\varphi_{n,j} = O_j(1/(n+2))$. Therefore

$$f(\varphi_{n,j}) = 1 + O_j\left(\frac{1}{(n+2)^2}\right), \qquad \frac{1}{f(\varphi_{n,j})} = 1 + O_j\left(\frac{1}{(n+2)^2}\right),$$

$$\frac{(n+2)}{2}\beta(\varphi_{n,j}) = \frac{\alpha_j + \delta_{n,j}}{2} + O_j\left(\frac{1}{(n+2)^2}\right),$$

$$\tanh\frac{(n+2)}{2}\beta(\varphi_{n,j}) = \tanh\frac{\alpha_j + \delta_{n,j}}{2} + O_j\left(\frac{1}{(n+2)^2}\right).$$

By the mean value theorem, there exist some numbers $\xi_{1,n,j}$ and $\xi_{2,n,j}$ between $\alpha_j/2$ and $(\alpha_j + \delta_{n,j})/2$ such that

$$\tanh\frac{\alpha_j + \delta_{n,j}}{2} - \tanh\frac{\alpha_j}{2} = \tanh'(\xi_{1,n,j})\frac{\delta_{n,j}}{2}$$

and

$$\tan\frac{\alpha_j + \delta_{n,j}}{2} - \tan\frac{\alpha_j}{2} = \tan'(\xi_{2,n,j})\frac{\delta_{n,j}}{2}.$$

After replacing x by $\varphi_{n,j}$, equation (6.1) takes the form

$$\begin{aligned}\tanh\frac{\alpha_j}{2} &+ \tanh'(\xi_{1,n,j})\frac{\delta_{n,j}}{2} + O_j\left(\frac{1}{(n+2)^2}\right)\\ &= (-1)^j\left(\tan\frac{\alpha_j}{2} + \tan'(\xi_{2,n,j})\frac{\delta_{n,j}}{2} + O_j\left(\frac{1}{(n+2)^2}\right)\right).\end{aligned}$$

Using the definition of α_j, this can be simplified to

$$\left(\tan'(\xi_{2,n,j}) + (-1)^{j-1}\tanh'(\xi_{1,n,j})\right)\delta_{n,j} = O_j\left(\frac{1}{(n+2)^2}\right).$$

The coefficient before $\delta_{n,j}$ is strictly positive and bounded away from zero. Indeed, for all x from the considered domain $(j\pi/2,(j+1)\pi/2)$ we have $\tan'(x) > 1$ and

$$\tanh'(x) = \frac{1}{1+x^2} < \frac{1}{1+\frac{\pi^2}{4}} < \frac{1}{2},$$

thus

$$\tan'(\xi_{2,n,j}) + (-1)^{j-1}\tanh'(\xi_{1,n,j}) > \frac{1}{2}.$$

Therefore $\delta_{n,j} = O_j(1/(n+2)^2)$, which is equivalent to (6.3). The function g has the following asymptotic expansion near the point 0:

$$g(x) = x^4 + O(x^6). \tag{6.5}$$

Using the formula $\lambda_{n,j} = g(\varphi_{n,j})$ and combining (6.3) with (6.5), we arrive at (6.4). □

Numerical test 6.2. Denote by $\varepsilon_{n,j}$ the absolute value of the residue in (6.4):

$$\varepsilon_{n,j} := \left|\lambda_{n,j} - \left(\frac{\alpha_j}{n+2}\right)^4\right|.$$

Similarly to Numerical test 5.4, the exact eigenvalues $\lambda_{n,j}$ and the coefficients α_j are computed in high-precision arithmetic with 100 decimal digits after the floating point. The next table shows $\varepsilon_{n,j}$ and $(n+2)^6\varepsilon_{n,j}$ corresponding to $j = 1, 2$ and to various values of n.

	$n = 64$	$n = 256$	$n = 1024$	$n = 4096$	$n = 16384$
$\varepsilon_{n,1}$	$6.3 \cdot 10^{-9}$	$1.8 \cdot 10^{-11}$	$4.5 \cdot 10^{-16}$	$1.1 \cdot 10^{-19}$	$2.7 \cdot 10^{-23}$
$(n+2)^6\varepsilon_{n,1}$	523.37	524.39	524.46	524.46	524.46
$\varepsilon_{n,2}$	$1.1 \cdot 10^{-7}$	$3.1 \cdot 10^{-11}$	$7.9 \cdot 10^{-15}$	$2.0 \cdot 10^{-18}$	$4.9 \cdot 10^{-22}$
$(n+2)^6\varepsilon_{n,2}$	9315.7	9266.9	9263.7	9263.5	9263.4

Moreover, numerical experiments show that

$$\max_{1\le n\le 100000}((n+2)^6\varepsilon_{n,1}) < 524.47.$$

Remark 6.3. Notice that formula (2.7) from [1] does not have the form (6.3) because the numerator $u_{1,j}$ in this formula depends on n in a complicated manner.

Remark 6.4. Proposition 6.1 has trivial corollaries about the norm of the inverse matrix and the condition number:

$$\|T_n^{-1}(g)\|_2 \sim \left(\frac{n+2}{\alpha_1}\right)^4, \quad \operatorname{cond}_2(T_n(g)) \sim 16\left(\frac{n+2}{\alpha_1}\right)^4, \quad \text{as} \quad n\to\infty.$$

Remark 6.5. Proposition 6.1 is not really new. Parter [12, 13] showed that if g_m is given by (2.1), then the corresponding eigenvalues satisfy

$$\lambda_{n,j} = \frac{\gamma_j(m)}{(n+2)^{2m}} + o\left(\frac{1}{(n+2)^{2m}}\right) \quad \text{as } n \to \infty \tag{6.6}$$

with some constant $\gamma_j(m)$ for each fixed j. Our proposition identifies $\gamma_1(2)$ as α_1^4 and improves the $o(1/(n+2)^4)$ to $O(1/(n+2)^6)$. Parter also had explicit formulas for $\gamma_j(2)$ in terms of the solutions of certain transcendental equations. Widom [15, 16] derived results like (6.6) by replacing matrices by integral operators with piecewise constant kernels and subsequently proving the convergence of the appropriately scaled integral operators. Widom's approach delivered the constants $\gamma_j(m)$ as the reciprocals of the eigenvalues of certain integral operators. More about these pioneering works can be found in [7, pp. 256–259] and in [8]. The proof of Proposition 6.1 given above is different from the ones by Parter and Widom.

Remark 6.6. If we pass to the denominator $n+1$ in formula (6.4), then it becomes more complicated:

$$\lambda_{n,j} = \frac{\alpha_j^4}{(n+1)^4} - \frac{4\alpha_j^4}{(n+1)^5} + O_j\left(\frac{1}{(n+2)^6}\right).$$

This reveals that the denominator $n+2$ is more convenient when studying the asymptotic behavior of the first eigenvalues in the example (1.1).

7. The regular four term asymptotic expansion for the example

Lemma 7.1. *Let* $g(x) = \left(2\sin\frac{x}{2}\right)^4$ *and let* $d_0, \dots, d_4$ *be the same functions as in Proposition* 5.3. *Then, as* $n \to \infty$, *we have the asymptotic expansions*

$$\sum_{k=0}^{3} \frac{d_k\left(\frac{j\pi}{n+2}\right)}{(n+2)^k} = \frac{(j\pi + \eta(0))^4 - \eta(0)^4}{(n+2)^4} + O\left(\frac{j^4}{(n+2)^5}\right), \tag{7.1}$$

$$\sum_{k=0}^{4} \frac{d_k\left(\frac{j\pi}{n+2}\right)}{(n+2)^k} = \left(\frac{j\pi + \eta(0)}{n+2}\right)^4 + O\left(\frac{j^4}{(n+2)^5}\right), \tag{7.2}$$

uniformly in j.

Proof. By (5.12), the function g and its derivatives admit the following asymptotic expansions near the point 0:

$$\begin{aligned} &g(x) = x^4 + O(x^6), \quad g'(x) = 4x^3 + O(x^5), \quad g''(x) = 12x^2 + O(x^4), \\ &g'''(x) = 24x + O(x^3), \quad g^{(4)}(x) = 24 + O(x^2) \qquad (x \to 0). \end{aligned} \tag{7.3}$$

Applying (5.13) and taking into account that η is smooth, we see that

$$c_0(x) = x, \qquad c_1(x) = \eta(x) = \eta(0) + O(x) \qquad (x \to 0) \tag{7.4}$$

and that the functions c_2, c_3, c_4 are bounded. Substituting (7.3) and (7.4) into the formulas (3.9), we get the following expansions of $d_0(x), \ldots, d_4(x)$, as $x \to 0$:

$$d_0(x) = x^4 + O(x^6), \quad d_1(x) = g'(x)c_1(x) = 4x^3\eta(0) + O(x^4),$$
$$d_2(x) = \frac{1}{2}g''(x)c_1^2(x) + O(x^3) = 6x^2\eta^2(0) + O(x^3),$$
$$d_3(x) = \frac{1}{6}g'''(x)c_1^3(x) + O(x^2) = 4x\eta^3(0) + O(x^2),$$
$$d_4(x) = \frac{1}{24}g^{(4)}(x)c_1^4(x) + O(x) = \eta^4(0) + O(x).$$

Using these formulas and the binomial theorem, we arrive at (7.2). Moving in (7.2) the summand with $k = 4$ to the right-hand side we obtain (7.1). □

The following proposition proves Theorem 1.2 (b).

Proposition 7.2. *Let* $g(x) = \left(2\sin\frac{x}{2}\right)^4$ *and* $d_0, \ldots, d_3 \colon [0, \pi] \to \mathbb{R}$ *be the functions from the proof of Proposition* 5.3. *Then there exists a* $C > 0$ *such that*

$$\left| \lambda_{n,j} - \sum_{k=0}^{3} \frac{d_k\left(\frac{j\pi}{n+2}\right)}{(n+2)^k} \right| \le \frac{C}{(n+2)^4} \tag{7.5}$$

for all $n \in \mathbb{N}$ *and all* $j \in \{1, \ldots, n\}$.

Proof. Thanks to Proposition 5.3 we are left with the case $j < 2\log(n+2)$. Using (5.4), the upper estimate (5.5), and the smoothness of η, we conclude that

$$\varphi_{n,j} = \frac{j\pi + \eta(0)}{n+2} + O\left(\frac{j}{(n+2)^2}\right) + O\left(\frac{e^{-2j}}{n+2}\right). \tag{7.6}$$

From (6.5) we therefore obtain that

$$\lambda_{n,j} = g(\varphi_{n,j}) = \varphi_{n,j}^4 + O\left(\frac{(\log(n+2))^6}{(n+2)^6}\right) = \varphi_{n,j}^4 + O\left(\frac{1}{(n+2)^4}\right).$$

Expanding $\varphi_{n,j}^4$ by the multinomial theorem and separating the main term, we get

$$\varphi_{n,j}^4 = \left(\frac{j\pi + \eta(0)}{n+2}\right)^4 + \sum_{\substack{p,q,r\ge 0\\ p+q+r=4\\ p<4}} O\left(\frac{(j\pi + \eta(0))^p j^q\, e^{-2jr}}{(n+2)^{p+2q+r}}\right).$$

The sum over p, q, r can be divided into the part with $q > 0$ and the part with $q = 0$ and estimated by

$$\sum_{\substack{p,q,r\ge 0\\ p+q+r=4\\ q>0}} O\left(\frac{(j\pi + \eta(0))^p j^q}{(n+2)^{4+q}}\right) + \sum_{\substack{p,r\ge 0\\ p+r=4\\ r>0}} O\left(\frac{(j\pi + \eta(0))^p e^{-2jr}}{(n+2)^4}\right) = O\left(\frac{1}{(n+2)^4}\right).$$

Consequently, the true asymptotic expansion of $\lambda_{n,j}$ under the condition $j < 2\log(n+2)$ is

$$\lambda_{n,j} = \left(\frac{j\pi + \eta(0)}{n+2}\right)^4 + O\left(\frac{1}{(n+2)^4}\right). \tag{7.7}$$

On the other hand, using (7.1) and the fact that $j^4 = O(n+2)$, we get

$$\sum_{k=0}^{3} \frac{d_k(u_{n,j})}{(n+2)^k} = \left(\frac{j\pi + \eta(0)}{n+2}\right)^4 + O\left(\frac{1}{(n+2)^4}\right). \tag{7.8}$$

Comparing (7.7) and (7.8), we obtain the required result. □

Numerical test 7.3. Denote by Δ_n the maximal error in (7.5):

$$\Delta_n := \max_{1\le j\le n} \left| \lambda_{n,j} - \sum_{k=0}^{3} \frac{d_k(u_{n,j})}{(n+2)^k} \right|.$$

The following table shows Δ_n and $(n+2)^4\Delta_n$ for various values of n.

	$n = 64$	$n = 256$	$n = 1024$	$n = 4096$	$n = 16384$
Δ_n	$7.6 \cdot 10^{-6}$	$3.2 \cdot 10^{-8}$	$1.3 \cdot 10^{-10}$	$5.1 \cdot 10^{-13}$	$2.0 \cdot 10^{-15}$
$(n+2)^4\Delta_n$	143.97	143.05	142.81	142.75	142.74

According to this table, the numbers Δ_n really behave like $O(1/(n+2)^4)$.

Remark 7.4. Let us again embark on the case $p = 3$ and thus on Theorem 1.2 (b) and the previous Numerical test 7.3. This test suggests that we could be satisfied by an error of 10^{-15} for $n = 16\,384$. However, as the first eigenvalues are also of order 10^{-15} we obtain nothing but an upper bound for them. In other words, the approximation of the first eigenvalues $\lambda_{n,j}$ by $\sum_{k=0}^{3} \frac{d_k(u_{n,j})}{(n+2)^k}$ is bad in the sense that the absolute error of this approximation is of the same order $O_j(1/(n+2)^4)$ as the eigenvalue $\lambda_{n,j}$ which we want to approximate! To state it in yet different terms, for each fixed j, the residues

$$\omega_{n,j} := \lambda_{n,j} - \sum_{k=0}^{3} \frac{d_k(\frac{j\pi}{n+2})}{(n+2)^k}$$

decay at the same rate $O_j(1/(n+2)^4)$ as the eigenvalues $\lambda_{n,j}$ and the distances between them, and the corresponding relative errors do not tend to zero:

$$\frac{\omega_{n,j}}{\lambda_{n,j}} \to \frac{\alpha_j^4 + \eta(0)^4 - (j\pi + \eta(0))^4}{\alpha_j^4} \neq 0,$$

$$\frac{\omega_{n,j}}{\lambda_{n,j+1} - \lambda_{n,j}} \to \frac{\alpha_j^4 + \eta(0)^4 - (j\pi + \eta(0))^4}{\alpha_{j+1}^4 - \alpha_j^4} \neq 0.$$

Compared to this, the residues of the asymptotic expansions for simple-loop symbols (see [2, 5]) can be bounded by $o(\frac{j\,(n+1-j)}{n^2}\,\frac{1}{n^p})$, where p is related with the smoothness of the symbols, and the expression $\frac{j\,(n+1-j)}{n^2}$ is in the simple-loop case always comparable with the distance $\lambda_{n,j+1} - \lambda_{n,j}$ between the consecutive eigenvalues, i.e., there exist $C_1 > 0$ and $C_2 > 0$ such that

$$C_1 \frac{j(n+1-j)}{n^2} \le \lambda_{n,j+1} - \lambda_{n,j} \le C_2 \frac{j(n+1-j)}{n^2}.$$

Clearly, the quotient $\frac{|\omega_{n,j}|}{\lambda_{n,j+1}-\lambda_{n,j}}$ is a more adequate measure of the quality of the approximation than just the absolute error $|\omega_{n,j}|$.

8. There is no regular five term asymptotic expansion for the example

As said, Ekström, Garoni, and Serra-Capizzano [10] conjectured that for every infinitely smooth 2π-periodic real-valued even function g, strictly increasing on $[0,\pi]$, the eigenvalues $\lambda_{n,j}$ of the corresponding Toeplitz matrices admit an asymptotic expansion of the regular form (1.2) for every order p.

We now show that for the symbol $g(x) = \left(2\sin\frac{x}{2}\right)^4$ an asymptotic expansion of the form (1.2) cannot be true for $p = 4$. This disproves Conjecture 1 from [10].

We remark that the following proposition is actually stronger than the third part of Theorem 1.2. Namely, Theorem 1.2(c) states that (1.6) cannot hold with the functions $d_1, \dots, d_4$ appearing in (1.4). The following proposition tells us that (1.6) is also impossible for any other choice of continuous functions $d_1, \dots, d_4$. The reason is of course Proposition 4.2.

Proposition 8.1. *Let $g(x) = \left(2\sin\frac{x}{2}\right)^4$. Denote by $\lambda_{n,1}, \dots, \lambda_{n,n}$ the eigenvalues of the Toeplitz matrices $T_n(g)$, written in the ascending order. Then there do not exist continuous functions $d_0, \dots, d_4 \colon [0,\pi] \to \mathbb{R}$ and numbers $C > 0$, $N \in \mathbb{N}$, such that for every $n \ge N$ and every $j \in \{1, \dots, n\}$*

$$\left| \lambda_{n,j} - \sum_{k=0}^{4} \frac{d_k\left(\frac{j\pi}{n+2}\right)}{(n+2)^k} \right| \le \frac{C}{(n+2)^5}. \tag{8.1}$$

Proof. Reasoning by contradiction, assume there exist functions $d_0, \dots, d_4$ and numbers C and N with the required properties. Put

$$J = \left\{(n,j) \in \mathbb{N}^2 \colon\ n \ge N,\ 2\log(n+2) \le j \le n\right\}.$$

Clearly, this set J asymptotically fills $[0,\pi]$ by quotients.

So, by Proposition 4.2, the functions $d_0, \dots, d_4$ from (8.1) must be the same as the functions $d_0, \dots, d_4$ from Proposition 5.3. In other words, the asymptotic expansion (5.11) from Proposition 5.3 holds for every pair (n,j) with n large enough and j in $\{1, \dots, n\}$, that is, without the restriction $j \ge 2\log(n+2)$.

Combining (8.1) with (7.2), we see that for each fixed j the eigenvalue $\lambda_{n,j}$ must have the asymptotic behavior

$$\lambda_{n,j} = \left(\frac{j\pi + \eta(0)}{n+2}\right)^4 + O_j\left(\frac{1}{(n+2)^5}\right). \tag{8.2}$$

Since $\eta(0) = 2\arctan(1) = \frac{\pi}{2}$, we obtain for $j = 1$ that

$$\lambda_{n,1} = \left(\frac{3\pi/2}{n+2}\right)^4 + O\left(\frac{1}{(n+2)^5}\right), \tag{8.3}$$

which contradicts Proposition 6.1 because $3\pi/2 \neq \alpha_1$. □

Remark 8.2. Here is an alternative way to finish the proof of Proposition 8.1. After having formula (8.2), we obtain the following hypothetical asymptotic relation between the two first eigenvalues:

$$\lim_{n\to\infty} \left((n+2)\left(\lambda_{n,2}^{1/4} - \lambda_{n,1}^{1/4}\right)\right) = (2\pi + \eta(0)) - (\pi + \eta(0)) = \pi.$$

But this contradicts Proposition 6.1, according to which

$$\lim_{n\to\infty} \left((n+2)\left(\lambda_{n,2}^{1/4} - \lambda_{n,1}^{1/4}\right)\right) = \alpha_2 - \alpha_1 < \pi.$$

In this reasoning we do not use the value $\eta(0)$.

Proof of Theorem 1.2. The existence of the asymptotic expansions (1.4) follows from Proposition 5.3, its uniqueness is a consequence of Proposition 4.2, formula (1.5) was established in Proposition 7.2, and the impossibility of (1.6) is just Proposition 8.1. □

Proof of Theorem 1.1. The functions $d_0, d_1, \ldots$ from Proposition 5.3 are infinitely smooth on $[0, \pi]$, and thus, by Remark 3.3, the expansion (5.11) with $p = 4$ can be rewritten in the form (1.3) with some infinitely smooth functions $f_0, \ldots, f_4$. So, (1.3) is true for all (n, j) satisfying $2\log(n+2) \leq j \leq n$.

Contrary to what we want, assume that there are $f_0, \ldots, f_4$, C, and N as in the statement of Theorem 1.1. Then, by Proposition 4.2, the functions $f_0, \ldots, f_4$ are the same as those in the previous paragraph. In particular, $f_0, \ldots, f_4$ must be infinitely smooth. In this case, the asymptotic expansion (1.3) can be rewritten in powers of $1/(n+2)$ and is true for all n and j with $n \geq N$ and $1 \leq j \leq n$. This contradicts Proposition 8.1. □

We conclude with a conjecture about the eigenvalues of Toeplitz matrices generated by (2.1).

Conjecture 8.3. *Let $g_m(x) = \left(2\sin\frac{x}{2}\right)^{2m}$ with an integer $m \geq 3$. If $p \leq 2m-1$, there are $N_p \in \mathbb{N}$ and $D_p > 0$ such that*

$$\left|\lambda_{n,j} - \sum_{k=0}^{p} \frac{d_k\left(\frac{j\pi}{n+2}\right)}{(n+2)^k}\right| \leq \frac{D_p}{(n+2)^{p+1}} \tag{8.4}$$

for all $n \geq N_p$ and all j in $\{1, \ldots, n\}$. For $p = 2m$, inequality (8.4) does not hold for all sufficiently large n and all $1 \leq j \leq n$, but it holds for for all sufficiently large n and all j not too close to 1, say, for $(\log(n+2))^2 \leq j \leq n$.

References

[1] M. Barrera and S.M. Grudsky, Asymptotics of eigenvalues for pentadiagonal symmetric Toeplitz matrices, *Operator Theory: Adv. and Appl.* **259**, Birkhäuser, 2017, 51–77. DOI: 10.1007/978-3-319-49182-0_7

[2] J.M. Bogoya, A. Böttcher, S.M. Grudsky, and E.A. Maximenko, Eigenvalues of Hermitian Toeplitz matrices with smooth simple-loop symbols, *J. Math. Analysis Appl.* **422** (2015), 1308–1334. DOI: 10.1016/j.jmaa.2014.09.057

[3] J.M. Bogoya, A. Böttcher, S.M. Grudsky, and E.A. Maximenko, Maximum norm versions of the Szegő and Avram–Parter theorems for Toeplitz matrices, *J. Approx. Theory* **196** (2015), 79–100. DOI: 10.1016/j.jat.2015.03.003

[4] J.M. Bogoya, A. Böttcher, and E.A. Maximenko, From convergence in distribution to uniform convergence, *Boletín de la Sociedad Matemática Mexicana* **22** (2016), no. 2, 695–710. DOI: 10.1007/s40590-016-0105-y

[5] J.M. Bogoya, S.M. Grudsky, and E.A. Maximenko, Eigenvalues of Hermitian Toeplitz matrices generated by simple-loop symbols with relaxed smoothness, *Operator Theory: Adv. and Appl.* **259**, Birkhäuser, 2017, 179–212. DOI: 10.1007/978-3-319-49182-0_11

[6] A. Böttcher, S.M. Grudsky, and E.A. Maksimenko, Inside the eigenvalues of certain Hermitian Toeplitz band matrices, *J. Comput. Appl. Math.* **233** (2010), 2245–2264. DOI: 10.1016/j.cam.2009.10.010

[7] A. Böttcher and S.M. Grudsky, *Spectral Properties of Banded Toeplitz Matrices*, SIAM, Philadelphia, 2005. DOI: 10.1137/1.9780898717853

[8] A. Böttcher and H. Widom, From Toeplitz eigenvalues through Green's kernels to higher-order Wirtinger-Sobolev inequalities, *Operator Theory: Adv. and Appl.* **171**, Birkhäuser, 2006, 73–87. DOI: 10.1007/978-3-7643-7980-3_4

[9] P. Deift, A. Its, and I. Krasovsky, Eigenvalues of Toeplitz matrices in the bulk of the spectrum, *Bull. Inst. Math. Acad. Sin. (N.S.)* **7** (2012), 437–461. URL: http://web.math.sinica.edu.tw/bulletin_ns/20124/2012401.pdf

[10] S.-E. Ekström, C. Garoni, and S. Serra-Capizzano, Are the eigenvalues of banded symmetric Toeplitz matrices known in almost closed form? *Experimental Mathematics*, 10 pp., 2017. DOI: 10.1080/10586458.2017.1320241

[11] M. Elouafi, On a relationship between Chebyshev polynomials and Toeplitz determinants, *Applied Mathematics and Computation* **229** (2014), 27–33. DOI: 10.1016/j.amc.2013.12.029

[12] S.V. Parter, Extreme eigenvalues of Toeplitz forms and applications to elliptic difference equations, *Trans. Amer. Math. Soc.* **99** (1961), 153–192. DOI: 10.2307/1993449

[13] S.V. Parter, On the extreme eigenvalues of truncated Toeplitz matrices, *Bull. Amer. Math. Soc.* **67** (1961), 191–196. DOI: 10.1090/S0002-9904-1961-10563-6

[14] W.F. Trench, Interlacement of the even and odd spectra of real symmetric Toeplitz matrices, *Linear Alg. Appl.* **195** (1993), 59–68. DOI: 10.1016/0024-3795(93)90256-N

[15] H. Widom, *Extreme eigenvalues of translation kernels, Trans. Amer. Math. Soc.* **88** (1958), 491–522. DOI: 10.1090/S0002-9947-1961-0138980-4

[16] H. Widom, Extreme eigenvalues of N-dimensional convolution operators, *Trans. Amer. Math. Soc.* **106** (1963), 391–414. DOI: 10.2307/1993750

Mauricio Barrera
CINVESTAV
Departamento de Matemáticas
Apartado Postal 07360
Ciudad de México
Mexico
e-mail: mabarrera@math.cinvestav.mx

Albrecht Böttcher
Technische Universität Chemnitz
Fakultät für Mathematik
09107 Chemnitz
Germany
e-mail: aboettch@mathematik.tu-chemnitz.de

Sergei M. Grudsky
CINVESTAV
Departamento de Matemáticas
Apartado Postal 07360
Ciudad de México
Mexico
e-mail: grudsky@math.cinvestav.mx

Egor A. Maximenko
Instituto Politécnico Nacional
Escuela Superior de Física y Matemáticas
Apartado Postal 07730
Ciudad de México
Mexico
e-mail: emaximenko@ipn.mx

Spectral regularity of a C^*-algebra generated by two-dimensional singular integral operators

Harm Bart, Torsten Ehrhardt and Bernd Silbermann

Abstract. Given a bounded simply connected domain $U \subset \mathbb{C}$ having a Lyapunov curve as its boundary, let $\mathcal{L}(L^2(U))$ stand for the C^*-algebra of all bounded linear operators acting on the Hilbert space $L^2(U)$ with Lebesgue area measure. We show that the smallest C^*-subalgebra $\mathcal{A}$ of $\mathcal{L}(L^2(U))$ containing the singular integral operator

$$(S_U f)(z) = -\frac{1}{\pi} \int_U \frac{f(w)}{(z-w)^2}\, dA(w),$$

along with its adjoint

$$(S_U^* f)(z) = -\frac{1}{\pi} \int_U \frac{f(w)}{(\overline{z}-\overline{w})^2}\, dA(w),$$

all multiplication operators $aI, a \in C(\overline{U})$, and all compact operators on $L^2(U)$, is spectrally regular. Roughly speaking the latter means the following: if the contour integral of the logarithmic derivative of an analytic $\mathcal{A}$-valued function f is vanishing (or is quasi-nilpotent), then f takes invertible values on the inner domain of the contour in question.

Mathematics Subject Classification (2010). Primary: 45E05, Secondary: 30G30, 47A53, 47L10.

Keywords. Analytic vector-valued function, logarithmic residue, spectral regularity, two-dimensional singular integral operator, C^*-algebra.

In memory of Georg Heinig

1. Introduction

Let $\mathcal{B}$ be a complex unital Banach algebra and let Δ be a bounded Cauchy domain in $\mathbb{C}$ (see [13, 19]). The positively oriented boundary of Δ will be denoted by $\partial\Delta$. Let $\mathcal{A}_\partial(\Delta, \mathcal{B})$ stand for the set of all $\mathcal{B}$-valued functions f which are defined and analytic on an open neighborhood of $\overline{\Delta} = \partial\Delta \cup \Delta$

© Springer International Publishing AG, part of Springer Nature 2018
A. Böttcher et al. (eds.), *The Diversity and Beauty of Applied Operator Theory*, Operator Theory: Advances and Applications 268, https://doi.org/10.1007/978-3-319-75996-8_3

and which take invertible values on all of $\partial\Delta$. For $f \in \mathcal{A}_\partial(\Delta, \mathcal{B})$ the contour integral

$$\mathrm{LR}(f;\Delta) = \frac{1}{2\pi i} \int_{\partial\Delta} f'(\lambda) f(\lambda)^{-1} \, d\lambda \,, \tag{1.1}$$

is well defined and called the *logarithmic residue* of f with respect to Δ.

In the scalar case $\mathcal{B} = \mathbb{C}$, the logarithmic residue is equal to the number of zeros of f in Δ, multiplicities taken into account. Thus, in this situation, the integral (1.1) vanishes if and only if f takes non-zero values, not only on $\partial\Delta$ (which has been assumed via the invertibility requirement in order to let (1.1) make sense) but on all of Δ.

This leads to the following question: *Given a unital Banach algebra $\mathcal{B}$ and a function $f \in \mathcal{A}_\partial(\Delta, \mathcal{B})$ with* $\mathrm{LR}(f;\Delta) = 0$, *does it follow that f takes invertible values on all of Δ?* In general the answer to this question is negative. The Banach algebra $\mathcal{B} = \mathcal{L}(\ell_2)$ of all bounded linear operators acting on ℓ_2 together with a suitably defined analytic function f provides a counterexample (see [2]). More counterexamples are presented in [8]. On the other hand, it turns out that for several classes of Banach algebras $\mathcal{B}$ the above conclusion is true for all functions $f \in \mathcal{A}_\partial(\Delta, \mathcal{B})$ and all Cauchy domains Δ. Early results of this character can be found in [1, 3, 4]. They became the starting point for a quest for more Banach algebras of the type in question.

A thorough investigation from a general point of view was carried out in [5]. It was recognized there that for all Banach algebras for which we were able to show that the above implication is true, also a stronger statement is true in which the assumption $\mathrm{LR}(f;\Delta) = 0$ is replaced by the weaker assumption that $\mathrm{LR}(f;\Delta)$ is a quasi-nilpotent element in $\mathcal{B}$. Recall that an element in a Banach algebra is quasi-nilpotent if its spectrum equals $\{0\}$. In line with this, the following notion of spectral regularity of a Banach algebra was defined. A unital complex Banach algebra $\mathcal{B}$ is called *spectrally regular* if for every Cauchy domain Δ and every function $f \in \mathcal{A}_\partial(\Delta, \mathcal{B})$ the property that $\mathrm{LR}(f;\Delta)$ is quasi-nilpotent implies that f takes invertible values on all of Δ.

We do not know whether the notion of spectral regularity and that in its weaker form, in which $\mathrm{LR}(f;\Delta)$ being quasi-nilpotent is replaced by $\mathrm{LR}(f;\Delta) = 0$, are always the same.

Among the spectrally regular Banach algebras are the commutative Banach algebras [1], the matrix algebras [4], the polynomial identity algebras [3], the approximately finite-dimensional Banach algebras [9], and the Banach algebras of quasi-triangular operators [11]. Additional classes, also relevant for the considerations in this papers can be found in [7, Sect. 8]).

Notice that expression (1.1) defines the left logarithmic residue. There is also a right version obtained by replacing the left logarithmic derivative $f'(\lambda)f(\lambda)^{-1}$ by the right logarithmic derivative $f(\lambda)^{-1}f'(\lambda)$. As far as the issues considered in the present paper are concerned, the results that can be obtained for the left and the right version of the logarithmic residue are analogous. Therefore we will only consider the left version.

The purpose of this paper is to prove that a certain Banach algebra that can be associated with two-dimensional singular operators is spectrally regular. Here is the description of the algebra in question.

Given a bounded simply connected domain $U \subset \mathbb{C}$ whose boundary ∂U is a Lyapunov curve, let S_U stand for the singular integral operator acting on the Hilbert space $L^2(U)$,

$$(S_U f)(z) = -\frac{1}{\pi} \int_U \frac{f(w)}{(z-w)^2}\, dA(w). \tag{1.2}$$

Here $dA = dx\, dy$ denotes the area measure. Note that the adjoint of S_U is given by

$$(S_U^* f)(z) = -\frac{1}{\pi} \int_U \frac{f(w)}{(\overline{z}-\overline{w})^2}\, dA(w). \tag{1.3}$$

Finally, multiplication operators cI with $c \in C(\overline{U})$ are defined by

$$cI : f(z) \in L^2(U) \mapsto c(z)f(z) \in L^2(U). \tag{1.4}$$

Note that these are bounded linear operators acting on $L^2(U)$.

In the following, $\mathcal{A}$ stands for the smallest closed subalgebra of $\mathcal{L}(L^2(U))$ which contains S_U, its adjoint S_U^*, all multiplication operators cI with $c \in C(\overline{U})$, and all compact operators on $L^2(U)$. It is the aim of this note to prove that the C^*-algebra $\mathcal{A}$ is spectrally regular.

Acknowledgement. The authors thank the anonymous referee for a helpful comment leading to an improvement of the paper.

2. Preliminaries

The verification of the spectral regularity of $\mathcal{A}$ is a nice application of a general principle developed in [5]. This principle allows us to prove the spectral regularity of a Banach algebra by establishing the existence of a family of associated Banach algebras for which spectral regularity has already been verified. These associated Banach algebras reflect in some way the structure of the given Banach algebra and are usually easier to analyze.

The general principle meant above is formulated in [5, Theorem 3.1]. However, for our purposes a simpler version (Corollary 3.3 of [5]) is sufficient.

Proposition 2.1. *Let $\mathcal{B}$ be a unital Banach algebra. For ω in a (non-empty) index set Ω, let $\mathcal{B}_\omega$ be a spectrally regular Banach algebra, and let $\phi_\omega : \mathcal{B} \to \mathcal{B}_\omega$ be a Banach algebra homomorphism. Suppose*

$$\bigcap_{\omega \in \Omega} \operatorname{Ker} \phi_\omega \subseteq \mathcal{R}(\mathcal{B}), \tag{2.1}$$

where $\mathcal{R}(\mathcal{B})$ stands for the radical of $\mathcal{B}$. Then $\mathcal{B}$ is spectrally regular.

A family $\{\phi_\omega : \mathcal{B} \to \mathcal{B}_\omega\}_{\omega\in\Omega}$ of Banach algebra homomorphisms for which (2.1) holds will be called *radical-separating*. If, in fact, the intersections of the kernels equals $\{0\}$, then the family is called *separating*. Notice that in the case of a C^*-algebra $\mathcal{B}$ both notions coincide since the radical is trivial.

Another notion is of importance. A family $\{\phi_\omega : \mathcal{B} \to \mathcal{B}_\omega\}_{\omega\in\Omega}$ of unital Banach algebra homomorphisms is called *sufficient* if for every $b \in \mathcal{B}$, the element b is invertible in $\mathcal{B}$ if (and only if) for every $\omega \in \Omega$ the element $\phi_\omega(b)$ is invertible in $\mathcal{B}_\omega$.

It is easy to verify that a sufficient family is automatically radical-separating (see [5, Proposition 3.4]), although the converse does not hold. For this as well as related notions characterizing families of homomorphisms see [6]. In any case, we can draw the following conclusion (see [5, Corollary 3.5]).

Corollary 2.2. *Let $\mathcal{B}$ be a unital Banach algebra, and let $\{\phi_\omega : \mathcal{B} \to \mathcal{B}_\omega\}_{\omega\in\Omega}$ be a sufficient family of homomorphisms into spectrally regular Banach algebras B_ω. Then $\mathcal{B}$ is spectrally regular, too.*

We will need another consequence of [5, Theorem 3.1], which is stated as [5, Corollary 4.13]. The expression $\mathcal{K}(X)$ is used to denote the set of all compact operators on a Banach space X.

Proposition 2.3. *Let X be an infinite-dimensional Banach space, and let $\mathcal{B}$ be a unital Banach subalgebra of $\mathcal{L}(X)$. If the quotient algebra $\mathcal{B}/(\mathcal{K}(X)\cap\mathcal{B})$ is spectrally regular, then so is $\mathcal{B}$.*

3. C^*-algebras generated by isometries and partial isometries

Before we turn our attention to the C^*-algebra $\mathcal{A}$ we need to show the spectral regularity of certain auxiliary C^*-algebras. The results in question were already presented in [7, Sect. 8]. For the convenience of the reader we recall them here, in fact with simplified arguments.

Let $\mathcal{B}$ be a C^*-algebra with unit element e. We say that $\mathcal{B}$ is generated by a *non-unitary isometry* v if $v^*v = e \neq vv^*$ and $\mathcal{B}$ coincides with the smallest C^*-subalgebra of $\mathcal{B}$ containing e, v, and v^*. Algebras of this type are universal, i.e., any two C^*-algebras generated by non-unitary isometries are isomorphic to each other as C^*-algebras (cf. [10, 12], [20, Sect. 4.23 to Sect. 4.25], or [14, Sect. XXXII.1]). A concrete example is the Toeplitz algebra $\mathcal{T}(C) \subseteq \mathcal{L}(\ell^2)$ which is generated by the forward shift

$$V : \{x_n\}_{n=0}^{\infty} \in \ell^2 \mapsto \{0, x_0, x_1, x_2, \dots\} \in \ell^2.$$

The following theorem is taken from [7, Sect. 8]. The reasoning given there relies on the quite complicated Fredholm theory for C^*-algebras. Here we present the result with a more direct proof.

Theorem 3.1. *Any C^*-algebra $\mathcal{B}$ generated by a non-unitary isometry is spectrally regular.*

Proof. Because of universality we can assume without loss of generality that $\mathcal{B}$ is the Toeplitz algebra $\mathcal{T}(C)$. It is known that the Toeplitz algebra contains the ideal $\mathcal{K}(\ell^2)$ of all compact operators, and that the quotient algebra $\mathcal{B}^\pi = \mathcal{T}(C)/\mathcal{K}(\ell^2)$ is commutative. In fact, it is isomorphic to the C^*-algebra $C(\mathbb{T})$ of all continuous complex functions on the unit circle $\mathbb{T}$. Since commutative Banach algebras are spectrally regular, we conclude by using Corollary 2.3 that $\mathcal{B}$ is spectrally regular as well. The spectral regularity of commutative Banach algebras can be derived from Proposition 2.1 or Corollary 2.2 by using Gelfand theory (cf. [1]). □

Next let us consider a special instance of a C^*-algebra $\mathcal{B}$ which is generated by a partial isometry. We do not know if every C^*-algebra generated by a partial isometry is spectrally regular.

Let $\mathcal{B}_1$ and $\mathcal{B}_2$ be two C^*-algebras with unit elements e_1 and e_2 generated by non-unitary isometries $v_1 \in \mathcal{B}_1$ and $v_2 \in \mathcal{B}_2$, respectively. Consider the C^*-algebra $\mathcal{B}_1 \times \mathcal{B}_2$ with component-wise algebraic operations and maximum norm. The unit element is (e_1, e_2). Now let $\mathcal{B}$ be the smallest C^*-subalgebra of $\mathcal{B}_1 \times \mathcal{B}_2$ containing the unit element, the element $w = (v_1, v_2^*)$ and its adjoint $w^* = (v_1^*, v_2)$.

Note that w (as well as w^*) is a partial isometry because $ww^*w = w$ and $w^*ww^* = w^*$. Thus $\mathcal{B}$ is a C^*-algebra generated by a partial isometry. The choice of the partial isometries is quite special, however. The C^*-algebra $\mathcal{B}$ can be described as a proper subalgebra of $\mathcal{B}_1 \times \mathcal{B}_2$ in a relatively explicit way (see [7, Theorem 8.4]). From the description given there it is clear that it is universal, i.e., up to isomorphy, it does not depend on the particular choice of v_1 and v_2. Spectral regularity was already observed in [7]. Here, again, we give a simpler argument.

Proposition 3.2. *The C^*-algebra $\mathcal{B}$ constructed above is spectrally regular.*

Proof. Consider the $*$-homomorphisms

$$\phi_i : (x_1, x_2) \in \mathcal{B} \mapsto x_i \in \mathcal{B}_i, \qquad i = 1, 2,$$

and apply Proposition 2.1. Notice that the family $\{\phi_1, \phi_2\}$ is separating. □

In [7, Sect. 8], two concrete realizations of such a C^*-algebra were given. One was related to sequences of finite sections of Toeplitz operators. The other is of interest to us. Details and background for what follows in the remainder of this section can be found in [22, Theorems 2.1 and 3.5], or [17, Sect. 2] (see also [7, Example 8.5]).

Let Π stand for the upper half plane in $\mathbb{C}$. We consider the two-dimensional singular integral operators S_Π and its adjoint S_Π^*, which are defined as in (1.2) and (1.3). These are bounded linear operators acting on $L^2(\Pi)$. It is known that $L^2(\Pi)$ is the orthogonal sum of two subspaces H and $\widetilde{H}$ such that both S_Π and S_Π^* have these two spaces as invariant subspaces, S_Π restricted on H is a non-unitary isometry on H, and S_Π^* restricted to $\widetilde{H}$ is a non-unitary isometry on $\widetilde{H}$.

Under the identification of $L^2(\Pi)$ with $H \dotplus \widetilde{H}$, the operators S_Π and S_Π^* take the form

$$S_\Pi = \begin{pmatrix} S_\Pi|_H & 0 \\ 0 & S_\Pi|_{\widetilde{H}} \end{pmatrix}, \qquad S_\Pi^* = \begin{pmatrix} S_\Pi^*|_H & 0 \\ 0 & S_\Pi^*|_{\widetilde{H}} \end{pmatrix}.$$

In other words, $S_\Pi = (v_1, v_2^*)$ and $S_\Pi^* = (v_1^*, v_2)$, where $v_1 = S_\Pi|_H \in \mathcal{L}(H)$ and $v_2 = S_\Pi^*|_{\widetilde{H}} \in \mathcal{L}(\widetilde{H})$ are both non-unitary isometries. Therefore, the smallest closed unital subalgebra of $\mathcal{L}(L^2(\Pi))$ containing S_Π and its adjoint is a C^*-algebra $\mathcal{B}$ of the type constructed above.

Corollary 3.3. *The C^*-algebra $\mathcal{B}_\Pi$ generated by S_Π, S_Π^* and the identity operator is spectrally regular.*

4. The spectral regularity of $\mathcal{A}$

In this section we will reach our goal: proving that the C^*-algebra $\mathcal{A}$ is spectrally regular. For this we need to draw on the Fredholm theory for operators in $\mathcal{A}$, which was developed originally by Vasilevski [21]. It can also be found in the paper [15], to which we refer for details. The Fredholm theory allows us to gain some insight into the structure of the Calkin algebra $\mathcal{A}/\mathcal{K}$. Here, and it what follows, $\mathcal{K} = \mathcal{K}(L^2(U))$ denotes the ideal of all compact linear operators acting on $L^2(U)$.

The paper [15] has defined the algebra $\mathcal{A}$ without explicitly including all compact operators. The results mentioned below remain valid since including $\mathcal{K}$ does not change the Fredholm theory (see also Section 5 below).

Let $S_{\mathbb{R}^2}$ and $S_{\mathbb{R}^2}^*$ stand for the singular integral operators on $\mathbb{R}^2 \cong \mathbb{C}$, which are also defined by (1.2) and (1.3) with $U = \mathbb{C}$. It is known (see, e.g., [18, Chapter X, p. 249] or [16, Sect. 2.3]) that

$$S_{\mathbb{R}^2} = \mathcal{F}^{-1}\,\psi(\xi)\,\mathcal{F},$$

where $\mathcal{F}$ is the two-dimensional Fourier transform on $\mathbb{R}^2 \cong \mathbb{C}$ and $\psi(\xi) = \bar{\xi}/\xi$. In particular, $S_{\mathbb{R}^2}$ is a unitary operator with spectrum equal to the unit circle. Clearly, the singular integral operator S_U on a domain U contained in $\mathbb{C}$ can be thought of as the compression of $S_{\mathbb{R}^2}$ onto U, i.e., $S_U = \chi_U S_{\mathbb{R}^2} \chi_U$ and $S_U^* = \chi_U S_{\mathbb{R}^2}^* \chi_U$.

The following auxiliary result is stated in [15] as Lemma 5.2. It is a consequence of that fact that S_U and S_U^* commute with the multiplication operator cI, $c \in C(\overline{U})$, modulo compact operators. This, on the other hand, follows from the fact that $S_{\mathbb{R}^2}$ and $S_{\mathbb{R}^2}^*$ commute with cI, $c \in C(\dot{\mathbb{R}}^2)$, modulo compact operators. The latter has been proved in [20, Chapter X, Theorem 7.1].

Lemma 4.1. *The commutator $cIA - AcI$ belongs to $\mathcal{K}$ for every function $c \in C(\overline{U})$ and every $A \in \mathcal{A}$.*

We will denote the Calkin algebra $\mathcal{A}/\mathcal{K}$ by $\mathcal{A}^\pi$. It is a C^*-subalgebra of $\mathcal{L}(L^2(U))/\mathcal{K}$. Writing A^π for the coset $A+\mathcal{K}$ in $\mathcal{A}/\mathcal{K}$, we have that

$$\pi : A \in \mathcal{A} \mapsto A^\pi \in \mathcal{A}^\pi$$

is the canonical homomorphism from $\mathcal{A}$ onto $\mathcal{A}^\pi$.

Lemma 4.2. *The set* $\mathcal{Z}^\pi = \{(cI)^\pi : c \in C(\overline{U})\}$ *is a central* C^**-subalgebra of* $\mathcal{A}^\pi$*, which is isomorphic to* $C(\overline{U})$*. Consequently, the maximal ideal space of* $\mathcal{Z}^\pi$ *is homeomorphic to* $\overline{U}$*, and for each* $t \in \overline{U}$*, the corresponding maximal ideal of* $\mathcal{Z}^\pi$ *is given by* $m_t = \{(cI)^\pi : c \in C(\overline{U}),\ c(t)=0\}$.

The isomorphy of $\mathcal{Z}^\pi$ with $C(\overline{U})$ can be obtained from [15]. Although is not explicitly shown there, it can be easily derived using an argument involving the strong limits considered in the proof of [15, Theorem 5.5].

For each $t \in \overline{U}$, let J_t^π stand for the smallest closed ideal of $\mathcal{A}^\pi$ which contains m_t. Note that J_t^π is $*$-ideal of $\mathcal{A}^\pi$. Furthermore, let $\pi_t : \mathcal{A}^\pi \to \mathcal{A}_t^\pi$ be the canonical homomorphism onto the quotient algebra $\mathcal{A}_t^\pi = \mathcal{A}^\pi / J_t^\pi$.

As a consequence of the local principle of Allan–Douglas, the following result is obtained in [15, Theorem 5.4].

Theorem 4.3. *Let* $A^\pi \in \mathcal{A}^\pi$*. Then* A^π *is invertible in* $\mathcal{A}^\pi$ *if and only if for every* $t \in \overline{U}$ *the coset* $A_t^\pi = A^\pi + J_t^\pi$ *is invertible in* $\mathcal{A}_t^\pi$.

The next step, the identification of the local algebras $\mathcal{A}_t^\pi$, has also been accomplished in [15, Theorem 5.5].

For a subset T of a Banach algebra $\mathcal{B}$, let $\operatorname{alg}_{\mathcal{B}} T$ stand for the smallest closed subalgebra of $\mathcal{B}$ containing T.

Theorem 4.4. *Let* U *be a bounded simply connected domain in* $\mathbb{C}$ *with Lyapunov boundary* ∂U*. Then the following statements hold:*

(i) *for* $t \in U$*, the local algebra* $\mathcal{A}_t^\pi$ *is* $*$*-isomorphic to*

$$\operatorname{alg}_{\mathcal{L}(L^2(\mathbb{R}^2))}\{I, S_{\mathbb{R}^2}, S_{\mathbb{R}^2}^*\}$$

where the $*$*-isomorphism is given by*

$$(cI)_t^\pi \mapsto c(t)I, \qquad (S_U)_t^\pi \mapsto S_{\mathbb{R}^2}, \qquad (S_U^*)_t^\pi \mapsto S_{\mathbb{R}^2}^*\,;$$

(ii) *for* $t \in \partial U$*, the local algebra* $\mathcal{A}_t^\pi$ *is* $*$*-isomorphic to*

$$\operatorname{alg}_{\mathcal{L}(L^2(\Pi))}\{I, S_\Pi, S_\Pi^*\}$$

where the $*$*-isomorphism is given by*

$$(cI)_t^\pi \mapsto c(t)I, \qquad (S_U)_t^\pi \mapsto S_\Pi, \qquad (S_U^*)_t^\pi \mapsto S_\Pi^*\,.$$

Now we are able to address the issue of spectral regularity.

Theorem 4.5. *For a bounded simply connected domain* U *in* $\mathbb{C}$ *with Lyapunov boundary, the algebras* $\mathcal{A}$ *and* $\mathcal{A}^\pi$ *are spectrally regular.*

Proof. Note that $\mathrm{alg}_{\mathcal{L}(L^2(\Pi))}\{I, S_\Pi, S_\Pi^*\} = \mathcal{B}_\Pi$ is spectrally regular by Corollary 3.3. On the other hand, as stated before, $S_{\mathbb{R}^2}$ is unitary. Therefore, the C^*-algebra $\mathrm{alg}_{\mathcal{L}(L^2(\mathbb{R}^2))}\{I, S_{\mathbb{R}^2}, S_{\mathbb{R}^2}^*\}$ is commutative, hence spectrally regular. As spectral regularity (obviously) is a notion which is invariant under Banach algebra isomorphisms, we conclude from Theorem 4.4 that all local algebras $\mathcal{A}_t^\pi$ $(t \in \overline{U})$ are spectrally regular.

The family of homomorphisms $\{\pi_t : \mathcal{A}^\pi \to \mathcal{A}_t^\pi\}_{t\in\overline{U}}$ is a sufficient family by Theorem 4.3. It follows from Corollary 2.2 that $\mathcal{A}^\pi$ is spectrally regular. Finally, we use Proposition 2.3 in order to conclude that $\mathcal{A}$ is spectrally regular, too. □

Theorem 4.4 is formulated under the assumption that U is a simply connected domain with Lyapunov boundary. This means in particular that U is homeomorphic to the unit disk. The authors conjecture that the assumption of simple connectedness is due to the method of proof in [15] and could probably be removed. We are able to establish the following modest extension of Theorem 4.5.

Theorem 4.6. *Let U be the finite union of bounded simply connected domains $U_1, \dots, U_N$ with Lyapunov boundaries such that $\overline{U}_j \cap \overline{U}_k = \emptyset$ for $j \neq k$. Then the algebras $\mathcal{A}$ and $\mathcal{A}^\pi$ are spectrally regular.*

Here $\mathcal{A}$ is defined in the same way as before but with the simple connectedness requirement on U removed.

Proof. We rely on the decomposition $L^2(U) = L^2(U_1) \dotplus L^2(U_2) \dotplus \cdots \dotplus L^2(U_N)$. Under this decomposition operators on $L^2(U)$ can be identified with block operators. In fact, for $c \in C(\overline{U})$, the multiplication operator cI can be identified with $\mathrm{diag}\,(c_1 I, c_2 I, \dots, c_N I)$ where $c_1, \dots, c_N$ are the restrictions of c to $\overline{U}_k$, respectively. Because the integral operators with kernel $1/(z-w)^2$ and $1/(\overline{z}-\overline{w})^2$, thought of as acting from $L^2(U_j)$ to $L^2(U_k)$, are compact for $j \neq k$, it follows that S_U can be identified with $\mathrm{diag}\,(S_{U_1}, S_{U_2}, \dots, S_{U_N})$ modulo a compact operator. Therefore, the operators in $\mathcal{A}$ are of block form with compact off-diagonal entries, while the entry on the k-th position on the diagonal belongs to $\mathcal{A}[U_k]$ given by

$$\mathcal{A}[U_k] = \mathrm{alg}_{\mathcal{L}(L^2(U_k))}\{S_{U_k}, S_{U_k}^*, c_k I : c_k \in C(\overline{U}_k)\} \cup \mathcal{K}(L^2(U_k)).$$

Passing to the quotients with respect to the compact operators, we see that $\mathcal{A}^\pi$ is isomorphic as a C^*-algebra to the finite direct product of the C^*-algebras $A[U_k]^\pi = \mathcal{A}[U_k]/\mathcal{K}(L^2(U_k))$. The latter algebras are spectrally regular by Theorem 4.5. Invoking Proposition 2.1 we conclude that $\mathcal{A}^\pi$ is spectrally regular. As before, this implies spectral regularity of $\mathcal{A}$ by Proposition 2.3. □

5. Concluding remarks

Let $\mathcal{A}_0$ be the smallest closed subalgebra of $\mathcal{L}(L^2(U))$ containing S_U, S_U^*, and all multiplication operators cI with $c \in C(\overline{U})$. Clearly, then $\mathcal{A}$ is the smallest closed subalgebra of $\mathcal{L}(L^2(U))$ which contains $\mathcal{A}_0$ and the ideal $\mathcal{K}$ of all compact operators. If U is a disk or a half-plane, then results of [16] (namely, Lemma 2.6 and statements made in the proof of Proposition 2.4) imply that $\mathcal{K} \subset \mathcal{A}_0$ (and thus $\mathcal{A}_0 = \mathcal{A}$). For other domains U, the question whether $\mathcal{K} \subset \mathcal{A}_0$ seems to be open.

Putting this issue aside, we observe that $\mathcal{A}_0$ is a C^*-subalgebra of $\mathcal{A}$. Now spectral regularity is a hereditary property which carries over to subalgebras. This was mentioned in [5] as a consequence of Corollary 4.1 there. (It can also be obtained from Proposition 2.1 above with the single homomorphism $\phi_0 : \mathcal{A}_0 \to \mathcal{A}$ being the embedding map.) We conclude that the spectral regularity of $\mathcal{A}$ implies that of $\mathcal{A}_0$. Therefore, for those domains considered in the previous two theorems, the spectral regularity of $\mathcal{A}_0$ is established, too.

References

[1] H. Bart, Spectral properties of locally holomorphic vector-valued functions, *Pacific J. Math.* **52** (1974), 321–329.

[2] H. Bart, T. Ehrhardt, and B. Silbermann, Zero sums of idempotents in Banach algebras, *Integral Equations Operator Theory* **19** (1994), 125–134.

[3] H. Bart, T. Ehrhardt, and B. Silbermann, Logarithmic residues in Banach algebras, *Integral Equations Operator Theory* **19** (1994), 135–152.

[4] H. Bart, T. Ehrhardt, and B. Silbermann, Logarithmic residues of Fredholm operator-valued functions and sums of finite rank operators, *Operator Theory: Advances and Applications* **130**, Birkhäuser, Basel, 2001, 83–106.

[5] H. Bart, T. Ehrhardt, and B. Silbermann, Spectral regularity of Banach algebras and non-commutative Gelfand theory, *Operator Theory: Advances and Applications* **218**, Birkhäuser, Basel, 2012, 123–153.

[6] H. Bart, T. Ehrhardt, and B. Silbermann, Families of homomorphisms in non-commutative Gelfand theory: comparisons and examples, *Operator Theory: Advances and Applications* **221**, Birkhäuser, Basel, 2012, 131–159.

[7] H. Bart, T. Ehrhardt, and B. Silbermann, Logarithmic residues, Rouché's theorem, and spectral regularity: The C^*-algebra case, *Indagationes Mathematicae* **23** (2012), 816–847.

[8] H. Bart, T. Ehrhardt, and B. Silbermann, Zero sums of idempotents and Banach algebras failing to be spectrally regular, *Operator Theory: Advances and Applications* **237**, Birkhäuser, Basel, 2013, 41–78.

[9] H. Bart, T. Ehrhardt, and B. Silbermann, Approximately finite-dimensional Banach algebras are spectrally regular, *Linear Algebra Appl.* **470** (2015), 185–199.

[10] L.A. Coburn, The C^*-algebra generated by an isometry, *Bull. Amer. Math. Soc.* **73** (1967), 722–726.

[11] T. Ehrhardt and B. Silbermann, Banach algebras of quasi-triangular operators are spectrally regular, *Linear Algebra Appl.* **439** (2013), 577–583.

[12] I. Gohberg, An application of the theory of normed rings to singular integral equations, *Uspeki Mat. Nauk (N.S.)* **7** (2(48)) (1952), 149–156 (in Russian).

[13] I. Gohberg, S. Goldberg, and M.A. Kaashoek, *Classes of Linear Operators, Vol. I*, Operator Theory: Advances and Applications **49**, Birkhäuser, Basel, 1990.

[14] I. Gohberg, S. Goldberg, and M.A. Kaashoek, *Classes of Linear Operators, Vol. II*, Operator Theory: Advances and Applications **63**, Birkhäuser, Basel, 1993.

[15] Y.I. Karlovich and V.A. Mozel, On nonlocal C^*-algebras of two-dimensional singular integral operators, *Operator Theory: Advances and Applications* **220**, Birkhäuser, Basel, 2012, 115–135.

[16] Y.I. Karlovich and L. Pessoa, Algebras generated by Bergman and anti-Bergman projections and by multiplications with piecewise continuous coefficients, *Integral Equations Operator Theory* **52** (2005), 219–270.

[17] Y.I. Karlovich and L. Pessoa, C^*-algebras of Bergman type operators with piecewise continuous coefficients, *Integral Equations Operator Theory* **57** (2007), 521–565.

[18] S.G. Michlin and S. Prössdorf, *Singular Integral Operators*, Springer-Verlag, Berlin, 1986.

[19] A.E. Taylor and D.C. Lay, *Introduction to Functional Analysis*, 2nd edition, Wiley, New York, 1980.

[20] S. Prössdorf and B. Silbermann, *Numerical Analysis for Integral and Related Operator Equations*, Operator Theory: Advances and Applications **52**, Birkhäuser, Basel, 1991.

[21] N.L. Vasilevski, *Multidimensional Singular Integral Operators with Discontinuous Classical Symbols*, Doctoral thesis, Odessa, 1985.

[22] N.L. Vasilevski, Poly-Bergman spaces and two-dimensional singular integral operators. *Operator Theory: Advances and Applications* **171**, Birkhäuser, Basel, 2007, 349–359.

Harm Bart
Econometric Institute
Erasmus University Rotterdam
3000 DR Rotterdam
The Netherlands
e-mail: `bart@ese.eur.nl`

Torsten Ehrhardt
Mathematics Department
University of California
Santa Cruz, CA-95064
USA
e-mail: `tehrhard@ucsc.edu`

Bernd Silbermann
Fakultät für Mathematik
Technische Universität Chemnitz
09107 Chemnitz
Germany
e-mail: `silbermn.toeplitz@googlemail.com`

A spectral shift function for Schrödinger operators with singular interactions

Jussi Behrndt, Fritz Gesztesy and Shu Nakamura

Abstract. For the pair $(-\Delta, -\Delta - \alpha\delta_{\mathcal{C}})$ of self-adjoint Schrödinger operators in $L^2(\mathbb{R}^n)$ a spectral shift function is determined in an explicit form with the help of (energy parameter dependent) Dirichlet-to-Neumann maps. Here $\delta_{\mathcal{C}}$ denotes a singular δ-potential which is supported on a smooth compact hypersurface $\mathcal{C} \subset \mathbb{R}^n$ and α is a real-valued function on $\mathcal{C}$.

Mathematics Subject Classification (2010). Primary 35J10; Secondary 47A40, 47A55, 47B25, 81Q10.

Keywords. δ-potential, singular interaction, boundary triple, trace formula, Weyl–Titchmarsh function, Dirichlet-to-Neumann map.

1. Introduction

The goal of this paper is to determine a spectral shift function for the pair $(H, H_{\delta,\alpha})$, where $H = -\Delta$ is the usual self-adjoint Laplacian in $L^2(\mathbb{R}^n)$, and $H_{\delta,\alpha} = -\Delta - \alpha\delta_{\mathcal{C}}$ is a singular perturbation of H by a δ-potential of variable real-valued strength $\alpha \in C^1(\mathcal{C})$ supported on some smooth, compact hypersurface $\mathcal{C}$ that splits $\mathbb{R}^n$, $n \geq 2$, into a bounded interior and an unbounded exterior domain. Schrödinger operators with δ-interactions are often used as idealized models of physical systems with short-range potentials; in the simplest case point interactions are considered, but in the last decades also interactions supported on curves and hypersurfaces have attracted a lot of attention, see the monographs [2, 4, 26], the review [22], and, for instance, [3, 5, 9, 12, 13, 18, 23, 24, 25, 27, 35] for a small selection of papers in this area.

It is known from [9] (see also [12]) that for an integer $m > (n/2) - 1$ the m-th power of the resolvents of H and $H_{\delta,\alpha}$ differs by a trace class operator,

$$\left[(H_{\delta,\alpha} - zI_{L^2(\mathbb{R}^n)})^{-m} - (H - zI_{L^2(\mathbb{R}^n)})^{-m}\right] \in \mathfrak{S}_1(L^2(\mathbb{R}^n)). \qquad (1.1)$$

Since H and $H_{\delta,\alpha}$ are bounded from below, [38, Theorem 8.9.1, p. 306–307] applies (upon replacing the pair $(H,H_{\delta,\alpha})$ by $(H+CI_{L^2(\mathbb{R}^n)}, H_{\delta,\alpha}+CI_{L^2(\mathbb{R}^n)})$

© Springer International Publishing AG, part of Springer Nature 2018

A. Böttcher et al. (eds.), *The Diversity and Beauty of Applied Operator Theory*, Operator Theory: Advances and Applications 268, https://doi.org/10.1007/978-3-319-75996-8_4

such that $H + CI_{L^2(\mathbb{R}^n)} \geq I_{L^2(\mathbb{R}^n)}$ and $H_{\delta,\alpha} + CI_{L^2(\mathbb{R}^n)} \geq I_{L^2(\mathbb{R}^n)}$ for some $C > 0$) and there exists a real-valued function $\xi \in L^1_{\text{loc}}(\mathbb{R})$ satisfying

$$\int_{\mathbb{R}} \frac{|\xi(\lambda)|\, d\lambda}{(1+|\lambda|)^{m+1}} < \infty$$

such that the trace formula

$$\operatorname{tr}_{L^2(\mathbb{R}^n)}\big((H_{\delta,\alpha} - zI_{L^2(\mathbb{R}^n)})^{-m} - (H - zI_{L^2(\mathbb{R}^n)})^{-m}\big) = -m \int_{\mathbb{R}} \frac{\xi(\lambda)\, d\lambda}{(\lambda - z)^{m+1}}$$

is valid for all $z \in \rho(H_{\delta,\alpha}) \cap \rho(H)$. The function ξ in the integrand on the right-hand side is called a *spectral shift function* of the pair $(H, H_{\delta,\alpha})$. For more details, the history and developments of the spectral shift function we refer the reader to the survey papers [14, 16, 17], the standard monographs [38, 40], the paper [39], and the original works [33, 34] by I. M. Lifshitz and [31, 32] by M. G. Krein.

Our approach in this note is based on techniques from extension theory of symmetric operators and relies on a recent representation result of the spectral shift function in terms of an abstract Weyl–Titchmarsh m-function from [6], which we recall in Section 3 for the convenience of the reader. In our situation this abstract Weyl–Titchmarsh m-function will turn out to be a combination of energy dependent Dirichlet-to-Neumann maps $\mathcal{D}_{\mathrm{i}}(z)$ and $\mathcal{D}_{\mathrm{e}}(z)$ associated to $-\Delta$ on the interior and exterior domain, respectively. More precisely, we shall interpret H and $H_{\delta,\alpha}$ as self-adjoint extensions of the densely defined closed symmetric operator

$$Sf = -\Delta f, \quad \operatorname{dom}(S) = \big\{f \in H^2(\mathbb{R}^n) \,\big|\, f\!\upharpoonright_{\mathcal{C}} = 0\big\},$$

and make use of the concept of so-called quasi boundary triples and their Weyl functions (see [7, 8]). It will then turn out in Theorem 4.5 that the trace class condition (1.1) is satisfied, and in the special case $\alpha(x) < 0$, $x \in \mathcal{C}$, the function

$$\xi(\lambda) = \sum_{j\in J} \lim_{\varepsilon\downarrow 0} \frac{1}{\pi} \\ \times \Big(\big(\operatorname{Im}\big(\log\big(\big(\mathcal{D}_{\mathrm{i}}(\lambda+i\varepsilon) + \mathcal{D}_{\mathrm{e}}(\lambda+i\varepsilon)\big)^{-1} - \alpha^{-1}\big)\big)\big)\varphi_j, \varphi_j\Big)_{L^2(\mathcal{C})}$$

for a.e. $\lambda \in \mathbb{R}$, is a spectral shift function for the pair $(H, H_{\delta,\alpha})$ such that $\xi(\lambda) = 0$ for $\lambda < 0$; here $(\varphi_j)_{j\in J}$ is an orthonormal basis in $L^2(\mathcal{C})$. For the case that no sign condition on the function α is assumed, a slightly more involved formula for the spectral shift function is provided in Theorem 4.3 and in Corollary 4.4.

Next, we briefly summarize the basic notation used in this paper. Let $\mathcal{G}$, $\mathfrak{H}$, etc., be separable complex Hilbert spaces, $(\cdot,\cdot)_{\mathfrak{H}}$ the scalar product in $\mathfrak{H}$ (linear in the first factor), and $I_{\mathfrak{H}}$ the identity operator in $\mathfrak{H}$. If T is a linear operator mapping (a subspace of) a Hilbert space into another, $\operatorname{dom}(T)$ denotes the domain and $\operatorname{ran}(T)$ is the range of T. The closure of a closable operator S is denoted by $\overline{S}$. The spectrum and resolvent set of a closed linear operator in $\mathfrak{H}$ will be denoted by $\sigma(\cdot)$ and $\rho(\cdot)$, respectively. The Banach space of bounded linear operators in $\mathfrak{H}$ is denoted by $\mathcal{L}(\mathfrak{H})$; in the context of two Hilbert spaces, $\mathfrak{H}_j$, $j = 1, 2$, we use the analogous abbreviation $\mathcal{L}(\mathfrak{H}_1, \mathfrak{H}_2)$. The p-th Schatten–von Neumann ideal consists of compact operators with singular values in ℓ^p, $p > 0$, and is denoted by $\mathfrak{S}_p(\mathfrak{H})$ and $\mathfrak{S}_p(\mathfrak{H}_1, \mathfrak{H}_2)$. For $\Omega \subseteq \mathbb{R}^n$ nonempty, $n \in \mathbb{N}$, we suppress the n-dimensional Lebesgue measure $d^n x$ and use the shorthand notation $L^2(\Omega) := L^2(\Omega; d^n x)$; similarly, if $\partial\Omega$ is sufficiently regular we write $L^2(\partial\Omega) := L^2(\partial\Omega; d^{n-1}\sigma)$, with $d^{n-1}\sigma$ the surface measure on $\partial\Omega$. We also abbreviate $\mathbb{C}_\pm := \{z \in \mathbb{C} \mid \operatorname{Im}(z) \gtrless 0\}$ and $\mathbb{N}_0 = \mathbb{N} \cup \{0\}$.

2. Quasi boundary triples and their Weyl functions

In this preliminary section we briefly recall the concept of quasi boundary triples and their Weyl functions from extension theory of symmetric operators, which will be used in the next sections. We refer to [7, 8] for more details on quasi boundary triples and to [19, 20, 21, 29, 37] for the closely related concepts of generalized and ordinary boundary triples.

Throughout this section let $\mathfrak{H}$ be a separable Hilbert space and let S be a densely defined closed symmetric operator in $\mathfrak{H}$.

Definition 2.1. Let $T \subset S^*$ be a linear operator in $\mathfrak{H}$ such that $\overline{T} = S^*$. A triple $\{\mathcal{G}, \Gamma_0, \Gamma_1\}$ is said to be a *quasi boundary triple* for $T \subset S^*$ if $\mathcal{G}$ is a Hilbert space and $\Gamma_0, \Gamma_1 : \operatorname{dom}(T) \to \mathcal{G}$ are linear mappings such that the following conditions (i)–(iii) are satisfied:

(i) The abstract Green's identity

$$(Tf, g)_{\mathfrak{H}} - (f, Tg)_{\mathfrak{H}} = (\Gamma_1 f, \Gamma_0 g)_{\mathcal{G}} - (\Gamma_0 f, \Gamma_1 g)_{\mathcal{G}}$$

holds for all $f, g \in \operatorname{dom}(T)$.

(ii) The range of the map $(\Gamma_0, \Gamma_1)^\top : \operatorname{dom}(T) \to \mathcal{G} \times \mathcal{G}$ is dense.

(iii) The operator $A_0 := T \upharpoonright \ker(\Gamma_0)$ is self-adjoint in $\mathfrak{H}$.

The next theorem from [7, 8] contains a sufficient condition for a triple $\{\mathcal{G}, \Gamma_0, \Gamma_1\}$ to be a quasi boundary triple. It will be used in the proof of Theorem 4.3.

Theorem 2.2. *Let $\mathfrak{H}$ and $\mathcal{G}$ be separable Hilbert spaces and let T be a linear operator in $\mathfrak{H}$. Assume that $\Gamma_0, \Gamma_1 : \operatorname{dom}(T) \to \mathcal{G}$ are linear mappings such that the following conditions* (i)–(iii) *hold:*

(i) *The abstract Green's identity*

$$(Tf, g)_{\mathfrak{H}} - (f, Tg)_{\mathfrak{H}} = (\Gamma_1 f, \Gamma_0 g)_{\mathcal{G}} - (\Gamma_0 f, \Gamma_1 g)_{\mathcal{G}}$$

holds for all $f, g \in \operatorname{dom}(T)$.

(ii) *The range of $(\Gamma_0, \Gamma_1)^\top : \operatorname{dom}(T) \to \mathcal{G} \times \mathcal{G}$ is dense and $\ker(\Gamma_0) \cap \ker(\Gamma_1)$ is dense in $\mathfrak{H}$.*

(iii) *$T \upharpoonright \ker(\Gamma_0)$ is an extension of a self-adjoint operator A_0.*

Then

$$S := T \upharpoonright \big(\ker(\Gamma_0) \cap \ker(\Gamma_1)\big)$$

is a densely defined closed symmetric operator in $\mathfrak{H}$ *such that* $\overline{T} = S^*$ *holds and* $\{\mathcal{G}, \Gamma_0, \Gamma_1\}$ *is a quasi boundary triple for* S^* *with* $A_0 = T \upharpoonright \ker(\Gamma_0)$.

Next, we recall the definition of the γ-field γ and Weyl function M associated to a quasi boundary triple, which is formally the same as in [20, 21] for the case of ordinary or generalized boundary triples. For this let $\{\mathcal{G}, \Gamma_0, \Gamma_1\}$ be a quasi boundary triple for $T \subset S^*$ with $A_0 = T \upharpoonright \ker(\Gamma_0)$. We note that the direct sum decomposition

$$\operatorname{dom}(T) = \operatorname{dom}(A_0) \dotplus \ker(T - zI_{\mathfrak{H}}) = \ker(\Gamma_0) \dotplus \ker(T - zI_{\mathfrak{H}})$$

of $\operatorname{dom}(T)$ holds for all $z \in \rho(A_0)$, and hence the mapping $\Gamma_0 \upharpoonright \ker(T - zI_{\mathfrak{H}})$ is injective for all $z \in \rho(A_0)$ and its range coincides with $\operatorname{ran}(\Gamma_0)$.

Definition 2.3. Let $T \subset S^*$ be a linear operator in $\mathfrak{H}$ such that $\overline{T} = S^*$ and let $\{\mathcal{G}, \Gamma_0, \Gamma_1\}$ be a quasi boundary triple for $T \subset S^*$ with $A_0 = T \upharpoonright \ker(\Gamma_0)$. The γ*-field* γ and the *Weyl function* M corresponding to $\{\mathcal{G}, \Gamma_0, \Gamma_1\}$ are operator-valued functions on $\rho(A_0)$ which are defined by

$$z \mapsto \gamma(z) := \big(\Gamma_0 \upharpoonright \ker(T - zI_{\mathfrak{H}})\big)^{-1} \text{ and } z \mapsto M(z) := \Gamma_1\big(\Gamma_0 \upharpoonright \ker(T - zI_{\mathfrak{H}})\big)^{-1}.$$

Various useful properties of the γ-field and Weyl function associated to a quasi boundary triple were provided in [7, 8, 11], see also [19, 20, 21, 37] for the special cases of ordinary and generalized boundary triples. In the following we only recall some properties important for our purposes. We first note that the values $\gamma(z)$, $z \in \rho(A_0)$, of the γ-field are operators defined on the dense subspace $\operatorname{ran}(\Gamma_0) \subset \mathcal{G}$ which map onto $\ker(T - zI_{\mathfrak{H}}) \subset \mathfrak{H}$. The operators $\gamma(z)$, $z \in \rho(A_0)$, are bounded and admit continuous extensions $\overline{\gamma(z)} \in \mathcal{L}(\mathcal{G}, \mathfrak{H})$, the function $z \mapsto \overline{\gamma(z)}$ is analytic on $\rho(A_0)$, and one has

$$\frac{d^k}{dz^k}\overline{\gamma(z)} = k!\,(A_0 - zI_{\mathfrak{H}})^{-k}\overline{\gamma(z)}, \qquad k \in \mathbb{N}_0,\ z \in \rho(A_0).$$

For the adjoint operators $\gamma(z)^* \in \mathcal{L}(\mathfrak{H}, \mathcal{G})$, $z \in \rho(A_0)$, it follows from the abstract Green's identity in Definition 2.1 (i) that

$$\gamma(z)^* = \Gamma_1(A_0 - \overline{z}I_{\mathfrak{H}})^{-1}, \quad z \in \rho(A_0), \tag{2.1}$$

and one has

$$\frac{d^k}{dz^k}\gamma(\overline{z})^* = k!\,\gamma(\overline{z})^*(A_0 - zI_{\mathfrak{H}})^{-k}, \qquad k \in \mathbb{N}_0,\ z \in \rho(A_0). \tag{2.2}$$

The values $M(z)$, $z \in \rho(A_0)$, of the Weyl function M associated to a quasi boundary triple are operators in $\mathcal{G}$ with $\operatorname{dom}(M(z)) = \operatorname{ran}(\Gamma_0)$ and $\operatorname{ran}(M(z)) \subseteq \operatorname{ran}(\Gamma_1)$ for all $z \in \rho(A_0)$. In general, $M(z)$ may be an unbounded operator, which is not necessarily closed, but closable. One can show that $z \mapsto M(z)\varphi$ is holomorphic on $\rho(A_0)$ for all $\varphi \in \operatorname{ran}(\Gamma_0)$ and in the case where the values $M(z)$ are densely defined bounded operators for some and hence for all $z \in \rho(A_0)$, one has

$$\frac{d^k}{dz^k}\overline{M(z)} = k!\,\gamma(\overline{z})^*(A_0 - zI_{\mathfrak{H}})^{-(k-1)}\overline{\gamma(z)}, \quad k \in \mathbb{N},\ z \in \rho(A_0). \tag{2.3}$$

3. A representation formula for the spectral shift function

Let A and B be self-adjoint operators in a separable Hilbert space $\mathfrak{H}$ and assume that the closed symmetric operator $S = A \cap B$, that is,

$$Sf = Af = Bf, \quad \operatorname{dom}(S) = \{f \in \operatorname{dom}(A) \cap \operatorname{dom}(B) \,|\, Af = Bf\}, \tag{3.1}$$

is densely defined. According to [6, Proposition 2.4] there exists a quasi boundary triple $\{\mathcal{G}, \Gamma_0, \Gamma_1\}$ with γ-field γ and Weyl function M such that

$$A = T \upharpoonright \ker(\Gamma_0) \text{ and } B = T \upharpoonright \ker(\Gamma_1), \tag{3.2}$$

and

$$(B - zI_{\mathfrak{H}})^{-1} - (A - zI_{\mathfrak{H}})^{-1} = -\gamma(z)M(z)^{-1}\gamma(\overline{z})^*, \quad z \in \rho(A) \cap \rho(B). \tag{3.3}$$

Next we recall the main result in the abstract part of [6], in which an explicit expression for a spectral shift function of the pair (A, B) in terms of the Weyl function M is found. We refer the reader to [6, Section 4] for a detailed discussion and the proof of Theorem 3.1. We shall use the logarithm of a boundedly invertible dissipative operator in the formula for the spectral shift function below. Here we define for $K \in \mathcal{L}(\mathcal{G})$ with $\operatorname{Im}(K) \geq 0$ and $0 \in \rho(K)$ the logarithm as

$$\log(K) := -i \int_0^\infty \left[(K + i\lambda I_{\mathcal{G}})^{-1} - (1 + i\lambda)^{-1} I_{\mathcal{G}}\right] d\lambda;$$

cf. [28, Section 2] for more details. We only mention that $\log(K) \in \mathcal{L}(\mathcal{G})$ by [28, Lemma 2.6].

Theorem 3.1. *Let A and B be self-adjoint operators in a separable Hilbert space $\mathfrak{H}$ and assume that for some $\zeta_0 \in \rho(A) \cap \rho(B) \cap \mathbb{R}$ the sign condition*

$$(A - \zeta_0 I_{\mathfrak{H}})^{-1} \geq (B - \zeta_0 I_{\mathfrak{H}})^{-1} \tag{3.4}$$

holds. Let the closed symmetric operator $S = A \cap B$ in (3.1) be densely defined and let $\{\mathcal{G}, \Gamma_0, \Gamma_1\}$ be a quasi boundary triple with γ-field γ and Weyl function M such that (3.2), and hence also (3.3), hold. Assume that $M(z_1)$, $M(z_2)^{-1}$ are bounded (not necessarily everywhere defined) operators in $\mathcal{G}$ for some $z_1, z_2 \in \rho(A) \cap \rho(B)$ and that for some $k \in \mathbb{N}_0$, all $p, q \in \mathbb{N}_0$, and all $z \in \rho(A) \cap \rho(B)$,

$$\left(\frac{d^p}{dz^p}\overline{\gamma(z)}\right)\frac{d^q}{dz^q}\left(M(z)^{-1}\gamma(\overline{z})^*\right) \in \mathfrak{S}_1(\mathfrak{H}), \quad p + q = 2k,$$

$$\left(\frac{d^q}{dz^q}\left(M(z)^{-1}\gamma(\overline{z})^*\right)\right)\frac{d^p}{dz^p}\overline{\gamma(z)} \in \mathfrak{S}_1(\mathcal{G}), \quad p + q = 2k,$$

and

$$\frac{d^j}{dz^j}\overline{M(z)} \in \mathfrak{S}_{(2k+1)/j}(\mathcal{G}), \quad j = 1, \ldots, 2k+1.$$

Then the following assertions (i) *and* (ii) *hold:*

(i) *The difference of the* $(2k+1)$*-th power of the resolvents of* A *and* B *is a trace class operator, that is,*

$$\big[(B - zI_{\mathfrak{H}})^{-(2k+1)} - (A - zI_{\mathfrak{H}})^{-(2k+1)}\big] \in \mathfrak{S}_1(\mathfrak{H})$$

holds for all $z \in \rho(A) \cap \rho(B)$.

(ii) *For any orthonormal basis* $\{\varphi_j\}_{j\in J}$ *in* $\mathcal{G}$ *the function*

$$\xi(\lambda) = \sum_{j\in J} \lim_{\varepsilon\downarrow 0} \frac{1}{\pi}\big(\operatorname{Im}\big(\log\big(\overline{M(\lambda + i\varepsilon)}\big)\big)\varphi_j, \varphi_j\big)_{\mathcal{G}} \quad \textit{for a.e. } \lambda \in \mathbb{R}$$

is a spectral shift function for the pair (A, B) *such that* $\xi(\lambda) = 0$ *in an open neighborhood of* ζ_0*; the function* ξ *does not depend on the choice of the orthonormal basis* $(\varphi_j)_{j\in J}$*. In particular, the trace formula*

$$\begin{aligned}&\operatorname{tr}_{\mathfrak{H}}\big((B - zI_{\mathfrak{H}})^{-(2k+1)} - (A - zI_{\mathfrak{H}})^{-(2k+1)}\big)\\ &\quad = -(2k+1)\int_{\mathbb{R}} \frac{\xi(\lambda)\,d\lambda}{(\lambda - z)^{2k+2}}, \quad z \in \rho(A) \cap \rho(B),\end{aligned}$$

holds.

In the special case $k = 0$ Theorem 3.1 can be reformulated and slightly improved; cf. [6, Corollary 4.2]. Here the essential feature is that the limit $\operatorname{Im}(\log(\overline{M(\lambda + i0)}))$ exists in $\mathfrak{S}_1(\mathcal{G})$ for a.e. $\lambda \in \mathbb{R}$.

Corollary 3.2. *Let* A *and* B *be self-adjoint operators in a separable Hilbert space* $\mathfrak{H}$ *and assume that for some* $\zeta_0 \in \rho(A) \cap \rho(B) \cap \mathbb{R}$ *the sign condition*

$$(A - \zeta_0 I_{\mathfrak{H}})^{-1} \geq (B - \zeta_0 I_{\mathfrak{H}})^{-1}$$

holds. Assume that the closed symmetric operator $S = A \cap B$ *in* (3.1) *is densely defined and let* $\{\mathcal{G}, \Gamma_0, \Gamma_1\}$ *be a quasi boundary triple with* γ*-field* γ *and Weyl function* M *such that* (3.2)*, and hence also* (3.3)*, hold. Assume that* $M(z_1)$, $M(z_2)^{-1}$ *are bounded* (*not necessarily everywhere defined*) *operators in* $\mathcal{G}$ *for some* $z_1, z_2 \in \rho(A)$ *and that* $\overline{\gamma(z_0)} \in \mathfrak{S}_2(\mathcal{G}, \mathfrak{H})$ *for some* $z_0 \in \rho(A)$*. Then the following assertions* (i)–(iii) *hold:*

(i) *The difference of the resolvents of* A *and* B *is a trace class operator, that is,*

$$\big[(B - zI_{\mathfrak{H}})^{-1} - (A - zI_{\mathfrak{H}})^{-1}\big] \in \mathfrak{S}_1(\mathfrak{H})$$

holds for all $z \in \rho(A) \cap \rho(B)$.

(ii) $\operatorname{Im}(\log(\overline{M(z)})) \in \mathfrak{S}_1(\mathcal{G})$ *for all* $z \in \mathbb{C}\backslash\mathbb{R}$ *and the limit*

$$\operatorname{Im}\big(\log\big(\overline{M(\lambda + i0)}\big)\big) := \lim_{\varepsilon\downarrow 0} \operatorname{Im}\big(\log\big(\overline{M(\lambda + i\varepsilon)}\big)\big)$$

exists for a.e. $\lambda \in \mathbb{R}$ *in* $\mathfrak{S}_1(\mathcal{G})$.

(iii) *The function*

$$\xi(\lambda) = \frac{1}{\pi} \operatorname{tr}_{\mathcal{G}}\big(\operatorname{Im}\big(\log\big(\overline{M(\lambda + i0)}\big)\big)\big) \quad \textit{for a.e. } \lambda \in \mathbb{R}$$

is a spectral shift function for the pair (A, B) *such that* $\xi(\lambda) = 0$ *in an open neighborhood of* ζ_0 *and the trace formula*

$$\operatorname{tr}_{\mathfrak{H}}\big((B - zI_{\mathfrak{H}})^{-1} - (A - zI_{\mathfrak{H}})^{-1}\big) = -\int_{\mathbb{R}} \frac{\xi(\lambda)\, d\lambda}{(\lambda - z)^2}$$

is valid for all $z \in \rho(A) \cap \rho(B)$.

We also recall from [6, Section 4] how the sign condition (3.4) in the assumptions in Theorem 3.1 can be replaced by some weaker comparability condition, which is satisfied in our main application in the next section. Again, let A and B be self-adjoint operators in a separable Hilbert space $\mathfrak{H}$ and assume that there exists a self-adjoint operator C in $\mathfrak{H}$ such that

$$(C - \zeta_A I_{\mathfrak{H}})^{-1} \geq (A - \zeta_A I_{\mathfrak{H}})^{-1} \quad \text{and} \quad (C - \zeta_B I_{\mathfrak{H}})^{-1} \geq (B - \zeta_B I_{\mathfrak{H}})^{-1}$$

for some $\zeta_A \in \rho(A) \cap \rho(C) \cap \mathbb{R}$ and some $\zeta_B \in \rho(B) \cap \rho(C) \cap \mathbb{R}$, respectively. Assume that the closed symmetric operators $S_A = A \cap C$ and $S_B = B \cap C$ are both densely defined and choose quasi boundary triples $\{\mathcal{G}_A, \Gamma_0^A, \Gamma_1^A\}$ and $\{\mathcal{G}_B, \Gamma_0^B, \Gamma_1^B\}$ with γ-fields γ_A, γ_B and Weyl functions M_A, M_B for

$$T_A = S_A^* \upharpoonright \big(\operatorname{dom}(A) + \operatorname{dom}(C)\big) \quad \text{and} \quad T_B = S_B^* \upharpoonright \big(\operatorname{dom}(B) + \operatorname{dom}(C)\big)$$

such that

$$C = T_A \upharpoonright \ker(\Gamma_0^A) = T_B \upharpoonright \ker(\Gamma_0^B)$$

and

$$A = T_A \upharpoonright \ker(\Gamma_1^A) \quad \text{and} \quad B = T_B \upharpoonright \ker(\Gamma_1^B)$$

(cf. [6, Proposition 2.4]). Next, assume that for some $k \in \mathbb{N}_0$, the conditions in Theorem 3.1 are satisfied for the γ-fields γ_A, γ_B and the Weyl functions M_A, M_B. Then the difference of the $(2k+1)$-th power of the resolvents of A and C, and the difference of the $(2k+1)$-th power of the resolvents of B and C are trace class operators, and for orthonormal bases $(\varphi_j)_{j\in J}$ in $\mathcal{G}_A$ and $(\psi_\ell)_{\ell\in L}$ in $\mathcal{G}_B$ ($J, L \subseteq \mathbb{N}$ appropriate index sets),

$$\xi_A(\lambda) = \sum_{j\in J} \lim_{\varepsilon\downarrow 0} \frac{1}{\pi} \big(\operatorname{Im}\big(\log\big(\overline{M_A(\lambda + i\varepsilon)}\big)\big)\varphi_j, \varphi_j\big)_{\mathcal{G}_A} \quad \text{for a.e. } \lambda \in \mathbb{R}$$

and

$$\xi_B(\lambda) = \sum_{\ell\in L} \lim_{\varepsilon\downarrow 0} \frac{1}{\pi} \big(\operatorname{Im}\big(\log\big(\overline{M_B(\lambda + i\varepsilon)}\big)\big)\psi_\ell, \psi_\ell\big)_{\mathcal{G}_B} \quad \text{for a.e. } \lambda \in \mathbb{R}$$

are spectral shift functions for the pairs (C, A) and (C, B), respectively.

It follows for $z \in \rho(A) \cap \rho(B) \cap \rho(C)$ that

$$\begin{aligned}
&\operatorname{tr}_{\mathfrak{H}}\big((B - zI_{\mathfrak{H}})^{-(2k+1)} - (A - zI_{\mathfrak{H}})^{-(2k+1)}\big)\\
&\quad = \operatorname{tr}_{\mathfrak{H}}\big((B - zI_{\mathfrak{H}})^{-(2k+1)} - (C - zI_{\mathfrak{H}})^{-(2k+1)}\big)\\
&\qquad - \operatorname{tr}_{\mathfrak{H}}\big((A - zI_{\mathfrak{H}})^{-(2k+1)} - (C - zI_{\mathfrak{H}})^{-(2k+1)}\big)\\
&\quad = -(2k+1)\int_{\mathbb{R}} \frac{[\xi_B(\lambda) - \xi_A(\lambda)]\, d\lambda}{(\lambda - z)^{2k+2}}
\end{aligned}$$

and

$$\int_{\mathbb{R}} \frac{|\xi_B(\lambda) - \xi_A(\lambda)|\, d\lambda}{(1 + |\lambda|)^{2m+2}} < \infty.$$

Therefore,

$$\xi(\lambda) = \xi_B(\lambda) - \xi_A(\lambda) \text{ for a.e. } \lambda \in \mathbb{R} \tag{3.5}$$

is a spectral shift function for the pair (A, B), and in the special case where $\mathcal{G}_A = \mathcal{G}_B := \mathcal{G}$ and $(\varphi_j)_{j\in J}$ is an orthonormal basis in $\mathcal{G}$, one infers that

$$\xi(\lambda) = \sum_{j\in J} \lim_{\varepsilon\downarrow 0} \frac{1}{\pi}\Big(\big(\operatorname{Im}\big(\log\big(\overline{M_B(\lambda + i\varepsilon)}\big) - \log\big(\overline{M_A(\lambda + i\varepsilon)}\big)\big)\varphi_j, \varphi_j\Big)_{\mathcal{G}} \tag{3.6}$$

for a.e. $\lambda \in \mathbb{R}$. We emphasize that in contrast to the spectral shift function in Theorem 3.1, here the spectral shift function ξ in (3.5) and (3.6) is not necessarily nonnegative.

4. Schrödinger operators with δ-potentials supported on hypersurfaces

The aim of this section is to determine a spectral shift function for the pair $(H, H_{\delta,\alpha})$, where $H = -\Delta$ is the usual self-adjoint Laplacian in $L^2(\mathbb{R}^n)$, and $H_{\delta,\alpha} = -\Delta - \alpha\delta_{\mathcal{C}}$ is a self-adjoint Schrödinger operator with δ-potential of strength α supported on a compact hypersurface $\mathcal{C}$ in $\mathbb{R}^n$ which splits $\mathbb{R}^n$ in a bounded interior domain and an unbounded exterior domain. Throughout this section we shall assume that the following hypothesis holds.

Hypothesis 4.1. *Let $n \in \mathbb{N}$, $n \geq 2$, and $\Omega_{\rm i}$ be a nonempty, open, bounded interior domain in $\mathbb{R}^n$ with a smooth boundary $\partial\Omega_{\rm i}$ and let $\Omega_{\rm e} = \mathbb{R}^n\backslash\overline{\Omega_{\rm i}}$ be the corresponding exterior domain. The common boundary of the interior domain $\Omega_{\rm i}$ and exterior domain $\Omega_{\rm e}$ will be denoted by $\mathcal{C} = \partial\Omega_{\rm e} = \partial\Omega_{\rm i}$. Furthermore, let $\alpha \in C^1(\mathcal{C})$ be a real-valued function on the boundary $\mathcal{C}$.*

We consider the self-adjoint operators in $L^2(\mathbb{R}^n)$ given by

$$Hf = -\Delta f, \quad \operatorname{dom}(H) = H^2(\mathbb{R}^n),$$

and

$$\begin{aligned}
&H_{\delta,\alpha} f = -\Delta f,\\
&\operatorname{dom}(H_{\delta,\alpha}) = \left\{ f = \begin{pmatrix} f_{\rm i} \\ f_{\rm e} \end{pmatrix} \in H^2(\Omega_{\rm i}) \times H^2(\Omega_{\rm e}) \,\middle|\, \begin{array}{l} \gamma_D^{\rm i} f_{\rm i} = \gamma_D^{\rm e} f_{\rm e}, \\ \alpha\gamma_D^{\rm i} f_{\rm i} = \gamma_N^{\rm i} f_{\rm i} + \gamma_N^{\rm e} f_{\rm e} \end{array} \right\}.
\end{aligned}$$

Here $f_{\rm i}$ and $f_{\rm e}$ denote the restrictions of a function f on $\mathbb{R}^n$ onto $\Omega_{\rm i}$ and $\Omega_{\rm e}$, and $\gamma_D^{\rm i}$, $\gamma_D^{\rm e}$ and $\gamma_N^{\rm i}$, $\gamma_N^{\rm e}$ are the Dirichlet and Neumann trace operators on $H^2(\Omega_{\rm i})$ and $H^2(\Omega_{\rm e})$, respectively. We note that $H_{\delta,\alpha}$ coincides with the self-adjoint operator associated to the quadratic form

$$\mathfrak{h}_{\delta,\alpha}[f,g] = (\nabla f, \nabla g)_{(L^2(\mathbb{R}^n))^n} - \int_{\mathcal{C}} \alpha(x) f(x)\overline{g(x)}\, d\sigma(x), \qquad f,g \in H^1(\mathbb{R}^n),$$

see [9, Proposition 3.7] and [18] for more details. For $c \in \mathbb{R}$ we shall also make use of the self-adjoint operator

$$H_{\delta,c} f = -\Delta f,$$

$$\operatorname{dom}(H_{\delta,c}) = \left\{ f = \begin{pmatrix} f_{\rm i} \\ f_{\rm e} \end{pmatrix} \in H^2(\Omega_{\rm i}) \times H^2(\Omega_{\rm e}) \,\middle|\, \begin{matrix} \gamma_D^{\rm i} f_{\rm i} = \gamma_D^{\rm e} f_{\rm e}, \\ c\gamma_D^{\rm i} f_{\rm i} = \gamma_N^{\rm i} f_{\rm i} + \gamma_N^{\rm e} f_{\rm e} \end{matrix} \right\}.$$

The following lemma will be useful for the $\mathfrak{S}_p$-estimates in the proof of Theorem 4.3 (cf. [10, Lemma 4.7]).

Lemma 4.2. *Let $X \in \mathcal{L}(L^2(\mathbb{R}^n), H^t(\mathcal{C}))$, and assume that* $\operatorname{ran}(X) \subseteq H^s(\mathcal{C})$ *for some $s > t \geq 0$. Then X is compact and*

$$X \in \mathfrak{S}_r\big(L^2(\mathbb{R}^n), H^t(\mathcal{C})\big) \ \textit{for all}\ r > (n-1)/(s-t).$$

Next we define interior and exterior Dirichlet-to-Neumann maps $\mathcal{D}_{\rm i}(z)$ and $\mathcal{D}_{\rm e}(\zeta)$ as operators in $L^2(\mathcal{C})$ for all $z,\zeta \in \mathbb{C}\backslash[0,\infty) = \rho(H)$. One notes that for $\varphi, \psi \in H^1(\mathcal{C})$ and $z,\zeta \in \mathbb{C}\backslash[0,\infty)$, the boundary value problems

$$-\Delta f_{{\rm i},z} = z f_{{\rm i},z}, \quad \gamma_D^{\rm i} f_{{\rm i},z} = \varphi, \tag{4.1}$$

and

$$-\Delta f_{{\rm e},\zeta} = \zeta f_{{\rm e},\zeta}, \quad \gamma_D^{\rm e} f_{{\rm e},\zeta} = \psi, \tag{4.2}$$

admit unique solutions $f_{{\rm i},z} \in H^{3/2}(\Omega_{\rm i})$ and $f_{{\rm e},\zeta} \in H^{3/2}(\Omega_{\rm e})$, respectively. The corresponding solution operators are denoted by

$$P_{\rm i}(z) : L^2(\mathcal{C}) \to L^2(\Omega_{\rm i}), \quad \varphi \mapsto f_{{\rm i},z},$$

and

$$P_{\rm e}(\zeta) : L^2(\mathcal{C}) \to L^2(\Omega_{\rm e}), \quad \psi \mapsto f_{{\rm e},\zeta}.$$

The *interior Dirichlet-to-Neumann map* in $L^2(\mathcal{C})$,

$$\mathcal{D}_{\rm i}(z) : H^1(\mathcal{C}) \to L^2(\mathcal{C}), \quad \varphi \mapsto \gamma_N^{\rm i} P_{\rm i}(z)\varphi, \tag{4.3}$$

maps Dirichlet boundary values $\gamma_D^{\rm i} f_{{\rm i},z}$ of the solutions $f_{{\rm i},z} \in H^{3/2}(\Omega_{\rm i})$ of (4.1) to the corresponding Neumann boundary values $\gamma_N^{\rm i} f_{{\rm i},z}$, and the *exterior Dirichlet-to-Neumann map* in $L^2(\mathcal{C})$,

$$\mathcal{D}_{\rm e}(\zeta) : H^1(\mathcal{C}) \to L^2(\mathcal{C}), \quad \psi \mapsto \gamma_N^{\rm e} P_{\rm e}(\zeta)\psi, \tag{4.4}$$

maps Dirichlet boundary values $\gamma_D^{\rm e} f_{{\rm e},\zeta}$ of the solutions $f_{{\rm e},\zeta} \in H^{3/2}(\Omega_{\rm e})$ of (4.2) to the corresponding Neumann boundary values $\gamma_N^{\rm e} f_{{\rm e},\zeta}$. The interior and exterior Dirichlet-to-Neumann maps are both closed unbounded operators in $L^2(\mathcal{C})$.

In the next theorem a spectral shift function for the pair $(H, H_{\delta,\alpha})$ is expressed in terms of the limits of the sum of the interior and exterior Dirichlet-to-Neumann map $\mathcal{D}_{\rm i}(z)$ and $\mathcal{D}_{\rm e}(z)$ and the function α. It will turn out that the operators $\mathcal{D}_{\rm i}(z)+\mathcal{D}_{\rm e}(z)$ are boundedly invertible for all $z\in\mathbb{C}\backslash[0,\infty)$ and for our purposes it is convenient to work with the function

$$z\mapsto \mathcal{E}(z)=\big(\mathcal{D}_{\rm i}(z)+\mathcal{D}_{\rm e}(z)\big)^{-1},\quad z\in\mathbb{C}\backslash[0,\infty). \tag{4.5}$$

It was shown in [9, Proposition 3.2 (iii) and Remark 3.3] that $\mathcal{E}(z)$ is a compact operator in $L^2(\mathcal{C})$ which extends the acoustic single layer potential for the Helmholtz equation, that is,

$$(\mathcal{E}(z)\varphi)(x)=\int_{\mathcal{C}}G(z,x,y)\varphi(y)d\sigma(y),\quad x\in\mathcal{C},\ \varphi\in C^\infty(\mathcal{C}),$$

where $G(z,\cdot\,,\cdot\,)$, $z\in\mathbb{C}\backslash[0,\infty)$, represents the integral kernel of the resolvent of H (cf. [36, Chapter 6] and [9, Remark 3.3]). Explicitly,

$$\begin{aligned} G(z,x,y)&=(i/4)\big(2\pi z^{-1/2}|x-y|\big)^{(2-n)/2}H^{(1)}_{(n-2)/2}\big(z^{1/2}|x-y|\big),\\ &\qquad z\in\mathbb{C}\backslash[0,\infty),\ \operatorname{Im}\big(z^{1/2}\big)>0,\ x,y\in\mathbb{R}^n,\ x\neq y,\ n\geqslant 2. \end{aligned}$$

Here $H^{(1)}_\nu(\cdot\,)$ denotes the Hankel function of the first kind with index $\nu\geq 0$ (cf. [1, Sect. 9.1]).

We mention that for the difference of the $(2k+1)$-th power of the resolvents in the next theorem the trace class property is known from [9] (see also [12]).

Theorem 4.3. *Assume Hypothesis* 4.1, *let* $\mathcal{E}(z)$ *be defined as in* (4.5), *let* $\alpha\in C^1(\mathcal{C})$ *be a real-valued function and fix* $c>0$ *such that* $\alpha(x)<c$ *for all* $x\in\mathcal{C}$. *Then the following assertions* (i) *and* (ii) *hold for* $k\in\mathbb{N}_0$ *such that* $k\geq(n-3)/4$:

(i) *The difference of the* $(2k+1)$*-th power of the resolvents of* H *and* $H_{\delta,\alpha}$ *is a trace class operator, that is,*

$$\big[(H_{\delta,\alpha}-zI_{L^2(\mathbb{R}^n)})^{-(2k+1)}-(H-zI_{L^2(\mathbb{R}^n)})^{-(2k+1)}\big]\in\mathfrak{S}_1\big(L^2(\mathbb{R}^n)\big)$$

holds for all $z\in\rho(H_{\delta,\alpha})=\rho(H)\cap\rho(H_{\delta,\alpha})$.

(ii) *For any orthonormal basis* $(\varphi_j)_{j\in J}$ *in* $L^2(\mathcal{C})$ *the function*

$$\xi(\lambda)=\sum_{j\in J}\lim_{\varepsilon\downarrow 0}\frac{1}{\pi}\Big(\big(\operatorname{Im}\big(\log(\mathcal{M}_\alpha(\lambda+i\varepsilon))-\log(\mathcal{M}_0(\lambda+i\varepsilon))\big)\big)\varphi_j,\varphi_j\Big)_{L^2(\mathcal{C})}$$

for a.e. $\lambda\in\mathbb{R}$ *with*

$$\mathcal{M}_0(z)=-c^{-1}\big(c\mathcal{E}(z)-I_{L^2(\mathcal{C})}\big)^{-1}, \tag{4.6}$$

$$\mathcal{M}_\alpha(z)=(c-\alpha)^{-1}\big(\alpha\mathcal{E}(z)-I_{L^2(\mathcal{C})}\big)\big(c\mathcal{E}(z)-I_{L^2(\mathcal{C})}\big)^{-1}, \tag{4.7}$$

for $z\in\mathbb{C}\backslash\mathbb{R}$, *is a spectral shift function for the pair* $(H,H_{\delta,\alpha})$ *such that* $\xi(\lambda)=0$ *for* $\lambda<\inf(\sigma(H_{\delta,c}))$ *and the trace formula*

$$\begin{aligned}&\operatorname{tr}_{L^2(\mathbb{R}^n)}\big((H_{\delta,\alpha}-zI_{L^2(\mathbb{R}^n)})^{-(2k+1)}-(H-zI_{L^2(\mathbb{R}^n)})^{-(2k+1)}\big)\\ &\quad=-(2k+1)\int_{\mathbb{R}}\frac{\xi(\lambda)\,d\lambda}{(\lambda-z)^{2k+2}}\end{aligned}$$

is valid for all $z\in\rho(H_{\delta,\alpha})=\rho(H)\cap\rho(H_{\delta,\alpha})$.

Proof. The structure and underlying idea of the proof of Theorem 4.3 is as follows. In the first two steps a suitable quasi boundary triple and its Weyl function are constructed. In the third step it is shown that the assumptions in Theorem 3.1 are satisfied.

Step 1. Since $c - \alpha(x) \neq 0$ for all $x \in \mathcal{C}$ by assumption, the closed symmetric operator $S = H_{\delta,c} \cap H_{\delta,\alpha}$ is given by

$$Sf = -\Delta f, \quad \mathrm{dom}(S) = \bigl\{ f \in H^2(\mathbb{R}^n) \,\big|\, \gamma_D^{\mathrm{i}} f_{\mathrm{i}} = \gamma_D^{\mathrm{e}} f_{\mathrm{e}} = 0 \bigr\}.$$

In this step we show that the operator

$$T = -\Delta, \quad \mathrm{dom}(T) = \left\{ f = \begin{pmatrix} f_{\mathrm{i}} \\ f_{\mathrm{e}} \end{pmatrix} \in H^2(\Omega_{\mathrm{i}}) \times H^2(\Omega_{\mathrm{e}}) \,\middle|\, \gamma_D^{\mathrm{i}} f_{\mathrm{i}} = \gamma_D^{\mathrm{e}} f_{\mathrm{e}} \right\},$$

satisfies $\overline{T} = S^*$ and that $\{L^2(\mathcal{C}), \Gamma_0, \Gamma_1\}$, where

$$\Gamma_0 f = c\gamma_D^{\mathrm{i}} f_{\mathrm{i}} - (\gamma_N^{\mathrm{i}} f_{\mathrm{i}} + \gamma_N^{\mathrm{e}} f_{\mathrm{e}}), \quad \mathrm{dom}(\Gamma_0) = \mathrm{dom}(T), \tag{4.8}$$

and

$$\Gamma_1 f = \frac{1}{c-\alpha}\bigl(\alpha\gamma_D^{\mathrm{i}} f_{\mathrm{i}} - (\gamma_N^{\mathrm{i}} f_{\mathrm{i}} + \gamma_N^{\mathrm{e}} f_{\mathrm{e}})\bigr), \quad \mathrm{dom}(\Gamma_1) = \mathrm{dom}(T), \tag{4.9}$$

is a quasi boundary triple for $T \subset S^*$ such that

$$H_{\delta,c} = T \upharpoonright \ker(\Gamma_0) \ \text{ and } \ H_{\delta,\alpha} = T \upharpoonright \ker(\Gamma_1). \tag{4.10}$$

For the proof of this fact we make use of Theorem 2.2 and verify next that assumptions (i)–(iii) in Theorem 2.2 are satisfied with the above choice of S, T and boundary maps Γ_0 and Γ_1. For $f, g \in \mathrm{dom}(T)$ one computes

$$\begin{aligned}
&(\Gamma_1 f, \Gamma_0 g)_{L^2(\mathcal{C})} - (\Gamma_0 f, \Gamma_1 g)_{L^2(\mathcal{C})} \\
&\quad = \Bigl(\tfrac{1}{c-\alpha}\bigl(\alpha\gamma_D^{\mathrm{i}} f_{\mathrm{i}} - (\gamma_N^{\mathrm{i}} f_{\mathrm{i}} + \gamma_N^{\mathrm{e}} f_{\mathrm{e}})\bigr), c\gamma_D^{\mathrm{i}} g_{\mathrm{i}} - (\gamma_N^{\mathrm{i}} g_{\mathrm{i}} + \gamma_N^{\mathrm{e}} g_{\mathrm{e}})\Bigr)_{L^2(\mathcal{C})} \\
&\qquad - \Bigl(c\gamma_D^{\mathrm{i}} f_{\mathrm{i}} - (\gamma_N^{\mathrm{i}} f_{\mathrm{i}} + \gamma_N^{\mathrm{e}} f_{\mathrm{e}}), \tfrac{1}{c-\alpha}\bigl(\alpha\gamma_D^{\mathrm{i}} g_{\mathrm{i}} - (\gamma_N^{\mathrm{i}} g_{\mathrm{i}} + \gamma_N^{\mathrm{e}} g_{\mathrm{e}})\bigr)\Bigr)_{L^2(\mathcal{C})} \\
&\quad = -\bigl(\tfrac{\alpha}{c-\alpha}\gamma_D^{\mathrm{i}} f_{\mathrm{i}}, \gamma_N^{\mathrm{i}} g_{\mathrm{i}} + \gamma_N^{\mathrm{e}} g_{\mathrm{e}}\bigr)_{L^2(\mathcal{C})} - \bigl(\gamma_N^{\mathrm{i}} f_{\mathrm{i}} + \gamma_N^{\mathrm{e}} f_{\mathrm{e}}, \tfrac{c}{c-\alpha}\gamma_D^{\mathrm{i}} g_{\mathrm{i}}\bigr)_{L^2(\mathcal{C})} \\
&\qquad + \bigl(\tfrac{c}{c-\alpha}\gamma_D^{\mathrm{i}} f_{\mathrm{i}}, \gamma_N^{\mathrm{i}} g_{\mathrm{i}} + \gamma_N^{\mathrm{e}} g_{\mathrm{e}}\bigr)_{L^2(\mathcal{C})} + \bigl(\gamma_N^{\mathrm{i}} f_{\mathrm{i}} + \gamma_N^{\mathrm{e}} f_{\mathrm{e}}, \tfrac{\alpha}{c-\alpha}\gamma_D^{\mathrm{i}} g_{\mathrm{i}}\bigr)_{L^2(\mathcal{C})} \\
&\quad = \bigl(\gamma_D^{\mathrm{i}} f_{\mathrm{i}}, \gamma_N^{\mathrm{i}} g_{\mathrm{i}} + \gamma_N^{\mathrm{e}} g_{\mathrm{e}}\bigr)_{L^2(\mathcal{C})} - \bigl(\gamma_N^{\mathrm{i}} f_{\mathrm{i}} + \gamma_N^{\mathrm{e}} f_{\mathrm{e}}, \gamma_D^{\mathrm{i}} g_{\mathrm{i}}\bigr)_{L^2(\mathcal{C})},
\end{aligned}$$

and on the other hand, Green's identity and $\gamma_D^{\mathrm{i}} f_{\mathrm{i}} = \gamma_D^{\mathrm{e}} f_{\mathrm{e}}$ and $\gamma_D^{\mathrm{i}} g_{\mathrm{i}} = \gamma_D^{\mathrm{e}} g_{\mathrm{e}}$ yield

$$\begin{aligned}
&(Tf, g)_{L^2(\mathbb{R}^n)} - (f, Tg)_{L^2(\mathbb{R}^n)} \\
&\quad = (-\Delta f_{\mathrm{i}}, g_{\mathrm{i}})_{L^2(\Omega_{\mathrm{i}})} - (f_{\mathrm{i}}, -\Delta g_{\mathrm{i}})_{L^2(\Omega_{\mathrm{i}})} \\
&\qquad + (-\Delta f_{\mathrm{e}}, g_{\mathrm{e}})_{L^2(\Omega_{\mathrm{e}})} - (f_{\mathrm{e}}, -\Delta g_{\mathrm{e}})_{L^2(\Omega_{\mathrm{e}})} \\
&\quad = (\gamma_D^{\mathrm{i}} f_{\mathrm{i}}, \gamma_N^{\mathrm{i}} g_{\mathrm{i}})_{L^2(\mathcal{C})} - (\gamma_N^{\mathrm{i}} f_{\mathrm{i}}, \gamma_D^{\mathrm{i}} g_{\mathrm{i}})_{L^2(\mathcal{C})} \\
&\qquad + (\gamma_D^{\mathrm{e}} f_{\mathrm{e}}, \gamma_N^{\mathrm{e}} g_{\mathrm{e}})_{L^2(\mathcal{C})} - (\gamma_N^{\mathrm{e}} f_{\mathrm{e}}, \gamma_D^{\mathrm{e}} g_{\mathrm{e}})_{L^2(\mathcal{C})} \\
&\quad = \bigl(\gamma_D^{\mathrm{i}} f_{\mathrm{i}}, \gamma_N^{\mathrm{i}} g_{\mathrm{i}} + \gamma_N^{\mathrm{e}} g_{\mathrm{e}}\bigr)_{L^2(\mathcal{C})} - \bigl(\gamma_N^{\mathrm{i}} f_{\mathrm{i}} + \gamma_N^{\mathrm{e}} f_{\mathrm{e}}, \gamma_D^{\mathrm{i}} g_{\mathrm{i}}\bigr)_{L^2(\mathcal{C})},
\end{aligned}$$

and hence condition (i) in Theorem 2.2 holds. Next, in order to show that $\operatorname{ran}(\Gamma_0, \Gamma_1)^\top$ is dense in $L^2(\mathcal{C})$ we recall that

$$\begin{pmatrix} \gamma_D^{\mathrm{i}} \\ \gamma_N^{\mathrm{i}} \end{pmatrix} : H^2(\Omega_{\mathrm{i}}) \to H^{3/2}(\mathcal{C}) \times H^{1/2}(\mathcal{C})$$

and

$$\begin{pmatrix} \gamma_D^{\mathrm{e}} \\ \gamma_N^{\mathrm{e}} \end{pmatrix} : H^2(\Omega_{\mathrm{e}}) \to H^{3/2}(\mathcal{C}) \times H^{1/2}(\mathcal{C})$$

are surjective mappings. It follows that also the mapping

$$\begin{pmatrix} \gamma_D^{\mathrm{i}} \\ \gamma_N^{\mathrm{i}} + \gamma_N^{\mathrm{e}} \end{pmatrix} : \operatorname{dom}(T) \to H^{3/2}(\mathcal{C}) \times H^{1/2}(\mathcal{C}) \tag{4.11}$$

is surjective, and since the 2×2-block operator matrix

$$\Theta := \begin{pmatrix} cI_{L^2(\mathcal{C})} & -I_{L^2(\mathcal{C})} \\ \frac{\alpha}{c-\alpha} I_{L^2(\mathcal{C})} & -\frac{1}{c-\alpha} I_{L^2(\mathcal{C})} \end{pmatrix}$$

is an isomorphism in $L^2(\mathcal{C}) \times L^2(\mathcal{C})$, it follows that the range of the mapping

$$\begin{pmatrix} \Gamma_0 \\ \Gamma_1 \end{pmatrix} = \Theta \begin{pmatrix} \gamma_D^{\mathrm{i}} \\ \gamma_N^{\mathrm{i}} + \gamma_N^{\mathrm{e}} \end{pmatrix} : \operatorname{dom}(T) \to L^2(\mathcal{C}) \times L^2(\mathcal{C}),$$

is dense. Furthermore, as $C_0^\infty(\Omega_{\mathrm{i}}) \times C_0^\infty(\Omega_{\mathrm{e}})$ is contained in $\ker(\Gamma_0) \cap \ker(\Gamma_1)$, it is clear that $\ker(\Gamma_0) \cap \ker(\Gamma_1)$ is dense in $L^2(\mathbb{R}^n)$. Hence one concludes that condition (ii) in Theorem 2.2 is satisfied. Condition (iii) in Theorem 2.2 is satisfied since (4.10) holds by construction and $H_{\delta,c}$ is self-adjoint. Thus, Theorem 2.2 implies that the closed symmetric operator

$$T \upharpoonright \big(\ker(\Gamma_0) \cap \ker(\Gamma_1)\big) = H_{\delta,c} \cap H_{\delta,\alpha} = S$$

is densely defined, its adjoint coincides with $\overline{T}$, and $\{L^2(\mathcal{C}), \Gamma_0, \Gamma_1\}$ is a quasi boundary triple for $T \subset S^*$ such that (4.10) holds.

Step 2. In this step we prove that for $z \in \rho(H_{\delta,c}) \cap \rho(H)$ the Weyl function corresponding to the quasi boundary triple $\{L^2(\mathcal{C}), \Gamma_0, \Gamma_1\}$ is given by

$$\begin{aligned} M(z) &= \frac{1}{c-\alpha}\big(\alpha\mathcal{E}_{1/2}(z) - I_{L^2(\mathcal{C})}\big)\big(c\mathcal{E}_{1/2}(z) - I_{L^2(\mathcal{C})}\big)^{-1}, \\ \operatorname{dom}(M(z)) &= H^{1/2}(\mathcal{C}), \end{aligned} \tag{4.12}$$

where $\mathcal{E}_{1/2}(z)$ denotes the restriction of the operator $\mathcal{E}(z)$ in (4.5) onto $H^{1/2}(\mathcal{C})$. Furthermore, we verify that $M(z_1)$ and $M(z_2)^{-1}$ are bounded for some $z_1, z_2 \in \mathbb{C}\backslash\mathbb{R}$, and we conclude that the closures of the operators $M(z)$, $z \in \mathbb{C}\backslash\mathbb{R}$, in $L^2(\mathcal{C})$ are given by the operators $\mathcal{M}_\alpha(z)$ in (4.6), (4.7).

It will first be shown that the operator $\mathcal{E}(z)$ and its restriction $\mathcal{E}_{1/2}(z)$ are well-defined for all $z \in \rho(H) = \mathbb{C}\backslash[0,\infty)$. For this fix $z \in \mathbb{C}\backslash[0,\infty)$, and let

$$f_z = \begin{pmatrix} f_{\mathrm{i},z} \\ f_{\mathrm{e},z} \end{pmatrix} \in H^{3/2}(\Omega_{\mathrm{i}}) \times H^{3/2}(\Omega_{\mathrm{e}}) \tag{4.13}$$

be such that $\gamma_D^{\mathrm{i}} f_{\mathrm{i},z} = \gamma_D^{\mathrm{e}} f_{\mathrm{e},z}$, and

$$-\Delta f_{\mathrm{i},z} = z f_{\mathrm{i},z} \text{ and } -\Delta f_{\mathrm{e},z} = z f_{\mathrm{e},z}.$$

From the definition of $\mathcal{D}_{\mathrm{i}}(z)$ and $\mathcal{D}_{\mathrm{e}}(z)$ in (4.3) and (4.4) one concludes that

$$\begin{aligned}\big(\mathcal{D}_{\mathrm{i}}(z) + \mathcal{D}_{\mathrm{e}}(z)\big)\gamma_D^{\mathrm{i}} f_{\mathrm{i},z} &= \mathcal{D}_{\mathrm{i}}(z)\gamma_D^{\mathrm{i}} f_{\mathrm{i},z} + \mathcal{D}_{\mathrm{e}}(z)\gamma_D^{\mathrm{e}} f_{\mathrm{e},z} \\ &= \gamma_N^{\mathrm{i}} f_{\mathrm{i},z} + \gamma_N^{\mathrm{e}} f_{\mathrm{e},z}.\end{aligned} \tag{4.14}$$

This also proves that $\mathcal{D}_{\mathrm{i}}(z) + \mathcal{D}_{\mathrm{e}}(z)$ is injective for $z \in \mathbb{C}\backslash[0,\infty)$. In fact, otherwise there would exist a function $f_z = (f_{\mathrm{i},z}, f_{\mathrm{e},z})^\top \neq 0$ as in (4.13) which would satisfy both conditions

$$\gamma_D^{\mathrm{i}} f_{\mathrm{i},z} = \gamma_D^{\mathrm{e}} f_{\mathrm{e},z} \text{ and } \gamma_N^{\mathrm{i}} f_{\mathrm{i},z} + \gamma_N^{\mathrm{e}} f_{\mathrm{e},z} = 0, \tag{4.15}$$

and hence for all $h \in \mathrm{dom}(H) = H^2(\mathbb{R}^n)$, Green's identity together with the conditions (4.15) would imply

$$\begin{aligned}&(Hh, f_z)_{L^2(\mathbb{R}^n)} - (h, zf_z)_{L^2(\mathbb{R}^n)} \\ &\quad = (-\Delta h_{\mathrm{i}}, f_{\mathrm{i},z})_{L^2(\Omega_{\mathrm{i}})} - (h_{\mathrm{i}}, -\Delta f_{\mathrm{i},z})_{L^2(\Omega_{\mathrm{i}})} \\ &\qquad + (-\Delta h_{\mathrm{e}}, f_{\mathrm{e},z})_{L^2(\Omega_{\mathrm{e}})} - (h_{\mathrm{e}}, -\Delta f_{\mathrm{e},z})_{L^2(\Omega_{\mathrm{e}})} \\ &\quad = (\gamma_D^{\mathrm{i}} h_{\mathrm{i}}, \gamma_N^{\mathrm{i}} f_{\mathrm{i},z})_{L^2(\mathcal{C})} - (\gamma_N^{\mathrm{i}} h_{\mathrm{i}}, \gamma_D^{\mathrm{i}} f_{\mathrm{i},z})_{L^2(\mathcal{C})} \\ &\qquad + (\gamma_D^{\mathrm{e}} h_{\mathrm{e}}, \gamma_N^{\mathrm{e}} f_{\mathrm{e},z})_{L^2(\mathcal{C})} - (\gamma_N^{\mathrm{e}} h_{\mathrm{e}}, \gamma_D^{\mathrm{e}} f_{\mathrm{e},z})_{L^2(\mathcal{C})} \\ &\quad = 0,\end{aligned} \tag{4.16}$$

that is, $f_z \in \mathrm{dom}(H)$ and $Hf_z = zf_z$; a contradiction since $z \in \rho(H)$. Hence,

$$\ker\big(\mathcal{D}_{\mathrm{i}}(z) + \mathcal{D}_{\mathrm{e}}(z)\big) = \{0\}, \quad z \in \mathbb{C}\backslash[0,\infty),$$

and if we denote the restrictions of $\mathcal{D}_{\mathrm{i}}(z)$ and $\mathcal{D}_{\mathrm{e}}(z)$ onto $H^{3/2}(\mathcal{C})$ by $\mathcal{D}_{\mathrm{i},3/2}(z)$ and $\mathcal{D}_{\mathrm{e},3/2}(z)$, respectively, then also $\ker(\mathcal{D}_{\mathrm{i},3/2}(z) + \mathcal{D}_{\mathrm{e},3/2}(z)) = \{0\}$ for $z \in \mathbb{C}\backslash[0,\infty)$. Thus, we have shown that $\mathcal{E}(z)$ and its restriction $\mathcal{E}_{1/2}(z)$ are well-defined for all $z \in \rho(H) = \mathbb{C}\backslash[0,\infty)$.

Furthermore, if the function f_z in (4.13) belongs to $H^2(\Omega_{\mathrm{i}}) \times H^2(\Omega_{\mathrm{e}})$, that is, $f_z \in \ker(T - zI_{L^2(\mathbb{R}^n)})$, then $\gamma_D^{\mathrm{i}} f_{\mathrm{i},z} = \gamma_D^{\mathrm{e}} f_{\mathrm{e},z} \in H^{3/2}(\mathcal{C})$ and hence besides (4.14) one also has

$$\big(\mathcal{D}_{\mathrm{i},3/2}(z) + \mathcal{D}_{\mathrm{e},3/2}(z)\big)\gamma_D^{\mathrm{i}} f_{\mathrm{i},z} = \gamma_N^{\mathrm{i}} f_{\mathrm{i},z} + \gamma_N^{\mathrm{e}} f_{\mathrm{e},z} \in H^{1/2}(\mathcal{C}). \tag{4.17}$$

One concludes from (4.17) that

$$\mathcal{E}_{1/2}(z)\big(\gamma_N^{\mathrm{i}} f_{\mathrm{i},z} + \gamma_N^{\mathrm{e}} f_{\mathrm{e},z}\big) = \gamma_D^{\mathrm{i}} f_{\mathrm{i},z},$$

and from (4.8) one then obtains

$$\begin{aligned}\big(c\mathcal{E}_{1/2}(z) - I_{L^2(\mathcal{C})}\big)\big(\gamma_N^{\mathrm{i}} f_{\mathrm{i},z} + \gamma_N^{\mathrm{e}} f_{\mathrm{e},z}\big) &= c\gamma_D^{\mathrm{i}} f_{\mathrm{i},z} - \big(\gamma_N^{\mathrm{i}} f_{\mathrm{i},z} + \gamma_N^{\mathrm{e}} f_{\mathrm{e},z}\big) \\ &= \Gamma_0 f_z,\end{aligned} \tag{4.18}$$

and

$$\big(\alpha\mathcal{E}_{1/2}(z) - I_{L^2(\mathcal{C})}\big)\big(\gamma_N^{\mathrm{i}} f_{\mathrm{i},z} + \gamma_N^{\mathrm{e}} f_{\mathrm{e},z}\big) = \alpha\gamma_D^{\mathrm{i}} f_{\mathrm{i},z} - \big(\gamma_N^{\mathrm{i}} f_{\mathrm{i},z} + \gamma_N^{\mathrm{e}} f_{\mathrm{e},z}\big). \tag{4.19}$$

For $z \in \rho(H_{\delta,c}) \cap \rho(H)$ one verifies $\ker(c\mathcal{E}_{1/2}(z) - I_{L^2(\mathcal{C})}) = \{0\}$ with the help of (4.18). Then (4.8) and (4.11) yield

$$\operatorname{ran}\big(c\mathcal{E}_{1/2}(z) - I_{L^2(\mathcal{C})}\big) = \operatorname{ran}(\Gamma_0) = H^{1/2}(\mathcal{C}).$$

Thus, it follows from (4.18), (4.19), and (4.9) that

$$\begin{aligned} &\frac{1}{c-\alpha}\big(\alpha\mathcal{E}_{1/2}(z) - I_{L^2(\mathcal{C})}\big)\big(c\mathcal{E}_{1/2}(z) - I_{L^2(\mathcal{C})}\big)^{-1}\Gamma_0 f_z \\ &\quad = \frac{1}{c-\alpha}\big(\alpha\mathcal{E}_{1/2}(z) - I_{L^2(\mathcal{C})}\big)\big(\gamma_N^{\rm i} f_{{\rm i},z} + \gamma_N^{\rm e} f_{{\rm e},z}\big) \\ &\quad = \frac{1}{c-\alpha}\Big(\alpha\gamma_D^{\rm i} f_{{\rm i},z} - \big(\gamma_N^{\rm i} f_{{\rm i},z} + \gamma_N^{\rm e} f_{{\rm e},z}\big)\Big) \\ &\quad = \Gamma_1 f_z \end{aligned}$$

holds for all $z \in \rho(H_{\delta,c}) \cap \rho(H)$. This proves that the Weyl function corresponding to the quasi boundary triple (4.8)–(4.9) is given by (4.12).

Next it will be shown that $M(z)$ and $M(z)^{-1}$ are bounded for $z \in \mathbb{C}\backslash\mathbb{R}$. For this it suffices to check that the operators

$$\alpha\mathcal{E}_{1/2}(z) - I_{L^2(\mathcal{C})} \quad \text{and} \quad c\mathcal{E}_{1/2}(z) - I_{L^2(\mathcal{C})} \tag{4.20}$$

are bounded and have bounded inverses. The argument is the same for both operators in (4.20) and hence we discuss $\alpha\mathcal{E}_{1/2}(z) - I_{L^2(\mathcal{C})}$ only. One recalls that

$$\mathcal{D}_{\rm i}(z) + \mathcal{D}_{\rm e}(z), \quad z \in \mathbb{C}\backslash\mathbb{R},$$

maps onto $L^2(\mathcal{C})$, is boundedly invertible, and its inverse $\mathcal{E}(z)$ in (4.5) is a compact operator in $L^2(\mathcal{C})$ with $\operatorname{ran}(\mathcal{E}(z)) = H^1(\mathcal{C})$ (see [9, Proposition 3.2 (iii)]). Hence also the restriction $\mathcal{E}_{1/2}(z)$ of $\mathcal{E}(z)$ onto $H^{1/2}(\mathcal{C})$ is bounded in $L^2(\mathcal{C})$. It follows that $\alpha\mathcal{E}_{1/2}(z) - I_{L^2(\mathcal{C})}$ is bounded, and its closure is given by

$$\overline{\alpha\mathcal{E}_{1/2}(z) - I_{L^2(\mathcal{C})}} = \alpha\mathcal{E}(z) - I_{L^2(\mathcal{C})} \in \mathcal{L}\big(L^2(\mathcal{C})\big), \quad z \in \mathbb{C}\backslash\mathbb{R}. \tag{4.21}$$

In order to show that the inverse $(\alpha\mathcal{E}_{1/2}(z) - I_{L^2(\mathcal{C})})^{-1}$ exists and is bounded for $z \in \mathbb{C}\backslash\mathbb{R}$ we first check that

$$\ker\big(\alpha\mathcal{E}(z) - I_{L^2(\mathcal{C})}\big) = \{0\}, \quad z \in \mathbb{C}\backslash\mathbb{R}. \tag{4.22}$$

In fact, assume that $z \in \mathbb{C}\backslash\mathbb{R}$ and $\varphi \in L^2(\mathcal{C})$ are such that $\alpha\mathcal{E}(z)\varphi = \varphi$. It follows from $\operatorname{dom}(\mathcal{E}(z)) = \operatorname{ran}(\mathcal{D}_{\rm i}(z) + \mathcal{D}_{\rm e}(z)) = L^2(\mathcal{C})$ that there exists $\psi \in H^1(\mathcal{C})$ such that

$$\varphi = \big(\mathcal{D}_{\rm i}(z) + \mathcal{D}_{\rm e}(z)\big)\psi, \tag{4.23}$$

and from (4.1)–(4.2) one concludes that there exists a unique

$$f_z = \begin{pmatrix} f_{{\rm i},z} \\ f_{{\rm e},z} \end{pmatrix} \in H^{3/2}(\Omega_{\rm i}) \times H^{3/2}(\Omega_{\rm e})$$

such that

$$\gamma_D^{\rm i} f_{{\rm i},z} = \gamma_D^{\rm e} f_{{\rm e},z} = \psi, \tag{4.24}$$

and

$$-\Delta f_{{\rm i},z} = z f_{{\rm i},z} \quad \text{and} \quad -\Delta f_{{\rm e},z} = z f_{{\rm e},z}.$$

Since $\varphi = \alpha\mathcal{E}(z)\varphi = \alpha\psi$ by (4.23), one obtains from (4.14), (4.24), and (4.23) that

$$\begin{aligned} \gamma_N^{\mathrm{i}} f_{\mathrm{i},z} + \gamma_N^{\mathrm{e}} f_{\mathrm{e},z} &= \big(\mathcal{D}_{\mathrm{i}}(z) + \mathcal{D}_{\mathrm{e}}(z)\big)\gamma_D^{\mathrm{i}} f_{\mathrm{i},z} \\ &= \big(\mathcal{D}_{\mathrm{i}}(z) + \mathcal{D}_{\mathrm{e}}(z)\big)\psi \\ &= \varphi \\ &= \alpha\psi \\ &= \alpha\gamma_D^{\mathrm{i}} f_{\mathrm{i},z}. \end{aligned} \tag{4.25}$$

For $h = (h_{\mathrm{i}}, h_{\mathrm{e}})^\top \in \operatorname{dom}(H_{\delta,\alpha})$ one has

$$\gamma_D^{\mathrm{i}} h_{\mathrm{i}} = \gamma_D^{\mathrm{e}} h_{\mathrm{e}} \text{ and } \gamma_N^{\mathrm{i}} h_{\mathrm{i}} + \gamma_N^{\mathrm{e}} h_{\mathrm{e}} = \alpha\gamma_D^{\mathrm{i}} h_{\mathrm{i}}, \tag{4.26}$$

and in a similar way as in (4.16), Green's identity together with (4.24), (4.25), and (4.26) imply

$$\begin{aligned} &(H_{\delta,\alpha} h, f_z)_{L^2(\mathbb{R}^n)} - (h, z f_z)_{L^2(\mathbb{R}^n)} \\ &\quad = (-\Delta h_{\mathrm{i}}, f_{\mathrm{i},z})_{L^2(\Omega_{\mathrm{i}})} - (h_{\mathrm{i}}, -\Delta f_{\mathrm{i},z})_{L^2(\Omega_{\mathrm{i}})} \\ &\qquad + (-\Delta h_{\mathrm{e}}, f_{\mathrm{e},z})_{L^2(\Omega_{\mathrm{e}})} - (h_{\mathrm{e}}, -\Delta f_{\mathrm{e},z})_{L^2(\Omega_{\mathrm{e}})} \\ &\quad = (\gamma_D^{\mathrm{i}} h_{\mathrm{i}}, \gamma_N^{\mathrm{i}} f_{\mathrm{i},z})_{L^2(\mathcal{C})} - (\gamma_N^{\mathrm{i}} h_{\mathrm{i}}, \gamma_D^{\mathrm{i}} f_{\mathrm{i},z})_{L^2(\mathcal{C})} \\ &\qquad + (\gamma_D^{\mathrm{e}} h_{\mathrm{e}}, \gamma_N^{\mathrm{e}} f_{\mathrm{e},z})_{L^2(\mathcal{C})} - (\gamma_N^{\mathrm{e}} h_{\mathrm{e}}, \gamma_D^{\mathrm{e}} f_{\mathrm{e},z})_{L^2(\mathcal{C})} \\ &\quad = \big(\gamma_D^{\mathrm{i}} h_{\mathrm{i}}, \gamma_N^{\mathrm{i}} f_{\mathrm{i},z} + \gamma_N^{\mathrm{e}} f_{\mathrm{e},z}\big)_{L^2(\mathcal{C})} - \big(\gamma_N^{\mathrm{i}} h_{\mathrm{i}} + \gamma_N^{\mathrm{e}} h_{\mathrm{e}}, \gamma_D^{\mathrm{i}} f_{\mathrm{i},z}\big)_{L^2(\mathcal{C})} \\ &\quad = \big(\gamma_D^{\mathrm{i}} h_{\mathrm{i}}, \alpha\gamma_D^{\mathrm{i}} f_{\mathrm{i},z}\big)_{L^2(\mathcal{C})} - \big(\alpha\gamma_D^{\mathrm{i}} h_{\mathrm{i}}, \gamma_D^{\mathrm{i}} f_{\mathrm{i},z}\big)_{L^2(\mathcal{C})} \\ &\quad = 0. \end{aligned}$$

As $H_{\delta,\alpha}$ is self-adjoint one concludes that $f_z \in \operatorname{dom}(H_{\delta,\alpha})$ and

$$f_z \in \ker(H_{\delta,\alpha} - z I_{L^2(\mathbb{R}^n)}).$$

Since $z \in \mathbb{C}\backslash\mathbb{R}$, this yields $f_z = 0$ and therefore, $\psi = \gamma_D^{\mathrm{i}} f_{\mathrm{i},z} = 0$ and hence $\varphi = 0$ by (4.23), implying (4.22).

Since $\mathcal{E}(z)$ is a compact operator in $L^2(\mathcal{C})$ (see [9, Proposition 3.2 (iii)]) also $\alpha\mathcal{E}(z)$ is compact and together with (4.22) one concludes that

$$(\alpha\mathcal{E}(z) - I_{L^2(\mathcal{C})})^{-1} \in \mathcal{L}\big(L^2(\mathcal{C})\big). \tag{4.27}$$

Hence also the restriction

$$\big(\alpha\mathcal{E}_{1/2}(z) - I_{L^2(\mathcal{C})}\big)^{-1}$$

is a bounded operator in $L^2(\mathcal{C})$. Summing up, we have shown that the operators in (4.20) are bounded and have bounded inverses for all $z \in \mathbb{C}\backslash\mathbb{R}$, and hence the values $M(z)$ of the Weyl function in (4.12) are bounded and have bounded inverses for all $z \in \mathbb{C}\backslash\mathbb{R}$. From (4.12), (4.21) and (4.27) it follows that that the closures of the operators $M(z)$, $z \in \mathbb{C}\backslash\mathbb{R}$, in $L^2(\mathcal{C})$ are given by the operators $\mathcal{M}_\alpha(z)$ in (4.6), (4.7).

Step 3. Now we check that the operators $\{H_{\delta,c}, H_{\delta,\alpha}\}$ and the Weyl function corresponding to the quasi boundary triple $\{L^2(\mathcal{C}), \Gamma_0, \Gamma_1\}$ in Step 1 satisfy the assumptions of Theorem 3.1 for $n \in \mathbb{N}$, $n \geq 2$, and all $k \geq (n-3)/4$.

In fact, the sign condition (3.4) follows from the assumption $\alpha(x) < c$ and the fact that the closed quadratic forms $\mathfrak{h}_{\delta,\alpha}$ and $\mathfrak{h}_{\delta,c}$ associated to $H_{\delta,\alpha}$ and $H_{\delta,c}$ satisfy the inequality $\mathfrak{h}_{\delta,c} \leq \mathfrak{h}_{\delta,\alpha}$. More precisely, the inequality for the quadratic forms yields $\inf(\sigma(H_{\delta,c})) \leq \inf(\sigma(H_{\delta,\alpha}))$, and for $\zeta < \inf(\sigma(H_{\delta,c}))$ the forms $\mathfrak{h}_{\delta,c} - \zeta$ and $\mathfrak{h}_{\delta,\alpha} - \zeta$ are both nonnegative, satisfy the inequality $\mathfrak{h}_{\delta,c} - \zeta \leq \mathfrak{h}_{\delta,\alpha} - \zeta$, and hence the resolvents of the corresponding nonnegative self-adjoint operators $H_{\delta,c} - \zeta I_{L^2(\mathbb{R}^n)}$ and $H_{\delta,\alpha} - \zeta I_{L^2(\mathbb{R}^n)}$ satisfy the inequality

$$(H_{\delta,c} - \zeta I_{L^2(\mathbb{R}^n)})^{-1} \geq (H_{\delta,\alpha} - \zeta I_{L^2(\mathbb{R}^n)})^{-1}, \quad \zeta < \inf(\sigma(H_{\delta,c}))$$

(see, e.g., [30, Chapter VI, § 2.6] or [15, Chapter 10, §2, Theorem 6]). Thus the sign condition (3.4) in the assumptions of Theorem 3.1 holds.

In order to verify the $\mathfrak{S}_p$-conditions

$$\overline{\gamma(z)}^{(p)} \big(M(z)^{-1}\gamma(\overline{z})^*\big)^{(q)} \in \mathfrak{S}_1\big(L^2(\mathbb{R}^n)\big), \quad p+q = 2k, \tag{4.28}$$

$$\big(M(z)^{-1}\gamma(\overline{z})^*\big)^{(q)}\overline{\gamma(z)}^{(p)} \in \mathfrak{S}_1\big(L^2(\mathcal{C})\big), \quad p+q = 2k, \tag{4.29}$$

and

$$\frac{d^j}{dz^j}\overline{M(z)} \in \mathfrak{S}_{(2k+1)/j}\big(L^2(\mathcal{C})\big), \quad j = 1,\dots,2k+1, \tag{4.30}$$

for all $z \in \rho(H_{\delta,c}) \cap \rho(H_{\delta,\alpha})$ in the assumptions of Theorem 3.1, one first recalls the smoothing property

$$(H_{\delta,c} - zI_{L^2(\mathbb{R}^n)})^{-1} f \in H^{k+2}(\Omega_{\rm i}) \times H^{k+2}(\Omega_{\rm e}) \tag{4.31}$$

for $f \in H^k(\Omega_{\rm i}) \times H^k(\Omega_{\rm e})$ and $k \in \mathbb{N}_0$, which follows, for instance, from [36, Theorem 4.20]. Next one observes that (2.1), (4.9), and the definition of $H_{\delta,c}$ imply

$$\begin{aligned}\gamma(\overline{z})^* f &= \Gamma_1 (H_{\delta,c} - zI_{L^2(\mathbb{R}^n)})^{-1} f \\ &= (c-\alpha)^{-1}\big(\alpha\gamma_D^{\rm i} - (\gamma_N^{\rm i} + \gamma_N^{\rm e})\big)(H_{\delta,c} - zI_{L^2(\mathbb{R}^n)})^{-1} f \\ &= (c-\alpha)^{-1}\big(c\gamma_D^{\rm i} - (\gamma_N^{\rm i} + \gamma_N^{\rm e}) + (\alpha - c)\gamma_D^{\rm i}\big)(H_{\delta,c} - zI_{L^2(\mathbb{R}^n)})^{-1} f,\end{aligned}$$

which yields

$$\gamma(\overline{z})^* f = -\gamma_D^{\rm i}(H_{\delta,c} - zI_{L^2(\mathbb{R}^n)})^{-1} f, \quad f \in L^2(\mathbb{R}^n). \tag{4.32}$$

Hence (2.2), (4.31), and Lemma 4.2 imply

$$\big(\gamma(\overline{z})^*\big)^{(q)} = -q!\,\gamma_D^{\rm i}(H_{\delta,c} - zI_{L^2(\mathbb{R}^n)})^{-(q+1)} \in \mathfrak{S}_r\big(L^2(\mathbb{R}^n), L^2(\mathcal{C})\big) \tag{4.33}$$

for $r > (n-1)/[2q + (3/2)]$, $z \in \rho(H_{\delta,c})$ and $q \in \mathbb{N}_0$ (cf. [12, Lemma 3.1] for the case $c = 0$). One also has

$$\overline{\gamma(z)}^{(p)} \in \mathfrak{S}_r\big(L^2(\mathcal{C}), L^2(\mathbb{R}^n)\big), \quad r > (n-1)/[2p + (3/2)], \tag{4.34}$$

for all $z \in \rho(H_{\delta,c})$ and $p \in \mathbb{N}_0$. Furthermore,

$$\frac{d^j}{dz^j}\overline{M(z)} = j!\,\gamma(\overline{z})^*(H_{\delta,c} - zI_{L^2(\mathbb{R}^n)})^{-(j-1)}\overline{\gamma(z)} \tag{4.35}$$

by (2.3) and with the help of (4.32) it follows that

$$\gamma(\overline{z})^*(H_{\delta,c}-zI_{L^2(\mathbb{R}^n)})^{-(j-1)} = -\gamma_D^{\mathrm{i}}(H_{\delta,c}-zI_{L^2(\mathbb{R}^n)})^{-j} \in \mathfrak{S}_x\big(L^2(\mathbb{R}^n), L^2(\mathcal{C})\big)$$

for $x > (n-1)/[2j-(1/2)]$. Moreover, we have $\overline{\gamma(z)} \in \mathfrak{S}_y(L^2(\mathcal{C}), L^2(\mathbb{R}^n))$ for $y > 2(n-1)/3$ by (4.34) and hence it follows from (4.35) and the well-known property $PQ \in \mathfrak{S}_w$ for $P \in \mathfrak{S}_x$, $Q \in \mathfrak{S}_y$, and $x^{-1} + y^{-1} = w^{-1}$, that

$$\frac{d^j}{dz^j}\overline{M(z)} \in \mathfrak{S}_w\big(L^2(\mathcal{C})\big), \quad w > (n-1)/(2j+1), \ z \in \rho(H_{\delta,c}), \ j \in \mathbb{N}. \tag{4.36}$$

One observes that

$$\frac{d}{dz}\big[\overline{M(z)}\big]^{-1} = -\big[\overline{M(z)}\big]^{-1}\left(\frac{d}{dz}\overline{M(z)}\right)\big[\overline{M(z)}\big]^{-1}, \quad z \in \rho(H_{\delta,c}) \cap \rho(H_{\delta,\alpha}),$$

that $\big[\overline{M(z)}\big]^{-1}$ is bounded, and by (4.36) that for $j \in \mathbb{N}$ also

$$\begin{aligned}\frac{d^j}{dz^j}\big[\overline{M(z)}\big]^{-1} \in \mathfrak{S}_w\big(L^2(\mathcal{C})\big), \quad & w > (n-1)/(2j+1), \\ & z \in \rho(H_{\delta,c}) \cap \rho(H_{\delta,\alpha});\end{aligned} \tag{4.37}$$

we leave the formal induction step to the reader. Therefore,

$$\begin{aligned}\big(M(z)^{-1}\gamma(\overline{z})^*\big)^{(q)} &= \big(\big[\overline{M(z)}\big]^{-1}\gamma(\overline{z})^*\big)^{(q)} \\ &= \sum_{\substack{p+m=q \\ p,m\geqslant 0}} \binom{q}{p} \big(\big[\overline{M(z)}\big]^{-1}\big)^{(p)} \big(\gamma(\overline{z})^*\big)^{(m)} \\ &= \big[\overline{M(z)}\big]^{-1}\big(\gamma(\overline{z})^*\big)^{(q)} + \sum_{\substack{p+m=q \\ p>0, m\geq 0}} \binom{q}{p} \big(\big[\overline{M(z)}\big]^{-1}\big)^{(p)} \big(\gamma(\overline{z})^*\big)^{(m)},\end{aligned} \tag{4.38}$$

and one has

$$\big[\overline{M(z)}\big]^{-1}(\gamma(\overline{z})^*)^{(q)} \in \mathfrak{S}_r\big(L^2(\mathbb{R}^n), L^2(\mathcal{C})\big)$$

for $r > (n-1)/[2q+(3/2)]$ by (4.33) and each summand (and hence also the finite sum) on the right-hand side in (4.38) is in $\mathfrak{S}_r(L^2(\mathbb{R}^n), L^2(\mathcal{C}))$ for $r > (n-1)/[2p+1+2m+(3/2)] = (n-1)/[2q+(5/2)]$, which follows from (4.37) and (4.33). Hence one has

$$\big(M(z)^{-1}\gamma(\overline{z})^*\big)^{(q)} \in \mathfrak{S}_r\big(L^2(\mathbb{R}^n), L^2(\mathcal{C})\big) \tag{4.39}$$

for $r > (n-1)/[2q+(3/2)]$ and $z \in \rho(H_{\delta,c}) \cap \rho(H_{\delta,\alpha})$. From (4.34) and (4.39) one then concludes

$$\overline{\gamma(z)}^{(p)}\big(M(z)^{-1}\gamma(\overline{z})^*\big)^{(q)} \in \mathfrak{S}_r\big(L^2(\mathbb{R}^n)\big)$$

for $r > (n-1)/[2(p+q)+3] = (n-1)/(4k+3)$, and since $k \geq (n-3)/4$, one has $1 > (n-1)/(4k+3)$, that is, the trace class condition (4.28) is satisfied. The same argument shows that (4.29) is satisfied. Finally, (4.30) follows from (4.36) and the fact that $k \geq (n-3)/4$ implies

$$\frac{2k+1}{j} \geq \frac{n-1}{2j} > \frac{n-1}{2j+1}, \quad j = 1, \ldots, 2k+1.$$

Hence the assumptions in Theorem 3.1 are satisfied with S in Step 1, the quasi boundary triple in (4.8)–(4.9), the corresponding γ-field, and Weyl function in (4.12). Therefore, Theorem 3.1 yields assertion (i) in Theorem 4.3 with H replaced by $H_{\delta,c}$. In addition, for any orthonormal basis $\{\varphi_j\}_{j\in J}$ in $L^2(\mathcal{C})$, the function

$$\xi_\alpha(\lambda) = \sum_{j\in J} \lim_{\varepsilon\downarrow 0} \frac{1}{\pi} \big(\mathrm{Im}\big(\log(\mathcal{M}_\alpha(\lambda + i\varepsilon))\big)\varphi_j, \varphi_j\big)_{L^2(\mathcal{C})} \quad \text{for a.e. } \lambda \in \mathbb{R}$$

is a spectral shift function for the pair $(H_{\delta,c}, H_{\delta,\alpha})$ such that $\xi_\alpha(\lambda) = 0$ for $\lambda < \inf(\sigma(H_{\delta,c})) \leq \inf(\sigma(H_{\delta,\alpha}))$ and the trace formula

$$\begin{aligned} &\mathrm{tr}_{L^2(\mathbb{R}^n)}\big((H_{\delta,\alpha} - zI_{L^2(\mathbb{R}^n)})^{-(2k+1)} - (H_{\delta,c} - zI_{L^2(\mathbb{R}^n)})^{-(2k+1)}\big) \\ &\quad = -(2k+1) \int_{\mathbb{R}} \frac{\xi_\alpha(\lambda)\, d\lambda}{(\lambda - z)^{2k+2}}, \qquad z \in \rho(H_{\delta,c}) \cap \rho(H_{\delta,\alpha}), \end{aligned}$$

holds.

The above considerations remain valid in the special case $\alpha = 0$ which corresponds to the pair $(H_{\delta,c}, H)$ and yields an analogous representation for a spectral shift function ξ_0. Finally it follows from the considerations in the end of Section 3 (see (3.5)) that

$$\begin{aligned} \xi(\lambda) &= \xi_\alpha(\lambda) - \xi_0(\lambda) \\ &= \sum_{j\in J} \lim_{\varepsilon\downarrow 0} \frac{1}{\pi} \Big(\big(\mathrm{Im}\big(\log(\mathcal{M}_\alpha(\lambda + i\varepsilon)) - \log(\mathcal{M}_0(\lambda + i\varepsilon))\big)\big)\varphi_j, \varphi_j\Big)_{L^2(\mathcal{C})} \end{aligned}$$

for a.e. $\lambda \in \mathbb{R}$ is a spectral shift function for the pair $(H, H_{\delta,\alpha})$ such that $\xi(\lambda) = 0$ for $\lambda < \inf(\sigma(H_{\delta,c})) \leq \inf\{\sigma(H), \sigma(H_{\delta,\alpha})\}$. This completes the proof of Theorem 4.3. □

In space dimensions $n = 2$ and $n = 3$ one can choose $k = 0$ in Theorem 4.3 and together with Corollary 3.2 one obtains the following result.

Corollary 4.4. *Let the assumptions and $\mathcal{M}_\alpha$ and $\mathcal{M}_0$ be as in Theorem 4.3, and suppose that $n = 2$ or $n = 3$. Then the following assertions* (i)–(iii) *hold:*

(i) *The difference of the resolvents of H and $H_{\delta,\alpha}$ is a trace class operator, that is, for all $z \in \rho(H_{\delta,\alpha}) = \rho(H) \cap \rho(H_{\delta,\alpha})$,*

$$\big[(H_{\delta,\alpha} - zI_{L^2(\mathbb{R}^n)})^{-1} - (H - zI_{L^2(\mathbb{R}^n)})^{-1}\big] \in \mathfrak{S}_1\big(L^2(\mathbb{R}^n)\big).$$

(ii) $\mathrm{Im}(\log(\mathcal{M}_\alpha(z))) \in \mathfrak{S}_1(L^2(\mathcal{C}))$ *and* $\mathrm{Im}(\log(\mathcal{M}_0(z))) \in \mathfrak{S}_1(L^2(\mathcal{C}))$ *for all $z \in \mathbb{C}\backslash\mathbb{R}$, and the limits*

$$\mathrm{Im}\big(\log(\mathcal{M}_\alpha(\lambda + i0))\big) := \lim_{\varepsilon\downarrow 0} \mathrm{Im}\big(\log(\mathcal{M}_\alpha(\lambda + i\varepsilon))\big)$$

and

$$\mathrm{Im}\big(\log(\mathcal{M}_0(\lambda + i0))\big) := \lim_{\varepsilon\downarrow 0} \mathrm{Im}\big(\log(\mathcal{M}_0(\lambda + i\varepsilon))\big)$$

exist for a.e. $\lambda \in \mathbb{R}$ in $\mathfrak{S}_1(L^2(\mathcal{C}))$.

(iii) *The function defined by*

$$\xi(\lambda) = \frac{1}{\pi} \operatorname{tr}_{L^2(\mathcal{C})}\big(\operatorname{Im}\big(\log(\mathcal{M}_\alpha(\lambda + i0)) - \log(\mathcal{M}_0(\lambda + i0))\big)\big)$$

for a.e. $\lambda \in \mathbb{R}$ is a spectral shift function for the pair $(H, H_{\delta,\alpha})$ such that $\xi(\lambda) = 0$ for $\lambda < \inf(\sigma(H_{\delta,c}))$ and the trace formula

$$\operatorname{tr}_{L^2(\mathbb{R}^n)}\big((H_{\delta,\alpha} - zI_{L^2(\mathbb{R}^n)})^{-1} - (H - zI_{L^2(\mathbb{R}^n)})^{-1}\big) = -\int_{\mathbb{R}} \frac{\xi(\lambda)\,d\lambda}{(\lambda - z)^2}$$

is valid for all $z \in \rho(H_{\delta,\alpha}) = \rho(H) \cap \rho(H_{\delta,\alpha})$.

In the special case $\alpha < 0$, Theorem 4.3 simplifies slightly since in that case the sign condition (3.4) in Theorem 3.1 is satisfied by the pair $(H, H_{\delta,\alpha})$. Hence it is not necessary to introduce the operator $H_{\delta,c}$ as a comparison operator in the proof of Theorem 4.3. Instead, one considers the operators S and T in Step 1 of the proof of Theorem 4.3, and defines the boundary maps by

$$\Gamma_0 f = -\gamma_N^{\mathrm{i}} f_{\mathrm{i}} - \gamma_N^{\mathrm{e}} f_{\mathrm{e}}, \quad \operatorname{dom}(\Gamma_0) = \operatorname{dom}(T),$$

and

$$\Gamma_1 f = -\gamma_D^{\mathrm{i}} f_{\mathrm{i}} + \frac{1}{\alpha}(\gamma_N^{\mathrm{i}} f_{\mathrm{i}} + \gamma_N^{\mathrm{e}} f_{\mathrm{e}})), \quad \operatorname{dom}(\Gamma_1) = \operatorname{dom}(T).$$

In this case the corresponding Weyl function is given by

$$M(z) = \mathcal{E}_{1/2}(z) - \alpha^{-1} I_{L^2(\mathcal{C})}, \quad z \in \mathbb{C}\backslash\mathbb{R},$$

and hence the next statement follows in the same way as Theorem 4.3 from our abstract result Theorem 3.1.

Theorem 4.5. *Assume Hypothesis* 4.1, *let $\mathcal{E}(z)$ be defined as in* (4.5), *and let $\alpha \in C^1(\mathcal{C})$ be a real-valued function such that $\alpha(x) < 0$ for all $x \in \mathcal{C}$. Then the following assertions* (i) *and* (ii) *hold for $k \in \mathbb{N}_0$ such that $k \geq (n-3)/4$:*

(i) *The difference of the $(2k+1)$-th power of the resolvents of H and $H_{\delta,\alpha}$ is a trace class operator, that is,*

$$\big[(H_{\delta,\alpha} - zI_{L^2(\mathbb{R}^n)})^{-(2k+1)} - (H - zI_{L^2(\mathbb{R}^n)})^{-(2k+1)}\big] \in \mathfrak{S}_1\big(L^2(\mathbb{R}^n)\big)$$

holds for all $z \in \rho(H_{\delta,\alpha}) = \rho(H) \cap \rho(H_{\delta,\alpha})$.

(ii) *For any orthonormal basis $(\varphi_j)_{j\in J}$ in $L^2(\mathcal{C})$ the function defined by*

$$\xi(\lambda) = \sum_{j\in J} \lim_{\varepsilon\downarrow 0} \frac{1}{\pi}\big(\operatorname{Im}\big(\log(\mathcal{E}(t + i\varepsilon) - \alpha^{-1} I_{L^2(\mathcal{C})})\big)\varphi_j, \varphi_j\big)_{L^2(\mathcal{C})}$$

for a.e. $\lambda \in \mathbb{R}$ is a spectral shift function for the pair $(H, H_{\delta,\alpha})$ such that $\xi(\lambda) = 0$ for $\lambda < 0$ and the trace formula

$$\begin{aligned}\operatorname{tr}_{L^2(\mathbb{R}^n)}&\big((H_{\delta,\alpha} - zI_{L^2(\mathbb{R}^n)})^{-(2k+1)} - (H - zI_{L^2(\mathbb{R}^n)})^{-(2k+1)}\big)\\ &= -(2k+1)\int_{\mathbb{R}} \frac{\xi(\lambda)\,d\lambda}{(\lambda - z)^{2k+2}}\end{aligned}$$

is valid for all $z \in \mathbb{C}\backslash[0, \infty)$.

The analog of Corollary 4.4 again holds in the special cases $n = 2$ and $n = 3$; we omit further details.

Acknowledgments. J.B. is most grateful for the stimulating research stay and the hospitality at the Graduate School of Mathematical Sciences of the University of Tokyo from April to July 2016, where parts of this paper were written. F.G. is indebted to all organizers of the IWOTA 2017 Conference for creating such a stimulating atmosphere and for the great hospitality in Chemnitz, Germany, August 14–18, 2017. The authors also wish to thank Hagen Neidhardt for fruitful discussions and helpful remarks. Finally, we thank the anonymous referee and Albrecht Böttcher for a careful reading of our manuscript and for very helpful comments.

This work is supported by International Relations and Mobility Programs of the TU Graz and the Austrian Science Fund (FWF), project P-25162-N26.

References

[1] M. Abramowitz and I.A. Stegun, *Handbook of Mathematical Functions*, Dover, New York, 1972.

[2] S. Albeverio, F. Gesztesy, R. Høegh-Krohn, and H. Holden, *Solvable Models in Quantum Mechanics*, 2nd edition. With an appendix by Pavel Exner. AMS Chelsea Publishing, Providence, RI, 2005.

[3] S. Albeverio, A. Kostenko, M.M. Malamud, and H. Neidhardt, Spherical Schrödinger operators with δ-type interactions, *J. Math. Phys.* **54** (2013), 052103.

[4] S. Albeverio and P. Kurasov, *Singular Perturbations of Differential Operators*, London Mathematical Society Lecture Note Series, Vol. **271**, Cambridge University Press, Cambridge, 2000.

[5] J.-P. Antoine, F. Gesztesy, and J. Shabani, Exactly solvable models of sphere interactions in quantum mechanics, *J. Phys. A* **20** (1987), 3687–3712.

[6] J. Behrndt, F. Gesztesy, and S. Nakamura, Spectral shift functions and Dirichlet-to-Neumann maps, *Math. Ann.*, DOI 10.1007/s00208-017-1593-4.

[7] J. Behrndt and M. Langer, Boundary value problems for elliptic partial differential operators on bounded domains, *J. Funct. Anal.* **243** (2007), 536–565.

[8] J. Behrndt and M. Langer, Elliptic operators, Dirichlet-to-Neumann maps and quasi boundary triples, in: *Operator Methods for Boundary Value Problems*, London Math. Soc. Lecture Note Series, Vol. **404**, 2012, pp. 121–160.

[9] J. Behrndt, M. Langer, and V. Lotoreichik, Schrödinger operators with δ and δ'-potentials supported on hypersurfaces, *Ann. Henri Poincaré* **14** (2013), 385–423.

[10] J. Behrndt, M. Langer, and V. Lotoreichik, Spectral estimates for resolvent differences of self-adjoint elliptic operators, *Integral Equations Operator Theory* **77** (2013), 1–37.

[11] J. Behrndt, M. Langer, and V. Lotoreichik, Trace formulae and singular values of resolvent power differences of self-adjoint elliptic operators, *J. London Math. Soc.* **88** (2013), 319–337.

[12] J. Behrndt, M. Langer, and V. Lotoreichik, Trace formulae for Schrödinger operators with singular interactions, in: *Functional Analysis and Operator Theory for Quantum Physics*, J. Dittrich, H. Kovarik, and A. Laptev (eds.), EMS Publishing House, EMS, ETH–Zürich, Switzerland, 2017, 129–152.

[13] J. Behrndt, M.M. Malamud, and H. Neidhardt, Scattering matrices and Dirichlet-to-Neumann maps, *J. Funct. Anal.* **273** (2017), 1970–2025.

[14] M.Sh. Birman and A.B. Pushnitski, Spectral shift function, amazing and multifaceted, *Integral Equations Operator Theory* **30** (1998), 191–199.

[15] M.Sh. Birman and M.Z. Solomjak, *Spectral Theory of Self-Adjoint Operators in Hilbert Spaces*, D. Reidel Publishing Co., Dordrecht, 1987.

[16] M.Sh. Birman and D.R. Yafaev, The spectral shift function. The papers of M.G. Krein and their further development, *Algebra i Analiz* **4** (1992), no. 5, 1–44; translation in *St. Petersburg Math. J.* **4** (1993), no. 5, 833–870.

[17] M.Sh. Birman and D.R. Yafaev, Spectral properties of the scattering matrix, *Algebra i Analiz* **4** (1992), no. 6, 1–27; translation in *St. Petersburg Math. J.* **4** (1993), no. 6, 1055–1079.

[18] J.F. Brasche, P. Exner, Yu.A. Kuperin, and P. Seba, Schrödinger operators with singular interactions, *J. Math. Anal. Appl.* **184** (1994), 112–139.

[19] J. Brüning, V. Geyler, and K. Pankrashkin, Spectra of self-adjoint extensions and applications to solvable Schrödinger operators, *Rev. Math. Phys.* **20** (2008), 1–70.

[20] V.A. Derkach and M.M. Malamud, Generalized resolvents and the boundary value problems for Hermitian operators with gaps, *J. Funct. Anal.* **95** (1991), 1–95.

[21] V.A. Derkach and M.M. Malamud, The extension theory of Hermitian operators and the moment problem, *J. Math. Sci. (NY)* **73** (1995), 141–242.

[22] P. Exner, Leaky quantum graphs: a review, *Proc. Symp. Pure Math.* **77** (2008), 523–564.

[23] P. Exner and T. Ichinose, Geometrically induced spectrum in curved leaky wires, *J. Phys. A* **34** (2001), 1439–1450.

[24] P. Exner and S. Kondej, Bound states due to a strong δ interaction supported by a curved surface, *J. Phys. A* **36** (2003), 443–457.

[25] P. Exner and S. Kondej, Scattering by local deformations of a straight leaky wire, *J. Phys. A* **38** (2005), 4865–4874.

[26] P. Exner and H. Kovařík, *Quantum Waveguides*, Springer, Cham, 2015.

[27] P. Exner and K. Yoshitomi, Asymptotics of eigenvalues of the Schrödinger operator with a strong δ-interaction on a loop, *J. Geom. Phys.* **41** (2002), 344–358.

[28] F. Gesztesy, K.A. Makarov, and S.N. Naboko, The spectral shift operator, *Operator Theory Advances Applications* **108**, Birkhäuser, 1999, 59–90.

[29] V.I. Gorbachuk and M.L. Gorbachuk, *Boundary Value Problems for Operator Differential Equations*, Kluwer Academic Publishers, Dordrecht, 1991.

[30] T. Kato, *Perturbation Theory for Linear Operators*, Grundlehren der mathematischen Wissenschaften, Vol. **132**, corr. printing of the 2nd ed., Springer, Berlin, 1980.

[31] M.G. Krein, On the trace formula in perturbation theory, *Mat. Sbornik* **33** (1953), 597–626.

[32] M.G. Krein, On perturbation determinants and a trace formula for unitary and self-adjoint operators, *Dokl. Akad. Nauk SSSR* **144** (1962), 268–271.

[33] I.M. Lifshits, On a problem of the theory of perturbations connected with quantum statistics, *Uspehi Matem. Nauk* **7** (1952), 171–180.

[34] I.M. Lifšic, Some problems of the dynamic theory of nonideal crystal lattices, *Nuovo Cimento Suppl.* **3** (Ser. X) (1956), 716–734.

[35] A. Mantile, A. Posilicano, and M. Sini, Self-adjoint elliptic operators with boundary conditions on not closed hypersurfaces, *J. Diff. Eq.* **261** (2016), 1–55.

[36] W. McLean, *Strongly Elliptic Systems and Boundary Integral Equations*, Cambridge University Press, Cambridge, 2000.

[37] K. Schmüdgen, *Unbounded Self-Adjoint Operators on Hilbert Space*, Springer, Dordrecht, 2012.

[38] D. R. Yafaev, *Mathematical Scattering Theory. General Theory*, Translations of Mathematical Monographs, Vol. **105**. Amer. Math. Soc., Providence, RI, 1992.

[39] D.R. Yafaev, A trace formula for the Dirac operator, *Bull. London Math. Soc.* **37** (2005), 908–918.

[40] D.R. Yafaev, *Mathematical Scattering Theory. Analytic Theory*, Mathematical Surveys and Monographs, Vol. **158**, Amer. Math. Soc., Providence, RI, 2010.

Jussi Behrndt
Institut für Angewandte Mathematik, Technische Universität Graz
Steyrergasse 30, 8010 Graz, Austria
e-mail: behrndt@tugraz.at
URL: http://www.math.tugraz.at/~behrndt/

Fritz Gesztesy
Department of Mathematics Baylor University
One Bear Place #97328, Waco, TX 76798-7328, USA
e-mail: Fritz_Gesztesy@baylor.edu
URL: http://www.baylor.edu/math/index.php?id=935340

Shu Nakamura
Graduate School of Mathematical Sciences, University of Tokyo
3-8-1, Komaba, Meguro-ku, Tokyo, Japan 153-8914
e-mail: shu@ms.u-tokyo.ac.jp
URL: http://www.ms.u-tokyo.ac.jp/~shu/

Quantum graph with the Dirac operator and resonance states completeness

Irina V. Blinova and Igor Y. Popov

Abstract. Quantum graphs with the Dirac operator at the edges are considered. Resonances (quasi-eigenvalues) and resonance states are found for certain star-like graphs and graphs with loops. Completeness of the resonance states on finite subgraphs is studied. Due to use of a functional model, the problem reduces to factorization of the characteristic matrix-function. The result is compared with the corresponding completeness theorem for the Schrödinger quantum graph.

Mathematics Subject Classification (2010). Primary 81U20; Secondary 46N50.

Keywords. Spectrum, resonance, completeness.

1. Introduction

The problems of resonances and resonance states have a long history. The oldest one in the field is the problem of resonance state description for the Helmholtz resonator posed by Rayleigh a century ago. The completeness problem is a part of this general task. Let us briefly describe it for the Helmholtz resonator. Consider a closed resonator. The Neumann (or Dirichlet) Laplacian for this domain has purely discrete spectrum with complete system of eigenfunctions. If we consider the analogous operator for the resonator coupled to the external domain through a small opening, the situation changes. Eigenvalues transform to quasi-eigenvalues (resonances) [1, 2, 3, 4]. Correspondingly, a natural question appears: is the system of quasi-eigenstates complete? This problem is related to the Sz.-Nagy functional model [5, 6, 7]. Starting with work [8], it is known that the scattering matrix is the same as the characteristic function from the functional model.

This work was partially financially supported by the Government of the Russian Federation (grant 074-U01), DFG Grant NE 1439/3-1, grant 16-11-10330 of Russian Science Foundation.

© Springer International Publishing AG, part of Springer Nature 2018
A. Böttcher et al. (eds.), *The Diversity and Beauty of Applied Operator Theory*, Operator Theory: Advances and Applications 268, https://doi.org/10.1007/978-3-319-75996-8_5

This observation allowed the establishment of many interesting relations in scattering theory. In particular, root vectors in the functional model correspond to resonance states in scattering theory. The problem of completeness of the system of root vectors is related to the factorization problem for the characteristic function. Correspondingly, one can study completeness using factorization. Particularly, for the finite-dimensional case, this approach gives one an effective completeness criterion [6].

The simplest model for an open resonator is based on a quantum graph. As for quantum mechanics, this problem is related to a particle in a quantum dot. If one assumes the Schrödinger operator at the graph edges then the model corresponds to a non-relativistic particle. If one deals with a relativistic particle, the operator should be replaced by the Dirac one (see, e.g., [9, 10] and references in [11]). The problem is finite-dimensional and we can use the above mentioned completeness criterion. In the present paper, we deal with quantum graphs with the Dirac operator at the edges. The following graphs are studied: 1) a segment attached to a line; 2) a loop attached to semi-infinite lines at two points; 3) a loop attached to a line at one point; 4) a loop attached to a line through a segment. We study the completeness of resonance states on the finite subgraphs for these systems. The corresponding non-relativistic case was considered in [12, 13, 14]

2. Model construction

2.1. Dirac operator

We consider the Dirac operator at the edges of a metric graph Γ (E is the set of edges, V is the set of vertices). The Dirac operator on the graph is defined in the conventional way (see, e.g., [15]).

Definition 2.1. The Dirac operator H on $L_2(\Gamma) \otimes \mathbb{C}^2$, where Γ is a metric graph, acts as follows:

$$H = i\hbar c \frac{d}{dx} \otimes \sigma_1 + mc^2 \otimes \sigma_3$$

where $\sigma_1 = \left(\begin{smallmatrix} 0 & 1 \\ 1 & 0 \end{smallmatrix}\right)$ and $\sigma_3 = \left(\begin{smallmatrix} 1 & 0 \\ 0 & -1 \end{smallmatrix}\right)$ are the Pauli matrices, $\hbar$ is the Planck constant, c is the speed of light, m is the particle mass. The domain is

$$\mathcal{D}(H) = \Bigg\{ \psi = \begin{pmatrix} \psi_1 \\ \psi_2 \end{pmatrix}, \psi \in W_2^1(\Gamma \setminus V) \otimes \mathbb{C}^2, \psi_1 \in AC_{\mathrm{loc}}(\Gamma),$$

$$\psi_2 \in AC_{\mathrm{loc}}(\Gamma \setminus V), \sum_j \pm \psi_2^j(v) = \frac{i\alpha}{c} \psi_1^j(v), v \in V \Bigg\}, \qquad (2.1)$$

where the summation is over all edges including the vertex v, sign "plus" is chosen for outgoing edges, sign "minus" for incoming edges, $\alpha \in \mathbb{R}$ characterizes the strength of the point-like potential at the vertex v, $AC_{\mathrm{loc}}(\Gamma)$ is the space of absolutely continuous functions, and W_2^1 is the Sobolev space.

The spectral problem reduces to the equation

$$\begin{pmatrix} mc^2 & -i\hbar c\frac{d}{dx} \\ -i\hbar c\frac{d}{dx} & -mc^2 \end{pmatrix}\begin{pmatrix}\psi_1\\ \psi_2\end{pmatrix} = \lambda\begin{pmatrix}\psi_1\\ \psi_2\end{pmatrix}$$

at the edges and the matching conditions (see (2.1)) at the graph vertices. The solution of the equation has the form

$$\begin{cases} \psi_1 = C_1e^{ikx} + C_2e^{-ikx}, \\ \psi_2 = \beta(C_1e^{ikx} - C_2e^{-ikx}). \end{cases} \tag{2.2}$$

Here and throughout below in the text, $k = \frac{\sqrt{\lambda^2-m^2c^4}}{\hbar c}$ is the wave number and $\beta = \operatorname{sign}(\lambda + mc^2)\sqrt{\frac{\lambda-mc^2}{\lambda+mc^2}}$.

2.2. Lax–Phillips approach and functional model

For our purposes, it is convenient to consider the scattering in the framework of the Lax–Phillips approach [16]. Let us briefly describe the method for the case of the simplest graph structure shown in Fig. 1.

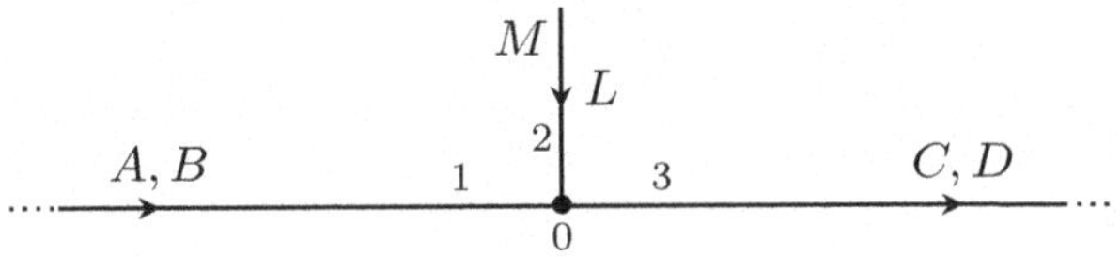

FIGURE 1. Graph structure: vertical segment as a model of a resonator. The arrows show the orientation of edges, the notations A, B, C, D, M, L will be explained below in the next section.

Consider the Cauchy problem for the time-dependent Dirac equation:

$$\begin{cases} i\hbar u_t' = Hu, \\ u(x,0) = u^0(x), \ x \in \Gamma. \end{cases} \tag{2.3}$$

Here $u(x,t) = \begin{pmatrix} u_1(x,t) \\ u_2(x,t)\end{pmatrix}$. Let $\mathcal{E}$ be the Hilbert space of two-component functions u on the graph Γ with the norm

$$\left\|\begin{pmatrix} u_1(x,t) \\ u_2(x,t)\end{pmatrix}\right\|_{\mathcal{E}}^2 = \int_\Gamma (|u_1(x,t)|^2 + |u_2(x,t)|^2)dx.$$

The solution for problem (2.3) is given by a continuous, one parameter, evolution unitary group $U(t)|_{t\in\mathbb{R}}$ of operators in $\mathcal{E}$:

$$U(t)\begin{pmatrix} u_1(x,0) \\ u_2(x,0)\end{pmatrix} = \begin{pmatrix} u_1(x,t) \\ u_2(x,t)\end{pmatrix}.$$

It is important for the Lax–Phillips scattering theory that there exist two orthogonal subspaces D_- and D_+ in $\mathcal{E}$, called, correspondingly, the incoming and outgoing subspaces, with the properties listed in the following definition.

Definition 2.2. The outgoing subspace D_+ is a subspace of $\mathcal{E}$ having the following properties:

(a) $U(t)D_+ \subset D_+$ for $t > 0$,

(b) $\cap_{t>0}U(t)D_+ = \{0\}$,

(c) $\overline{\cup_{t<0}U(t)D_+} = \mathcal{E}$.

Remark 2.3. D_- is defined analogously (with the natural replacement $t < 0$ instead of $t > 0$). The subspace D_- corresponds to incoming waves which do not interact with the target (scatterer) prior to $t = 0$ while the subspace D_+ corresponds to outgoing waves which do not interact with the target after $t = 0$.

Let P_- be the orthogonal projection of $\mathcal{E}$ onto the orthogonal complement of D_- and P_+ be the orthogonal projection of $\mathcal{E}$ onto the orthogonal complement of D_+. Consider the family $\{Z(t)\}|_{t\geq 0}$ of operators on $\mathcal{E}$ (known as the Lax–Phillips semigroup) defined by

$$Z(t) = P_+U(t)P_-, \quad t \geq 0.$$

Lax and Phillips proved the following theorem [16].

Theorem 2.4. *The operators $\{Z(t)\}|_{t\geq 0}$ annihilate D_+ and D_-, map the orthogonal complement subspace $K = \mathcal{E} \ominus (D_- \oplus D_+)$ into itself and form a strongly continuous semigroup (i.e., $Z(t_1)Z(t_2) = Z(t_1 + t_2)$ for $t_1, t_2 \geq 0$) of contraction operators on K. Furthermore, we have* $\text{s-lim}_{t\to\infty} Z(t) = 0$.

The following theorem was proved in [17].

Theorem 2.5. *If D_+ is an outgoing subspace with respect to the unitary group $\{U(t)\}_{t\in\mathbb{R}}$ defined on a Hilbert space $\mathcal{E}$, then $\mathcal{E}$ can be represented isometrically as the Hilbert space of functions $L_2(\mathbb{R}, N)$ for some Hilbert space N (called the auxiliary Hilbert space) in such a way that $U(t)$ goes to translation to the right by t units and D_+ is mapped onto $L_2(\mathbb{R}_+, N)$. This representation is unique up to an isomorphism of N.*

Such a representation is called an outgoing translation representation. Analogously, one can obtain an incoming translation representation, i.e., if D_- is an incoming subspace with respect to the group $\{U(t)\}_{t\in\mathbb{R}}$ then there is a representation in which $\mathcal{E}$ is mapped isometrically onto $L_2(\mathbb{R}, N)$, $U(t)$ goes to translation to the right by t units and D_- is mapped onto $L_2(\mathbb{R}_-, N)$.

The Lax–Phillips scattering operator $\tilde{S}$ is defined as follows. Suppose $W_+ : \mathcal{E} \to L_2(\mathbb{R}, N)$ and $W_- : \mathcal{E} \to L_2(\mathbb{R}, N)$ are the mappings of $\mathcal{E}$ onto the outgoing and incoming translation representations, respectively. The map $\tilde{S} : L_2(\mathbb{R}, N) \to L_2(\mathbb{R}, N)$ is defined by the formula

$$\tilde{S} = W_+(W_-)^{-1}.$$

Lax and Phillips proved that this definition is equivalent to the standard definition of the scattering operator. For most purposes it is more convenient to work with the Fourier transforms of the incoming and outgoing translation representations, respectively, called the incoming spectral representation and outgoing spectral representation. According to the Paley–Wiener theorem [18], in the incoming spectral representation, D_- is represented by $H^2_+(\mathbb{R}, N)$, i.e., by the space of boundary values on $\mathbb{R}$ of functions in the Hardy space $H^2(\mathbb{C}^+, N)$ of vector-valued functions (with values in N) defined in the upper half-plane $\mathbb{C}^+$. Correspondingly, the same theorem gives one a symmetric result: in the outgoing spectral representation, D_+ is represented by $H^2_-(\mathbb{R}, N)$ where $H^2_-(\mathbb{R}, N)$ is the space of boundary values on $\mathbb{R}$ of functions in the Hardy space $H^2(\mathbb{C}^-, N)$ of vector-valued functions (with values in N) defined in the lower half-plane $\mathbb{C}^-$. Accordingly, the scattering operator $\tilde{S}$ in the spectral representation is transformed to

$$S = F\tilde{S}F^{-1},$$

where F is the Fourier transform operator. The operator S is realized in the spectral representation as the operator of multiplication by the operator-valued function $S(\cdot) : \mathbb{R} \to B(N)$, where $B(N)$ is the space of all bounded linear operators on N. $S(\cdot)$ is called the Lax–Phillips S-matrix. The following theorem ([16]) presents the main properties of S.

Theorem 2.6. (a) $S(\cdot)$ *is the boundary value on* $\mathbb{R}$ *of an operator-valued function* $S(\cdot) : \mathbb{C}^+ \to B(N)$ *analytic in* $\mathbb{C}^+$.
(b) $\|S(z)\| \leq 1$ *for every* $z \in \mathbb{C}^+$.
(c) $S(E)$, $E \in \mathbb{R}$, *is, pointwise, a unitary operator on* N.

The analytic continuation of $S(\cdot)$ from the upper half-plane to the lower half-plane is constructed in a conventional manner:

$$S(z) = (S^*(\overline{z}))^{-1}, \quad \Im z < 0.$$

Thus, $S(\cdot)$ is a meromorphic operator-valued function on the whole complex plane. Let B be the generator of the semigroup $Z(t) : Z(t) = \exp i\mathsf{B}t, t > 0$. The eigenvalues of B are called resonances and the corresponding eigenvectors are the resonance states. There is a relation between the eigenvalues of B and the poles of the S-matrix. It is described in the following theorem from [16].

Theorem 2.7. *If* $\Im k < 0$, *then* k *belongs to the point spectrum of* B *if and only if* $S^*(\overline{k})$ *has a non-trivial null space.*

Remark 2.8. The theorem shows that a pole of the Lax–Phillips S-matrix at a point k in the lower half-plane is associated with an eigenvalue k of the generator of the Lax–Phillips semigroup. In other words, resonance poles of the Lax–Phillips S-matrix correspond to eigenvalues of the Lax–Phillips semigroup with well defined eigenvectors belonging to the so-called the resonance subspace $K = \mathcal{E} \ominus (D_- \oplus D_+)$.

Let us return to the problem of the Dirac quantum graph. In this case, analogously to the Schrödinger graph, one can construct $D_\pm$ and the spectral representations explicitly. Accordingly, the following lemmas take place analogously to the corresponding lemmas in [12].

Lemma 2.9. *There is a pair of isometric maps* $T_\pm : \mathcal{E} \to L_2(\mathbb{R}, \mathbb{C}^2)$ *(the outgoing and incoming spectral representations) having the following properties:*

$$T_\pm U(t) = e^{ikt} T_\pm, \quad T_+ D_+ = H_+^2(\mathbb{C}^2), \quad T_- D_- = H_-^2(\mathbb{C}^2),$$

where $H_\pm^2$ *is the Hardy space of the upper (lower) half-plane.*

Lemma 2.10. *We have*

$$T_- D_- = H_-^2(\mathbb{C}^2), \quad T_- D_+ = S H_+^2(\mathbb{C}^2), \quad T_- U(t) = e^{ikt} T_-,$$

the matrix-function S *is an inner function in* $\mathbb{C}_+$*, and*

$$K_- = T_- K = H_+^2 \ominus S H_+^2, \quad T_- Z(t)|_K = P_{K_-} e^{ikt} T_-|_{K_-}.$$

As an inner function, S can be represented in the form $S = \Pi\Theta$, where Π is a Blaschke–Potapov product and Θ is a singular inner function [5, 6, 7]. We are interested in the completeness of the system of resonance states. It is related to the factorization of the scattering matrix. The next theorem shows this relation (we use here the notations described above).

Theorem 2.11 (Completeness criterion from [6]). *The following statements are equivalent:*

1. *The operator* B *is complete;*
2. *The operator* B^* *is complete;*
3. S *is a Blaschke–Potapov product.*

Remark 2.12. The auxiliary space N in our case is $\mathbb{C}^2$. The operator is said to be complete if it has a complete set of the root vectors.

There is a simple criterion for the absence of the singular inner factor in the case $\dim N < \infty$ (in the general operator case there is no such simple criterion).

Theorem 2.13 ([6]). *Let* $\dim N < \infty$*. The following statements are equivalent:*

1. S *is a Blaschke–Potapov product;*
2. *we have*

$$\lim_{r\to 1} \int_{C_r} \ln|\det S(k)| \frac{2i}{(k+i)^2} dk = 0, \tag{2.4}$$

where C_r *is the image of* $|\zeta| = r$ *under the inverse Cayley transform.*

The integration curve can be parameterized as $C_r = \{\mathsf{R}(r)e^{it} + i\mathsf{C}(r) \mid t \in [0, 2\pi)\}$ (see (2.6) below). For brevity, we define

$$s(k) = |\det S(k)|,$$

and after throwing away constants which are irrelevant for convergence, we obtain the final form of the criterion (2.4), which is convenient for us and will be used afterwards:

$$\lim_{r\to 1} \int_0^{2\pi} \frac{\mathsf{R}(r) \ln(s(\mathsf{R}(r)e^{it} + i\mathsf{C}(r)))}{(\mathsf{R}(r)e^{it} + i\mathsf{C}(r) + i)^2} dt = 0, \tag{2.5}$$

where

$$\mathsf{C}(r) = \frac{1+r^2}{1-r^2}, \quad \mathsf{R}(r) = \frac{2r}{1-r^2}. \tag{2.6}$$

It should be noted that $\mathsf{R} \to \infty$ corresponds to $r \to 1$.

2.3. Scattering matrix

Let us return to the Dirac operator and consider a system consisting of a subgraph playing the role of the resonator and two semi-infinite wires Ω_j, Ω_k. The wave functions for Ω_j are denoted by $\psi_1^{(j)}$ and $\psi_2^{(j)}$ with the corresponding coefficients A and B, while the wave functions for Ω_k are denoted by $\psi_1^{(k)}$ and $\psi_2^{(k)}$ with the corresponding coefficients C and D. The matrix S gives us the following relation between A, B, C, D:

$$\begin{pmatrix} B \\ C \end{pmatrix} = S \begin{pmatrix} A \\ D \end{pmatrix}.$$

Let $A = 1$, $D = 0$. Then $B = R$, $C = T$. The scattering matrix has the form

$$S = \begin{pmatrix} R & T \\ T & R \end{pmatrix}.$$

3. Resonance states completeness for graphs of different structures

3.1. A line with attached segment

Consider a segment as a model of a resonator (Fig. 1). The wave function at each edge has the form

$$\begin{cases} \psi_1^{(1)} = Ae^{ikx} + Be^{-ikx}, & \psi_2^{(1)} = \beta(Ae^{ikx} - Be^{-ikx}), \\ \psi_1^{(2)} = iM \sin kx, & \psi_2^{(2)} = \beta M \cos kx, \\ \psi_1^{(3)} = Ce^{ikx} + De^{-ikx}, & \psi_2^{(3)} = \beta(Ce^{ikx} - De^{-ikx}), \\ k = \dfrac{\sqrt{\lambda^2 - m^2c^4}}{\hbar c}. \end{cases} \tag{3.1}$$

The boundary condition at the internal vertex is as follows:

$$\begin{cases} \psi_1^{(1)}(0) = \psi_1^{(2)}(L) = \psi_1^{(3)}(0), \\ -\psi_2^{(1)}(0) - \psi_2^{(2)}(L) + \psi_2^{(3)}(0) = \frac{i\alpha}{c}\psi_1^{(1)}(0). \end{cases} \tag{3.2}$$

Using (3.1) and (3.2) for the case $A = 1$, $B = R$, $C = T$, $D = 0$, one obtains

$$\begin{cases} R = \dfrac{\gamma - i\cot kL}{2 - \gamma + i\cot kL}, \\ T = \dfrac{2}{2 - \gamma + i\cot kL}. \end{cases}$$

Here and below in the text $\gamma = \frac{i\alpha}{c\beta}$. Correspondingly, one has the following expression for $s(k)$:

$$s(k) = |R^2 - T^2| = \left|\frac{2+\gamma - i\cot kL}{2-\gamma + i\cot kL}\right|.$$

If $\gamma = 0$, then $s(k) = \left|\frac{-2\sin kL + i\cos kL}{2\sin kL + i\cos kL}\right|$.

Let us prove the completeness using the criterion (2.5). We have to estimate the integral

$$\int_0^{2\pi} F(t)dt = \int_0^{2\pi} \frac{\mathsf{R}(r)\ln(s(\mathsf{R}(r)e^{it} + i\mathsf{C}(r)))}{(\mathsf{R}(r)e^{it} + i\mathsf{C}(r) + i)^2} dt.$$

Here C, R are given by (2.6), s is the function

$$s(k) = \left|\frac{(3+\gamma)e^{ix}e^{-y} - (1+\gamma)e^{-ix}e^{y}}{(3-\gamma)e^{ix}e^{-y} - (1-\gamma)e^{-ix}e^{y}}\right|,$$

where $k = x + iy$, $L = 1$, $x = \mathsf{R}\cos t$, $y = \mathsf{R}\sin t + \mathsf{C}$. The integration curve is divided into several parts. The first part is that inside a strip $0 < y < \delta$. Taking into account that at the real axis ($y = 0$) one has $s(k) = 1$, one obtains $|\ln s(\mathsf{R}e^{it} + \mathsf{C}i)| < \delta$. The length of the corresponding part of the circle is of order $\sqrt{2\mathsf{R}\delta}$. As a result, the integral over this part of the curve is $o(1/\sqrt{\mathsf{R}})$ and tends to zero if $\mathsf{R} \to \infty$.

The second part of the integral is related to the singularities of F, i.e., the roots of $s(k)$ (resonances). These values are roots of an analytic function. Correspondingly, the number of roots at the integration curve is finite. Let t_0 be the value of a parameter corresponding to a resonance. Let us take a vicinity $(t_0 - \delta_1', t_0 + \delta_1)$ such that outside it we have

$$|\ln s(\mathsf{R}e^{it} + \mathsf{C}i)| < c_1. \tag{3.3}$$

One can find such δ_1', δ_1, because if $e^{2y} > \frac{3+\gamma}{1+\gamma}$ then $s(k)$ has no roots. Let us take δ_1', δ_1 such that $e^{2y} > 4\frac{3+\gamma}{1+\gamma}$ outside the interval and, correspondingly, $|s(k)| \le c_3$.

Inside the interval, we have

$$|F| \le c_2 \mathsf{R}^{-1} \ln t.$$

The corresponding integral is estimated as

$$I_2 = \left|\int_{t_0-\delta_1'}^{t_0+\delta_1} F(t)dt\right| \le c_2 \mathsf{R}^{-1}\delta_1 \ln \delta_1.$$

On the remaining part of the integration curve we have $|F| \le c_1 \mathsf{R}^{-1}$, and the length of the integration interval is not greater than 2π.

Thus, the procedure of estimation is as follows. Choose δ_1', δ_1 to separate the root (or roots) of $s(k)$. If $t_0 - \delta_1 > 0$ then consider $(0, t_0 - \delta_1]$ separately (for the second semi-circle $\pi \le t < 2\pi$ the consideration is analogous). For this part of the curve with small t (i.e. small y), the estimate of the integral is $O(1/\sqrt{\mathsf{R}})$. For the part of the curve outside these intervals, the estimate of the integral is $O(1/\mathsf{R})$. Consequently, the full integral is estimated as $O(1/\sqrt{\mathsf{R}})$, i.e., the integral tends to zero if $\mathsf{R} \to \infty$. In accordance with the completeness criterion we come to the following theorem.

Theorem 3.1. *The system of resonance states is complete in* $L_2(\Omega_2)$.

3.2. A loop with two semi-infinite lines attached

Consider another graph-type model of a resonator - a loop with two edges of lengths L_2, L_3 with $L_2 < L_3$, and two semi-infinite wires attached (Fig. 2).

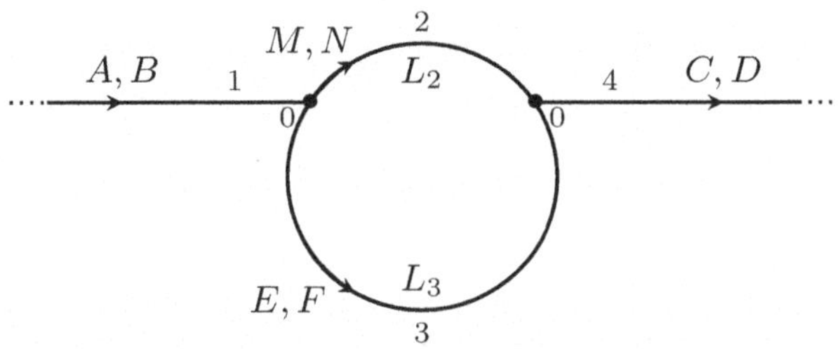

FIGURE 2. Graph structure: A loop with two semi-infinite lines attached. L_2, L_3 are the lengths of the corresponding edges, the coefficients A, B, C, D, E, F, M, N are explained in the text.

The wave function at each edge is as follows:

$$\begin{cases} \psi_1^{(1)} = Ae^{ikx} + Be^{-ikx}, & \psi_2^{(1)} = \beta(Ae^{ikx} - Be^{-ikx}), \\ \psi_1^{(2)} = Me^{ikx} + Ne^{-ikx}, & \psi_2^{(2)} = \beta(Me^{ikx} - Ne^{-ikx}), \\ \psi_1^{(3)} = Ee^{ikx} + Fe^{-ikx}, & \psi_2^{(3)} = \beta(Ee^{ikx} - Fe^{-ikx}) \\ \psi_1^{(4)} = Ce^{ikx} + De^{-ikx}, & \psi_2^{(4)} = \beta(Ce^{ikx} - De^{-ikx}), \\ k = \dfrac{\sqrt{\lambda^2 - m^2c^4}}{\hbar c}. \end{cases}$$

The boundary condition at two vertices are given by

$$\begin{cases} \psi_1^{(1)}(0) = \psi_1^{(2)}(0) = \psi_1^{(3)}(0), \\ -\psi_2^{(1)}(0) + \psi_2^{(2)}(0) + \psi_2^{(3)}(0) = \frac{i\alpha}{c}\psi_1^{(1)}(0), \\ \psi_1^{(2)}(L_2) = \psi_1^{(3)}(L_3) = \psi_1^{(4)}(0), \\ -\psi_2^{(2)}(L_2) - \psi_2^{(3)}(L_3) + \psi_2^{(4)}(0) = \frac{i\alpha}{c}\psi_1^{(4)}(0). \end{cases}$$

The reflection and the transmission coefficients are

$$\begin{cases} R = -1 - \dfrac{2\beta_1}{\beta_2^2 - \beta_1^2}, \\ T = \dfrac{2\beta_2}{\beta_2^2 - \beta_1^2}, \end{cases}$$

where $\beta_1 = i\cot kL_2 + i\cot kL_3 + 1 - \gamma$, $\beta_2 = \frac{1}{i\sin kL_2} + \frac{1}{i\sin kL_3}$. Correspondingly, the absolute value of the determinant of the matrix S is

$$s(k) = |R^2 - T^2| = \left|1 + \frac{4(\beta_1 - 1)}{\beta_2^2 - \beta_1^2}\right|.$$

Thus, we arrive at the final expression

$$s(k) = \left|\frac{-3 - 2\gamma - \gamma^2 - \beta_3 + 2i(1+\gamma)(\cot kL_2 + \cot kL_3)}{-3 + 2\gamma - \gamma^2 - \beta_3 - 2i(1-\gamma)(\cot kL_2 + \cot kL_3)}\right|,$$
$$\beta_3 = \tan\frac{kL_2}{2}\cot\frac{kL_3}{2} + \tan\frac{kL_3}{2}\cot\frac{kL_2}{2}.$$

If $\gamma = 0$, then

$$s(k) = \left|\frac{-3 - \tan\frac{kL_2}{2}\cot\frac{kL_3}{2} - \tan\frac{kL_3}{2}\cot\frac{kL_2}{2} + 2i(\cot kL_2 + \cot kL_3)}{-3 - \tan\frac{kL_2}{2}\cot\frac{kL_3}{2} - \tan\frac{kL_3}{2}\cot\frac{kL_2}{2} - 2i(\cot kL_2 + \cot kL_3)}\right|.$$

For equal edges $L_2 = L_3 = L$, one has

$$s(k) = \left|\frac{-5 + 4i\cot kL}{-5 - 4i\cot kL}\right| = \left|\frac{4i\cos kL - 5\sin kL}{-4i\cos kL - 5\sin kL}\right|.$$

The investigation of the integral from the completeness criterion is analogous to the previous section. The result is presented in the following theorem.

Theorem 3.2. *The system of resonance states is complete in* $L_2(\Omega_2 \cup \Omega_3)$.

3.3. A loop touched by a line

Consider a loop coupled to a line at one point (Fig. 3).

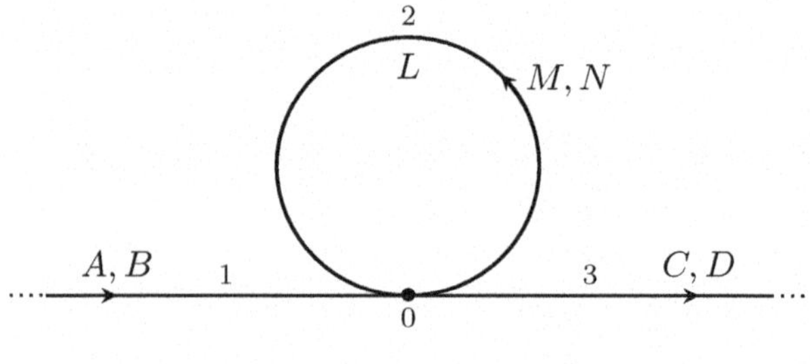

FIGURE 3. Graph structure: Loop of length L coupled to a line at one point.

The wave function at the edges has the form

$$\begin{cases} \psi_1^{(1)} = Ae^{ikx} + Be^{-ikx}, & \psi_2^{(1)} = \beta(Ae^{ikx} - Be^{-ikx}), \\ \psi_1^{(2)} = Me^{ikx} + Ne^{-ikx}, & \psi_2^{(2)} = \beta(Me^{ikx} - Ne^{-ikx}), \\ \psi_1^{(3)} = Ce^{ikx} + De^{-ikx}, & \psi_2^{(3)} = \beta(Ce^{ikx} - De^{-ikx}), \\ k = \dfrac{\sqrt{\lambda^2 - m^2c^4}}{\hbar c}. \end{cases}$$

The conditions at the vertex are as follows:

$$\begin{cases} \psi_1^{(1)}(0) = \psi_1^{(2)}(0) = \psi_1^{(2)}(L) = \psi_1^{(3)}(0), \\ -\psi_2^{(1)}(0) + \psi_2^{(2)}(0) - \psi_2^{(2)}(L) + \psi_2^{(3)}(0) = \frac{i\alpha}{c}\psi_1^{(1)}(0). \end{cases}$$

This yields the transmission and reflection coefficients

$$\begin{cases} R = \dfrac{2e^{ikL} - 2 + \gamma(1 + e^{ikL})}{4 - \gamma(1 + e^{ikL})}, \\ T = \dfrac{2(1 + e^{ikL})}{4 - \gamma(1 + e^{ikL})}. \end{cases}$$

The S-matrix determinant for this case takes the form

$$s(k) = \left| \frac{4e^{ikL} + \gamma(1 + e^{ikL})}{4 - \gamma(1 + e^{ikL})} \right|.$$

If $\gamma \neq 0$, then the integral estimation is similar to the previous section. If $\gamma = 0$, then $s(k) = |e^{ikL}|$. In this case, the result differs from the previous one. It is clear that $\ln s(k)$ has linear growth in the upper half-plane, and the corresponding integral does not tend to zero for $\mathsf{R} \to \infty$ (moreover, the presence of the singular inner factor is clear directly from the expression for $s(k)$). We so come to the following theorem.

Theorem 3.3. *If $\gamma \neq 0$, then the system of resonance states is complete in the space $L_2(\Omega_2)$, whereas if $\gamma = 0$, then the system of resonance states is not complete in $L_2(\Omega_2)$.*

3.4. A loop coupled to a line through a segment

To study the completeness/incompleteness situation in more detail, let us consider a small perturbation of the system: a graph with a connection through a segment (may be, arbitrarily small) between a loop and a line (Fig. 4).

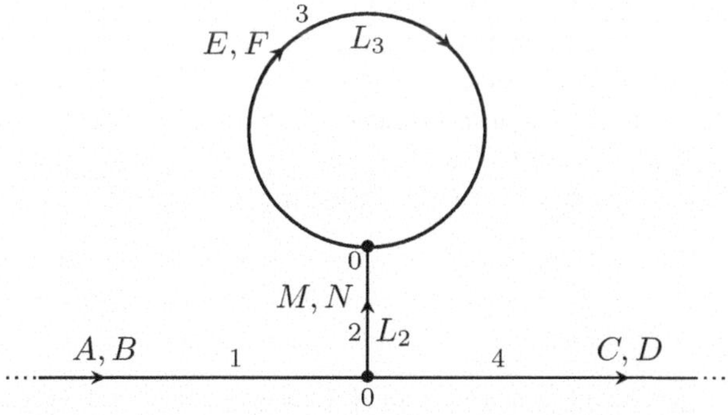

FIGURE 4. Graph structure: A loop coupled to a line through a segment.

The wave functions are

$$\begin{cases} \psi_1^{(1)} = Ae^{ikx} + Be^{-ikx}, & \psi_2^{(1)} = \beta(Ae^{ikx} - Be^{-ikx}), \\ \psi_1^{(2)} = Me^{ikx} + Ne^{-ikx}, & \psi_2^{(2)} = \beta(Me^{ikx} - Ne^{-ikx}), \\ \psi_1^{(3)} = Ee^{ikx} + Fe^{-ikx}, & \psi_2^{(3)} = \beta(Ee^{ikx} - Fe^{-ikx}), \\ \psi_1^{(4)} = Ce^{ikx} + De^{-ikx}, & \psi_2^{(4)} = \beta(Ce^{ikx} - De^{-ikx}), \\ k = \dfrac{\sqrt{\lambda^2 - m^2c^4}}{\hbar c} & \end{cases}$$

and the boundary conditions at the two graph vertices read

$$\begin{cases} \psi_1^{(1)}(0) = \psi_1^{(2)}(0) = \psi_1^{(4)}(0), \\ -\psi_2^{(1)}(0) + \psi_2^{(2)}(0) + \psi_2^{(4)}(0) = \frac{i\alpha}{c}\psi_1^{(1)}(0), \\ \psi_1^{(2)}(L_2) = \psi_1^{(3)}(0) = \psi_1^{(3)}(L_3), \\ -\psi_2^{(2)}(L_2) + \psi_2^{(3)}(0) - \psi_2^{(3)}(L_3) = \frac{i\alpha}{c}\psi_1^{(3)}(0). \end{cases}$$

This leads to the following expressions for the reflection and transmission coefficients:

$$T = 2\frac{-e^{ikL_2}(\beta_6 - 1) + e^{-ikL_2}(\beta_6 + 1)}{e^{ikL_2}(\gamma - 1)(\beta_6 - 1) + e^{-ikL_2}(3 - \gamma)(\beta_6 + 1)}, \qquad R = T - 1,$$

where $\beta_6 = 2\frac{1-e^{ikL_3}}{1+e^{ikL_3}} - \gamma$. Correspondingly,

$$s(k) = \left| \frac{e^{ikL_2}(\gamma + 3)\beta_4 - e^{-ikL_2}(\gamma + 1)\beta_5}{e^{ikL_2}(\gamma - 1)\beta_4 + e^{-ikL_2}(3 - \gamma)\beta_5} \right|$$

with $\beta_4 = 1 - 3e^{ikL_3} - \gamma(1 + e^{ikL_3})$ and $\beta_5 = 3 - e^{ikL_3} - \gamma(1 + e^{ikL_3})$. If $\gamma = 0$ then

$$s(k) = \left| \frac{3e^{ikL_2}(1 - 3e^{ikL_3}) - e^{-ikL_2}(3 - e^{ikL_3})}{-e^{ikL_2}(1 - 3e^{ikL_3}) + 3e^{-ikL_2}(3 - e^{ikL_3})} \right|.$$

If $L_2 = 0$ then one has the natural answer $s(k) = |e^{ikL_3}|$ as in the previous section.

The integral estimation is analogous to the cases considered above. We have completeness of the resonance states in $L_2(\Omega_3)$. Thus, only the case $L_2 = 0$, $\gamma = 0$ leads to incompleteness. Any perturbation (small coupling segment or point-like potential at the vertex, i.e., $\gamma \neq 0$) restores the completeness. This is summarized in the following theorem.

Theorem 3.4. *If $\gamma \neq 0$ or $L_2 \neq 0$, then the system of resonance states is complete in $L_2(\Omega_3)$. If $L_2 = 0$ and $\gamma = 0$, then the system of resonance states is not complete in $L_2(\Omega_3)$.*

Remark 3.5. The results obtained in Theorems 3.1 – 3.4 can be compared with those for Schrödinger quantum graphs [13, 14]. For each of the graphs considered, there is no difference between the completeness of the resonance states in the Dirac and Schrödinger cases.

References

[1] P. Exner, V. Lotoreichik, and M. Tater, On resonances and bound states of Smilansky Hamiltonian, *Nanosystems: Phys. Chem. Math.* **7** (2016), 789–802.

[2] A. Aslanyan, L. Parnovski, and D. Vassiliev, Complex resonances in acoustic waveguides, *Q. J. Mech. Appl. Math.* **53** (2000), 429–447.

[3] P. Duclos, P. Exner, and B. Meller, Open quantum dots: Resonances from perturbed symmetry and bound states in strong magnetic fields, *Rep. Math. Phys.* **47** (2001), 253–267.

[4] J. Edward, On the resonances of the Laplacian on waveguides, *J. Math. Anal. Appl.* **272** (2002), 89–116.

[5] B. Sz.-Nagy, C. Foias, H. Bercovici, and L. Kerchy, *Harmonic Analysis of Operators on Hilbert Space*, 2nd ed., Springer, Berlin, 2010.

[6] N. Nikol'skii, *Treatise on the Shift Operator: Spectral Function Theory*, Springer Science & Business Media, Berlin, 2012.

[7] S.V. Khrushchev, N.K. Nikol'skii, and B.S. Pavlov, Unconditional bases of exponentials and of reproducing kernels, Complex Analysis and Spectral Theory (Leningrad, 1979/1980), *Lecture Notes in Math.* **864**, 214-335, Springer, 1981.

[8] V.M. Adamyan and D.Z. Arov, On a class of scattering operators and characteristic operator-functions of contractions, *Dokl. Akad. Nauk SSSR* **160** (1965), 9–12 (in Russian).

[9] F. Gesztesy and P. Šeba, New analytically solvable models of relativistic point interactions, *Lett. Math. Phys.* **13** (1987), 345–358.

[10] S. Benvegnù and L. Dabrowski, Relativistic point interaction, *Lett. Math. Phys.* **30** (1994), 159–167.

[11] I.Y. Popov, P.A. Kurasov, S.N. Naboko, A.A. Kiselev, A.E. Ryzhkov, A.M. Yafyasov, G.P. Miroshnichenko, Yu.E. Karpeshina, V.I. Kruglov, T.F. Pankratova, and A.I. Popov, A distinguished mathematical physicist Boris S. Pavlov, *Nanosystems: Phys. Chem. Math.* **7** (2016), 782–788.

[12] I.Y. Popov and A.I. Popov, Quantum dot with attached wires: Resonant states completeness, *Rep. on Math. Phys.* **80** (2017), 1–10.

[13] D.A. Gerasimov and I.Y. Popov, Completeness of resonance states for quantum graph with two semi-infinite edges, *Complex Variables and Elliptic Equations* **62** (2017). DOI: 10.1080/17476933.2017.1289517.

[14] I.Y. Popov and A.I. Popov, Line with attached segment as a model of Helmholtz resonator: Resonant states completeness, *Journal of King Saud University - Science* **29** (2017), 133–136.

[15] G. Berkolaiko and P. Kuchment, *Introduction to Quantum Graphs*, AMS, Providence, 2012.

[16] P.D. Lax and R.S. Phillips, *Scattering Theory*, Academic Press, New York, 1967.

[17] I.P. Cornfield, S.V. Fomin and Ya.G. Sinai, *Ergodic Theory*, Springer, Berlin, 1982.

[18] R.E.A.C. Paley and N. Wiener, *Fourier Transforms in the Complex Domain*, Amer. Math. Soc. Colloq. Pub. **19**, New York, 1934.

Irina V. Blinova and Igor Y. Popov
ITMO University
Kronverkskiy, 49
St. Petersburg, 197101
Russia
e-mail: irin-a@yandex.ru
popov1955@gmail.com

Robert Sheckley's Answerer for two orthogonal projections

Albrecht Böttcher and Ilya M. Spitkovsky

Abstract. The meta theorem of this paper is that Halmos' two projections theorem is something like Robert Sheckley's Answerer: no question about the W^*- and C^*-algebras generated by two orthogonal projections will go unanswered, provided the question is not foolish. An alternative approach to questions about two orthogonal projections makes use of the supersymmetry equality introduced by Avron, Seiler, and Simon. A noteworthy insight of the paper reveals that the supersymmetric approach is nothing but Halmos in different language and hence an equivalent Answerer.

Mathematics Subject Classification (2010). Primary 47L15; Secondary 47A53, 47A60, 47B15, 47C15.

Keywords. Orthogonal projection, C^*-algebra, W^*-algebra, Drazin inverse, Fredholm operator, trace-class operator.

1. Introduction

One of the books which had a great influence on us when we just started studying Functional Analysis was Glazman and Lyubich's [12]. In particular, we always remembered Glazman's famous "And how does this look in the two-dimensional case?" question when someone was describing to him an elaborate infinite-dimensional construction, and the claim that "quite frequently this shocking question helped to better understand the gist of the matter". The topic of this paper is a striking example of the validity of Glazman's approach.

So, let us start with a pair of orthogonal projections P, Q acting on $\mathbb{C}^2$. If one of them, say P, is the zero or the identity operator, we may diagonalize Q by a unitary similarity to $\operatorname{diag}[0,0]$, $\operatorname{diag}[1,0]$, or $\operatorname{diag}[1,1]$, while P remains equal to $\operatorname{diag}[0,0]$ or $\operatorname{diag}[1,1]$ under this unitary similarity.

The second author was supported in part by Faculty Research funding from the Division of Science and Mathematics, New York University Abu Dhabi.

© Springer International Publishing AG, part of Springer Nature 2018
A. Böttcher et al. (eds.), *The Diversity and Beauty of Applied Operator Theory*, Operator Theory: Advances and Applications 268, https://doi.org/10.1007/978-3-319-75996-8_6

Thus suppose P, Q both have rank one. A unitary similarity can then be used to put P in the diagonal form diag$[1, 0]$. The matrix of Q in the respective basis is Hermitian, with zero determinant and the trace equal to one. An additional (diagonal) unitary similarity, while leaving the representation of P unchanged, allows us to make the off-diagonal entries of this matrix equal and non-negative, without changing its diagonal entries. It is thus bound to equal

$$\begin{bmatrix} x & \sqrt{x(1-x)} \\ \sqrt{x(1-x)} & 1-x \end{bmatrix}$$

with $x \in (0, 1)$ (the values $x = 0, 1$ are excluded because otherwise Q would commute with P).

2. Canonical representation

This picture extends to the general Hilbert space setting in the most natural and direct way. Namely, according to Halmos' paper [13], for a pair of orthogonal projections acting on a Hilbert space $\mathcal{H}$ there exists an orthogonal decomposition

$$\mathcal{H} = \mathcal{M}_{00} \oplus \mathcal{M}_{01} \oplus \mathcal{M}_{10} \oplus \mathcal{M}_{11} \oplus (\mathcal{M} \oplus \mathcal{M}'), \tag{1}$$

with respect to which

$$\begin{aligned} P &= I \oplus I \oplus 0 \oplus 0 \oplus \boldsymbol{W}^* \begin{bmatrix} I & 0 \\ 0 & 0 \end{bmatrix} \boldsymbol{W}, \\ Q &= I \oplus 0 \oplus I \oplus 0 \oplus \boldsymbol{W}^* \begin{bmatrix} H & \sqrt{H(I-H)} \\ \sqrt{H(I-H)} & I-H \end{bmatrix} \boldsymbol{W}, \end{aligned} \tag{2}$$

where

$$\boldsymbol{W} = \begin{bmatrix} I & 0 \\ 0 & W \end{bmatrix},$$

$W : \mathcal{M}' \longrightarrow \mathcal{M}$ is unitary, and H is the compression of Q to $\mathcal{M}$. The operator H is selfadjoint with spectrum $\sigma(H) \subset [0, 1]$ and $0, 1$ not being its eigenvalues. We refer to [6, 16] for more on the history of this representation before and after Halmos, for full proofs, and for related topics. One more proof will be given in Section 12.

Of course,

$$\begin{aligned} \mathcal{M}_{00} &= \operatorname{Im} P \cap \operatorname{Im} Q, & \mathcal{M}_{01} &= \operatorname{Im} P \cap \operatorname{Ker} Q, \\ \mathcal{M}_{10} &= \operatorname{Ker} P \cap \operatorname{Im} Q, & \mathcal{M}_{11} &= \operatorname{Ker} P \cap \operatorname{Ker} Q, \end{aligned}$$

and so

$$\mathcal{M} = \operatorname{Im} P \ominus (\mathcal{M}_{00} \oplus \mathcal{M}_{01}), \tag{3}$$

while

$$\mathcal{M}' = \operatorname{Ker} P \ominus (\mathcal{M}_{10} \oplus \mathcal{M}_{11}).$$

It is an implicit consequence of (2) that $\dim \mathcal{M}' = \dim \mathcal{M}$. In what follows, for simplicity of notation we will identify $\mathcal{M}'$ with $\mathcal{M}$ via their isomorphism W. In other words, we will drop the factors $\boldsymbol{W}, \boldsymbol{W}^*$ in (2).

The operators P and Q commute if and only if the last summand in (2) is missing, that is, $\mathcal{M}(=\mathcal{M}')=\{0\}$. This P,Q configuration is of course not very interesting, though should be accounted for. Another extreme is $\mathcal{M}_{ij}=\{0\}$ for all $i,j=0,1$. If this is the case, P and Q are said to be in the *generic position.*

3. Algebras

Based on (2), a description of the von Neumann algebra $\mathcal{A}(P,Q)$ generated by P and Q was obtained in [11]. The elements of $\mathcal{A}(P,Q)$ are all the operators of the form

$$\left(\bigoplus a_{ij}I_{\mathcal{M}_{ij}}\right)\oplus\begin{bmatrix}\phi_{00}(H) & \phi_{01}(H)\\ \phi_{10}(H) & \phi_{11}(H)\end{bmatrix}, \tag{4}$$

where $a_{ij}\in\mathbb{C}$, the direct sum in the parentheses is taken with respect to $i,j=0,1$ for which $\mathcal{M}_{ij}\neq\{0\}$ and the functions ϕ_{ij} are Borel-measurable and essentially (with respect to the spectral measure of H) bounded on $[0,1]$.

With the notation $\Phi=\begin{bmatrix}\phi_{00} & \phi_{01}\\ \phi_{10} & \phi_{11}\end{bmatrix}$, we can (and sometimes will) abbreviate (4) to

$$(\oplus a_{ij}I_{\mathcal{M}_{ij}})\oplus\Phi(H). \tag{5}$$

Invoking the spectral representation

$$H=\int_{\sigma(H)}\lambda\, dE(\lambda)$$

of H, we can also rewrite (4) as

$$\left(\bigoplus a_{ij}I_{\mathcal{M}_{ij}}\right)\oplus\int_{\sigma(H)}\Phi(\lambda)dE(\lambda).$$

The elements of the C^*-algebra $\mathcal{B}(P,Q)$ generated by P and Q are distinguished among those of the form (4) by the following [15, 21] additional properties[1]:

(i) The functions ϕ_{ij} are continuous on $[0,1]$, not just measurable;
(ii) If $0\in\sigma(H)$, then $\phi_{01}(0)=\phi_{10}(0)=0$, $a_{00}=\phi_{11}(0)$, $a_{11}=\phi_{00}(0)$;
(iii) If $1\in\sigma(H)$, then $\phi_{01}(1)=\phi_{10}(1)=0$, $a_{01}=\phi_{11}(1)$, $a_{10}=\phi_{00}(1)$.

In the finite-dimensional setting the algebras $\mathcal{A}(P,Q)$ and $\mathcal{B}(P,Q)$ of course coincide, and their elements are (up to a unitary similarity which we agreed to ignore) of the form

$$\left(\bigoplus a_{ij}I_{\mathcal{M}_{ij}}\right)\oplus\left(\bigoplus_{\lambda_j\in\sigma(H)}\Phi(\lambda_j)\right). \tag{6}$$

[1] Of course, conditions on a_{ij} below are meaningful only if the respective subspaces $\mathcal{M}_{ij}$ are non-zero.

4. The Answerer

Independently of whether $\mathcal{H}$ is finite- or infinite-dimensional, the representations (4)–(6) allow us to settle any meaningful question about operators from the algebras generated by the pair P, Q. The real challenge is to ask the right questions, and this brings us to Robert Sheckley's famous short story "Ask a foolish question", written in 1953.

In that story, we encounter an Answerer, a machine built a long time ago by a race and left back on a planet after the race disappeared. "He [the Answerer] knew the nature of things, and why things are as they are, and what they are, and what it all means. Answerer could answer anything, provided it was a legitimate question." For example, he could not give an answer to the question "Is the universe expanding?" What he replied was " 'Expansion' is a term inapplicable to the situation. Universe, as the Questioner views it, is an illusory concept." Another drastic passage in the story says "Imagine a bushman walking up to a physicist and asking him why he can't shoot his arrow into the sun. The scientist can explain it only in his own terms. What would happen?" – "The scientist wouldn't even attempt it, ... he would know the limitations of the questioner." – "How do you explain the earth's rotation to a bushman? Or better, how do you explain relativity to him, maintaining scientific rigor in your explanation at all times, of course." – "We're bushmen. But the gap is much greater here. Worm and super-man, perhaps. The worm desires to know the nature of dirt, and why there's so much of it." The quintessence of the story is that

"In order to ask a question you must already know most of the answer."

In what follows we embark on some questions about two orthogonal projections we consider as meaningful and will show what kind of answer Halmos' theorem will give.

5. Routine

Some necessary bookkeeping was performed in [18]. An explicit, though somewhat cumbersome, description was provided there for the kernels and ranges of operators $A \in \mathcal{A}(P, Q)$. Based on those, Fredholmness and invertibility criteria, formulas for spectra and essential spectra, norms, and the Moore–Penrose inverse $A^\dagger$ (when it exists) were derived.

To give a taste of these results, here is the description of $\operatorname{Ker} A$ for A given by (4). Let $\mathcal{M}^{(r)}$ be the spectral subspace of H corresponding to the subset Δ_r of $\sigma(H)$ on which $\Phi(t)$ has rank $r \in \{0, 1, 2\}$. Let also

$$\phi = \sum_{i,j=0,1} |\phi_{ij}|^2, \quad \chi_i = \sqrt{\frac{|\phi_{0i}|^2 + |\phi_{1i}|^2}{\phi}}, \quad i = 0, 1,$$

and

$$u = \exp\left(i \arg(\phi_{01}\overline{\phi_{00}} + \phi_{11}\overline{\phi_{10}})\right).$$

Then

$$\operatorname{Ker} A = \left(\bigoplus_{a_{ij}=0} \mathcal{M}_{ij} \right) \oplus \left(\mathcal{M}^{(0)} \oplus \mathcal{M}^{(0)} \right) \oplus \begin{bmatrix} u(H)\chi_1(H) \\ -\chi_0(H) \end{bmatrix} (\mathcal{M}^{(1)}). \quad (7)$$

For example, let $A = I - Q$. In its representation (4) we then have

$$a_{00} = a_{10} = 0, \quad a_{01} = a_{11} = 1, \quad (8)$$

$$\phi_{00}(t) = 1 - t, \quad \phi_{11}(t) = t, \quad \phi_{01}(t) = \phi_{10}(t) = -\sqrt{t(1-t)}.$$

Consequently,

$$\phi = 1, \quad u = -1, \quad \chi_1(t) = \sqrt{t}, \quad \chi_0(t) = \sqrt{1-t}. \quad (9)$$

Plugging (8), (9) into (7) we see that

$$\operatorname{Im} Q \; (= \operatorname{Ker}(I - Q)) = \mathcal{M}_{00} \oplus \mathcal{M}_{10} \oplus \begin{bmatrix} \sqrt{H} \\ \sqrt{I-H} \end{bmatrix} (\mathcal{M}). \quad (10)$$

Since A^* belongs to $\mathcal{A}(P, Q)$ along with A, the description of $\operatorname{Ker} A^*$ follows from (7) via a simple change of notation. The closures of $\operatorname{Im} A^*$ and $\operatorname{Im} A$ can then be obtained as the respective orthogonal complements. Note however that [18] provides the description of these ranges themselves, not just their closures. In particular, $\operatorname{Im} A$ and $\operatorname{Im} A^*$ are closed if and only if

$$\det \Phi \text{ and } \phi \text{ are separated from 0 on } \Delta_2 \text{ and } \Delta_1 \text{ respectively,} \quad (11)$$

so (11) is also a criterion for $A^\dagger$ to exist. In its turn, A is invertible if and only if $\det \Phi$ is separated from zero on the whole Δ and, in addition, $a_{ij} \neq 0$ whenever $\mathcal{M}_{ij} \neq \{0\}$.

Example. Consider $A = P - Q$. Its representation (5) has the form

$$0_{\mathcal{M}_{00}} \oplus I_{\mathcal{M}_{01}} \oplus (-I)_{\mathcal{M}_{10}} \oplus 0_{\mathcal{M}_{11}} \oplus \begin{bmatrix} I - H & -\sqrt{H(I-H)} \\ -\sqrt{H(I-H)} & H - I \end{bmatrix}, \quad (12)$$

and so the respective matrix Φ is

$$\Phi_{P-Q}(t) = \begin{bmatrix} 1 - t & -\sqrt{t(1-t)} \\ -\sqrt{t(1-t)} & t - 1 \end{bmatrix} \quad (13)$$

with the characteristic polynomial $\lambda^2 + t - 1$. It immediately follows that

$$\sigma(P - Q) = \left\{ \pm\sqrt{1-t} \colon t \in \sigma(H) \right\}, \quad (14)$$

which is a subset of $[-1, 1]$ that is symmetric about the origin, with the additional eigenvalues $1, -1$ or 0 materializing if and only if the respective subspace $\mathcal{M}_{01}$, $\mathcal{M}_{10}$, or $\mathcal{M}_{00} \oplus \mathcal{M}_{11}$ is non-trivial.

6. Anticommutators

To provide yet another example of how easily the considerations of Section 5 generate some nice formulas, we turn to the anticommutator $PQ + QP$ of P, Q. For simplicity, take P and Q in generic position. Then

$$PQ + QP - \lambda I = \begin{bmatrix} 2H - \lambda I & \sqrt{H(I-H)} \\ \sqrt{H(I-H)} & -\lambda I \end{bmatrix}. \tag{15}$$

Since the entries of the operator matrix on the right-hand side of (15) commute pairwise, according to [14, Problem 70] it is invertible only simultaneously with its formal determinant

$$\begin{aligned}\lambda^2 I - 2\lambda H - H + H^2 &= (\lambda I - H)^2 - H \\ &= (\lambda I - H + \sqrt{H})(\lambda I - H - \sqrt{H}).\end{aligned}$$

Consequently,

$$\sigma(PQ + QP) = \{\lambda \pm \sqrt{\lambda} \colon \lambda \in \sigma(H)\}.$$

In particular, $PQ+QP$ is invertible if and only if $0, 1 \notin \sigma(H)$. Note that this is always the case if $\dim \mathcal{H} < \infty$.

On the other hand, in our setting H is simply the operator PQP considered on $\operatorname{Im} P$, and thus $\sigma(H) \cup \{0\} = \sigma(PQP)$. We therefore conclude that the spectrum of the anticommutator $PQ+QP$ is the set $\{\lambda \pm \sqrt{\lambda} \colon \lambda \in \sigma(PQP)\}$ from which the origin should be removed if $0, 1 \notin \sigma(H)$. This covers the result of [9].

Moreover, since $PQ + QP$ is a positive semi-definite operator, its norm coincides with the maximum of its spectrum. Therefore,

$$\|PQ + QP\| = \max\{\lambda + \sqrt{\lambda} \colon \lambda \in \sigma(PQP)\} = \|PQP\| + \|PQP\|^{1/2}. \tag{16}$$

In its turn, $\|PQP\| = \|PQ(PQ)^*\| = \|PQ\|^2$, and (16) can be rewritten as

$$\|PQ + QP\| = \|PQ\|^2 + \|PQ\|.$$

The latter formula was the main subject of Walters' [22].

7. Drazin invertibility

Recall that an operator A acting on a Hilbert (or even a Banach) space is *Drazin invertible* if and only if the sequences $\operatorname{Im} A^j$ and $\operatorname{Ker} A^j$ stabilize. If this is the case, and k is the smallest non-negative integer for which $\operatorname{Ker} A^k = \operatorname{Ker} A^{k+1}$ and $\operatorname{Im} A^k = \operatorname{Im} A^{k+1}$, the *Drazin inverse* X of A is defined uniquely by the properties

$$A^{k+1}X = A^k, \quad XAX = X, \quad AX = XA.$$

A criterion for Drazin invertibility of operators $A \in \mathcal{A}(P, Q)$ and a formula for their Drazin inverse A^D was found in [5]. Setting $\Delta_{11} := \{t \in \Delta_1 \colon \operatorname{trace} \Phi(t) \neq 0\}$, we have that A is Drazin invertible if and only if

$$\det \Phi|\Delta_2 \text{ and } \operatorname{trace} \Phi|\Delta_{11} \text{ are separated from } 0. \tag{17}$$

Note that the first parts of conditions (11), (17) are the same, while the second requirement of (17) implies that ϕ is separated from zero on Δ_{11} though not necessarily on the whole Δ_1. So, if $\Delta_{10} := \Delta_1 \setminus \Delta_{11} \neq \emptyset$, a Drazin invertible operator A may or may not have closed range and thus be Moore–Penrose invertible or not (and, even if it is, $A^D \neq A^\dagger$). This is exactly the case when $k = 2$. On the other hand, if $\Delta_{10} = \emptyset$, then condition (17) implies (11). So, A is Moore–Penrose invertible with $A^D = A^\dagger$ and k is either zero (in which case A is invertible in the usual sense) or $k = 1$.

If A is a polynomial in P and Q, the functions $\det \Phi, \operatorname{trace} \Phi$ are also polynomial. This allows us to simplify (17) accordingly. To illustrate things, consider a linear combination $A = aP + bQ$. In that case

$$\Phi(t) = \begin{bmatrix} a+bt & b\sqrt{t(1-t)} \\ b\sqrt{t(1-t)} & b(1-t) \end{bmatrix},$$

implying $\det \Phi(t) = ab(1-t)$ and $\operatorname{trace} \Phi(t) = a + b$. So, this particular A is Drazin invertible if and only if $a = 0$ or $b = 0$ or $1 \notin \sigma(H)$. Indeed, if $a = b = 0$, then $A = 0$ is Drazin invertible. If $a = 0$ and b is different from 0, then Δ_2 is empty and the trace is separated from zero, so (17) holds. Analogously for $b = 0$ and a different from zero. Finally, if ab is different from zero, then Δ_2 is the whole spectrum with 1 deleted, and in order for $\det \Phi$ to be separated from zero on it it is necessary and sufficient that the spectrum is separated from the point 1. But this is exactly the condition that 1 is not in $\sigma(H)$. Note that in all these cases, A is also Moore–Penrose invertible. We remark that the differences $P - Q$, along with some other simple polynomials in P, Q, were treated by Deng [7], prompting the considerations of [5].

8. Compatible ranges

As in [8], we will say that an operator A acting on $\mathcal{H}$ has the *compatible range* (CoR) *property* if A and A^* coincide on $(\operatorname{Ker} A + \operatorname{Ker} A^*)^\perp$. It is easy to see (and was also observed in [8]) that all the products $P, PQ, PQP, \dots$ have this property. Those containing an odd number of factors are Hermitian, which of course implies CoR. On the other hand, the product of $n = 2k$ interlacing Ps and Qs is $A = (PQ)^k$. So, $\operatorname{Ker} A \supset \operatorname{Ker} Q$, $\operatorname{Ker} A^* \supset \operatorname{Ker} P$, and in the notation of (2) we have $(\operatorname{Ker} A + \operatorname{Ker} A^*)^\perp = \mathcal{M}_{00}$. It remains to observe that the restrictions of both A and A^* to this subspace are equal to the identity operator.

A somewhat tedious but straightforward computation of

$$(\operatorname{Ker} A + \operatorname{Ker} A^*)^\perp$$

with the use of (7) and its analogue for A^* leads to the CoR criterion for arbitrary $A \in \mathcal{A}(P, Q)$ obtained in [19]. Namely:

The operator (4) has the CoR property if and only if $a_{ij} \in \mathbb{R}$ when $\mathcal{M}_{ij} \neq \{0\}$ and for (almost) every $t \in \Delta$ the matrix $\Phi(t)$ is either (i) Hermitian or (ii) singular but not normal.

9. A distance formula

Along with P, Q, let us introduce the involution $U = 2Q - I$. If R is an orthogonal projection, then, following [23], UR is called the symmetry of R (with respect to U) and R is said to be *orthogonal to its symmetry* if $RUR = 0$. Denote by $\mathcal{Q}_U$ the set of all orthogonal projections R satisfying the orthogonality equation $RUR = 0$.

It was shown in [23] that if P is "nearly orthogonal to its symmetry" (quantitatively, $x := \|PUP\| < \xi \approx 0.455$), then

$$\operatorname{dist}(P, \mathcal{Q}_U) \leq \frac{1}{2}x + 4x^2. \tag{18}$$

In fact, concentrating on $\mathcal{Q}_U^0 := \mathcal{Q}_U \cap \mathcal{A}(P, Q)$ and computing the norms along the lines of Section 5 we arrive at the following result established in [20]: if $\mathcal{M}_{00} = \mathcal{M}_{01} = \{0\}$ in (1), then

$$\operatorname{dist}(P, \mathcal{Q}_U^0) = \sqrt{\frac{1}{2}\left(1 - \sqrt{1 - x^2}\right)} = \frac{1}{2}x + \frac{1}{16}x^3 + \cdots, \tag{19}$$

and $\operatorname{dist}(P, \mathcal{Q}_U^0) = 1$ otherwise. Note that the latter case is only possible if $\|PUP\| = 1$ and note also that there are no a priori restrictions on $\|PUP\|$ in order for (19) to hold.

The distance (19) is actually attained and, if $\|PUP\| < 1$, the respective element of $\mathcal{Q}_U^0$ lies in $\mathcal{B}(P, Q)$.

10. Index and trace

According to [3], (P, Q) is a *Fredholm pair* if the operator

$$C := QP\colon \operatorname{Im} P \longrightarrow \operatorname{Im} Q \tag{20}$$

is Fredholm, and the index $\operatorname{ind}(P, Q)$ of the pair (P, Q) is by definition the index of C. Using (3) and (10), we can rewrite (20) in a more detailed form:

$$C\colon \mathcal{M}_{00} \oplus \mathcal{M}_{01} \oplus \mathcal{M} \longrightarrow \mathcal{M}_{00} \oplus \mathcal{M}_{10} \oplus \mathcal{N},$$

where $\mathcal{N} = \begin{bmatrix} \sqrt{H} \\ \sqrt{I - H} \end{bmatrix} (\mathcal{M})$.

Now observe that C acts as the identity on $\mathcal{M}_{00}$, the zero on $\mathcal{M}_{01}$, while its action on $\mathcal{M}$ is the composition of the unitary operator

$$\begin{bmatrix} \sqrt{H} \\ \sqrt{I - H} \end{bmatrix} : \mathcal{M} \to \mathcal{N}$$

with $\operatorname{diag}[\sqrt{H}, \sqrt{H}]$. We conclude that $\operatorname{Ker} C = \mathcal{M}_{01}$ while $\operatorname{Im} C$ is the orthogonal sum of $\mathcal{M}_{00}$ with a dense subspace of $\mathcal{N}$ which is closed if and only if the operator H is invertible. In particular, $(\operatorname{Im} C)^\perp = \mathcal{M}_{10}$.

So, the pair (P, Q) is Fredholm if and only if $\mathcal{M}_{01}, \mathcal{M}_{10}$ are finite-dimensional and H is invertible. Moreover, if these conditions hold, then

$$\operatorname{ind}(P, Q) = \dim \mathcal{M}_{01} - \dim \mathcal{M}_{10}.$$

This result can be recast in terms of the difference $P-Q$. Namely, the operator H is invertible if and only if ± 1 are at most isolated points of $\sigma(P-Q)$ (see formula (14) and the explanations following it), while $\mathcal{M}_{01}$ and $\mathcal{M}_{10}$ are simply the eigenspaces of $P-Q$ corresponding to ± 1, due to (12). We thus arrive at Proposition 3.1 of [3], which says that the pair (P,Q) is Fredholm if and only if ± 1 are (at most) isolated points of $\sigma(P-Q)$ having finite multiplicity and that under these conditions

$$\operatorname{ind}(P,Q) = \dim \operatorname{Ker}(P-Q-I) - \dim \operatorname{Ker}(P-Q+I). \tag{21}$$

Because $\sigma(P-Q) \subset [-1,1]$, we see in particular that if P,Q are in generic position, then the pair is Fredholm if and only if

$$\|P-Q\| < 1, \tag{22}$$

and then $\operatorname{ind}(P,Q) = 0$. This was pointed out in [1].

Let us now consider powers of $P-Q$. Since (13) may be rewritten as

$$\Phi_{P-Q} = \sqrt{1-t}\begin{bmatrix} \sqrt{1-t} & -\sqrt{t} \\ -\sqrt{t} & -\sqrt{1-t} \end{bmatrix},$$

with the matrix factor on the right-hand side being an involution, it is easy to see that (12) implies that, for every even $k = 2n$,

$$(P-Q)^k = 0_{\mathcal{M}_{00}} \oplus I_{\mathcal{M}_{01}} \oplus I_{\mathcal{M}_{10}} \oplus 0_{\mathcal{M}_{11}} \oplus \operatorname{diag}[(I-H)^n, (I-H)^n].$$

Consequently, for odd powers $k = 2n+1$,

$$(P-Q)^k = 0_{\mathcal{M}_{00}} \oplus I_{\mathcal{M}_{01}} \oplus (-I)_{\mathcal{M}_{10}} \oplus 0_{\mathcal{M}_{11}} \oplus (I-H)^{n+1/2}\begin{bmatrix} \sqrt{I-H} & -\sqrt{H} \\ -\sqrt{H} & -\sqrt{I-H} \end{bmatrix}. \tag{23}$$

Suppose now that for some m the m-th power of $P-Q$ is a trace class operator. Then $\mathcal{M}_{01}, \mathcal{M}_{10}$ are finite-dimensional, and for every $k \geq m$ the last direct summand in (23) is a zero-trace operator. We thus have

$$\operatorname{trace}(P-Q)^k = \dim \operatorname{Ker}(P-Q-I) - \dim \operatorname{Ker}(P-Q+I) \tag{24}$$

independently of k.

Note also that $(P-Q)^k$ being a trace class operator implies that $P-Q$, and therefore $I-H$, is compact. Then H, as a Fredholm operator with zero index and (by its construction) satisfying $\operatorname{Ker} H = \{0\}$ is in fact invertible. As stated above, the pair (P,Q) is thus Fredholm, and (21) holds. Comparing (21) with (24), we arrive at the formula

$$\operatorname{trace}(P-Q)^k = \operatorname{ind}(P,Q)$$

valid for any odd $k \geq m$ provided that $(P-Q)^m$ is trace class. This is [3, Theorem 4.1]. In relation to their physics applications, the results of this section are also treated in [2].

11. Intertwining

In the early 1950s, Kato (unpublished) found a unitary operator U satisfying $UP = QU$ provided that (22) holds. In [3] it was established that, under the same condition (22), the unitary U can be constructed to satisfy the two equations

$$UP = QU \text{ and } UQ = PU; \tag{25}$$

we will say that such a U *intertwines* P with Q.

A necessary and sufficient condition for such U to exist is that in (1)

$$\dim \mathcal{M}_{01} = \dim \mathcal{M}_{10}, \tag{26}$$

see [24, Theorem 6]. Note that (22) implies

$$\mathcal{M}_{01} = \mathcal{M}_{10} = \{0\}, \tag{27}$$

so that (26) holds in a trivial way.

A description of all U satisfying (25) was provided in [10]. In the notation (2) it looks as follows [4]:

$$U = U_0 \oplus \begin{bmatrix} 0 & U_{10} \\ U_{01} & 0 \end{bmatrix} \oplus U_1 \oplus \mathbf{W}^* \begin{bmatrix} V & 0 \\ 0 & V \end{bmatrix} \begin{bmatrix} \sqrt{H} & \sqrt{I-H} \\ \sqrt{I-H} & -\sqrt{H} \end{bmatrix} \mathbf{W}. \tag{28}$$

Here U_j, U_{ij} are arbitrary unitary operators acting on $\mathcal{M}_{jj}$ and from $\mathcal{M}_{ji}$ onto $\mathcal{M}_{ij}$, respectively, and V is an arbitrary unitary operator acting on $\mathcal{M}$ and commuting with H.

Invoking (4), it was also observed in [4] that operators U intertwining P and Q can be chosen in $\mathcal{A}(P,Q)$ only if instead of (26) the stronger condition (27) is imposed. All such operators U are then given by

$$U = a_0 I_{\mathcal{M}_{00}} \oplus a_1 I_{\mathcal{M}_{11}} \oplus \mathbf{W}^* \begin{bmatrix} \phi(H) & 0 \\ 0 & \phi(H) \end{bmatrix} \begin{bmatrix} \sqrt{H} & \sqrt{I-H} \\ \sqrt{I-H} & -\sqrt{H} \end{bmatrix} \mathbf{W}, \tag{29}$$

where $|a_0| = |a_1| = 1$ and ϕ is a Borel-measurable unimodular function on $[0,1]$.

In its turn, such U lie in $\mathcal{B}(P,Q)$ if and only if the unimodular function ϕ is continuous on $[0,1]$, not just measurable. Finally, if the pair P,Q is in generic position and the spectrum of PQP is simple, then all operators satisfying (25) lie in $\mathcal{A}(P,Q)$.

12. The supersymmetric approach

The pertinent results of Sections 10 and 11 were obtained in [2, 3] solely based on the simple (and directly verifiable) observation that for any two orthogonal projections P,Q the (selfadjoint) operators

$$A = P - Q, \quad B = I - P - Q \tag{30}$$

satisfy

$$A^2 + B^2 = I, \quad AB + BA = 0. \tag{31}$$

Because of the second formula in (31), it is natural to speak of the supersymmetric approach.

The approach of [1, 9, 10] and [24] was geometrical, using either (2) or its equivalents. In [17] a point was made to derive the existence criterion for the intertwining unitary U via the supersymmetric approach. For the description of all such U, this was done in [4, Section 4]. Here we would like to show how the reasoning of the latter, with some modifications, can be used to derive Halmos' canonical representation (1),(2) for

$$P = \frac{1}{2}(I + A - B), \quad Q = \frac{1}{2}(I - A - B) \tag{32}$$

directly from (31).

The first of the formulas (31) implies that the restriction of B to the eigenspaces of A corresponding to the eigenvalues ± 1 equals zero. Denote these eigenspaces by $\mathcal{M}_{01}$ and $\mathcal{M}_{10}$, respectively, and let $\langle \cdot, \cdot \rangle$ be the scalar product in $\mathcal{H}$. Then for a unit vector $x \in \mathcal{M}_{01}$ we have $\langle (P-Q)x, x\rangle = 1$. But both $\langle Px, x\rangle$ and $\langle Qx, x\rangle$ take their values in $[0,1]$, leaving us with the only option $\langle Px, x\rangle = 1$, $\langle Qx, x\rangle = 0$. This in turn implies $Px = x$ and $Qx = 0$, i.e., $P|\mathcal{M}_{01} = I$, $Q|\mathcal{M}_{01} = 0$. Similarly, $P|\mathcal{M}_{10} = 0$, $Q|\mathcal{M}_{10} = I$. This agrees with (2) and allows us to consider now the restrictions of A, B to the orthogonal complement $\mathcal{H}'$ of $\mathcal{M}_{01} \oplus \mathcal{M}_{10}$.

Representing $\mathcal{H}'$ as the orthogonal sum of $\operatorname{Ker} A$ and the spectral subspaces of A corresponding to the positive (resp., negative) parts of its spectrum, we can write $A|\mathcal{H}'$ and $B|\mathcal{H}'$ as $A' = \operatorname{diag}[0, A_+, -A_-]$ and

$$B' = \begin{bmatrix} B_{00} & B_{01} & B_{02} \\ B_{01}^* & B_{11} & B_{12} \\ B_{02}^* & B_{12}^* & B_{22} \end{bmatrix},$$

with $A_\pm$ being positive definite operators.

Since formulas (31) carry over to A', B', we have in particular

$$A_+B_{11} + B_{11}A_+ = 0.$$

Thus, the operator A_+B_{11} has zero Hermitian part, and so its spectrum is purely imaginary. Since for any two operators X, Y we have $\sigma(XY) \cup \{0\}$ and $\sigma(YX) \cup \{0\}$ being equal, the spectrum of $A_+^{1/2}B_{11}A^{1/2}$ also is purely imaginary. On the other hand, the latter operator is selfadjoint, and hence its spectrum is real. Combining these two observations we conclude that the selfadjoint operator $A_+^{1/2}B_{11}A^{1/2}$ has zero spectrum and thus itself is zero. From the injectivity of A_+ we conclude that $B_{11} = 0$. Similarly,

$$A_-B_{22} + B_{22}A_- = 0$$

implies that $B_{22} = 0$.

With these simplifications in mind, the second part of (31) is now equivalent to

$$B_{01}A_+ = 0, \quad B_{02}A_- = 0, \tag{33}$$

and

$$A_+B_{12} = B_{12}A_-. \tag{34}$$

Invoking the injectivity of $A_\pm$ again, we see from (33) that the blocks B_{01}, B_{02} are also equal to zero, and so B' takes the form

$$B_{00} \oplus \begin{bmatrix} 0 & B_{12} \\ B_{12}^* & 0 \end{bmatrix}.$$

In particular, $\operatorname{Ker} A$ is an invariant subspace of B. According to the first formula in (31), the restriction B_{00} of B to $\operatorname{Ker} A$ is a (selfadjoint) involution. Consequently, $\operatorname{Ker} A$ splits into the orthogonal sum of the eigenspaces of B corresponding to the eigenvalues ± 1. Denoting them by $\mathcal{M}_{00}$ and $\mathcal{M}_{11}$ and using (32), we find ourselves in agreement with (2) again.

With a slight abuse of notation, we are now left with the following. Let A, B be given by

$$A = \begin{bmatrix} A_+ & 0 \\ 0 & -A_- \end{bmatrix}, \quad B = \begin{bmatrix} 0 & B_{12} \\ B_{12}^* & 0 \end{bmatrix},$$

with $A_\pm$ positive definite and not having 1 as an eigenvalue. Let also (34) hold and suppose

$$A_+^2 + B_{12}B_{12}^* = I, \quad A_-^2 + B_{12}^*B_{12} = I. \tag{35}$$

Our task is to show that then the pair (32) is in generic position and admits the respective representation (2).

Of course, (35) is simply the first condition in (31) written block-wise.

Since 1 is not an eigenvalue of $A_\pm$, equalities (35) imply that B_{12} has zero kernel and dense range. In its polar representation

$$B_{12} = CV, \quad C = \sqrt{B_{12}^*B_{12}}$$

the operator V is an isometry between the domains of $A_\pm$, implying in particular that these domains have equal dimensions. The unitary similarity $\operatorname{diag}[I, V]$ allows us to replace the pair (A, B) by

$$\begin{bmatrix} A_+ & 0 \\ 0 & -VA_-V^* \end{bmatrix}, \quad B = \begin{bmatrix} 0 & C \\ C & 0 \end{bmatrix},$$

for which (35) turns into $A_+^2 + C^2 = (VA_-V^*)^2 + C^2 = I$.

But A_+ and VA_-V^* are both positive definite. So, the latter equality defines them uniquely as

$$A_+ = VA_-V^* = \sqrt{I - C^2} := S.$$

We have thus found a unitary similarity under which P, Q become

$$P = \frac{1}{2}\begin{bmatrix} I+S & -C \\ -C & I-S \end{bmatrix}, \quad Q = \frac{1}{2}\begin{bmatrix} I-S & -C \\ -C & I+S \end{bmatrix}. \tag{36}$$

A side note: the representation (36), being more "balanced", has some advantages over the generic portion of (2). In particular, the existence of an intertwining U becomes obvious: the permutation $\left[\begin{smallmatrix} 0 & I \\ I & 0 \end{smallmatrix}\right]$ does the job.

For the task at hand, however, one more unitary similarity is needed, one which reduces P from (36) to the form $\mathrm{diag}[I, 0]$. To this end, let us introduce the selfadjoint involution

$$J = \frac{\sqrt{2}}{2}(I+S)^{-1/2}\begin{bmatrix} C & -(I+S) \\ -(I+S) & -C \end{bmatrix}.$$

A direct computation shows that then indeed $JPJ = \mathrm{diag}[I, 0]$, while

$$JQJ = \begin{bmatrix} S^2 & CS \\ CS & C^2 \end{bmatrix}.$$

It remains to relabel $C^2 = H$.

So, it is not surprising that any result pertinent to pairs of orthogonal projections can be derived from scratch just by using the purely algebraic relations (31). The supersymmetric approach is an Answerer that can rival with Halmos' theorem.

References

[1] W.O. Amrein and K.B. Sinha, On pairs of projections in a Hilbert space, *Linear Algebra Appl.* **208/209** (1994), 425–435.

[2] J.E. Avron, R. Seiler, and B. Simon, Charge deficency, charge transport and comparison of dimensions, *Comm. Math. Phys.* **159** (1994), 399–422.

[3] J.E. Avron, R. Seiler, and B. Simon, The index of a pair of projections, *J. Functional Analysis* **120** (1994), 220–237.

[4] A. Böttcher, B. Simon, and I. Spitkovsky, Similarity between two projections, *Integral Equations and Operator Theory* **89** (2017), 507–518.

[5] A. Böttcher and I. Spitkovsky, Drazin inversion in the von Neumann algebra generated by two orthogonal projections, *J. Math. Anal. Appl.* **358** (2009), 403–409.

[6] A. Böttcher and I. M. Spitkovsky, A gentle guide to the basics of two projections theory, *Linear Algebra Appl.* **432** (2010), 1412–1459.

[7] C.Y. Deng, The Drazin inverses of products and differences of orthogonal projections, *J. Math. Anal. Appl.* **335** (2007), 64–171.

[8] M.S. Djikić, Operators with compatible ranges, *Filomat* **31** (2017), 4579–4585.

[9] Y.-N. Dou, H.-K. Du, and Y.Q. Wang, Spectra of anticommutator for two orthogonal projections, arXiv.math.SP/1705.05866v1, 1–5, 2017.

[10] Y.-N. Dou, W.-J. Shi, M.-M. Cui, and H.-K. Du, General explicit descriptions for intertwining operators and direct rotations of two orthogonal projections, *Linear Algebra Appl.* **531** (2017), 575–591.

[11] R. Giles and H. Kummer, A matrix representation of a pair of projections in a Hilbert space, *Canad. Math. Bull.* **14** (1971), 35–44.

[12] I.M. Glazman and Yu.I. Lyubich, *Finite-Dimensional Analysis in Problems* (in Russian), Nauka, Moscow, 1969.

[13] P.R. Halmos, Two subspaces, *Trans. Amer. Math. Soc.* **144** (1969), 381–389.

[14] P.R. Halmos, *A Hilbert Space Problem Book,* 2nd ed., Graduate Texts in Mathematics **19**, Encyclopedia of Mathematics and its Applications **17**, Springer-Verlag, New York and Berlin, 1982.

[15] G.K. Pedersen, Measure theory for C^*-algebras. II, *Math. Scand.* **22** (1968), 63–74.

[16] S. Roch, P.A. Santos, and B. Silbermann, *Non-Commutative Gelfand Theories,* Springer-Verlag, London, 2011.

[17] B. Simon, Unitaries permuting two orthogonal projections, *Linear Algebra Appl.* **528** (2017), 436–441.

[18] I.M. Spitkovsky, Once more on algebras generated by two projections, *Linear Algebra Appl.* **208/209** (1994), 377–395.

[19] I.M. Spitkovsky, Operators with compatible ranges in an algebra generated by two orthogonal projections, *Advances in Operator Theory* **3** (2018), 117–122.

[20] I.M. Spitkovsky, A distance formula related to a family of projections orthogonal to their symmetries, *Operator Theory: Advances and Applications* (to appear).

[21] N. Vasilevsky and I. Spitkovsky, On the algebra generated by two projections (in Russian), *Doklady Akad. Nauk Ukrain. SSR, Ser. A* **8** (1981), 10–13.

[22] S. Walters, Anticommutator norm formula for projection operators, arXiv. math.FA/1604.00699v1, 1–9, 2016.

[23] S. Walters, Projection operators nearly orthogonal to their symmetries, *J. Math. Anal. Appl.* **446** (2017), 1356–1361.

[24] Y.Q. Wang, H.K. Du, and Y.N. Dou, On the index of Fredholm pairs of idempotents, *Acta Math. Sin. (Engl. Ser.)* **25** (2009), 679–686.

Albrecht Böttcher
Fakultät für Mathematik
TU Chemnitz
D-09107 Chemnitz
Germany
e-mail: `aboettch@math.tu-chemnitz.de`

Ilya M. Spitkovsky
Division of Science
New York University Abu Dhabi (NYUAD)
Saadiyat Island
P.O. Box 129188 Abu Dhabi
UAE
e-mail: `ims2@nyu.edu, imspitkovsky@gmail.com`

Toeplitz kernels and model spaces

M. Cristina Câmara and Jonathan R. Partington

Abstract. We review some classical and more recent results concerning kernels of Toeplitz operators and their relations with model spaces, which are themselves Toeplitz kernels of a special kind. We highlight the fundamental role played by the existence of maximal vectors for every nontrivial Toeplitz kernel.

Mathematics Subject Classification (2010). Primary 47B35; Secondary 30H10.

Keywords. Toeplitz kernel, model space, nearly-invariant subspace, minimal kernel, multiplier, Carleson measure.

1. Introduction

We shall mostly be discussing Toeplitz operators on the familiar Hardy space $H^2 = H^2(\mathbb{D})$ of the unit disc $\mathbb{D}$, which embeds isometrically as a closed subspace of $L^2(\mathbb{T})$, where $\mathbb{T}$ is the unit circle, by means of non-tangential limits. These are standard facts that can be found in many places, such as [14, 30].

For a symbol $g \in L^\infty(\mathbb{T})$ the Toeplitz operator $T_g : H^2 \to H^2$ is defined by

$$T_g f = P_{H^2}(g \cdot f) \qquad (f \in H^2),$$

where P_{H^2} denotes the orthogonal projection from $L^2(\mathbb{T})$ onto H^2.

Similarly we may define Toeplitz operators on the Hardy space $H^2(\mathbb{C}^+)$ of the upper half-plane, which embeds as a closed subspace of $L^2(\mathbb{R})$, and we shall use the same notation, since the context should always be clear, writing

$$T_g f = P_{H^2(\mathbb{C}^+)}(g \cdot f) \qquad (f \in H^2(\mathbb{C}^+)),$$

where $P_{H^2(\mathbb{C}^+)}$ is the orthogonal projection from $L^2(\mathbb{R})$ onto $H^2(\mathbb{C}^+)$.

The kernels of such operators have been a subject of serious study for at least fifty years, and one particular example here is the class of model spaces. Let $\theta \in H^\infty = H^\infty(\mathbb{D})$ be an inner function, that is $|\theta(t)| = 1$ almost

This work was partially supported by FCT/Portugal through the grant UID/MAT/04459/2013.

© Springer International Publishing AG, part of Springer Nature 2018

A. Böttcher et al. (eds.), *The Diversity and Beauty of Applied Operator Theory*, Operator Theory: Advances and Applications 268, https://doi.org/10.1007/978-3-319-75996-8_7

everywhere on $\mathbb{T}$, and consider the Toeplitz operator $T_{\overline{\theta}}$. It is easily verified that its kernel is the space

$$K_\theta := H^2 \ominus \theta H^2 = H^2 \cap \theta \overline{H_0^2},$$

where $\overline{H_0^2}$ denotes the orthogonal complement of H^2 in $L^2(\mathbb{R})$. It follows from Beurling's theorem that these spaces K_θ are the nontrivial closed invariant subspaces of the backward shift operator $S^* = T_{\bar{z}}$, defined by

$$S^* f(z) = \frac{f(z) - f(0)}{z} \qquad (f \in H^2, \ z \in \mathbb{D}).$$

They include the spaces of polynomials of degree at most n for $n = 0, 1, 2, \ldots$ (take $\theta(z) = z^{n+1}$), as well as the finite-dimensional spaces consisting of rational functions (each such n-dimensional space corresponds to taking θ to be a Blaschke product of degree n). For a good recent book on model spaces, see [19].

Another example, which has applications in systems and control theory, is the space corresponding to the inner function $\theta_T(s) = e^{isT}$ in $H^\infty(\mathbb{C}^+)$, for a fixed $T > 0$. For by the Paley–Wiener theorem, the Fourier transform establishes a canonical isometric isomorphism between $L^2(0, \infty)$ and $H^2(\mathbb{C}^+)$, mapping the subspace $L^2(0, T)$ onto K_{θ_T}.

As we shall now see, the class of Toeplitz kernels, which includes the class of model spaces, can itself be described in terms of model spaces. Most of the results we present are valid (with suitable modifications) in H^p for $1 < p < \infty$, as well as in Hardy spaces on the half-plane. The interested reader may refer back to the original sources.

We recall first one classical result of Coburn [10], that for $g \in L^\infty(\mathbb{T})$ not almost everywhere 0, either $\ker T_g = \{0\}$ or $\ker T_g^* = \{0\}$ (note that $T_g^* = T_{\overline{g}}$). This was proved as an intermediate step towards showing that the Weyl spectrum of a Toeplitz operator coincides with its essential spectrum.

2. Background results

2.1. The 1980s

The papers of Nakazi [29], Hayashi [22, 23], Hitt [25], and Sarason [31] were all published within a short space of time.

Nakazi's paper is mostly concerned with finite-dimensional Toeplitz kernels, but does explore the role of *rigid* functions in the context of Toeplitz kernels. He uses the term *p-strong* for an outer function $f \in H^p$ with the property that if $kf \in H^p$ for some measurable k with $k \geq 0$ a.e., then k is constant, although nowadays the term *rigid* is generally adopted. He then shows that $\dim \ker T_g = n$, a non-zero integer, if and only if $\ker T_g = uP_{n-1}$, where $u \in H^2$ with u^2 rigid, and P_{n-1} is the space of polynomials of degree at most $n-1$. Nakazi's work also bears on extremal problems and the properties of Hankel operators.

In fact, a function $f \in H^1$ with $\|f\| = 1$ is rigid if and only if it is an *exposed point* of the ball of H^1; that is, if and only if there is a functional

$\phi \in (H^1)^*$ such that $\phi(f) = \|\phi\| = \|f\| = 1$, and such that if $\phi(g) = 1$ for some g with $\|g\| = 1$, then $g = f$. Chapter 6 of [18] contains a useful discussion of this result.

Meanwhile, Hayashi [23] showed that the kernel of a Toeplitz operator T_g can be written as uK_θ, where u is outer and θ is inner with $\theta(0) = 0$, and u multiplies the model space K_θ isometrically onto $\ker T_g$. Every closed subspace M of H^2 possesses a reproducing kernel $k_w \in M$ (where $w \in \mathbb{D}$), such that $\langle f, k_w \rangle = f(w)$ for $f \in M$, and, as an application of his main result, Hayashi gave an expression for the reproducing kernel corresponding to a Toeplitz kernel, namely,

$$k_w(z) = \overline{u(w)}u(z)\frac{1 - \overline{\theta(w)}\theta(z)}{1 - \overline{w}z},$$

for $w, z \in \mathbb{D}$, where $\ker T_g = uK_\theta$. Hayashi also noted in [22] that every nontrivial Toeplitz kernel T_g is equal to $\ker T_{\overline{h}/h}$ for some outer function h, a significant simplification in the analysis of Toeplitz kernels. Moreover, in the representation uK_θ, we have that u^2 is rigid.

Hitt's work was mostly concerned with the Hardy space $H^2(\mathbb{A})$ of the annulus $\mathbb{A} = \{z \in \mathbb{C} : 1 < |z| < R\}$ for some $R > 1$, and in classifying those closed subspaces of $H^2(\mathbb{A})$ invariant under $Sf(z) = zf(z)$. To do this he made a study of subspaces M of $H^2(\mathbb{D})$ that are *nearly invariant* under the backwards shift S^*, i.e., $f \in M$ and $f(0) = 0$ implies that $S^*f \in M$. (Again, his original terminology, *weakly invariant*, has been superseded.)

It is easy to see that a Toeplitz kernel is nearly S^*-invariant, for if $f \in \ker T_g$ with $f(0) = 0$, then $gf \in \overline{H_0^2}$ and so $g(\overline{z}f) \in \overline{H_0^2}$ also, with $\overline{z}f \in H^2$, which means that $\overline{z}f \in \ker T_g$ too. Indeed, a similar argument shows that we may divide out each inner factor while remaining in the kernel.

Thus Hitt proved the following result.

Theorem 2.1. *The nearly S^*-invariant subspaces have the form $M = uK$, with $u \in M$ of unit norm, $u(0) > 0$, and u orthogonal to all elements of M vanishing at the origin, K an S^*-invariant subspace, and the operator of multiplication by u is isometric from K into H^2.*

Note that K may be H^2 itself, as for example θH^2 is nearly S^*-invariant if θ is an inner function with $\theta(0) \neq 0$. This case is often overlooked, but these spaces θH^2 are not Toeplitz kernels, since they are not invariant under dividing by θ. The case we are most interested in is $K = K_\theta$, with θ inner.

The link with $H^2(\mathbb{A})$ is that if M is an invariant subspace of $H^2(\mathbb{A})$, then under the change of variable $s = 1/z$, the subspace $M \cap H^2(\mathbb{C} \setminus \overline{\mathbb{D}})$ corresponds to a nearly S^*-invariant subspace.

Sarason gave a new proof of Hitt's theorem using the de Branges–Rovnyak spaces studied in [12]. He further showed that the inner function θ in the representation $\ker T_g = uK_\theta$ divides $(F - 1)/(F + 1)$, where F is the Herglotz integral of $|u|^2$.

2.2. The 1990s

Hayashi [24] and Sarason [32] continued to examine the nearly S^*-invariant subspaces which are kernels of Toeplitz operators.

Hayashi gave a complete characterization of such uK_θ, as follows. Let $u \in H^2$ be outer with $u(0) > 0$, let F be the Herglotz integral of $|u|^2$, and $b = (F-1)/(F+1)$. Let a be the outer function with $a(0) > 0$ such that a.e. $|a|^2 + |b|^2 = 1$. We have $a = 2f/(F+1)$ and $f = a/(1-b)$, and we write $u_\theta = a/(1-\theta b)$.

Theorem 2.2. *Let $M = uK_\theta$ as in Theorem 2.1. Then M is the kernel of a Toeplitz operator if and only if u is outer and $a/(1-z\overline{\theta}b))^2$ is an exposed point of the unit ball of H^1.*

Another way of writing this is to say that as follows.

Theorem 2.3. *The nontrivial kernels of Toeplitz operators are the subspaces of the form $M = u_\theta K_{z\theta}$, where θ is inner and $u \in H^2$ is outer with $u(0) > 0$ and u^2 an exposed point of the unit ball of H^1.*

Sarason gave an alternative proof of Hayashi's result, and a further discussion of rigid functions (for example the 1-dimensional Toeplitz kernels are spanned by functions u with u^2 rigid, and an outer function u is rigid if and only if $\ker T_{\overline{u}/u} = \{0\})$.

2.3. The 2000s and 2010s

Dyakonov [15] took an alternative approach to Toeplitz kernels, using Bourgain's factorization for a unimodular function ψ [1, 5], namely that there is a triple (B, b, g) such that $\psi = b\overline{g}/(Bg)$, where b and B are Blaschke products and g is an invertible element in H^∞.

As a result he showed the following result (in fact he showed a similar result in H^p for $p > 1$).

Theorem 2.4. *For every $\psi \in L^\infty \setminus \{0\}$, there exists a triple (B, b, g) such that* $\ker T_\psi = gb^{-1}(K_B \cap bH^2)$.

Then Makarov and Poltoratski [27], working in the upper half-plane $\mathbb{C}^+$, considered uniqueness sets. A Blaschke set $\Lambda \subset \mathbb{C}^+$ is said to be a uniqueness set for K_θ if every function in K_θ that vanishes on Λ vanishes identically. This property is equivalent to the injectivity property for Toeplitz operators, i.e., $\ker T_{\overline{\Theta}B} = \{0\}$, where B is the Blaschke product with zero set Λ. Using these ideas they gave a necessary and sufficient condition for the injectivity of a Toeplitz operator with the symbol $U = e^{i\gamma}$ where γ is a real-analytic real function.

Before describing more recent work, we mention the survey article of Hartmann and Mitkovski [21] and the book of Fricain and Mashreghi [18], which give good treatments of the material we have discussed above. Then the theory of model spaces and their operators (including composition operators, multipliers, restricted shifts and indeed more general truncated Toeplitz operators) forms the subject of a monograph [19].

3. Near invariance and minimal kernels

Toeplitz kernels form one of the most important classes of nearly S^*-invariant subspaces. One may look at this property as meaning that if there is an element of a Toeplitz kernel K of the form zf_+ with $f_+ \in H^2$, then $f_+ \in K$. In particular, one cannot have a one-dimensional Toeplitz kernel whose elements all vanish at 0. It is easy to see that an analogous property holds when z is replaced by the inverse of a function $\eta \in \overline{H^\infty}$, as, for instance, an inner function. More generally, if η is a complex-valued function defined a.e. on $\mathbb{T}$, we say that a proper closed subspace $\mathcal{E}$ of H^2 is *nearly η-invariant* if, for all $f_+ \in \mathcal{E}$, $\eta f_+ \in H^2$ implies that $\eta f_+ \in \mathcal{E}$. Thus, saying that $\mathcal{E}$ is nearly S^*-invariant is equivalent to saying that $\mathcal{E}$ is nearly $\overline{z}$-invariant.

It can be shown [6] that if $\eta \in H^\infty$ and η is not constant, then no finite-dimensional kernel is nearly η-invariant. However, one can characterize a vast class of functions η, besides those in $\overline{H^\infty}$, for which all Toeplitz kernels are nearly η-invariant. Let $\mathcal{N}_2$ denote the class of all such functions. We have the following.

Theorem 3.1 ([6]). *If $\eta : X \to \mathbb{C}$, measurable and defined on a set $X \subset \mathbb{T}$ such that $\mathbb{T} \setminus X$ has measure zero, satisfies*

$$L^2(\mathbb{T}) \cap \eta \overline{H_0^2} \subset \overline{H_0^2},$$

then every Toeplitz kernel is nearly η-invariant, i.e., $\eta \in \mathcal{N}_2$.

Note that the class described in this theorem is rather large, including various well-known classes of functions, not necessarily bounded [6], in particular all rational functions whose poles are in the closed disc $\overline{\mathbb{D}}$ and all functions belonging to $\overline{H_0^2}$, as for instance those in $\overline{\theta} K_\theta = \overline{z K_\theta}$ for some inner function θ. We conclude therefore that if $\ker T_g \neq \{0\}$ (with $g \in L^\infty(\mathbb{T})$), then, for each η in that class, all H^2 functions that can be obtained from $f_+ \in \ker T_g$ by factoring out η^{-1} must also belong to $\ker T_g$. This establishes some sort of "lower bound" for the Toeplitz kernel. For example, we have the following.

Theorem 3.2 ([6]). *A Toeplitz kernel that contains an element of the form $\phi_+ = R f_+$, where $f_+ \in H^2$ and $R \in H^\infty$ is a rational function of the form $R = p_1/p_2$, with p_1 and p_2 polynomials with no common zeroes, and $\deg p_1 \leq \deg p_2$, has dimension at least $d := P - Z + 1$, where P is the number of poles of R, and Z is the number of zeroes of R in the exterior of $\mathbb{D}$ (including ∞).*

As another example, we have that if an inner function θ belongs to a Toeplitz kernel K, then $K \supset K_\theta$ [6]. Thus, if θ is a singular inner function, then K must be infinite-dimensional.

These lower bounds imply that, if $f_+ \in H^2$ has a non-constant inner factor, then $\operatorname{span}\{f_+\}$ cannot be a Toeplitz kernel. On the other hand, it is easy to see that there always exists a Toeplitz kernel containing f_+, namely $\ker T_{\overline{z}\overline{f_+}/f_+}$, where the symbol is unimodular. We are thus led to the question whether there is some "smaller" Toeplitz kernel containing f_+. Or, in finite-dimensional language, is there a minimum dimension for a Toeplitz kernel containing f_+? And can there be two different Toeplitz kernels with that minimum dimension, such that f_+ is contained in both? The answer to the first question is affirmative, while the second question has a negative answer. We have the following result.

Theorem 3.3 ([6]). *Let $f_+ \in H^2 \setminus \{0\}$ and let $f_+ = IO_+$ be its inner–outer factorization. Then there exists a minimal Toeplitz kernel containing* $\operatorname{span}\{f_+\}$, *written* $\mathrm{K}_{\min}(f_+)$, *such that every Toeplitz kernel K with $f_+ \in K$ contains* $\mathrm{K}_{\min}(f_+)$, *and we have*

$$\mathrm{K}_{\min}(f_+) = \ker T_{\overline{zIO_+}/O_+}. \tag{3.1}$$

For example, given an inner function θ, every kernel containing θ must contain K_θ, as mentioned before; the minimum kernel for θ is

$$\mathrm{K}_{\min}(\theta) = \ker T_{\overline{z\theta}} = K_\theta \oplus \operatorname{span}\{\theta\} = K_\theta \oplus \theta K_z.$$

If a Toeplitz kernel is the minimal kernel for $f_+ \in H^2$, we say that f_+ is a *maximal function* or *maximal vector* for K. Since every Toeplitz kernel is the kernel of an operator $T_{\overline{zIO_+}/O_+}$ for some inner function I and outer function $O_+ \in H^2$ [32] we conclude the following.

Corollary 3.4. *Every Toeplitz kernel has a maximal function.*

Note that this implies that every Toeplitz kernel K contains an outer function, since, with the notation above, if $IO_+ \in K$, then $O_+ \in K$ by near invariance.

One may ask when $\mathrm{K}_{\min}(f_+) = \operatorname{span}\{f_+\}$, i.e., it is one-dimensional. There is a close connection between one-dimensional Toeplitz kernels in H^2 and rigid functions in H^2. It is easy to see that every rigid function is outer, and every rigid function in H^1 is the square of an outer function in H^2. We have the following.

Theorem 3.5 ([32]). *If $f_+ \in H^2 \setminus \{0\}$, then $\mathcal{E} = \operatorname{span}\{f_+\}$ is a Toeplitz kernel if and only if f_+ is outer and f_+^2 is rigid in H^2. In that case $\mathcal{E} = \ker T_{\overline{zf_+}/f_+}$.*

4. Maximal functions in model spaces

The maximal vectors for a given Toeplitz kernel can be characterized as follows.

Theorem 4.1 ([8]). *Let $g \in L^\infty \setminus \{0\}$ be such that $\ker T_g$ is nontrivial. Then k_+ is a maximal vector for $\ker T_g$ if and only if $k_+ \in H^2$ and $k_+ = g^{-1}\overline{zp_+}$, where $p_+ \in H^2$ is outer.*

Since model spaces are Toeplitz kernels ($K_\theta = \ker T_{\overline{\theta}}$), the maximal vectors are the functions $k_+ \in H^2$ of the form

$$k_+ = \theta \overline{zp_+} \qquad (p_+ \in H^2,\ \text{outer}),$$

i.e., such that $\theta\overline{z k_+}$ is an outer function. Thus, the reproducing kernel function, defined for each $w \in \mathbb{D}$ by

$$k_w^\theta(z) := \frac{1-\overline{\theta(w)}\theta(z)}{1-\overline{w}z}, \qquad (z \in \mathbb{T}),$$

is not in general a maximal vector for K_θ, since

$$\theta\overline{z k_w^\theta} = \frac{\theta - \theta(w)}{z-w},$$

which is not outer in general. On the other hand, we have that

$$\widetilde{k_w^\theta}(z) := \frac{\theta(z)-\theta(w)}{z-w}$$

is a maximal vector for K_θ, for every $w \in \mathbb{D}$.

Other maximal vectors for the model space K_θ can be found using the result that follows. We use the notation $\mathcal{G}H^\infty$ for the set of invertible elements of the algebra H^∞.

Theorem 4.2. *If f_+ is a maximal vector for* $\ker T_g$*, where $g \in L^\infty(\mathbb{T})$, then $\theta h_+^{-1} f_+$ is a maximal vector for* $\ker T_{h_-\overline{\theta}gh_+}$*, for every inner function θ and every $h_+ \in \mathcal{G}H^\infty$, $h_- \in \mathcal{G}\overline{H^\infty}$.*

Proof. From Theorem 4.1, if $\mathrm{K}_{\min}(f_+) = \ker T_g$, then $gf_+ = \overline{zp_+}$, where $p_+ \in H^2$ is outer. Therefore $\theta h_+^{-1} f_+ \in H^2$ is such that

$$h_-\overline{\theta}h_+g(\theta h_+^{-1}f_+) = h_-gf_+ = \overline{z}(h_-\overline{p_+}),$$

and using Theorem 4.1 again, we see that $\mathrm{K}_{\min}(\theta h_+^{-1}f_+) = \ker T_{h_-\overline{\theta}h_+g}$. □

If the inner function is a finite Blaschke product B, with $B(z_0) = 0$ for some $z_0 \in \mathbb{D}$, then it is easy to see from Theorem 3.3 that

$$\mathrm{K}_{\min}\left(\frac{B}{z-z_0}\right) = \ker T_{\overline{B}} = K_B.$$

Now each inner function θ can be factorized as

$$\theta = h_-Bh_+,$$

where $B = \frac{\theta - a}{1-\overline{a}\theta}$ with $|a| < 1$ is a Blaschke product and $h_- = 1+a\overline{B} \in \mathcal{G}\overline{H^\infty}$, and $h_+ = \frac{1}{1+\overline{a}B} = \overline{h_-^{-1}} \in \mathcal{G}H^\infty$ [30]; thus it follows from Theorem 4.2 that

$$\phi_+^\theta := \overline{h_-^{-1}}\frac{B}{z-z_0} = h_+\frac{B}{z-z_0} = \frac{h_-^{-1}\theta}{z-z_0} \tag{4.1}$$

is a maximal vector for $K_\theta = \ker T_{\overline{\theta}}$.

Note that, from (4.1), we can express θ in terms of these maximal vectors for K_θ, using the same notation as above:

$$\theta = (z-z_0)h_-\phi_+^\theta. \tag{4.2}$$

From Theorem 4.2, applied to Toeplitz kernels that are model spaces, we also obtain the following.

Theorem 4.3 ([9]). *Let θ and θ_1 be inner functions. If k_{1+} is a maximal vector for K_{θ_1}, then θk_{1+} is a maximal vector for $K_{\theta\theta_1} = K_{\theta_1} \oplus \theta_1 K_\theta$.*

Thus if $\mathrm{K}_{\min}(k_{1+})$ is a model space K_{θ_1}, then $\mathrm{K}_{\min}(\theta k_{1+})$ is also a model space, $K_{\theta\theta_1}$ for all inner functions θ.

More generally, one can consider the minimal kernel containing a given set of functions. In particular, when these functions are maximal vectors for model spaces, we obtain the following generalization of the previous result.

Theorem 4.4 ([9]). *Let $k_{1+}, k_{2+}, \dots, k_{n_+} \in H^2$ be maximal vectors for the spaces $K_{\theta_1}, K_{\theta_2}, \dots, K_{\theta_n}$, respectively, where every θ_j is an inner function, for $j = 1, 2, \dots, n$. Then there exists a minimal kernel containing the vectors $\{k_{j+} : j = 1, 2, \dots, n\}$, and for $\theta = \mathrm{LCM}(\theta_1, \theta_2, \dots, \theta_n)$ we have*

$$K = K_\theta = \mathrm{clos}_{H^2}(K_{\theta_1} + K_{\theta_2} + \cdots + K_{\theta_n}) = K_{\theta_j} \oplus \theta_j K_{\theta\overline{\theta_j}},$$

for each $j = 1, 2, \dots, n$.

5. On the relations between $\ker T_g$ and $\ker T_{\theta g}$

Direct sum decompositions of the form $K_{\theta\theta_1} = K_{\theta_1} \oplus \theta_1 K_\theta$ can also be expressed in terms of maximal functions, using (4.2) with θ replaced by θ_1:

$$K_{\theta\theta_1} = K_{\theta_1} \oplus (z - z_0) h_- \phi_+^{\theta_1} K_\theta. \tag{5.1}$$

For $g = \overline{\theta\theta_1}$ the identity (5.1) is equivalent to

$$\ker T_g = \ker T_{\theta g} \oplus (z - z_0) h_- \phi_+^{\theta g} K_\theta, \tag{5.2}$$

where $\phi_+^{\theta g}$ is a maximal vector for $\ker T_{\theta g}$ and $h_- = 1$ if θ is a Blaschke product with $\theta(z_0) = 0$. This relation can be extended for general $g \in L^\infty(\mathbb{T})$ when θ is a finite Blaschke product, in terms of maximal functions and model spaces.

Indeed for every $g \in L^\infty(\mathbb{T})$ and every non-constant inner function θ, we have

$$\ker T_{\theta g} \subsetneq \ker T_g,$$

whenever $\ker T_g \neq \{0\}$.

If θ is not a finite Blaschke product and $\dim \ker T_g < \infty$, then actually $\ker T_{\theta g} = \{0\}$; while, if $\ker T_g$ is infinite-dimensional, then $\ker T_{\theta g}$ may or may not be finite-dimensional, and in particular it can be $\{0\}$ — as it happens, for instance, when $\overline{g}$ is an inner function dividing θ, or in the case of the following example.

Example ([8, 9]). For $\theta(z) = \exp(\frac{z+1}{z-1})$ and $\psi(z) = \exp(\frac{z-1}{z+1})$, we have $\ker T_{\overline{z}\theta\overline{\psi}} = \{0\}$.

For finite Blaschke products θ we have the following.

Theorem 5.1 ([9]). *If $g \in L^\infty(\mathbb{T})$ and θ is a finite Blaschke product, then*

$$\dim\ker T_g < \infty \quad \textit{if and only if} \quad \dim\ker T_{\theta g} < \infty,$$

and $\ker T_g$ *is finite-dimensional if and only if there exists a* $k_0 \in \mathbb{Z}$ *such that* $\ker T_{z^{k_0}g} = \{0\}$*; in that case* $\dim\ker T_g \leq \max\{0, k_0\}$*. Moreover, if* $\dim\ker T_g < \infty$*, we have*

$$\dim\ker T_{\theta g} = \max\{0, \dim\ker T_g - k\}, \tag{5.3}$$

where k is the number of zeroes of θ counting their multiplicity.

Thus, in particular, if $\dim\ker T_g = d < \infty$ and θ is a finite Blaschke product such that $\dim K_\theta = k \leq d$, then

$$\dim\ker T_{\theta g} = \dim\ker T_g - k. \tag{5.4}$$

Of course, when $\ker T_g$ is infinite-dimensional and the same happens with $\ker T_{\theta g}$, it is not possible to relate their dimension as in (5.4). We can, however, use maximal functions to present an alternative relation, analogous to (5.2), which not only generalizes Theorem 5.1 but moreover sheds new light on the meaning of (5.3) when $k < \dim\ker T_g < \infty$.

Theorem 5.2 ([9]). *Let $g \in L^\infty(\mathbb{T})$ and let B be a finite Blaschke product of degree k. If* $\dim\ker T_g \leq k$*, then* $\ker T_{Bg} = \{0\}$*; if* $\dim\ker T_g > k$*, then*

$$\ker T_g = \ker T_{Bg} \oplus (z - z_0)\phi_+ K_B,$$

where z_0 is a zero of B and ϕ_+ is a maximal function for $\ker T_{Bg}$*.*

6. Injective Toeplitz operators

Clearly, the existence of maximal functions and the results of the previous section are closely connected with the question of injectivity of Toeplitz operators, which in turn is equivalent to the question whether the Riemann–Hilbert problem $gf_+ = f_-$, with $f_+ \in H^2$ and $f_- \in \overline{H_0^2}$, has a nontrivial solution.

It is well known that various properties of a Toeplitz operator, and in particular of its kernel, can be described in terms of an appropriate factorization of its symbol ([4, 13, 20, 26, 28]). For instance, the so-called L^2-factorization is a representation of the symbol $g \in L^\infty(\mathbb{T})$ as a product

$$g = g_- d g_+^{-1}, \tag{6.1}$$

where $g_+^{\pm 1} \in H^2$, $g_-^{\pm 1} \in \overline{H^2}$ and $d(z) = z^k$ for some $k \in \mathbb{Z}$. If g is invertible in $L^\infty(\mathbb{T})$ and admits an L^2-factorization, then $\dim\ker T_g = |k|$ if $k \leq 0$, and $\dim\ker T_g^* = k$ if $k > 0$. The factorization (6.1) is called a bounded factorization when $g_+^{\pm 1}, \overline{g_-^{\pm 1}} \in H^\infty$. In various subalgebras of $L^\infty(\mathbb{T})$, every invertible element admits a factorization of the form (6.1), where the middle factor d is an inner function. This is the case in the Wiener algebra on $\mathbb{T}$ and in the analogous algebra APW of almost-periodic functions on the real line $\mathbb{R}$.

In the latter case d may be a singular inner function, $d(\xi) = \exp(-i\lambda\xi)$ with $\lambda \in \mathbb{R}$, and we have that if $g \in APW$ is invertible in $L^\infty(\mathbb{R})$ then $\ker T_g$ is either trivial or isomorphic to an infinite-dimensional model space K_θ with $\theta(\xi) = \exp(i\lambda\xi)$, depending on whether $\lambda \le 0$ or $\lambda > 0$. For more details see [8] and [3, Sec. 8.3].

For $g_1, g_2 \in L^\infty(\mathbb{T})$, we say that $g_1 \sim g_2$ if and only if there are functions $h_+ \in \mathcal{G}H^\infty$, $h_- \in \mathcal{G}\overline{H^\infty}$ such that $g_1 = h_- g_2 h_+$, and in that case we have $\ker T_{g_1} = h_+^{-1} \ker T_{g_2}$ (which we write as $\ker T_{g_1} \sim \ker T_{g_2}$). Thus if (6.1) is a bounded factorization, we have $g \sim z^k$ and $\ker T_g = \{0\}$ if $k \ge 0$, and $\ker T_g \sim K_{z^{|k|}}$ if $k < 0$.

L^2 factorizations are a particular case of factorizations of the form

$$g = g_- \theta^{-N} g_+^{-1}, \qquad g_- \in \overline{H^2}, \quad g_+ \in H^2, \tag{6.2}$$

where θ is an inner function and $N \in \mathbb{Z}$. We have the following.

Theorem 6.1 **([7, 8]).** *If $g \in L^\infty(\mathbb{T})$ admits a factorization* (6.2), *where $\overline{g_-}$ and g_+ are outer functions in H^2, with g_+^2 rigid in H^1, then*

$$\ker T_g \neq \{0\} \quad \textit{if and only if} \quad N > 0.$$

If $N > 0$ and θ is a finite Blaschke product of degree k, then $\dim \ker T_g = kN$; *if θ is not a finite Blaschke product, then* $\dim \ker T_g = \infty$.

We also have the following.

Theorem 6.2 **([7, 29]).** *For $g \in L^\infty(\mathbb{T})$, $\ker T_g$ is nontrivial of finite dimension if and only if, for some $N \in \mathbb{N}$, g admits a factorization $g = g_- z^{-N} g_+^{-1}$, where $g_- \in \overline{H_0^2}$ is outer, and $g_+ \in H^2$ is outer with g_+^2 rigid in H^1. In that case* $\ker T_g = \ker T_{z^{-N}\overline{g_+}/g_+}$, *and* $\dim \ker T_g = N$.

Some other results regarding conditions for injectivity or non-injectivity of Toeplitz operators will be mentioned in the next section.

7. Multipliers between Toeplitz kernels

The existence of maximal vectors for every non-zero Toeplitz kernel also provides test functions for various properties of these spaces.

In [11] Crofoot characterized the multipliers from a model space *onto* another. Partly motivated by that work, Fricain, Hartmann and Ross addressed in [17] the question of which holomorphic functions w multiply a model space K_θ into another model space K_ϕ. Their main result shows that w multiplies K_θ into K_ϕ (written $w \in \mathcal{M}(K_\theta, K_\phi)$) if and only if

(i) w multiplies the function $S^*\theta = \widetilde{k_0^\theta}$ into K_ϕ, and
(ii) w multiplies K_θ into H^2, which can be expressed by saying that $|w|^2\, dm$ is a Carleson measure for K_θ.

Model spaces being a particular type of Toeplitz kernel, that question may be posed more generally for the latter. We may also ask whether more general test functions can be used, other than $S^*\theta$.

In this more general setting, one immediately notices that, unlike multipliers between model spaces, multipliers between general Toeplitz kernels need not lie in H^2. In fact, for model spaces, we must have $w \in H^2$ if $w \in \mathcal{M}(K_\theta, K_\phi)$, because we must then have $wk_0^\theta \in K_\phi \subset H^2$, and $1/k_0^\theta \in H^\infty$; but the function $w(z) = (z-1)^{-1/2}$ multiplies $\ker T_g$, with $g(z) = z^{-3/2}$ and $\arg z \in [0, 2\pi)$ for $z \in \mathbb{T}$, onto the model space $K_z = \ker T_{\overline{z}}$ consisting of the constant functions, even though $w \notin H^2$.

One can characterize all multipliers from one Toeplitz kernel into another as follows. We denote by $\mathcal{C}(\ker T_g)$ the class of all w such that $|w|^2\, dm$ is a Carleson measure for $\ker T_g$, i.e., $w \ker T_g \subset L^2(\mathbb{T})$, and by $\mathcal{N}_+$ the Smirnov class.

Theorem 7.1 **([8]).** *Let $g, h \in L^\infty(\mathbb{T}) \setminus \{0\}$ be such that $\ker T_g$ and $\ker T_h$ are nontrivial. Then the following are equivalent:*

(i) *$w \in \mathcal{M}(\ker T_g, \ker T_h)$;*

(ii) *$w \in \mathcal{C}(\ker T_g)$ and $wk_+ \in \ker T_h$ for some (and hence all) maximal vectors k_+ of $\ker T_g$;*

(iii) *$w \in \mathcal{C}(\ker T_g)$ and $hg^{-1}w \in \overline{\mathcal{N}_+}$.*

Note that if k_+ is not a maximal vector for $\ker T_g$, then k_+ cannot be used as a test function; for example, the function $w = 1$ is not a multiplier from $\ker T_g$ into $\mathrm{K}_{\min}(k_+)$, even though $wk_+ \in \mathrm{K}_{\min}(k_+)$.

Corollary 7.2 **([8]).** *With the same assumptions as in Theorem 7.1, and assuming moreover that $hg^{-1} \in L^\infty(\mathbb{T})$, one has*

$$w \in \mathcal{M}(\ker T_g, \ker T_h) \quad \textit{if and only if} \quad w \in \mathcal{C}(\ker T_g) \cap \ker T_{\overline{z}gh^{-1}}.$$

By considering the special case $g = \overline{\theta}$, where θ is inner, we obtain the following result.

Corollary 7.3 **([8, 17]).** *Let θ be inner and let $h \in L^\infty(\mathbb{T}) \setminus \{0\}$ be such that $\ker T_h$ is nontrivial. Then the following are equivalent:*

(i) *$w \in \mathcal{M}(K_\theta, \ker T_h)$;*

(ii) *$w \in \mathcal{C}(K_\theta)$ and $wS^*\theta \in \ker T_h$;*

(iii) *$w \in \mathcal{C}(K_\theta) \cap \ker T_{\overline{z}\theta h}$.*

The last two corollaries also bring out a close connection between the existence of non-zero multipliers in $L^2(\mathbb{T})$ and their description, on the one hand, and the question of injectivity of an associated Toeplitz operator and the characterization of its kernel (discussed in Sections 5 and 6), on the other hand. Thus, for instance, the result of Example 5 implies that, since $T_{\overline{z}\theta\overline{\psi}}$ is injective in that case, we have $\mathcal{M}(K_\theta, K_\phi) = \{0\}$. Another example is the following.

Example. Let θ, ϕ be two inner functions with $\phi \preceq \theta$, i.e., $K_\phi \subset K_\theta$. Then $\dim \ker T_{\overline{z}\theta\overline{\phi}} \leq 1$, as $\theta\overline{\phi} \in H^\infty$ and $\ker T_{\theta\overline{\phi}} = \{0\}$ ([2]). We have $\ker T_{\overline{z}\theta\overline{\phi}} = \mathbb{C}$ if $\phi = a\theta$ with $a \in \mathbb{C}$, $|a| = 1$, and we have $\ker T_{\overline{z}\theta\overline{\phi}} = \{0\}$ if $\phi \prec \theta$; therefore $\mathcal{M}(K_\theta, K_\phi) \neq \{0\}$ if and only if $K_\theta = K_\phi$, in which case $\mathcal{M}(K_\theta, K_\phi) = \mathbb{C}$.

The class of bounded multipliers,

$$\mathcal{M}_\infty(\ker T_g, \ker T_h) = \mathcal{M}(\ker T_g, \ker T_h) \cap H^\infty,$$

is of great importance. For instance, the question whether $w = 1$ is a multiplier from $\ker T_g$ into $\ker T_h$ is equivalent to asking whether $\ker T_g \subset \ker T_h$. Noting that the Carleson measure condition is redundant for bounded w, we obtain the following characterization from Theorem 7.1.

Theorem 7.4. *Let $g, h \in L^\infty(\mathbb{T}) \setminus \{0\}$ be such that $\ker T_g$ and $\ker T_h$ are nontrivial. Then the following are equivalent:*

(i) *$w \in \mathcal{M}_\infty(\ker T_g, \ker T_h)$;*
(ii) *$w \in H^\infty$ and $wk_+ \in \ker T_h$ for some (and hence all) maximal vectors k_+ of $\ker T_g$;*
(iii) *$w \in H^\infty$ and $hg^{-1}w \in \overline{H^\infty}$ (assuming that $hg^{-1} \in L^\infty(\mathbb{T})$).*

For model spaces, we thus recover the main theorem on bounded multipliers from [17].

Corollary 7.5. *Let θ and ϕ be inner functions, and let $w \in H^2$. Then*

$$w \in \mathcal{M}_\infty(K_\theta, K_\phi) \iff w \in \ker T_{\overline{z\theta}\phi} \cap H^\infty \iff wS^*\theta \in K_\phi \cap H^\infty$$
$$\iff w \in H^\infty \quad \text{and} \quad \theta\overline{\phi}w \in \overline{H^\infty}.$$

Applying the results of Theorem 7.4 to $w = 1$ we obtain moreover the following results.

Corollary 7.6. *Under the same assumptions as in Theorem 7.4, the following conditions are equivalent:*

(i) *$\ker T_g \subset \ker T_h$;*
(ii) *$hg^{-1} \in \overline{\mathcal{N}_+}$;*
(iii) *there exists a maximal function k_+ for $\ker T_g$ such that $k_+ \in \ker T_h$.*

If, moreover, $\ker T_g$ contains a maximal vector k_+ with $k_+, k_+^{-1} \in L^\infty(\mathbb{T})$, then each of the above conditions is equivalent to

(iv) *$k_+ \in \ker T_h \cap H^\infty$.*

Corollary 7.7. *Under the same assumptions as in Theorem 7.4, if hg^{-1} is in $\mathcal{G}L^\infty(\mathbb{T})$, then*

$$\ker T_g \subset \ker T_h \quad \text{if and only if} \quad hg^{-1} \in \overline{H^\infty}.$$

This last result implies in particular that, assuming that $hg^{-1} \in L^\infty(\mathbb{T})$, a Toeplitz kernel is contained in another Toeplitz kernel if and only if they take the form $\ker T_g$ and $\ker T_{\theta g}$ for some inner function θ and $g \in L^\infty(\mathbb{T})$ (cf. Section 5).

Corollary 7.8. *Under the same assumptions as in Theorem 7.4, we have $\ker T_g = \ker T_h$ if and only if $g/h = \overline{p_+/q_+}$ with $p_+, q_+ \in H^2$ outer. If moreover $hg^{-1} \in \mathcal{G}L^\infty(\mathbb{T})$, then we have*

$$\ker T_g = \ker T_h \quad \text{if and only if} \quad hg^{-1} \in \mathcal{G}\overline{H^\infty}.$$

We can draw several interesting conclusions from these results:

1. First, we can characterize the Toeplitz kernels that are contained in a given model space K_θ ($\ker T_g = \ker T_{\overline{\theta}\alpha}$, with α inner), and those that contain K_θ ($\ker T_g$ with $g \in \overline{\theta H^\infty}$), assuming that the symbols are in $\mathcal{G}L^\infty(\mathbb{T})$.

2. Second, while (3.1) provides an expression for a (unimodular) symbol g such that $\ker T_g$ is the minimal kernel for a given function with inner–outer factorization $\phi_+ = IO_+$, it is not claimed that all Toeplitz operators with that kernel have the same symbol. Indeed, from Corollary 7.8, we have that if $\ker T_g = \mathrm{K}_{\min}(\phi_+)$ with $\phi_+ = IO_+$, then

$$g = \frac{\overline{p_+}}{\overline{q_+}} \frac{\overline{IO_+}}{O_+} \qquad \text{with} \;\; p_+, q_+ \in H^2 \text{ outer};$$

if, moreover, $g \in \mathcal{G}L^\infty(\mathbb{T})$, then $g = h_- \overline{IO_+}/\overline{O}_+$, with $h_- \in \mathcal{G}\overline{H^\infty}$.

3. Clearly, a Toeplitz operator with unimodular symbol u is non-injective if and only if it has a maximal vector, i.e., there exist an inner function I and an outer function $O_+ \in H^2$ such that $\ker T_u = \mathrm{K}_{\min}(IO_+) = \ker T_{\overline{IO_+}/O_+}$, which is equivalent, as shown in point 2, to having

$$u = \overline{z} \frac{\overline{IO_+}}{O_+} h_-, \qquad \text{with} \;\; h_- \in \mathcal{G}\overline{H^\infty}.$$

Since $|h_-| = 1$ a.e. on $\mathbb{T}$, we conclude that h_- must be a unimodular constant, and therefore T_u is non-injective if and only if

$$u = \overline{z}\overline{IO_+}/O_+,$$

thus recovering a result by Makarov and Poltoratski [27, Lem. 3.2].

4. Since there are different maximal functions for each Toeplitz kernel with dimension greater than 1, one may ask how they can be related. Again, from Corollary 7.8, we see that if $\mathrm{K}_{\min}(f_{1+}) = \mathrm{K}_{\min}(f_{2+})$, where $f_{1+} = I_1 O_{1+}$ and $f_{2+} = I_2 O_{2+}$ with I_1, I_2 inner and $O_{1+}, O_{2+} \in H^2$ outer, then

$$\frac{\overline{I_1 O_{1+}}}{O_{1+}} = \frac{\overline{I_2 O_{2+}}}{O_{2+}} h_-,$$

where $h_- \in \mathcal{G}\overline{H^\infty}$, $|h_-| = 1$, and so h_- is constant. Thus, finally f_{1+} and f_{2+} are related by

$$f_{2+} = f_{1+} \frac{\overline{O_{2+}}}{\overline{O_{1+}}}.$$

References

[1] S. Barclay, A solution to the Douglas-Rudin problem for matrix-valued functions, *Proc. Lond. Math. Soc.* (3) **99** (2009), no. 3, 757–786.

[2] C. Benhida, M.C. Câmara, and C. Diogo, Some properties of the kernel and the cokernel of Toeplitz operators with matrix symbols, *Linear Algebra Appl.* **432** (2010), no. 1, 307–317.

[3] A. Böttcher, Y.I. Karlovich, and I.M. Spitkovsky, *Convolution Operators and Factorization of Almost Periodic Matrix Functions,* Operator Theory: Advances and Applications **131**, Birkhäuser Verlag, Basel, 2002.

[4] A. Böttcher and B. Silbermann, *Analysis of Toeplitz Operators*, Springer-Verlag, Berlin, 1990.

[5] J. Bourgain, A problem of Douglas and Rudin on factorization, *Pacific J. Math.* **121** (1986), no. 1, 47–50.

[6] M.C. Câmara and J.R. Partington, Near invariance and kernels of Toeplitz operators, *J. Anal. Math.* **124** (2014), 235–260.

[7] M.C. Câmara and J.R. Partington, Finite-dimensional Toeplitz kernels and nearly-invariant subspaces, *J. Operator Theory* **75** (2016), no. 1, 75–90.

[8] M.C. Câmara and J.R. Partington, Multipliers and equivalences between Toeplitz kernels, `https://arxiv.org/abs/1611.08429`

[9] M.C. Câmara, M.T. Malheiro, and J.R. Partington, Model spaces and Toeplitz kernels in reflexive Hardy space, *Oper. Matrices* **10** (2016), no. 1, 127–148.

[10] L.A. Coburn, Weyl's theorem for nonnormal operators. *Michigan Math. J.* **13** (1966), 285–288.

[11] R.B. Crofoot, Multipliers between invariant subspaces of the backward shift, *Pacific J. Math.* **166** (1994), no. 2, 225–246.

[12] L. de Branges and J. Rovnyak, *Square Summable Power Series*, Holt, Rinehart and Winston, New York–Toronto–London, 1966.

[13] R. Duduchava, *Integral Equations in Convolution with Discontinuous Presymbols. Singular Integral Equations with Fixed Singularities, and their Applications to some Problems of Mechanics*, Teubner-Texte zur Mathematik, Leipzig, 1979.

[14] P.L. Duren, *Theory of H^p Spaces*, Dover, New York, 2000.

[15] K.M. Dyakonov, Kernels of Toeplitz operators via Bourgain's factorization theorem. *J. Funct. Anal.* **170** (2000), no. 1, 93–106.

[16] I.A. Feldman and I.C. Gohberg, Wiener-Hopf integro-difference equations. *Dokl. Akad. Nauk SSSR* **183** (1968), 25–28. English translation: *Soviet Math. Dokl.* **9** (1968), 1312–1316.

[17] E. Fricain, A. Hartmann and W.T. Ross, Multipliers between model spaces, *Studia Math.*, to appear. `http://arxiv.org/abs/1605.07418`.

[18] E. Fricain and J. Mashreghi, *The Theory of $\mathcal{H}(b)$ Spaces*, Vol. 1, New Mathematical Monographs **20**, Cambridge University Press, Cambridge, 2016.

[19] S.R. Garcia, J. Mashreghi, and W.T. Ross, *Introduction to Model Spaces and their Operators*, Cambridge Studies in Advanced Mathematics **148**, Cambridge University Press, Cambridge, 2016.

[20] I. Gohberg and N. Krupnik, *One-dimensional Linear Singular Integral Equations*, Vols. I and II, Birkhäuser Verlag, Basel, 1992.

[21] A. Hartmann and M. Mitkovski, Kernels of Toeplitz operators, *Recent progress on operator theory and approximation in spaces of analytic functions*, 147–177, Contemp. Math. **679**, Amer. Math. Soc., Providence, RI, 2016.

[22] E. Hayashi, The solution sets of extremal problems in H^1, *Proc. Amer. Math. Soc.* **93** (1985), no. 4, 690–696.

[23] E. Hayashi, The kernel of a Toeplitz operator, *Integral Equations Operator Theory* **9** (1986), no. 4, 588–591.

[24] E. Hayashi, Classification of nearly invariant subspaces of the backward shift. *Proc. Amer. Math. Soc.* **110** (1990), no. 2, 441–448.

[25] D. Hitt, Invariant subspaces of $\mathcal{H}^2$ of an annulus, *Pacific J. Math.* **134** (1988), no. 1, 101–120.

[26] G.S. Litvinchuk and I.M. Spitkovsky, *Factorization of Measurable Matrix Functions*, Birkhäuser Verlag, Basel and Boston, 1987.

[27] N. Makarov and A. Poltoratski, Meromorphic inner functions, Toeplitz kernels and the uncertainty principle, *Perspectives in analysis*, 185–252, Math. Phys. Stud. **27**, Springer, Berlin, 2005.

[28] S.G. Mikhlin and S. Prössdorf, *Singular Integral Operators*, Translated from the German by Albrecht Böttcher and Reinhard Lehmann, Springer-Verlag, Berlin, 1986.

[29] T. Nakazi, Kernels of Toeplitz operators, *J. Math. Soc. Japan* **38** (1986), no. 4, 607–616.

[30] N.K. Nikolski, *Operators, Functions, and Systems: an Easy Reading, Vol. 1, Hardy, Hankel, and Toeplitz*, Translated from the French by Andreas Hartmann, Mathematical Surveys and Monographs **92**, American Mathematical Society, Providence, RI, 2002.

[31] D. Sarason, Nearly invariant subspaces of the backward shift, *Contributions to operator theory and its applications* (Mesa, AZ, 1987), 481–493, Oper. Theory Adv. Appl. **35**, Birkhäuser, Basel, 1988.

[32] D. Sarason, Kernels of Toeplitz operators, *Toeplitz operators and related topics (Santa Cruz, CA, 1992)*, 153–164, Oper. Theory Adv. Appl. **71**, Birkhäuser, Basel, 1994.

M. Cristina Câmara
Center for Mathematical Analysis
Geometry and Dynamical Systems
Instituto Superior Técnico, Universidade de Lisboa
Av. Rovisco Pais, 1049-001 Lisboa
Portugal.
e-mail: `ccamara@math.ist.utl.pt`

Jonathan R. Partington
School of Mathematics
University of Leeds
Leeds LS2 9JT
U.K.
e-mail: `j.r.partington@leeds.ac.uk`

Frames, operator representations, and open problems

Ole Christensen and Marzieh Hasannasab

Abstract. A frame in a Hilbert space $\mathcal{H}$ is a countable collection of elements in $\mathcal{H}$ that allows each $f \in \mathcal{H}$ to be expanded as an (infinite) linear combination of the frame elements. Frames generalize the well-known orthonormal bases, but provide much more flexibility and can often be constructed with properties that are not possible for orthonormal bases. We will present the basic facts in frame theory with focus on their operator theoretical characterizations and discuss open problems concerning representations of frames in terms of iterations of a fixed operator. These problems come up in the context of dynamical sampling, a topic that has recently attracted considerably interest within harmonic analysis. The goal of the paper is twofold, namely, that experts in operator theory will explore the potential of frames, and that frame theory will benefit from insight provided by the operator theory community.

Mathematics Subject Classification (2010). 42C15.

Keywords. Frames, dual frames, dynamical sampling, operator theory.

1. Introduction and motivation

A coherent state is a (typically overcomplete) system of vectors in a Hilbert space $\mathcal{H}$. In general it is given by the action of a class of linear operators on a single element in the underlying Hilbert space. In particular, it could be given by iterated action of a fixed operator on a single element, i.e., as $\{T^n\varphi\}_{n=0}^{\infty}$ for some $\varphi \in \mathcal{H}$ and a linear operator $T : \mathcal{H} \to \mathcal{H}$. Coherent states play an important role in mathematical physics [20, 22], operator theory, and modern applied harmonic analysis [14, 7]. In particular, a Gabor system (see the definition below) is a coherent state.

Systems of vectors on the form $\{T^n\varphi\}_{n=0}^{\infty}$ also appear in the more recent context of dynamical sampling [1, 2, 3, 23]. Key questions in this context are whether $\{T^n\varphi\}_{n=0}^{\infty}$ can form a basis or a frame when the operator T belongs to a certain class of operators, e.g., normal operators or self-adjoint operators.

© Springer International Publishing AG, part of Springer Nature 2018
A. Böttcher et al. (eds.), *The Diversity and Beauty of Applied Operator Theory*, Operator Theory: Advances and Applications 268, https://doi.org/10.1007/978-3-319-75996-8_8

We will postpone the formal definition of a frame to Section 2 and just mention that a frame is a collection of vectors in $\mathcal{H}$ that allows each $f \in \mathcal{H}$ to be expanded as an (infinite) linear combination of the frame elements. Frames are getting increasingly popular in applied harmonic analysis and signal processing, mainly due to the fact that they are much more flexible and easier to construct with prescribed properties than the classical orthonormal bases.

A different approach was taken in the papers [9, 10, 12]: here, the starting point is a frame and the question is when and how it has a representation of the form

$$\{T^n\varphi\}_{n=0}^{\infty} \text{ for some } \varphi \in \mathcal{H} \text{ and a bounded linear operator } T : \mathcal{H} \to \mathcal{H}. \quad (1.1)$$

We will give a short survey of some of the most important results concerning frame representations of the form (1.1). As inspiration for the reader we will also state a number of open problems.

The above questions will also be analyzed with a different indexing, i.e., considering systems on the form $\{T^n\varphi\}_{n=-\infty}^{\infty}$ instead of $\{T^n\varphi\}_{n=0}^{\infty}$. The indexing in terms of $\mathbb{Z}$ is natural for several well-known classes of frames, and the theoretical conditions for a frame having such a representation with a bounded operator T are similar to the ones for systems indexed by $\mathbb{N}_0$. The change in indexing gives an interesting twist on the problem. For example, a shift-invariant system (see Section 4 for the definition) always has a representation $\{T^n\varphi\}_{n=-\infty}^{\infty}$ with a bounded operator T, but it does not have a representation with the indexing in (1.1). This observation indeed leads to one of the open problems, stated for the so-called Gabor frames in $L^2(\mathbb{R})$.

The paper is organized as follows. In Section 2 we will provide a very short survey on frame theory with focus on operator theoretical characterizations. Section 3 discusses representations of frames via iterated systems of operators acting on a single element in the underlying Hilbert space. Finally, in Section 4 we consider concrete classes of frames, namely, shift-invariant frames and Gabor frames, and discuss a number of open problems related to operator representations.

2. Frame theory

Let $\mathcal{H}$ denote a separable Hilbert space. A sequence $\{f_k\}_{k=1}^{\infty}$ in $\mathcal{H}$ is a *frame* for $\mathcal{H}$ if there exist constants $A, B > 0$ such that

$$A\,||f||^2 \leq \sum_{k=1}^{\infty} |\langle f, f_k\rangle|^2 \leq B\,||f||^2 \quad \forall f \in \mathcal{H};$$

it is a *frame sequence* if the stated inequalities hold for all $f \in \overline{\text{span}}\{f_k\}_{k=1}^{\infty}$. The sequence $\{f_k\}_{k=1}^{\infty}$ is a *Bessel sequence* if at least the upper frame condition holds. Also, $\{f_k\}_{k=1}^{\infty}$ is called a *Riesz sequence* if there exist constants $A, B > 0$ such that

$$A \sum |c_k|^2 \leq \left|\left|\sum c_k f_k\right|\right|^2 \leq B \sum |c_k|^2$$

for all finite scalar sequences $\{c_k\}_{k=1}^{\infty}$. A *Riesz basis* is a Riesz sequence $\{f_k\}_{k=1}^{\infty}$ for which $\overline{\text{span}}\{f_k\}_{k=1}^{\infty} = \mathcal{H}$.

It is well-known that the above concepts have operator theoretical characterizations, see, e.g., [7].

Theorem 2.1. *Consider a sequence $\{f_k\}_{k=1}^{\infty}$ in a separable Hilbert space $\mathcal{H}$. Then the followings hold:*

(i) *$\{f_k\}_{k=1}^{\infty}$ is a Bessel sequence if and only if $U : \{c_k\}_{k=1}^{\infty} \mapsto \sum_{k=1}^{\infty} c_k f_k$ is a well-defined mapping from $\ell^2(\mathbb{N})$ to $\mathcal{H}$, i.e., the infinite series is convergent for all $\{c_k\}_{k=1}^{\infty} \in \ell^2(\mathbb{N})$; in the affirmative case the operator U is linear and bounded.*

(ii) *$\{f_k\}_{k=1}^{\infty}$ is a frame if and only if the mapping $\{c_k\}_{k=1}^{\infty} \mapsto \sum_{k=1}^{\infty} c_k f_k$ is well-defined from $\ell^2(\mathbb{N})$ to $\mathcal{H}$ and surjective.*

(iii) *$\{f_k\}_{k=1}^{\infty}$ is a Riesz basis if and only if the mapping $\{c_k\}_{k=1}^{\infty} \mapsto \sum_{k=1}^{\infty} c_k f_k$ is well-defined from $\ell^2(\mathbb{N})$ to $\mathcal{H}$ and bijective.*

Theorem 2.1 tells us that if $\{f_k\}_{k=1}^{\infty}$ is a Bessel sequence, then the *synthesis operator* defined by

$$U : \ell^2(\mathbb{N}) \to \mathcal{H}, \quad U\{c_k\}_{k=1}^{\infty} := \sum_{k=1}^{\infty} c_k f_k \tag{2.1}$$

is well-defined and bounded. A central role will be played by the kernel of the operator U, i.e., the subset of $\ell^2(\mathbb{N})$ given by

$$N_U = \left\{ \{c_k\}_{k=1}^{\infty} \in \ell^2(\mathbb{N}) \;\middle|\; \sum_{k=1}^{\infty} c_k f_k = 0 \right\}. \tag{2.2}$$

We will now state one more characterization of frames, in terms of a condition that is very similar to the Riesz basis condition, except that it only takes place on sequences in the orthogonal complement of the kernel of the operator U.

Lemma 2.2. *A sequence $\{f_k\}_{k=1}^{\infty}$ in $\mathcal{H}$ is a frame for $\mathcal{H}$ with bounds A, B if and only if the following conditions are satisfied:*

(i) *$\{f_k\}_{k=1}^{\infty}$ is complete in $\mathcal{H}$.*

(ii) *The synthesis operator U is well defined on $\ell^2(\mathbb{N})$ and*

$$A \sum_{k=1}^{\infty} |c_k|^2 \leq ||U\{c_k\}_{k=1}^{\infty}||^2 \leq B \sum_{k=1}^{\infty} |c_k|^2 \quad \forall \{c_k\}_{k=1}^{\infty} \in N_U^{\perp}. \tag{2.3}$$

The *excess* of a frame is the number of elements that can be removed yet leaving a frame. It is well-known that the excess equals $\dim(N_U)$; see [4].

Given a Bessel sequence $\{f_k\}_{k=1}^{\infty}$, the *frame operator* is defined by

$$S : \mathcal{H} \to \mathcal{H}, \quad Sf := UU^* f = \sum_{k=1}^{\infty} \langle f, f_k \rangle f_k.$$

For a frame $\{f_k\}_{k=1}^{\infty}$, the frame operator is bounded, bijective, and self-adjoint; these properties immediately lead to the important *frame decomposition*

$$f = SS^{-1}f = \sum_{k=1}^{\infty} \langle f, S^{-1}f_k \rangle f_k \quad \forall f \in \mathcal{H}. \tag{2.4}$$

The sequence $\{S^{-1}f_k\}_{k=1}^{\infty}$ is also a frame; it is called the *canonical dual frame.*

One of the most striking properties of frames is that they can be *over-complete*; the intuitive interpretation of this is that a frame might consist of more elements than necessary to span the Hilbert space. In practice, this means that if $\{f_k\}_{k=1}^{\infty}$ is a frame but not a basis, then there exists $\{g_k\}_{k=1}^{\infty} \neq \{S^{-1}f_k\}_{k=1}^{\infty}$ such that

$$f = \sum_{k=1}^{\infty} \langle f, g_k \rangle f_k \quad \forall f \in \mathcal{H}. \tag{2.5}$$

Any frame $\{g_k\}_{k=1}^{\infty}$ satisfying (2.5) for a given frame $\{f_k\}_{k=1}^{\infty}$ is called a *dual frame* of $\{f_k\}_{k=1}^{\infty}$. Note that if $\{f_k\}_{k=1}^{\infty}$ and $\{g_k\}_{k=1}^{\infty}$ are Bessel sequences with synthesis operators U, V, respectively, then (2.5) means precisely that

$$UV^* = I.$$

We refer to [7] and [18] for more information about frames and Riesz bases.

3. Operator representations of frames

Formulated in purely mathematical terms, dynamical sampling in a Hilbert space $\mathcal{H}$ deals with frame properties of sequences in $\mathcal{H}$ of the form $\{T^n\varphi\}_{n=0}^{\infty}$, where $\varphi \in \mathcal{H}$ and $T : \mathcal{H} \to \mathcal{H}$ is a linear operator. The purpose of this section is to give an introduction to the topic that allows to discuss a number of important open problems as well.

Considering a class $\mathcal{A}$ of operators $T : \mathcal{H} \to \mathcal{H}$, typical questions in dynamical sampling are as follows:

- Can $\{T^n\varphi\}_{n=0}^{\infty}$ be a basis for $\mathcal{H}$ for some $T \in \mathcal{A}, \varphi \in \mathcal{H}$?
- Can $\{T^n\varphi\}_{n=0}^{\infty}$ be a frame for $\mathcal{H}$ for some $T \in \mathcal{A}, \varphi \in \mathcal{H}$?

Unfortunately, the theory of dynamical sampling is full of no-go results. Let us state some of them here.

Theorem 3.1. *Consider a bounded operator $T : \mathcal{H} \to \mathcal{H}$, and let $\varphi \in \mathcal{H}$. Then the followings hold:*

(i) *If T is normal, then $\{T^n\varphi\}_{n=0}^{\infty}$ can not be a basis* [2];
(ii) *If T is unitary, then $\{T^n\varphi\}_{n=0}^{\infty}$ can not be a frame* [3];
(iii) *If T is compact, then $\{T^n\varphi\}_{n=0}^{\infty}$ can not be a frame* [12].

The first construction of a frame of the form $\{T^n\varphi\}_{n=0}^\infty$ was obtained in [1] and further discussed in [2, 3]. Since it deals with a class of diagonalizable operators it is natural to formulate it for the Hilbert space $\ell^2(\mathbb{N})$, where the operator T can be considered as a matrix.

Theorem 3.2. *Consider a diagonal matrix $T = [a_{jk}]_{j,k\in\mathbb{N}}$ given by $a_{kk} = \lambda_k$, $a_{jk} = 0$, $j \neq k$, and a sequence $\varphi = \{\varphi(k)\}_{k\in\mathbb{N}} \in \ell^2(\mathbb{N})$. Then the system $\{T^n\varphi\}_{n=0}^\infty$ is a frame for $\ell^2(\mathbb{N})$ if and only if*

(i) $|\lambda_k| < 1$ *for all* $k \in \mathbb{N}$;
(ii) $|\lambda_k| \to 1$ *as* $k \to \infty$;
(iii) *The sequence* $\{\lambda_k\}_{k=1}^\infty$ *satisfies the Carleson condition, i.e.,*

$$\inf_n \prod_{n\neq k} \frac{|\lambda_k - \lambda_n|}{|1 - \lambda_k \overline{\lambda_n}|} > 0; \tag{3.1}$$

(iv) $\varphi(k) = m_k\sqrt{1 - |\lambda_k|^2}$ *for a scalar-sequence* $\{m_k\}_{k=1}^\infty$ *that is bounded below away from zero and above.*

In the affirmative case $\{T^n\varphi\}_{n=0}^\infty$ *is overcomplete, i.e., not a basis.*

The fact that the Carleson condition comes in as the key condition in Theorem 3.2 indicates that the result is based on deep results concerning interpolation theory in spaces of analytic functions. Note that a self-contained proof (just based on a single result in [24] and standard frame theory) was given later in [13].

The construction in Theorem 3.2 can be extended to diagonalizable matrices, but very little is known in the literature about how and when one can construct frames for general matrices. Let us formulate this as an open problem.

Problem 1. Identify a class of non-diagonalizable matrices $T = [a_{jk}]_{j,k\in\mathbb{N}}$ for which

(i) T is a bounded operator on $\ell^2(\mathbb{N})$;
(ii) $\{T^n\varphi\}_{n=0}^\infty$ is a frame for $\ell^2(\mathbb{N})$ for some $\varphi \in \ell^2(\mathbb{N})$.

The class of frames that are known to be representable on the form $\{T^n\varphi\}_{n=0}^\infty$ for a bounded operator T is indeed limited: except for the construction in Theorem 3.2, such a representation has only been obtained for non-redundant frames, i.e., Riesz bases [9].

Note that the way the key questions in dynamical sampling are formulated in the introduction puts the operator in the central spot: we start with an operator and ask for frame properties of the associated iterated system. The opposite approach was taken in the papers [9, 10, 12], where we considered a given frame $\{f_k\}_{k=1}^\infty$ and ask for the existence of a representation of the form

$$\{f_k\}_{k=1}^\infty = \{T^n f_1\}_{n=0}^\infty, \tag{3.2}$$

where $T : \text{span}\{f_k\}_{k=1}^\infty \to \mathcal{H}$ is a bounded linear operator.

The following result collects a generalization of results from the papers [9, 10, 12]. Indeed, for reasons that will become clear in Section 4, we will generalize (3.2) and ask for representations of the form

$$\{f_k\}_{k=1}^{\infty} = \{a_n T^n f_1\}_{n=0}^{\infty}, \tag{3.3}$$

for some scalars $a_n \neq 0$ with $\sup_n |\frac{a_n}{a_{n+1}}| < \infty$ and a bounded linear operator $T : \text{span}\{f_k\}_{k=1}^{\infty} \to \mathcal{H}$. We will see in the subsequent example that the technical condition on the scalars a_n is necessary. Define the weighted right-shift operator on $\ell^2(\mathbb{N}_0)$ by

$$\mathcal{T}_\omega : \ell^2(\mathbb{N}_0) \to \ell^2(\mathbb{N}_0), \quad \mathcal{T}_\omega(c_0, c_1, \cdots) = \left(0, \frac{a_0}{a_1}c_0, \frac{a_1}{a_2}c_1, \cdots\right). \tag{3.4}$$

Theorem 3.3. *Consider a frame $\{f_k\}_{k=1}^{\infty}$ for an infinite-dimensional Hilbert space $\mathcal{H}$, and a sequence of non-zero scalars $\{a_n\}_{n=0}^{\infty}$ with $\sup_n |\frac{a_n}{a_{n+1}}| < \infty$. Then the followings hold:*

(i) *$\{f_k\}_{k=1}^{\infty}$ is linearly independent if and only if there exists a linear operator $T : \text{span}\{f_k\}_{k=1}^{\infty} \to \mathcal{H}$ such that (3.3) is satisfied.*
(ii) *Assume that $\{f_k\}_{k=1}^{\infty}$ is linearly independent. Then the operator T in (3.3) is bounded if and only if the kernel N_U of the synthesis operator is invariant under weighted right-shifts given as in (3.4); in particular T is bounded if $\{f_k\}_{k=1}^{\infty}$ is a Riesz basis.*
(iii) *Assume that $\{f_k\}_{k=1}^{\infty}$ is linearly independent and overcomplete. If the operator T in (3.3) is bounded, then $\{f_k\}_{k=1}^{\infty}$ has infinite excess.*

Proof. The proof of (i) is similar to the proof of the non-weighted case given in [9]. In order to prove (ii), consider now the representation (3.3). Let us write the synthesis operator for the frame $\{a_n T^n f_1\}_{n=0}^{\infty}$ as

$$U : \ell^2(\mathbb{N}_0) \to \mathcal{H}, \quad U\{c_n\}_{n=0}^{\infty} = \sum_{n=0}^{\infty} c_n a_n T^n f_1.$$

Now, assume first that T is bounded and $\{c_n\}_{n=0}^{\infty} \in N_U$. Then

$$\begin{aligned} U\mathcal{T}_\omega\{c_n\}_{n=0}^{\infty} = \sum_{n=1}^{\infty} \frac{a_{n-1}}{a_n} c_{n-1} a_n T^n f_1 &= \sum_{n=0}^{\infty} c_n a_n T^{n+1} f_1 \\ &= T \sum_{n=0}^{\infty} c_n a_n T^n f_1 = 0. \end{aligned}$$

Therefore $\mathcal{T}_\omega\{c_n\}_{n=0}^{\infty} \in N_U$. Conversely, assume that N_U is invariant under the weighted right-shift operator. We want to prove that T is bounded. Consider an element $f \in \mathcal{H}$ that has a finite expansion in terms of the frame $\{a_n T^n f_1\}_{n=0}^{\infty}$, i.e., $f = \sum_{n=0}^{N} c_n a_n T^n f_1$ for some $N \in \mathbb{N}, c_n \in \mathbb{C}$. Letting $c_n = 0$ for $n > N$, we consider $\{c_n\}_{n=0}^{\infty}$ as a sequence in $\ell^2(\mathbb{N}_0)$. Choose $\{d_n\}_{n=0}^{\infty} \in N_U$ and $\{r_n\}_{n=0}^{\infty} \in N_U^{\perp}$ such that $\{c_n\}_{n=0}^{\infty} = \{d_n\}_{n=0}^{\infty} + \{r_n\}_{n=0}^{\infty}$.

Letting A, B denote frame bounds for $\{f_k\}_{k=1}^\infty$, Lemma 2.2 implies that

$$\begin{aligned}
\|Tf\|^2 &= \left\| T\sum_{n=0}^{N} c_n a_n T^n f_1 \right\|^2 \\
&= \left\| T\sum_{n=0}^{\infty} r_n a_n T^n f_1 \right\|^2 = \left\| \sum_{n=0}^{\infty} r_n a_n T^{n+1} f_1 \right\|^2 \\
&= \left\| \sum_{n=0}^{\infty} r_n \frac{a_n}{a_{n+1}} a_{n+1} T^{n+1} f_1 \right\|^2 \le B \sup_n \left| \frac{a_n}{a_{n+1}} \right|^2 \sum_{n=0}^{\infty} |r_n|^2 \\
&\le BA^{-1} \sup_n \left| \frac{a_n}{a_{n+1}} \right|^2 \left\| \sum_{n=0}^{\infty} r_n a_n T^n f_1 \right\|^2 = BA^{-1} \sup_n \left| \frac{a_n}{a_{n+1}} \right|^2 \|f\|^2.
\end{aligned}$$

Thus T is bounded, as claimed.
For (iii), note that if $\{f_k\}_{k=1}^\infty = \{a_n T^n f_1\}_{n=0}^\infty$, then for $k \ge 1$,

$$f_{k+1} = a_k T^k f_1 = \frac{a_k}{a_{k-1}} T(a_{k-1} T^{k-1} f_1) = \frac{a_k}{a_{k-1}} T f_k. \tag{3.5}$$

Now, if $\{f_k\}_{k=1}^\infty$ has finite excess, then there exists some $K \in \mathbb{N} \cup \{0\}$ such that $\{f_k\}_{k=K}^\infty$ is an overcomplete frame sequence and $\{f_k\}_{k=K+1}^\infty$ is a Riesz sequence. Therefore there exists a non-zero sequence $\{c_k\}_{k=K}^\infty$ such that $\sum_{k=K}^\infty c_k f_k = 0$. Since T is bounded, we have

$$0 = \sum_{k=K}^{\infty} c_k T f_k = \sum_{k=K}^{\infty} c_k \frac{a_{k-1}}{a_k} f_{k+1} = \sum_{k=K+1}^{\infty} c_{k-1} \frac{a_{k-2}}{a_{k-1}} f_k.$$

Therefore $c_k = 0$ for $k \in \{K, K+1, \cdots\}$, which is a contradiction. So T can not be bounded. □

Example. Consider a Riesz basis $\{f_k\}_{k=1}^\infty$ for $\mathcal{H}$, with frame bounds $A, B > 0$. Theorem 3.3 implies that for any sequence of non-zero scalars $\{a_n\}_{n=0}^\infty$ such that

$$\sup_n \left| \frac{a_n}{a_{n+1}} \right| < \infty, \tag{3.6}$$

there is a bounded operator $T : \mathcal{H} \to \mathcal{H}$ such that $\{f_k\}_{k=1}^\infty = \{a_n T^n f_1\}_{n=0}^\infty$. The condition (3.6) is indeed necessary for this conclusion to hold. To see this, note first that if $\{f_k\}_{k=1}^\infty = \{a_n T^n f_1\}_{n=0}^\infty$, then as we saw in (3.5) we have $Tf_k = \frac{a_{k-1}}{a_k} f_{k+1}$. Using that $\sqrt{A} \le ||f_k|| \le \sqrt{B}$ for all $k \in \mathbb{N}$, it follows that

$$||Tf_k|| = \left\| \frac{a_{k-1}}{a_k} f_{k+1} \right\| \ge \left| \frac{a_{k-1}}{a_k} \right| \sqrt{A} \ge \left| \frac{a_{k-1}}{a_k} \right| \sqrt{\frac{A}{B}}\, ||f_k||,$$

which implies that T is unbounded if (3.6) is violated.

Cyclic vectors and hypercyclic vectors

We would like to point out that the frame condition on an iterated system $\{T^n\varphi\}_{n=0}^{\infty}$ is indeed very different from the conditions that are typically considered in operator theory for such systems. First, consider a linear operator $T : \mathcal{H} \to \mathcal{H}$ and recall that a vector $\varphi \in \mathcal{H}$ is said to be *cyclic* with respect to T if $\overline{\text{span}}\{T^n\varphi\}_{n=0}^{\infty} = \mathcal{H}$. By (2.4) this condition is satisfied if $\{T^n\varphi\}_{n=0}^{\infty}$ is a frame for $\mathcal{H}$. However, the frame condition is significantly stronger than the condition of φ being cyclic. In order to illustrate this, let $\{e_k\}_{k=1}^{\infty}$ denote an orthonormal basis for $\mathcal{H}$ and consider the family $\{f_k\}_{k=1}^{\infty} := \{e_k + e_{k+1}\}_{k=1}^{\infty}$. Define the operator T by $Te_k := e_{k+1}$; then T can be extended to a bounded linear operator on $\mathcal{H}$ and $\{f_k\}_{k=1}^{\infty} = \{T^n(e_1 + e_2)\}_{n=0}^{\infty}$. It is known (see Example 5.4.6 in [7]) that $\overline{\text{span}}\{f_k\}_{k=1}^{\infty} = \mathcal{H}$, which implies that the vector $\varphi := e_1 + e_2$ is cyclic with respect to the operator T; however, the same example in [7] shows that $\{T^n(e_1 + e_2)\}_{n=0}^{\infty}$ is not a frame.

Recall also that a vector $\varphi \in \mathcal{H}$ is *hypercyclic* with respect to the operator T if $\{T^n\varphi\}_{n=0}^{\infty}$ is dense in $\mathcal{H}$. This condition is way too strong in the context of frames, as it implies that $\{T^n\varphi\}_{n=0}^{\infty}$ does not satisfy the Bessel condition.

4. Shift-invariant systems and Gabor frames

In this section we will consider some classes of explicitly given frames in the Hilbert space $L^2(\mathbb{R})$ and open problems related to dynamical sampling. For our purpose the central class of frames is the so-called Gabor frames, but it is natural also to consider shift-invariant systems. Both systems are defined in terms of certain classes of operators on $L^2(\mathbb{R})$.

For $a \in \mathbb{R}$, define the *translation operator*

$$T_a : L^2(\mathbb{R}) \to L^2(\mathbb{R}), \quad T_a f(x) := f(x - a)$$

and the *modulation operator*

$$E_a : L^2(\mathbb{R}) \to L^2(\mathbb{R}), \quad E_a f(x) := e^{2\pi i a x} f(x).$$

The translation operators and the modulation operators are unitary. We define the Fourier transform of $f \in L^1(\mathbb{R})$ by

$$\widehat{f}(\gamma) = \mathcal{F}f(\gamma) = \int_{-\infty}^{\infty} f(x) e^{-2\pi i \gamma x}\, dx$$

and extend it in the standard way to a unitary operator on $L^2(\mathbb{R})$.

Given a function $\varphi \in L^2(\mathbb{R})$ and some $b > 0$, the associated *shift-invariant system* is given by $\{T_{kb}\varphi\}_{k\in\mathbb{Z}}$. The frame properties of such systems are well understood. The following proposition collects some of the main results. Given $\varphi \in L^2(\mathbb{R})$ and some $b > 0$, consider the function

$$\Phi(\gamma) := \sum_{k\in\mathbb{Z}} \left| \widehat{\varphi}\left(\frac{\gamma + k}{b} \right) \right|^2, \ \gamma \in \mathbb{R}. \tag{4.1}$$

Proposition 4.1. *Let $\varphi \in L^2(\mathbb{R}) \setminus \{0\}$ and $b > 0$ be given. Then the followings hold:*

(i) *$\{T_{kb}\varphi\}_{k\in\mathbb{Z}}$ is linearly independent.*
(ii) *$\{T_{kb}\varphi\}_{k\in\mathbb{Z}}$ is a Riesz basis if and only if there exist $A, B > 0$ such that $A \le \Phi(\gamma) \le B$ for almost all $\gamma \in [0,1]$.*
(iii) *$\{T_{kb}\varphi\}_{k\in\mathbb{Z}}$ is a frame sequence if and only if there exist $A, B > 0$ such that $A \le \Phi(\gamma) \le B$ for almost all $\gamma \in [0,1] \setminus \{\gamma \in [0,1] \mid \Phi(\gamma) = 0\}$.*
(iv) *$\{T_{kb}\varphi\}_{k\in\mathbb{Z}}$ can at most be a frame for a proper subspace of $L^2(\mathbb{R})$.*
(v) *If $\{T_{kb}\varphi\}_{k\in\mathbb{Z}}$ is an overcomplete frame sequence, it has infinite excess.*
(vi) *$\{T_{kb}\varphi\}_{k\in\mathbb{Z}} = \{(T_b)^k\varphi\}_{k\in\mathbb{Z}}$, i.e., the system $\{T_{kb}\varphi\}_{k\in\mathbb{Z}}$ has the form of an iterated system indexed by $\mathbb{Z}$.*

The result in (i) is well-known, and (ii) & (iii) are proved in [5]; (iv) is proved in [8], (v) is proved in [4, 10], and (vi) is clear.

Note that the representation of a shift-invariant system as an iterated system in Proposition 4.1 (v) differs from the ones appearing in Section 3 in terms of the index set $\mathbb{Z}$. It was recently shown in [11] that even though a shift-invariant frame $\{T_{kb}\varphi\}_{k\in\mathbb{Z}}$ can be re-indexed and be represented in the form (1.1) for a linear operator $T : \text{span}\{T_{kb}\varphi\}_{k\in\mathbb{Z}} \to L^2(\mathbb{R})$, the operator T can only be bounded if $\{T_{kb}\varphi\}_{k\in\mathbb{Z}}$ is a basis.

Let us now introduce the Gabor systems. Given some $a, b > 0$ and a function $g \in L^2(\mathbb{R})$, the associated *Gabor system* is the collection of functions given by

$$\{E_{mb}T_{na}g\}_{m,n\in\mathbb{Z}} = \left\{e^{2\pi i m b x} g(x - na)\right\}_{m,n\in\mathbb{Z}}.$$

Gabor systems play an important role in time-frequency analysis; we will just state the properties that are necessary for the flow of the current paper, and refer to [17, 15, 16, 7] for much more information.

Proposition 4.2. *Let $g \in L^2(\mathbb{R}) \setminus \{0\}$. Then the followings hold:*

(i) *$\{E_{mb}T_{na}g\}_{m,n\in\mathbb{Z}}$ is linearly independent.*
(ii) *If $\{E_{mb}T_{na}g\}_{m,n\in\mathbb{Z}}$ is a frame for $L^2(\mathbb{R})$, then $ab \le 1$.*
(iii) *If $\{E_{mb}T_{na}g\}_{m,n\in\mathbb{Z}}$ is a frame for $L^2(\mathbb{R})$, then $\{E_{mb}T_{na}g\}_{m,n\in\mathbb{Z}}$ is a Riesz basis if and only if $ab = 1$.*
(iv) *If $\{E_{mb}T_{na}g\}_{m,n\in\mathbb{Z}}$ is an overcomplete frame for $L^2(\mathbb{R})$, then it has infinite excess.*

The result in (i) was proved in [21] (hereby confirming a conjecture stated in [19]); (ii) & (iii) are classical results [17, 7], and (iv) is proved in [4].

Note that since a Gabor frame $\{E_{mb}T_{na}g\}_{m,n\in\mathbb{Z}}$ is linearly independent by Proposition 4.2 (i), Proposition 3.3 shows that any reordering $\{f_k\}_{k=1}^{\infty}$ can be represented on the form $\{T^n\varphi\}_{n=0}^{\infty}$ for $T : L^2(\mathbb{R}) \to L^2(\mathbb{R})$ some linear operator and some $\varphi \in L^2(\mathbb{R})$. However, it was recently shown in [11] that the operator T always is unbounded, except in the case where the Gabor frame is a Riesz basis, i.e., if $ab = 1$. This is indeed the reason that we considered the more general representations (3.3) in Section 3:

the hope is that the possibility of choosing appropriate coefficients a_n allows to find a bounded operator T. Let us formulate the key question as an open problem.

Problem 2. Do there exist overcomplete Gabor frames $\{E_{mb}T_{na}g\}_{m,n\in\mathbb{Z}}$ such that an appropriate ordering $\{f_k\}_{k=1}^{\infty}$ of the frame elements has a representation

$$\{f_k\}_{k=1}^{\infty} = \{a_n T^n \varphi\}_{n=0}^{\infty}, \tag{4.2}$$

for some scalars $a_n \neq 0$, a bounded operator $T : L^2(\mathbb{R}) \to L^2(\mathbb{R})$, and some $\varphi \in L^2(\mathbb{R})$?

Note that if the Gabor frame $\{E_{mb}T_{na}g\}_{m,n\in\mathbb{Z}}$ is generated by a function g for which $||g|| = 1$ and we assume that $a_n > 0$ for all $n \in \mathbb{N}_0$, the representation (4.2) implies that $a_n = ||T^n\varphi||^{-1}$, and thus the representation takes the form

$$\{f_k\}_{k=1}^{\infty} = \left\{ \frac{T^n \varphi}{||T^n \varphi||} \right\}_{n=0}^{\infty}. \tag{4.3}$$

For shift-invariant frames $\{T_{kb}\varphi\}_{k\in\mathbb{Z}}$, we just saw that they are indeed "born" having the structure of an iterated system, indexed by $\mathbb{Z}$. Thus, it is natural to ask whether such a representation is possible for Gabor frames as well. Let us formulate this as the final problem.

Problem 3. Do there exist overcomplete Gabor frames $\{E_{mb}T_{na}g\}_{m,n\in\mathbb{Z}}$ such that an appropriate ordering $\{f_k\}_{k=-\infty}^{\infty}$ of the frame elements has a representation

$$\{f_k\}_{k=-\infty}^{\infty} = \{T^n \varphi\}_{n=-\infty}^{\infty},$$

for some bounded operator $T : L^2(\mathbb{R}) \to L^2(\mathbb{R})$ and some $\varphi \in L^2(\mathbb{R})$?

Acknowledgment. The authors would like to thank Albrecht Böttcher for many useful comments to the manuscript.

References

[1] A. Aldroubi, C. Cabrelli, U. Molter, and S. Tang, Dynamical sampling, *Appl. Harm. Anal. Appl.* **42** (2017), no. 3, 378–401.

[2] A. Aldroubi, C. Cabrelli, A.F. Çakmak, U. Molter, and A. Petrosyan, Iterative actions of normal operators, *J. Funct. Anal.* **272** (2017), no. 3, 1121–1146.

[3] A. Aldroubi, and A. Petrosyan, Dynamical sampling and systems from iterative actions of operators, in: *Frames and Other Bases in Abstract and Function Spaces*, eds. H. Mhaskar, I. Pesenson, D.X. Zhou, Q.T. Le Gia, and A. Mayeli, Birkhäuser, Boston, 2017.

[4] R. Balan, P. Casazza, C. Heil, and Z. Landau, Deficits and excesses of frames, *Adv. Comp. Math.* **18** (2002), 93–116.

[5] J. Benedetto and S. Li, The theory of multiresolution analysis frames and applications to filter banks, *Appl. Comp. Harm. Anal.* **5** (1998), 389–427.

[6] C. Cabrelli, U. Molter, V. Paternostro, and F. Philipp, Dynamical sampling on finite index sets, Preprint, 2017.

[7] O. Christensen, *An introduction to frames and Riesz bases,* 2nd expanded ed., Birkhäuser, Boston, 2016.

[8] O. Christensen, B. Deng, and C. Heil, Density of Gabor frames, *Appl. Comp. Harm. Anal.* **7** (1999), 292–304.

[9] O. Christensen, and M. Hasannasab, Frame properties of systems arising via iterative actions of operators, To appear in *Appl. Comp. Harm. Anal.*.

[10] O. Christensen, and M. Hasannasab, Operator representations of frames: boundedness, duality, and stability, *Integral Equations and Operator Theory* **88** (2017), no. 4, 483–499.

[11] O. Christensen, M. Hasannasab, and F. Philipp, Frame properties of operator orbits, Submitted, 2018.

[12] O. Christensen, M. Hasannasab, and E. Rashidi, Dynamical sampling and frame representations with bounded operators, Preprint, 2017.

[13] O. Christensen, M. Hasannasab, and D.T. Stoeva, Operator representations of sequences and dynamical sampling, Preprint, 2017.

[14] I. Daubechies, The wavelet transformation, time-frequency localization and signal analysis, *IEEE Trans. Inform. Theory* **36** (1990), 961–1005.

[15] H.G. Feichtinger and T. Strohmer (eds.), *Gabor Analysis and Algorithms: Theory and Applications,* Birkhäuser, Boston, 1998.

[16] H.G. Feichtinger and T. Strohmer (eds.), *Advances in Gabor Analysis,* Birkhäuser, Boston, 2002.

[17] K. Gröchenig, *Foundations of time-frequency analysis,* Birkhäuser, Boston, 2000.

[18] C. Heil, *A basis theory primer,* Expanded ed., Applied and Numerical Harmonic Analysis, Birkhäuser, New York, 2011.

[19] C. Heil, J. Ramanathan, and P. Topiwala, Linear independence of time-frequency translates, *Proc. Amer. Math. Soc.* **124** (1996), 2787–2795.

[20] J. Klauder and B. Skagerstam, *Coherent states. Applications in physics and mathematical physics,* World Scientific, Singapore, 1985.

[21] P. Linnell, Von Neumann algebras and linear independence of translates, *Proc. Amer. Math. Soc.* **127** (1999), no. 11, 3269–3277.

[22] A. Perelomov, *Generalized coherent states and their applications,* Springer, New York, 2012.

[23] F. Philipp, Bessel orbits of normal operators, *J. Math. Anal. Appl.* **448** (2017), 767–785.

[24] H.S. Shapiro and A.L. Shields, On some interpolation problems for analytic functions, *American Journal of Mathematics* **83** (1961), no. 3, 513–532.

Ole Christensen and Marzieh Hasannasab
Technical University of Denmark
DTU Compute, Building 303, 2800 Lyngby
Denmark
e-mail: ochr@dtu.dk
mhas@dtu.dk

A survey on solvable sesquilinear forms

Rosario Corso

Abstract. The aim of this paper is to present a unified theory of many Kato type representation theorems in terms of solvable forms on a Hilbert space $(\mathcal{H}, \langle \cdot, \cdot \rangle)$. In particular, for some sesquilinear forms Ω on a dense domain $\mathcal{D} \subseteq \mathcal{H}$ one looks for a representation $\Omega(\xi, \eta) = \langle T\xi, \eta \rangle$ $(\xi \in D(T), \eta \in \mathcal{D})$, where T is a densely defined closed operator with domain $D(T) \subseteq \mathcal{D}$. There are two characteristic aspects of a solvable form on $\mathcal{H}$. One is that the domain of the form can be turned into a reflexive Banach space that need not be a Hilbert space. The second one is that representation theorems hold after perturbing the form by a bounded form that is not necessarily a multiple of the inner product of $\mathcal{H}$.

Mathematics Subject Classification (2010). Primary 47A07; Secondary 47A10, 47A12.

Keywords. Kato's representation theorems, q-closed and solvable sesquilinear forms.

1. Introduction

Let $\mathcal{H}$ be a Hilbert space with inner product $\langle \cdot, \cdot \rangle$. Bounded linear operators and bounded sesquilinear forms are related by the formula

$$\Omega(\xi, \eta) = \langle T\xi, \eta \rangle \quad \forall \xi, \eta \in \mathcal{H},$$

which holds for every bounded sesquilinear form Ω and for some bounded linear operator T by Riesz's classical representation theorem. The situation in the unbounded case is more complicated. One of the earliest results on this topic is formulated by Kato in [8].

Kato's first representation theorem. *Let Ω be a densely defined closed sectorial form with domain $\mathcal{D} \subseteq \mathcal{H}$. Then there exists a unique m-sectorial operator T, with domain $D(T) \subseteq \mathcal{D}$, such that*

$$\Omega(\xi, \eta) = \langle T\xi, \eta \rangle \quad \forall \xi \in D(T), \ \eta \in \mathcal{D}. \tag{1.1}$$

© Springer International Publishing AG, part of Springer Nature 2018

A. Böttcher et al. (eds.), *The Diversity and Beauty of Applied Operator Theory*, Operator Theory: Advances and Applications 268, https://doi.org/10.1007/978-3-319-75996-8_9

Here there are some differences compared to the bounded case. For example, representation (1.1) does not necessarily hold on the whole $\mathcal{D}$ because in general $D(T)$ is smaller than $\mathcal{D}$. However, $D(T)$ is not a 'small' subspace since it is dense in $\mathcal{H}$. It is worth mentioning that a representation like (1.1) can be given for any sesquilinear form Ω considering the operator defined by

$$D(T) = \{\xi \in \mathcal{D} : \exists \chi \in \mathcal{H},\ \Omega(\xi, \eta) = \langle \chi, \eta \rangle\ \forall \eta \in \mathcal{D}\} \tag{1.2}$$

and $T\xi = \chi$ for all $\xi \in D(T)$ and χ as in (1.2). Note that this T, the operator *associated* to Ω, is the maximal operator that satisfies (1.1). However, usually one is looking for operators T satisfying some additional requirements, such as closedness or properties of the resolvent set, like in Kato's theorem.

A bijection between densely defined closed sectorial forms and their associated operators (i.e. m-sectorial operators) is valid. But this bijection is not preserved when we consider a larger class of sesquilinear forms. Indeed, there exist many sesquilinear forms with the same associated operator (see Proposition 4.2 of [6]). Although in the unbounded case the representation on the whole domain and the correspondence between forms and operators are lost, we have the following strong result (see [8]).

Kato's second representation theorem. *Let Ω be a densely defined closed non-negative sesquilinear form with domain $\mathcal{D}$ and let T be its positive self-adjoint associated operator. Then $\mathcal{D} = D(T^{\frac{1}{2}})$ and*

$$\Omega(\xi, \eta) = \langle T^{\frac{1}{2}}\xi, T^{\frac{1}{2}}\eta \rangle \quad \forall \xi, \eta \in \mathcal{D}. \tag{1.3}$$

We stress that in (1.3) the representation is well-defined in $\mathcal{D}$, which is also the domain of a positive self-adjoint operator. Nevertheless, this last theorem does not have direct generalizations without the condition of positivity. Indeed, Example 2.11, Proposition 4.2 of [6], and Example 5.4 of [4] show sesquilinear forms that satisfy the first type of representation but not the second.

Kato's theorems lead to several applications. These include, for instance, a way to define the Friedrichs extension of densely defined sectorial operators [8, Ch. VI.2.3], a proof of von Neumann's theorem about the operator T^*T when T is densely defined and closed [8, Example VI.2.13], and a way to prove that some operators are m-sectorial or self-adjoint (see [8, Ch. VI] and also [5] for some generalizations). There are cases in which it is simpler to handle forms rather than operators. Indeed, the sum of two operators might be defined in a small subspace, but with closed forms one can define a special sum that has a dense domain (see [13] for the concrete example of the so-called *form sum* of the operators $Af = -f''$ and $\delta f = f(0)$ with $f \in C_0^\infty(\mathbb{R})$).

Recently, the first representation theorem has been generalized to the context of *q-closed* and *solvable forms* in [1] and, successively, in [2] (see Theorem 2.3 below). The second one has been extended to solvable forms in [3] (see Theorem 2.5 below). Solvable forms constitute a unified theory of many representation theorems (for example [4, 6, 9, 10, 11, 14, 15]).

The new aspects of solvable forms, compared to the ones in the works mentioned above, are the following (see Definitions 2.1 and 2.2). First, the structure of a *reflexive Banach space* on the domain of the form need not be the structure of a Hilbert space. Secondly, representation theorems hold for perturbations of the form by a *bounded form* instead by a multiple of the inner product of the Hilbert space. These conditions are stressed in Example 7.3 of [2] and Example 2.9 in Section 2, respectively.

This paper is organized as follows. In Section 2 we give the definition of solvable forms and their representation theorems. We show in Section 3 some properties of these forms in terms of the numerical range. Section 4 provides an exposition of particular cases of solvable forms known in the literature. In the final section we discuss another representation, the *Radon–Nikodym-like representation.*

2. The representation theorems

Throughout this paper we will use the following notations. $\mathcal{H}$ is a Hilbert space with inner product $\langle\cdot,\cdot\rangle$ and norm $\|\cdot\|$, $\mathcal{D}$ is a dense subspace of $\mathcal{H}$, $D(T), R(T)$, and $\rho(T)$ are the *domain*, *range*, and *resolvent set* of an operator T on $\mathcal{H}$, respectively, $\mathcal{B}(\mathcal{H})$ is the set of *bounded* linear operators defined everywhere on $\mathcal{H}$, $\Re B$ and $\Im B$ are the *real* and *imaginary parts* of an operator $B \in \mathcal{B}(\mathcal{H})$, respectively,

$$\mathfrak{n}_T := \{\langle T\xi,\xi\rangle : \xi \in D(T),\ \|\xi\| = 1\}$$

is the *numerical range* of T, and l_p with $p > 1$ is the classic Banach space with the usual norm.

We will consider sesquilinear forms defined on $\mathcal{D}$, i.e., maps $\mathcal{D}\times\mathcal{D}\to\mathbb{C}$ which are linear in the first component and anti-linear in the second one. If Ω is a sesquilinear form defined on $\mathcal{D}$, then the *adjoint form* Ω^* of Ω is given by $\Omega^*(\xi,\eta) = \overline{\Omega(\eta,\xi)}$ for $\xi,\eta\in\mathcal{D}$. The *real* and *imaginary* parts $\Re\Omega$ and $\Im\Omega$ are $\Re\Omega = \frac{1}{2}(\Omega+\Omega^*)$ and $\Im\Omega = \frac{1}{2i}(\Omega-\Omega^*)$, respectively. The *numerical range* of Ω is

$$\mathfrak{n}_\Omega := \{\Omega(\xi,\xi) : \xi\in\mathcal{D},\ \|\xi\| = 1\}.$$

Ω is said to be *symmetric* if $\Omega = \Omega^*$ (i.e., $\mathfrak{n}_\Omega \subseteq \mathbb{R}$) and, in particular, Ω is *non-negative* if $\mathfrak{n}_\Omega \subseteq [0,+\infty)$. We will denote by ι the sesquilinear form $\iota(\xi,\eta) = \langle\xi,\eta\rangle$ $(\xi,\eta\in\mathcal{H})$ and by ϑ the null form on $\mathcal{H}$. The following definition of q-closed forms is taken from [2, Proposition 3.2].

Definition 2.1. A sesquilinear form Ω on $\mathcal{D}$ is called *q-closed with respect to a norm on* $\mathcal{D}$ which is denoted by $\|\cdot\|_\Omega$ if

1. $\mathcal{E}_\Omega := \mathcal{D}[\|\cdot\|_\Omega]$ is a reflexive Banach space;
2. the embedding $\mathcal{E}_\Omega \hookrightarrow \mathcal{H}$ is continuous;
3. there exists a $\beta \geq 0$ such that $|\Omega(\xi,\eta)| \leq \beta\|\xi\|_\Omega\|\eta\|_\Omega$ for all $\xi,\eta\in\mathcal{D}$; i.e., Ω is bounded on $\mathcal{E}_\Omega$.

If $\mathcal{E}_\Omega$ is a Hilbert space, then Ω is called *q-closed with respect to the inner product of* $\mathcal{E}_\Omega$.

Let Ω be a q-closed sesquilinear form on $\mathcal{D}$ w.r.t. $\|\cdot\|_\Omega$. We denote by $\mathcal{E}_\Omega^\times$ the conjugate dual space of $\mathcal{E}_\Omega := \mathcal{D}[\|\cdot\|_\Omega]$ and by $\langle \Lambda, \xi \rangle$ the action of the conjugate linear functional $\Lambda \in \mathcal{E}_\Omega^\times$ on an element $\xi \in \mathcal{D}$. The reason why we use the symbol $\langle \cdot, \cdot \rangle$ also here is that $\mathcal{H}$ is continuously embedded into $\mathcal{E}_\Omega^\times$ and the action of elements of $\mathcal{E}_\Omega^\times$ is an extension of the inner product of $\mathcal{H}$ (see [2, Sect. 4]). Let $\mathfrak{P}(\Omega)$ be the set of bounded sesquilinear forms Υ on $\mathcal{H}$ such that

1. if $(\Omega + \Upsilon)(\xi, \eta) = 0$ for all $\eta \in \mathcal{D}$, then $\xi = 0$;
2. for all $\Lambda \in \mathcal{E}_\Omega^\times$ there exists a $\xi \in \mathcal{D}$ such that $\langle \Lambda, \eta \rangle = (\Omega + \Upsilon)(\xi, \eta)$ for all $\eta \in \mathcal{D}$.

Definition 2.2. If the set $\mathfrak{P}(\Omega)$ is not empty, then Ω is said to be *solvable w.r.t.* $\|\cdot\|_\Omega$ (if moreover $\|\cdot\|_\Omega$ is a Hilbert norm, then Ω is also said to be *solvable w.r.t. the inner product* induced by $\|\cdot\|_\Omega$).

Solvable forms are q-closed forms characterized by the existence of a bounded sesquilinear form Υ on $\mathcal{H}$ such that the operator $X_\Upsilon : \mathcal{E}_\Omega \to \mathcal{E}_\Omega^\times$ is bijective, where $\langle X_\Upsilon \xi, \eta \rangle = \Omega(\xi, \eta) + \Upsilon(\xi, \eta)$ for all $\eta \in \mathcal{E}_\Omega$ (see [1, Lemma 5.6]). Therefore, the set $\mathfrak{P}(\Omega)$ denotes perturbations of Ω with bounded forms which induce a bijection of $\mathcal{E}_\Omega$ onto $\mathcal{E}_\Omega^\times$. Equivalent characterizations of solvable forms are provided by [2, Lemma 5.1].

The next theorem generalizes Kato's first representation theorem to solvable forms (for a proof see [2, Theorem 4.6] and also [3, Theorem 2.7]).

Theorem 2.3. *Let Ω be a solvable sesquilinear form on $\mathcal{D}$ w.r.t. a norm $\|\cdot\|_\Omega$. Then there exists a closed operator T, with dense domain $D(T) \subseteq \mathcal{D}$ in $\mathcal{H}$, such that the following statements hold.*

1. *$\Omega(\xi, \eta) = \langle T\xi, \eta \rangle$ for all $\xi \in D(T), \eta \in \mathcal{D}$.*
2. *$D(T)$ is dense in $\mathcal{D}[\|\cdot\|_\Omega]$.*
3. *A bounded form $\Upsilon(\cdot, \cdot) = \langle B\cdot, \cdot \rangle$ belongs to $\mathfrak{P}(\Omega)$ if and only if 0 is in $\rho(T + B)$. In particular, if $\Upsilon = -\lambda\iota$, with $\lambda \in \mathbb{C}$, then $\Upsilon \in \mathfrak{P}(\Omega)$ if and only if $\lambda \in \rho(T)$.*

The operator T is uniquely determined by the following condition. Let ξ, χ be in $\mathcal{H}$. Then $\xi \in D(T)$ and $T\xi = \chi$ if and only if $\xi \in \mathcal{D}$ and $\Omega(\xi, \eta) = \langle \chi, \eta \rangle$ for all η belonging to a dense subset of $\mathcal{D}[\|\cdot\|_\Omega]$.

Kato's second theorem is generalized in Theorem 2.5 below for the special class of hyper-solvable forms defined in the following way (see [3, Lemma 4.14, Theorem 4.17]).

Definition 2.4. A solvable sesquilinear form on $\mathcal{D}$ with associated operator T is said to be *hyper-solvable* if $\mathcal{D} = D(|T|^{\frac{1}{2}})$.

Theorem 2.5. *Let Ω be a hyper-solvable sesquilinear form on $\mathcal{D}$ w.r.t. a norm $\|\cdot\|_\Omega$ and with associated operator T. Then $\mathcal{D} = D(|T^*|^{\frac{1}{2}})$ and*

$$\Omega(\xi, \eta) = \langle U|T|^{\frac{1}{2}}\xi, |T^*|^{\frac{1}{2}}\eta \rangle \quad \forall \xi, \eta \in \mathcal{D},$$

$$\Omega(\xi, \eta) = \langle |T^*|^{\frac{1}{2}}U\xi, |T^*|^{\frac{1}{2}}\eta \rangle \quad \forall \xi, \eta \in \mathcal{D},$$

where $T = U|T| = |T^*|U$ is the polar decomposition of T, and $\|\cdot\|_\Omega$ is equivalent to the graph norms of $|T|^{\frac{1}{2}}$ and of $|T^*|^{\frac{1}{2}}$.

Remark 2.6. According to Theorem 2.3 and since the resolvent set of a closed operator is open we obtain the following property: if Ω is a solvable sesquilinear form and $\Upsilon \in \mathfrak{P}(\Omega)$, then there exists a $\delta > 0$ such that $(\Upsilon + \mu\iota) \in \mathfrak{P}(\Omega)$, for all $|\mu| < \delta$.

We mention some other features of a q-closed and solvable form Ω w.r.t. $\|\cdot\|_\Omega$:

- the same property of being q-closed or solvable holds for the adjoint Ω^* ([2, Theorem 4.11]) (this implies also that $\Re\Omega$ and $\Im\Omega$ are q-closed forms);
- the operators associated to Ω and to Ω^* are the adjoints of each other ([2, Theorem 4.11]);
- the peculiarity of symmetric solvable forms to have self-adjoint associated operators ([2, Corollary 4.14]);
- all norms w.r.t. which Ω is q-closed or solvable are equivalent to $\|\cdot\|_\Omega$ ([2, Theorems 3.8 and 4.4]);
- different hyper-solvable forms have different associated operators ([3, Theorem 5.3]).

Remark 2.7. Let Ω_1, Ω_2 be two q-closed sesquilinear forms on $\mathcal{D}$ w.r.t. $\|\cdot\|_1$ and $\|\cdot\|_2$, respectively, and let $c \in \mathbb{C}$. Then the two norms are equivalent by [2, Theorem 2.5] and the sesquilinear forms $c\Omega_1, \Omega_1^*, \Re\Omega_1, \Im\Omega_1, \Omega_1 + \Omega_2$ are q-closed w.r.t. both $\|\cdot\|_1$ and $\|\cdot\|_2$.

We conclude this section by presenting some examples of solvable forms (cf. [2, Example 4.16] and [3, Example 4.5]).

Example 2.8. Let $\underline{\alpha} := \{\alpha_n\}$ be a sequence of complex numbers and

$$\Omega_{\underline{\alpha}}(\{\xi_n\}, \{\eta_n\}) = \sum_{n=1}^{\infty} \alpha_n \xi_n \overline{\eta_n}$$

with domain $\mathcal{D} = \{\{\xi_n\} \in l_2 : \sum_{n=1}^{\infty} |\alpha_n||\xi_n|^2 < \infty\}$. The form $\Omega_{\underline{\alpha}}$ is hyper-solvable w.r.t. the norm given by

$$\|\{\xi_n\}\|_{\Omega_{\underline{\alpha}}} = \left(\sum_{n=1}^{\infty} |\xi_n|^2 + \sum_{n=1}^{\infty} |\alpha_n||\xi_n|^2\right)^{\frac{1}{2}}.$$

Moreover,

1. if $\overline{\{\alpha_n : n \in \mathbb{N}\}} \neq \mathbb{C}$, then $-\lambda\iota \in \mathfrak{P}(\Omega_{\underline{\alpha}})$ for all $\lambda \notin \overline{\{\alpha_n : n \in \mathbb{N}\}}$;
2. if in the general case[1] we define the sequence $\underline{\beta} = \{\beta_n\}$ by $\beta_n = -\alpha_n + 1$ for $|\alpha_n| \leq 1$ and $\beta_n = 0$ for $|\alpha_n| > 1$, then the form $\Omega_{\underline{\beta}}$ is bounded and $0 \notin \overline{\{\alpha_n + \beta_n : n \in \mathbb{N}\}}$, and, by the previous case, $\Omega_{\underline{\alpha}} + \Omega_{\underline{\beta}}$ is solvable, and $\Omega_{\underline{\beta}} \in \mathfrak{P}(\Omega_{\underline{\alpha}})$.

[1]The case $\overline{\{\alpha_n : n \in \mathbb{N}\}} = \mathbb{C}$ is not considered in [1, 2].

The operator associated to $\Omega_{\underline{\alpha}}$ is the operator $\mathscr{M}_{\underline{\alpha}}$ of multiplication by $\underline{\alpha}$ with domain

$$D(\mathscr{M}_{\underline{\alpha}}) = \left\{ \{\xi_n\} \in l_2 : \sum_{n=1}^{\infty} |\alpha_n|^2 |\xi_n|^2 < \infty \right\},$$

which is given by $\mathscr{M}_{\underline{\alpha}}\{\xi_n\} = \{\alpha_n \xi_n\}$ for $\{\xi_n\} \in D(\mathscr{M}_{\underline{\alpha}})$.

The next is a new example of a solvable sesquilinear form.

Example 2.9. Let $1 < p < 2$ and q be such that $\frac{1}{p} + \frac{1}{q} = 1$. For convenience we denote by $\underline{\xi} = \{\xi_n\}$ the generic element of the space l_r with $r > 1$. Let moreover $\mathcal{D} = l_p \oplus l_q$, which is a reflexive Banach space if it is endowed with the norm $\|(\underline{\xi}, \underline{\eta})\|_{\mathcal{D}} = \|\underline{\xi}\|_p + \|\underline{\eta}\|_q$ (as usual, $\|\cdot\|_p$ and $\|\cdot\|_q$ are the classical norms on l_p and l_q, respectively). The Banach space $\mathcal{D}[\|\cdot\|_{\mathcal{D}}]$ will be denoted by $\mathcal{E}$. Observe that $\mathcal{E}$ is isomorphic to its conjugate dual space $\mathcal{E}^\times$. Indeed, we have the isomorphism (we identify $\mathcal{E}^\times$ with $l_q \oplus l_p$)

$$\begin{aligned} X : \quad \mathcal{E} \quad &\to \quad \mathcal{E}^\times \\ (\underline{\xi}, \underline{\eta}) &\mapsto (\underline{\eta}, \underline{\xi}). \end{aligned}$$

The action of X is given by

$$\langle X(\underline{\xi}, \underline{\eta}), (\underline{\xi}', \underline{\eta}') \rangle = \sum_{n=1}^{\infty} (\eta_n \overline{\xi_n'} + \xi_n \overline{\eta_n'}) \tag{2.1}$$

for all $(\underline{\xi}, \underline{\eta}), (\underline{\xi}', \underline{\eta}') \in \mathcal{E}$. Now we define a sesquilinear form Ω on $\mathcal{D}$ exactly by (2.1); i.e., for $(\underline{\xi}, \underline{\eta}), (\underline{\xi}', \underline{\eta}') \in \mathcal{E}$,

$$\Omega((\underline{\xi}, \underline{\eta}), (\underline{\xi}', \underline{\eta}')) := \sum_{n=1}^{\infty} (\eta_n \overline{\xi_n'} + \xi_n \overline{\eta_n'}).$$

Then Ω is bounded on $\mathcal{E}$. Indeed, an easy computation shows that

$$\begin{aligned} \left| \Omega((\underline{\xi}, \underline{\eta}), (\underline{\xi}', \underline{\eta}')) \right| &\leq \left| \sum_{n=1}^{\infty} \eta_n \overline{\xi_n'} \right| + \left| \sum_{n=1}^{\infty} \xi_n \overline{\eta_n'} \right| \\ &\leq \|\underline{\xi}'\|_p \|\underline{\eta}\|_q + \|\underline{\xi}\|_p \|\underline{\eta}'\|_q \leq \|(\underline{\xi}, \underline{\eta})\|_{\mathcal{D}} \|(\underline{\xi}', \underline{\eta}')\|_{\mathcal{D}}. \end{aligned}$$

Our goal is to show that Ω is solvable w.r.t. $\|\cdot\|_{\mathcal{E}}$. The first thing we need is a Hilbert space in which $\mathcal{E}$ can be continuously embedded with dense range. To this end, we list the following observations:

- l_p is continuously embedded in the Hilbert space $\mathcal{H}_1 := l_2$ with dense range.
- An inner product on l_q can be given by

  $$[\underline{\eta}, \underline{\eta}'] = \sum_{n=1}^{\infty} 2^{-n} \eta_n \overline{\eta_n'}, \quad \underline{\eta}, \underline{\eta}' \in l_q.$$

  This is well defined since

  $$|[\underline{\eta}, \underline{\eta}']| \leq \sum_{n=1}^{\infty} 2^{-n} |\eta_n||\eta_n'| \leq \sum_{m=1}^{\infty} 2^{-m} \|\underline{\eta}\|_q \|\underline{\eta}'\|_q = \|\underline{\eta}\|_q \|\underline{\eta}'\|_q. \tag{2.2}$$

Let $(\mathcal{H}_2, [\cdot,\cdot])$ be the completion of the pre-Hilbert space $(l_q, [\cdot,\cdot])$. Moreover, by (2.2) we obtain $[\eta,\eta]^{\frac{1}{2}} \leq \|\eta\|_q$ for every $\eta \in l_q$. Hence, l_q is continuously embedded into $\mathcal{H}_2$, and of course the range is dense.
- $\mathcal{E}$ is continuously embedded into $\mathcal{H} := \mathcal{H}_1 \oplus \mathcal{H}_2$ and the range is dense.

These observations prove that Ω is q-closed w.r.t. $\|\cdot\|_{\mathcal{D}}$. Moreover, by [1, Lemma 5.6] Ω is solvable w.r.t. $\|\cdot\|_{\mathcal{D}}$ (in fact the operator X_ϑ coincides with X, which is bijective).

However, Ω is not solvable w.r.t. any inner product. Indeed, presuming the contrary, $\mathcal{D}$ would be a Hilbert space with the same topology as that of $\mathcal{E}$ by [2, Theorem 3.8]. The subspace $l_p \oplus \{0\}$ is closed in $\mathcal{E}$, therefore l_p would be a Hilbert space with the same topology as the one induced by $\|\cdot\|_p$. But we know that l_p is not isomorphic to a Hilbert space (for example, it is a consequence of [12]).

Moreover, this form is not hyper-solvable by [3, Corollary 4.4].

3. Numerical range

The purpose of this short section is to show that the numerical range of a q-closed form plays a special role for the property of being solvable.

Lemma 3.1.[2] *Let Ω be a solvable sesquilinear form and let $\mathfrak{n}_\Omega \neq \mathbb{C}$ be its numerical range. Assume that $\mathfrak{m}$ is a connected component of $\overline{\mathfrak{n}_\Omega}^c$, the complementary set of the closure of $\mathfrak{n}_\Omega$. Then the following statements are equivalent:*

1. $-\lambda\iota \in \mathfrak{P}(\Omega)$ *for some* $\lambda \in \mathfrak{m}$;
2. $-\mu\iota \in \mathfrak{P}(\Omega)$ *for all* $\mu \in \mathfrak{m}$.

Proof. This is an immediate consequence of statement 3 of Theorem 2.3 and of the fact that the defect numbers of the associated operator are constant on $\mathfrak{m}$ (see [8, Theorem V.3.2]). □

Let Ω be a q-closed sesquilinear form on $\mathcal{D}$ w.r.t. a norm $\|\cdot\|_\Omega$ with numerical range $\mathfrak{n}_\Omega \neq \mathbb{C}$. Let Υ be a bounded form such that $\mathfrak{n}_\Omega \cap \mathfrak{n}_{-\Upsilon} = \varnothing$, where $\mathfrak{n}_{-\Upsilon}$ is the numerical range of $-\Upsilon$ (we may in particular take $\Upsilon = -\lambda\iota$ with $\lambda \notin \mathfrak{n}_\Omega$). Theorem 5.2 of [2] gives an equivalent condition for Υ to be in $\mathfrak{P}(\Omega)$, and Theorem 5.4 of [2] contains a sufficient condition. By statement 2 of Corollary 5.3 of [2], $\Upsilon \in \mathfrak{P}(\Omega)$ if and only if the map

$$\xi \mapsto \sup_{\|\eta\|_\Omega = 1} |(\Omega + \Upsilon)(\xi,\eta)|$$

defines a norm on $\mathcal{D}$ that is equivalent to $\|\cdot\|_\Omega$.

We also mention that the numerical range of the operator associated to a solvable sesquilinear form is a dense subset of the numerical range of the form ([2, Proposition 4.13]).

[2] A particular case of this result for symmetric forms is [3, Corollary 2.8].

4. Special cases

Many representation theorems in the literature for sesquilinear forms are particular cases of Theorem 2.3. The following list witnesses this assertion.

Lemma 4.1. *Let Ω be a sesquilinear form on $\mathcal{D}$.*

1. Ω *satisfies* [4, Theorem 3.3] *if and only if* Ω *is solvable w.r.t. an inner product,* $\mathfrak{n}_\Omega \subseteq \mathbb{R}$ *and* $-\lambda\iota \in \mathfrak{P}(\Omega)$ *for some* $\lambda \in \mathbb{R}$.
2. Ω *satisfies* [6, Theorem 2.3] *if and only if* Ω *is solvable w.r.t. an inner product,* $\mathfrak{n}_\Omega \subseteq \mathbb{R}$ *and* $\vartheta \in \mathfrak{P}(\Omega)$.
3. Ω *is symmetric and satisfies Kato's first theorem if and only if* Ω *is solvable and* $\mathfrak{n}_\Omega$ *is contained in the half-line* $[\omega, +\infty)$ *for some* $\omega \in \mathbb{R}$.
4. Ω *satisfies Kato's first theorem if and only if* Ω *is solvable,* $\mathfrak{n}_\Omega$ *is contained in a sector* $\mathcal{S} := \{\lambda \in \mathbb{C} : \arg(\lambda - \gamma) \leq \theta\}$*, where* $\gamma \in \mathbb{R}$ *and* $0 \leq \theta < \frac{\pi}{2}$*, and* $-\lambda\iota \in \mathfrak{P}(\Omega)$ *for some* $\lambda \notin \mathcal{S}^c$.
5. Ω *satisfies* [10, Theorem 3.1] *if and only if* Ω *is solvable w.r.t. an inner product,* $\mathfrak{n}_\Omega$ *is contained in the half-plane* $\{\lambda \in \mathbb{C} : \Re\lambda \geq 0\}$*, and* $-i\iota \in \mathfrak{P}(\Omega)$.
6. Ω *satisfies* [11, Proposition 2.1] *if and only if* Ω *is solvable w.r.t. an inner product and* $-\lambda\iota \in \mathfrak{P}(\Omega)$ *for some* $\lambda \in \mathbb{C}$.
7. *If* Ω *satisfies* [14, Theorem 2.3]*, then* Ω *is solvable w.r.t. an inner product,* $\mathfrak{n}_\Omega \subseteq \mathbb{R}$*, and* $\mathfrak{P}(\Omega)$ *contains a bounded form which may not be a multiple of the inner product* $\langle \cdot, \cdot \rangle$.
8. *If* Ω *satisfies* [15, Theorem 11.3]*, then* Ω *is solvable w.r.t. an inner product,* $\mathfrak{n}_\Omega$ *is contained in a half-plane which excludes* 0*, and* $\vartheta \in \mathfrak{P}(\Omega)$.

Proof. Item 3 is proved in [3, Proposition 2.9]. Item 4 is a consequence of [2, Proposition 7.1] and Lemma 3.1. The other results are contained in [2, Section 7]. Note that if Ω satisfies [15, Theorem 11.3] (see also [9]), then $|\Omega(\xi,\xi)| \geq \omega\|\xi\|^2$ for all $\xi \in \mathcal{D}$ and some constant $\omega > 0$. Therefore, $0 \notin \overline{\mathfrak{n}_\Omega}$, and since $\mathfrak{n}_\Omega$ is convex, it is contained in a half-plane which excludes 0. □

Remark 4.2. The sesquilinear forms in Example 2.8 with $\overline{\{\alpha_n : n \in \mathbb{N}\}} = \mathbb{C}$, in Example 2.9, and in Example 7.3 of [2] satisfy Theorem 2.3. But the representation theorems listed in Lemma 4.1 cannot be used for these forms.

As for Kato's second representation theorem, Theorem 2.5 generalizes also Theorem 4.2 of [4], Theorem 2.10 of [6], and Theorem 3.1 of [14].

5. Radon–Nikodym-like representation

Theorem 3.8 of [3] provides another representation of sesquilinear forms, one with weaker hypothesis. In particular, a sesquilinear form Ω on $\mathcal{D}$ is q-closed w.r.t. an inner product if and only if

$$\Omega(\xi,\eta) = \langle QH\xi, H\eta\rangle \quad \forall \xi,\eta \in \mathcal{D}, \tag{5.1}$$

where H is a positive self-adjoint operator with domain $D(H) = \mathcal{D}$ and $0 \in \rho(H)$, and $Q \in \mathcal{B}(\mathcal{H})$.

We call a representation as in (5.1) a *Radon–Nikodym-like representation* of Ω. It is never unique (since we can act on Q, H by multiplication with scalars) and Lemma 3.7 of [3] gives a way to obtain this type of representation.

Remark 5.1. (a) Actually, Theorem 3.8 of [3] is another generalization of Kato's second theorem. Indeed, (1.3) is equal to (5.1), with Q being the identity operator and $H = T^{\frac{1}{2}}$.

(b) If Ω is a closed sectorial form with vertex 0 then a Radon–Nikodym-like representation is given by formula (3.5) of [8, Chapter VI].

(c) In [6, 19] the authors dealt with sesquilinear forms like (5.1). In particular, in [6] the operator Q is symmetric and $0 \in \rho(Q)$.

(d) The motivation for the name 'Radon–Nikodym-like' comes from a more general context (see Theorem 3.6 and Example 6.2 of [1]). Previous works on Radon–Nikodym style theorems, in the non-negative case, are [16, 18], which concern the Lebesgue decomposition of non-negative forms (see also [17]). We mention that Theorem 2.2 of [18] and Theorem 3 of [16], with the so-called singular part null, are Kato's second version theorems in a framework with two non-negative sesquilinear forms. However, in this paper by 'Radon–Nikodym-like representation' we mean also that $D(H) = \mathcal{D}$ in (5.1).

Let $\mathcal{S}$ be the family of all q-closed sesquilinear forms on $\mathcal{D}$ w.r.t. to some inner product and let $\mathcal{F}$ be the family of all positive self-adjoint operators H with $D(H) = \mathcal{D}$ and $0 \in \rho(H)$. By [3, Lemma 3.7] we can define a map using the Radon–Nikodym-like representation as

$$\mathfrak{b} : \mathcal{S} \times \mathcal{F} \to \mathcal{B}(\mathcal{H}), \quad \mathfrak{b}(\Omega, H) = Q,$$

where Q is the operator in (5.1). For a fixed $H \in \mathcal{F}$ we can also define a map by

$$\mathfrak{b}_H : \mathcal{S} \to \mathcal{B}(\mathcal{H}), \quad \mathfrak{b}_H(\Omega) = \mathfrak{b}(\Omega, H) = Q.$$

The following proposition is an immediate consequence of [3, Lemma 3.7, Theorem 3.8, and Proposition 3.12].

Proposition 5.2. *For every $H \in \mathcal{F}$, $\mathfrak{b}_H$ establishes an isomorphism between the vector spaces $\mathcal{S}$ and $\mathcal{B}(\mathcal{H})$. Moreover, for every $\Omega \in \mathcal{S}$,*

$$\mathfrak{b}_H(\Omega^*) = \mathfrak{b}_H(\Omega)^*, \quad \mathfrak{b}_H(\Re\Omega) = \Re\mathfrak{b}_H(\Omega), \quad \mathfrak{b}_H(\Im\Omega) = \Im\mathfrak{b}_H(\Omega).$$

Remark 5.3. Let $\mathfrak{n}$ be one of the sets $(0, +\infty)$, $[0, +\infty)$, $\mathbb{R}$, $\{\lambda \in \mathbb{C} : \Re\lambda \geq 0\}$, or $\{\lambda \in \mathbb{C} : \arg(\lambda) \leq \theta\}$ with $0 \leq \theta < \frac{\pi}{2}$. Clearly, $\mathfrak{n}_Q \subseteq \mathfrak{n}$ if, and only if, $\mathfrak{n}_\Omega \subseteq \mathfrak{n}$.

Corollary 5.4. *Suppose that Ω is a q-closed sesquilinear form on $\mathcal{D}$ with Radon–Nikodym-like representation* (5.1).

1. *If $0 \notin \overline{\mathfrak{n}_Q}$, then Ω is solvable and $\vartheta \in \mathfrak{P}(\Omega)$.*
2. *If $\mathfrak{n}_{\Re Q} \subseteq [\gamma, +\infty)$ with $\gamma > 0$, then Ω is a closed sectorial form in Kato's sense.*

Proof. Suppose $0 \notin \overline{\mathfrak{n}_Q}$. Then $0 \in \rho(Q)$ (see [7, Problem 214]) and Ω is solvable with $\vartheta \in \mathfrak{P}(\Omega)$ by [3, Theorem 3.8]. In particular, if $\mathfrak{n}_{\Re Q} \subseteq [\gamma, +\infty)$ with $\gamma > 0$, then Ω is solvable with $\vartheta \in \mathfrak{P}(\Re\Omega)$. Moreover, taking into account that $\mathfrak{n}_Q$ is a bounded subset, we conclude that $\mathfrak{n}_Q$ is contained in a sector $\mathcal{S} = \{\lambda \in \mathbb{C} : \arg(\lambda) \leq \theta\}$ with $0 \leq \theta < \frac{\pi}{2}$. As mentioned in Remark 5.3, Ω has numerical range in $\mathcal{S}$, and there exists a $\lambda < 0$ such that $-\lambda\iota \in \mathfrak{P}(\Re\Omega)$ by Remark 2.6. Finally, Theorem 4.1 implies that Ω is sectorial closed in Kato's sense. □

Acknowledgements

The author gratefully acknowledges the helpful remarks and suggestions of the referees. This work was supported by the Gruppo Nazionale per l'Analisi Matematica, la Probabilità e le loro Applicazioni (GNAMPA) of the Istituto Nazionale di Alta Matematica (INdAM) (project 'Problemi spettrali e di rappresentazione in quasi *-algebre di operatori' 2017).

References

[1] S. Di Bella and C. Trapani, Some representation theorems for sesquilinear forms, *J. Math. Anal. Appl.* **451** (2017), 64–83.

[2] R. Corso and C. Trapani, Representation theorems for solvable sesquilinear forms, *Integral Equ. Oper. Theory* **89** (2017), 43–68.

[3] R. Corso, A Kato's second type representation theorem for solvable sesquilinear forms, arXiv:1707.05073v2, math.FA, 2017; to appear in *J. Math. Anal. Appl.*, DOI 10.1016/j.jmaa.2017.12.058.

[4] A. Fleige, S. Hassi, and H.S.V. de Snoo, A Krein space approach to representation theorems and generalized Friedrichs extensions, *Acta Sci. Math. (Szeged)* **66** (2000), 633–650.

[5] A. Fleige, S. Hassi, H.S.V. de Snoo, and H. Winkler, Sesquilinear forms corresponding to a non-semibounded Sturm-Liouville operator, *Proc. Roy. Soc. Edinburgh* **140A** (2010), 291–318.

[6] L. Grubišić, V. Kostrykin, K.A. Makarov, and K. Veselić, Representation theorems for indefinite quadratic forms revisited, *Mathematika* **59**, (2013), 169—189.

[7] P.R. Halmos, *A Hilbert Space Problem Book*, 2nd ed., Springer, New York, 1982.

[8] T. Kato, *Perturbation Theory for Linear Operators*, Springer, Berlin, 1966.

[9] J.L. Lions, *Équations différentielles opérationnelles et problèmes aux limites*, Springer, Berlin, Göttingen, Heidelberg, 1961.

[10] A. McIntosh, Representation of bilinear forms in Hilbert space by linear operators, *Trans. Amer. Math. Soc.* **131** (1968), 365–377.

[11] A. McIntosh, Hermitian bilinear forms which are not semibounded, *Bull. Amer. Math. Soc.* **76** (1970), 732—737.

[12] F.J. Murray, On complementary manifolds and projections in spaces L_p and l_p, *Trans. Amer. Math. Soc.* **41** (1937), 138–152.

[13] M. Reed and B. Simon, *Methods of Modern Mathematical Physics I. Functional Analysis*, Academic Press, New York, 1972.

[14] S. Schmitz, Representation theorems for indefinite quadratic forms without spectral gap, *Integral Equ. Oper. Theory* **83** (2015), 73—94.

[15] K. Schmüdgen, *Unbounded Self-adjoint Operators on Hilbert Space*, Springer, Dordrecht, 2012.

[16] Z. Sebestyén and T. Titkos, A Radon–Nikodym type theorem for forms, *Positivity* **17** (2013), 863—873.

[17] B. Simon, A canonical decomposition for quadratic forms with applications to monotone convergence theorems, *J. Funct. Anal.* **28** (1978), 371—385.

[18] Zs. Tarcsay, Radon-Nikodym theorems for nonnegative forms, measures and representable functionals, *Complex Anal. Oper. Theory* **10** (2016), 479—494.

[19] K. Veselić, Spectral perturbation bounds for sefadjoint operators. I, *Oper. Matrices* **2** (2008), 307—339.

Rosario Corso
Dipartimento di Matematica e Informatica
Università degli Studi di Palermo
Via Archirafi 34
I-90123 Palermo
Italy
e-mail: `rosario.corso@studium.unict.it`

An application of limiting interpolation to Fourier series theory

Leo R.Ya. Doktorski

Abstract. The limiting real interpolation method is applied to describe the behavior of the Fourier coefficients of functions that belong to spaces which are "very close" to L_2. The Fourier coefficients are taken with respect to bounded orthonormal systems.

Mathematics Subject Classification (2010). Primary 42A16; Secondary 42A24, 43A15, 46B70.

Keywords. Orthonormal bounded system, Fourier coefficients, Real interpolation method, Limiting reiteration theorems.

1. Introduction

For simplicity, we consider (equivalence classes of) complex-valued measurable functions on $(0, 1)$. Let $\{\varphi_n\}$ ($n \in \mathbb{N}$ or $n \in \mathbb{Z}$) be an orthonormal system in L_2 bounded in L_∞,

$$\sup ||\varphi_n||_{L_\infty(0,1)} = M \ (< \infty)\,. \tag{1.1}$$

Everywhere below we denote by $\{c_n(f)\}$ the Fourier coefficients of a function f with respect to the system $\{\varphi_n\}$,

$$c_n \equiv c_n(f) := \int_0^1 f(x)\,\overline{\varphi}_n(x)\,dx.$$

We write $\mathcal{F}$ for the Fourier series map assigning the sequence of Fourier coefficients to a function f, i.e., $\mathcal{F}(f) = \{c_n(f)\}$. It is known [35, 39] that $\mathcal{F}$ is a bounded linear operator from L_2 to l_2 and from L_1 to l_∞ with the norms equal to 1 and M, respectively:

$$||\mathcal{F}|L_2 \to l_2|| = 1, \tag{1.2}$$

$$||\mathcal{F}|L_1 \to l_\infty|| = M. \tag{1.3}$$

The results of Hausdorff, Young, and Paley describe the behavior of the Fourier coefficients of functions that belong to the Lebesgue spaces L_p ($1 < p < 2$). For further information about classical results dealing with the

© Springer International Publishing AG, part of Springer Nature 2018
A. Böttcher et al. (eds.), *The Diversity and Beauty of Applied Operator Theory*, Operator Theory: Advances and Applications 268, https://doi.org/10.1007/978-3-319-75996-8_10

Fourier series map, we refer to [2, 3, 27, 28, 39]. Interpolation between (1.2) and (1.3) by the classical real interpolation method $(*,*)_{\theta,q}$ $(0<\theta<1)$ provides such a description for the Lorentz spaces $L_{p,q}$ [1, 35], namely,

$$\|\mathcal{F}|L_{p,q}\to l_{r,q}\|\le C_{p,q}M^{\frac{2}{p}-1}.$$

Application of the real interpolation functor $(*,*)_{\theta,q,\alpha}$ $(0<\theta<1,\ \alpha\in\mathbb{R})$ involving a logarithmic factor shows that

$$\mathcal{F}:L_{p,q}\,(\log L)_\alpha\to l_{r,q}\,(\log L)_\alpha$$

holds for the Lorentz–Zygmund spaces $L_{p,q}\,(\log L)_\alpha$ [2, 14, 15]. In both formulas $1<p<2$, $\frac{1}{r}=1-\frac{1}{p}$, $0<q\le\infty$. Note that in the scale $L_{p,q}\,(\log L)_\alpha$ this result is optimal (see [34, Theorem 5.3]). Similar results were obtained for $p=1$ in [1, 7, 14]. For $1<p<2$, other approaches can be found in [26, 33, 34].

But for spaces which are "very close" to L_2, all these methods do not work. Instead, two other approaches can be applied. One of them is a "direct way" based on estimates of a K-functional [5, 6, 29, 30, 32]. The other one is based on limiting real interpolation methods [7, 11, 14, 36].

Comparable results are also known for the inverse transformation

$$\mathcal{F}^{-1}:\{c_n\}\mapsto\sum c_n\varphi_n\,(x)\,,$$

applied to complex sequences $\{c_n\}$. It is known [35, 39] that

$$\left\|\mathcal{F}^{-1}|l_2\to L_2\right\|=1,\tag{1.4}$$

$$\left\|F^{-1}|l_1\to L_\infty\right\|\le M.\tag{1.5}$$

Interpolation between (1.4) and (1.5) by the real interpolation methods

$$(*,*)_{\theta,q}\ \text{ or }\ (*,*)_{\theta,q,\alpha}\quad(0<\theta<1)$$

leads also to results for Lorentz and Lorentz–Zygmund sequence spaces with main parameter between 1 and 2 [14, 15, 35]. However, for the spaces which are "very close" to l_2 all these methods do not work either.

The main objective of this work is to study the Fourier series map and its inverse for the spaces which are "very close" to L_2 or l_2 via limiting interpolation methods as well as to estimate the norms of these operators. To be more precise, we prove the following assertions.

Theorem 1.1. *Let* $0<q\le\infty$ *and* $\alpha<-\frac{1}{q}$. *Then*

$$\left(\sum_{k=1}^{\infty}\left((1+\log k)^\alpha\left(\sum_{i=1}^{k}c_i^{*2}\,(f)\right)^{\frac{1}{2}}\right)^q\frac{1}{k}\right)^{\frac{1}{q}}$$
$$\le C\min\left(M,(1+\log M)^{|\alpha|}\right)$$
$$\times\left(\int_0^1\left((1-\log t)^\alpha\left(\int_t^1 f^{*2}\,(u)\,du\right)^{\frac{1}{2}}\right)^q\frac{dt}{t}\right)^{\frac{1}{q}}$$

for some constant C which depends only on q and α. (As usual, here and below the integral and the sum should be replaced by the supremum when $q = \infty$. Moreover, f^* and $\{c_k^*\}$ ($k \in \mathbb{N}$) are the non-increasing rearrangements of the function $|f|$ and the sequence $\{|c_n|\}$, respectively.)

Theorem 1.2. *Let $0 < q \leq \infty$ and $\alpha > -\frac{1}{q}$. Then for any $\varepsilon > 0$,*

$$\left(\int_0^1 \left((1-\log t)^\alpha \left(\int_0^t f^{*2}(u)\,du\right)^{\frac{1}{2}}\right)^q \frac{dt}{t}\right)^{\frac{1}{q}}$$
$$\leq C \min\left(M, (1+\log M)^{|\alpha|+\varepsilon+\frac{1}{q}}\right)$$
$$\times \left(\sum_{k=1}^\infty \left((1+\log k)^\alpha \left(\sum_{i=k}^\infty c_i^{*2}(f)\right)^{\frac{1}{2}}\right)^q \frac{1}{k}\right)^{\frac{1}{q}}$$

with some constant C which depends only on q, α, and ε. In particular, this means that if the expression on the right-hand side of the estimate above exists, then the series $\sum c_n \varphi_n(x)$ converges in the (quasi-) norm determined by the expression on the left-hand side.

Remark 1.3. (i) The orthonormality of the system $\{\varphi_n\}$ and (1.1) imply that always $M \geq 1$.

(ii) Theorems 1.1 and 1.2 remain true for systems $\{\varphi_n\}$ that are bounded in L_Q for some $Q \in (2, \infty]$ [28]. The estimates do not depend on Q.

This paper is organized as follows. Section 2 contains necessary notations, definitions, and auxiliary results. Theorems 1.1 and 1.2 will be proved and discussed in Sections 3 and 4, respectively.

2. Notations, definitions, and auxiliary results

If X is a (quasi-) Banach space and $x \in X$, then its (quasi-) norm is denoted by $\|x\|_X$. We write $X \cong Y$ if the spaces X and Y are isomorphic. Throughout the paper, $L_p(a,b)$ ($0 < p \leq \infty, -\infty \leq a < b \leq \infty$) is the usual quasinormed Lebesgue space L_p on the interval (a,b). For $p \geq 1$, it is a Banach space. By C we designate different positive constants which are independent on all significant arguments. If f and g are positive functions, we will write $f \prec g$ if $f \leq Cg$ and $f \simeq g$ if $f \prec g$ and $g \prec f$.

2.1. Interpolation spaces

Let $\overline{X} \equiv (X_0, X_1)$ be a compatible couple of (quasi-) Banach spaces. The Peetre K-functional is defined by

$$\begin{aligned} K(t,x) &\equiv K\left(t,x;\overline{X}\right) \\ &:= \inf_{x=x_0+x_1;\, x_i \in X_i} \left(\|x_0\|_{X_0} + t\|x_1\|_{X_1}\right) \quad (x \in X_0 + X_1,\ t > 0). \end{aligned}$$

For further information about properties of the K-functional and the real interpolation method we refer to [3, 4, 27, 38]. For our purposes, it is enough to consider only ordered couples $X_0 \supset X_1$ with the norm of the embedding equal to 1. This will be denoted as $X_0 \overset{1}{\supset} X$. In this case $K(t,x) \simeq \|x\|_{X_0}$ for $t > 1$; see [4].

Definition 2.1. Let $X_0 \overset{1}{\supset} X_1$, $0 \le \theta \le 1$, $0 < q \le \infty$, and $\alpha \in \mathbb{R}$. We define the (quasi-) Banach space $\overline{X}_{\theta,q,\alpha} \equiv (X_0, X_1)_{\theta,q,\alpha}$ as

$$\left\{ x \in X_0 + X_1 \mid \|x\|_{\overline{X}_{\theta,q,\alpha}} := \left\| t^{-\theta-\frac{1}{q}} \left(1 + |\log t|\right)^{\alpha} K(t,x) \right\|_{L_q(0,1)} < \infty \right\}.$$

It makes only sense to consider the spaces $\overline{X}_{\theta,q,\alpha}$ on the set

$$\left\{ (\theta,q,\alpha) \in [0,1] \times (0,\infty] \times \mathbb{R} \mid 0 < \theta < 1, \text{ or } \theta = 0,\ q \le \infty,\ \alpha \ge -\frac{1}{q}, \right.$$
$$\left. \text{or } \theta = 1,\ q \le \infty,\ \alpha < -\frac{1}{q}, \text{ or } \theta = 1,\ q = \infty,\ \alpha = 0 \right\}.$$

Note that the functors $(X_0, X_1)_{0,q,\alpha}$ and $(X_0, X_1)_{1,q,\alpha}$ produce spaces which are "very close" to X_0 and X_1 respectively. This definition can be found in a lot of papers (see, e.g., [10, 12, 15, 16, 17, 19, 21, 36]). Different versions of the next lemma can be found in the literature, see [12, Theorem 2.5], [8, Theorem 4.9], [9, Eq. (2.3) and (2.4)], [18], and [22, Theorem 3.5].

Lemma 2.2. *Let $X_0 \overset{1}{\supset} X_1$ and $Y_0 \overset{1}{\supset} Y_1$ be (quasi-) Banach spaces and let T be a (quasi-) linear bounded operator, $T : X_j \to Y_j$ with norms $M_j := \|T|X_j \to Y_j\| > 0$ $(j = 0, 1)$. Additionally suppose that $0 < q \le \infty$.*

(a) *If $M_0 \ge M_1$ and $\alpha < -\frac{1}{q}$, then T is bounded from $\overline{X}_{1,q,\alpha}$ to $\overline{Y}_{1,q,\alpha}$ and*

$$\left\| T | \overline{X}_{1,q,\alpha} \to \overline{Y}_{1,q,\alpha} \right\| \le \min \left(M_0, \left(1 + \log \left(\frac{M_0}{M_1} \right)^{|\alpha|} \right) M_1 \right).$$

(b) *If $M_0 \le M_1$ and $\alpha \ge -\frac{1}{q}$, then T is bounded from $\overline{X}_{0,q,\alpha}$ to $\overline{Y}_{0,q,\alpha}$ and for every $\varepsilon > 0$,*

$$\left\| T | \overline{X}_{0,q,\alpha} \to \overline{Y}_{0,q,\alpha} \right\| \prec \min \left(M_1, \left(1 + \log \left(\frac{M_1}{M_0} \right) \right)^{|\alpha| + \varepsilon + \frac{1}{q}} M_0 \right).$$

Proof. First, notice that if $x \in X_0 + X_1$, then

$$K\left(t, Tx; \overline{Y}\right) \le \max\left(M_0, M_1\right) K\left(t, x; \overline{X}\right). \tag{2.1}$$

Moreover,

$$K\left(t, Tx; \overline{Y}\right) \le M_0 K\left(\frac{M_1}{M_0} t, x; \overline{X}\right). \tag{2.2}$$

We begin with the assertion (a). It is not difficult to show that

$$\left(1 + |\log(uv)|\right)^{\alpha} \le \left(1 + |\log(u)|\right)^{\alpha} \left(1 + |\log(v)|\right)^{|\alpha|} \qquad (u, v > 0).$$

By means of this inequality and of (2.2), because $\frac{M_1}{M_0} \leq 1$, and using the change of variable $u = \frac{M_1}{M_0}t$, we obtain

$$\begin{aligned}\|Tx\|_{\overline{Y}_{1,q,\alpha}} &\leq M_0 \left\| t^{-1-\frac{1}{q}} \left(1 + |\log t|\right)^{\alpha} K\left(\frac{M_1}{M_0}t, x; \overline{X}\right) \right\|_{L_q(0,1)} \\ &= M_1 \left\| u^{-1-\frac{1}{q}} \left(1 + \left|\log\left(\frac{M_0}{M_1}u\right)\right|\right)^{\alpha} K\left(u, x; \overline{X}\right) \right\|_{L_q\left(0, \frac{M_1}{M_0}\right)} \\ &\leq M_1 \left(1 + \log\left(\frac{M_0}{M_1}\right)\right)^{|\alpha|} \left\| u^{-1-\frac{1}{q}} \left(1 + |\log u|\right)^{\alpha} K\left(u, x; \overline{X}\right) \right\|_{L_q(0,1)} \\ &= M_1 \left(1 + \log\left(\frac{M_0}{M_1}\right)\right)^{|\alpha|} \|x\|_{\overline{X}_{1,q,\alpha}}.\end{aligned}$$

Combining this with (2.1) we get assertion (a).

Next we consider the case (b). For real numbers α_0 und α_∞ we put, as usual,

$$l^{(\alpha_0,\alpha_\infty)}(t) := \begin{cases} (1 - \log t)^{\alpha_0}, & \text{if } 0 < t \leq 1, \\ (1 + \log t)^{\alpha_\infty}, & \text{if } 1 < t < \infty. \end{cases}$$

Now we show that for every $\varepsilon > 0$,

$$\begin{aligned}&\left\| t^{-\frac{1}{q}} \left(1 + |\log t|\right)^{\alpha} K\left(t, x; \overline{X}\right) \right\|_{L_q(0,1)} \\ &\qquad \simeq \left\| t^{-\frac{1}{q}} l^{\left(\alpha, -\varepsilon - \frac{1}{q}\right)}(t) K\left(t, x; \overline{X}\right) \right\|_{L_q(0,\infty)}.\end{aligned} \tag{2.3}$$

It is enough to check that

$$\begin{aligned}&\left\| t^{-\frac{1}{q}} \left(1 + |\log t|\right)^{-\left(\varepsilon + \frac{1}{q}\right)} K\left(t, x; \overline{X}\right) \right\|_{L_q(1,\infty)} \\ &\qquad \prec \left\| t^{-\frac{1}{q}} \left(1 + |\log t|\right)^{\alpha} K\left(t, x; \overline{X}\right) \right\|_{L_q(0,1)}.\end{aligned}$$

Observe first that $\left\| t^{1-\frac{1}{q}} \left(1 + |\log t|\right)^{\alpha} \right\|_{L_q(0,1)} < \infty$ and if $\varepsilon > 0$, then

$$\left\| t^{-\frac{1}{q}} \left(1 + |\log t|\right)^{-\left(\varepsilon + \frac{1}{q}\right)} \right\|_{L_q(1,\infty)} < \infty.$$

Using that $t^{-1}K\left(t, x; \overline{X}\right)$ is non-increasing and $K\left(t, x; \overline{X}\right) \simeq K\left(1, x; \overline{X}\right)$ for $t > 1$, we obtain

$$\begin{aligned}&\left\| t^{-\frac{1}{q}} \left(1 + |\log t|\right)^{-\left(\varepsilon + \frac{1}{q}\right)} K\left(t, x; \overline{X}\right) \right\|_{L_q(1,\infty)} \\ &\quad \simeq K\left(1, x; \overline{X}\right) \left\| t^{-\frac{1}{q}} \left(1 + |\log t|\right)^{-\left(\varepsilon + \frac{1}{q}\right)} \right\|_{L_q(1,\infty)} \\ &\quad \simeq K\left(1, x; \overline{X}\right) \simeq K\left(1, x; \overline{X}\right) \left\| t^{1-\frac{1}{q}} \left(1 + |\log t|\right)^{\alpha} \right\|_{L_q(0,1)} \\ &\quad \leq \left\| t^{-\frac{1}{q}} \left(1 + |\log t|\right)^{\alpha} K\left(t, x; \overline{X}\right) \right\|_{L_q(0,1)}.\end{aligned}$$

So, (2.3) is proved. It can be shown that (cf. [9], p. 169.)

$$l^{(\alpha_0,\alpha_\infty)}(uv) \le l^{(\alpha_0,\alpha_\infty)}(u)\,(1+|\log v|)^{|\alpha_0|+|\alpha_\infty|} \qquad (u,v>0).$$

Therefore,

$$l^{\left(\alpha,-\varepsilon-\frac{1}{q}\right)}(uv) \le l^{\left(\alpha,-\varepsilon-\frac{1}{q}\right)}(u)\,(1+|\log v|)^{|\alpha|+\varepsilon+\frac{1}{q}}.$$

Since $M_0 \le M_1$, using (2.2) and (2.3), and by means of the change of variable $u = \frac{M_1}{M_0}t$, we obtain

$$\begin{aligned}
\|Tx\|_{\overline{Y}_{0,q,\alpha}} &\simeq \left\| t^{-\frac{1}{q}} l^{\left(\alpha,-\varepsilon-\frac{1}{q}\right)}(t)\,K\left(t,Tx;\overline{Y}\right)\right\|_{L_q(0,\infty)} \\
&\le M_0 \left\| t^{-\frac{1}{q}} l^{\left(\alpha,-\varepsilon-\frac{1}{q}\right)}(t)\,K\left(\frac{M_1}{M_0}t,x;\overline{X}\right)\right\|_{L_q(0,\infty)} \\
&= M_0 \left\| u^{-\frac{1}{q}} l^{\left(\alpha,-\varepsilon-\frac{1}{q}\right)}\left(\frac{M_0}{M_1}u\right)\,K\left(u,x;\overline{X}\right)\right\|_{L_q(0,\infty)} \\
&\prec M_0\left(1+\log\left(\frac{M_1}{M_0}\right)\right)^{|\alpha|+\varepsilon+\frac{1}{q}} \left\| u^{-\frac{1}{q}} l^{\left(\alpha,-\varepsilon-\frac{1}{q}\right)}(u)\,K\left(u,x;\overline{X}\right)\right\|_{L_q(0,\infty)} \\
&\simeq M_0\left(1+\log\left(\frac{M_1}{M_0}\right)\right)^{|\alpha|+\varepsilon+\frac{1}{q}} \left\| u^{-\frac{1}{q}} (1-\log u)^{\alpha}\,K\left(u,x;\overline{X}\right)\right\|_{L_q(0,1)} \\
&= M_0\left(1+\log\left(\frac{M_1}{M_0}\right)\right)^{|\alpha|+\varepsilon+\frac{1}{q}} \|x\|_{\overline{X}_{0,q,\alpha}}.
\end{aligned}$$

Combining this with (2.1), we complete the proof. □

Remark 2.3. Due to Lemma 2.2, we have in Theorems 1.1 and 1.2 an expression of the form

$$G(M,\gamma) := \min\left(M,(1+\log M)^{\gamma}\right)$$

with $M \ge 1$ and $\gamma \ge 0$. It is clear that

$$G(M,\gamma) = \begin{cases} (1+\log M)^{\gamma} & \text{if } \gamma \le \log M/\log(1+\log M), \\ M & \text{otherwise.} \end{cases}$$

In particular, $G(M,\gamma) = (1+\log M)^{\gamma}$ if $\gamma \le 1$.

Next we consider the following limiting interpolation spaces. These spaces allow us to formulate reiteration theorems in the limiting cases $\theta = 0$ and $\theta = 1$. In more general form, these spaces were introduced and investigated in [14, 16, 19, 23, 25].

Definition 2.4. Let $X_0 \overset{1}{\supset} X_1$, $0 < q, r \le \infty$, and $\alpha \in \mathbb{R}$. We denote by $\overline{X}^{\mathcal{L}}_{\theta,q,\alpha,r}$ $(0 \le \theta < 1)$ and $\overline{X}^{\mathcal{R}}_{\theta,q,\alpha,r}$ $(0 < \theta \le 1)$ the sets of elements $x \in X_0$ for which the expressions

$$\|x\|_{\overline{X}^{\mathcal{L}}_{\theta,q,\alpha,r}} := \left\| t^{-\frac{1}{q}}(1+|\log t|)^{\alpha} \left\| u^{-\theta-1/r}K(u,x)\right\|_{L_r(0,t)}\right\|_{L_q(0,1)}$$

and

$$\|x\|_{\overline{X}^{\mathcal{R}}_{\theta,q,\alpha,r}} := \left\| t^{-\frac{1}{q}} \left(1+|\log t|\right)^{\alpha} \left\| u^{-\theta-1/r} K\left(u,x\right) \right\|_{L_r(t,1)} \right\|_{L_q(0,1)},$$

respectively, are finite.

The next lemma follows from [14, Lemma 6.2] and [16, Lemma 4]; see also [19].

Lemma 2.5. *Let* $0 < q, r \leq \infty$. *Additionally suppose that* $0 \leq \theta < 1$ *and* $\alpha > -\frac{1}{q}$ (*or* $0 < \theta \leq 1$ *and* $\alpha < -\frac{1}{q}$). *Then*

$$\overline{X}_{\theta,q,\alpha+\frac{1}{\min(r,q)}} \subset \overline{X}^{\mathcal{L}}_{\theta,q,\alpha,r} \text{ (or } \overline{X}^{\mathcal{R}}_{\theta,q,\alpha,r} \text{ respectively)}$$
$$\subset \overline{X}_{\theta,q,\alpha+\frac{1}{\max(r,q)}} \cap \overline{X}_{\theta,\max(r,q),\alpha+\frac{1}{q}}.$$

2.2. Function spaces

In this section, we give the necessary definitions of function and sequence spaces. We consider (equivalence classes of) complex-valued measurable functions on $(0,1)$ and bounded complex-valued sequences $\{c_n\}$. As usual, f^* and $\{c_k^*\}$ $(k \in \mathbb{N})$ are the non-increasing rearrangements of a function $|f|$ and of a sequence $\{|c_n|\}$, respectively. The Lorentz–Zygmund spaces can be defined as follows.

Definition 2.6. Let $0 < p, q \leq \infty$ and $\alpha \in \mathbb{R}$. Put

$$L_{p,q}\left(\log L\right)_{\alpha} := \left\{ f \mid \|f\|_{L_{p,q}(\log L)_{\alpha}} := \left\| t^{\frac{1}{p}-\frac{1}{q}} \left(1+|\log t|\right)^{\alpha} f^*\left(t\right) \right\|_{L_q(0,1)} < \infty \right\}.$$

Analogously

$$l_{p,q}\left(\log l\right)_{\alpha} := \left\{ \{c_k\} \mid \|\{c_k\}\|_{l_{p,q}(\log l)_{\alpha}} := \left\| k^{\frac{1}{p}-\frac{1}{q}} \left(1+\log k\right)^{\alpha} c_k^* \right\|_{l_q} < \infty \right\}.$$

These spaces are studied in [2, 13, 15, 16, 17, 19, 20, 21, 31]. See also [3] and [25]. Note that $L_{p,q} = L_{p,q}\left(\log L\right)_0$, $L_p = L_{p,p}$, $l_{p,q} = l_{p,q}\left(\log l\right)_0$, and $l_p = l_{p,p}$. Concerning the next definition, we refer to [14, 19, 25].

Definition 2.7. Let $0 < q \leq \infty$ and $\alpha \in \mathbb{R}$. Put

$$L^{\mathcal{L}}_{2,q,\alpha,2} := \left\{ f \mid \|f\|_{L^{\mathcal{L}}_{2,q,\alpha,2}} := \left\| t^{-\frac{1}{q}} \left(1+|\log t|\right)^{\alpha} \|f^*\left(u\right)\|_{L_2(0,t)} \right\|_{L_q(0,1)} < \infty \right\},$$

$$L^{\mathcal{R}}_{2,q,\alpha,2} := \left\{ f \mid \|f\|_{L^{\mathcal{R}}_{2,q,\alpha,2}} := \left\| t^{-\frac{1}{q}} \left(1+|\log t|\right)^{\alpha} \|f^*\left(u\right)\|_{L_2(t,1)} \right\|_{L_q(0,1)} < \infty \right\}.$$

Analogously,

$$l^{\mathcal{L}}_{2,q,\alpha,2} :=$$

$$\left\{ \{c_k\} \mid \|\{c_k\}\|_{l^{\mathcal{L}}_{2,q,\alpha,2}} := \left(\sum_{k=1}^{\infty} \left((1+\log k)^{\alpha} \left(\sum_{i=1}^{k} c_i^{*2} \right)^{\frac{1}{2}} \right)^{q} \frac{1}{k} \right)^{\frac{1}{q}} < \infty \right\},$$

$$l^{\mathcal{R}}_{2,q,\alpha,2} :=$$

$$\left\{ \{c_k\} \mid \|\{c_k\}\|_{l^{\mathcal{R}}_{2,q,\alpha,2}} := \left(\sum_{k=1}^{\infty} \left((1+\log k)^{\alpha} \left(\sum_{i=k}^{\infty} c_i^{*2} \right)^{\frac{1}{2}} \right)^{q} \frac{1}{k} \right)^{\frac{1}{q}} < \infty \right\}.$$

Note that in the terminology of [24], $L^{\mathcal{R}}_{2,\infty,\alpha,2}$ is the generalized grand Lorentz space $L^{2),2}_{(1+|\log t|)^{\alpha}}$ and $l^{\mathcal{L}}_{2,\infty,\alpha,2}$ is the generalized grand Lorentz space of sequences $l^{2),2}_{(1+|\log t|)^{\alpha}}$. The following result is a consequence of [14, Corollaries 7.3 and 7.9], [19, Theorem 8.9], and [25, Theorem 5.7].

Lemma 2.8. *Let $0 < q \leq \infty$. If $\alpha < -\frac{1}{q}$, then*

$$(L_1, L_2)_{1,q,\alpha} \cong L^{\mathcal{R}}_{2,q,\alpha,2}, \qquad (l_\infty, l_2)_{1,q,\alpha} \cong l^{\mathcal{L}}_{2,q,\alpha,2}.$$

If $\alpha > -\frac{1}{q}$, then

$$(L_2, L_\infty)_{0,q,\alpha} \cong L^{\mathcal{L}}_{2,q,\alpha,2}, \qquad (l_2, l_1)_{0,q,\alpha} \cong l^{\mathcal{R}}_{2,q,\alpha,2}.$$

By Lemmas 2.5 and 2.8, we get the following embeddings (cf. [14, Corollaries 7.4 and 7.8]).

Lemma 2.9. *Let $0 < q \leq \infty$. If $\alpha < -\frac{1}{q}$ (or $\alpha > -\frac{1}{q}$), then*

$$\begin{aligned} L_{2,q}(\log L)_{\alpha+\frac{1}{\min(q,2)}} &\subset L^{\mathcal{R}}_{2,q,\alpha,2} \quad (\textit{or } L^{\mathcal{L}}_{2,q,\alpha,2} \textit{ respectively}) \\ &\subset L_{2,q}(\log L)_{\alpha+\frac{1}{\max(q,2)}} \cap L_{2,\max(q,2)}(\log L)_{\alpha+\frac{1}{q}}, \end{aligned} \tag{2.4}$$

$$\begin{aligned} l_{2,q}(\log l)_{\alpha+\frac{1}{\min(q,2)}} &\subset l^{\mathcal{L}}_{2,q,\alpha,2} \quad (\textit{or } l^{\mathcal{R}}_{2,q,\alpha,2} \textit{ respectively}) \\ &\subset l_{2,q}(\log l)_{\alpha+\frac{1}{\max(q,2)}} \cap l_{2,\max(q,2)}(\log l)_{\alpha+\frac{1}{q}}. \end{aligned} \tag{2.5}$$

Remark 2.10. For $0 < q < 2$, the spaces $L_{2,q}(\log L)_{\alpha+\frac{1}{2}}$ and $L_{2,2}(\log L)_{\alpha+\frac{1}{q}}$ are incomparable [37].

Corollary 2.11. *If $\alpha < -\frac{1}{2}$ (or $\alpha > -\frac{1}{2}$), we have the isomorphisms*

$$\begin{aligned} &L^{\mathcal{R}}_{2,2,\alpha,2} \ (\textit{or } L^{\mathcal{L}}_{2,2,\alpha,2} \textit{ respectively}) \cong L_{2,2}(\log L)_{\alpha+\frac{1}{2}}, \\ &l^{\mathcal{L}}_{2,2,\alpha,2} \ (\textit{or } l^{\mathcal{R}}_{2,2,\alpha,2} \textit{ respectively}) \cong l_{2,2}(\log l)_{\alpha+\frac{1}{2}}. \end{aligned} \tag{2.6}$$

For the scale $l^{\mathcal{L}}_{2,q,\alpha,2}$, we also need the following inclusion.

Lemma 2.12. *If $0 < q < \infty$ and $\alpha < -\frac{1}{q}$, then $l^{\mathcal{L}}_{2,q,\alpha,2} \subset l^{\mathcal{L}}_{2,\infty,\alpha+\frac{1}{q},2}$.*

Proof. Let $\{c_k\} \in l^{\mathcal{L}}_{2,q,\alpha,2}$. The sequence $\{b_k\} := \{(\sum_{i=1}^{k} c_i^{*2})^{q/2}\}$ is non-negative and non-decreasing. Because $0 < q < \infty$ and $q\alpha < -1$, we obtain for $m \geq 1$ that

$$(1+\log m)^{1+q\alpha} \simeq \int_m^\infty (1+\log x)^{1+q\alpha}\frac{dx}{x} \simeq \sum_{k=m}^\infty (1+\log k)^{q\alpha}\frac{1}{k}.$$

Therefore,

$$(1+\log m)^{1+q\alpha} b_m \simeq b_m \sum_{k=m}^\infty (1+\log k)^{q\alpha}\frac{1}{k} \leq \sum_{k=m}^\infty (1+\log k)^{q\alpha} b_k \frac{1}{k}.$$

Thus, for all $m \geq 1$, we have

$$(1+\log m)^{\alpha+\frac{1}{q}} \left(\sum_{i=1}^m c_i^{*2}\right)^{\frac{1}{2}}$$

$$\prec \left(\sum_{k=m}^\infty \left((1+\log k)^\alpha \left(\sum_{i=1}^k c_i^{*2}\right)^{\frac{1}{2}}\right)^q \frac{1}{k}\right)^{\frac{1}{q}} \leq \|\{c_k\}\|_{l^{\mathcal{L}}_{2,q,\alpha,2}},$$

and finally

$$\|\{c_k\}\|_{l^{\mathcal{L}}_{2,\infty,\alpha+\frac{1}{q},2}} = \sup_{m\geq 1}(1+\log m)^{\alpha+\frac{1}{q}}\left(\sum_{i=1}^m c_i^{*2}\right)^{\frac{1}{2}} \prec \|\{c_k\}\|_{l^{\mathcal{L}}_{2,q,\alpha,2}}. \qquad \square$$

3. Proof of Theorem 1.1, corollaries, and remarks

Proof of Theorem 1.1. The assertion of Theorem 1.1 can be reformulated as follows: if $0 < q \leq \infty$ and $\alpha < -\frac{1}{q}$, then the Fourier series map $\mathcal{F}$ is bounded from $L^{\mathcal{R}}_{2,q,\alpha,2}$ to $l^{\mathcal{L}}_{2,q,\alpha,2}$ and

$$\left\|\mathcal{F} \mid L^{\mathcal{R}}_{2,q,\alpha,2} \to l^{\mathcal{L}}_{2,q,\alpha,2}\right\| \prec \min\left(M, (1+\log M)^{|\alpha|}\right). \qquad (3.1)$$

By Lemma 2.8, we have $(L_1, L_2)_{1,q,\alpha} \cong L^{\mathcal{R}}_{2,q,\alpha,2}$ and $(l_\infty, l_2)_{1,q,\alpha} \cong l^{\mathcal{L}}_{2,q,\alpha,2}$. Now due to (1.2) and (1.3), it only remains to apply Lemma 2.2. $\square$

Remark 3.1. Theorem 8.2 (b) in [14] is a special case of Theorem 1.1 for the system $\{e^{i2\pi nx}\}$.

Using the isomorphisms (2.6), we get the following corollary.

Corollary 3.2. *If $\alpha < -\frac{1}{2}$, then*

$$\left\|\mathcal{F} \mid L_{2,2}(\log L)_{\alpha+\frac{1}{2}} \to l_{2,2}(\log L)_{\alpha+\frac{1}{2}}\right\| \prec \min\left(M, (1+\log M)^{|\alpha|}\right).$$

Remark 3.3. Writing down Corollary 3.2 in the special case $\varphi_n(x) = e^{i2\pi nx}$ and $\alpha = -1$, we obtain [11, Theorem 8.5] (see also [36, Theorem 3.25]).

The following result is a consequence of (3.1) and the first of the inclusions (2.4).

Corollary 3.4. *If* $2 \le q \le \infty$ *and* $\alpha < \frac{1}{2} - \frac{1}{q}$, *then*

$$\left\| \mathcal{F} \mid L_{2,q} (\log L)_\alpha \to l^{\mathcal{L}}_{2,q,\alpha-\frac{1}{2},2} \right\| \prec \min \left(M, (1+\log M)^{\left|\alpha - \frac{1}{2}\right|} \right).$$

In particular, due to Remark 2.3, *if* $2 < q \le \infty$, *then*

$$\left\| \mathcal{F} \mid L_{2,q} \to l^{\mathcal{L}}_{2,q,-\frac{1}{2},2} \right\| \prec (1+\log M)^{\frac{1}{2}}.$$

Remark 3.5. S. V. Bochkarev [5, 6] proved the following estimate (see also [29]):

$$\left\| \mathcal{F} \mid L_{2,q} \to l^{\mathcal{L}}_{2,\infty,\frac{1}{q}-\frac{1}{2},2} \right\| \prec M \quad (2 < q \le \infty). \tag{3.2}$$

Note that in the case $q = \infty$, (3.2) was proved in [32]. In [30], Bochkarev's inequality was improved in Lorentz–Zygmund spaces, namely,

$$\left\| \mathcal{F} \mid L_{2,q} (\log L)_\alpha \to l^{\mathcal{L}}_{2,\infty,\alpha+\frac{1}{q}-\frac{1}{2},2} \right\| \prec M \quad \left(2 < q \le \infty, \alpha < \tfrac{1}{2} - \tfrac{1}{q}\right). \tag{3.3}$$

Due to Lemma 2.12, if $2 < q < \infty$ and $\alpha < \frac{1}{2} - \frac{1}{q}$, we have the inclusion $l^{\mathcal{L}}_{2,q,\alpha-\frac{1}{2},2} \subset l^{\mathcal{L}}_{2,\infty,\alpha+\frac{1}{q}-\frac{1}{2},2}$. So, Corollary 3.4 improves both inequalities (3.2) and (3.3).

Corollary 3.6. *Let* $0 < q \le \infty$ *and* $\alpha < -\frac{1}{q}$. *Then*

$$\left\| \mathcal{F} \mid L_{2,q} (\log L)_{\alpha+\frac{1}{\min(q,2)}} \to l_{2,q} (\log l)_{\alpha+\frac{1}{\max(q,2)}} \cap l_{2,\max(q,2)} (\log l)_{\alpha+\frac{1}{q}} \right\|$$
$$\prec \min \left(M, (1+\log M)^{|\alpha|} \right).$$

Remark 3.7. In the special case $\{\varphi_n(x)\} = \{e^{i2\pi nx}\}$, Corollary 3.6 yields [7, Theorem 5.3].

4. Proof of Theorem 1.2, corollaries, and remarks

Proof of Theorem 1.2. The assertion of Theorem 1.2 can be restated as follows: if $0 < q \le \infty$ and $\alpha > -\frac{1}{q}$, then the operator $\mathcal{F}^{-1}$ is bounded from $l^{\mathcal{R}}_{2,q,\alpha,2}$ to $L^{\mathcal{L}}_{2,q,\alpha,2}$ and, for every $\varepsilon > 0$,

$$\left\| \mathcal{F}^{-1} \mid l^{\mathcal{R}}_{2,q,\alpha,2} \to L^{\mathcal{L}}_{2,q,\alpha,2} \right\| \prec \min \left(M, (1+\log M)^{|\alpha|+\frac{1}{q}+\varepsilon} \right). \tag{4.1}$$

By Lemma 2.8, we have $(l_2, l_1)_{0,q,\alpha} \cong l^{\mathcal{R}}_{2,q,\alpha,2}$ and $(L_2, L_\infty)_{0,q,\alpha} \cong L^{\mathcal{L}}_{2,q,\alpha,2}$. Thus, due to (1.4) and (1.5), the assertion follows from Lemma 2.2. □

The following result is a consequence of (4.1) and the first of the inclusions (2.5).

Corollary 4.1. *If* $0 < q \le \infty$ *and* $\alpha > -\frac{1}{q}$, *then, for every* $\varepsilon > 0$,

$$\left\| \mathcal{F}^{-1} \mid l_{2,q} (\log L)_{\alpha+\frac{1}{\min(q,2)}} \to L^{\mathcal{L}}_{2,q,\alpha,2} \right\| \prec \min \left(M, (1+\log M)^{|\alpha|+\frac{1}{q}+\varepsilon} \right).$$

Applying Lemma 2.9 to both spaces in (4.1), we get the next corollary.

Corollary 4.2. *Suppose* $0 < q \leq \infty$ *and* $\alpha > -\frac{1}{q}$. *Then, for every* $\varepsilon > 0$,

$$\left\| \mathcal{F}^{-1} \mid l_{2,q} (\log L)_{\alpha + \frac{1}{\min(q,2)}} \to L_{2,q} (\log L)_{\alpha + \frac{1}{\max(q,2)}} \cap L_{2,\max(q,2)} (\log L)_{\alpha + \frac{1}{q}} \right\|$$

$$\prec \min \left(M, (1 + \log M)^{|\alpha| + \frac{1}{q} + \varepsilon} \right).$$

Remark 4.3. Writing down Theorem 1.2 and Corollary 4.2 in the special case of the system $\{e^{i2\pi nx}\}$, we obtain [14, Theorem 8.2 (g) and Lemma 8.4], respectively.

References

[1] C. Bennett, Banach function spaces and interpolation methods III. Hausdorff-Young estimates, *J. Approx. Theory* **13** (1975), 267–275.

[2] C. Bennett and K. Rudnick, On Lorentz-Zygmund spaces, *Diss. Math.* **175** (1980), 5–67.

[3] C. Bennett and R. Sharpley, *Interpolation of Operators,* Academic Press, Boston, 1988.

[4] J. Bergh and J. Löfström, *Interpolation Spaces. An Introduction,* Springer, Berlin, 1976.

[5] S.V. Bochkarev, The Hausdorff-Young-Riesz theorem in Lorentz spaces and multiplicative inequalities, *Proc. Steklov Inst. Math.* **219** (1997), 103–114.

[6] S.V. Bochkarev, Estimation of the Fourier coefficients of functions from Lorentz spaces, *Soviet Math. Dokl.* **57** (1998), no. 3, 454–457.

[7] F. Cobos and O. Domíngues, Approximation spaces, limiting interpolation and Besov spaces, *J. Approx. Theory* **189** (2015), 43–66.

[8] F. Cobos, L.M. Fernández-Cabrera, Th. Kühn, and T. Ullrich, On an extreme class of real interpolation spaces, *J. Func. Anal.* **256** (2009), 2321–2366.

[9] F. Cobos, L.M. Fernández-Cabrera, and A. Martínez, On a paper of Edmunds and Opic on limiting interpolation of compact operators between L_p spaces, *Math. Nachr.* **288** (2015), 167–175.

[10] F. Cobos and Th. Kühn, Equivalence of K- and J-methods for limiting real interpolation spaces, *J. Func. Anal.* **261** (2011), no. 12, 3696–3722.

[11] F. Cobos and A. Segurado, Limiting real interpolation methods for arbitrary Banach couples, *Studia Math.* **213** (2012), no. 3, 243–273.

[12] R.Ya. Doktorskii, A multiparametric real interpolation method I. Multiparametric interpolation functors (Russian), Manuscript No. **6120-B88**, deposited at VINITI (1988).

[13] R.Ya. Doktorskii, A multiparametric real interpolation method III. Lorentz-Zygmund spaces (Russian), Manuscript No. **6070-B88**, deposited at VINITI (1988).

[14] R.Ya. Doktorskii, A multiparametric real interpolation method IV. Reiteration relations for "limiting" cases $\eta = 0$ and $\eta = 1$, application to the Fourier series theory (Russian), Manuscript No. **4637-B89**, deposited at VINITI (1989).

[15] R.Ya. Doktorskii, Description of the scale of the Lorentz-Zygmund spaces via a multiparametric modification of the real interpolation method (Russian), in: *Differential and integral equations and their applications*, Kalmytsk. Gos. Univ, Elista, 1990, 38–48.

[16] R.Ya. Doktorskii, Reiteration relations of the real interpolation method, *Soviet. Math. Dokl.* **44** (1992), no. 3, 665–669.

[17] L.R.Ya. Doktorski, Limiting reiteration for real interpolation with logarithmic functions, *Bol. Soc. Mat. Mex.* **22** (2016), no. 2, 679–693.

[18] D.E. Edmunds and B. Opic, Limiting variants of Krasnosel'skiĭ's compact interpolation theorem, *J. Funct. Anal.* **266** (2014), 3265–3285.

[19] W.D. Evans and B. Opic, Real interpolation with logarithmic functors and reiteration, *Canad. J. Math.* **52** (2000), no. 5, 920–960.

[20] W.D. Evans, B. Opic, and L. Pick, Interpolation of operators on scales of generalized Lorentz-Zygmund spaces, *Math. Nachr.* **182** (1996), 127–181.

[21] W.D. Evans, B. Opic, and L. Pick, Real interpolation with logarithmic functors, *J. of Inequal. & Appl.* **7** (2002), no. 2, 187–269.

[22] P. Fernández-Martínez, A. Segurado, and T. Signes, Compactness results for a class of limiting interpolation methods, *Mediterr. J. Math.* **13** (2016), no. 5, 2959–2979.

[23] P. Fernández-Martínez, and T. Signes, Limit cases of reiteration theorems, *Math. Nachr.* **288** (2015), no. 1, 25–47.

[24] A. Fiorenza and G.E. Karadzhov, Grand and small Lebesgue spaces and their analogs, *J. Anal. Appl.* **23** (2004), no. 4, 657–681.

[25] A. Gogatishvili, B. Opic, and W. Trebels, Limiting reiteration for real interpolation with slowly varying functions, *Math. Nachr.* **278** (2005), no. 1–2, 86–107.

[26] A. Kopezhanova and L.-E. Persson, On summability of the Fourier coefficients in bounded orthonormal systems for functions from some Lorentz type spaces, *Eurasian Mathematical Journal* **1** (2010), no. 2, 76–85.

[27] S.G. Krein, Ju.I. Petunin, and E.M. Semenov, *Interpolation of Linear Operators,* Transl. Math. Monogr. **54**, Amer. Math. Soc., Providence, RI, 1982.

[28] J. Marcinkiewicz and A. Zygmund, Some theorems on orthogonal systems, *Fund. math.* **28** (1937), 309–335.

[29] G.K. Mussabayeva, Inequality type Bochkarev, *KazNU Bull. Math., Mech., Comp. Science Series* **3** (2014), no. 82, 12–17.

[30] H. Oba, E. Sato, and Y. Sato, A note on Lorentz-Zygmund spaces, *Georgian Math. J.* **18** (2011), 533–548.

[31] B. Opic and L. Pick, On generalized Lorentz-Zygmund spaces, *Math. Inequal. & Appl.* **2** (1999), no. 3, 391–467.

[32] V.I. Ovchinnikov, V.D. Raspopova, and V.A. Rodin, Sharp estimates of the Fourier coefficients of summable functions and K-functionals, *Math. Notes* **32** (1982), no. 3, 627–631.

[33] L.-E. Persson, Relations between summability of functions and their Fourier series, *Acta Math. Acad. Sci. Hungar* **27** (1976), 267–280.

[34] J. Rastegari and G. Sinnamon, Fourier series in weigted Lorentz spaces, *J. Fourier Anal. Appl.* **22** (2016), no. 5, 1192–1223.

[35] Y. Sagher, Interpolation of r-Banach spaces, *Stud. Math.* **41** (1972), no. 1, 45–70.

[36] A. Segurado, *Limiting interpolation methods,* Ph.D. Thesis, Madrid, 2015.

[37] R. Sharpley, Counterexamples for classical operators in Lorentz-Zygmund spaces, *Studia Math.* **68** (1980), 141–158.

[38] H. Triebel, *Interpolation Theory, Function Spaces, Differential Operators,* North-Holland, Amsterdam, 1978.

[39] A. Zygmund, *Trigonometric Series,* 2nd ed., Cambridge Univ. Press, New York, 1968.

Leo R.Ya. Doktorski
Department Object Recognition
Fraunhofer Institute of Optronics
System Technologies
and Image Exploitation IOSB
Gutleuthausstr. 1
76275 Ettlingen
Germany
e-mail: leo.doktorski@iosb.fraunhofer.de

Isomorphisms of $AC(\sigma)$ spaces for countable sets

Ian Doust and Shaymaa Al-shakarchi

Abstract. It is known that the classical Banach–Stone theorem does not extend to the class of $AC(\sigma)$ spaces of absolutely continuous functions defined on compact subsets of the complex plane. On the other hand, if σ is restricted to the set of compact polygons, then all the corresponding $AC(\sigma)$ spaces are isomorphic (as algebras). In this paper we examine the case where σ is the spectrum of a compact operator, and show that in this case one can obtain an infinite family of homeomorphic sets for which the corresponding function spaces are not isomorphic.

Mathematics Subject Classification (2010). Primary: 46J10; Secondary: 46J35, 47B40, 26B30.

Keywords. $AC(\sigma)$ spaces, functions of bounded variation, compact operators.

1. Introduction

Well-bounded operators are one generalization of self-adjoint operators to the Banach space setting. A bounded linear operator T on a Banach space is said to be well-bounded if there is a compact interval $[a, b] \subseteq \mathbb{R}$ such that T admits a bounded $AC[a, b]$ functional calculus. At least on reflexive spaces such operators possess a type of spectral decomposition theory similar to that for self-adjoint operators, but one which allows conditionally rather than unconditionally convergent spectral expansions.

Even on a general Banach space, every compact well-bounded operator admits a diagonal representation of the form

$$T = \sum_{j=1}^{\infty} \lambda_j P_j, \tag{1.1}$$

where $\{\lambda_j\}$ is the set of nonzero eigenvalues of T and $\{P_j\}$ is the corresponding set of Riesz projections onto the eigenspaces. Conversely, under suitable conditions on $\{\lambda_j\}$ and $\{P_j\}$, any operator formed in this way is compact and

© Springer International Publishing AG, part of Springer Nature 2018

A. Böttcher et al. (eds.), *The Diversity and Beauty of Applied Operator Theory*, Operator Theory: Advances and Applications 268, https://doi.org/10.1007/978-3-319-75996-8_11

well-bounded [7]. For example, if $\{Q_j\}_{j=0}^{\infty}$ are the projections associated with a Schauder decomposition of the space X and $P_j = Q_j - Q_{j-1}$, $j = 1, 2, \ldots$, then $\sum_{j=1}^{\infty} \lambda_j P_j$ is compact and well-bounded for any decreasing sequence $\{\lambda_j\}$ of positive reals converging to zero.

An obstruction to extending this theory to provide an analogue of normal operators was the lack of a good replacement for the function algebra $AC[a, b]$. Ideally the functional calculus for an operator T should only depend on the values of the function on $\sigma(T)$, and so one would like a suitable algebra $AC(\sigma)$ where σ is any nonempty compact subset of $\mathbb{C}$. Over the years a number of papers addressed this problem (see for example, [10, 5, 6]) without providing a fully satisfactory theory.

For use in spectral theory, any proposed definition of a Banach algebra $AC(\sigma)$ of 'absolutely continuous functions' on a compact set $\sigma \subseteq \mathbb{C}$ should have the properties that:

1. it should agree with the usual definition if σ is an interval in $\mathbb{R}$;
2. $AC(\sigma)$ should contain all sufficiently well-behaved functions;
3. if $\alpha, \beta \in \mathbb{C}$ with $\alpha \neq 0$, then the space $AC(\alpha\sigma + \beta)$ should be isometrically isomorphic to $AC(\sigma)$.

The final condition is capturing the fact that the spectral decomposition and functional calculus for $\alpha T + \beta$ should essentially match that of T.

Since none of the existing concepts of absolute continuity for functions defined on subsets of the plane satisfied these conditions, a new definition (which does satisfy properties 1, 2, and 3) was introduced in [1], and these are the spaces $AC(\sigma)$ which we consider in this paper. We briefly outline the appropriate definitions in Section 2.

An $AC(\sigma)$ operator is defined to be one which admits a bounded $AC(\sigma)$ functional calculus. This class includes not only all normal Hilbert space operators, but also the classes of scalar-type spectral operators, well-bounded operators, and trigonometrically well-bounded operators acting on any Banach space [2]. Compact $AC(\sigma)$ operators admit a spectral decomposition as a possibly conditionally convergent sum of the form (1.1) (see [3]). Conversely, Theorem 5.1 of [3] shows how to construct large families of compact $AC(\sigma)$ operators on a Banach space X from conditional decompositions of X. In particular, αT is a compact $AC(\alpha\sigma(T))$ operator whenever $\alpha \in \mathbb{C}$ and $T \in B(X)$ is compact and well-bounded.

For a normal operator T on a Hilbert space, the C^*-algebra $C^*(T)$ generated by T is isometrically isomorphic to $C(\sigma(T))$. The Banach–Stone and Gelfand–Kolmogorov theorems say that if Ω_1 and Ω_2 are compact Hausdorff spaces then the following conditions are equivalent:

1. Ω_1 and Ω_2 are homeomorphic,
2. $C(\Omega_1)$ and $C(\Omega_2)$ are linearly isometric as Banach spaces,
3. $C(\Omega_1)$ and $C(\Omega_2)$ are isomorphic as algebras (or as C^*-algebras).

This greatly limits the structure of the algebras $C^*(T)$ for compact normal operators.

The link between the Banach algebra

$$\mathcal{B}_T = \mathrm{cl}\{f(T) : f \in AC(\sigma)\}$$

generated by an $AC(\sigma)$ operator and the function algebra $AC(\sigma)$ is less direct (even in the case that $\sigma = \sigma(T)$). Nonetheless, it is natural to ask about the extent to which a Banach–Stone type theorem might apply in this setting. In particular, we are predominantly interested in determining conditions under which two such spaces $AC(\sigma_1)$, $AC(\sigma_2)$ are isomorphic as Banach algebras.

To be definite, if $\mathcal{A}$ and $\mathcal{B}$ are algebras, we shall say that $\mathcal{A}$ is isomorphic as an algebra to $\mathcal{B}$ if there exists an algebra isomorphism (that is, a linear and multiplicative bijection) $\Phi : \mathcal{A} \to \mathcal{B}$. If $\mathcal{A}$ and $\mathcal{B}$ are Banach algebras, then we shall say that they are isomorphic as Banach algebras, and write $\mathcal{A} \simeq \mathcal{B}$ if there is an algebra isomorphism $\Phi : \mathcal{A} \to \mathcal{B}$ such that Φ and Φ^{-1} are continuous. Since this will generally be the context in this paper, unless otherwise specified, the term isomorphic should be taken to mean isomorphic as Banach algebras.

In [9] it was shown that if $AC(\sigma_1)$ and $AC(\sigma_2)$ are isomorphic as algebras, then any algebra isomorphism is necessarily bicontinuous and hence the spaces are in fact isomorphic as Banach algebras. However, unlike the case for the $C(\Omega)$ spaces, the algebra isomorphisms need not be isometric.

One could also consider the question as to when $AC(\sigma_1)$ and $AC(\sigma_2)$ are isomorphic as Banach spaces. Although we will not pursue this here, we will see in Section 6 that it is easy to construct examples where two of these spaces are linearly isomorphic, but not isomorphic as algebras.

One direction of the Gelfand–Kolmogorov theorem carries over to the current setting. If $AC(\sigma_1)$ and $AC(\sigma_2)$ are isomorphic as algebras, then the sets σ_1 and σ_2 must be homeomorphic subsets of $\mathbb{C}$ (see [9, Theorem 2.6]). The converse direction was shown to fail since the spaces of absolutely continuous functions on the closed disc and the closed square are not isomorphic. In a positive direction, the spaces for any two closed polygons are necessarily isomorphic Banach algebras. (Indeed, this result can be extended to a more general class of sets consisting of polygonal regions with polygonal holes.)

The aim of this paper is to examine the situation where σ is the spectrum of a compact operator, and more particularly where σ is a compact countable subset of the plane with a single limit point. All sets in this latter category are of course homeomorphic, but we shall show that one can get infinitely many non-isomorphic $AC(\sigma)$ spaces from such sets. On the other hand, if we require that these sets are subsets of $\mathbb{R}$, then there are exactly two such spaces up to isomorphism.

The $AC(\sigma)$ spaces are defined as subalgebras of spaces of functions of bounded variation, denoted by $BV(\sigma)$. For general sets σ the isomorphisms between $BV(\sigma)$ spaces have been much less studied. Certainly isomorphisms of such spaces need not be associated with homeomorphisms of the domains of the function spaces (see Example 3.4). On the other hand, for the limited range of sets we are working with here, $BV(\sigma)$ is not much bigger than $AC(\sigma)$ and many proofs can be adapted to give a result about these larger spaces.

2. Preliminaries

In this section we shall briefly outline the definition of the spaces $AC(\sigma)$ and $BV(\sigma)$. Throughout, σ, σ_1 and σ_2 will denote nonempty compact subsets of the plane, which, for notational convenience, we shall often identify as $\mathbb{R}^2$. We shall work throughout with algebras of complex-valued functions.

Suppose that $f : \sigma \to \mathbb{C}$. Let $S = \left[\boldsymbol{x}_0, \boldsymbol{x}_1, \dots, \boldsymbol{x}_n\right]$ be a finite ordered list of elements of σ, where, for the moment, we shall assume that $n \geq 1$. Let γ_S denote the piecewise linear curve joining the points of S in order. Note that the elements of such a list do not need to be distinct.

The *curve variation of f on the set S* is defined to be

$$\operatorname{cvar}(f, S) = \sum_{i=1}^{n} \left| f(\boldsymbol{x}_i) - f(\boldsymbol{x}_{i-1}) \right|. \tag{2.1}$$

Unless f is constant, this quantity can be made arbitrarily large by taking S to consist of a repeating sequence of points on which f differs. To deal with this we associate to each list S a variation factor $\operatorname{vf}(S)$. Loosely speaking, this is the greatest number of times that γ_S crosses any line in the plane. To make this more precise we need the concept of a crossing segment.

Definition 2.1. Suppose that ℓ is a line in the plane. We say that $\overline{\boldsymbol{x}_i\, \boldsymbol{x}_{i+1}}$, the line segment joining $\boldsymbol{x}_i$ to $\boldsymbol{x}_{i+1}$, is a *crossing segment* of $S = \left[\boldsymbol{x}_0, \boldsymbol{x}_1, \dots, \boldsymbol{x}_n\right]$ on ℓ if any one of the following holds:

(i) $\boldsymbol{x}_i$ and $\boldsymbol{x}_{i+1}$ lie on (strictly) opposite sides of ℓ.
(ii) $i = 0$ and $\boldsymbol{x}_i \in \ell$.
(iii) $i > 0$, $\boldsymbol{x}_i \in \ell$ and $\boldsymbol{x}_{i-1} \notin \ell$.
(iv) $i = n-1$, $\boldsymbol{x}_i \notin \ell$ and $\boldsymbol{x}_{i+1} \in \ell$.

In this case we shall write $\overline{\boldsymbol{x}_i\, \boldsymbol{x}_{i+1}} \in X(S, \ell)$.

Definition 2.2. Let $\operatorname{vf}(S, \ell)$ denote the number of crossing segments of S on ℓ. The *variation factor* of S is defined to be

$$\operatorname{vf}(S) = \max_{\ell} \operatorname{vf}(S, \ell).$$

Clearly $1 \leq \operatorname{vf}(S) \leq n$. For completeness, in the case that $S = \left[\boldsymbol{x}_0\right]$ we set $\operatorname{cvar}(f, \left[\boldsymbol{x}_0\right]) = 0$ and let $\operatorname{vf}(\left[\boldsymbol{x}_0\right], \ell) = 1$ whenever $\boldsymbol{x}_0 \in \ell$.

Definition 2.3. The *two-dimensional variation* of a function $f : \sigma \to \mathbb{C}$ is defined to be

$$\operatorname{var}(f, \sigma) = \sup_{S} \frac{\operatorname{cvar}(f, S)}{\operatorname{vf}(S)}, \tag{2.2}$$

where the supremum is taken over all finite ordered lists of elements of σ.

The *variation norm* of such a function is

$$\|f\|_{BV(\sigma)} = \|f\|_\infty + \operatorname{var}(f, \sigma),$$

and the set of functions of bounded variation on σ is

$$BV(\sigma) = \{f : \sigma \to \mathbb{C} : \|f\|_{BV(\sigma)} < \infty\}.$$

The space $BV(\sigma)$ is a Banach algebra under pointwise operations [1, Theorem 3.8]. If $\sigma = [a,b] \subseteq \mathbb{R}$ then the above definition is equivalent to the more classical one.

Let $\mathcal{P}_2$ denote the space of polynomials in two real variables of the form $p(x,y) = \sum_{n,m} c_{nm} x^n y^m$, and let $\mathcal{P}_2(\sigma)$ denote the restrictions of elements on $\mathcal{P}_2$ to σ. The algebra $\mathcal{P}_2(\sigma)$ is always a subalgebra of $BV(\sigma)$ [1, Corollary 3.14].

Definition 2.4. The set of *absolutely continuous* functions on σ, denoted by $AC(\sigma)$, is the closure of $\mathcal{P}_2(\sigma)$ in $BV(\sigma)$.

The set $AC(\sigma)$ forms a closed subalgebra of $BV(\sigma)$ and hence is a Banach algebra. Again, if $\sigma = [a,b]$, this definition reduces to the classical definition.

More generally, we always have that $C^1(\sigma) \subseteq AC(\sigma) \subseteq C(\sigma)$, where one interprets $C^1(\sigma)$ as consisting of all functions for which there is a C^1 extension to an open neighbourhood of σ (see [8]).

3. Locally piecewise affine maps

It is essentially a consequence of the classical Banach–Stone theorem that any algebra isomorphism between two $AC(\sigma)$ spaces must take the form of a composition operator determined by a homeomorphism. (Note that this is not true for the $BV(\sigma)$ spaces.)

Theorem 3.1 ([9, Theorem 2.6]). *Suppose that σ_1 and σ_2 are nonempty compact subsets of the plane. If $\Phi : AC(\sigma_1) \to AC(\sigma_2)$ is an isomorphism, then there exists a homeomorphism $h : \sigma_1 \to \sigma_2$ such that $\Phi(f) = f \circ h^{-1}$ for all $f \in AC(\sigma_1)$.*

Not all homeomorphisms $h : \sigma_1 \to \sigma_2$ produce algebra isomorphisms, but a large class of suitable maps can be obtained by taking compositions of what are known as locally piecewise affine maps.

Let $\alpha : \mathbb{R}^2 \to \mathbb{R}^2$ be an invertible affine map, and let C be a convex n-gon. Then $\alpha(C)$ is also a convex n-gon. Denote the sides of C by $s_1, \dots, s_n$. Suppose that $\boldsymbol{x}_0 \in \operatorname{int}(C)$. The point $\boldsymbol{x}_0$ determines a triangulation $T_1, \dots, T_n$ of C, where T_j is the (closed) triangle with side s_j and vertex $\boldsymbol{x}_0$. A point $\boldsymbol{y}_0 \in \operatorname{int}(\alpha(C))$ determines a similar triangularization $\hat{T}_1, \dots, \hat{T}_n$ of $\alpha(C)$, where the numbering is such that $\alpha(s_j) \subseteq \hat{T}_j$.

Lemma 3.2. *With the notation as above, there is a unique map $h : \mathbb{R}^2 \to \mathbb{R}^2$ such that*

1. *$h(\boldsymbol{x}) = \alpha(\boldsymbol{x})$ for $\boldsymbol{x} \notin \operatorname{int}(C)$;*
2. *h maps T_j onto $\hat{T}_j$, for $1 \le j \le n$;*
3. *$\alpha_j = h|T_j$ is affine, for $1 \le j \le n$;*
4. *$h(\boldsymbol{x}_0) = \boldsymbol{y}_0$.*

We shall say that h is the *locally piecewise affine* map determined by $(C, \alpha, \boldsymbol{x}_0, \boldsymbol{y}_0)$.

The important property of locally piecewise affine maps is that they preserve the isomorphism class of $AC(\sigma)$ spaces. (Explicit bounds on the norms of the isomorphisms are given in [9], but we shall not need these here. In any case, the known bounds are unlikely to be sharp.)

Theorem 3.3 ([3, Theorem 5.5]). *Suppose that σ is a nonempty compact subset of the plane, and that h is a locally piecewise affine map. Then we have $BV(\sigma) \simeq BV(h(\sigma))$ and $AC(\sigma) \simeq AC(h(\sigma))$.*

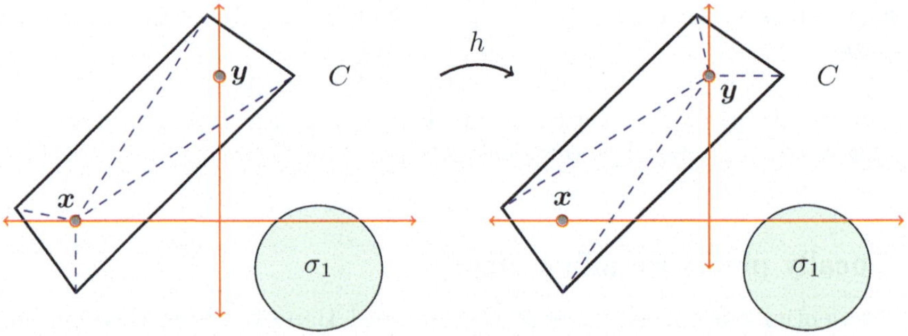

FIGURE 1. A locally piecewise affine map h moving $\boldsymbol{x}$ to $\boldsymbol{y}$.

For most applications it suffices to restrict one's attention to locally piecewise affine maps where the map α is the identity. This allows you to move certain parts of σ while leaving other parts fixed. In particular, if σ_1 is a compact set and $\boldsymbol{x}$ and $\boldsymbol{y}$ are points in the complement of σ_1 which can be joined by a polygonal path which avoids σ_1, then $BV(\sigma_1 \cup \{\boldsymbol{x}\}) \simeq BV(\sigma_1 \cup \{\boldsymbol{y}\})$ and $AC(\sigma_1 \cup \{\boldsymbol{x}\}) \simeq AC(\sigma_1 \cup \{\boldsymbol{y}\})$ (see Figure 1). This would be sufficient to prove our main theorem in Section 5, but in the next section we shall prove a more general result which removes the requirement that there be a path from $\boldsymbol{x}$ to $\boldsymbol{y}$.

The following example shows that there are isomorphisms of $BV(\sigma)$ spaces which are not induced by homeomorphisms of the domains σ_1 and σ_2.

Example 3.4. Let $\sigma_1 = \sigma_2 = \{0\} \cup \{\frac{1}{n}\}_{n=1}^{\infty}$. Define $h : \sigma_1 \to \sigma_2$ by

$$h(\boldsymbol{x}) = \begin{cases} 1, & \boldsymbol{x} = 0, \\ 0, & \boldsymbol{x} = 1, \\ \boldsymbol{x}, & \text{otherwise.} \end{cases}$$

and for $f \in BV(\sigma_1)$ let $\Phi(f) : \sigma_2 \to \mathbb{C}$ be $\Phi(f) = f \circ h^{-1}$. A simple calculation shows that $\frac{1}{3}\operatorname{var}(f, \sigma_1) \le \operatorname{var}(\Phi(f), \sigma_2) \le 3\operatorname{var}(f, \sigma_1)$, and so Φ is a Banach algebra isomorphism from $BV(\sigma_1)$ to $BV(\sigma_2)$. The map h is of course not a homeomorphism.

On the other hand, as in the example, all isomorphisms of $BV(\sigma)$ spaces do come from composition with a bijection of the two domains.

Theorem 3.5. *Suppose that σ_1 and σ_2 are nonempty compact subsets of the plane. If $\Phi : BV(\sigma_1) \to BV(\sigma_2)$ is an isomorphism, then there exists a bijection $h : \sigma_1 \to \sigma_2$ such that $\Phi(f) = f \circ h^{-1}$ for all $f \in BV(\sigma_1)$.*

Proof. Since Φ is an algebra isomorphism, it must map idempotents to idempotents. Note that for all $z \in \sigma_1$, the function $f_z = \chi_{\{z\}}$ lies in $BV(\sigma_1)$ and hence $g_z = \Phi(f_z)$ is an idempotent in $BV(\sigma_2)$. Since Φ is one-to-one, g_z is not the zero function and hence the support of g_z is a nonempty set $\tau \subseteq \sigma_2$. If τ is more than a singleton then we can choose $w \in \tau$ and write $g_z = \chi_{\{w\}} + \chi_{S_T \setminus \{w\}}$ as a sum of two nonzero idempotents in $BV(\sigma_2)$. But then $f_z = \Phi^{-1}(\chi_{\{w\}}) + \Phi^{-1}(\chi_{S \setminus \{w\}})$ is the sum of two nonzero idempotents in $BV(\sigma_1)$ which is impossible. It follows that g_z is the characteristic function of a singleton set, and this clearly induces a map $h : \sigma_1 \to \sigma_2$ so that $\Phi(f_z) = \chi_{\{h(z)\}}$. Indeed, by considering Φ^{-1} it is clear that h must be a bijection between the two sets. □

4. Isolated points

In general, calculating $\|f\|_{BV(\sigma)}$, or indeed checking that a function f is in $AC(\sigma)$ can be challenging. One way to simplify things is to break σ into smaller pieces and then deal with the restrictions of f to these pieces. If σ_1 is a compact subset of σ and $f \in AC(\sigma)$, then it is easy to check that $f|\sigma_1 \in AC(\sigma_1)$ and $\|f|\sigma_1\|_{BV(\sigma_1)} \le \|f\|_{BV(\sigma)}$. However there are simple examples (see, e.g., [8, Example 3.3]) where $\sigma = \sigma_1 \cup \sigma_2$, $f|\sigma_1 \in AC(\sigma_1)$, $f|\sigma_2 \in AC(\sigma_2)$, but $f \notin BV(\sigma)$.

If one has disjoint sets σ_1 and σ_2, then the situation is rather better. Writing $\sigma = \sigma_1 \cup \sigma_2$ one essentially has that $AC(\sigma) = AC(\sigma_1) \oplus AC(\sigma_2)$; see [8, Corollary 5.3]. To formally make sense of this one needs to identify $AC(\sigma_1)$ with the set $\{f \in AC(\sigma) : \operatorname{supp}(f) \subseteq \sigma_1\}$. This requires that if one extends a function $g \in AC(\sigma_1)$ to all of σ by making it zero on σ_2, then the extended function is absolutely continuous. While this is indeed always true, the constant C_{σ_1,σ_2} such that $\|f\|_{BV(\sigma)} \le C_{\sigma_1,\sigma_2} \|f|\sigma_1\|_{BV(\sigma_1)}$ depends on the geometric configuration of the two sets, and is not bounded by any universal constant.

For what we need later in the paper, we shall just need to consider the special case where σ_2 is an isolated singleton point. For the remainder of this section then assume that σ_1 is a nonempty compact subset of $\mathbb{C}$, that $\boldsymbol{z} \notin \sigma_1$ and that $\sigma = \sigma_1 \cup \{\boldsymbol{z}\}$. It is worth noting (using Proposition 4.4 of [1] for example) that $\chi_{\{\boldsymbol{z}\}}$ is always an element of $AC(\sigma)$.

For $f \in BV(\sigma)$ let

$$\|f\|_D = \|f\|_{D(\sigma_1,\boldsymbol{z})} = \|f|\sigma_1\|_{BV(\sigma_1)} + |f(\boldsymbol{z})|.$$

(To prevent the notation from becoming too cumbersome we will usually just write $\|f\|_{BV(\sigma_1)}$ rather than $\|f|\sigma_1\|_{BV(\sigma_1)}$ unless there is some risk of confusion.)

Proposition 4.1. *The norm $\|\cdot\|_D$ is equivalent to the usual norm $\|\cdot\|_{BV(\sigma)}$ on $BV(\sigma)$.*

Proof. We first remark that it is clear that $\|\cdot\|_D$ is a norm on $BV(\sigma)$. Also, noting the above remarks, $\|f\|_D \le 2\|f\|_{BV(\sigma)}$ so we just need to find a suitable lower bound for $\|f\|_D$.

Suppose then that $f \in BV(\sigma)$. Let $S = [\boldsymbol{x}_0, \boldsymbol{x}_1, \dots, \boldsymbol{x}_n]$ be an ordered list of points in σ and let $S' = [\boldsymbol{y}_0, \dots, \boldsymbol{y}_m]$ be the list S with those points equal to $\boldsymbol{z}$ omitted. Our aim is to compare $\operatorname{cvar}(f, S)$ with $\operatorname{cvar}(f, S')$. In calculating $\operatorname{cvar}(f, S)$ we may assume that no two consecutive points in this list are both equal to $\boldsymbol{z}$, and that S' is nonempty. Let N be the number of times that the point $\boldsymbol{z}$ occurs in the list S.

Now if $\boldsymbol{x}_k = \boldsymbol{z}$ for some $0 < k < n$, then

$$\begin{aligned}|f(\boldsymbol{x}_k) - f(\boldsymbol{x}_{k-1})| &+ |f(\boldsymbol{x}_{k+1}) - f(\boldsymbol{x}_k)| \\ &\le 2\|f|\sigma_1\|_\infty + 2|f(\boldsymbol{z})| \\ &\le |f(\boldsymbol{x}_{k+1}) - f(\boldsymbol{x}_{k-1})| + 2\|f|\sigma_1\|_\infty + 2|f(\boldsymbol{z})|.\end{aligned}$$

If $\boldsymbol{x}_0 = \boldsymbol{z}$, then $|f(\boldsymbol{x}_1) - f(\boldsymbol{x}_0)| \le \|f|\sigma_1\|_\infty + |f(\boldsymbol{z})|$, and a similar estimate applies if $\boldsymbol{x}_n = \boldsymbol{z}$. Putting these together shows that

$$\begin{aligned}\operatorname{cvar}(f, S) &= \sum_{k=1}^{n} |f(\boldsymbol{x}_k) - f(\boldsymbol{x}_{k-1})| \\ &\le \sum_{k=1}^{m} |f(\boldsymbol{y}_k) - f(\boldsymbol{y}_{k-1})| + 2N(\|f|\sigma_1\|_\infty + |f(\boldsymbol{z})|).\end{aligned}$$

Let ℓ be any line through $\boldsymbol{z}$ which doesn't intersect any other points of S. Checking Definition 2.1, one sees that we get a crossing segment of S on ℓ for each time that $\boldsymbol{x}_k = \boldsymbol{z}$ and so $\operatorname{vf}(S) \ge \operatorname{vf}(S, \ell) \ge N$. By [8, Proposition 3.5] we also have that $\operatorname{vf}(S) \ge \operatorname{vf}(S')$. Thus

$$\begin{aligned}\frac{\operatorname{cvar}(f, S)}{\operatorname{vf}(S)} &\le \frac{\operatorname{cvar}(f, S') + 2N(\|f|\sigma_1\|_\infty + |f(\boldsymbol{z})|)}{\operatorname{vf}(S)} \\ &\le \frac{\operatorname{cvar}(f, S')}{\operatorname{vf}(S')} + \frac{2N(\|f|\sigma_1\|_\infty + |f(\boldsymbol{z})|)}{N} \\ &\le \operatorname{var}(f, \sigma_1) + 2(\|f|\sigma_1\|_\infty + |f(\boldsymbol{z})|) \\ &\le 2\|f\|_D.\end{aligned}$$

Taking the supremum over all lists S then shows that $\operatorname{var}(f, \sigma) \le 2\|f\|_D$ and hence that

$$\|f\|_{BV(\sigma)} = \|f\|_\infty + \operatorname{var}(f, \sigma) \le 3\|f\|_D,$$

which completes the proof. □

The constants obtained in the proof of Proposition 4.1 are in fact sharp. Suppose that $\sigma_1 = \{-1, 1\}$, $\boldsymbol{z} = 0$ and $\sigma = \sigma_1 \cup \{z\}$. Then $\|\chi_{\{0\}}\|_D = 1$ while $\|\chi_{\{0\}}\|_{BV(\sigma)} = 3$. On the other hand, if f is the constant function 1, then $\|f\|_D = 2$ while $\|f\|_{BV(\sigma)} = 1$.

Proposition 4.2. *$f \in AC(\sigma)$ if and only if $f|\sigma_1 \in AC(\sigma_1)$.*

Proof. Rather than using the heavy machinery of [8, Section 5], we give a more direct proof using the definition of absolute continuity. As noted above, one just needs to show that if $f|\sigma_1 \in AC(\sigma_1)$, then $f \in AC(\sigma)$. Suppose then that $f|\sigma_1 \in AC(\sigma_1)$. Given $\epsilon > 0$, there exists a polynomial $p \in \mathcal{P}_2$ such that $\|f - p\|_{BV(\sigma_1)} < \epsilon/3$. Define $g : \sigma \to \mathbb{C}$ by $g = p + (f(\boldsymbol{z}) - p(\boldsymbol{z}))\chi_{\{z\}}$. Then we have $g \in AC(\sigma)$ and $\|f - g\|_{BV(\sigma)} \le 3\|f - g\|_D = 3\|f - p\|_{BV(\sigma_1)} < \epsilon$ since $\chi_{\{z\}} \in AC(\sigma)$. As $AC(\sigma)$ is closed, this shows that $f \in AC(\sigma)$. ☐

Corollary 4.3. *Suppose that σ_1 is a nonempty compact subset of $\mathbb{C}$ and that $\boldsymbol{x}$ and $\boldsymbol{y}$ are points in the complement of σ_1. Then $BV(\sigma_1 \cup \{\boldsymbol{x}\}) \simeq BV(\sigma_1 \cup \{\boldsymbol{y}\})$ and $AC(\sigma_1 \cup \{\boldsymbol{x}\}) \simeq AC(\sigma_1 \cup \{\boldsymbol{y}\})$.*

Proof. Let $h : \sigma_1 \cup \{\boldsymbol{x}\} \to \sigma_1 \cup \{\boldsymbol{y}\}$ be the natural homeomorphism, which is the identity on σ_1 and which maps $\boldsymbol{x}$ to $\boldsymbol{y}$, and for $f \in BV(\sigma_1 \cup \{\boldsymbol{x}\})$ let $\Phi(f) = f \circ h^{-1}$. Then Φ is an algebra isomorphism of $BV(\sigma_1 \cup \{\boldsymbol{x}\})$ onto $BV(\sigma_1 \cup \{\boldsymbol{y}\})$ which is isometric under the norms $\|\cdot\|_{D(\sigma_1,\boldsymbol{x})}$ and $\|\cdot\|_{D(\sigma_1,\boldsymbol{y})}$, and hence it is certainly bicontinuous under the respective BV norms.

It follows immediately from Proposition 4.2 that Φ preserves absolute continuity as well. ☐

More generally of course, this result says that one can move any finite number of isolated points around the complex plane without altering the isomorphism class of these spaces.

5. C-sets

The spectrum of a compact operator is either finite or else a countable set with limit point 0. If σ has n elements, then $AC(\sigma)$ is an n-dimensional algebra and consequently for finite sets, one has a trivial Banach–Stone type theorem: $AC(\sigma_1) \simeq AC(\sigma_2)$ if and only if σ_1 and σ_2 have the same number of elements. (Of course the same result is also true for the $BV(\sigma)$ spaces.)

The case where σ is a countable set is more complicated however.

Definition 5.1. We shall say that a subset $\sigma \subseteq \mathbb{C}$ is a *C-set* if it is a countably infinite compact set with unique limit point 0. If further $\sigma \subseteq \mathbb{R}$, we shall say that σ is a *real C-set.*

Any two C-sets are homeomorphic, but as we shall see, they can produce an infinite number of non-isomorphic spaces of absolutely continuous functions. In most of what follows, it is not particularly important that the limit point of the set is 0 since one can apply a simple translation of the domain σ

to achieve this and any such translation induces an isometric isomorphism of the corresponding function spaces.

The easiest C-sets to deal with are what were called spoke sets in [3], that is, sets which are contained in a finite number of rays emanating from the origin. To state our main theorem, we shall need a slight variant of this concept. For $\theta \in [0, 2\pi)$ let R_θ denote the ray $\{te^{i\theta} : t \ge 0\}$.

Definition 5.2. Suppose that k is a positive integer. We shall say that a C-set σ is a *k-ray set* if there are k distinct rays $R_{\theta_1}, \dots, R_{\theta_k}$ such that

1. $\sigma_j := \sigma \cap R_{\theta_j}$ is infinite for each j,
2. $\sigma_0 := \sigma \setminus (\sigma_1 \cup \cdots \cup \sigma_k)$ is finite.

If σ_0 is empty, then we shall say that σ is a *strict k-ray set.*

Although in general the calculation of norms in $BV(\sigma)$ can be difficult, if σ is a strict k-ray set, then we can pass to a much more tractable equivalent norm, called the spoke norm in [3].

Definition 5.3. Suppose that σ is a strict k-ray set. The *k-spoke norm* on $BV(\sigma)$ is (using the notation of Definition 5.2)

$$\|f\|_{Sp(k)} = |f(0)| + \sum_{j=1}^{k} \|f - f(0)\|_{BV(\sigma_j)} .$$

Since each of the subsets σ_j is contained in a straight line, the calculation of the variation over these is straightforward. If we write $\sigma_j = \{0\} \cup \{\lambda_{j,i}\}_{i=1}^{\infty}$ with $|\lambda_{j,1}| > |\lambda_{j,2}| > \cdots$, then

$$\|f - f(0)\|_{BV(\sigma_j)} = \sup_i |f(\lambda_{j,i}) - f(0)| + \sum_{i=1}^{\infty} |f(\lambda_{j,i}) - f(\lambda_{j,i+1})|.$$

Proposition 5.4 ([3, Proposition 4.3]). *Suppose that σ is a strict k-ray set. Then for all $f \in BV(\sigma)$,*

$$\frac{1}{2k+1} \|f\|_{Sp(k)} \le \|f\|_{BV(\sigma)} \le 3 \|f\|_{Sp(k)} .$$

One property which significantly simplifies the analysis for such spaces is that for a k-ray set σ, one always has $AC(\sigma) = BV(\sigma) \cap C(\sigma)$. In particular a function of bounded variation on such a set σ is absolutely continuous if and only if it is continuous at the origin.

Proposition 5.5. *If σ is a k-ray set then $AC(\sigma) = BV(\sigma) \cap C(\sigma)$.*

Proof. Since $AC(\sigma)$ is always a subset of $BV(\sigma) \cap C(\sigma)$ we just need to prove the reverse inclusion.

Suppose first that σ is a strict k-ray set and $f \in BV(\sigma) \cap C(\sigma)$. For $n = 1, 2, \dots$, define $g_n : \sigma \to \mathbb{C}$ by

$$g_n(z) = \begin{Bmatrix} f(z), & \text{if } |z| \ge \frac{1}{n}, \\ f(0), & \text{if } |z| < \frac{1}{n} \end{Bmatrix} = f(0) + \sum_{|z| \ge 1/n} (f(z) - f(0)) \chi_{\{z\}}.$$

Since $\chi_{\{z\}} \in AC(\sigma)$ for all nonzero points in σ, clearly $g_n \in AC(\sigma)$. Now

$$\|f - g_n\|_{Sp(k)} = \sum_{j=1}^{k} \|f - g_n\|_{BV(\sigma_j)} \cdot \tag{5.1}$$

Fix j and label the elements of σ_j as above. Then, for all n there exists an index $I_{j,n}$ such that $|\lambda_{j,i}| < \frac{1}{n}$ if and only if $i \geq I_{j,n}$. Thus

$$\begin{aligned} &\|f - g_n\|_{BV(\sigma_j)} \\ &= \sup_{i \geq I_{j,n}} |f(\lambda_{j,i}) - f(0)| + \sum_{i \geq I_{j,n}} |f(\lambda_{j,i}) - f(\lambda_{j,i+1})| + |f(\lambda_{j,I_{j,n}} - f(0)|. \end{aligned}$$

The first and last of these terms converge to zero since $f \in C(\sigma)$. The middle term also converges to zero since it is the tail of a convergent sum.

Since we can make each of the k terms in (5.1) as small as we like, $\|f - g_n\|_{Sp(k)} \to 0$ and hence $g_n \to f$ in $BV(\sigma)$. Thus $f \in AC(\sigma)$.

Suppose finally that σ is not a strict k-ray set, that is $\sigma_0 \neq \emptyset$. Let $\sigma' = \sigma \setminus \sigma_0$. If $f \in BV(\sigma) \cap C(\sigma)$, then $f|\sigma' \in BV(\sigma') \cap C(\sigma')$. By the above, $f|\sigma' \in AC(\sigma')$. Repeated use of Proposition 4.2 then shows $f \in AC(\sigma)$. □

It would be interesting to know whether Theorem 5.5 holds for more general C-sets.

Corollary 5.6. *Suppose that σ is a strict k-ray set and that $f : \sigma \to \mathbb{C}$. For $j = 1, \ldots, k$, let f_j denote the restriction of f to σ_j. Then $f \in AC(\sigma)$ if and only if $f_j \in AC(\sigma_j)$ for all j.*

Proof. By Lemma 4.5 of [1], if $f \in AC(\sigma)$, then the restriction of f to any compact subset is also absolutely continuous. If each $f_j \in AC(\sigma_j)$, then certainly $f \in C(\sigma)$. Furthermore $\|f\|_{Sp(k)}$ is finite and hence $f \in BV(\sigma)$. Thus, by Proposition 5.5, $f \in AC(\sigma)$. □

Theorem 5.7. *Suppose that σ is a k-ray set and that τ is an ℓ-ray set. Then $AC(\sigma) \simeq AC(\tau)$ if and only if $k = \ell$.*

Proof. Write $\sigma = \cup_{j=0}^{k} \sigma_j$ and $\tau = \cup_{j=0}^{\ell} \tau_j$ as in Definition 5.2. It follows from Corollary 4.3 that by moving the finite number of points in σ_0 onto one of the rays containing a set σ_j, that $AC(\sigma)$ is isomorphic to $AC(\sigma')$ for some strict k-ray set. To prove the theorem then, it suffices therefore to assume that σ and τ are strict k and ℓ-ray sets.

Suppose first that $k > \ell$ and that there is a Banach algebra isomorphism Φ from $AC(\sigma)$ to $AC(\tau)$. By Theorem 3.1, $\Phi(f) = f \circ h^{-1}$ for some homeomorphism $h : \sigma \to \tau$.

By the pigeonhole principle there exists $L \in \{1, \ldots, \ell\}$ so that $h(\sigma_j) \cap \tau_L$ is infinite for (at least) two distinct sets values of j. Without loss of generality we will assume that this is true for $j = 1$ and $j = 2$. Indeed, since rotations produce isometric isomorphisms of these spaces, we may also assume that $\tau_L \subset [0, \infty)$. Let $\sigma_j = \{0\} \cup \{\lambda_{j,i}\}_{i=1}^{\infty}$, where the points are labelled so that $|\lambda_{j,1}| > |\lambda_{j,2}| > \cdots$. There must then be two increasing sequences $i_1 < i_2 < \cdots$ and $k_1 < k_2 < \cdots$ such that

$$h(\lambda_{1,i_1}) > h(\lambda_{2,k_1}) > h(\lambda_{1,i_2}) > h(\lambda_{2,k_2}) > \cdots .$$

For $n = 1, 2, \ldots$ define $f_n \in AC(\sigma)$ by

$$f_n(z) = \begin{cases} 1, & z \in \{\lambda_{1,1}, \ldots, \lambda_{1,n}\}, \\ 0, & \text{otherwise.} \end{cases}$$

Then $\|f_n\|_{Sp(k)} = 2$ for all n, but $\|\Phi(f_n)\|_{Sp(\ell)} \geq 2n$. Using Proposition 5.4, this means that Φ must be unbounded, which is impossible. Hence no such isomorphism can exist.

Finally, suppose that $k = \ell$. For each $j = 1, 2, \ldots, k$ order the elements of σ_j and τ_j by decreasing modulus and let h_j be the unique homeomorphism from σ_j to τ_j which preserves this ordering. Let h be the homeomorphism whose restriction to each σ_j is h_j and let $\Phi(f) = f \circ h^{-1}$. Then Φ is an isometric isomorphism from $(BV(\sigma), \|\cdot\|_{Sp(k)})$ to $(BV(\tau), \|\cdot\|_{Sp(k)})$, and hence is a Banach algebra isomorphism between these spaces under their usual BV norms. Since Φ is also an isomorphism from $C(\sigma)$ to $C(\tau)$, the result now follows from Proposition 5.5. □

Corollary 5.8. *There are infinitely many mutually non-isomorphic $AC(\sigma)$ spaces with σ a C-set.*

Clearly, any real C-set is either a 1-ray set, or a 2-ray set.

Corollary 5.9. *There are exactly two isomorphism classes of $AC(\sigma)$ spaces with σ a real C-set.*

We should point out at this point that Theorem 5.7 is far from a characterization of the sets τ for which $AC(\tau)$ is isomorphic to $AC(\sigma)$ where σ is some k-ray set.

Example 5.10. Let $\tau = \{0\} \cup \{\frac{1}{j} + \frac{i}{j^2}\}_{j=1}^{\infty}$ and let $\sigma = \{0\} \cup \{\frac{1}{j}\}_{j=1}^{\infty}$. Clearly, τ is not a k-ray set for any k.

For $f \in BV(\sigma)$ let $\Phi(f)(t + it^2) = f(t)$, $t \in \sigma$. It follows from [1, Lemma 3.12] that $\|\Phi(f)\|_{BV(\tau)} \leq \|f\|_{BV(\sigma)}$. For the other direction, suppose that $\lambda_0 \leq \lambda_1 \leq \cdots \leq \lambda_n$ are points in σ and let $S = [\lambda_0 + i\lambda_0^2, \ldots, \lambda_n + i\lambda_n^2]$ be the corresponding list of points in τ. It is easy to see that $\mathrm{vf}(S)$ is 2 if $n > 1$ (and is 1 if $n = 1$). Then

$$\begin{aligned} \sum_{j=1}^{n} |f(\lambda_j)) - f(\lambda_{j-1})| &= \sum_{j=1}^{n} |\Phi(f)(\lambda_j + i\lambda_j^2)) - \Phi(f)(\lambda_{j-1} + i\lambda_j^2)| \\ &\leq 2 \frac{\mathrm{cvar}(\Phi(f), S)}{\mathrm{vf}(S)} \\ &\leq 2\,\mathrm{var}(\Phi(f), \tau). \end{aligned}$$

Since the variation of f is given by the supremum of such sums over all such ordered subsets of σ, we have $\operatorname{var}(f,\sigma) \le 2\operatorname{var}(\Phi(f),\tau)$ and hence $\|f\|_{BV(\sigma)} \le 2\,\|\Phi(f)\|_{BV(\tau)}$. This shows that $BV(\sigma) \simeq BV(\tau)$.

Proposition 4.4 of [1] ensures that if $f \in AC(\sigma)$, then $\Phi(f) \in AC(\tau)$. Conversely, if $g = \Phi(f) \in AC(\tau)$, then certainly $g \in C(\tau)$ and consequently $f \in C(\sigma)$. By the previous paragraph $f \in BV(\sigma)$ too and hence, by Proposition 5.5, $f \in AC(\sigma)$. Thus $AC(\sigma) \simeq AC(\tau)$.

Example 5.11. Let $\sigma = \{0\} \cup \{\frac{e^{i/m}}{n} : n, m \in \mathbb{Z}^+\} \cup \{\frac{1}{n} : n \in \mathbb{Z}^+\}$ (where $\mathbb{Z}^+$ denotes the set of positive integers) and let τ be an ℓ-ray set. Repeating the proof of Theorem 5.7, one sees that there can be no Banach algebra isomorphism from $AC(\sigma)$ to $AC(\tau)$, so even among C-sets there are more isomorphism classes than those captured by Theorem 5.7.

The corresponding result for the $BV(\sigma)$ spaces also holds.

Corollary 5.12. *Suppose that σ is a k-ray set and that τ is an ℓ-ray set. Then $BV(\sigma) \simeq BV(\tau)$ if and only if $k = \ell$.*

Proof. The proof is more or less identical to that of Theorem 5.7. In showing that $AC(\sigma) \not\simeq AC(\tau)$ for $k \neq \ell$ we used the fact that any isomorphism between these spaces is of the form $\Phi(f) = f \circ h^{-1}$. In showing that such a map cannot be bounded, the continuity of h was not used, only the fact that h must be a bijection, and so one may use Theorem 3.5 in place of Theorem 3.1 in this case.

The fact that $BV(\sigma) \simeq BV(\tau)$ for $k = \ell$ is already noted in the above proof. □

6. Operator algebras

If $\sigma = \{0\} \cup \{\frac{1}{n}\}_{n=1}^{\infty}$, the map $\Psi : AC(\sigma) \to \ell^1$,

$$\Psi(f) = \big(f(1), f(\tfrac{1}{2}) - f(1), f(\tfrac{1}{3}) - f(\tfrac{1}{2}), \dots\big)$$

is a Banach space isomorphism. Indeed it is not hard to see that Proposition 5.4 implies that if σ is a strict k-ray set, then, as Banach spaces, $AC(\sigma)$ is isomorphic to $\oplus_{j=1}^{k} \ell^1$, which in turn is isomorphic to ℓ^1, and consequently all such $AC(\sigma)$ spaces are Banach space isomorphic.

Given any nonempty compact set $\sigma \subseteq \mathbb{C}$, the operator $Tg(z) = zg(z)$ acting on $AC(\sigma)$ is an $AC(\sigma)$ operator. Indeed the functional calculus for T is given by $f(T)g = fg$ for $f \in AC(\sigma)$, from which one can deduce that $\|f(T)\| = \|f\|_{BV(\sigma)}$, and therefore the Banach algebra generated by the functional calculus for T is isomorphic to $AC(\sigma)$. Proposition 6.1 of [3] shows that if σ is a C-set, then any such operator T is a compact $AC(\sigma)$ operator.

Combining these observations, together with Corollary 5.8, shows that on ℓ^1 there are infinitely many non-isomorphic Banach subalgebras of $B(\ell^1)$ which are generated by (non-finite rank) compact $AC(\sigma)$ operators on ℓ^1, so things are rather different to the known situation for compact normal operators on ℓ^2.

References

[1] B. Ashton and I. Doust, Functions of bounded variation on compact subsets of the plane, *Studia Math.* **169** (2005), 163–188.

[2] B. Ashton and I. Doust, A comparison of algebras of functions of bounded variation, *Proc. Edin. Math. Soc.* **49** (2006), 575–591.

[3] B. Ashton and I. Doust, Compact $AC(\sigma)$ operators, *Integral Equations Operator Theory* **63** (2009), 459–472.

[4] B. Ashton and I. Doust, $AC(\sigma)$ operators, *J. Operator Theory* **65** (2011), 255–279.

[5] E. Berkson and T.A. Gillespie, Absolutely continuous functions of two variables and well-bounded operators, *J. London Math. Soc. (2)* **30** (1984), 305–321.

[6] E. Berkson and T.A. Gillespie, *AC* functions on the circle and spectral families, *J. Operator Theory* **13** (1985), 33–47.

[7] Q. Cheng and I. Doust, Compact well-bounded operators, *Glasg. Math. J.* **43** (2001), 467–475.

[8] I. Doust and M. Leinert, Approximation in $AC(\sigma)$, arXiv:1312.1806v1, 2013.

[9] I. Doust and M. Leinert, Isomorphisms of $AC(\sigma)$ spaces, *Studia Math.* **228** (2015), 7–31.

[10] J.R. Ringrose, On well-bounded operators II, *Proc. London. Math. Soc. (3)* **13** (1963), 613–638.

Ian Doust and Shaymaa Al-shakarchi
School of Mathematics and Statistics
University of New South Wales
UNSW Sydney 2052
Australia
e-mail: i.doust@unsw.edu.au

Restricted inversion of split-Bezoutians

Torsten Ehrhardt and Karla Rost

In memory of Georg Heinig

Abstract. The main aim of the present paper is to compute inverses of split-Bezoutians considered as linear operators restricted to subspaces of symmetric or skewsymmetric vectors. Such results are important, e.g., for the inversion of nonsingular, centrosymmetric or centroskewsymmetric Toeplitz-plus-Hankel Bezoutians B of order n. To realize this inversion we present algorithms with $O(n^2)$ computational complexity, which involves an explicit representation of B^{-1} as a sum of a Toeplitz and a Hankel matrix. Based on different ideas such inversion formulas have already been proved in previous papers by the authors. Here we focus on the occurring splitting parts since they are of interest also in a more general context. The main key is the solution of the converse problem: the inversion of Toeplitz-plus-Hankel matrices. An advantage of this approach is that all appearing special cases can be dealt with in the same, relatively straightforward way without any additional assumptions.

Mathematics Subject Classification (2010). Primary 15A09; Secondary 15B05, 65F05.

Keywords. Bezoutian matrix, Toeplitz matrix, Hankel matrix, Toeplitz-plus-Hankel matrix, matrix inversion.

1. Introduction

In the present paper we deal with special types of structured matrices, the so-called split-Bezoutians. The entries of all vectors and matrices considered here belong to a field $\mathbb{F}$ with characteristic not equal to 2. The subject of split-Bezoutians is inspired by a series of papers dedicated to the inversion of Toeplitz-plus-Hankel Bezoutians. The starting point was [2], where inversion algorithms for Toeplitz Bezoutians and Hankel Bezoutians were established. They relied on the result of [1] that the nullspace of generalized resultant matrices can be described by means of solutions of Bezout equations. In [3], [4], and [5] the inversion of Toeplitz-plus-Hankel Bezoutians was discussed for the first time.

© Springer International Publishing AG, part of Springer Nature 2018
A. Böttcher et al. (eds.), *The Diversity and Beauty of Applied Operator Theory*, Operator Theory: Advances and Applications 268, https://doi.org/10.1007/978-3-319-75996-8_12

Let us recall some underlying notions. A matrix $B = [b_{ij}]_{i,j=0}^{n-1}$ is called a *Toeplitz-plus-Hankel Bezoutian* (briefly, *$T+H$-Bezoutian*) if in polynomial language

$$B(t,s) := \sum_{i,j=0}^{n-1} b_{ij} t^i s^j = \frac{\sum_{i=1}^{4} \mathbf{f}_i(t)\mathbf{g}_i(s)}{(t-s)(1-ts)}$$

for certain polynomials $\mathbf{f}_i(t), \mathbf{g}_i(t)$, $i = 1, \dots, 4$. In the papers [3], [4], and [5], the $T+H$-Bezoutians are assumed to be centrosymmetric or centroskewsymmetric. Recall that an $n \times n$ matrix A is called *centrosymmetric* if $J_n A J_n = A$ and *centroskewsymmetric* if $J_n A J_n = -A$, where J_n denotes the flip matrix of order n,

$$J_n := \begin{bmatrix} 0 & & 1 \\ & \cdot^{\cdot^{\cdot}} & \\ 1 & & 0 \end{bmatrix}. \tag{1.1}$$

In both cases a splitting of the $T + H$-Bezoutian into two special matrices, called *split-Bezoutians*, is possible. In fact, four kinds of split-Bezoutians occur. For centrosymmetric $T + H$-Bezoutians we have a splitting

$$B = B_{++} + B_{--},$$

while in the centroskewsymmtric case we have

$$B = B_{+-} + B_{-+}.$$

The split-Bezoutians $B_{\pm\pm}$ can be thought of as acting between the spaces $\mathbb{F}^n_\pm$ of symmetric or skewsymmetric vectors (i.e., vectors satisfying $\mathbf{x} = \pm J_n \mathbf{x}$). The type of the split-Bezoutians is indicated by '$\pm$'-signs. Split-Bezoutians have a structure which is simpler than that of general $T + H$-Bezoutians. They are given by a pair of vectors $\{\mathbf{f}, \mathbf{g}\}$. For instance, split-Bezoutians of $(++)$type are of the form

$$B_{++}(t,s) = \frac{\mathbf{f}(t)\mathbf{g}(s) - \mathbf{g}(t)\mathbf{f}(s)}{(t-s)(1-ts)}$$

with polynomials $\mathbf{f}(t)$ and $\mathbf{g}(t)$ whose coefficients are the entries of symmetric vectors $\mathbf{f}$ and $\mathbf{g}$.

In the centrosymmetric case these splittings were discovered in Section 8 of [12] (see also Section 5 of [13] for the centro(skew)symmetric cases and more details).

A few words about the history. Bezoutians were considered first in connection with elimination theory by Euler in 1748, Bezout in 1764 and Cayley in 1857 (see, e.g., [18]). Much later, in 1974, their importance for the inversion of Hankel and Toeplitz matrices was discovered by Lander [15]. He observed that the inverse of a nonsingular Hankel (Toeplitz) matrix is a Hankel (Toeplitz) Bezoutian and vice versa. In 1988 Heinig and Rost [11] discovered that inverses of $T + H$-Bezoutians are *Toeplitz-plus-Hankel matrices* (briefly, *$T+H$ matrices*), i.e., matrices $T_n(\mathbf{a}) + H_n(\mathbf{s})$ which are the sum of a Toeplitz matrix $T_n(\mathbf{a}) = [\,a_{i-j}\,]_{i,j=0}^{n-1}$ and a Hankel matrix $H_n(\mathbf{s}) = [\,s_{i+j}\,]_{i,j=0}^{n-1}$. The converse is also true.

Let us mention that there is a vast literature dedicated to the inversion of Toeplitz, Hankel, and also Toeplitz-plus-Hankel matrices, which started with the papers [17], [6], [16], and [10]. On the other hand, the converse problem – the inversion of Bezoutians – has received little attention in the past (see [9], [8], [7]).

The main aim of the present paper is to investigate the invertibility of split-Bezoutians of order $n\,(> 2)$ in more detail. These matrices are always singular, but, under certain assumptions, restrictedly invertible. By this we mean the following. For example, if B is an $n \times n$ split-Bezoutian of $(++)$type, then it maps $\mathbb{F}_+^n$ into itself. We call B *restrictedly invertible* if this map from $\mathbb{F}_+^n$ to $\mathbb{F}_+^n$ is bijective. The inverse map is called the *restricted inverse* and denoted by $B^{(-1)}$.

It turns out that the restricted inverse of a split-Bezoutian is a (particular) $T + H$ matrix. Generally, it is given by only one vector. For instance, the restricted inverse of a split-Bezoutian of $(++)$type is a $T + H$ matrix of the form

$$T_n(\mathbf{c})(I + J_n), \tag{1.2}$$

where $\mathbf{c}$ is a symmetric vector called *symbol* of the Toeplitz matrix $T_n(\mathbf{c})$. Note that $T_n(\mathbf{c})J_n$ is a Hankel matrix.

Our task will be the following. Given a pair of vectors $\{\mathbf{f}, \mathbf{g}\}$ which determines a split-Bezoutian B, and assuming that the split-Bezoutian is restrictedly invertible, compute the vector $\mathbf{c}$ which describes the restricted inverse of B, a $T + H$ matrix (1.2).

Our considerations are based on the results of [13] concerning the inversion of centro(skew)symmetric Toeplitz-plus-Hankel matrices by means of four special solutions of (pure) Toeplitz equations. These solutions allow an explicit description of the Toeplitz-plus-Hankel matrix inverses, which are $T+H$-Bezoutians and which decompose into split-Bezoutians. In other words, the solutions of certain Toeplitz systems give rise to split-Bezoutians, which in our case are given. So we have to reinterpret these linear systems. The entries of the Toeplitz matrices are now the unknowns, whereas the former solution vectors are given. The reinterpretation leads to *inhomogeneous resultant equations* the solutions of which are just the symbols of the $T + H$ matrices we are looking for.

The paper is organized as follows. In Section 2 we introduce notational conventions. The notion of splitting and the restricted invertibility of general centro(skew)symmetric matrices is discussed in Section 3. In Section 4 some basic properties of centro(skew)symmetric $T + H$ matrices are considered. Section 5 is dedicated to $T + H$-Bezoutians and split-Bezoutians of different types denoted by $(++)$, $(--)$, $(+-)$, and $(-+)$. In particular, criteria for their restricted invertibility are established, the reduction of split-Bezoutians of $(--)$, $(+-)$, and $(-+)$type to split-Bezoutians of $(++)$type is presented, and examples are discussed.

In Section 6 the results of [13] on inversion of $T+H$ matrices by special solutions of related Toeplitz equations are presented. Morever, their connection to split-Bezoutians is revealed. A first step towards the afore-mentioned reinterpretation is done in that the restricted invertibility of (special) $T+H$ matrices is examined. Their restricted inverses are the split-Bezoutians.

In Section 7 the second step of the reinterpretation is realized. Here inhomogeneous resultant equations occur, which are solvable if and only if the corresponding split-Bezoutians are restrictedly invertible. The solution of these resultant equations determines the symbol of the $T+H$ matrices connected with restricted inverses of the different types of split-Bezoutians. To summarize, at this point, we have reduced the problem of inverting split-Bezoutians of order n to the problem of solving inhomogeneous resultant equations. The kernels of the corresponding resultant matrices are also determined.

The next step is the construction of particular solutions to these resultant equations. Here the consideration of six cases is necessary. This is because in the $(\pm\pm)$ cases we have to distinguish between n even and n odd. In the $(\pm\mp)$ cases only n even is of interest. This is done in Section 8. The resultant equations are interpreted as polynomial equations, and solutions are constructed by solving Bezout equations. Note that generalized resultant equations and Bezout equations already occurred in connection with the inversion of Toeplitz (or Hankel) Bezoutians [2].

Section 9 is the extract of the previous considerations and presents the main results, the restricted inversion of the split-Bezoutians in all six cases. The resulting algorithms use only the extended Euclidian algorithm (in order to solve Bezout equations) and polynomial multiplications. Hence their computational complexity is $O(n^2)$. Finally, in Section 10, we illustrate the algorithms with some simple examples.

The results obtained here can be applied to the inversion of centro-(skew)symmetric $T+H$-Bezoutians. They confirm the results of [4] and [5]. Notice that the proof of the results of [5] required an artificial extra-condition, which is now shown to be unnecessary. Moreover, the approach presented here is unified and simpler than that one of [5].

2. Preliminaries

Throughout this paper we consider vectors or matrices whose entries belong to a field $\mathbb{F}$ with a characteristic not equal to 2. By $\mathbb{F}^n$ we denote the linear space of all vectors of length n, I_n denotes the identity matrix in $\mathbb{F}^{n\times n}$ and J_n denotes the flip matrix introduced in (1.1).

In what follows we often use polynomial language. We associate with a matrix $A = [\,a_{ij}\,]_{i,j=0}^{n-1} \in \mathbb{F}^{n\times n}$ the bivariate polynomial

$$A(t,s) := \sum_{i,j=0}^{n-1} a_{ij}\, t^i s^j \,, \tag{2.1}$$

and call it the *generating polynomial of A*.

Similarly, with a vector $\mathbf{x} = (x_j)_{j=0}^{m-1} \in \mathbb{F}^m$ we associate the polynomial

$$\mathbf{x}(t) := \sum_{j=0}^{m-1} x_j t^j \in \mathbb{F}^m[t], \tag{2.2}$$

where $\mathbb{F}^m[t]$ denotes the linear space of all polynomials in t of degree less than m with coefficients in $\mathbb{F}$. For a vector $\mathbf{x} \in \mathbb{F}^n$ we put

$$\mathbf{x}^J := J_n \mathbf{x}.$$

In polynomial language this means

$$\mathbf{x}^J(t) = \mathbf{x}(t^{-1})t^{n-1}.$$

A vector $\mathbf{x} \in \mathbb{F}^n$, or its corresponding polynomial $\mathbf{x}(t) \in \mathbb{F}^n[t]$, is said to be *symmetric* if $\mathbf{x} = \mathbf{x}^J$ and *skewsymmetric* if $\mathbf{x} = -\mathbf{x}^J$. The subspaces of $\mathbb{F}^n$ consisting of all symmetric or skewsymmetric vectors, respectively, are denoted by

$$\mathbb{F}^n_{\pm} := \{ \mathbf{x} \in \mathbb{F}^n \;:\; \mathbf{x}^J = \pm \mathbf{x} \}. \tag{2.3}$$

The corresponding subspaces of polynomials are abbreviated by $\mathbb{F}^n_{\pm}[t]$. The matrices

$$P_{\pm} := \tfrac{1}{2}(I_n \pm J_n) \tag{2.4}$$

are the projections from $\mathbb{F}^n$ onto $\mathbb{F}^n_{\pm}$ with kernel $\mathbb{F}^n_{\mp}$. Note that here we use the assumption that the characteristic of $\mathbb{F}$ is not equal to 2. For later reference, let us recall the following relations between the spaces of polynomials:

$$\begin{aligned} \mathbb{F}^{2\ell}_{\pm}[t] &= \{ \; (t \pm 1)\mathbf{x}(t) \;:\; \mathbf{x}(t) \in \mathbb{F}^{2\ell-1}_{+}[t] \; \}, \\ \mathbb{F}^{2\ell+1}_{-}[t] &= \{ \; (t^2 - 1)\mathbf{x}(t) \;:\; \mathbf{x}(t) \in \mathbb{F}^{2\ell-1}_{+}[t] \; \}. \end{aligned} \tag{2.5}$$

To introduce the matrices under considerations let us define Toeplitz and Hankel matrices. The $n \times n$ *Hankel matrix* generated by $\mathbf{s} = (s_i)_{i=0}^{2n-2} \in \mathbb{F}^{2n-1}$ is the matrix

$$H_n(\mathbf{s}) = [\, s_{i+j} \,]_{i,j=0}^{n-1}.$$

Clearly, $H_n(\mathbf{s})$ is symmetric. The $n \times n$ *Toeplitz matrix* generated by the vector $\mathbf{a} = (a_i)_{i=-n+1}^{n-1} \in \mathbb{F}^{2n-1}$ is the matrix

$$T_n(\mathbf{a}) = [\, a_{i-j} \,]_{i,j=0}^{n-1}.$$

The vector $\mathbf{a}$ (or $\mathbf{s}$) is called the *symbol* of the Toeplitz (or the Hankel) matrix.

For a Toeplitz matrix we have $T_n(\mathbf{a})^T = J_n T_n(\mathbf{a}) J_n$, which means that $T_n(\mathbf{a})$ is *persymmetric*. Note that

$$T_n(\mathbf{a}^J) = T_n(\mathbf{a})^T = J_n T_n(\mathbf{a}) J_n \text{ with } \mathbf{a}^J = J_{2n-1}\mathbf{a}. \tag{2.6}$$

Therefore, a Toeplitz matrix is symmetric if and only if it is centrosymmetric, or, equivalently, if its symbol $\mathbf{a}$ is a symmetric vector. Likewise, a Toeplitz matrix is skewsymmetric if and only if it is centroskewsymmetric, or, if its symbol is skewsymmetric.

Toeplitz-plus-Hankel matrices (shortly, $T + H$ *matrices*) are matrices which are a sum of a Toeplitz and a Hankel matrix. Since $T_n(\mathbf{b})J_n$ is a

Hankel matrix it is possible to represent any $T+H$ matrix by means of two Toeplitz matrices,

$$R_n = T_n(\mathbf{a}) + T_n(\mathbf{b})J_n \qquad (\mathbf{a},\mathbf{b} \in \mathbb{F}^{2n-1}). \tag{2.7}$$

Related to this representation is another one, using the projections (2.4) and the symbols $\mathbf{c} = \mathbf{a}+\mathbf{b}$ and $\mathbf{d} = \mathbf{a}-\mathbf{b}$, namely

$$R_n = T_n(\mathbf{c})P_+ + T_n(\mathbf{d})P_- . \tag{2.8}$$

3. Centro(skew)symmetric matrices: splitting and restricted invertibility

We are now going to discuss some very basic concepts, which facilitates the presentation of the results in the subsequent sections.

As before, let $\mathbb{F}$ stand for a field of characteristic different from two, and let $A \in \mathbb{F}^{n\times n}$. Recall that A is called *centrosymmetric* if $A = J_n A J_n$, and A is called *centroskewsymmetric* if $A = -J_n A J_n$. Alternative characterizations can be given in terms of P_+ and P_-. Indeed, A is centrosymmetric if and only if

$$P_- A P_+ = P_+ A P_- = 0, \tag{3.1}$$

and A is centroskewsymmetric if and only if

$$P_- A P_- = P_+ A P_+ = 0. \tag{3.2}$$

We also note that A is centrosymmetric if and only if A has both $\mathbb{F}^n_+$ and $\mathbb{F}^n_-$ as invariant subspaces, while A is centroskewsymmetric if and only if A maps $\mathbb{F}^n_+$ into $\mathbb{F}^n_-$ and maps $\mathbb{F}^n_-$ into $\mathbb{F}^n_+$. Hence $AP_\pm = P_\pm A P_\pm$ if A is centrosymmetric, and $AP_\pm = P_\mp A P_\pm$ if A is centroskewsymmetric.

We will observe in a few moments that centro(skew)symmetric matrices allow a unique splitting into two matrices of a more particular structure, and this splitting allows us to reduce the inversion of such matrices to two other invertibility problems.

We need to prepare with the following definitions. A matrix A said to be a *matrix of* $(++)$*type* or *of* $(--)$*type* if

$$A = P_+ A P_+ \quad \text{or} \quad A = P_- A P_-,$$

respectively. Furthermore, a matrix A is said to be of $(+-)$type or $(-+)$type if

$$A = P_+ A P_- \quad \text{or} \quad A = P_- A P_+,$$

respectively. Note that, for instance, a matrix of $(+-)$type has symmetric columns and skewsymmetric rows, and similar statements hold for matrices of the other types as well. Furthermore, matrices of $(++)$type or $(--)$type are centrosymmetric, and matrices of $(+-)$type or $(+-)$type are centroskewsymmetric.

Let $s_1, s_2 \in \{+,-\}$, and let A be a matrix of $(s_1 s_2)$type. We can view such a matrix as a linear map for $\mathbb{F}^n_{s_2}$ into $\mathbb{F}^n_{s_1}$. If this map is invertible, we call the matrix A *restrictedly invertible* and denote the corresponding inverse by

$A^{(-1)}$. The *restricted inverse* is a map from $\mathbb{F}^n_{s_1}$ to $\mathbb{F}^n_{s_2}$ and can be identified with a matrix of (s_2s_1)type. More specifically, if A is a matrix of (s_1s_2)type, then its restricted inverse $A^{(-1)}$ is a matrix of (s_2s_1)type such that

$$A^{(-1)}A = P_{s_2} \quad \text{and} \quad AA^{(-1)} = P_{s_1}.$$

The relationship between centro(skew)symmetric matrices and the matrices of the four types is now clarified in the following proposition.

Proposition 3.1.

(a) *Each centrosymmetric matrix A admits a unique splitting*

$$A = A_{++} + A_{--}$$

into matrices of $(++)$type and $(--)$type. Moreover, A is invertible if and only if both A_{++} and A_{--} are restrictedly invertible. In this case,

$$A^{-1} = A_{++}^{(-1)} + A_{--}^{(-1)}$$

and $A_{++}^{(-1)} = P_+A^{-1}P_+$ and $A_{--}^{(-1)} = P_-A^{-1}P_-$.

(b) *Each centroskewsymmetric matrix A admits a unique splitting*

$$A = A_{-+} + A_{+-}$$

into matrices of $(-+)$type and $(+-)$type. Moreover, A is invertible if and only if both A_{-+} and A_{+-} are restrictedly invertible. In this case,

$$A^{-1} = A_{-+}^{(-1)} + A_{+-}^{(-1)}$$

and $A_{-+}^{(-1)} = P_+A^{-1}P_-$ and $A_{+-}^{(-1)} = P_-A^{-1}P_+$.

Indeed, the splitting parts of A can be obtained by $A_{\pm\pm} = P_\pm AP_\pm$ and $A_{\pm\mp} = P_\pm AP_\mp$, respectively (see also (3.1) and (3.2)).

We note that in the centroskewsymmetric case invertibility and restricted invertibility can only occur if n is even. Indeed, only for even n the dimensions of $\mathbb{F}^n_+$ and $\mathbb{F}^n_-$ are the same.

4. Centro(skew)symmetric $T+H$ matrices

The goal of this section is to recall how centrosymmetric and centroskewsymmetric $T+H$ matrices can be represented. Furthermore, the connection with $T+H$ matrices of $(++)$, $(--)$, $(+-)$, and $(-+)$type is made.

Let us start with the observation, already mentioned above, that a general $T+H$ matrix can be written in the form

$$T_n(\mathbf{c})P_+ + T_n(\mathbf{d})P_- .$$

The choice of $\mathbf{c}$ and $\mathbf{d}$ is in general not unique. The reason is that there exist matrices of "checkerboard" structure which are both Toeplitz and Hankel (see [13, Sect. 2]).

The following basic result shows under which conditions a $T+H$ matrix is centrosymmetric or centroskewsymmetric. It also addresses the non-uniqueness of the symbol. For the details of the proof we refer to [3, Prop. 3.3] and [4, Prop. 3.1].

Proposition 4.1.

(a) *A $T+H$ matrix*

$$R_n = T_n(\mathbf{c})P_+ + T_n(\mathbf{d})P_- \tag{4.1}$$

is centrosymmetric if and only if $\mathbf{c}$ *and* $\mathbf{d}$ *both are symmetric, i.e.,* $\mathbf{c},\mathbf{d} \in \mathbb{F}_+^{2n-1}$.

(b) *We have* $T_n(\mathbf{c})P_+ + T_n(\mathbf{d})P_- = \mathbf{0}$ *if and only if*

$$\mathbf{c} = \begin{cases} \mathbf{0} & \text{if } n \text{ is odd,} \\ \mathbf{e}_{\alpha,-\alpha} & \text{if } n \text{ is even,} \end{cases} \qquad \mathbf{d} = \begin{cases} \mathbf{e}_{\alpha,\beta} & \text{if } n \text{ is odd,} \\ \mathbf{e}_{\alpha,\alpha} & \text{if } n \text{ is even,} \end{cases} \tag{4.2}$$

where $\mathbf{e}_{\alpha,\beta} = (\alpha,\beta,\alpha,\beta,\ldots,\beta,\alpha)^T \in \mathbb{F}_+^{2n-1}$.

(c) *An $n\times n$ matrix R_n is a centroskewsymmetric $T+H$ matrix if and only if it can be written in the form*

$$R_n = T_n(\mathbf{c})P_+ + T_n(\mathbf{d})P_- \tag{4.3}$$

with skewsymmetric $\mathbf{c},\mathbf{d} \in \mathbb{F}_-^{2n-1}$. *The choice of these vectors is unique.*

Concerning general $T+H$ matrices, (b) tells us that the choice of the symbols $\mathbf{c}$ and $\mathbf{d}$ is not unique. This is equally true in the centrosymmetric case. In the centroskewsymmetric case, non-uniqueness can also occur, but only if one allows for non-skewsymmetric vectors. As stated in (c), the restriction to skewsymmetric vectors makes the choice of the symbols unique.

Parts (a) and (c) of the previous proposition imply that centrosymmetric $T+H$ matrices can also be written as

$$R_n = P_+T_n(\mathbf{c})P_+ + P_-T_n(\mathbf{d})P_-$$

with symmetric $\mathbf{c},\mathbf{d} \in \mathbb{F}_+^{2n-1}$, while centroskewsymmetric $T+H$ matrices can be written as

$$R_n = P_-T_n(\mathbf{c})P_+ + P_+T_n(\mathbf{d})P_-$$

with skewsymmetric $\mathbf{c},\mathbf{d} \in \mathbb{F}_-^{2n-1}$. Indeed, to see this recall (3.1) and (3.2) and take the remarks made after formula (2.6) into account. These two formulas represent the splitting of centro(skew)symmetric $T+H$ matrices into $T+H$ matrices of $(++)$, $(--)$, $(-+)$, and $(+-)$type, respectively.

In particular, we can characterize $T+H$ matrices of the four types. Firstly, a $T+H$ matrix of $(++)$type or $(--)$type is given by

$$P_+T_n(\mathbf{c})P_+ \quad \text{and} \quad P_-T_n(\mathbf{d})P_-\,,$$

respectively, with symmetric $\mathbf{c},\mathbf{d} \in \mathbb{F}_+^{2n-1}$. The symbols are not necessarily unique, but the only modifications are those described in (4.2). Secondly, a $T+H$ matrix of $(-+)$type or $(+-)$type is given by

$$P_-T_n(\mathbf{c})P_+ \quad \text{and} \quad P_+T_n(\mathbf{d})P_-\,,$$

respectively, with skewsymmetric symbol $\mathbf{c}, \mathbf{d} \in \mathbb{F}_-^{2n-1}$. Within the class of skewsymmetric vectors, these symbols are unique.

At this point let us mention that there is a relationship between the restricted invertibility of the $T + H$ matrices of the various types and the invertibility of a Toeplitz matrix.

Proposition 4.2. *Let $\mathbf{c} \in \mathbb{F}_-^{2n-1}$. Then the following are equivalent:*

(a) *$T_n(\mathbf{c})$ is invertible;*
(b) *$P_+ T_n(\mathbf{c}) P_-$ is restrictedly invertible;*
(c) *$P_- T_n(\mathbf{c}) P_+$ is restrictedly invertible.*

Proof. Note that we have $P_+ T_n(\mathbf{c}) P_+ = P_- T_n(\mathbf{c}) P_- = \mathbf{0}$ and therefore $T_n(\mathbf{c}) = P_+ T_n(\mathbf{c}) P_- + P_- T_n(\mathbf{c}) P_+$. The equivalence of (a) with (b) and (c) follows from Proposition 3.1.

The equivalence of (b) with (c) becomes clear by passing to the transpose, $(P_+ T_n(\mathbf{c}) P_-)^T = P_- T_n(\mathbf{c}^J) P_+ = -P_- T_n(\mathbf{c}) P_+$. □

Proposition 4.3. *Let $\mathbf{c} \in \mathbb{F}_+^{2n-1}$. Then $T_n(\mathbf{c})$ is invertible if and only if both matrices $P_+ T_n(\mathbf{c}) P_+$ and $P_- T_n(\mathbf{c}) P_-$ are restrictedly invertible.*

Proof. It suffices to note that $P_+ T_n(\mathbf{c}) P_- = P_- T_n(\mathbf{c}) P_+ = \mathbf{0}$ and hence $T_n(\mathbf{c}) = P_+ T_n(\mathbf{c}) P_+ + P_- T_n(\mathbf{c}) P_-$. Proposition 3.1 gives the assertion. □

Note that the restricted invertibility of $P_+ T_n(\mathbf{c}) P_+$ is not equivalent to the restricted invertibility of $P_- T_n(\mathbf{c}) P_-$. For example, if $\mathbf{c} = (0, 1, 0, 1, 0)^T$, then $P_+ T_3(\mathbf{c}) P_+$ is restrictedly invertible, but $P_- T_3(\mathbf{c}) P_- = \mathbf{0}$ is not. On the other hand, if we take $\mathbf{d} = (-1, 0, 1, 0, -1)^T$, then $P_+ T_3(\mathbf{d}) P_+$ is not restrictedly invertible, while $P_- T_3(\mathbf{d}) P_-$ is. Clearly, $T_3(\mathbf{c})$ and $T_3(\mathbf{d})$ are not invertible.

5. Centro(skew)symmetric $T+H$-Bezoutians and split-Bezoutians

Recall that a $T + H$-Bezoutian is an $n \times n$ matrix B for which there exist eight vectors $\mathbf{f}_i, \mathbf{g}_i \in \mathbb{F}^{n+2}$ $(i = 1, 2, 3, 4)$ such that, in polynomial language,

$$B(t, s) = \frac{\sum_{i=1}^4 \mathbf{g}_i(t) \mathbf{f}_i(s)}{(t - s)(1 - ts)}.$$

The relationship between $T + H$-Bezoutians and $T + H$ matrices is shown in the following important theorem, which was established in [11].

Theorem 5.1. *The inverse of an invertible $T + H$-Bezoutian is an invertible $T + H$ matrix and vice versa.*

We will see in a few moments that if an invertible $T + H$-Bezoutian is centrosymmetric or centroskewsymmetric, then it admits a decomposition into a sum of two particular $T + H$-Bezoutians, called split-Bezoutians. The definition of these two split-Bezoutians is different in the centrosymmetric and the centroskewsymmetric case.

5.1. The centrosymmetric case

An $n \times n$ matrix B is called a *split-Bezoutian of* $(++)$*type* or $(--)$*type*, if

$$B(t,s) = \frac{\mathbf{f}(t)\mathbf{g}(s) - \mathbf{g}(t)\mathbf{f}(s)}{(t-s)(1-ts)} \tag{5.1}$$

with symmetric $\mathbf{f}, \mathbf{g} \in \mathbb{F}_+^{n+2}$ or skewsymmetric $\mathbf{f}, \mathbf{g} \in \mathbb{F}_-^{n+2}$, respectively. In this case we will use the notation

$$B = \mathrm{Bez}_{\mathrm{sp}}(\mathbf{f}, \mathbf{g}).$$

It is not difficult to verify that the split-Bezoutians are well defined for every pair of (skew)symmetric vectors $\mathbf{f}$ and $\mathbf{g}$. Indeed, it suffices to notice that (5.1) is a polynomial in t and s.

These split-Bezoutians are matrices of $(++)$type and $(--)$type, respectively, in the sense of Section 3. In particular, they are centrosymmetric matrices. Furthermore, the sum of a split-Bezoutian of $(++)$type and a split-Bezoutian of $(--)$type is a centrosymmetric $T+H$-Bezoutian.

The converse of this statement holds under the additional assumption of invertibility. This is made precise in the following result, which is taken from Theorem 5.12 of [13] and which discusses the splitting of *invertible* centrosymmetric $T+H$-Bezoutians.

Theorem 5.2. *Each invertible, centrosymmetric $n \times n$ $T+H$-Bezoutian B allows the splitting*

$$B = B_{++} + B_{--}\,, \tag{5.2}$$

where $B_{\pm\pm} = \mathrm{Bez}_{\mathrm{sp}}(\mathbf{f}_\pm, \mathbf{g}_\pm)$ *are split-Bezoutians of* $(\pm\pm)$*type with vectors* $\mathbf{f}_\pm, \mathbf{g}_\pm \in \mathbb{F}_\pm^{n+2}$ *satisfying the conditions*

$$\gcd(\mathbf{f}_+, \mathbf{g}_+) = \begin{cases} 1 & \text{if } n \text{ is odd,} \\ t+1 & \text{if } n \text{ is even,} \end{cases} \tag{5.3}$$

$$\gcd(\mathbf{f}_-, \mathbf{g}_-) = \begin{cases} t^2-1 & \text{if } n \text{ is odd,} \\ t-1 & \text{if } n \text{ is even.} \end{cases} \tag{5.4}$$

Conversely, the conditions (5.3) *and* (5.4) *imply that the $T+H$-Bezoutian defined by* (5.2) *is invertible.*

Here we use the abbreviation $\gcd(\mathbf{f}, \mathbf{g})$ for the greatest common divisor of the polynomials $\mathbf{f}(t)$ and $\mathbf{g}(t)$. As already noticed in (2.5), any polynomial $\mathbf{p}_\pm(t) \in \mathbb{F}_\pm^n[t]$ has always a zero at ∓ 1 in the case where n is even. In case n is odd we have $\mathbf{p}_-(\pm 1) = 0$. Consequently, the conditions (5.3) and (5.4) mean that *the greatest common divisors are minimal.*

As a consequence of the previous theorem and of Theorem 5.1 we obtain a relationship between split-Bezoutians of $(\pm\pm)$type and $T+H$ matrices of $(\pm\pm)$type.

Corollary 5.3. *The restricted inverse of a restrictedly invertible split-Bezoutian of $(\pm\pm)$type is a restrictedly invertible $T+H$ matrix of $(\pm\pm)$type, and vice versa. In particular, a split-Bezoutian* $\mathrm{Bez}_{\mathrm{sp}}(\mathbf{f}_\pm, \mathbf{g}_\pm)$ *of* $(\pm\pm)$*type is* *restrictedly invertible if and only if the condition* (5.3) *or* (5.4), *respectively, is satisfied.*

Proof. We first need to establish the rather simple fact that there exist restrictedly invertible split-Bezoutians and $T + H$ matrices of both $(++)$ and $(--)$type in every dimension n. Indeed, we can consider the identity matrix I_n, which is a centrosymmetric $T + H$ matrix and a centrosymmetric $T + H$-Bezoutian, and has the corresponding splitting $I_n = P_+ + P_-$. Therefore, P_+ and P_- are split-Bezoutians as well as $T+H$ matrices of $(++)$ and $(--)$type, respectively. Since they are projections, they are their own restricted inverses. More specifically, $P_\pm = \mathrm{Bez}_{\mathrm{sp}}(\mathbf{f}_\pm^{(0)}, \mathbf{g}_\pm^{(0)})$ with

$$\mathbf{f}_\pm^{(0)}(t) = -\frac{1}{2}\left(t^{n+1} \pm 1\right), \qquad \mathbf{g}_\pm^{(0)}(t) = t^n \pm t,$$

and it can be verified by direct computation that $\{\mathbf{f}_\pm^{(0)}, \mathbf{g}_\pm^{(0)}\}$ satisfy (5.3) and (5.4), respectively. Note that this also follows from the splitting $I_n = P_+ + P_-$ by applying Theorem 5.2.

Now let B_{++} be a split-Bezoutian of $(++)$type. Then

$$B = B_{++} + P_-$$

is a centrosymmetric $T+H$-Bezoutian. Because of Proposition 3.1 the matrix B is invertible if and only if B_{++} is restrictedly invertible. Using the previous theorem we see that this is the case precisely if $\{\mathbf{f}_+(t), \mathbf{g}_+(t)\}$ satisfies the condition (5.3).

Let us proceed to show that the restricted inverse of a split-Bezoutian is a $T + H$ matrix of corresponding type and vice versa. First assume that B_{++} is restrictedly invertible, i.e., B is invertible. Then, by Theorem 5.1, the inverse of B is a (centrosymmetric) $T+H$ matrix. Again, by Proposition 3.1, the restricted inverse is given by

$$B_{++}^{(-1)} = P_+ B^{-1} P_+ .$$

In view of (4.1) it is a $T + H$ matrix of $(++)$type.

Now let us show the converse. Assume that A_{++} is a restrictedly invertible $T + H$ matrix of $(++)$type. Consider $A = A_{++} + P_-$, which is an invertible centrosymmetric $T + H$ matrix. By Theorem 5.1, its inverse is a centrosymmetric $T + H$-Bezoutian $B = A^{-1}$, which can be split into a sum $B_{++} + B_{--}$. Now, by Proposition 3.1 we have

$$A_{++}^{(-1)} = P_+ A^{-1} P_+ = B_{++}.$$

Hence the restricted inverse of a $T+H$ matrix of $(++)$type is a split-Bezoutian of $(++)$type.

Finally, remark that the case of split-Bezoutians and $T+H$ matrices of $(--)$type can be proved similarly by changing the roles of P_+ and P_-. □

We proceed with the observation that the two kinds of split-Bezoutians considered here can be reduced to split-Bezoutians of $(++)$type and of odd order. This was shown in [13, Thm. 5.12]. It is a simple consequence of the fact that depending on whether n is even or odd and whether the split-Bezoutian is of $(++)$ or $(--)$type, the polynomials $\mathbf{f}(t)$ and $\mathbf{g}(t)$ may have common factors $(t-1)$ or $(t+1)$ (see Theorem 5.2 and formulas (2.5)), which one can pull out in the representation (5.1).

To that aim let us introduce the $n \times (n-1)$ matrices $M_{n-1}^{t\pm 1}$ and the $n \times (n-2)$ matrix $M_{n-2}^{t^2-1}$,

$$M_{n-1}^{t\pm 1} = \begin{bmatrix} \pm 1 & & & 0 \\ 1 & \pm 1 & & \\ & 1 & \ddots & \\ & & \ddots & \pm 1 \\ 0 & & & 1 \end{bmatrix}, \quad M_{n-2}^{t^2-1} = \begin{bmatrix} -1 & & & 0 \\ 0 & -1 & & \\ 1 & 0 & \ddots & \\ & 1 & \ddots & -1 \\ & & \ddots & 0 \\ 0 & & & 1 \end{bmatrix}. \tag{5.5}$$

These are matrices of multiplication operators (in the corresponding polynomial spaces with respect to the canonical bases) with the polynomials being indicated in their superscripts.

Finally, taking Theorem 5.2 and Corollary 5.3 into account we obtain the following.

Theorem 5.4. *Let B_{++} and B_{--} be $n \times n$ split-Bezoutians of $(++)$ and $(--)$type, respectively.*

(a) *If n is even, then there exist $\mathbf{f}_i, \mathbf{g}_i \in \mathbb{F}_+^{n+1}$, $i = 1, 2$, such that*

$$\begin{aligned} B_{++} &= M_{n-1}^{t+1} \operatorname{Bez}_{\mathrm{sp}}(\mathbf{f}_1, \mathbf{g}_1)(M_{n-1}^{t+1})^T, \\ B_{--} &= M_{n-1}^{t-1} \operatorname{Bez}_{\mathrm{sp}}(\mathbf{f}_2, \mathbf{g}_2)(M_{n-1}^{t-1})^T. \end{aligned} \tag{5.6}$$

Moreover, the split-Bezoutians are restrictedly invertible if and only if the pair $\{\mathbf{f}_1(t), \mathbf{g}_1(t)\}$ or $\{\mathbf{f}_2(t), \mathbf{g}_2(t)\}$, respectively, is coprime.

(b) *If n is odd, then there exist $\mathbf{f}_1, \mathbf{g}_1 \in \mathbb{F}_+^{n+2}$ and $\mathbf{f}_2, \mathbf{g}_2 \in \mathbb{F}_+^n$ such that*

$$\begin{aligned} B_{++} &= \operatorname{Bez}_{\mathrm{sp}}(\mathbf{f}_1, \mathbf{g}_1), \\ B_{--} &= M_{n-2}^{t^2-1} \operatorname{Bez}_{\mathrm{sp}}(\mathbf{f}_2, \mathbf{g}_2)(M_{n-2}^{t^2-1})^T. \end{aligned} \tag{5.7}$$

Moreover, the split-Bezoutians are restrictedly invertible if and only if the pair $\{\mathbf{f}_1(t), \mathbf{g}_1(t)\}$ or $\{\mathbf{f}_2(t), \mathbf{g}_2(t)\}$, respectively, is coprime.

The formulas (5.6) and (5.7) can be expressed in polynomial language. Then they read

$$\begin{aligned} B_{++}(t,s) &= (t+1)\frac{\mathbf{f}_1(t)\mathbf{g}_1(s) - \mathbf{g}_1(t)\mathbf{f}_1(s)}{(t-s)(1-ts)}(s+1), \\ B_{--}(t,s) &= (t-1)\frac{\mathbf{f}_2(t)\mathbf{g}_2(s) - \mathbf{g}_2(t)\mathbf{f}_2(s)}{(t-s)(1-ts)}(s-1) \end{aligned} \tag{5.8}$$

for n even, while

$$\begin{aligned} B_{++}(t,s) &= \frac{\mathbf{f}_1(t)\mathbf{g}_1(s) - \mathbf{g}_1(t)\mathbf{f}_1(s)}{(t-s)(1-ts)}, \\ B_{--}(t,s) &= (t^2-1)\frac{\mathbf{f}_2(t)\mathbf{g}_2(s) - \mathbf{g}_2(t)\mathbf{f}_2(s)}{(t-s)(1-ts)}(s^2-1) \end{aligned} \tag{5.9}$$

for n odd. Note that in all cases of (5.6) and (5.7) the terms in the middle, $\mathrm{Bez_{sp}}(\mathbf{f}_i, \mathbf{g}_i)$, are split-Bezoutians of odd order and $(++)$type.

5.2. The centroskewsymmetric case

Let us now consider centroskewsymmetric $T+H$-Bezoutians. We will assume that the order n is even since centroskewsymmetric matrices of odd order are singular.

We start with the following splitting theorem about centroskewsymmetric $T+H$-Bezoutians, which was proved in [13, Thm. 5.13].

Theorem 5.5. *Each invertible, centroskewsymmetric $T+H$-Bezoutian B of even order n allows the splitting $B = B_{+-} + B_{-+}$ with*

$$B_{+-}(t,s) = (t+1)\frac{\mathbf{f}_1(t)\mathbf{g}_1(s) - \mathbf{g}_1(t)\mathbf{f}_1(s)}{(t-s)(1-ts)}(s-1), \tag{5.10}$$

$$B_{-+}(t,s) = (t-1)\frac{\mathbf{f}_2(t)\mathbf{g}_2(s) - \mathbf{g}_2(t)\mathbf{f}_2(s)}{(t-s)(1-ts)}(s+1), \tag{5.11}$$

where $\mathbf{f}_i, \mathbf{g}_i \in \mathbb{F}_+^{n+1}$ are such that $\{\mathbf{f}_i(t), \mathbf{g}_i(t)\}$ are coprime $(i = 1,2)$. Conversely, each centroskewsymmetric $T+H$-Bezoutian B defined in this way with coprime $\{\mathbf{f}_i(t), \mathbf{g}_i(t)\}$, $i = 1,2$, is invertible.

In matrix form, the formulas (5.10) and (5.11) read

$$B_{+-} = M_{n-1}^{t+1}\mathrm{Bez_{sp}}(\mathbf{f}_1, \mathbf{g}_1)(M_{n-1}^{t-1})^T, \tag{5.12}$$

$$B_{-+} = M_{n-1}^{t-1}\mathrm{Bez_{sp}}(\mathbf{f}_2, \mathbf{g}_2)(M_{n-1}^{t+1})^T. \tag{5.13}$$

Again, as before, the terms in the middle, $\mathrm{Bez_{sp}}(\mathbf{f}_i, \mathbf{g}_i)$, are split-Bezoutians of $(++)$type and odd order $n-1$.

We will refer to matrices of the kind (5.10) and (5.11) as *split-Bezoutians of $(+-)$type* and *of $(-+)$type*, respectively. These split-Bezoutians are matrices of $(+-)$type and $(-+)$type, respectively, in the sense of Section 3. Hence they are centroskewsymmetric $T+H$-Bezoutians. In fact, a sum of a split-Bezoutian of $(+-)$type and a split-Bezoutian of $(-+)$type is a centroskewsymmetric $T+H$-Bezoutian.

In contrast to the centrosymmetric case, the factors on the left and the right hand side in (5.10) and (5.11) are different, so that these representations cannot be brought into a simpler form (5.1) involving only two polynomials.

However, one can obtain three equivalent representations by means of simple substitutions. Given $\mathbf{f}_1, \mathbf{g}_1, \mathbf{f}_2, \mathbf{g}_2 \in \mathbb{F}_+^{n+1}$, introduce $\mathbf{f}_\pm, \mathbf{g}_\pm, \hat{\mathbf{f}}_\pm, \hat{\mathbf{g}}_\pm \in \mathbb{F}_\pm^{n+2}$ by

$$\mathbf{f}_1(t) = \frac{\mathbf{f}_+(t)}{t+1} = \frac{\hat{\mathbf{f}}_-(t)}{t-1}, \qquad \mathbf{g}_1(t) = \frac{\mathbf{g}_+(t)}{t+1} = \frac{\hat{\mathbf{g}}_-(t)}{t-1},$$
$$\mathbf{f}_2(t) = \frac{\mathbf{f}_-(t)}{t-1} = \frac{\hat{\mathbf{f}}_+(t)}{t+1}, \qquad \mathbf{g}_2(t) = \frac{\mathbf{g}_-(t)}{t-1} = \frac{\hat{\mathbf{g}}_+(t)}{t+1}.$$

Then we obtain the following representations:

$$\begin{aligned} B_{\pm\mp}(t,s) &= \left(\frac{\mathbf{f}_\pm(t)\mathbf{g}_\pm(s) - \mathbf{g}_\pm(t)\mathbf{f}_\pm(s)}{(t-s)(1-ts)} \right) \frac{s\mp 1}{s\pm 1} \\ &= \frac{t\pm 1}{t\mp 1} \left(\frac{\hat{\mathbf{f}}_\mp(t)\hat{\mathbf{g}}_\mp(s) - \hat{\mathbf{g}}_\mp(t)\hat{\mathbf{f}}_\mp(s)}{(t-s)(1-ts)} \right) \\ &= \frac{\mathbf{f}_\pm(t)\hat{\mathbf{g}}_\mp(s) - \mathbf{g}_\pm(t)\hat{\mathbf{f}}_\mp(s)}{(t-s)(1-ts)}. \end{aligned}$$

Notice that the last representation comes closest to resembling (5.1), but it involves four polynomials. The other two representations are 'adjoint' to each other. One of them was already used in [13].

From the previous theorem it is possible to derive the following result, which examines the restricted invertibility of split-Bezoutians of $(\pm\mp)$type and its relationship to $T+H$ matrices of $(\mp\pm)$type.

Corollary 5.6. *Let n be even. The restricted inverse of a restrictedly invertible split-Bezoutian of $(\pm\mp)$type is a restrictedly invertible $T+H$ matrix of $(\mp\pm)$type, and vice versa. In particular, a split-Bezoutian $B_{\pm\mp}$ of $(\pm\mp)$type given by* (5.10)–(5.11) (*or* (5.12)–(5.13)) *is restrictedly invertible if and only if the pair of polynomials $\{\mathbf{f}_1(t), \mathbf{g}_1(t)\}$ or $\{\mathbf{f}_2(t), \mathbf{g}_2(t)\}$, respectively, is coprime.*

Proof. As before, we need to show that there exist restrictedly invertible split-Bezoutians of $(\pm\mp)$type in even dimension n. Let

$$C_{+-} = M_{n-1}^{t+1} P_+ (M_{n-1}^{t-1})^T, \qquad C_{-+} = M_{n-1}^{t-1} P_+ (M_{n-1}^{t+1})^T,$$

where P_+ is of order $n-1$. Since $P_+ = \mathrm{Bez}_{\mathrm{sp}}(\mathbf{f}, \mathbf{g})$ with two polynomials $\mathbf{f}(t) = -\frac{1}{2}(t^n+1)$, $\mathbf{g}(t) = t^{n-1} + t$, which are coprime, it follows from the previous theorem that $C = C_{+-} + C_{-+}$ is an invertible $T+H$-Bezoutian. Consequently, the splitting parts $C_{\pm\mp}$ are restrictedly invertible split-Bezoutians of $(\pm\mp)$type.

Now we can proceed similarly as in the proof of Corollary 5.3. For instance, given a split-Bezoutian B_{+-} of $(+-)$type, we define a centroskewsymmetric $T+H$-Bezoutian $B = B_{+-} + C_{-+}$. For the converse, if we are given a $T+H$ matrix A_{-+} of $(-+)$type, we are going to consider $A = A_{-+} + C_{-+}^{(-1)}$, which is a centroskewsymmetric $T+H$ matrix. We leave the details to the reader. □

5.3. The uniqueness of the split-Bezoutians

In the last two subsections, we obtained six different kinds of representations for the split-Bezoutians of the various types, namely (5.6), (5.7), (5.12), and (5.13). Note that one has corresponding polynomial representations (5.8), (5.9), (5.10), and (5.11). In these polynomial representations, only the terms on the left and on the right are different, but the term in the middle,

$$\frac{\mathbf{f}(t)\mathbf{g}(s) - \mathbf{g}(t)\mathbf{f}(s)}{(t-s)(1-st)},$$

is always of the same form.

Later we need the following simple fact about split-Bezoutians. It is known for Hankel- and Toeplitz-Bezoutians (see Corollaries 2.3 and 2.10 in [14]) and split-Bezoutians of $(++)$type (see Remark 4.3 in [5]), but holds for split-Bezoutians of $(--)$ or $(\pm\mp)$type as well.

Lemma 5.7. *Two nonzero split-Bezoutians (of $(\pm\pm)$type or $(\pm\mp)$type) given by pairs of vectors $\{\mathbf{f},\mathbf{g}\}$ and $\{\mathbf{u},\mathbf{v}\}$, respectively, are equal if and only if there is $\Phi \in \mathbb{F}^{2\times 2}$ with $\det\Phi = 1$ such that*

$$[\,\mathbf{u},\mathbf{v}\,] = [\,\mathbf{f},\mathbf{g}\,]\,\Phi\,. \tag{5.14}$$

Proof. In view of the polynomial representations of the two Bezoutians, we can cancel possible factors on the left and the right as well as remove the denominator in the middle term in order to get

$$\mathbf{f}(t)\mathbf{g}(s) - \mathbf{g}(t)\mathbf{f}(s) = \mathbf{u}(t)\mathbf{v}(s) - \mathbf{v}(t)\mathbf{u}(s)\,.$$

Translated into matrix language this means that

$$[\,\mathbf{f},\ \mathbf{g}\,]\begin{bmatrix} 0 & -1 \\ 1 & 0 \end{bmatrix}\begin{bmatrix} \mathbf{f}^T \\ \mathbf{g}^T \end{bmatrix} = [\,\mathbf{u},\ \mathbf{v}\,]\begin{bmatrix} 0 & -1 \\ 1 & 0 \end{bmatrix}\begin{bmatrix} \mathbf{u}^T \\ \mathbf{v}^T \end{bmatrix}.$$

Note that $[\mathbf{f},\ \mathbf{g}]$ has rank two because otherwise the two vectors are linearly dependent, which would imply that the Bezoutian vanishes. The same holds for $[\mathbf{u},\ \mathbf{v}]$. It follows that there exists an invertible Φ such that (5.14) holds. Now plug this relation into the above formula in order to conclude that $\det\Phi = 1$. □

5.4. Split-Bezoutians versus $T+H$-Bezoutians

Let us make some further comments on the relationship between the split-Bezoutians of the various types and centro(skew)symmetric $T+H$-Bezoutians. These remarks are added for illustration, but are not needed subsequently.

Clearly, each split-Bezoutian of $(\pm\pm)$type is a (centrosymmetric) $T+H$-Bezoutian of $(\pm\pm)$type. However, not every $T+H$-Bezoutian of $(\pm\pm)$type is a split-Bezoutian. For example, let B_1 and B_2 be two $n\times n$ split-Bezoutians of $(++)$type given by

$$B_i(t,s) = \frac{\mathbf{f}_i(t)\mathbf{g}_i(s) - \mathbf{g}_i(t)\mathbf{f}_i(s)}{(t-s)(1-ts)}, \quad (i=1,2),$$

where $\mathbf{f}_i, \mathbf{g}_i \in \mathbb{F}_+^{n+2}$. Then, clearly, $B = B_1 + B_2$ is a $T+H$-Bezoutian and it is a matrix of $(++)$type. But, for general choices of $\{\mathbf{f}_1,\mathbf{f}_2,\mathbf{g}_1,\mathbf{g}_2\}$ and if $n\geq 3$, the matrix B is not a split-Bezoutian. The same construction works in the case of split-Bezoutians of $(--)$type if $n \geq 4$.

Now we are going to provide examples of $T+H$-Bezoutians of $(\pm\pm)$type which are not split-Bezoutians. Note that each such matrix is a non-invertible centrosymmetric $T+H$-Bezoutian which cannot be written as the sum of a split-Bezoutian of $(++)$type and a split-Bezoutian of $(--)$type. Hence the statement of Theorem 5.2 is not true without the invertibility assumption.

Example 5.8. The following matrices B_+ and B_- describe two $n \times n$ centrosymmetric $T+H$-Bezoutians of $(++)$type and $(--)$type. Assume that the order is $n \geq 3$ and $n \geq 4$, respectively.

$$B_+ = \begin{bmatrix} 1 & 1 & & & & 1 & 1 \\ & 1 & & & & 1 & \\ & & 1 & & 1 & & \\ & & & \ddots\!\!\!\!\iddots & & & \\ & & 1 & & 1 & & \\ & 1 & & & & 1 & \\ 1 & 1 & & & & 1 & 1 \end{bmatrix}, \quad B_- = \begin{bmatrix} 1 & 1 & & & & -1 & -1 \\ & 1 & & & & -1 & \\ & & 1 & & -1 & & \\ & & & \ddots\!\!\!\!\iddots & & & \\ & & -1 & & 1 & & \\ & -1 & & & & 1 & \\ -1 & -1 & & & & 1 & 1 \end{bmatrix}.$$

It can be verified straightforwardly that both are restrictedly invertible. On the other hand, neither of them is a split-Bezoutian.

In the case of centroskewsymmetric $T+H$-Bezoutians of $(\pm\mp)$type, similar statements hold, and corresponding examples can be given as well.

6. Inversion results for $T+H$ matrices

In this section we consider the invertibility of centrosymmetric and centroskewsymmetric $T+H$ matrices. In [13] necessary and sufficient conditions for the invertibility of centro(skew)symmetric $T+H$ matrices were given. Using these results we derive necessary and sufficient conditions for the restricted invertibility of $T+H$ matrices of $(\pm\pm)$ and $(\pm\mp)$type. These results are used later on for studying the (restricted) invertibility of split-Bezoutians of the various types.

To begin with, recall from Section 4 that $T+H$ matrices can be written in the form

$$R_n = T_n(\mathbf{c})P_+ + T_n(\mathbf{d})P_-$$

where the symbols $\mathbf{c}$ and $\mathbf{d}$ are both symmetric or skewsymmetric vectors if the matrix R_n is centrosymmetric or centroskewsymmetric, respectively.

6.1. The centrosymmetric case

We first start with the issue of invertibility of a centrosymmetric $T+H$ matrix R_n. As it will turn out, it is related to the solvability of two systems of linear equations being of Toeplitz form, which we are going to state now.

For given symmetric $\mathbf{c}, \mathbf{d} \in \mathbb{F}_+^{2n-1}$, we consider the systems

$$T_n(\mathbf{c})(u_j^+)_{j=1}^n = 2P_+\mathbf{e}_1\,, \qquad T_n(\mathbf{c})(v_j^+)_{j=1}^n = -2P_+(c_j)_{j=1}^n\,, \tag{6.1}$$

$$T_n(\mathbf{d})(u_j^-)_{j=1}^n = 2P_-\mathbf{e}_1\,, \qquad T_n(\mathbf{d})(v_j^-)_{j=1}^n = -2P_-(d_j)_{j=1}^n\,. \tag{6.2}$$

Here we choose arbitrary constants $c_n = c_{-n}$, and $d_n = d_{-n}$. The solvability of these systems is independent of the choice of these constants. Indeed, if the systems have solutions $(u_j^\pm)$, $(v_j^\pm)$ for some specific choice of $c_{\pm n}$ or $d_{\pm n}$, then for another choice a solution is given by $(u_j^\pm)$, $(v_j^\pm) + \lambda_\pm \cdot (u_j^\pm)$, where $\lambda_\pm \in \mathbb{F}$ is a constant depending on the new (and the old) choices.

Furthermore, if these systems have a solution, then they also have a solution where the vectors $(u_j^\pm)_{j=1}^n$ and $(v_j^\pm)_{j=1}^n$ belong to $\mathbb{F}_\pm^n$. In fact, using that $\mathbf{c}$ and $\mathbf{d}$ are symmetric, one can replace $(u_j^\pm)$ by $P_\pm(u_j^\pm)$ and $(v_j^\pm)$ by $P_\pm(v_j^\pm)$. Henceforth we will only be interested in symmetric and skewsymmetric solutions.

From the solutions of (6.1) and (6.2) we will pass to augmented vectors,

$$\mathbf{u}_\pm = \begin{bmatrix} 0 \\ u_1^\pm \\ \vdots \\ u_n^\pm \\ 0 \end{bmatrix} \in \mathbb{F}_\pm^{n+2}, \quad \mathbf{v}_\pm = \begin{bmatrix} 1 \\ v_1^\pm \\ \vdots \\ v_n^\pm \\ \pm 1 \end{bmatrix} \in \mathbb{F}_\pm^{n+2}. \tag{6.3}$$

In terms of the augmented vectors, the systems (6.1) and (6.2) are equivalent to

$$\begin{bmatrix} c_1 & c_0 & c_1 & \dots & c_n \\ \vdots & \ddots & \ddots & \ddots & \vdots \\ c_n & \dots & c_1 & c_0 & c_1 \end{bmatrix} [\,\mathbf{u}_+,\ \mathbf{v}_+\,] = [\,2P_+\mathbf{e}_1\,,\ \mathbf{0}\,] \tag{6.4}$$

and

$$\begin{bmatrix} d_1 & d_0 & d_1 & \dots & d_n \\ \vdots & \ddots & \ddots & \ddots & \vdots \\ d_n & \dots & d_1 & d_0 & d_1 \end{bmatrix} [\,\mathbf{u}_-,\ \mathbf{v}_-\,] = [\,2P_-\mathbf{e}_1\,,\ \mathbf{0}\,]\,, \tag{6.5}$$

respectively.

The following theorem was proved in [13, Thm. 3.3].

Theorem 6.1. *Let* $\mathbf{c}, \mathbf{d} \in \mathbb{F}_+^{2n-1}$ *and* $R_n = P_+T_n(\mathbf{c})P_+ + P_-T_n(\mathbf{d})P_-$. *Then* R_n *is invertible if and only if the Toeplitz equations* (6.1) *and* (6.2) *have* (*symmetric respectively skewsymmetric*) *solutions. In this case, the inverse of* R_n *is given by the corresponding augmented vectors* (6.3) *of these solutions,*

$$R_n^{-1} = \frac{1}{2}\Big(B_{++} + B_{--}\Big), \tag{6.6}$$

where $B_{\pm\pm} = \mathrm{Bez_{sp}}(\mathbf{u}_\pm, \mathbf{v}_\pm)$, *i.e.,*

$$B_{\pm\pm}(t,s) = \frac{\mathbf{u}_\pm(t)\mathbf{v}_\pm(s) - \mathbf{v}_\pm(t)\mathbf{u}_\pm(s)}{(t-s)(1-ts)}. \tag{6.7}$$

Note that the invertibility of the Toeplitz matrices $T_n(\mathbf{c})$ or $T_n(\mathbf{d})$ is, of course, sufficient for the (unique) solvability of the systems (6.1) and (6.2), while it is not always necessary. Therefore, in the case of invertible R_n given as in Theorem 6.1, the solutions to (6.1) and (6.2) need not be unique. However, one can show that the solutions are unique within the set of symmetric and skewsymmetric vectors.

Using the same kind of argument as in the proof of Corollary 5.3 one can derive from the previous theorem a corresponding result for $T+H$ matrices of $(\pm\pm)$type.

Corollary 6.2.

(a) *Let* $\mathbf{c} \in \mathbb{F}_+^{2n-1}$. *Then* $P_+T_n(\mathbf{c})P_+$ *is restrictedly invertible if and only if the Toeplitz equations* (6.1) *have symmetric solutions. In this case*

$$(P_+T_n(\mathbf{c})P_+)^{(-1)} = \frac{1}{2}\mathrm{Bez}_{\mathrm{sp}}(\mathbf{u}_+, \mathbf{v}_+),$$

where $\mathbf{u}_+, \mathbf{v}_+$ *are defined in* (6.3).

(b) *Let* $\mathbf{d} \in \mathbb{F}_+^{2n-1}$. *Then* $P_-T_n(\mathbf{d})P_-$ *is restrictedly invertible if and only if the Toeplitz equations* (6.2) *have skewsymmetric solutions. In this case*

$$(P_-T_n(\mathbf{d})P_-)^{(-1)} = \frac{1}{2}\mathrm{Bez}_{\mathrm{sp}}(\mathbf{u}_-, \mathbf{v}_-),$$

where $\mathbf{u}_-, \mathbf{v}_-$ *are defined in* (6.3).

The example of vectors $\mathbf{c}$ and $\mathbf{d}$ given at the end of Section 4 show that it is possible that $P_+T_n(\mathbf{c})P_+$ is restrictedly invertible, but $T_n(\mathbf{c})$ is not invertible. Likewise, $P_-T_n(\mathbf{d})P_-$ is restrictedly invertible whereas $T_n(\mathbf{d})$ is not invertible. While the systems (6.1) and (6.2) have solutions, we see that in these cases the solutions are not unique. However, as already mentioned above, if we restrict ourselves to symmetric or skewsymmetric solutions, respectively, then uniqueness is guaranteed.

6.2. The centroskewsymmetric case

In this subsection we consider centroskewsymmetric $T+H$ matrices. We will assume that the order n is even because this is a necessary condition for the invertibility of any centroskewsymmetric matrix. We start again from the representation (2.8) of general $T+H$ matrices.

The invertibility of centroskewsymmetric $T+H$ matrices R_n is again connected with systems of Toeplitz form. For given $\mathbf{c}, \mathbf{d} \in \mathbb{F}_-^{2n-1}$ consider

$$T_n(\mathbf{c})(u_j^+)_{j=1}^n = 2P_-\mathbf{e}_1, \qquad T_n(\mathbf{c})(v_j^+)_{j=1}^n = -2P_-(c_j)_{j=1}^n, \tag{6.8}$$

$$T_n(\mathbf{d})(u_j^-)_{j=1}^n = 2P_+\mathbf{e}_1, \qquad T_n(\mathbf{d})(v_j^-)_{j=1}^n = -2P_+(d_j)_{j=1}^n, \tag{6.9}$$

where we choose arbitrary constants $c_n = -c_{-n}$, $d_n = -d_{-n}$. As before, the solvability of (6.8) or (6.9), respectively, is independent of the choice of these constants. This can be seen in the same way as in the centrosymmetric case.

As before, if these systems have a solution, then there is also a solution in which $(u_j^\pm)_{j=1}^n$ and $(v_j^\pm)_{j=1}^n$ belong to $\mathbb{F}_\pm^n$. Furthermore, given these solutions we will pass to augmented vectors (6.3) as we did in the centrosymmetric case.

In terms of the augmented vectors, the system (6.8) is equivalent to

$$\begin{bmatrix} c_1 & 0 & -c_1 & \dots & -c_n \\ \vdots & \ddots & \ddots & \ddots & \vdots \\ c_n & \dots & c_1 & 0 & -c_1 \end{bmatrix} [\,\mathbf{u}_+,\ \mathbf{v}_+\,] = [\,2P_-\mathbf{e}_1\,,\ \mathbf{0}\,]\,, \tag{6.10}$$

and the system (6.9) is equivalent to

$$\begin{bmatrix} d_1 & 0 & -d_1 & \dots & -d_n \\ \vdots & \ddots & \ddots & \ddots & \vdots \\ d_n & \dots & d_1 & 0 & -d_1 \end{bmatrix} [\,\mathbf{u}_-,\ \mathbf{v}_-\,] = [\,2P_+\mathbf{e}_1\,,\ \mathbf{0}\,]\,. \tag{6.11}$$

The following theorem was proved in [13, Thm. 3.6].

Theorem 6.3. *Let $\mathbf{c}, \mathbf{d} \in \mathbb{F}_-^{2n-1}$ and let $R_n = P_-T_n(\mathbf{c})P_+ + P_+T_n(\mathbf{d})P_-$ with n being even. Then R_n is invertible if and only if the Toeplitz equations* (6.8) *and* (6.9) *have (symmetric respectively skewsymmetric) solutions. In this case, the inverse of R_n is given by the corresponding augmented vectors* (6.3) *of these solutions,*

$$R_n^{-1} = \frac{1}{2}\Big(B_{+-} + B_{-+}\Big),$$

where

$$B_{\pm\mp}(t,s) = \frac{\mathbf{u}_\pm(t)\mathbf{v}_\pm(s) - \mathbf{v}_\pm(t)\mathbf{u}_\pm(s)}{(t-s)(1-ts)}\,\frac{s\mp 1}{s\pm 1}. \tag{6.12}$$

Contrary to the centrosymmetric case, it is known that in the centroskewsymmetric case $R_n = P_-T_n(\mathbf{c})P_+ + P_+T_n(\mathbf{d})P_-$ is invertible if and only if both $T_n(\mathbf{c})$ and $T_n(\mathbf{d})$ are invertible. Indeed, this follows from Propositions 3.1 and 4.2 (see also [13, Corollary 3.7]). As a consequence, if the Toeplitz equations (6.8) or (6.9) have a solution, then the solution is unique and it is necessarily symmetric or skewsymmetric, respectively.

Finally, as in the symmetric case, we are able to conclude a corresponding restricted invertibility result for $T + H$ matrices of $(\pm\mp)$type.

Corollary 6.4.

(a) *Let $\mathbf{c} \in \mathbb{F}_-^{2n-1}$. Then $P_-T_n(\mathbf{c})P_+$ is restrictedly invertible if and only if the Toeplitz equations* (6.8) *have (symmetric) solutions. In this case*

$$(P_-T_n(\mathbf{c})P_+)^{(-1)} = B_{+-},$$

where B_{+-} is defined in (6.12) *with the augmented vectors of the solutions.*

(b) *Let $\mathbf{d} \in \mathbb{F}_-^{2n-1}$. Then $P_+T_n(\mathbf{d})P_-$ is restrictedly invertible if and only if the Toeplitz equations* (6.9) *have (skewsymmetric) solutions. In this case*

$$(P_+T_n(\mathbf{d})P_-)^{(-1)} = B_{-+},$$

where B_{-+} is defined in (6.12) *with the augmented vectors of the solutions.*

Note that by Proposition 4.2, $P_-T_n(\mathbf{c})P_+$ is restrictedly invertible if and only if $T_n(\mathbf{c})$ is invertible. Likewise, $P_+T_n(\mathbf{d})P_-$ is restrictedly invertible if and only if $T_n(\mathbf{d})$ is invertible. In this case, the solutions to (6.8) and (6.9), respectively, are unique.

7. Resultant equations

In the previous section we arrived at the equations (6.1) and (6.2) in the centrosymmetric case, which are equivalent to equations (6.4) and (6.5). In the centroskewsymmetric case we obtained the equations (6.8) and (6.9), which are equivalent to (6.10) and (6.11).

Our goal is now to express these equations as resultant equations. At the same time, we relate these equations to the restricted invertibility of the split-Bezoutians of the various types rather than the restricted invertibility of $T + H$ matrices.

7.1. Resultant matrices

Hereafter we need the following $m \times (m+k)$ matrix which is associated with a vector $\mathbf{w} = (w_i)_{i=0}^k \in \mathbb{F}^{k+1}$,

$$D_{m,m+k}(\mathbf{w}) = \begin{bmatrix} w_0 & w_1 & \dots & w_k & & & 0 \\ & w_0 & w_1 & \dots & w_k & & \\ & & \ddots & \ddots & & \ddots & \\ 0 & & & w_0 & w_1 & \dots & w_k \end{bmatrix}. \tag{7.1}$$

The equation $D_{m,m+k}(\mathbf{w})\mathbf{x} = \mathbf{y}$ can be interpreted in the language of rational functions as follows:

$$\mathbf{w}(t^{-1})\mathbf{x}(t) \equiv \mathbf{y}(t) \mod \{\dots, t^{-2}, t^{-1}, t^m, t^{m+1}, \dots\}. \tag{7.2}$$

Indeed, write

$$\mathbf{w}(t^{-1}) = w_0 + w_1 t^{-1} + \dots + w_k t^{-k} \text{ and } \mathbf{x}(t) = x_0 + x_1 t + \dots + x_{m+k-1} t^{m+k-1}.$$

The coefficients of the powers $1, t, \dots, t^{m-1}$ in the product $\mathbf{w}(t^{-1})\mathbf{x}(t)$ have to coincide with those of the polynomial $\mathbf{y}(t) = y_0 + y_1 t + \dots + y_{m-1} t^{m-1}$.

Let $\mathbf{u} \in \mathbb{F}^{m_1+1}, \mathbf{v} \in \mathbb{F}^{m_2+1}$ be nonzero vectors, $p < \min\{m_1, m_2\}$, allowing p to be negative, and let $m_1, m_2 \geq 0$. We introduce the *resultant matrix* $\mathrm{Res}^p(\mathbf{u}, \mathbf{v})$ of $\mathbf{u}$ and $\mathbf{v}$,

$$\mathrm{Res}^p(\mathbf{u}, \mathbf{v}) = \begin{bmatrix} D_{m_2-p, m_1+m_2-p}(\mathbf{u}) \\ D_{m_1-p, m_1+m_2-p}(\mathbf{v}) \end{bmatrix}, \tag{7.3}$$

i.e.,

$$\mathrm{Res}^p(\mathbf{u},\mathbf{v}) = \underbrace{\left[\begin{array}{cccccccc} u_0 & u_1 & \dots & u_{m_1} & & & & 0 \\ & u_0 & u_1 & \dots & u_{m_1} & & & \\ & & \ddots & \ddots & & & \ddots & \\ 0 & & & u_0 & u_1 & & \dots & u_{m_1} \\ v_0 & v_1 & \dots & v_{m_2} & & & & 0 \\ & v_0 & v_1 & \dots & v_{m_2} & & & \\ & & \ddots & \ddots & & & \ddots & \\ 0 & & & v_0 & v_1 & & \dots & v_{m_2} \end{array}\right]}_{m_1+m_2-p} \begin{array}{l} \left.\vphantom{\begin{array}{c}1\\1\\1\\1\end{array}}\right\} m_2-p \\ \left.\vphantom{\begin{array}{c}1\\1\\1\\1\end{array}}\right\} m_1-p \end{array}.$$

In the square case $p = 0$ we have the classical Sylvester resultant matrix. Define

$$\nu = \deg\gcd(\mathbf{u}(t),\mathbf{v}(t)) + \nu_\infty$$

where $\nu_\infty = \min\{m_1 - \deg\mathbf{u}(t), m_2 - \deg\mathbf{v}(t)\}$. More specifically (see [1, Sect. 3]) it can be shown that the dimension of the nullspace of the transpose of $\mathrm{Res}^p(\mathbf{u},\mathbf{v})$ is

$$\dim\ker(\mathrm{Res}^p(\mathbf{u},\mathbf{v}))^T = \max\{0, \nu - p\},$$

and thus

$$\dim\ker\mathrm{Res}^p(\mathbf{u},\mathbf{v}) = \max\{p,\nu\}. \tag{7.4}$$

7.2. The centrosymmetric case

In the centrosymmetric case (see Corollary 6.2), we arrive at the equations (6.4) and (6.5) with $\mathbf{u}_\pm, \mathbf{v}_\pm \in \mathbb{F}_\pm^{n+2}$ given in augmented form (6.3). We can equivalently rewrite these equations in the following way. Equation (6.4) is equivalent to

$$\left[\begin{array}{cccccccccc} 0 & u_1^+ & \dots & u_n^+ & 0 & & & & & 0 \\ & 0 & u_1^+ & \dots & u_n^+ & 0 & & & & \\ & & \ddots & \ddots & & \ddots & \ddots & & & \\ 0 & & & 0 & u_1^+ & \dots & u_n^+ & 0 & & \\ \hline 1 & v_1^+ & \dots & v_n^+ & 1 & & & & & 0 \\ & 1 & v_1^+ & \dots & v_n^+ & 1 & & & & \\ & & \ddots & \ddots & & \ddots & \ddots & & & \\ 0 & & & 1 & v_1^+ & \dots & v_n^+ & 1 & & \end{array}\right] \left[\begin{array}{c} c_n \\ \vdots \\ c_1 \\ c_0 \\ c_1 \\ \vdots \\ c_n \end{array}\right] = \left[\begin{array}{c} 1 \\ 0 \\ \vdots \\ 0 \\ 1 \\ \hline 0 \\ \vdots \\ 0 \end{array}\right],$$

and equation (6.5) is equivalent to

$$\left[\begin{array}{ccccccccc}
0 & u_1^- & \dots & u_n^- & 0 & & & & 0 \\
 & 0 & u_1^- & \dots & u_n^- & 0 & & & \\
 & & \ddots & \ddots & & \ddots & \ddots & & \\
0 & & & 0 & u_1^- & & \dots & u_n^- & 0 \\
\hline
1 & v_1^- & \dots & v_n^- & -1 & & & & 0 \\
 & 1 & v_1^- & \dots & v_n^- & -1 & & & \\
 & & \ddots & \ddots & & \ddots & \ddots & & \\
0 & & & 1 & v_1^- & & \dots & v_n^- & -1
\end{array}\right]
\left[\begin{array}{c} d_n \\ \vdots \\ d_1 \\ d_0 \\ d_1 \\ \vdots \\ d_n \end{array}\right]
=
\left[\begin{array}{c} -1 \\ 0 \\ \vdots \\ 0 \\ 1 \\ \hline 0 \\ \vdots \\ 0 \end{array}\right].$$

Therein the underlying matrices are of size $2n \times (2n+1)$, where we have n rows containing the $u_j^\pm$ and n rows containing the $v_j^\pm$.

Note that each of the variables c_n and d_n, respectively, occurs only in two equations, and these two equations are the same because we assume that $(v_j^+)_{j=1}^n$ is symmetric and $(v_j^-)_{j=1}^n$ is skewsymmetric. The equations read

$$c_n + v_1^+ c_{n-1} + \dots v_n^+ c_0 + c_1 = 0$$

and

$$d_n + v_1^- d_{n-1} + \dots v_n^- d_0 - d_1 = 0,$$

respectively. This means we can eliminate c_n and d_n from the systems by deleting the first and the last equation of the subsystems of equations involving the $v_j^\pm$. After this, the variables c_n and d_n are annihilated by matrix-vector multiplication, and therefore we can delete the first and the last column of the matrix.

As a result, we obtain systems with the following underlying matrices:

$$\underbrace{\left[\begin{array}{cccccccccc}
u_1^\pm & u_2^\pm & \dots & u_{n-1}^\pm & u_n^\pm & & & & & 0 \\
 & u_1^\pm & u_2^\pm & \dots & u_{n-1}^\pm & u_n^\pm & & & & \\
 & & \ddots & \ddots & & \ddots & \ddots & & & \\
 & & & u_1^\pm & u_2^\pm & \dots & u_{n-1}^\pm & u_n^\pm & & \\
0 & & & & u_1^\pm & u_2^\pm & \dots & u_{n-1}^\pm & u_n^\pm & \\
\hline
1 & v_1^\pm & v_2^\pm & \dots & v_{n-1}^\pm & v_n^\pm & \pm 1 & & & 0 \\
 & \ddots & \ddots & \ddots & & \ddots & \ddots & \ddots & & \\
0 & & 1 & v_1^\pm & v_2^\pm & \dots & v_{n-1}^\pm & v_n^\pm & \pm 1 &
\end{array}\right]}_{2n-1}
\begin{array}{l} \left.\vphantom{\begin{array}{c}1\\1\\1\\1\\1\end{array}}\right\} n \\ \left.\vphantom{\begin{array}{c}1\\1\\1\end{array}}\right\} n-2 \end{array}$$

These are the resultant matrices $\mathrm{Res}^1(\mathring{\mathbf{u}}_\pm\,,\,\mathbf{v}_\pm)$. Here

$$\mathring{\mathbf{u}}_\pm = (u_j^\pm)_{j=1}^n \tag{7.5}$$

denote the *reduced vectors*, which are associated with the vectors $\mathbf{u}_\pm$ in augmented form (6.3). As a consequence, we obtain the following lemma.

Lemma 7.1. *Let $\mathbf{c}, \mathbf{d} \in \mathbb{F}_+^{2n-1}$, let $\mathbf{u}_\pm, \mathbf{v}_\pm \in \mathbb{F}_\pm^{n+2}$ be of the augmented form (6.3), and let $\mathring{\mathbf{u}}_\pm \in \mathbb{F}_\pm^n$ the reduced vectors (7.5) associated with $\mathbf{u}_\pm$.*

(a) *There exists $c_n = c_{-n} \in \mathbb{F}$ such that (6.4) is true if and only if*

$$\mathrm{Res}^1(\mathring{\mathbf{u}}_+\,,\,\mathbf{v}_+)\,\mathbf{c} = \begin{bmatrix} 2P_+\mathbf{e}_1 \\ \mathbf{0} \end{bmatrix}. \tag{7.6}$$

(b) *There exists $d_n = d_{-n} \in \mathbb{F}$ such that (6.5) is true if and only if*

$$\mathrm{Res}^1(\mathring{\mathbf{u}}_-\,,\,\mathbf{v}_-)\,\mathbf{d} = \begin{bmatrix} -2P_-\mathbf{e}_1 \\ \mathbf{0} \end{bmatrix}. \tag{7.7}$$

Theorem 7.2. *Let $\mathbf{u}_\pm, \mathbf{v}_\pm \in \mathbb{F}_\pm^{n+2}$ be of the form (6.3), and let*

$$B_{\pm\pm} = \mathrm{Bez}_{\mathrm{sp}}(\mathbf{u}_\pm, \mathbf{v}_\pm).$$

(a) *B_{++} is restrictedly invertible if and only if equation (7.6) has a solution $\mathbf{c} \in \mathbb{F}_+^{2n+1}$. In this case,*

$$B_{++}^{(-1)} = \frac{1}{2} P_+ T_n(\mathbf{c}) P_+.$$

(b) *B_{--} is restrictedly invertible if and only if equation (7.7) has a solution $\mathbf{d} \in \mathbb{F}_-^{2n+1}$. In this case,*

$$B_{--}^{(-1)} = \frac{1}{2} P_- T_n(\mathbf{d}) P_-.$$

Proof. $\Leftarrow$: Let us first assume that the systems (7.6) or (7.7), respectively, have a solution. We are going to show that $B_{\pm\pm}$ are restrictedly invertible and that their restricted inverses are given as above.

Since $\mathbf{u}_\pm, \mathbf{v}_\pm$ with the corresponding reduced vector $\mathring{\mathbf{u}}_\pm$ satisfy the systems (7.6) or (7.7), Lemma 7.1 implies that there exist $c_n = c_{-n}$ and $d_n = d_{-n}$ such that the Toeplitz systems (6.4) or (6.5) are satisfied. These Toeplitz systems are in turn equivalent to the systems (6.1) or (6.2). Hence the system (6.1) or (6.2) is solvable, and the pair of vectors $\{\mathbf{u}_\pm, \mathbf{v}_\pm\}$ represents a solution.

Now Corollary 6.2 implies that

$$R_+ := P_+ T_n(\mathbf{c}) P_+ \quad \text{or} \quad R_- := P_- T_n(\mathbf{d}) P_-$$

is restrictedly invertible and that

$$R_\pm^{(-1)} = \frac{1}{2} B_{\pm\pm}$$

with the split-Bezoutian $B_{\pm\pm} = \mathrm{Bez}_{\mathrm{sp}}(\mathbf{u}_\pm, \mathbf{v}_\pm)$. Hence $B_{\pm\pm}$ is restrictedly invertible, and the above formula is true.

$\Rightarrow$: Now assume that $B_{\pm\pm}$ is restrictedly invertible. Then the restricted inverse is of the form

$$\frac{1}{2}B_{++}^{(-1)} = P_+T_n(\mathbf{c})P_+, \qquad \frac{1}{2}B_{--}^{(-1)} = P_-T_n(\mathbf{d})P_-$$

with certain $\mathbf{c},\mathbf{d} \in \mathbb{F}_+^{2n-1}$ (see Corollary 5.3). As $P_+T_n(\mathbf{c})P_+$ or $P_-T_n(\mathbf{d})P_-$ is restrictedly invertible, too, Corollary 6.2 implies that there are $\tilde{\mathbf{u}}_\pm, \tilde{\mathbf{v}}_\pm$ (possibly different from $\mathbf{u}_\pm, \mathbf{v}_\pm$) such that (6.1) or (6.2), respectively, holds. Moreover,

$$\frac{1}{2}(P_+T_n(\mathbf{c})P_+)^{(-1)} = B_{++} = \mathrm{Bez}_{\mathrm{sp}}(\tilde{\mathbf{u}}_+, \tilde{\mathbf{v}}_+),$$
$$\frac{1}{2}(P_-T_n(\mathbf{d})P_-)^{(-1)} = B_{--} = \mathrm{Bez}_{\mathrm{sp}}(\tilde{\mathbf{u}}_-, \tilde{\mathbf{v}}_-).$$

Consequently, $\mathrm{Bez}_{\mathrm{sp}}(\tilde{\mathbf{u}}_\pm, \tilde{\mathbf{v}}_\pm) = \mathrm{Bez}_{\mathrm{sp}}(\mathbf{u}_\pm, \mathbf{v}_\pm)$. From Lemma 5.7 it follows that there exist $\Phi_\pm \in \mathbb{F}^{2\times 2}$ such that $\det \Phi_\pm = 1$ and

$$\begin{bmatrix} \mathbf{u}_\pm, \mathbf{v}_\pm \end{bmatrix} = \begin{bmatrix} \tilde{\mathbf{u}}_\pm, \tilde{\mathbf{v}}_\pm \end{bmatrix} \Phi_\pm .$$

Since both underlying pairs of vectors are of augmented form, we conclude that

$$\Phi_\pm = \begin{bmatrix} 1 & \lambda_\pm \\ 0 & 1 \end{bmatrix}$$

for some $\lambda_\pm \in \mathbb{F}$. From the comments made in connection with (6.1) or (6.2) it follows that $\mathbf{u}_\pm, \mathbf{v}_\pm$ also satisfy the system (6.1) or (6.2) with possibly different $c_{\pm n}, d_{\pm n}$. As a consequence, systems (7.6) and (7.7) are satisfied. In other words, for the given $\mathbf{u}_\pm, \mathbf{v}_\pm$ these systems have the solutions $\mathbf{c}, \mathbf{d}$. $\square$

7.3. The centroskewsymmetric case

In the centroskewsymmetric case (see Corollary 6.4), we are led to the equations (6.10) and (6.11) with $\mathbf{u}_\pm, \mathbf{v}_\pm \in \mathbb{F}_\pm^{n+2}$ given in augmented form (6.3). We can equivalently rewrite these equations as follows. Equation (6.10) is equivalent to

$$\left[\begin{array}{cccccccc} 0 & u_1^+ & \dots & u_n^+ & 0 & & & 0 \\ & 0 & u_1^+ & \dots & u_n^+ & 0 & & \\ & & \ddots & \ddots & & \ddots & \ddots & \\ 0 & & & 0 & u_1^+ & \dots & u_n^+ & 0 \\ \hline 1 & v_1^+ & \dots & v_n^+ & 1 & & & 0 \\ & 1 & v_1^+ & \dots & v_n^+ & 1 & & \\ & & \ddots & \ddots & & \ddots & \ddots & \\ 0 & & & 1 & v_1^+ & \dots & v_n^+ & 1 \end{array}\right] \begin{bmatrix} -c_n \\ \vdots \\ -c_1 \\ 0 \\ c_1 \\ \vdots \\ c_n \end{bmatrix} = \left[\begin{array}{c} 1 \\ 0 \\ \vdots \\ 0 \\ -1 \\ \hline 0 \\ \vdots \\ 0 \end{array}\right],$$

and equation (6.11) is equivalent to

$$\left[\begin{array}{ccccccccc} 0 & u_1^- & \dots & u_n^- & 0 & & & & 0 \\ & 0 & u_1^- & \dots & u_n^- & 0 & & & \\ & & \ddots & \ddots & & \ddots & \ddots & & \\ 0 & & & 0 & u_1^- & \dots & u_n^- & & 0 \\ \hline 1 & v_1^- & \dots & v_n^- & -1 & & & & 0 \\ & 1 & v_1^- & \dots & v_n^- & -1 & & & \\ & & \ddots & \ddots & & \ddots & \ddots & & \\ 0 & & & 1 & v_1^- & \dots & v_n^- & & -1 \end{array}\right] \begin{bmatrix} -d_n \\ \vdots \\ -d_1 \\ 0 \\ d_1 \\ \vdots \\ d_n \end{bmatrix} = \left[\begin{array}{c} -1 \\ 0 \\ \vdots \\ 0 \\ -1 \\ \hline 0 \\ \vdots \\ 0 \end{array}\right].$$

As before, the underlying matrices are of size $2n \times (2n+1)$, where we have n rows containing the $u_j^\pm$ and n rows containing the $v_j^\pm$.

We can also eliminate $c_n = -c_{-n}$ and $d_n = -d_{-n}$ in the same way as in Subsection 7.2. This leads to the following lemma.

Lemma 7.3. *Let* $\mathbf{c}, \mathbf{d} \in \mathbb{F}_-^{2n-1}$*, let* $\mathbf{u}_\pm, \mathbf{v}_\pm \in \mathbb{F}_\pm^{n+2}$ *be of the augmented form* (6.3)*, let* $\mathring{\mathbf{u}}_\pm \in \mathbb{F}_\pm^n$ *be the reduced vector* (7.5) *associated with* $\mathbf{u}_\pm$*.*

(a) *There exists* $c_n = -c_{-n} \in \mathbb{F}$ *such that* (6.10) *is true if and only if*

$$\operatorname{Res}^1(\mathring{\mathbf{u}}_+ , \mathbf{v}_+)\, \mathbf{c} = \begin{bmatrix} 2P_-\mathbf{e}_1 \\ \mathbf{0} \end{bmatrix}. \tag{7.8}$$

(b) *There exists* $d_n = -d_{-n} \in \mathbb{F}$ *such that* (6.11) *is true if and only if*

$$\operatorname{Res}^1(\mathring{\mathbf{u}}_- , \mathbf{v}_-)\, \mathbf{d} = \begin{bmatrix} -2P_+\mathbf{e}_1 \\ \mathbf{0} \end{bmatrix}. \tag{7.9}$$

The proof of the following theorem is completely analogous to that of Theorem 7.2.

Theorem 7.4. *Let* $\mathbf{u}_\pm, \mathbf{v}_\pm \in \mathbb{F}_\pm^{n+2}$ *be of the form* (6.3)*, and let* B_{+-} *and* B_{-+} *be given by* (6.12)*.*

(a) B_{+-} *is restrictedly invertible if and only if equation* (7.8) *has a solution* $\mathbf{c} \in \mathbb{F}_+^{2n+1}$*. In this case,*

$$B_{+-}^{(-1)} = \frac{1}{2} P_- T_n(\mathbf{c}) P_+.$$

(b) B_{-+} *is restrictedly invertible if and only if equation* (7.9) *has a solution* $\mathbf{d} \in \mathbb{F}_+^{2n+1}$*. In this case,*

$$B_{-+}^{(-1)} = \frac{1}{2} P_+ T_n(\mathbf{d}) P_-.$$

7.4. Kernels of resultant matrices

Theorems 7.2 and 7.4 tell us that if we want to invert split-Bezoutians of the various types, we are led to linear systems, namely (7.6), (7.7), (7.8), and (7.9), whose coefficient matrices are the $(2n-2)\times(2n-1)$ resultant matrices $\operatorname{Res}^1(\mathring{\mathbf{u}}_\pm, \mathbf{v}_\pm)$. Here $\mathbf{u}_\pm, \mathbf{v}_\pm \in \mathbb{F}_\pm^{n+2}$ are of the augmented form (6.3), and $\mathring{\mathbf{u}}_\pm \in \mathbb{F}_\pm^n$ are the reduced vectors (7.5). Note that in terms of the notation of Section 7.1, we are in the setting of

$$p = 1, \quad m_1 = n-1, \quad m_2 = n+1.$$

The afore-mentioned systems are inhomogeneous, and their solution vectors $\mathbf{c}$ and $\mathbf{d}$ are sought in $\mathbb{F}_+^{2n-1}$ (in the centrosymmetric case) or in $\mathbb{F}_-^{2n-1}$ (in the centroskewsymmetric case).

It is worthwhile to look at the kernels of these resultant matrices, which are non-trivial. According to Theorems 7.2 and 7.4, any solution to the afore-mentioned systems should give rise to the same inverse of the split-Bezoutian even though $\mathbf{c}$ and $\mathbf{d}$ may not be unique. To see this directly, we need the following lemma.

Lemma 7.5.

(a) *Let n be even, $\mathbf{u}_+, \mathbf{v}_+ \in \mathbb{F}_+^{n+2}$ and assume that* $\gcd(\mathbf{u}_+, \mathbf{v}_+) = t+1$. *Then*

$$\ker \operatorname{Res}^1(\mathring{\mathbf{u}}_+, \mathbf{v}_+) = \operatorname{lin}\left\{(1,-1,1,\dots,1,-1,1)^T\right\}.$$

(b) *Let n be even, $\mathbf{u}_-, \mathbf{v}_- \in \mathbb{F}_-^{n+2}$ and assume that* $\gcd(\mathbf{u}_-, \mathbf{v}_-) = t-1$. *Then*

$$\ker \operatorname{Res}^1(\mathring{\mathbf{u}}_-, \mathbf{v}_-) = \operatorname{lin}\left\{(1,1,1,\dots,1,1,1)^T\right\}.$$

(c) *Let n be odd, $\mathbf{u}_+, \mathbf{v}_+ \in \mathbb{F}_+^{n+2}$ and assume that* $\gcd(\mathbf{u}_+, \mathbf{v}_+) = 1$. *Then*

$$\ker \operatorname{Res}^1(\mathring{\mathbf{u}}_+, \mathbf{v}_+) = \operatorname{lin}\{\mathbf{q}\},$$

where

$$\mathbf{q}(t) = \boldsymbol{\alpha}(t)\boldsymbol{\beta}^J(t) - \boldsymbol{\beta}(t)\boldsymbol{\alpha}^J(t) \in \mathbb{F}_-^{2n-1}$$

with $\mathring{\mathbf{u}}_+(t)\boldsymbol{\alpha}(t) + \mathbf{v}_+(t)\boldsymbol{\beta}(t) = 1$, $\boldsymbol{\alpha} \in \mathbb{F}^{n+1}$, *and* $\boldsymbol{\beta} \in \mathbb{F}^{n-1}$.

(d) *Let n be odd, $\mathbf{u}_-, \mathbf{v}_- \in \mathbb{F}_-^{n+2}$ and assume that* $\gcd(\mathbf{u}_-, \mathbf{v}_-) = t^2-1$. *Then*

$$\ker \operatorname{Res}^1(\mathring{\mathbf{u}}_-, \mathbf{v}_-) = \operatorname{lin}\left\{(1,-1,1,\dots,1,-1,1)^T,\ (1,1,1,\dots,1,1,1)^T\right\}.$$

Proof. Since $\mathbf{v}_\pm(t)$ has no zero at $t=0$ and at $t=\infty$ (the latter means that the polynomial $\mathbf{v}_\pm(t)$ has the maximal degree $n+1$) it follows that $\gcd(\mathring{\mathbf{u}}_\pm, \mathbf{v}_\pm) = \gcd(\mathbf{u}_\pm, \mathbf{v}_\pm)$. Furthermore, with the notation of Section 7.1, $\nu_\infty = 0$ in all cases. Hence, we have

$$\nu = 1, \quad \nu = 1, \quad \nu = 0, \quad \nu = 2$$

in the cases (a), (b), (c), (d), respectively. As $p = 1$ in all cases, we get that the dimension of the kernel equals 1 in the cases (a), (b), (c), and 2 in the case (d) (see formula (7.4)).

In the cases (a), (b), (d) the form of the kernel can be found by direct inspection.

In the case (c) the kernel can be obtained as follows (see [1, Thm. 3.3]). Solve the Bezout equations

$$\mathring{\mathbf{u}}_+(t)\boldsymbol{\alpha}(t) + \mathbf{v}_+(t)\boldsymbol{\beta}(t) = 1, \quad \mathring{\mathbf{u}}_+^J(t)\boldsymbol{\gamma}^J(t) + \mathbf{v}_+^J(t)\boldsymbol{\delta}^J(t) = 1,$$

with $\boldsymbol{\alpha}, \boldsymbol{\gamma} \in \mathbb{F}^{n+1}$and $\boldsymbol{\beta}, \boldsymbol{\delta} \in \mathbb{F}^{n-1}$, and put $\mathbf{q}(t) = \boldsymbol{\alpha}(t)\boldsymbol{\delta}(t) - \boldsymbol{\beta}(t)\boldsymbol{\gamma}(t)$. Since $\mathring{\mathbf{u}}_+(t)$ and $\mathbf{v}_+(t)$ are generalized coprime, the Bezout equations have unique solutions, and because $\mathring{\mathbf{u}}_+$ and $\mathbf{v}_+$ are symmetric, it follows that $\boldsymbol{\alpha} = \boldsymbol{\gamma}^J$ and $\boldsymbol{\beta} = \boldsymbol{\delta}^J$. Then the one-dimensional kernel is spanned by the vector $\mathbf{q} \subset \mathbb{F}_-^{2n-1}$. □

Note that in case (c) the kernel is spanned by a skewsymmetric vector, while in cases (a), (b), (d), the kernel consists of symmetric vectors.

To justify the above claim that two different solutions to the systems considered in Theorems 7.2 and 7.4 lead to the same $T + H$ matrices (of the various kinds), it suffices to consider the intersections of the corresponding kernels with the spaces of symmetric and skewsymmetric vectors, respectively, and compare them with Proposition 4.1. With the notation of Proposition 4.1 we have

$$\ker \operatorname{Res}^1(\mathring{\mathbf{u}}_\pm, \mathbf{v}_\pm) \cap \mathbb{F}_+^{2n-1} = \{\, \mathbf{e}_{\alpha,\pm\alpha} \;:\; \alpha \in \mathbb{F}\,\}$$

in the case n even, and

$$\ker \operatorname{Res}^1(\mathring{\mathbf{u}}_\pm, \mathbf{v}_\pm) \cap \mathbb{F}_+^{2n-1} = \begin{cases} \{\mathbf{0}\} & \text{in the } + \text{ case}, \\ \{\mathbf{e}_{\alpha,\beta} \;:\alpha, \beta \in \mathbb{F}\} & \text{in the } - \text{ case}, \end{cases}$$

in the case n odd. Furthermore,

$$\ker \operatorname{Res}^1(\mathring{\mathbf{u}}_\pm, \mathbf{v}_\pm) \cap \mathbb{F}_-^{2n-1} = \{\mathbf{0}\}$$

in the case n even.

8. Solution of the resultant equations

The goal of this section is to find solutions to the equations (7.6), (7.7), (7.8), and (7.9). In the centrosymmetric case, we have to distinguish between n even and odd, while in the centroskewsymmetric case we only consider n even.

The basic observation is that these equations can be written in the language of polynomials (in t and t^{-1}). Observe that $\mathbf{u}_\pm, \mathbf{v}_\pm$ are given in augmented form (6.3), and that the corresponding reduced vector is given by $\mathring{\mathbf{u}}_\pm(t) = \frac{1}{t}\mathbf{u}_\pm(t)$. Using (7.2) we can reformulate (7.6) in the form

$$\begin{aligned} t\mathbf{u}_+(\tfrac{1}{t})\mathbf{c}(t) &\equiv 1 + t^{n-1} \quad \operatorname{mod} \{\ldots, t^{-3}, t^{-2}, t^{-1}, t^n, t^{n+1}, t^{n+2}, \ldots\}, \\ \mathbf{v}_+(\tfrac{1}{t})\mathbf{c}(t) &\equiv \mathbf{0} \qquad\qquad \operatorname{mod} \{\ldots, t^{-3}, t^{-2}, t^{-1}, t^{n-2}, t^{n-1}, t^n, \ldots\}, \end{aligned} \tag{8.1}$$

while the resultant equation (7.7) is equivalent to

$$\begin{aligned} t\mathbf{u}_-(\tfrac{1}{t})\mathbf{d}(t) &\equiv -1 + t^{n-1} \quad \bmod \{\dots, t^{-3}, t^{-2}, t^{-1}, t^n, t^{n+1}, t^{n+2}, \dots\}, \\ \mathbf{v}_-(\tfrac{1}{t})\mathbf{d}(t) &\equiv \mathbf{0} \quad \bmod \{\dots, t^{-3}, t^{-2}, t^{-1}, t^{n-2}, t^{n-1}, t^n, \dots\}. \end{aligned} \tag{8.2}$$

Regarding the centroskewsymmetric cases, the resultant equation (7.8) is equivalent to

$$\begin{aligned} t\mathbf{u}_+(\tfrac{1}{t})\mathbf{c}(t) &\equiv 1 - t^{n-1} \quad \bmod \{\dots, t^{-3}, t^{-2}, t^{-1}, t^n, t^{n+1}, t^{n+2}, \dots\}, \\ \mathbf{v}_+(\tfrac{1}{t})\mathbf{c}(t) &\equiv \mathbf{0} \quad \bmod \{\dots, t^{-3}, t^{-2}, t^{-1}, t^{n-2}, t^{n-1}, t^n, \dots\}, \end{aligned} \tag{8.3}$$

and the resultant equation (7.9) is equivalent to

$$\begin{aligned} t\mathbf{u}_-(\tfrac{1}{t})\mathbf{d}(t) &\equiv -1 - t^{n-1} \quad \bmod \{\dots, t^{-3}, t^{-2}, t^{-1}, t^n, t^{n+1}, t^{n+2}, \dots\}, \\ \mathbf{v}_-(\tfrac{1}{t})\mathbf{d}(t) &\equiv \mathbf{0} \quad \bmod \{\dots, t^{-3}, t^{-2}, t^{-1}, t^{n-2}, t^{n-1}, t^n, \dots\}. \end{aligned} \tag{8.4}$$

8.1. The centrosymmetric case

We start with considering the systems (8.1) and (8.2), the solution of which relates to the restricted inverses of split-Bezoutians $B_{\pm\pm}$ (see Theorem 7.2).

Note that the restricted invertibility of these split-Bezoutians requires that the g.c.d. of $\mathbf{u}_\pm(t)$ and $\mathbf{v}_\pm(t)$ is minimal (see Corollary 5.3 and formulas (5.3) and (5.4)). Hence it is natural to assume here the corresponding conditions of coprimeness.

As we have to distinguish between n even and odd, we end up with four cases to consider. The proofs are similar to each other in certain respects. Hence we will give the most detailed arguments in the first case, while we confine ourselves to the differences in the other cases.

Let us start with finding the solution of (8.1) in the case n is odd. Suppose $\mathbf{u}_+, \mathbf{v}_+ \in \mathbb{F}_+^{n+2}$ are of the form (6.3) and assume $\gcd(\mathbf{u}_+, \mathbf{v}_+) = 1$. Then the Bezout equation

$$\mathbf{u}_+(t)\boldsymbol{\alpha}_0(t) + \mathbf{v}_+(t)\boldsymbol{\beta}_0(t) = 1 \tag{8.5}$$

has unique solutions $\boldsymbol{\alpha}_0, \boldsymbol{\beta}_0 \in \mathbb{F}^{n+1}$. Recall that the solution $\mathbf{c}$ of (8.1) is related to the restricted inverse of the split-Bezoutian $B_{++} = \mathrm{Bez}_{\mathrm{sp}}(\mathbf{u}_+, \mathbf{v}_+)$ of size $n \times n$.

Theorem 8.1. *Let $\mathbf{z}(t)$ be the polynomial*

$$\mathbf{z}(t) = \boldsymbol{\alpha}_0(t)\boldsymbol{\beta}_0^J(t) - \boldsymbol{\beta}_0(t)\boldsymbol{\alpha}_0^J(t) \in \mathbb{F}_-^{2n+1}[t], \tag{8.6}$$

and let $\mathbf{y}(t) = (t^2 - 1)\mathbf{z}(t) \in \mathbb{F}_+^{2n+3}[t]$. Then

$$\mathbf{c} = \begin{bmatrix} 0 & 0 & 1 & & 0 & 0 & 0 \\ \vdots & \vdots & & \ddots & & \vdots & \vdots \\ 0 & 0 & 0 & & 1 & 0 & 0 \end{bmatrix} \mathbf{y} \in \mathbb{F}_+^{2n-1}$$

is a solution of (8.1).

Note that the vector $\mathbf{c}$ is obtained from the vector $\mathbf{y}$ by dropping the first and the last two entries.

Proof. We first notice that $\mathbf{z} = (z_i)_{i=0}^{2n} \in \mathbb{F}_-^{2n+1}$, and therefore we can write $\mathbf{z}(t) = z_0 + z_1 t + \cdots - z_1 t^{2n-1} - z_0 t^{2n}$. It follows that the polynomial $\mathbf{y}(t)$ is given by

$$\mathbf{y}(t) = -z_0 - z_1 t + \cdots - z_1 t^{2n+1} - z_0 t^{2n+2},$$

where we note only the first and last two coefficients. The vector $\mathbf{c}$ is obtained from $\mathbf{y}$ by dropping these coefficients, i.e.,

$$\begin{aligned}\mathbf{c}(t) &= t^{-2}\left(\mathbf{y}(t) + z_0 t^{2n+2} + z_1 t^{2n+1} + z_1 t + z_0\right)\\ &= \left(1 - \frac{1}{t^2}\right)\mathbf{z}(t) + z_0 t^{2n} + z_1 t^{2n-1} + z_1 \frac{1}{t} + z_0 \frac{1}{t^2}.\end{aligned}$$

Let us now verify the first equation in (8.1). Taking into account that $\mathbf{u}_+(0) = 0$ and $\deg \mathbf{u}_+(t) \le n$ (see (6.3)) we get

$$\begin{aligned}t\mathbf{u}_+\left(\frac{1}{t}\right)\mathbf{c}(t) &= \mathbf{u}_+\left(\frac{1}{t}\right)\left(\left(t - \frac{1}{t}\right)\mathbf{z}(t) + z_0 t^{2n+1} + z_1 t^{2n} + z_1 + z_0 \frac{1}{t}\right)\\ &\equiv \mathbf{u}_+\left(\frac{1}{t}\right)\left(t - \frac{1}{t}\right)\mathbf{z}(t).\end{aligned}$$

Note that here "$\equiv$" signals that equality holds up to linear combinations of t^k with $k < 0$ or $k \ge n$. In what follows we are going to apply formulas (8.5) and (8.6), and use the symmetry of $\mathbf{u}_+, \mathbf{v}_+ \in \mathbb{F}^{n+2}$, i.e.,

$$\mathbf{u}_+(t) = t^{n+1}\mathbf{u}_+\left(\frac{1}{t}\right) \quad \text{and} \quad \mathbf{v}_+(t) = t^{n+1}\mathbf{v}_+\left(\frac{1}{t}\right).$$

Furthermore, note that $\boldsymbol{\alpha}_0^J(t) = t^n \boldsymbol{\alpha}_0(t^{-1})$ and $\boldsymbol{\beta}_0^J(t) = t^n \boldsymbol{\beta}_0(t^{-1})$. It follows that

$$\begin{aligned}&t\mathbf{u}_+\left(\frac{1}{t}\right)\mathbf{c}(t)\\ &\equiv \mathbf{u}_+\left(\frac{1}{t}\right)\left(t - \frac{1}{t}\right)\left(\boldsymbol{\alpha}_0(t)t^n\boldsymbol{\beta}_0\left(\frac{1}{t}\right) - \boldsymbol{\beta}_0(t)t^n\boldsymbol{\alpha}_0\left(\frac{1}{t}\right)\right)\\ &= \left(t - \frac{1}{t}\right)\left(\frac{1}{t}\boldsymbol{\alpha}_0(t)\mathbf{u}_+(t)\boldsymbol{\beta}_0\left(\frac{1}{t}\right) - t^n\boldsymbol{\alpha}_0\left(\frac{1}{t}\right)\mathbf{u}_+\left(\frac{1}{t}\right)\boldsymbol{\beta}_0(t)\right)\\ &= \left(t - \frac{1}{t}\right)\left(\frac{1}{t}\left(1 - \boldsymbol{\beta}_0(t)\mathbf{v}_+(t)\right)\boldsymbol{\beta}_0\left(\frac{1}{t}\right) - t^n\left(1 - \boldsymbol{\beta}_0\left(\frac{1}{t}\right)\mathbf{v}_+\left(\frac{1}{t}\right)\right)\boldsymbol{\beta}_0(t)\right)\\ &= \left(t - \frac{1}{t}\right)\left(\frac{1}{t}\boldsymbol{\beta}_0\left(\frac{1}{t}\right) - t^n\boldsymbol{\beta}_0(t)\right)\\ &\equiv \boldsymbol{\beta}_0\left(\frac{1}{t}\right) + t^{n-1}\boldsymbol{\beta}_0(t).\end{aligned}$$

Now consider equation (8.5) and use that $\mathbf{u}_+(0) = 0$ and $\mathbf{v}_+(0) = 1$ (see (6.3)). We conclude that $\boldsymbol{\beta}_0(0) = 1$. From this it follows that

$$\boldsymbol{\beta}_0\left(\frac{1}{t}\right) + t^{n-1}\boldsymbol{\beta}_0(t) \equiv 1 + t^{n-1}.$$

Hence the first equation in (8.1) is satisfied.

Now let us verify the second equation in (8.1). Proceeding similar as before, but with $\deg \mathbf{v}_+(t) \le n+1$, we obtain first

$$\begin{aligned}\mathbf{v}_+\Big(\frac{1}{t}\Big)\mathbf{c}(t) &= \mathbf{v}_+\Big(\frac{1}{t}\Big)\Big(\Big(1-\frac{1}{t^2}\Big)\mathbf{z}(t) + z_0 t^{2n} + z_1 t^{2n-1} + z_1\frac{1}{t} + z_0\frac{1}{t^2}\Big)\\ &\equiv \mathbf{v}_+\Big(\frac{1}{t}\Big)\Big(1-\frac{1}{t^2}\Big)\mathbf{z}(t).\end{aligned}$$

Here "$\equiv$" signals that equality holds up to linear combinations of t^k with $k<0$ or $k \ge n-2$. Furthermore,

$$\begin{aligned}&\mathbf{v}_+\Big(\frac{1}{t}\Big)\mathbf{c}(t)\\ &\equiv \mathbf{v}_+\Big(\frac{1}{t}\Big)\Big(1-\frac{1}{t^2}\Big)\Big(\boldsymbol{\alpha}_0(t)t^n\boldsymbol{\beta}_0\Big(\frac{1}{t}\Big) - \boldsymbol{\beta}_0(t)t^n\boldsymbol{\alpha}_0\Big(\frac{1}{t}\Big)\Big)\\ &= \Big(1-\frac{1}{t^2}\Big)\Big(t^n\boldsymbol{\alpha}_0(t)\mathbf{v}_+\Big(\frac{1}{t}\Big)\boldsymbol{\beta}_0\Big(\frac{1}{t}\Big) - \frac{1}{t}\boldsymbol{\alpha}_0\Big(\frac{1}{t}\Big)\mathbf{v}_+(t)\boldsymbol{\beta}_0(t)\Big)\\ &= \Big(1-\frac{1}{t^2}\Big)\Big(t^n\boldsymbol{\alpha}_0(t)\Big(1-\mathbf{u}_+\Big(\frac{1}{t}\Big)\boldsymbol{\alpha}_0\Big(\frac{1}{t}\Big)\Big) - \frac{1}{t}\boldsymbol{\alpha}_0\Big(\frac{1}{t}\Big)(1-\mathbf{u}_+(t)\boldsymbol{\alpha}_0(t))\Big)\\ &= \Big(1-\frac{1}{t^2}\Big)\Big(t^n\boldsymbol{\alpha}_0(t) - \frac{1}{t}\boldsymbol{\alpha}_0\Big(\frac{1}{t}\Big)\Big) \equiv 0.\end{aligned}$$

Hence the second equation in (8.1) is satisfied, too. □

Next, let us find the solution of (8.2) in the case n is odd. This solution $\mathbf{d}$ corresponds to the restricted inverse of a split-Bezoutian, namely $B_{--} = \mathrm{Bez}_{\mathrm{sp}}(\mathbf{u}_-, \mathbf{v}_-)$, where $\mathbf{u}_-, \mathbf{v}_- \in \mathbb{F}_-^{n+2}$ are of the form (6.3). Due to (2.5) we can always write

$$\mathbf{u}_-(t) = (1-t^2)\mathbf{u}_0(t), \quad \mathbf{v}_-(t) = (1-t^2)\mathbf{v}_0(t),$$

where $\mathbf{u}_0, \mathbf{v}_0 \in \mathbb{F}_+^n$. It is natural to assume that $\gcd(\mathbf{u}_-, \mathbf{v}_-)$ is minimal (see (5.4)), and therefore we require that $\gcd(\mathbf{u}_0, \mathbf{v}_0) = 1$. Then the Bezout equation

$$\mathbf{u}_0(t)\boldsymbol{\alpha}_0(t) + \mathbf{v}_0(t)\boldsymbol{\beta}_0(t) = 1 \tag{8.7}$$

has unique solutions $\boldsymbol{\alpha}_0, \boldsymbol{\beta}_0 \in \mathbb{F}^{n-1}$.

Theorem 8.2. *Let $\mathbf{z}(t)$ be the polynomial*

$$\mathbf{z}(t) = \boldsymbol{\alpha}_0(t)\boldsymbol{\beta}_0^J(t) - \boldsymbol{\beta}_0(t)\boldsymbol{\alpha}_0^J(t) \in \mathbb{F}_-^{2n-3}[t]. \tag{8.8}$$

Then

$$\mathbf{d}(t) = \frac{t^2}{1-t^2}\,\mathbf{z}(t) \in \mathbb{F}_+^{2n-1}[t]$$

is a solution of (8.2).

Note that the vector $\mathbf{d}$ is obtained by adding two zeros at the beginning and two zeros at the end of the vector $\mathbf{y}$ where $\mathbf{y}(t) := \frac{1}{1-t^2}\mathbf{z}(t) \in \mathbb{F}_+^{2n-5}[t]$. We remark that $\mathbf{y}$ is well defined due to (2.5).

Proof. We start with observing that $\mathbf{u}_0(0) = 0, \mathbf{v}_0(0) = 1$, and

$$\frac{t^2}{t^2-1}\mathbf{u}_-\Big(\frac{1}{t}\Big) = \mathbf{u}_0\Big(\frac{1}{t}\Big) = t^{-(n-1)}\mathbf{u}_0(t).$$

A corresponding formula holds for $\mathbf{v}_-$ and $\mathbf{v}_0$ as well.

Let us consider the first equation in (8.2). Taking (8.7) and (8.8) into account, we get

$$\begin{aligned} t\mathbf{u}_-\Big(\frac{1}{t}\Big)\mathbf{d}(t) &= \mathbf{u}_-\Big(\frac{1}{t}\Big)\frac{t^3}{1-t^2}\left(\boldsymbol{\alpha}_0(t)t^{n-2}\boldsymbol{\beta}_0\Big(\frac{1}{t}\Big) - \boldsymbol{\beta}_0(t)t^{n-2}\boldsymbol{\alpha}_0\Big(\frac{1}{t}\Big)\right) \\ &= -\boldsymbol{\alpha}_0(t)\mathbf{u}_0(t)\boldsymbol{\beta}_0\Big(\frac{1}{t}\Big) + t^{n-1}\boldsymbol{\alpha}_0\Big(\frac{1}{t}\Big)\mathbf{u}_0\Big(\frac{1}{t}\Big)\boldsymbol{\beta}_0(t) \\ &= -(1-\boldsymbol{\beta}_0(t)\mathbf{v}_0(t))\boldsymbol{\beta}_0\Big(\frac{1}{t}\Big) + t^{n-1}\Big(1-\boldsymbol{\beta}_0\Big(\frac{1}{t}\Big)\mathbf{v}_0\Big(\frac{1}{t}\Big)\Big)\boldsymbol{\beta}_0(t) \\ &= -\boldsymbol{\beta}_0\Big(\frac{1}{t}\Big) + t^{n-1}\boldsymbol{\beta}_0(t) \equiv -1 + t^{n-1}. \end{aligned}$$

Here $\boldsymbol{\beta}_0(0) = 1$. Along the same lines we obtain for the second equation in (8.2),

$$\begin{aligned} \mathbf{v}_-\Big(\frac{1}{t}\Big)\mathbf{d}(t) &= \mathbf{v}_-\Big(\frac{1}{t}\Big)\frac{t^2}{1-t^2}\left(\boldsymbol{\alpha}_0(t)t^{n-2}\boldsymbol{\beta}_0\Big(\frac{1}{t}\Big) - \boldsymbol{\beta}_0(t)t^{n-2}\boldsymbol{\alpha}_0(\frac{1}{t})\right) \\ &= -t^{n-2}\boldsymbol{\alpha}_0(t)\mathbf{v}_0\Big(\frac{1}{t}\Big)\boldsymbol{\beta}_0\Big(\frac{1}{t}\Big) + \frac{1}{t}\boldsymbol{\alpha}_0\Big(\frac{1}{t}\Big)\mathbf{v}_0(t)\boldsymbol{\beta}_0(t) \\ &= -t^{n-2}\boldsymbol{\alpha}_0(t)\Big(1-\mathbf{u}_0\Big(\frac{1}{t}\Big)\boldsymbol{\alpha}_0\Big(\frac{1}{t}\Big)\Big) + \frac{1}{t}\boldsymbol{\alpha}_0\Big(\frac{1}{t}\Big)(1-\mathbf{u}_0(t)\boldsymbol{\alpha}_0(t)) \\ &= -t^{n-2}\boldsymbol{\alpha}_0(t) + \frac{1}{t}\boldsymbol{\alpha}_0\Big(\frac{1}{t}\Big) \equiv 0, \end{aligned}$$

which completes the proof. □

Now we turn to the case of n even. We first want to find a solution of equation (8.1). Note that the solution $\mathbf{c}$ is related to the restricted inverse of a split-Bezoutian of (++)type $B_{++} = \mathrm{Bez}_{\mathrm{sp}}(\mathbf{u}_+, \mathbf{v}_+)$ with $\mathbf{u}_+, \mathbf{v}_+ \in \mathbb{F}_+^{n+2}$ of the form (6.3). Using (2.5) we can write

$$\mathbf{u}_+(t) = (1+t)\mathbf{u}_0(t), \quad \mathbf{v}_+(t) = (1+t)\mathbf{v}_0(t).$$

with $\mathbf{u}_0, \mathbf{v}_0 \in \mathbb{F}_+^{n+1}$. The natural assumption is that $\gcd(\mathbf{u}_+, \mathbf{v}_+) = 1 + t$. In other words, we will assume $\gcd(\mathbf{u}_0, \mathbf{v}_0) = 1$. Then the Bezout equation

$$\mathbf{u}_0(t)\boldsymbol{\alpha}_0(t) + \mathbf{v}_0(t)\boldsymbol{\beta}_0(t) = 1 \tag{8.9}$$

has unique solutions $\boldsymbol{\alpha}_0, \boldsymbol{\beta}_0 \in \mathbb{F}^n$.

Theorem 8.3. *Let* $\mathbf{z}(t)$ *be the polynomial*

$$\mathbf{z}(t) = \boldsymbol{\alpha}_0(t)\boldsymbol{\beta}_0^J(t) - \boldsymbol{\beta}_0(t)\boldsymbol{\alpha}_0^J(t) \in \mathbb{F}_-^{2n-1}[t]. \tag{8.10}$$

Then

$$\mathbf{c}(t) = \frac{t-1}{t+1}\mathbf{z}(t) \in \mathbb{F}_+^{2n-1}[t]$$

is a solution of (8.1).

Proof. First note that $\mathbf{u}_0(0) = 1, \mathbf{v}_0(0) = 1$, and

$$\frac{t}{t+1}\mathbf{u}_+\left(\frac{1}{t}\right) = \mathbf{u}_0\left(\frac{1}{t}\right) = t^{-n}\mathbf{u}_0(t)\,.$$

For the first equation in (8.1) we obtain using (8.9),

$$\begin{aligned} t\mathbf{u}_+\left(\frac{1}{t}\right)\mathbf{c}(t) &= \mathbf{u}_+\left(\frac{1}{t}\right)\frac{t-1}{t+1}\left(\boldsymbol{\alpha}_0(t)t^n\boldsymbol{\beta}_0\left(\frac{1}{t}\right) - \boldsymbol{\beta}_0(t)t^n\boldsymbol{\alpha}_0\left(\frac{1}{t}\right)\right) \\ &= \frac{t-1}{t}\left(\mathbf{u}_0(t)\boldsymbol{\alpha}_0(t)\boldsymbol{\beta}_0\left(\frac{1}{t}\right) - t^n\mathbf{u}_0\left(\frac{1}{t}\right)\boldsymbol{\alpha}_0\left(\frac{1}{t}\right)\boldsymbol{\beta}_0(t)\right) \\ &= \left(1-\frac{1}{t}\right)\left(\boldsymbol{\beta}_0\left(\frac{1}{t}\right) - t^n\boldsymbol{\beta}_0(t)\right) \\ &\equiv \boldsymbol{\beta}_0\left(\frac{1}{t}\right) + t^{n-1}\boldsymbol{\beta}_0(t) \equiv 1 + t^{n-1}. \end{aligned}$$

Here $\boldsymbol{\beta}_0(0) = 1$. For the second equation in (8.1) we have

$$\begin{aligned} \mathbf{v}_+\left(\frac{1}{t}\right)\mathbf{c}(t) &= \mathbf{v}_+\left(\frac{1}{t}\right)\frac{t-1}{t+1}\left(\boldsymbol{\alpha}_0(t)t^{n-1}\boldsymbol{\beta}_0\left(\frac{1}{t}\right) - \boldsymbol{\alpha}_0\left(\frac{1}{t}\right)t^{n-1}\boldsymbol{\beta}_0(t)\right) \\ &= \frac{t-1}{t^2}\left(t^n\mathbf{v}_0\left(\frac{1}{t}\right)\boldsymbol{\beta}_0\left(\frac{1}{t}\right)\boldsymbol{\alpha}_0(t) - \mathbf{v}_0(t)\boldsymbol{\beta}_0(t)\boldsymbol{\alpha}_0\left(\frac{1}{t}\right)\right) \\ &= \frac{t-1}{t^2}\left(t^n\boldsymbol{\alpha}_0(t) - \boldsymbol{\alpha}_0\left(\frac{1}{t}\right)\right) \equiv 0. \end{aligned}$$

Hence the second equation in (8.1) is fulfilled. □

It remains to find the solution of equation (8.2) in the case n is even. Here the solution $\mathbf{d}$ corresponds to a split-Bezoutian $B_{--} = \mathrm{Bez}_{\mathrm{sp}}(\mathbf{u}_-, \mathbf{v}_-)$ of $(--)$type with $\mathbf{u}_-, \mathbf{v}_- \in \mathbb{F}_-^{n+2}$ of the form (6.3). The natural assumption that $\gcd(\mathbf{u}_-, \mathbf{v}_-) = 1 - t$ implies that we can write

$$\mathbf{u}_-(t) = (1-t)\mathbf{u}_0(t), \quad \mathbf{v}_-(t) = (1-t)\mathbf{v}_0(t)$$

with $\gcd(\mathbf{u}_0, \mathbf{v}_0) = 1$ and $\mathbf{u}_0, \mathbf{v}_0 \in \mathbb{F}_+^{n+1}$. Then the Bezout equation

$$\mathbf{u}_0(t)\boldsymbol{\alpha}_0(t) + \mathbf{v}_0(t)\boldsymbol{\beta}_0(t) = 1 \tag{8.11}$$

has unique solutions $\boldsymbol{\alpha}_0, \boldsymbol{\beta}_0 \in \mathbb{F}^n$.

Theorem 8.4. *Let* $\mathbf{z}(t)$ *be the polynomial*

$$\mathbf{z}(t) = \boldsymbol{\alpha}_0(t)\boldsymbol{\beta}_0^J(t) - \boldsymbol{\beta}_0(t)\boldsymbol{\alpha}_0^J(t) \in \mathbb{F}_-^{2n-1}[t]. \tag{8.12}$$

Then

$$\mathbf{d}(t) = \frac{1+t}{1-t}\mathbf{z}(t) \in \mathbb{F}_+^{2n-1}[t]$$

is a solution of (8.2).

Proof. As

$$\frac{t}{t-1}\mathbf{u}_-\left(\frac{1}{t}\right) = \mathbf{u}_0\left(\frac{1}{t}\right) = t^{-n}\mathbf{u}_0(t)$$

we obtain for the first equation

$$\begin{aligned} t\mathbf{u}_-\Big(\frac{1}{t}\Big)\mathbf{d}(t) &= \mathbf{u}_-\Big(\frac{1}{t}\Big)\frac{1+t}{1-t}\Big(\boldsymbol{\alpha}_0(t)t^n\boldsymbol{\beta}_0\Big(\frac{1}{t}\Big) - \boldsymbol{\beta}_0(t)t^n\boldsymbol{\alpha}_0\Big(\frac{1}{t}\Big)\Big) \\ &= -\frac{1+t}{t}\Big(\mathbf{u}_0(t)\boldsymbol{\alpha}_0(t)\boldsymbol{\beta}_0\Big(\frac{1}{t}\Big) - t^n\mathbf{u}_0\Big(\frac{1}{t}\Big)\boldsymbol{\alpha}_0\Big(\frac{1}{t}\Big)\boldsymbol{\beta}_0(t)\Big) \\ &= -\Big(1+\frac{1}{t}\Big)\Big(\boldsymbol{\beta}_0\Big(\frac{1}{t}\Big) - t^n\boldsymbol{\beta}_0(t)\Big) \\ &\equiv -\boldsymbol{\beta}_0\Big(\frac{1}{t}\Big) + t^{n-1}\boldsymbol{\beta}_0(t) \equiv -1 + t^{n-1} \end{aligned}$$

as $\boldsymbol{\beta}_0(0) = 1$. The verification of the second equation is analogous. □

8.2. The centroskewsymmetric case

Let us now discuss the cases related to the split-Bezoutians of $(\pm\mp)$type (see Theorem 7.4). Here n is even, and our goal is to find solutions to the equations (8.3) and (8.4).

We start with equation (8.3), the solution of which corresponds to the restricted inverse of a split-Bezoutian B_{+-} defined in (6.12) with $\mathbf{u}_+, \mathbf{v}_+ \in \mathbb{F}_+^{n+2}$ of the form (6.3). Due to (2.5) we can write

$$\mathbf{u}_+(t) = (1+t)\mathbf{u}_0(t), \;\; \mathbf{v}_+(t) = (1+t)\mathbf{v}_0(t)$$

with $\mathbf{u}_0, \mathbf{v}_0 \in \mathbb{F}_+^{n+1}$. The results of Section 5.2 (see Corollary 5.6 and the various representations of the split-Bezoutians) imply that B_{+-} is restrictedly invertible if and only if $\gcd(\mathbf{u}_0, \mathbf{v}_0) = 1$, which we will therefore assume. Then the Bezout equation

$$\mathbf{u}_0(t)\boldsymbol{\alpha}_0(t) + \mathbf{v}_0(t)\boldsymbol{\beta}_0(t) = 1 \tag{8.13}$$

has unique solutions $\boldsymbol{\alpha}_0, \boldsymbol{\beta}_0 \in \mathbb{F}^n$.

Theorem 8.5. *Let $\mathbf{c}(t)$ be the polynomial*

$$\mathbf{c}(t) = \boldsymbol{\alpha}_0(t)\boldsymbol{\beta}_0^J(t) - \boldsymbol{\beta}_0(t)\boldsymbol{\alpha}_0^J(t) \in \mathbb{F}_-^{2n-1}[t]. \tag{8.14}$$

Then $\mathbf{c}$ is a solution of (8.3).

Proof. For the first equation in (8.3) we obtain

$$\begin{aligned} t\mathbf{u}_+\Big(\frac{1}{t}\Big)\mathbf{c}(t) &= (1+t)\mathbf{u}_0\Big(\frac{1}{t}\Big)\Big(\boldsymbol{\alpha}_0(t)t^{n-1}\boldsymbol{\beta}_0\Big(\frac{1}{t}\Big) - \boldsymbol{\beta}_0(t)t^{n-1}\boldsymbol{\alpha}_0\Big(\frac{1}{t}\Big)\Big) \\ &= (1+t)\Big(\frac{1}{t}\mathbf{u}_0(t)\boldsymbol{\alpha}_0(t)\boldsymbol{\beta}_0\Big(\frac{1}{t}\Big) - t^{n-1}\mathbf{u}_0\Big(\frac{1}{t}\Big)\boldsymbol{\alpha}_0\Big(\frac{1}{t}\Big)\boldsymbol{\beta}_0(t)\Big) \\ &= (1+t)\Big(\frac{1}{t}\boldsymbol{\beta}_0\Big(\frac{1}{t}\Big) - t^{n-1}\boldsymbol{\beta}_0(t)\Big) \\ &\equiv \boldsymbol{\beta}_0\Big(\frac{1}{t}\Big) - t^{n-1}\boldsymbol{\beta}_0(t) \equiv 1 - t^{n-1}. \end{aligned}$$

Here $\boldsymbol{\beta}_0(0) = 1$ since $\mathbf{u}_+(0) = 0$ and $\mathbf{v}_+(0) = 1$. For the second equation we have

$$\begin{aligned}\mathbf{v}_+\Big(\frac{1}{t}\Big)\mathbf{c}(t) &= \Big(1+\frac{1}{t}\Big)\mathbf{v}_0\Big(\frac{1}{t}\Big)\Big(\boldsymbol{\alpha}_0(t)t^{n-1}\boldsymbol{\beta}_0\Big(\frac{1}{t}\Big) - \boldsymbol{\beta}_0(t)t^{n-1}\boldsymbol{\alpha}_0\Big(\frac{1}{t}\Big)\Big)\\ &= \Big(1+\frac{1}{t}\Big)\Big(t^{n-1}\mathbf{v}_0\Big(\frac{1}{t}\Big)\boldsymbol{\beta}_0\Big(\frac{1}{t}\Big)\boldsymbol{\alpha}_0(t) - \frac{1}{t}\mathbf{v}_0(t)\boldsymbol{\beta}_0(t)\boldsymbol{\alpha}_0\Big(\frac{1}{t}\Big)\Big)\\ &= \Big(1+\frac{1}{t}\Big)\Big(t^{n-1}\boldsymbol{\alpha}_0(t) - \frac{1}{t}\boldsymbol{\alpha}_0\Big(\frac{1}{t}\Big)\Big) \equiv 0.\end{aligned}$$

Hence the second equation in (8.3) is fulfilled. □

It remains to consider equation (8.4), whose solution corresponds to a split-Bezoutian B_{-+} defined in (6.12) with $\mathbf{u}_-, \mathbf{v}_- \in \mathbb{F}_-^{n+2}$ being of the form (6.3). Here we can write

$$\mathbf{u}_-(t) = (1-t)\mathbf{u}_0(t), \quad \mathbf{v}_-(t) = (1-t)\mathbf{v}_0(t)$$

with $\mathbf{u}_0, \mathbf{v}_0 \in \mathbb{F}_+^{n+1}$. We will assume $\gcd(\mathbf{u}_0, \mathbf{v}_0) = 1$. Then the Bezout equation

$$\mathbf{u}_0(t)\boldsymbol{\alpha}_0(t) + \mathbf{v}_0(t)\boldsymbol{\beta}_0(t) = 1 \tag{8.15}$$

has unique solutions $\boldsymbol{\alpha}_0, \boldsymbol{\beta}_0 \in \mathbb{F}^n$.

Theorem 8.6. *Let* $\mathbf{d}(t)$ *be the polynomial*

$$\mathbf{d}(t) = -\Big(\boldsymbol{\alpha}_0(t)\boldsymbol{\beta}_0^J(t) - \boldsymbol{\beta}_0(t)\boldsymbol{\alpha}_0^J(t)\Big) \in \mathbb{F}_-^{2n-1}[t]. \tag{8.16}$$

Then $\mathbf{d}$ *is a solution of* (8.4).

Proof. For the first equation of (8.4) we obtain

$$\begin{aligned}t\mathbf{u}_-\Big(\frac{1}{t}\Big)\mathbf{d}(t) &= (1-t)\mathbf{u}_0\Big(\frac{1}{t}\Big)\Big(\boldsymbol{\alpha}_0(t)t^{n-1}\boldsymbol{\beta}_0\Big(\frac{1}{t}\Big) - \boldsymbol{\beta}_0(t)t^{n-1}\boldsymbol{\alpha}_0\Big(\frac{1}{t}\Big)\Big)\\ &= (1-t)\Big(\frac{1}{t}\mathbf{u}_0(t)\boldsymbol{\alpha}_0(t)\boldsymbol{\beta}_0\Big(\frac{1}{t}\Big) - t^{n-1}\mathbf{u}_0\Big(\frac{1}{t}\Big)\boldsymbol{\alpha}_0\Big(\frac{1}{t}\Big)\boldsymbol{\beta}_0(t)\Big)\\ &= (1-t)\Big(\frac{1}{t}\boldsymbol{\beta}_0\Big(\frac{1}{t}\Big) - t^{n-1}\boldsymbol{\beta}_0(t)\Big)\\ &\equiv -\boldsymbol{\beta}_0\Big(\frac{1}{t}\Big) - t^{n-1}\boldsymbol{\beta}_0(t)\\ &\equiv -1 - t^{n-1}.\end{aligned}$$

Here $\boldsymbol{\beta}_0(0) = 1$. This proves the first equation. The verification of the second equation is analogous. □

9. Restricted inversion of split-Bezoutians

Using the results of the previous section and of Section 7 (Theorems 7.2 and 7.4), we can establish the following results concerning the restricted invertibility of split-Bezoutians. For sake of presentation we split the results into the cases n even and odd, and consider the different types of split-Bezoutians as subcases therein.

For all cases we rely on the following preparatory constructions. Given $\mathbf{f}, \mathbf{g} \in \mathbb{F}_+^{n+1+i}$ with $i \in \{-1, 0, 1\}$, assume that the pair $\{\mathbf{f}(t), \mathbf{g}(t)\}$ is coprime. We are going to define $\mathbf{z} \in \mathbb{F}_-^{2n-1+2i}$ as follows. First determine the (unique) solutions of the Bezout equation

$$\mathbf{f}(t)\boldsymbol{\alpha}(t) + \mathbf{g}(t)\boldsymbol{\beta}(t) = 1$$

with $\boldsymbol{\alpha}, \boldsymbol{\beta} \in \mathbb{F}^{n+i}$, and then define

$$\mathbf{z}(t) = \boldsymbol{\alpha}(t)\boldsymbol{\beta}^J(t) - \boldsymbol{\beta}(t)\boldsymbol{\alpha}^J(t) \in \mathbb{F}_-^{2n-1+2i}[t]\,. \tag{9.1}$$

Theorem 9.1. *Let n be even and $\mathbf{f}, \mathbf{g} \in \mathbb{F}_+^{n+1}$ be such that $\{\mathbf{f}(t), \mathbf{g}(t)\}$ is coprime. Define $\mathbf{z} \in \mathbb{F}_-^{2n-1}$ by (9.1).*

(a) *Then $B_{++} = M_{n-1}^{t+1}\mathrm{Bez}_{\mathrm{sp}}(\mathbf{f}, \mathbf{g})(M_{n-1}^{t+1})^T$ is restrictedly invertible and*

$$B_{++}^{(-1)} = \frac{1}{2}P_+T_n(\mathbf{c})P_+$$

with $\mathbf{c}(t) = \frac{t-1}{t+1}\mathbf{z}(t) \in \mathbb{F}_+^{2n-1}[t]$.

(b) *Then $B_{--} = M_{n-1}^{t-1}\mathrm{Bez}_{\mathrm{sp}}(\mathbf{f}, \mathbf{g})(M_{n-1}^{t-1})^T$ is restrictedly invertible and*

$$B_{--}^{(-1)} = \frac{1}{2}P_-T_n(\mathbf{d})P_-$$

with $\mathbf{d}(t) = \frac{1+t}{1-t}\mathbf{z}(t) \in \mathbb{F}_+^{2n-1}[t]$.

(c) *Then $B_{+-} = M_{n-1}^{t+1}\mathrm{Bez}_{\mathrm{sp}}(\mathbf{f}, \mathbf{g})(M_{n-1}^{t-1})^T$ is restrictedly invertible and*

$$B_{+-}^{(-1)} = \frac{1}{2}P_-T_n(\mathbf{c})P_+$$

with $\mathbf{c}(t) = \mathbf{z}(t) \in \mathbb{F}_-^{2n-1}[t]$.

(d) *Then $B_{-+} = M_{n-1}^{t-1}\mathrm{Bez}_{\mathrm{sp}}(\mathbf{f}, \mathbf{g})(M_{n-1}^{t+1})^T$ is restrictedly invertible and*

$$B_{-+}^{(-1)} = \frac{1}{2}P_+T_n(\mathbf{d})P_-$$

with $\mathbf{d}(t) = -\mathbf{z}(t) \in \mathbb{F}_-^{2n-1}[t]$.

Proof. For sake of definiteness, consider the case (a). The treatment of the other cases is analogous.

Let $B_{++} = M_{n-1}^{t+1}\mathrm{Bez}_{\mathrm{sp}}(\mathbf{f}, \mathbf{g})(M_{n-1}^{t+1})^T$ be given with coprime polynomials $\{\mathbf{f}(t), \mathbf{g}(t)\}$. Using the symmetry of these polynomials and that they cannot have a common zero (otherwise they are not coprime), it follows that there exists a $\Phi \in \mathbb{F}^{2\times 2}$ with $\det \Phi = 1$ such that the vectors $\mathbf{u}_0, \mathbf{v}_0$ defined by

$$[\,\mathbf{u}_0,\, \mathbf{v}_0\,] := [\,\mathbf{f},\, \mathbf{g}\,]\,\Phi$$

are in augmented form (6.3). So $\mathrm{Bez}_{\mathrm{sp}}(\mathbf{f}, \mathbf{g}) = \mathrm{Bez}_{\mathrm{sp}}(\mathbf{u}_0, \mathbf{v}_0)$ by Lemma 5.7, and we obtain $B_{++} = M_{n-1}^{t+1}\mathrm{Bez}_{\mathrm{sp}}(\mathbf{u}_0, \mathbf{v}_0)(M_{n-1}^{t+1})^T$, where $\{\mathbf{u}_0(t), \mathbf{v}_0(t)\}$ are coprime as well. Define $\mathbf{u}_+$ and $\mathbf{v}_+$ by

$$\mathbf{u}_+(t) = (t+1)\mathbf{u}_0(t), \qquad \mathbf{v}_+(t) = (t+1)\mathbf{v}_0(t).$$

Then $\mathbf{u}_+$ and $\mathbf{v}_+$ are also in augmented form, and $B_{++} = \mathrm{Bez}_{\mathrm{sp}}(\mathbf{u}_+, \mathbf{v}_+)$. Now use Theorem 8.3 to see that the vector $\mathbf{c}$ defined there is a solution of (8.1). To distinguish that vector there from the vector $\mathbf{c}$ considered in this theorem denote it by $\hat{\mathbf{c}}$.

Let us show that $\mathbf{c} = \hat{\mathbf{c}}$. Indeed, recalling the definitions we have

$$\hat{\mathbf{c}}(t) = \frac{t-1}{t+1}\hat{\mathbf{z}}(t), \qquad \mathbf{c}(t) = \frac{t-1}{t+1}\mathbf{z}(t)$$

with

$$\hat{\mathbf{z}}(t) = \boldsymbol{\alpha}_0(t)\boldsymbol{\beta}_0^J(t) - \boldsymbol{\beta}_0(t)\boldsymbol{\alpha}_0^J(t), \qquad \mathbf{z}(t) = \boldsymbol{\alpha}(t)\boldsymbol{\beta}^J(t) - \boldsymbol{\beta}(t)\boldsymbol{\alpha}^J(t),$$

and

$$\mathbf{u}_0(t)\boldsymbol{\alpha}_0(t) + \mathbf{v}_0(t)\boldsymbol{\beta}_0(t) = 1, \qquad \mathbf{f}(t)\boldsymbol{\alpha}(t) + \mathbf{g}(t)\boldsymbol{\beta}(t) = 1.$$

Since $[\mathbf{u}_0,\, \mathbf{v}_0] = [\mathbf{f},\, \mathbf{g},]\Phi$ the uniqueness of the Bezout equations implies that

$$[\,\boldsymbol{\alpha}_0,\, \boldsymbol{\beta}_0\,] = [\boldsymbol{\alpha},\, \boldsymbol{\beta}]\, \Phi^{-T},$$

where Φ^{-T} stands for the inverse of the transpose matrix. Since

$$\Phi^T \begin{bmatrix} 0 & -1 \\ 1 & 0 \end{bmatrix} \Phi = \begin{bmatrix} 0 & -1 \\ 1 & 0 \end{bmatrix}$$

as $\det \Phi = 1$, it follows that $\hat{\mathbf{z}} = \mathbf{z}$ via straightforward considerations (compare the proof of Theorem 8.1 of [5]). Therefore $\hat{\mathbf{c}} = \mathbf{c}$ as desired.

It follows that the vector $\mathbf{c}$ defined in the present theorem satisfies (8.1). As noted at the beginning of Section 8, this means it satisfies (7.6).

Now recall Theorem 7.2 (a) to conclude that B_{++} is restrictedly invertible and that

$$B_{++}^{(-1)} = \frac{1}{2} P_+ T_n(\mathbf{c}) P_+.$$

This proves statement (a).

For the remaining statements notice that we use Theorem 8.4 in case (b). In cases (c) and (d) we take Theorems 8.5 and 8.6 into account. Furthermore, we utilize Theorem 7.2 (b) in case (b) and Theorem 7.4 in cases (c) and (d). □

The proof of the result for the odd case is also completely analogous to the previous proof. Here we employ Theorems 8.1 and 8.2 together with Theorem 7.2 (a), (b).

Theorem 9.2. *Let n be odd.*

(a) *Let $\mathbf{f}, \mathbf{g} \in \mathbb{F}_+^{n+2}$ be such that $\{\mathbf{f}(t), \mathbf{g}(t)\}$ is coprime. Define $\mathbf{z} \in \mathbb{F}_-^{2n+1}$ by (9.1). Then $B_{++} = \mathrm{Bez}_{\mathrm{sp}}(\mathbf{f}, \mathbf{g})$ is restrictedly invertible and*

$$B_{++}^{(-1)} = \frac{1}{2} P_+ T_n(\mathbf{c}) P_+\,,$$

where

$$\mathbf{c} = \begin{bmatrix} 0 & 0 & 1 & & 0 & 0 & 0 \\ \vdots & \vdots & & \ddots & & \vdots & \vdots \\ 0 & 0 & 0 & & 1 & 0 & 0 \end{bmatrix} \mathbf{y} \in \mathbb{F}_+^{2n-1}$$

with $\mathbf{y}(t) = (t^2-1)\mathbf{z}(t) \in \mathbb{F}_+^{2n+3}[t]$.

(b) *Let* $\mathbf{f}, \mathbf{g} \in \mathbb{F}_+^n$ *be such that* $\{\mathbf{f}(t), \mathbf{g}(t)\}$ *is coprime. Define* $\mathbf{z} \in \mathbb{F}_-^{2n-3}$ *by* (9.1). *Then* $B_{--} = M_{n-2}^{t^2-1} \mathrm{Bez}_{\mathrm{sp}}(\mathbf{f}, \mathbf{g})(M_{n-2}^{t^2-1})^T$ *is restrictedly invertible and*

$$B_{--}^{(-1)} = \frac{1}{2} P_- T_n(\mathbf{d}) P_- ,$$

where

$$\mathbf{d}(t) = \frac{t^2}{1-t^2} \mathbf{z}(t) \in \mathbb{F}_+^{2n-1}[t].$$

The algorithms resulting from Theorems 9.1 and 9.2 have a computational complexity of $O(n^2)$. They require to apply the extended Euclidian algorithm to solve Bezout equations and polynomial multiplications. The polynomial division by factors $(t \pm 1)$ can be done with linear complexity. For more details see Remark 9.7 of [5].

By Proposition 3.1 the formulas in Theorems 9.1 and 9.2 lead to inversion formulas for centro(skew)symmetric $T+H$-Bezoutians. For instance, let n be even, and B be a centrosymmetric $T+H$-Bezoutian of order n given by its splitting $B = B_{++} + B_{--}$, where $B_{\pm\pm}$ are defined in (5.6) with $\{\mathbf{f}_i(t), \mathbf{g}_i(t)\}$ coprime polynomials ($i = 1, 2$). Then

$$B^{-1} = B_{++}^{(-1)} + B_{--}^{(-1)}$$

where $B_{\pm\pm}^{(-1)}$ can be computed as in Theorem 9.1, part (a) with $\{\mathbf{f}_1, \mathbf{g}_1\}$ and part (b) with $\{\mathbf{f}_2, \mathbf{g}_2\}$.

In case n is odd, we use Theorem 9.2. For centroskewsymmetric $T+H$-Bezoutians we apply Theorem 9.1 (c) and (d).

The formulas of Theorems 9.1 and 9.2 coincide with the formulas established in Section 9 of [5] (centrosymmetric case) and Section 6 of [4] (centroskewsymmetric case). In fact, we have shown here that they are true even without the additional assumptions required in [5].

10. Examples

We are going to illustrate the algorithms obtained in the previous section with some simple examples. We first focus on the four cases considered in Theorem 9.1 with $n = 4$. In all four cases we choose the vectors $\mathbf{f}, \mathbf{g} \in \mathbb{F}_+^5$ given by

$$\mathbf{f}(t) = t + t^2 + t^3, \qquad \mathbf{g}(t) = 1 + 2t^2 + t^4.$$

According to the definition (5.1) we get

$$\mathrm{Bez}_{\mathrm{sp}}(\mathbf{f}, \mathbf{g}) = \begin{bmatrix} 1 & 1 & 1 \\ 1 & 0 & 1 \\ 1 & 1 & 1 \end{bmatrix}.$$

The four split-Bezoutians considered in Theorem 9.1 evaluate now to

$$B_{++} = \begin{bmatrix} 1 & 2 & 2 & 1 \\ 2 & 3 & 3 & 3 \\ 2 & 3 & 3 & 3 \\ 1 & 2 & 2 & 1 \end{bmatrix}, \qquad B_{--} = \begin{bmatrix} 1 & 0 & 0 & -1 \\ 0 & -1 & 1 & 0 \\ 0 & 1 & -1 & 0 \\ -1 & 0 & 0 & 1 \end{bmatrix},$$

$$B_{+-} = \begin{bmatrix} -1 & 0 & 0 & 1 \\ -2 & 1 & -1 & 2 \\ -2 & 1 & -1 & 2 \\ -1 & 0 & 0 & 1 \end{bmatrix}, \qquad B_{-+} = \begin{bmatrix} -1 & -2 & -2 & -1 \\ 0 & 1 & 1 & 0 \\ 0 & -1 & -1 & 0 \\ 1 & 2 & 2 & 1 \end{bmatrix}.$$

To obtain the restricted inverses of these matrices, solve the Bezout equation $\mathbf{f}(t)\boldsymbol{\alpha}(t) + \mathbf{g}(t)\boldsymbol{\beta}(t) = 1$, $\boldsymbol{\alpha}, \boldsymbol{\beta} \in \mathbb{F}^4$ to get $\boldsymbol{\alpha}(t) = -2 - t - t^2 - t^3$ and $\boldsymbol{\beta}(t) = 1+2t+t^2$. Then by (9.1) we obtain $\mathbf{z}(t) = 1+t-t^2+t^4-t^5-t^6 \in \mathbb{F}^7_-[t]$. For the centrosymmetric cases it follows that

$$\mathbf{c}(t) = \frac{t-1}{t+1}\mathbf{z}(t) = -1 + t + t^2 - t^3 + t^4 + t^5 - t^6,$$
$$\mathbf{d}(t) = \frac{1+t}{1-t}\mathbf{z}(t) = 1 + 3t + 3t^2 + 2t^3 + 3t^4 + 4t^5 + t^6,$$

and thus $B_{++}^{(-1)} = \frac{1}{4}T_n(\mathbf{c})(I + J_n)$ and $B_{--}^{(-1)} = \frac{1}{4}T_n(\mathbf{d})(I - J_n)$ evaluate to

$$B_{++}^{(-1)} = \frac{1}{4}\begin{bmatrix} -3 & 2 & 2 & -3 \\ 2 & -1 & -1 & 2 \\ 2 & -1 & -1 & 2 \\ -3 & 2 & 2 & -3 \end{bmatrix}, \qquad B_{--}^{(-1)} = \frac{1}{4}\begin{bmatrix} 1 & 0 & 0 & -1 \\ 0 & -1 & 1 & 0 \\ 0 & 1 & -1 & 0 \\ -1 & 0 & 0 & 1 \end{bmatrix}.$$

For the centroskewsymmetric cases we have $\mathbf{c} = \mathbf{z}$ and $\mathbf{d} = -\mathbf{z}$. Here then $B_{+-}^{(-1)} = \frac{1}{4}T_n(\mathbf{c})(I + J_n)$ and $B_{-+}^{(-1)} = \frac{1}{4}T_n(\mathbf{d})(I - J_n)$ evaluate to

$$B_{+-}^{(-1)} = \frac{1}{4}\begin{bmatrix} -1 & 0 & 0 & -1 \\ -2 & 1 & 1 & -2 \\ 2 & -1 & -1 & 2 \\ 1 & 0 & 0 & 1 \end{bmatrix}, \qquad B_{-+}^{(-1)} = \frac{1}{4}\begin{bmatrix} -1 & -2 & 2 & 1 \\ 0 & 1 & -1 & 0 \\ 0 & 1 & -1 & 0 \\ -1 & -2 & 2 & 1 \end{bmatrix}.$$

It can be verified straightforwardly that these are indeed the restricted inverses.

Let us now illustrate Theorem 9.2. We consider $n = 5$ and start with case (b), where we use the same $\mathbf{f}, \mathbf{g} \in \mathbb{F}^5$ as before. Then the split-Bezoutian of $(--)$type is

$$B_{--} = \begin{bmatrix} 1 & 1 & 0 & -1 & -1 \\ 1 & 0 & 0 & 0 & -1 \\ 0 & 0 & 0 & 0 & 0 \\ -1 & 0 & 0 & 0 & 1 \\ -1 & -1 & 0 & 1 & 1 \end{bmatrix}.$$

The polynomial $\mathbf{z}(t)$ is the same as above, and

$$\frac{\mathbf{z}(t)}{1-t^2} = 1 + t + t^3 + t^4 \in \mathbb{F}^7_+[t],$$

which gives the vector $\mathbf{d} = (0,0,1,1,0,1,1,0,0) \in \mathbb{F}^9_+$ by adding zeros. The corresponding restricted inverse is $\frac{1}{4}T_n(\mathbf{d})(I - J_n)$, i.e.,

$$B_{--}^{(-1)} = \frac{1}{4}\begin{bmatrix} 0 & 1 & 0 & -1 & 0 \\ 1 & -1 & 0 & 1 & -1 \\ 0 & 0 & 0 & 0 & 0 \\ -1 & 1 & 0 & -1 & 1 \\ 0 & -1 & 0 & 1 & 0 \end{bmatrix}.$$

For the case (a) consider the following $\mathbf{f}, \mathbf{g} \in \mathbb{F}^7_+$,

$$\mathbf{f}(t) = t + t^2 + t^3 + t^4 + t^5, \qquad \mathbf{g}(t) = 1 + 3t^2 + 3t^4 + t^6.$$

The Bezoutian $B_{++} = \mathrm{Bez}_{\mathrm{sp}}(\mathbf{f}, \mathbf{g})$ of $(++)$type is

$$B_{++} = \begin{bmatrix} 1 & 1 & 1 & 1 & 1 \\ 1 & -1 & 2 & -1 & 1 \\ 1 & 2 & 0 & 2 & 1 \\ 1 & -1 & 2 & -1 & 1 \\ 1 & 1 & 1 & 1 & 1 \end{bmatrix}.$$

The solutions $\boldsymbol{\alpha}, \boldsymbol{\beta} \in \mathbb{F}^6$ of the Bezout equation are now

$$\boldsymbol{\alpha}(t) = -3 - 4t - 5t^2 - 4t^3 - 2t^4 - t^5, \qquad \boldsymbol{\beta}(t) = 1 + 3t + 4t^2 + 3t^3 + t^4.$$

It follows that

$$\mathbf{z}(t) = 1 + 2t + t^2 - t^3 - 2t^4 + 2t^6 + t^7 - t^8 - 2t^9 - t^{10} \in \mathbb{F}^{11}_-[t],$$
$$\mathbf{y}(t) = -1 - 2t + 3t^3 + 3t^4 - t^5 - t^6 - t^7 + 3t^8 + 3t^9 - 2t^{11} - t^{12} \in \mathbb{F}^{13}_+[t],$$

since $\mathbf{y}(t) = (t^2 - 1)\mathbf{z}(t)$. Thus $\mathbf{c} = (0,3,3,-1-4,-1,3,3,0) \in \mathbb{F}^9_+$ by chopping off four entries of $\mathbf{y}$. The corresponding restricted inverse is

$$B_{++}^{(-1)} = \frac{1}{4}\begin{bmatrix} -4 & 2 & 6 & 2 & -4 \\ 2 & -1 & -2 & -1 & 2 \\ 6 & -2 & -8 & -2 & 6 \\ 2 & -1 & -2 & -1 & 2 \\ -4 & 2 & 6 & 2 & -4 \end{bmatrix},$$

which is obtained from $\frac{1}{4}T_n(\mathbf{c})(I + J_n)$.

References

[1] T. Ehrhardt and K. Rost, On the kernel structure of generalized resultant matrices, *Indagationes Mathematicae* **23** (2012), 1053–1069.

[2] T. Ehrhardt and K. Rost, Resultant matrices and inversion of Bezoutians, *Linear Algebra Appl.* **439** (2013), 621–639.

[3] T. Ehrhardt and K. Rost, Inversion of centrosymmetric Toeplitz-plus-Hankel Bezoutians, *Electron. Trans. Numer. Anal.* **42** (2014), 106–135.

[4] T. Ehrhardt and K. Rost, Inversion of centroskewsymmetric Toeplitz-plus-Hankel Bezoutians, *Electron. J. Linear Algebra* **30** (2015), 336–359.

[5] T. Ehrhardt and K. Rost, Fast inversion of centrosymmetric Toeplitz-plus-Hankel Bezoutians, *Oper. Theory Adv. Appl.* **259**, 267–300, Birkhäuser, 2017.

[6] I.C. Gohberg and A.A. Semencul, The inversion of finite Toeplitz matrices and their continuous analogues (in Russian), *Mat. Issled.* **7** (1972), no. 2(24), 201–223, 290.

[7] M.C. Gouveia, Group and Moore-Penrose invertibility of Bezoutians, *Linear Algebra Appl.* **197/198** (1994), 495–509.

[8] G. Heinig and U. Jungnickel, Hankel matrices generated by the Markov parameters of rational functions, *Linear Algebra Appl.* **76** (1986), 121–135.

[9] G. Heinig and K. Rost, *Algebraic methods for Toeplitz-like matrices and operators, Operator Theory: Advances and Applications* **13**, Birkhäuser Verlag, Basel, 1984.

[10] G. Heinig and K. Rost, Fast inversion of Toeplitz-plus-Hankel matrices, *Wiss. Z. Tech. Hochsch. Karl-Marx-Stadt* **27** (1985), no. 1, 66–71.

[11] G. Heinig and K. Rost, On the inverses of Toeplitz-plus-Hankel matrices, *Linear Algebra Appl.* **106** (1988), 39–52.

[12] G. Heinig and K. Rost, Hartley transform representations of inverses of real Toeplitz-plus-Hankel matrices, In: *Proceedings of the International Conference on Fourier Analysis and Applications (Kuwait, 1998), Numer. Funct. Anal. Optim.* **21** (2000), 175–189.

[13] G. Heinig and K. Rost, Centrosymmetric and centro-skewsymmetric Toeplitz-plus-Hankel matrices and Bezoutians, *Linear Algebra Appl.* **366** (2003), 257–281.

[14] G. Heinig and K. Rost, Introduction to Bezoutians, *Oper. Theory Adv. Appl.* **199**, 25–118, Birkhäuser, 2010.

[15] F.I. Lander, The Bezoutian and the inversion of Hankel and Toeplitz matrices (in Russian), *Mat. Issled.* **9** (1974), no. 2(32), 69–87, 249–250.

[16] A.B. Nersesyan and A.A. Papoyan, Construction of a matrix inverse to the sum of Toeplitz and Hankel matrices (in Russian), *Izv. Akad. Nauk Armyan. SSR Ser. Mat.* **18(2)** (1983), 150–160.

[17] W.F. Trench, An algorithm for the inversion of finite Toeplitz matrices, *J. Soc. Indust. Appl. Math.* **12** (1964), 515–522.

[18] H.K. Wimmer, On the history of the Bezoutian and the resultant matrix, *Linear Algebra Appl.* **128** (1990), 27–34.

Torsten Ehrhardt
Mathematics Department, University of California
Santa Cruz
CA-95064, U.S.A.
e-mail: tehrhard@ucsc.edu

Karla Rost
Faculty of Mathematics, Technische Universität Chemnitz
Reichenhainer Straße 39
D-09126 Chemnitz, Germany
e-mail: karla.rost@mathematik.tu-chemnitz.de

Generalized backward shift operators on the ring $\mathbb{Z}[[x]]$, Cramer's rule for infinite linear systems, and p-adic integers

Sergey Gefter and Anna Goncharuk

Abstract. Let A be a generalized backward shift operator on $\mathbb{Z}[[x]]$ and $f(x)$ be a formal power series with integer coefficients. A criterion for the existence of a solution of the linear equation $(Ay)(x) + f(x) = y(x)$ in $\mathbb{Z}[[x]]$ is obtained. An explicit formula for its unique solution in $\mathbb{Z}[[x]]$ is found as well. The main results are based on using the p-adic topology on $\mathbb{Z}$ and on using a formal version of Cramer's rule for solving infinite linear systems.

Mathematics Subject Classification (2010). Primary 13J05; Secondary 35C10, 35E05, 44A35.

Keywords. Formal power series, generalized backward shift operator, Cramer's rule, p-adic integers, convolution.

1. Introduction

Let $a = (a_1, a_2, a_3, \ldots)$ be a sequence of positive integers, such that the following conditions hold:

1. infinitely many of a_i are greater than 1;
2. for each prime p, either p does not divide any of the numbers a_i or p divides an infinite number of the numbers a_i.

We define the operator A on the ring of formal power series with integer coefficients by

$$A(y_0 + y_1x + y_2x^2 + \cdots) = a_1y_1 + a_2y_2x + a_3y_3x^2 + \cdots$$

and consider the equation

$$(Ay)(x) + f(x) = y(x), \tag{1.1}$$

where $f(x) = f_0 + f_1x + f_2x^2 + \cdots \in \mathbb{Z}[[x]]$.

This work was completed with the support by the Akhiezer Fond.

© Springer International Publishing AG, part of Springer Nature 2018

A. Böttcher et al. (eds.), *The Diversity and Beauty of Applied Operator Theory*, Operator Theory: Advances and Applications 268, https://doi.org/10.1007/978-3-319-75996-8_13

It will be shown that this equation has at most one solution, however, it may have no solutions (see the examples in Section 2). We also obtain a criterion for the existence of a solution and find an explicit formula for this solution (see Theorem 4.2 and formula (4.1)).

Moreover we study the relation of equation (1.1) with an infinite linear system, which naturally arises from this equation. Infinite linear systems are studied in numerous works (see, for example, [1]–[3]).

We show that if a solution exists, then it can be found using Cramer's rule, where, in order to introduce determinants of infinite matrices, we use convergence in the p-adic and related topologies and on the integers $\mathbb{Z}$ (see Sections 3 and 4).

Notice, that if $a = (1, 2, 3, 4, \ldots)$, then conditions 1 and 2 hold and A is the differentiation operator. Then equation (1.1) becomes $y'(x)+f(x) = y(x)$. In the general case, the operator A can be regarded as an analogue of the Gelfond–Leontjev generalized differentiation operator (see [4], [5]).

Besides, if $a = (b, b, b, b, \ldots)$, $b \neq 1$, then conditions 1 and 2 hold as well and $A = b \cdot S^*$, where S^* is the backward shift operator. This operator is widely used in the theory of functions, the theory of operators, and the algebraic analysis (see, for example, [6]–[8]).

In Section 5 we introduce a special convolution generalizing the construction of the Hurwitz product for Laurent series (see [9, §1, Section 1.5] and [10]). Using this convolution we rewrite the solution of equation (1.1) in a form which is close to the usual one in linear differential equation theory (see Theorem 5.12). We can consider this result as an algebraic analogue of Theorem 4.4 and Corollary 4.5 from [11]. In Section 5 we also consider some concrete examples of equation (1.1).

2. Linear equations with a generalized backward shift operator

Lemma 2.1. *Equation* (1.1) *has at most one solution in* $\mathbb{Z}[[x]]$.

Proof. It suffices to prove that the homogeneous equation $Ay = y$ has only the trivial solution. Let $y(x) = y_0 + y_1 x + y_2 x^2 + \cdots$ be its solution. Then

$$a_1 y_1 + a_2 y_2 x + a_3 y_3 x^2 + \cdots = y_0 + y_1 x + y_2 x^2 + \cdots .$$

We have $y_n = a_{n+1} y_{n+1}$ for all n. Thus, $y_n = \frac{y_0}{a_1 a_2 \cdots a_n}$. Recall that all the coefficients of the formal series $y(x)$ are integers. Therefore, $y_n = 0$, so $y(x) = 0$. □

Remark 2.2. There are infinitely many solutions of (1.1) in $\mathbb{Z}[[x]]$ if $a_n = 1$ for all n greater than some n_0. Actually, solving the homogeneous equation we obtain $y_n = \frac{y_0}{a_1 a_2 \cdots a_n} = \frac{y_0}{a_1 a_2 \cdots a_{n_0}}$ for all $n \geq n_0$. So, we get different solutions for all y_0 divisible by $a_1 a_2 \cdots a_{n_0}$. That is, condition 1 is essential for uniqueness.

The next result is on a simple situation where the solution of the inhomogeneous equation is unique.

Lemma 2.3. *Equation* (1.1) *with* $f(x) = f_0 + f_1 x + \cdots + f_n x^n \in \mathbb{Z}[x]$ *being a polynomial has exactly one solution in* $\mathbb{Z}[x]$.

Proof. It is not difficult to check that the polynomial

$$f(x) + (Af)(x) + (A^2 f)(x) + (A^3 f)(x) + \cdots .$$

is a solution. □

From Lemmas 2.1 and 2.3 we obtain the following statement.

Theorem 2.4. *The set of formal power series* $f(x)$ *with integer coefficients such that equation* (1.1) *has a solution with integer coefficients is uncountable. Moreover, it is a dense submodule of the space of formal power series with integer coefficients equipped with the Krull topology.*

Proof. For the Krull topology, see [12, Ch. 1, §3, Sec. 4]. Actually, we can construct the inhomogeneity if we know a solution: $f(x) = y(x) - (Ay)(x)$, and we have an uncountable set of the power series with integer coefficients which could be a solution. There is a solution for all polynomial inhomogeneities, and as the set of polynomials is dense in $\mathbb{Z}[[x]]$, so the set of formal power series with integer coefficients $f(x)$ such that equation (1.1) has a solution with integer coefficients is dense in the Krull topology. □

Example. Consider (1.1) with $a = (2, 2, \ldots)$ and $f(x) = 1 + x + x^2 + x^3 + \cdots$, that is,

$$2(S^* y)(x) + 1 + x + x^2 + x^3 + \cdots = y(x). \tag{2.1}$$

It is easy to verify that this equation has $y(x) = -1 - x - x^2 - x^3 - \cdots$ as the unique solution from $\mathbb{Z}[[x]]$.

The next examples show that equation (1.1) not necessarily has a solution from $\mathbb{Z}[[x]]$.

Example. Consider (1.1) with $a = (3, 3, \ldots)$ and $f(x) = 1 + x + x^2 + x^3 + \cdots$, that is, the equation

$$3(S^* y)(x) + 1 + x + x^2 + x^3 + \cdots = y(x). \tag{2.2}$$

Let us prove that there are no solutions in $\mathbb{Z}[[x]]$. If $y(x) = y_0 + y_1 x + y_2 x^2 + \cdots$ is a solution, then the following equality holds:

$$3y_1 + 3y_2 x + 3y_3 x^2 + \cdots + 1 + x + x^2 + x^3 + \cdots = y_0 + y_1 x + y_2 x^2 + \cdots .$$

Therefore, there is a sequence $\{y_n\} \subset \mathbb{Z}$ so that the equalities $3y_{n+1} + 1 = y_n$ hold for all n. Notice, that the sequence $\{y_{n+1}\}$ also satisfies these equalities. As the solution is unique the sequences $\{y_n\}$ and $\{y_{n+1}\}$ coincide. Because $y_n = y_{n+1}$ for all n, then $y_n = -\frac{1}{2}$, so there is no integer solution of this equation.

Example. Consider equation (1.1) where we choose $a = (1, 2, 3, 4, \ldots)$ and $f(x) = 1 + x + x^2 + x^3 + \cdots$, that is,

$$y'(x) + 1 + x + x^2 + x^3 + \cdots = y(x). \tag{2.3}$$

Let us prove that there are no solutions in $\mathbb{Z}[[x]]$. If $y(x) = y_0+y_1x+y_2x^2+\cdots$ is a solution, we have

$$y_1+2y_2x+3y_3x^2+4y_4x^3+\cdots+1+x+x^2+x^3+\cdots = y_0+y_1x+y_2x^2+\cdots.$$

Therefore, there is a sequence $\{y_n\}$, $y_n \in \mathbb{Z}$ such that for all n the equalities $ny_n + 1 = y_{n-1}$ hold. Then, $y_{n+1} = \frac{y_n - 1}{n+1}$.

If $y_0 = 0$, then $y_1 = -1$, $y_2 = -1$, $y_3 = -\frac{2}{3}$. But we supposed the coefficients to be integers. Obviously, if $y_n = 0$, then $y_{n+1} = -\frac{1}{n+1}$, which is not an integer if $n \geq 1$. Now suppose that $y_n \neq 0$ for all n. Considering $|y_n|$ if $n > 2$ we get

$$|y_n| \leq \frac{1}{n}(|y_{n-1}| + 1) = \frac{1}{n}|y_{n-1}| + \frac{1}{n} < \frac{1}{n}|y_{n-1}| + \frac{n-1}{n}|y_{n-1}| = |y_{n-1}|.$$

Here we used that $y_n \in \mathbb{Z}$ and $y_n \neq 0$, so $|y_{n-1}| > \frac{1}{n-1}$ if $n > 2$ and $\frac{n-1}{n}|y_{n-1}| > \frac{1}{n}$. So, we get $|y_n| < |y_{n-1}|$ for all n. This is impossible for the integer coefficients.

Remark 2.5. There are infinitely many solutions of (1.1) in the ring $\mathbb{Q}[[x]]$: we can find a unique solution of this equation for each initial condition $y(0) \in \mathbb{Q}$.

3. Cramer's rule

As we saw in Section 2, equation (1.1) has at most one solution in $\mathbb{Z}[[x]]$, but it may have no solution. Let us try to find the solution of (1.1) assuming that it exists.

Substituting the solution $y(x) = y_0 + y_1x + y_2x^2 + \cdots$ into the equation $Ay + f(x) = y$ we get

$$a_1y_1+a_2y_2x+a_3y_3x^2+\cdots+f_0+f_1x+f_2x^2+f_3x^3+\cdots = y_0+y_1x+y_2x^2+\cdots,$$

which leads to a recurrence formula for the coefficients of $y(x)$, namely $a_{n+1}y_{n+1} + f_n = y_n$. It can be rewritten as an infinite system of linear equations:

$$By = f, \text{ where } B = \begin{pmatrix} 1 & -a_1 & 0 & 0 & \cdots \\ 0 & 1 & -a_2 & 0 & \cdots \\ 0 & 0 & 1 & -a_3 & \cdots \\ 0 & 0 & 0 & 1 & \cdots \\ \vdots & \vdots & \vdots & \vdots & \ddots \end{pmatrix}, \quad f = \begin{pmatrix} f_0 \\ f_1 \\ f_2 \\ f_3 \\ \vdots \end{pmatrix}. \tag{3.1}$$

Let us try to solve this system using the "formal" Cramer rule. Let B_i be the matrix formed by replacing the i-th column of B by f. Then

$$B_0 = \begin{pmatrix} f_0 & -a_1 & 0 & 0 & \cdots \\ f_1 & 1 & -a_2 & 0 & \cdots \\ f_2 & 0 & 1 & -a_3 & \cdots \\ f_3 & 0 & 0 & 1 & \cdots \\ \vdots & \vdots & \vdots & \vdots & \ddots \end{pmatrix}, \quad B_1 = \begin{pmatrix} 1 & f_0 & 0 & 0 & \cdots \\ 0 & f_1 & -a_2 & 0 & \cdots \\ 0 & f_2 & 1 & -a_3 & \cdots \\ 0 & f_3 & 0 & 1 & \cdots \\ \vdots & \vdots & \vdots & \vdots & \ddots \end{pmatrix}, \ldots.$$

By the formal Cramer rule, the "solution" of the system (3.1) is the sequence $\{y_n\}$ given by $y_n = \frac{\det B_n}{\det B}$. We are going to give meaning to the expressions $\det B_n$ and $\det B$. At first, let us denote the principal minor of order k of a matrix M as $\det M^{(k)}$. In Section 4 we will show that $\det B_n$ and $\det B$ can be regarded as certain limits: $\det B_n = \lim_{k\to\infty} \det B_n^{(k)}$, $\det B = \lim_{k\to\infty} \det B^{(k)}$.

Let us find explicit expressions for $\det B_n^{(k)}$. Obviously,

$$\det B_n^{(n+1)} = \det \begin{pmatrix} 1 & -a_1 & 0 & \cdots & f_0 \\ 0 & 1 & -a_2 & \cdots & f_1 \\ 0 & 0 & 1 & \cdots & f_2 \\ 0 & 0 & 0 & \cdots & f_3 \\ \vdots & \vdots & \vdots & \ddots & \vdots \\ 0 & 0 & 0 & \cdots & f_n \end{pmatrix} = f_n.$$

If $k > n$, then

$$\det B_n^{(k+1)} = \det \begin{pmatrix} 1 & -a_1 & \cdots & f_0 & 0 & \cdots & 0 \\ 0 & 1 & \cdots & f_1 & 0 & \cdots & 0 \\ \vdots & \vdots & \vdots & \vdots & \vdots & \ddots & \vdots \\ 0 & 0 & \cdots & f_n & 0 & \cdots & 0 \\ \vdots & \vdots & \ddots & \vdots & \vdots & \ddots & \vdots \\ 0 & 0 & \cdots & f_{k-1} & 0 & \cdots & -a_k \\ 0 & 0 & \cdots & f_k & 0 & \cdots & 1 \end{pmatrix}$$
$$= \det B_n^{(k)} + f_k a_{n+1} a_{n+2} \cdots a_k.$$

Hence, $\det B_n^{(k)} = f_n + \sum_{s=n+1}^{k} f_s a_{n+1} \cdots a_s = \sum_{s=n}^{k} f_s a_{n+1} a_{n+2} \cdots a_s$. Thus, we have the formal equalities

$$\det B = 1 \qquad \text{and} \tag{3.2}$$

$$y_n = \frac{\det B_n}{\det B} = \sum_{j=0}^{\infty} f_{n+j} a_{n+1} a_{n+2} \cdots a_{n+j}. \tag{3.3}$$

In the next section these equalities will be justified.

Remark 3.1. Let us solve system (3.1) finding the inverse matrix of B. We can construct it "algebraically" as the adjugate matrix of B. This gives

$$B^{-1} = \begin{pmatrix} 1 & a_1 & a_1a_2 & a_1a_2a_3 & a_1a_2a_3a_4 & \cdots \\ 0 & 1 & a_2 & a_2a_3 & a_2a_3a_4 & \cdots \\ 0 & 0 & 1 & a_3 & a_3a_4 & \cdots \\ 0 & 0 & 0 & 1 & a_4 & \cdots \\ 0 & 0 & 0 & 0 & 1 & \cdots \\ \vdots & \vdots & \ddots & \vdots & \vdots & \ddots \end{pmatrix}.$$

Now the infinite sums (3.3) of integers appear when multiplying this matrix by the vector f.

4. Using p-adic and related topologies

We denote by $\hat{a}$ the sequence obtained from $a = (a_1, a_2, a_3, \ldots)$ by deleting all a_k which are equal to 1. By condition 1 of Section 1, $\hat{a}$ is an infinite sequence of integers (all greater than 1). Let $\mathbb{Z}_{\hat{a}}$ be the ring of $\hat{a}$-adic integers with the standard topology (see [13, Chapter 2, §10]). For example, if p is a prime and $a = (p, p, p, \ldots)$, then $\mathbb{Z}_{\hat{a}}$ is the ring of p-adic integers $\mathbb{Z}_p$ (see [14, §3]). Below we use the following fact: the series $\sum_{n=0}^{\infty} x_n$ with $x_n \in \mathbb{Z}_{\hat{a}}$ converges in $\mathbb{Z}_{\hat{a}}$ if and only if $x_n \to 0$ in $\mathbb{Z}_{\hat{a}}$.

We first consider the question about the solvability of $Ay + f(x) = y$ in the ring $\mathbb{Z}_{\hat{a}}[[x]]$.

Theorem 4.1. *Let $f \in \mathbb{Z}[[x]]$. Equation* (1.1) *has the following unique solution in $\mathbb{Z}_{\hat{a}}[[x]]$:*

$$y(x) = f(x) + (Af)(x) + (A^2 f)(x) + (A^3 f)(x) + \cdots, \tag{4.1}$$

where the sum on the right hand side converges in $\mathbb{Z}_{\hat{a}}[[x]]$ in the topology of coefficientwise convergence.

Proof. First we prove the uniqueness. Solving the homogeneous equation, we obtain the recurrence formula $y_n = a_{n+1} y_{n+1}$. Hence $y_0 = a_1 \cdots a_n y_n$. Notice that the sequence $a_1 \cdots a_n$ tends to zero in $\mathbb{Z}_{\hat{a}}$. So $y_0 = 0$. Similarly, we obtain that $y_n = 0$ for $n \geq 1$. Thus, $y = 0$.

Let $f(x) = f_0 + f_1 x + f_2 x^2 + \cdots \in \mathbb{Z}[[x]]$. Consider the formal sum

$$f(x) + (Af)(x) + (A^2 f)(x) + (A^3 f)(x) + \cdots.$$

Notice that every formal power series $g(x)$ may be expressed in the following way:

$$g(x) = g(0) + \frac{(Ag)(0)}{a_1} x + \frac{(A^2 g)(0)}{a_1 a_2} x^2 + \frac{(A^3 g)(0)}{a_1 a_2 a_3} x^3 + \cdots.$$

It is a certain analogue of the Taylor series expansion. We expand each term of the series in the similar way and regroup terms, obtaining

$$\begin{aligned}
&f(x) + (Af)(x) + (A^2 f)(x) + (A^3 f)(x) + \cdots \\
&\quad = \left(f(0) + \frac{Af(0)}{a_1} x + \frac{A^2 f(0)}{a_1 a_2} x^2 + \cdots \right) \\
&\quad\quad + \left(Af(0) + \frac{A^2 f(0)}{a_1} x + \frac{A^3 f(0)}{a_1 a_2} x^2 + \cdots \right) \\
&\quad\quad + \left(A^2 f(0) + \frac{A^3 f(0)}{a_1} x + \frac{A^4 f(0)}{a_1 a_2} x^2 + \cdots \right) + \cdots \\
&\quad = (f(0) + Af(0) + A^2 f(0) + \cdots) \\
&\quad\quad + \left(\frac{Af(0)}{a_1} + \frac{A^2 f(0)}{a_1} + \frac{A^3 f(0)}{a_1} + \cdots \right) x \\
&\quad\quad + \left(\frac{A^2 f(0)}{a_1 a_2} + \frac{A^3 f(0)}{a_1 a_2} + \frac{A^4 f(0)}{a_1 a_2} + \cdots \right) x^2 + \cdots.
\end{aligned}$$

We first show that each of the series that occurs as the coefficient of a power of x converges in $\mathbb{Z}_{\hat{a}}$. Indeed, the n-th term in the coefficient of x^m has the form

$$\frac{(A^{m+n}f)(0)}{a_1 \cdots a_m} = a_{m+1} \cdots a_{m+n} f_{m+n}.$$

Therefore it tends to zero in $\mathbb{Z}_{\hat{a}}$ as $n \to \infty$ (actually, for all j there exist n such that $a_{m+1} \cdots a_{m+n}$ is divisible by a_j, because each prime divisor of a_j divides an infinite number of a_i due to condition 2 from Section 1). Since each term tends to zero, the series converges. Thus the coefficient of x^m belongs to $\mathbb{Z}_{\hat{a}}$ for every m. So, the series in the right-hand side of (4.1) is well defined as an element of $\mathbb{Z}_{\hat{a}}[[x]]$. It is easy to check that this series is a solution of (1.1). □

The next theorem is a criterion for equation (1.1) to have a solution in the ring $\mathbb{Z}[[x]]$.

Theorem 4.2. *Let $f \in \mathbb{Z}[[x]]$, $f(x) = f_0 + f_1 x + f_2 x^2 + \cdots$. The following statements are equivalent:*

(1) $f_0 + a_1 f_1 + a_1 a_2 f_2 + a_1 a_2 a_3 f_3 + a_1 a_2 a_3 a_4 f_4 + \cdots \in \mathbb{Z}$ *in* $\mathbb{Z}_{\hat{a}}$.
(2) *The equation $Ay + f(x) = y$ has a solution in $\mathbb{Z}[[x]]$.*

Proof. By Theorem 4.1, the unique solution of our equation is of the form (4.1). Let us prove that, if the statement (1) holds, then it belongs to $\mathbb{Z}[[x]]$. In other words, let us prove that the coefficient of x^n is an integer for every n. The proof is by induction. For y_0, note that

$$y_0 = f_0 + a_1 f_1 + a_1 a_2 f_2 + a_1 a_2 a_3 f_3 + \cdots .$$

This is an integer due to our assumption. Now suppose that the k-th coefficient

$$y_k = f_k + a_{k+1} f_{k+1} + a_{k+1} a_{k+2} f_{k+2} + \cdots$$

is an integer. We need to prove that the $k+1$-st coefficient

$$y_{k+1} = f_{k+1} + a_{k+2} f_{k+2} + a_{k+2} a_{k+3} f_{k+3} + \cdots$$

is an integer as well. We have $y_{k+1} = \frac{y_k - f_k}{a_{k+1}}$. Since $f_k \in \mathbb{Z}$ we conclude that $a_{k+1} f_{k+1} + a_{k+1} a_{k+2} f_{k+2} + a_{k+1} a_{k+2} a_{k+3} f_{k+3} + \cdots$ is an integer. We prove that the sum of this series is divisible by a_{k+1}. To this end, we make use of the following lemma.

Lemma 4.3. *Let $\{r_n\}$ be a sequence from $\mathbb{Z}$. If $a_k r_n \to l$ in $\mathbb{Z}_{\hat{a}}$ as $n \to \infty$, then $a_k | l$.*

Proof. Since $a_k r_n \to l$, it follows that $\sigma(a_k r_n, l)_a \to 0$, where σ is the standard metric on $\mathbb{Z}_{\hat{a}}$ (see [13, Chapter 2, §10]). Then there exists an n_0 such that $a_k r_n - l$ is divisible by a_k for all $n > n_0$. Thus, l is divisible by a_k, and $r_n \to \frac{l}{a_k}$. □

Therefore, if each term of a series, considered over $\mathbb{Z}_{\hat{a}}$, is divisible by a_k, then the sum of the series is also divisible by a_k. Moreover, if $\sum_{n=1}^{\infty} a_k r_n = l$, then $\sum_{n=1}^{\infty} r_n = \frac{l}{a_k}$.

Let us now return to the proof of the theorem. We have that all terms of the series $a_{k+1}f_{k+1} + a_{k+1}a_{k+2}f_{k+2} + a_{k+1}a_{k+2}a_{k+3}f_{k+3} + \cdots$, whose sum is equal to $y_k - f_k$, are divisible by a_{k+1}. Using the lemma, we get that the sum of the series is also divisible by a_{k+1}. Thus, the sum of the series $f_{k+1} + a_{k+2}f_{k+2} + a_{k+2}a_{k+3}f_{k+3} + \cdots = \frac{y_k - f_k}{a_{k+1}}$ is an integer. □

Now we can give meaning to the formal designs from Section 3.

Corollary 4.4. *If equation (1.1) has a solution in $\mathbb{Z}[[x]]$, then it is unique and its coefficients may be found, using Cramer's rule, as $y_k = \frac{\det B_k}{\det B}$, where the determinants of B, B_n are the limits of the principal minors of these matrices in $\mathbb{Z}_{\hat{a}}$, i.e.,*

$$\det B = \lim_{k\to\infty} B^{(k)}, \ \det B_n = \lim_{k\to\infty} B_n^{(k)}.$$

5. Some particular cases

5.1. Differential equations

If $a = (b, 2b, 3b, 4b, 5b, \ldots)$, then (1.1) is the differential equation

$$by'(x) + f(x) = y(x). \tag{5.1}$$

By Theorem 4.2 the solution of this equation can be written as

$$y(x) = f(x) + bf'(x) + b^2 f''(x) + b^3 f'''(x) + \cdots \tag{5.2}$$

in $\mathbb{Z}_{\hat{a}}[[x]]$, and the equation has the solution in $\mathbb{Z}[[x]]$ if and only if

$$f_0 + bf_1 + 2!b^2 f_2 + 3!b^3 f_3 + 4!b^4 f_4 + \cdots \in \mathbb{Z}.$$

Besides, in this case we can formulate another condition for the existence of a solution of this equation. To this end let us prove the following lemma.

Lemma 5.1. *Suppose $c \in \mathbb{Z}$ and $\{r_n\}$ is a sequence from $\mathbb{Z}$. Then the following statements are equivalent:*

(1) *$r_n \to c$ in $\mathbb{Z}_{\hat{a}}$, where $a = (b, 2b, 3b, 4b, \ldots)$.*
(2) *$r_n \to c$ in $\mathbb{Z}_p$ for all prime p.*

Proof. The statement (1) means that $\sigma(r_n, c)_a \to 0$. Then for each m there exists an n_0 such that $r_n - c$ is divisible by $a_0 a_1 a_2 \cdots a_m = m! b^{m+1}$ if $n > n_0$. So for every power p^k of a prime p there exists an n_0 such that $r_n - c$ is divisible by p^k if $n > n_0$. Therefore, $||r_n - c||_p \to 0$, where $||\cdot||_p$ is the standard norm in $\mathbb{Z}_p$. The reverse can be proved similarly, because $m!b^{m+1}$ can be written as a product of primes. □

Using this lemma and Theorem 4.2, we get the following theorem.

Theorem 5.2. *The following statements are equivalent:*

(1) *There exist $c \in \mathbb{Z}$ such that $f_0 + 1!bf_1 + 2!b^2 f_2 + \cdots = c$ in $\mathbb{Z}_p$ for all primes p.*

(2) *The equation $by' + f(x) = y$ has a solution from $\mathbb{Z}[[x]]$.*

Example. Let us return to the example from Section 2 and consider the equation (2.3).

Using the previous theorem we obtain, that this equation has a solution if and only if $1 + 1! + 2! + 3! + 4! + \cdots$ is integer in $\mathbb{Z}_p$ for all p. As the sum of the factorials is not integer in $\mathbb{Z}_p$, we conclude once more that this equation has no solution in $\mathbb{Z}[[x]]$.

Example. Consider the equation $y'(x) + x + 2x^2 + 3x^3 + \cdots = y(x)$.

One easily proves $1 \cdot 1! + 2 \cdot 2! + 3 \cdot 3! + 4 \cdot 4! + \cdots + n \cdot n! = (n+1)! - 1$. In $\mathbb{Z}_p$ the sequence $(n+1)!$ tends to zero, so $(n+1)! - 1$ tends to -1. Therefore, $1 \cdot 1! + 2 \cdot 2! + 3 \cdot 3! + 4 \cdot 4! + \cdots = -1$ is in $\mathbb{Z}_p$ for all p. Consequently, this equation has a solution. It is not difficult to check that the following series is its solution: $y(x) = -1 - x - x^2 - x^3 - \cdots$.

Now we are going to rewrite the solution (5.2) of equation (5.1) in the form that is usual in the theory of linear differential equations, namely, as a convolution of a fundamental solution with the inhomogeneity. So, we need to introduce the notion of a special convolution of an element from $\frac{1}{x}\mathbb{Z}[[\frac{1}{x}]]$ and an element from $\mathbb{Z}[[x]]$.

First let us consider the convolution of a formal Laurent series with only negative powers on $\mathbb{Z}$ and a polynomial on $\mathbb{Z}$ (see [15, Section 3]).

Definition 5.3. Suppose $Q(x) = \frac{q_1}{x} + \frac{q_2}{x^2} + \frac{q_3}{x^3} + \cdots \in \frac{1}{x}\mathbb{Z}[[\frac{1}{x}]]$ and $f(x) \in \mathbb{Z}[x]$. By definition, put

$$(Q * f)(x) = \operatorname{Res}_y(Q(y)f(x-y)), \tag{5.3}$$

where Res_y is the formal residue (see [16, Section 2.1]).

Here we consider $f(x-y)$ as an element of $\mathbb{Z}[x][y]$, i.e., as a polynomial in y the coefficients of which are polynomials in x,

$$f(x-y) = f(x) - \frac{f'(x)}{1!}y + \frac{f''(x)}{2!}y^2 - \frac{f'''(x)}{3!}y^3 + \frac{f^{(4)}(x)}{4!}y^4 - \cdots.$$

This is well defined; all the coefficients of $f^{(k)}(x)$ are divisible by $k!$.

The product $Q(y)f(x-y)$ is an element from $\mathbb{Z}[x][[y, \frac{1}{y}]]$, i.e., a two-sided formal Laurent series in the variable y with coefficients in $\mathbb{Z}[x]$.

From the definition of this convolution we obtain an explicit formula for the convolution:

$$(Q*f)(x) = q_1 f(x) - \frac{q_2}{1!}f'(x) + \frac{q_3}{2!}f''(x) - \frac{q_4}{3!}f'''(x) + \frac{q_5}{4!}f^{(4)}(x) - \cdots. \tag{5.4}$$

If we try to define the convolution of an element from $\frac{1}{x}\mathbb{Z}[[\frac{1}{x}]]$ and an element from $\mathbb{Z}[[x]]$ similarly using the formula (5.3), we will obtain sums of an infinite number of integers. Actually, if $f(x) = \sum_{k=0}^{\infty} f_k x^k \in \mathbb{Z}[[x]]$, then $f(x-y)$ is a power series in y with coefficients which are power series in x.

Then the formal product $Q(y)f(x-y)$ has infinite sums of integers as "coefficients", and these do not necessarily exist in a p-adic sense.

The "coefficient" of y^n in (5.3) is for all $n>0$ given by

$$\sum_{k=1}^{\infty} \frac{(-1)^{n+k} q_k f^{(n+k)}(x)}{(n+k)!}. \tag{5.5}$$

Then the "coefficient" of x^m in the "series" (5.5) is, again for all $n>0$, equal to

$$\sum_{k=1}^{\infty} (-1)^{n+k} \frac{(n+m+k)!}{(n+k)!m!} f_{n+m+k} q_k. \tag{5.6}$$

For all $n>0$, the "coefficient" in (5.3) of $\frac{1}{y^n}$ is

$$\sum_{k=0}^{\infty} (-1)^k \frac{q_{n+k} f^{(k)}(x)}{k!}. \tag{5.7}$$

Then, again for all $n>0$, the "coefficient" of x^m in the "series" (5.7) equals

$$\sum_{k=0}^{\infty} (-1)^k \frac{(m+k)!}{m!k!} f_{m+k} q_{n+k}. \tag{5.8}$$

Lemma 5.4. *Let* $Q(x) = \frac{q_1}{x} + \frac{q_2}{x^2} + \cdots + \frac{q_n}{x^n} + \cdots$ *be from* $\frac{1}{x}\mathbb{Z}[[\frac{1}{x}]]$ *and let* $f(x) = \sum_{i=0}^{\infty} f_k x^k$ *be from* $\mathbb{Z}[[x]]$, *and suppose* q_i *tends to zero in* $\mathbb{Z}_p$ *as* $i \to \infty$. *Then the sequences* $\frac{(n+m+i)!}{(n+i)!m!} f_{n+m+i} q_i$ *and* $\frac{(m+i)!}{m!i!} f_{m+i} q_{n+i}$ *tend to zero in* $\mathbb{Z}_p$ *as* $i \to \infty$, *so the series* (5.6) *and* (5.8) *converge in* $\mathbb{Z}_p$. *Therefore,* $Q(y)f(x-y) \in \mathbb{Z}_p[[x]][[y, \frac{1}{y}]]$.

Using Lemma 5.4 we can define a convolution of the two, $Q \in \frac{1}{x}\mathbb{Z}[[\frac{1}{x}]]$ and $f \in \mathbb{Z}[[x]]$.

Definition 5.5. Let $q_i \to 0$ in $\mathbb{Z}_p$, $Q = \sum_{k=1}^{\infty} \frac{q_k}{x^k} \in \frac{1}{x}\mathbb{Z}[[\frac{1}{x}]]$, and $f \in \mathbb{Z}[[x]]$. By definition, put

$$(Q * f)(x) = \operatorname{Res}_y(Q(y)f(x-y)), \tag{5.9}$$

where Res_y is the formal residue.

As a result of the convolution we get an element in $\mathbb{Z}_p[[x]]$ and it can be represented by the formula (5.4).

From Theorem 4.2 we now obtain the following.

Theorem 5.6. *Suppose* (5.1) *has a solution in* $\mathbb{Z}[[x]]$. *Then this solution has the form of the convolution* $\mathcal{E}_b * f$, *where* $\mathcal{E}_b(x) = \frac{1}{x} - \frac{1!b}{x^2} + \frac{2!b^2}{x^3} - \frac{3!b^3}{x^4} + \cdots$ *is the Euler series.*

That means we can regard the Euler series as a fundamental solution of equation (5.1) (see [17, Section 4.4]).

5.2. The backward shift operator

Let $b \in \mathbb{N}$ and $a = (b, b, b, b, \ldots)$. Then $A = bS^*$ where S^* is the backward shift operator, and equation (1.1) can be written as

$$b(S^*y)(x) + f(x) = y(x). \tag{5.10}$$

Notice that in the case $b = 1$ equation (5.10) has an infinite number of solutions in $\mathbb{Z}[[x]]$, while in all other cases it has at most one solution. In what follows we assume that $b > 1$, since otherwise condition 1 of Section 1 is violated. Then $\hat{a}$ matches with a, so $\mathbb{Z}_{\hat{a}} = \mathbb{Z}_a$.

By Theorem 4.2 the equation has a solution from $\mathbb{Z}[[x]]$ if and only if $f_0 + bf_1 + b^2 f_2 + b^3 f_3 + \cdots \in \mathbb{Z}$ in $\mathbb{Z}_a$, and the solution can then be written as

$$y(x) = f(x) + bS^*(f)(x) + b^2(S^*)^2(f)(x) + b^3(S^*)^3(f)(x) + \cdots .$$

In this case we can reformulate Theorem 4.2 using the next lemma.

Lemma 5.7. *Suppose $c \in \mathbb{Z}$ and $\{r_n\}$ is a sequence from $\mathbb{Z}$. Then the following statements are equivalent:*

(1) *$r_n \to c$ in $\mathbb{Z}_a$, where $a = (b, b, b, \ldots)$.*
(2) *$r_n \to c$ in $\mathbb{Z}_p$ for all prime divisors of b.*

Proof. The statement (1) means that $\sigma(r_n, c)_a \to 0$. Then for each m there exists an n_0 such that $r_n - c$ is divisible by $a_0 a_1 a_2 \cdots a_m = b^m$ if $n > n_0$. If p is a divisor of b, then $r_n - c$ is divisible by p^m. So for every power p^k of a prime $p|b$ there exists an n_0 such that $r_n - c$ is divisible by p^k if $n > n_0$. Therefore, $||r_n - c||_p \to 0$. The reverse can be proved similarly, because b^m can be written as the product of the prime divisors of b. □

From this we get the following theorem.

Theorem 5.8. *The following statements are equivalent:*

(1) *There exists a $c \in \mathbb{Z}$ such that $f_0 + bf_1 + b^2 f_2 + b^3 f_3 + \cdots = c$ in $\mathbb{Z}_p$ for all prime divisors of b.*
(2) *The equation $b(S^*y) + f(x) = y$ has a solution in $\mathbb{Z}[[x]]$.*

Example. Let us consider the equation (2.1). In this case

$$f_0 + bf_1 + b^2 f_2 + b^3 f_3 + \cdots = 1 + 2 + 2^2 + 2^3 + \cdots .$$

Note that $1 + 2 + 2^2 + 2^3 + \cdots + 2^k = 2^{k+1} - 1 \to -1$ when $k \to \infty$, so this sum is equal to -1.

Example. Now consider (2.2). In this case $f_0 + bf_1 + b^2 f_2 + b^3 f_3 + \cdots$ becomes $1+3+3^2+3^3+\cdots \notin \mathbb{Z}$ in $\mathbb{Z}_3$. Note that $1+3+3^2+3^3+\cdots+3^k = \frac{3^{k+1}-1}{2} \to -\frac{1}{2}$ as $k \to \infty$, so this sum is equal to $-\frac{1}{2}$ in $\mathbb{Z}_3$. So we once more see that this equation has no solution in $\mathbb{Z}[[x]]$.

Now we are going to rewrite the solution of equation (5.10) in the form of a convolution of a fundamental solution with the inhomogeneity. We need to introduce a convolution of an element from $\frac{1}{x}\mathbb{Z}[[\frac{1}{x}]]$ and an element from $\mathbb{Z}[[x]]$ for equation (5.10) similarly to formula (5.4).

Definition 5.9. Let p be a prime. Suppose $Q(x) = \frac{q_1}{x} + \frac{q_2}{x^2} + \frac{q_3}{x^3} + \cdots$, where q_i tends to zero in $\mathbb{Z}_p$ and $f(x) = f_0 + f_1 x + f_2 x^2 + \cdots$. By definition, put

$$(Q * f)(x) = q_1 f(x) - q_2 S^* f(x) + q_3 (S^*)^2 f(x) - q_4 (S^*)^3 f(x) + \cdots. \quad (5.11)$$

The coefficient of x^n in (5.11) is $q_1 f_n - q_2 f_{n+1} + q_3 f_{n+2} - q_4 f_{n+3} + \cdots$. Obviously $q_{i+1} f_{n+i}$ tends to zero as $i \to \infty$ in $\mathbb{Z}_p$. Then the series (5.11) converges, so the convolution is well defined and its result is included in $\mathbb{Z}_p$.

Similarly to Theorem 5.6, we obtain the following statement.

Theorem 5.10. *Suppose there exists a solution of equation* (5.10) *that belongs to* $\mathbb{Z}[[x]]$. *Then this solution has the form of the convolution* $\Delta_b * f$, *where* $\Delta_b(x) = \frac{1}{x} - \frac{b}{x^2} + \frac{b^2}{x^3} - \frac{b^3}{x^4} + \cdots$.

5.3. A convolution over $\mathbb{Z}_a$

To define the convolution of a Laurent series $Q(x) = \frac{q_1}{x} + \frac{q_2}{x^2} + \frac{q_3}{x_3} \cdots \in \frac{1}{x}\mathbb{Z}[[\frac{1}{x}]]$ with a power series $f(x) = f_0 + f_1 x + f_2 x^2 + \cdots \in \mathbb{Z}[[x]]$ over $\mathbb{Z}_a$, we need conditions that are stronger than those in the previous cases: suppose q_i is divisible by $a_1 a_2 \cdots a_{i-1}$.

Definition 5.11. Similarly to the formula (5.4), put, by definition,

$$(Q * f)_A = q_1 f - \frac{q_2 A(f)}{a_1} + \frac{q_3 A^2(f)}{a_1 a_2} - \frac{q_4 A^3(f)}{a_1 a_2 a_3} + \cdots. \quad (5.12)$$

The coefficient of x^n in (5.12) is

$$q_1 f_n - \frac{q_2 a_{n+1} f_{n+1}}{a_1} + \frac{q_3 a_{n+1} a_{n+2} f_{n+2}}{a_1 a_2} - \frac{q_4 a_{n+1} a_{n+2} a_{n+3} f_{n+3}}{a_1 a_2 a_3} + \cdots.$$

Because $\frac{q_i}{a_1 \cdots a_{i-1}}$ is an integer and $a_{n+1} a_{n+2} \cdots a_{n+i-1}$ tends to zero in $\mathbb{Z}_{\hat{a}}$ as $i \to \infty$, this series converges in $\mathbb{Z}_{\hat{a}}$. Therefore the convolution belongs to $\mathbb{Z}_{\hat{a}}[[x]]$.

From Theorem 4.2 we obtain the following result.

Theorem 5.12. *Suppose there exists a solution of equation* (1.1) *belonging to* $\mathbb{Z}[[x]]$. *Then this solution has the form of the convolution* $\mathcal{E} * f$, *where we have* $\mathcal{E}(x) = \frac{1}{x} - \frac{a_1}{x^2} + \frac{a_1 a_2}{x^3} - \frac{a_1 a_2 a_3}{x^4} + \cdots$.

Then we can regard the series $\mathcal{E}(x) = \frac{1}{x} - \frac{a_1}{x^2} + \frac{a_1 a_2}{x^3} - \frac{a_1 a_2 a_3}{x^4} + \cdots$ as a fundamental solution of the equation $(Ay)(x) + f(x) = y(x)$.

References

[1] J. Combes, Sur la résolution de certains systémes infinis d'équations linéaires, *Annales de la Faculté des sciences de Toulouse: Mathématiques* **28** (1964), no. 1, 149–159.

[2] R.G. Cooke, *Infinite Matrices and Sequence Spaces*, Dover, 1955.

[3] P.N. Shivakumar and K.C. Shivakumar, A rewiew of infinite matrices and their applications, *Linear Algebra and its Applications* **430** (2009), 976–998.

[4] A.O. Gel'fond and A.F. Leont'ev, On a generalization of Fourier series, *Mat. Sb. (N.S.)* **29(71)** (1951), no. 3, 477–500.

[5] Ju.F. Korobeĭnik, Compound operator equations in generalized derivatives and their applications to Appell sequences, *Mathematics of the USSR-Sbornik* **31** (1977), no. 4, 425–443.

[6] R.G. Douglas, H.S. Shapiro, and A.L. Shields, On cyclic vectors of the backward shift, *Bull. Amer. Math. Soc.* **73** (1967), no. 1, 156–159.

[7] N.K. Nikol'skii, *Treatise on the Shift Operator. Spectral Function Theory,* Spinger-Verlag, 1986.

[8] D. Przeworska-Rolewicz, *Logarithms and Antilogarithms. An Algebraic Analysis Approach*, Kluwer Academic Publishers, 1998.

[9] L. Bieberbach, *Analytische Fortsetzung,* Springer, 1955.

[10] A. Hurwitz, Sur un théoreme de M. Hadamard, *C.R. Acad. Sci.* **128** (1899), 350–353.

[11] S. Gefter and T. Stulova, On some vector differential operators of infinite order, *Operator Theory: Advances and Applications* **236**, 193–203, Birkhäuser, 2014.

[12] H. Grauert and R. Remmert, *Analytische Stellenalgebren*, Springer-Verlag, 1971.

[13] E. Hewitt and K.A. Ross, *Abstract Harmonic Analysis: Volume I, Structure of Topological Groups Integration Theory Group Representations*, Springer-Verlag, 1963.

[14] Z.I. Borevich and I.R. Shafarevich, *Number Theory*, Academic Press Inc.,1966.

[15] S. Gefter and A. Goncharuk, Fundamental solution of an implicit linear inhomogeneous first order differential equation over an arbitrary ring, *Journal of Mathematical Sciences* **219** (2016), no. 6, 922–935.

[16] V.G. Kac, *Vertex Algebras for Beginners*, 2nd ed., University Lecture Series **10**, American Mathematical Society, 1998.

[17] L. Hörmander, *The Analysis of Linear Partial Differential Operators I: Distribution Theory and Fourier Analysis*, Springer-Verlag, 1983.

Sergey Gefter and Anna Goncharuk
V.N. Karazin Kharkiv National University
School of Mathematics and Computer Sciences
4 sq. Svoboda
61022 Kharkiv
Ukraine
e-mail: `gefter@karazin.ua`
`angoncharuk@ukr.net`

Feynman path integral regularization using Fourier Integral Operator ζ-functions

Tobias Hartung

Abstract. We will have a closer look at a regularized path integral definition based on Fourier Integral Operator ζ-functions and the generalized Kontsevich–Vishik trace, as well as physical examples.

Using Feynman's path integral formulation of quantum mechanics, it is possible to formally write partition functions and expectations of observables in terms of operator traces. More precisely, Let U be the wave propagator (a Fourier Integral Operator of order 0) and Ω an observable (a pseudo-differential operator), then the expectation $\langle\Omega\rangle$ can formally be expressed as $\langle\Omega\rangle = \frac{\operatorname{tr}(U\Omega)}{\operatorname{tr} U}$. Unfortunately, the operators U and $U\Omega$ are not of trace-class in general. Hence, "regularizing the path integral" can be understood as "defining these traces." In particular, the traces should extend the classical trace on trace-class operators.

We therefore consider the generalized Kontsevich–Vishik trace (i.e., Fourier Integral Operator ζ-functions) since its restriction to pseudo-differential operators (obtained through Wick rotations if they are possible) is the unique extension of the classical trace. Applying the construction of the generalized Kontsevich–Vishik trace yields a new definition of the Feynman path integral whose predictions coincide with a number of well-known physical examples.

Mathematics Subject Classification (2010). Primary 58J40; Secondary 81Q30, 46F10.

Keywords. Feynman path integral, operator ζ-functions, Fourier Integral Operators.

Introduction

Quantum mechanics and Quantum Field Theories (QFTs) have had quite a unique history. Not many theories can claim to have been developed independently from two very different points of view; yet quantum mechanics has Schrödinger's differential equation and Heisenberg's matrix algebra formalism. As these two mathematical descriptions seem to be very distinct,

© Springer International Publishing AG, part of Springer Nature 2018

A. Böttcher et al. (eds.), *The Diversity and Beauty of Applied Operator Theory*, Operator Theory: Advances and Applications 268, https://doi.org/10.1007/978-3-319-75996-8_14

Dirac's transformation theory indeed proved their equivalence. Dirac [5] also introduced the first QFT that he called "Quantum Electrodynamics" (QED), however emerging infinities rendered progress of developing the theoretical frameworks of QFTs rather slow.

Based on some of Dirac's observations about the classical action in quantum mechanics, Feynman [7] added a third important mathematical formulation of quantum mechanics—the Feynman path integral—which could handle these infinities through means of a process called renormalization. In memory of the two independent formulations of quantum mechanics, Schwinger independently introduced an operator based formalism of renormalization, which was also independently developed by Tomonaga. Finally, Dyson showed that these formalisms of QFT are, in fact, equivalent and could be described in a field-theoretical framework.

In this paper, we will consider Feynman's formulation of a path integral. In particular, we want to address the problem of the path integral being a notoriously elusive object. In fact, only for quantum mechanics an analytically well-defined path integral construction is known. In most other cases, the path integral can merely be evaluated "formally", e.g., by means of a formal power series in the physical variables [18]. These developments gave rise to perturbation theoretical approaches to QFT. On the other hand, the quantum mechanical path integral formalism (as described in Section 1) can be defined by means of an inductive limit of the discretized system [35]. Wilson [37] further developed this idea for QFTs since the path integral of a quantum mechanical system in discretized space-time is always well-defined. Wilson's approach (Lattice QFT; LQFT) is, thus, non-perturbative and can describe physics beyond perturbation theory of QFTs. In order to study path integrals in LQFTs computationally, an additional transformation (Wick rotation) from Minkowski space-time to Euclidean space-time is introduced [4, 9, 21, 22, 23, 26].

Furthermore, this Wick rotation allows us to study non-discretized path integrals in Euclidean space-time within the framework of classical pseudo-differential operators and their traces and determinants [29]. Using operator ζ-functions (Section 2) these traces and determinants can be defined in terms of the Kontsevich–Vishik trace [19, 20]. Long before the Kontsevich–Vishik trace was developed, however, Hawking [16] studied the path integral with a curved space-time background applying spectral ζ-functions on the quadratic term of the power series expansion of the physical action. Gibbons, Hawking, and Perry [10] then studied convergence properties of the one-loop approximation of the path integral in this ζ-regularized setting.

Recent developments [14, 15] (Section 3) in the field of Fourier Integral Operator ζ-functions enable us to consider that path integral (discretized and non-discretized) in Minkowski space-time. This option is particularly interesting for LQFT as some observables, e.g., parton distribution functions of a nucleon, require light-cone dynamics [36] and, hence, are not directly accessible on a Euclidean lattice.

Using the path integral formulation obtained by the operator ζ-function setting (Section 4), we will consider a number of physical examples (Sections 6–10), as well as its application to LQFT (Section 11).

Our results are originally reported in [13]. The main focus of this paper is to have a more in-depth look at the operator theory underlying the approach to regularize Feynman's path integral. From this regularization, we obtain a new definition of the path integral. This definition will be given in Section 5, but we anticipate it already here.

Definition 5.1 Let H be the Hamiltonian, $(\mathfrak{G}(T,z))_{T\in\mathbb{R}_{>0},z\in\mathbb{C}}$ a gauged family of operators with

$$\forall T \in \mathbb{R}_{>0}: \ \mathfrak{G}(T,0) = \exp\left(\frac{-i}{\hbar}\int_0^T H(s)ds\right),$$

and Ω an observable. Then we define the *ζ-regularized expectation value* $\langle\Omega\rangle_{\mathfrak{G}}$ of Ω with respect to $\mathfrak{G}$ as

$$\begin{aligned}\langle\Omega\rangle_{\mathfrak{G}} &:= \lim_{T\to\infty}\lim_{z\to 0}\frac{\zeta\left(\mathfrak{G}(T,\cdot)\Omega\right)(z)}{\zeta\left(\mathfrak{G}(T,\cdot)\right)(z)}\\ &= \lim_{T\to\infty}\lim_{z\to 0}\frac{\left(\mathbb{C}_{\Re(\cdot)\ll 0}\ni s\mapsto \operatorname{tr}\left(\mathfrak{G}(T,s)\Omega\right)\in\mathbb{C}\right)\Big|_{\text{mer.}}(z)}{\left(\mathbb{C}_{\Re(\cdot)\ll 0}\ni s\mapsto \operatorname{tr}\mathfrak{G}(T,s)\in\mathbb{C}\right)\Big|_{\text{mer.}}(z)},\end{aligned}$$

where $f|_{\text{mer.}}$ denotes the meromorphic extension of a function f with maximal open and connected domain (provided f has open and connected domain).

Thus, we want to shine a light onto the mathematical background of this definition and discuss why this is a definition to consider. Finally, we quickly consider examples (Sections 6–10) and application to LQFT (Section 11)—which can both be found in more detail in [13]—to show that the results obtained using this new definition of the path integral coincide with physics.

1. The Feynman path integral

Feynman's path integral is a reformulation of quantum mechanics from a "particle point of view." To illustrate this, let us consider a particle moving through a double slit:

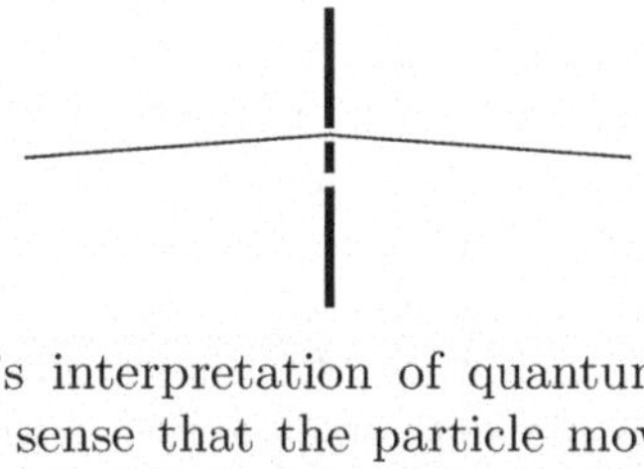

According to Feynman's interpretation of quantum mechanics, we should interpret physics in the sense that the particle moves not through one but both slits at the same time. To do this, we attribute a so-called quantum mechanical amplitude Φ_1 to the event of the particle moving through the top slit, and a quantum mechanical amplitude Φ_2 to the event of the particle moving through the bottom slit. Using the superposition principle, the overall probability density of detecting the particle on the right is then given by $P = |\Phi_1 + \Phi_2|^2$.

While computing such an example with a few multi-slits is seldom more difficult than cumbersome, this approach becomes rapidly unfeasible as the number of slits grows, since the number of possible paths, and hence amplitudes Φ_p in $P = \left|\sum_{p\in\text{paths}} \Phi_p\right|^2$ that need to be computed after quantization, grows even more rapidly:

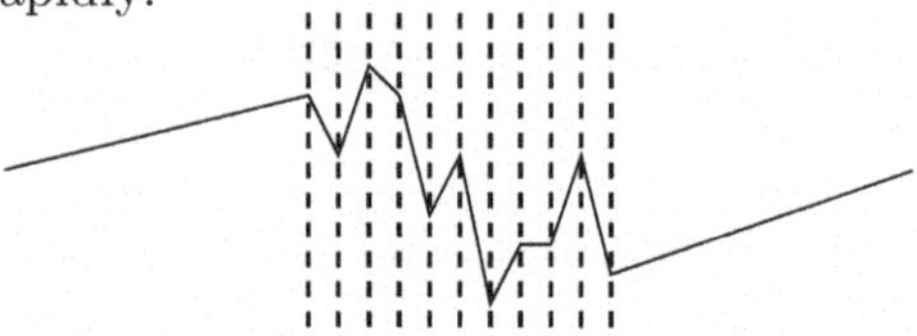

It is precisely this point at which Dirac's observations about the classical action before and after quantization are crucial. Feynman realized that

$$\forall p \in \text{paths} : \; \Phi_p = \exp\left(\frac{i}{\hbar} S_{\text{cl}}(p)\right),$$

where S_{cl} is the classical action, renders the computation of the limit "dense set of infinitesimal slits" possible. In order to compute this limit, we need to assume that the classical action can be evaluated at a given path. Thus, we need to assume that paths are continuous and have at least some regularity with respect to (weak) derivatives. Therefore, the set of paths is usually modeled as a set of sections of some vector bundle over the given space-time. Taking the inductive limit, Feynman formally replaced the "sum over all paths" by an "integral over all paths" to denote that the inductive limit has taken place and obtained the so-called Feynman propagator

$$K(t_1, x_1; t_0, x_0) = \int_{p\in\text{paths}((t_0,x_0)\to(t_1,x_1))} \exp\left(\frac{i}{\hbar} S_{\text{cl}}(p)\right) \mathcal{D}p,$$

which is the probability density of a particle measured at (t_0, x_0) to move to (t_1, x_1). While this formulation is well-defined in quantum mechanics and some very simple QFTs by means of this inductive limits of the discretized system, i.e., the multi-slit picture, this formula cannot be interpreted in general as no measure $\mathcal{D}p$ is known to exist (in fact, the "canonical choices" are known not to exist).

Perturbation theory: Since certain classes of free theories allow the inductive limit to be performed, the first methods of studying the Feynman propagator were based on perturbation theory. In more precise terms, physicists applied stationary phase approximation to the propagator which gave rise to Feynman diagrams. Considering Quantum Chromodynamics (QCD) for instance, these perturbation theoretical methods excel in high energy situations (e.g., quark-gluon plasma) or at length scales that are significantly smaller than 1 fm. However, they cannot produce effects like instantons, and, hence, they cannot accurately describe physics at the 1 fm scale or above (e.g., if you want to capture a proton in your theory).

Lattice QFT and non-perturbative methods: In order to study physics at length scales not reachable by perturbation theory, a number of non-perturbative methods were developed. Most notably, in Lattice QFT the Feynman propagator is still a well-defined object as we still live in the discretized multi-slit picture. Thus, the unknown inductive limit can be computed through extrapolation of discretized computations.

Probably the most prominent non-perturbative method (that does not require extrapolation of a family of discretized QFTs) is topological QFT (TQFT). A TQFT is a QFT that computes topological invariants and is closely related to knot theory, algebraic topology, and algebraic geometry. While a TQFT is inherently a non-perturbative description of physics, most methods known to compute interesting quantities are unfortunately based on perturbative frameworks.

ζ-functions: As such, a non-perturbative understanding of the Feynman propagator in the continuum is still an open problem that we want to address using operator ζ-functions. However, before we can do that, we need to reformulate the Feynman propagator in terms that are well-defined.

Axiomatically, it has to be possible to construct a Hilbert space $\mathcal{H}$, elements $\delta_x \in \mathcal{H}$, and operators $U(t,t')$ on $\mathcal{H}$ such that the Feynman propagator has the representation

$$K(t_1,x_1;t_0,x_0) = \langle \delta_{x_1}, U(t_1,t_0)\delta_{x_0}\rangle_{\mathcal{H}}\,.$$

The family U is the time-evolution semigroup (also known as the wave propagator) and satisfies the "Schrödinger equation"

$$\partial_{t_1} K(t_1,x_1;t_0,x_0) = \left\langle \delta_{x_1}, \frac{-i}{\hbar} H(t_1)U(t_1,t_0)\delta_{x_0}\right\rangle_{\mathcal{H}},$$

where the family H is called the Hamiltonian. This, in turn, induces the Hamiltonian formulation of our QFT,

$$\psi(t_1) = U(t_1,t_0)\psi(t_0) = \exp\left(-\frac{i}{\hbar}\int_{t_0}^{t_1} H(t)dt\right)\psi(t_0),$$

where ψ is the state vector of the "universe" and exp is the time-ordered exponential if the $H(t)$ don't commute.

Following Feynman's approach [7] (cf. "Some Remarks on Mathematical Rigor" in chapter 4-3 [8] and [3], as well), we will change the physics slightly and introduce a (flat) time torus of length T, i.e.,

$$\psi(T) = \exp\left(-\frac{i}{\hbar}\int_0^T H(t)dt\right)\psi(0) = \psi(0).$$

Locality of physical theories implies that the limit $T \to \infty$ lets us recover the QFT we started out with. On the other hand, the time torus allows us to introduce the partition function

$$Z_T \text{ “=” } \operatorname{tr}\exp\left(-\frac{i}{\hbar}\int_0^T H(t)dt\right),$$

which captures the statistical properties of the thermal equilibrium of the universe. Given an observable Ω, we define its expectation value $\langle\Omega\rangle_T$ on the time torus as

$$\langle\Omega\rangle_T \text{ “=” } \frac{\operatorname{tr}\exp\left(-\frac{i}{\hbar}\int_0^T H(t)dt\right)\Omega}{Z_T} = \frac{\operatorname{tr}\exp\left(-\frac{i}{\hbar}\int_0^T H(t)dt\right)\Omega}{\operatorname{tr}\exp\left(-\frac{i}{\hbar}\int_0^T H(t)dt\right)},$$

and we recover the actual expectation value $\langle\Omega\rangle$ in our QFT through the thermal limit

$$\langle\Omega\rangle := \lim_{T\to\infty}\langle\Omega\rangle_T.$$

In this reformulation, we can clearly see why the path integral is ill-defined; the operators whose traces we need to compute are not of trace-class, in general. Thus, these traces need to be constructed.

2. Operator ζ-functions

Operator ζ-functions were introduced by Ray and Singer [30, 31] as a trace construction. Given an operator A with purely discrete spectrum, Ray and Singer introduced the spectral ζ-function of A as

$$\zeta_\sigma(A)(z) := \sum_{\lambda\in\sigma(A)\setminus\{0\}} \mu_\lambda \lambda^z,$$

where μ_λ denotes the multiplicity of the eigenvalue λ. If $\zeta_\sigma(A)$ exists in some open subset of $\mathbb{C}$ and extends analytically to a neighborhood of 1, then the idea is to interpret $\zeta_\sigma(A)(1)$ as a generalized trace of A; very much in the same way that $\sum_{n\in\mathbb{N}} n^2$ can be interpreted as $\zeta_R(-2)$ where ζ_R is the Riemann ζ-function. This is precisely the notion of ζ-regularization Hawking [16] used in his work on path integrals with a curved space-time background.

If the operator A allows for the construction of complex powers (e.g., if it has a spectral cut) and the complex powers A^z are of trace-class in some open subset of $\mathbb{C}$, then the spectral ζ-function can be written as the analytic extension of

$$\zeta_\sigma(A)(z) = \operatorname{tr} A^z.$$

Such complex powers of operators and spectral ζ-functions were extensively studied based on pioneering work by Seeley [33]; in particular, with respect to classical pseudo-differential operators.

In the context of classical pseudo-differential operators, it is natural to consider a more general version of the ζ-function since taking complex powers is not always possible. Hence, rather than studying the holomorphic family A^z, one considers a more general holomorphic family of classical pseudo-differential operators $(A(z))_{z\in\mathbb{C}}$ where each symbol $\sigma(z)$ of $A(z)$ has an asymptotic expansion $\sum_{j\in\mathbb{N}_0} a_{m+qz-j}(z)$ with $a_{m+qz-j}(z)$ being homogeneous of degree $m+qz-j$ and $q\neq 0$.

Given a classical pseudo-differential operator A_0, a common construction is $A(z) := A_0 Q^z$ where Q is a classical pseudo-differential operator whose complex powers exist (e.g., a Laplacian), and q is the order of Q. Then it can be shown that $A(z)$ is of trace-class if $\Re(z)$ is sufficiently small (that is, sufficiently negative) and the map

$$\zeta(A_0, Q)(z) := \operatorname{tr} A_0 Q^z$$

has a meromorphic extension to $\mathbb{C}$ with at most isolated first-order poles. Furthermore, the Laurent expansion of $\zeta(A_0, Q)$ near zero is given by

$$\zeta(A_0, Q)(z) = \frac{\operatorname{res} A_0}{qz} + \sum_{k \in \mathbb{N}_0} c_k(A_0, Q) z^k,$$

where $\operatorname{res} A_0$ is the Wodzicki residue [38, 39]. The Wodzicki residue is independent of Q, in fact it only depends on the term with homogeneity $-\dim M$, and is, in general, the projectively unique continuous trace on the algebra of classical pseudo-differential operators on a compact Riemannian C^∞-manifold M without boundary. Hence, any ζ-function $\zeta(A_0, Q)$ is holomorphic in a neighborhood of zero if A_0 has no term with homogeneity $-\dim M$ in its asymptotic expansion of the symbol (e.g., for all non-integer-order classical pseudo-differential operators). Therefore, the degree $-\dim M$ is called the "critical degree of homogeneity."

If we are given an operator A_0 that has no critical degree of homogeneity, then the constant Laurent coefficient $c_0(A_0, Q)$ is independent of Q and tracial.[1] This (unbounded) trace is called the Kontsevich–Vishik trace [19, 20], and it can be shown [25] that the Kontsevich–Vishik trace is the only trace on the algebra of classical pseudo-differential operators that coincides with the canonical trace of trace-class operators in $L(L_2(M))$.

While the operator traces we would like to construct are not traces of classical pseudo-differential operators, they are very closely related to Fourier Integral Operators, which contain pseudo-differential operators as a special case. Since the Kontsevich–Vishik trace has been generalized to Fourier Integral Operators [14, 15], this generalized version of the Kontsevich–Vishik trace is a natural starting point for our trace construction in the path integral case. However, we will have to consider a slightly more generalized notion of ζ-functions.

Definition 2.1. Let $\mathcal{A}$ be an algebra of operators (usually containing many unbounded operators), $\mathcal{A}_0 \subseteq \mathcal{A}$ a subalgebra, $\tau : \mathcal{A}_0 \to \mathbb{C}$ tracial, $A : \mathbb{C} \to \mathcal{A}$ holomorphic, and $\Omega \subseteq \mathbb{C}$ open and connected such that $\forall z \in \Omega : A(z) \in \mathcal{A}_0$. We then define the *(generalized) ζ-function* $\zeta(A)$ of the family A with respect to τ as the maximal holomorphic extension of $\tau \circ A$ with open and connected domain $\Omega' \supseteq \Omega$.

[1] A functional τ on an algebra $\mathcal{A}$ is called tracial if and only if $\forall x, y \in \mathcal{A} : (xy, yx \in D(\tau) \Rightarrow \tau(xy) = \tau(yx))$.

Given such a (generalized) ζ-function $\zeta(A)$, we want to consider $\zeta(A)(0)$ as a regularized notion of $\tau(A(0))$. However, in order to do this, there are a number of questions to be answered. First and foremost, does the domain Ω' of $\zeta(A)$ contain zero for all suitable families A with $A(0) = A_0$ for a given A_0? If so, does $\zeta(A)(0)$ only depend on $A(0)$, or will different families A and B with $A(0) = B(0) = A_0$ yield $\zeta(A)(0) \neq \zeta(B)(0)$? Finally, if $A_0 \mapsto (A(z))_{z\in\mathbb{C}} \mapsto \zeta(A)(0) =: \tau^{\text{reg}}(A_0)$ is well-defined, is it tracial?

While the former two questions necessarily need to have a positive answer, it should be noted that traciality of τ^{reg} is not absolutely crucial. Though the physical interpretation of an anomalous trace might be difficult, trace anomalies can be very informative and useful. For instance, trace anomalies determine the energy momentum tensor of a two-dimensional black hole in a box and, for a four-dimensional black hole, it determines the energy momentum tensor up to a function of position [2].

3. Fourier Integral Operator ζ-functions

In order to answer these questions, we need to study our class of operators in more detail. More precisely, we will reduce the trace construction for the path integral to a trace construction problem for Fourier Integral Operators. Hence, in this section, we will give an overview of Fourier Integral Operators and their ζ-functions. For more details, please refer to [14, 15].

Throughout this section, we will consider a compact, orientable, connected, finite-dimensional Riemannian C^∞-manifold X without boundary and a closed conic Lagrangian submanifold Λ of $T^*X^2 \setminus 0$, where 0 denotes the zero-section of T^*X^2. Then X and Λ define classes of Fourier Integral Operators.

Definition 3.1. Let E be a (complex) vector bundle over X and Y a C^∞ submanifold of X without boundary. Then the space $I^m(X,Y;E)$ of distribution sections of E that are conormal to Y and of order less than or equal to m is the set of all distributions $u \in C_c^\infty(X,E)'$ such that

$$L_1 \dots L_N u \in B_{2,\infty,\text{loc}}^{-m-\frac{\dim X}{4}}(X,E)$$

for all $N \in \mathbb{N}_0$ and all first-order differential operators L_j between distribution sections of E whose coefficients are C^∞ tangential to Y.

In the definition above, $B_{p,q}^s(\mathbb{R}^n)$ denotes the usual Besov space, and, for $U \subseteq \mathbb{R}^n$ open, we define $B_{p,q,\text{loc}}^s(U)$ as the set of distributions $u \in C_c^\infty(U)'$ such that $\forall\varphi \in C_c^\infty(U) : \ \varphi u \in B_{p,q}^s(\mathbb{R}^n)$. This definition can then be lifted to manifolds in the usual manner. Furthermore, the definition of conormality can be extended to pseudo-differential operators from E to E whose principal symbol vanishes on Y. This lift and extension to pseudo-differential operators allows us to define Lagrangian distributions and Fourier Integral Operators in the sense of Hörmander (cf. Chapter 25 in [17]).

Definition 3.2. (i) Let E be a (complex) vector bundle over X. Then the space $I^m(X,\Lambda;E)$ of Lagrangian distribution sections of E of order less than or equal to m is the set of all distributions $u \in C_c^\infty(X,E)'$ such that $L_1 \dots L_N u \in B_{2,\infty,\mathrm{loc}}^{-m-\frac{\dim X}{4}}(X,E)$ for all $N \in \mathbb{N}_0$ and all properly supported first-order pseudo-differential operators $L_j \in \Psi^1(X;E,E)$ whose principal symbols vanish on Λ.

(ii) An integral operator $A : C_c^\infty(X,E) \to C_c^\infty(X,E)'$ is called Fourier Integral Operator of order m if and only if its kernel is an element of the space $I^m(X^2,\Lambda;E)$ for some Λ.

While Λ is geometrically the object defining classes of Fourier Integral Operators, it is often algebraically useful to introduce a "twist" and consider the canonical relation $\Gamma \subseteq (T^*X \setminus 0)^2$ (cf., e.g., Chapter 1 in [14]) which satisfies

$$\Lambda = \Gamma' := \left\{ ((x,\xi),(y,\eta)) \in (T^*X \setminus 0)^2 ;\ ((x,\xi),(y,-\eta)) \in \Gamma \right\}.$$

Γ is the object that describes whether or not the composition of given Fourier Integral Operators makes sense. In particular, for a homogeneous canonical relation Γ (cf., e.g., Chapter 1 in [14], Theorem 2.4.1 in [6], and Example 1 in [12]), the set of operators $\mathcal{A}_\Gamma$ with kernels in $\bigcup_{m\in\mathbb{R}} I^m(X^2;\Gamma')$ is an associative operator algebra which has non-trivial intersection with the set of trace-class operators in $L(L_2(X))$ (cf. Lemmata 1.12 and 1.13 in [14]). In fact, if the kernel k is in $I^m(X^2;\Gamma')$ with m sufficiently small, then the associated Fourier Integral Operator A is of trace-class and

$$\operatorname{tr} A = \int_X k(x,x) d\mathrm{vol}_X(x).$$

As we are interested in extending this trace to non-trace-class elements of $\mathcal{A}_\Gamma$, we need to consider holomorphic families $A \in C^\omega(\mathbb{C},\mathcal{A}_\Gamma)$ with kernels $k \in C^\omega\left(\mathbb{C}, I^m(X^2;\Gamma')\right)$ such that $\forall z \in \mathbb{C}:\ k(z) \in I^{m+\Re(z)}(X;\Gamma')$. We then define $\zeta(A)$ to be the maximal meromorphic extension of

$$\zeta(A)(z) := \operatorname{tr} A(z) = \int_X k(z)(x,x) d\mathrm{vol}_X(x),$$

which is well-defined for $\Re(z)$ sufficiently small, that is, $\Re(z) \ll 0$.

In order to show existence of a meromorphic extension to $\mathbb{C}$, we need to consider special cases of such families. The holomorphic families we will utilize to regularize the path integral are called "gauged Fourier Integral Operators with log-polyhomogeneous amplitudes." Gauged Fourier Integral Operators are holomorphic families of Fourier Integral Operators whose kernels can locally be written as

$$\forall z \in \mathbb{C}\ \forall x,y \in X:\ k(z)(x,y) = \int_{\mathbb{R}^N} e^{i\vartheta(x,y,\xi)} a(z)(x,y,\xi) d\xi, \tag{1}$$

where ϑ is a phase function,

$$\forall x, y \in X\ \forall \xi \in \mathbb{R}^N \setminus \{0\} :\ \vartheta(x, y, \xi) = \vartheta\left(x, y, \frac{\xi}{\|\xi\|_{\ell_2(N)}}\right) \|\xi\|_{\ell_2(N)}, \tag{2}$$

and

$$a(z)(x, y, \xi) = a_0(z)(x, y, \xi) + \sum_{\iota \in I} a_\iota(z)(x, y, \xi) \tag{3}$$

with $a_0(z) \in L_1(X \times X \times \mathbb{R}^N)$ and

$$a_\iota(z)(x, y, \xi) = \|\xi\|_{\ell_2(N)}^{d_\iota + z} \left(\ln \|\xi\|_{\ell_2(N)}\right)^{l_\iota} \tilde{a}_\iota(z) \left(x, y, \frac{\xi}{\|\xi\|_{\ell_2(N)}}\right). \tag{4}$$

Here, the d_ι are called the "degree of homogeneity of a_ι" and l_ι the "logarithmic order." Furthermore, if all $l_\iota = 0$, we call the amplitude homogeneous. For the ζ-function to exist as a meromorphic function on $\mathbb{C}$, we need a few more technical assumptions which we will omit at this point (cf. Chapter 2 in [14]).

Choosing a suitable gauge is a difficult task in many cases, especially if one requires the gauged kernels to define global densities. Locally, a gauge can always be written down fairly easily. For instance, the term $a_\iota(x, y, \xi)$ can be gauged locally to $a_\iota(x, y, \xi)\hat{a}(z)\big(x, \frac{\xi}{\|\xi\|_{\ell_2(N)}}\big) \|\xi\|_{\ell_2(N)}^z$, where $\hat{a}$ is any holomorphic section of the $(N-1)$-sphere bundle. However, in general, the associated kernels will depend on local coordinates and not be globally defined for all z. Per se, this will not introduce any difficulties as the entire ζ-function calculus works locally, but, if geometric properties of the manifold X are relevant, globally gauged kernels might be a necessity. In this situation, it is possible to consider a properly supported first-order pseudo-differential operator P for which complex powers P^z can be constructed. Supposing that we consider an operator with kernel k, then $P^z k$ defines a family of globally defined kernels and the associated family of operators is gauged.

The existence of ζ-functions of gauged Fourier Integral Operators with polyhomogeneous amplitudes as meromorphic functions on $\mathbb{C}$ was proven by Guillemin [11, 12], who also showed that they have at most isolated simple poles and studied the residues. The author [14, 15] has extended Guillemin's work to gauged Fourier Integral Operators with log-polyhomogeneous amplitudes by introducing the notion of gauged poly-log-homogeneous distributions. Using gauged poly-log-homogeneous distributions, it can be shown that ζ-functions of gauged Fourier Integral Operators with log-polyhomogeneous amplitudes have at most isolated poles of finite order located at $z = -N - d_\iota$ and the order is bounded by $l_\iota + 1$. Furthermore, the Laurent expansion can be computed.

Guillemin [12] studied the residues since they yield the (projectively) unique (bounded) trace on most algebras $\mathcal{A}_\Gamma$. However, we are looking for an extension of the classical trace (the residue trace vanishes on trace-class operators), that is, we are interested in the constant Laurent coefficient which is $\zeta(A)(0)$ if $\forall \iota \in I: \ d_\iota \neq -N$ (the non-critical case). In fact, in the non-critical case, $\zeta(A)$ is holomorphic in a neighborhood of zero and $\zeta(A)(0)$ depends only on $A(0)$. This can be used to prove that

$$A_0 \in \mathcal{A}_\Gamma \ \rightsquigarrow \ A \in C^\omega(\mathbb{C},\mathcal{A}_\Gamma) \text{ gauged with } A(0) = A_0 \ \rightsquigarrow \ \zeta(A)(0) \in \mathbb{C}$$

defines an unbounded tracial functional on $\mathcal{A}_\Gamma$ (cf. Chapter 7 in [14]); the generalized Kontsevich–Vishik trace.

To be more precise, let us consider a gauged Fourier Integral Operator A with kernel k as described in equations (1)–(4). Using Theorem 25.1.3 in [17] (cf. Chapter 4 [14] or [11]), it is possible to show that the distributional trace $\int_X k(z)(x,x)d\mathrm{vol}_X(x)$ can locally be written as a sum of a holomorphic function τ and a gauged poly-log-homogeneous distribution

$$\int_{\mathbb{R}_{\geq 1}\times\partial B_{\mathbb{R}^N}} \alpha(z)(r,\xi)d\mathrm{vol}_{\mathbb{R}_{\geq 1}\times\partial B_{\mathbb{R}^N}}(r,\xi).$$

The family α has an expansion $\alpha = \alpha_0 + \sum_{\iota\in I}\alpha_\iota$ where $I \subseteq \mathbb{N}$, $\forall z \in \Omega_0$: $\alpha_0(z) \in L_1(\mathbb{R}_{\geq 1}\times\partial B_{\mathbb{R}^N})$ and each $\alpha_\iota(z)$ is log-homogeneous with degree of homogeneity $d_\iota + z$ and logarithmic order $l_\iota \in \mathbb{N}_0$, i.e., there exists an $\tilde{\alpha}_\iota \in C^\infty(\mathbb{C}, L_1(\partial B_{\mathbb{R}^N}))$ such that

$$\forall r \in \mathbb{R}_{\geq 1}\ \forall \xi \in \partial B_{\mathbb{R}^N}: \ \alpha_\iota(z)(r,\xi) = r^{d_\iota+z}(\ln r)^{l_\iota}\tilde{\alpha}_\iota(z)(\xi).$$

Furthermore, we assume the following.

- The family $(\Re(d_\iota))_{\iota\in I}$ is bounded from above. (Note, we do not require $\Re(d_\iota) \to -\infty$. In fact, $\forall \iota \in I: \ \Re(d_\iota) = 42$ is entirely possible.)
- The map $I \ni \iota \mapsto (d_\iota, l_\iota) \in \mathbb{C}\times\mathbb{N}_0$ is injective.
- There are only finitely many ι satisfying $d_\iota = d$ for any given $d \in \mathbb{C}$.
- The family $((d_\iota - \delta)^{-1})_{\iota\in I}$ is in $\ell_2(I)$ for any $\delta \in \mathbb{C}\setminus\{d_\iota;\ \iota\in I\}$.
- Each $\sum_{\iota\in I}\tilde{\alpha}_\iota(z)$ converges unconditionally in $L_1(\partial B_{\mathbb{R}^N})$.[2]

Note that I is finite if a is a classical symbol, i.e., all the assumptions above are satisfied and Ω_0 can be chosen to be $\mathbb{C}_{\Re(\cdot)<R}$ for any $R \in \mathbb{R}_{>0}$. Furthermore, if A is a gauged pseudo-differential operator with symbol a, then (locally)

$$\alpha(z)(r,\xi) = \int_X a(z)(x,r\xi)d\mathrm{vol}_X(x).$$

Under the conditions above, we obtain

$$\begin{aligned}\zeta(\alpha)(z) &:= \int_{\mathbb{R}_{\geq 1}\times\partial B_{\mathbb{R}^N}} \alpha(z)d\mathrm{vol}_{\mathbb{R}_{\geq 1}\times\partial B_{\mathbb{R}^N}} \\ &= \underbrace{\int_{\mathbb{R}_{\geq 1}\times\partial B_{\mathbb{R}^N}} \alpha_0(z)d\mathrm{vol}_{\mathbb{R}_{\geq 1}\times\partial B_{\mathbb{R}^N}}}_{=:\tau_0(z)} + \sum_{\iota\in I}\int_{\mathbb{R}_{\geq 1}\times\partial B_{\mathbb{R}^N}} \alpha_\iota(z)d\mathrm{vol}_{\mathbb{R}_{\geq 1}\times\partial B}\end{aligned}$$

[2]Unconditional convergence of $\sum_{\iota\in I}\tilde{\alpha}_\iota(z)$ in $L_1(M)$ may also be replaced by the slightly weaker, though more artificial, condition $\sum_{\iota\in I}\|\tilde{\alpha}_\iota(z)\|^2_{L_1(M)} < \infty$. However, we need at least conditional convergence or $\sum_{\iota\in I}\alpha_\iota$ would not make sense, and having only conditional convergence (rather than unconditional convergence) would give rise to complications as we want to split off critical terms and treat them separately.

$$= \tau_0(z) + \sum_{\iota\in I} \int_{\mathbb{R}_{\geq 1}} \int_{\partial B_{\mathbb{R}^N}} r^{d_\iota+z+N-1}(\ln r)^{l_\iota} \tilde{\alpha}_\iota(z)(\xi) d\mathrm{vol}_{\partial B_{\mathbb{R}^N}}(\xi) dr$$

$$= \tau_0(z) + \sum_{\iota\in I} \frac{(-1)^{l_\iota+1} l_\iota! \int_{\partial B_{\mathbb{R}^N}} \tilde{\alpha}_\iota(z) d\mathrm{vol}_{\partial B_{\mathbb{R}^N}}}{(N+d_\iota+z)^{l_\iota+1}},$$

where $\sum_{\iota\in I} (-1)^{l_\iota+1} l_\iota! \int_{\partial B_{\mathbb{R}^N}} \tilde{\alpha}_\iota(z) d\mathrm{vol}_{\partial B_{\mathbb{R}^N}} / (N+d_\iota+z)^{l_\iota+1}$ is absolutely convergent (Lemma 2.2 in [14]). It is then possible to compute the Laurent expansion of $\zeta(\alpha)$ and transform back to the Fourier Integral Operator picture which yields the generalized Kontsevich–Vishik trace

$$\begin{aligned}\zeta(A)(0) = &\int_X \int_{B_{\mathbb{R}^N}(0,1)} e^{i\vartheta(x,x,\xi)} a(0)(x,x,\xi)\, d\xi\, d\mathrm{vol}_X(x)\\ &+ \int_{\mathbb{R}_{\geq 1}\times\partial B_{\mathbb{R}^N}} \int_X e^{i\vartheta(x,x,\xi)} a_0(0)(x,x,\xi)\, d\mathrm{vol}_X(x)\, d\mathrm{vol}_{\mathbb{R}_{\geq 1}\times\partial B}(\xi)\\ &+ \sum_{\iota\in I} \frac{(-1)^{l_\iota+1} l_\iota! \int_{X\times\partial B_{\mathbb{R}^N}} e^{i\vartheta(x,x,\xi)} \tilde{a}_\iota(0)(x,x,\xi)\, d\mathrm{vol}_{X\times\partial B}(x,\xi)}{(N+d_\iota)^{l_\iota+1}}\end{aligned}$$

provided that none of the d_ι are $-N$. Furthermore, by construction the generalized Kontsevich–Vishik trace coincides with the trace tr on trace-class operators.

It should also be noted that the (classical) Kontsevich–Vishik trace (that is the case of $\mathcal{A}_\Gamma$ being the algebra of classical pseudo-differential operators Ψ^{cl}) is the only trace on Ψ^{cl} that extends tr in $L(L_2(X))$ [25].

4. The path integral in the ζ-regularized setting

Recall that we are trying to make sense of the equality

$$\langle\Omega\rangle_T \text{ “=” } \frac{\operatorname{tr}\exp\left(-\frac{i}{\hbar}\int_0^T H(t)dt\right)\Omega}{Z_T} = \frac{\operatorname{tr}\exp\left(-\frac{i}{\hbar}\int_0^T H(t)dt\right)\Omega}{\operatorname{tr}\exp\left(-\frac{i}{\hbar}\int_0^T H(t)dt\right)}.$$

In order to use Fourier Integral Operator ζ-functions, we need to examine the operators in questions a little more closely.

The operator H is the Hamiltonian of our QFT. Thus, we may assume that H is a pseudo-differential operator with symbol (amplitude)

$$\sigma_H(t,x,r\xi) := h_2(t,x,\xi)r^2 + h_1(t,x,\xi)r + h_0(t,x,r,\xi),$$

where $\|\xi\|_{\ell_2} = 1$, the h_j are continuous, and $h_0(t,x,r,\xi)$ has an asymptotic expansion $\sum_{j\in\mathbb{N}_0} r^{-j} a_{-j}(t,x,\xi)$. Thus, $\exp(\frac{-i}{\hbar}\int_0^T H(s)ds)$ is an operator with kernel

$$\frac{1}{(2\pi)^{\dim X}} \int_{\mathbb{R}^{\dim X}} e^{i\langle x-y,\xi\rangle_{\ell_2(\dim X)}} \sigma_{\exp(\frac{-i}{\hbar}\int_0^T H(s)ds)}(x,\xi) d\xi.$$

If the symbol of H is independent of x, then the symbol $\sigma_{\exp\left(\frac{-i}{\hbar}\int_0^T H(s)ds\right)}$ satisfies

$$\sigma_{\exp\left(\frac{-i}{\hbar}\int_0^T H(s)ds\right)} = e^{iH_2}e^{iH_1}e^{\frac{-i}{\hbar}\int_0^T h_0(s,\cdot)ds}$$

with

$$H_2(r\xi) := \frac{-1}{\hbar}r^2\int_0^T h_2(s,\xi)ds$$

and

$$H_1(r\xi) := \frac{-1}{\hbar}r\int_0^T h_1(s,\xi)ds.$$

Since $h_0(t,r,\xi)$ has an asymptotic expansion $\sum_{j\in\mathbb{N}_0} r^{-j}a_{-j}(t,\xi)$, we obtain

$$\begin{aligned} e^{\frac{-i}{\hbar}\int_0^T h_0(s,r,\xi)ds} &= \sum_{k\in\mathbb{N}_0}\frac{\left(-\frac{i}{\hbar}\right)^k}{k!}\left(\int_0^T h_0(s,r,\xi)ds\right)^k \\ &\sim \sum_{k\in\mathbb{N}_0}\frac{\left(-\frac{i}{\hbar}\right)^k}{k!}\left(\sum_{j\in\mathbb{N}_0} r^{-j}\int_0^T a_{-j}(s,\xi)ds\right)^k, \end{aligned}$$

and, using

$$\left(\sum_{k\in\mathbb{N}_0} a_k X^k\right)^n = \sum_{m\in\mathbb{N}_0} c_m X^m$$

with $c_0 = a_0^n$ and $c_m = \frac{1}{ma_0}\sum_{k=1}^m (kn-m+k)a_k c_{m-k}$, we conclude that $e^{\frac{-i}{\hbar}\int_0^T h_0(s,r,\xi)ds}$ has an asymptotic expansion

$$e^{\frac{-i}{\hbar}\int_0^T h_0(s,r,\xi)ds} = b(r\xi) \sim \sum_{j\in\mathbb{N}_0} r^{-j}b_{-j}(\xi).$$

These formulae are more complicated if the symbol of H depends on x, but we still obtain a symbol of the form $\sigma_{\exp\left(\frac{-i}{\hbar}\int_0^T H(s)ds\right)} = e^{iH_2}e^{iH_1}e^{iH_0}$ with H_2 and H_1 homogeneous of degrees 2 and 1, and e^{iH_0} having an asymptotic expansion.

This shows that the denominator is very closely related to Fourier Integral Operators as discussed in Section 3 if it weren't for the second-order phase function. However, before we address this fact, let us consider the numerator $\exp\left(\frac{-i}{\hbar}\int_0^T H(s)ds\right)\Omega$. Using the Fourier transform $\mathcal{F}$, we observe for all $u,\varphi\in C_c^\infty(X)$

$$\begin{aligned} \langle\mathcal{F}\Omega\varphi,\mathcal{F}u\rangle &= \langle\varphi,\Omega^*u\rangle \\ &= \int_{\mathbb{R}^N}\varphi(x)\left(\int_{\mathbb{R}^N} e^{i\langle x,\xi\rangle}\sigma_{\Omega^*}(x,\xi)\mathcal{F}u(\xi)d\xi\right)^* dx \\ &= \int_{\mathbb{R}^N}\int_{\mathbb{R}^N}\varphi(x)e^{-i\langle x,\xi\rangle}\sigma_{\Omega^*}(x,\xi)^*\mathcal{F}u(\xi)^*d\xi dx, \end{aligned}$$

which implies

$$\mathcal{F}\Omega\varphi(\xi) = \int_{\mathbb{R}^N} e^{-i\langle x,\xi\rangle}\sigma_{\Omega^*}(x,\xi)^*\varphi(x)dx$$

and, thus,

$$\begin{aligned}\exp&\left(\frac{-i}{\hbar}\int_0^T H(s)ds\right)\Omega\varphi(x)\\ &= \int_{\mathbb{R}^N} e^{iH_2(x,\xi)}e^{iH_1(x,\xi)}b(x,\xi)e^{i\langle x,\xi\rangle}\mathcal{F}\Omega\varphi(\xi)d\xi\\ &= \int_{\mathbb{R}^N}\int_{\mathbb{R}^N} e^{iH_2(x,\xi)}e^{iH_1(x,\xi)}b(x,\xi)e^{i\langle x,\xi\rangle}e^{-i\langle y,\xi\rangle}\sigma_{\Omega^*}(y,\xi)^*\varphi(y)dyd\xi\\ &= \int_{\mathbb{R}^N}\int_{\mathbb{R}^N} e^{iH_2(x,\xi)}e^{iH_1(x,\xi)}b(x,\xi)e^{i\langle x,\xi\rangle}e^{-i\langle y,\xi\rangle}\sigma_{\Omega}(y,\xi)\varphi(y)dyd\xi,\end{aligned}$$

where we used the identity $\sigma_{A^*}(x,y,\xi) = \sigma_A(y,x,\xi)^*$ which holds for any pseudo-differential operator A. In other words,

$$\sigma_{\exp\left(\frac{-i}{\hbar}\int_0^T H(s)ds\right)\Omega}(x,y,\xi) = e^{iH_2(x,\xi)}e^{iH_1(x,\xi)}b(x,\xi)\sigma_\Omega(y,\xi).$$

Hence, both numerator and denominator have kernels of the form

$$\int_{\mathbb{R}^N} e^{i\langle x-y,\xi\rangle}e^{iH_2(x,\xi)}e^{iH_1(x,\xi)}a(x,\xi)d\xi,$$

where a is poly-(log)-homogeneous provided that h_0 and σ_Ω are poly-(log)-homogeneous.

For the ζ-function regularization, we now need to gauge these integrals. The choice of a reasonable gauge usually depends a lot on the specific operators. However, it is often convenient to choose an $\mathcal{M}$-gauge (or Mellin-gauge; cf. Definition 2.10 in [14])

$$\int_{\mathbb{R}^N} e^{i\langle x-y,\xi\rangle}e^{iH_2(x,\xi)}e^{iH_1(x,\xi)}a(x,\xi)\,\|\xi\|_{\ell_2(N)}^z\,d\xi.$$

Theorem 4.1. *Let X be a compact, orientable, N-dimensional Riemannian C^∞-manifold without boundary, σ_Ω polyhomogeneous, and*

$$Z = \int_X\int_{\mathbb{R}^N} e^{-i\sigma_H(x,\xi)}\sigma_\Omega(x,\xi)\ d\xi\ \mathrm{dvol}_X(x)$$

with

$$\forall(x,r,\eta)\in X\times\mathbb{R}_{\geq 0}\times\partial B_{\mathbb{R}^N}:\sigma_H(x,r\eta) = h_2(x,\eta)r^2 + h_1(x,\eta)r + h_0(x,r\eta),$$

where $h_2, h_1 \in C(X\times\partial B_{\mathbb{R}^N})$, h_0 polyhomogeneous, and

- (i) *either $h_2 = 0$ and $\vartheta(x,\xi) := h_1\left(x,\frac{\xi}{\|\xi\|_{\ell_2}}\right)\|\xi\|_{\ell_2}$ is a non-degenerate phase function*
- (ii) *or $\forall x\in X\ \forall\eta\in\partial B_{\mathbb{R}^N}:\ |h_2(x,\eta)| > 0$.*

Then Z can be regularized using Fourier Integral Operator ζ-functions.

Proof. We can absorb e^{ih_0} into the amplitude σ, and obtain

$$Z = \int_X \int_{\partial B_{\mathbb{R}^N}} \int_{\mathbb{R}_{>0}} e^{-i\left(h_2(x,\eta)r^2+h_1(x,\eta)r\right)}\sigma(x,r\eta)\ dr\ \mathrm{dvol}_{\partial B_{\mathbb{R}^N}}(\eta)\ \mathrm{dvol}_X(x).$$

For (i), suppose $h_2 = 0$ and $\vartheta(x,\xi) := h_1(x, \frac{\xi}{\|\xi\|_{\ell_2(N)}})\ \|\xi\|_{\ell_2(N)}$ is non-degenerate. Then Z is a Fourier Integral Operator trace already and the entire known theory (Section 3) is applicable.

For (ii), suppose $\forall x \in X\ \forall \eta \in \partial B_{\mathbb{R}^N}:\ |h_2(x,\eta)| > 0$ and

$$R := 1 + \max\left\{ \left|\frac{h_1(x,\eta)}{2h_2(x,\eta)}\right| \in \mathbb{R};\ (x,\eta) \in X \times \partial B_{\mathbb{R}^N} \right\}.$$

Then we can split Z into two parts

$$Z_1 := \int_{X\times\partial B_{\mathbb{R}^N}\times[0,R]} e^{-i\left(h_2(x,\eta)r^2+h_1(x,\eta)r\right)}\sigma(x,r\eta)\ \mathrm{dvol}_{X\times\partial B_{\mathbb{R}^N}\times[0,R]}(x,\eta,r)$$

and

$$Z_2 := \int_{X\times\partial B_{\mathbb{R}^N}\times\mathbb{R}_{\geq R}} e^{-i\left(h_2(x,\eta)r^2+h_1(x,\eta)r\right)}\sigma(x,r\eta)\ \mathrm{dvol}_{X\times\partial B_{\mathbb{R}^N}\times\mathbb{R}_{\geq R}}(x,\eta,r).$$

Regarding Z_1, we observe that $X \times \partial B_{\mathbb{R}^N} \times [0,R]$ is compact. Hence, Z_1 is well-defined and will yield a holomorphic function after gauging.

Regarding Z_2, we recall the assumption $|h_2| > 0$ and obtain

$$h_2(x,\eta)r^2 + h_1(x,\eta)r = h_2(x,\eta)\left(r + \frac{h_1(x,\eta)}{2h_2(x,\eta)}\right)^2 - \frac{h_1(x,\eta)^2}{4h_2(x,\eta)}.$$

Absorbing $e^{i\frac{h_1(x,\eta)^2}{4h_2(x,\eta)}}$ into σ and setting $\sigma_1(x,s,\eta) = \sigma\left(x, s - \frac{h_1(x,\eta)}{2h_2(x,\eta)}, \eta\right)$ yields

$$\begin{aligned} Z_2 &= \int_{X\times\partial B_{\mathbb{R}^N}} \int_{\mathbb{R}_{\geq R}} e^{-ih_2(x,\eta)\left(r+\frac{h_1(x,\eta)}{2h_2(x,\eta)}\right)^2}\sigma(x,r\eta)\ dr\ \mathrm{dvol}_{X\times\partial B_{\mathbb{R}^N}}(x,\eta) \\ &= \int_{X\times\partial B_{\mathbb{R}^N}} \int_{\mathbb{R}_{\geq R+\frac{h_1(x,\eta)}{2h_2(x,\eta)}}} e^{-ih_2(x,\eta)s^2}\sigma_1(x,s,\eta)\ ds\ \mathrm{dvol}_{X\times\partial B_{\mathbb{R}^N}}(x,\eta)) \\ &= \int_{X\times\partial B_{\mathbb{R}^N}} \int_{\mathbb{R}_{\geq\left(R+\frac{h_1(x,\eta)}{2h_2(x,\eta)}\right)^2}} e^{-ih_2(x,\eta)t}\sigma_1(x,\sqrt{t},\eta)\ \frac{dt}{2\sqrt{t}}\ \mathrm{dvol}_{X\times\partial B}(x,\eta). \end{aligned}$$

This shows there exists a Fourier Integral Operator with polyhomogeneous amplitude whose trace coincides with Z_2 (up to another term similar to Z_1) and gauging Z_2 gauges the Fourier Integral Operator, as well. In other words, existence of the ζ-function of that gauged Fourier Integral Operator implies that Z can be gauged, exists for $\Re(z)$ sufficiently small, and has a meromorphic extension to $\mathbb{C}$. □

Theorem 4.1 implies that we can write

$$\langle\Omega\rangle_T(z) = \frac{N(T,z)}{D(T,z)},$$

where both $N(T,\cdot)$ and $D(T,\cdot)$ are ζ-functions of gauged traces, i.e., meromorphic. Hence, point evaluation at $z=0$ might not be possible. Furthermore, gauge dependence of $\langle\Omega\rangle_T(0)$ needs to be addressed provided the point evaluation is possible.

Let us first assume that both $N(T,\cdot)$ and $D(T,\cdot)$ have no critical degree of homogeneity (that is $-N$ in the $h_2=0$ case and $1-2N$ in the $|H_2|>0$ case) and at most one of them has a zero in zero. Then gauge independence of the constant Laurent coefficient of the Fourier Integral Operator ζ-function (cf. Lemma 2.6 in [14]) implies that both $N(T,0)$ and $D(T,0)$ are gauge independent, as well. In particular, evaluation of $\langle\Omega\rangle_T(0)$ is possible if and only if $D(T,0)\neq 0$. Furthermore, if $D(T,0)\neq 0$, then $\langle\Omega\rangle_T(0)$ is independent of the gauges chosen in $N(T,\cdot)$ and $D(T,\cdot)$. Unfortunately, there is no reason to assume $D(T,0)\neq 0$ in general; in fact, the free relativistic fermion as described in Section 8 is an example of $N(T,0)=D(T,0)=0$.

In the case $N(T,0)=D(T,0)=0$, evaluation of $\langle\Omega\rangle_T(0)$ depends on the first non-zero Laurent coefficient, and we obtain the same situation as above but with some higher-order Laurent coefficient. Supposing we can evaluate $\langle\Omega\rangle_T(0)$, it is important to note that the higher-order Laurent coefficients are gauge dependent and, thus, so is $\langle\Omega\rangle_T(0)$. In other words, we have to choose our gauges in a physically meaningful way (cf. Section 5).

On the other hand, if critical degrees of homogeneity are present, then we have to expect $N(T,\cdot)$ and $D(T,\cdot)$ to have first-order poles in zero. If at least one of the residues does not vanish, then we are in exactly the same situation as in the non-critical case with at least one of the constant Laurent coefficients not vanishing; namely, evaluation of $\langle\Omega\rangle_T(0)$ is possible if and only if $\operatorname{res}_0 D(T,\cdot)\neq 0$ and both $\operatorname{res}_0 D(T,\cdot)$ and $\operatorname{res}_0 N(T,\cdot)$ are independent of the chosen gauges (cf. Lemma 2.5 in [14]). Hence, existence of $\langle\Omega\rangle_T(0)$ implies gauge independence in the critical case. Similarly, if both residues vanish, we have a quotient of higher-order Laurent coefficients which depend on the gauge (the constant Laurent coefficient is gauge dependent in the critical case), and the choice of gauge becomes important again.

5. Physical choices of gauges

At the end of the last section, we have noticed that the expectation value $\langle\Omega\rangle_T$ of an observable Ω on the time torus may depend on the gauges chosen for the numerator and denominator.

Recall that we have obtained the denominator $D(T,\cdot)$ by meromorphic extension of a function $\operatorname{tr}\mathfrak{G}(T,\cdot)$ where $\mathfrak{G}(T,\cdot)$ is a gauged family of operators such that $\mathfrak{G}(T,z)$ is of trace-class for $\Re(z)\ll 0$ and $\mathfrak{G}(T,0)$ coincides with the time-evolution operator $\exp(\frac{-i}{\hbar}\int_0^T H(s)ds)$ of the QFT we want to study. Thus, we may interpret $\mathfrak{G}(T,\cdot)$ as the holomorphic family time-evolution operators of a "holomorphic family $\mathfrak{Q}$ of QFTs." In that sense, $\operatorname{tr}\mathfrak{G}(T,z)$ is precisely the partition function of $\mathfrak{Q}(z)$ given $\Re(z)\ll 0$. However, since the path integral in $\mathfrak{Q}(z)$ with $\Re(z)\ll 0$ is well-defined, there is only one reasonable choice of gauge for the numerator, since Feynman's construction in $\mathfrak{Q}(z)$ implies that the expectation $\langle\Omega\rangle_{\mathfrak{Q}(z)}$ of Ω in $\mathfrak{Q}(z)$ satisfies

$$\forall z \in \mathbb{C}_{\Re(\cdot)\ll 0}: \ \langle\Omega\rangle_T(z) = \langle\Omega\rangle_{\mathfrak{Q}(z)} = \frac{\operatorname{tr}(\mathfrak{G}(T,z)\Omega)}{\operatorname{tr}\mathfrak{G}(T,z)}.$$

These observations lead to the following definition of a ζ-regularized expectation of Ω.

Definition 5.1. Let H be the Hamiltonian, $(\mathfrak{G}(T,z))_{T\in\mathbb{R}_{>0},z\in\mathbb{C}}$ a gauged family of operators with

$$\forall T \in \mathbb{R}_{>0}: \ \mathfrak{G}(T,0) = \exp\left(\frac{-i}{\hbar}\int_0^T H(s)ds\right),$$

and Ω an observable. Then we define the ζ-regularized expectation value $\langle\Omega\rangle_{\mathfrak{G}}$ of Ω with respect to $\mathfrak{G}$ as

$$\begin{aligned}\langle\Omega\rangle_{\mathfrak{G}} &:= \lim_{T\to\infty}\lim_{z\to 0}\frac{\zeta\left(\mathfrak{G}(T,\cdot)\Omega\right)(z)}{\zeta\left(\mathfrak{G}(T,\cdot)\right)(z)}\\ &= \lim_{T\to\infty}\lim_{z\to 0}\frac{\left(\mathbb{C}_{\Re(\cdot)\ll 0}\ni s\mapsto \operatorname{tr}(\mathfrak{G}(T,s)\Omega)\in\mathbb{C}\right)\Big|_{\text{mer.}}(z)}{\left(\mathbb{C}_{\Re(\cdot)\ll 0}\ni s\mapsto \operatorname{tr}\mathfrak{G}(T,s)\in\mathbb{C}\right)\Big|_{\text{mer.}}(z)},\end{aligned}$$

where $f|_{\text{mer.}}$ denotes the meromorphic extension of a function f with maximal open and connected domain (provided f has open and connected domain).

Given that this definition is rather esoteric, we will proceed with a number of physical examples to convince ourselves that this new way of interpreting the path integral is physically useful. It should be noted that the ζ-regularized expectation coincides with Feynman's definition whenever Feynman's notion is well-defined in the Hamiltonian formulation (i.e., the operators are of trace-class). As such, the ζ-definition is an extension of Feynman's definition and, since going forward we will not always explicitly define the time-evolution family $\mathfrak{G}$, we will drop the index $\mathfrak{G}$ from $\langle\Omega\rangle_{\mathfrak{G}}$ in the upcoming examples.

6. The topological oscillator

The first example we would like to consider is the topological oscillator (a.k.a. quantum rotor). It describes a particle of mass M moving on a circle with radius R. Parametrizing the circle as $(x,y) = (R\cos\varphi, R\sin\varphi)$ with free coordinate φ, we obtain the Lagrangian

$$\mathcal{L} = \frac{M}{2}(\dot{x}^2+\dot{y}^2) = \frac{J}{2}\dot{\varphi}^2,$$

where $J = MR^2$ is the moment of inertia. This yields the generalized momentum

$$p = \partial_{\dot{\varphi}}\mathcal{L} = J\dot{\varphi}$$

and Hamiltonian

$$H = \dot{\varphi}p - \mathcal{L} = \frac{p^2}{J} - \frac{p^2}{2J} = \frac{p^2}{2J}.$$

In other words, the symbol σ_H of H is

$$\sigma_H(x, r\xi) = \frac{r^2}{2J}.$$

As an observable, we will consider the topological charge

$$Q = \frac{1}{2\pi}\int_0^T \dot{\varphi}(t)dt = \frac{1}{2\pi}\int_0^T \frac{p(t)}{J}dt \quad \Rightarrow \quad \sigma_Q(x, r\xi) = \frac{1}{2\pi}\int_0^T \frac{r\xi}{J}dt = \frac{Tr\xi}{2\pi J},$$

which is the winding number of the path of the particle "over the entire lifetime of the universe." The topological charge is interesting since it allows us to compute the topological charge χ_{top} of the system

$$\chi_{\text{top}} = \lim_{T\to\infty}\left\langle \frac{Q^2}{-iT} \right\rangle_T$$

which is related to the energy gap

$$\Delta E = 2\pi^2\chi_{\text{top}}$$

between the ground state and first excited state of the particle. Using the Python 2.7 implementation [13], we obtain the physically correct values (in natural units $c = \hbar = 1$)

$$\chi_{\text{top}} = \frac{1}{4\pi^2 J} \qquad \text{and} \qquad \Delta E = \frac{1}{2J}.$$

7. The free massive Schwinger model

Our next example is the free massive Schwinger model [32]. The massive Schwinger model is a theory of quantum electrodynamics (QED) in two spacetime dimensions with Hamiltonian (in the zero-momentum frame; cf., e.g., equation (2.2) in [1])

$$H_m = \begin{pmatrix} mc^2 & -i\partial \\ -i\partial & mc^2 \end{pmatrix} = mc^2 + \underbrace{\begin{pmatrix} 0 & -i\partial \\ -i\partial & 0 \end{pmatrix}}_{=:H_0}.$$

Thus,

$$\sigma_{H_0}(x, r\xi) = \begin{pmatrix} 0 & -r\xi \\ -r\xi & 0 \end{pmatrix}$$

implies $\sigma_{H_0^{2k}}(x, r\xi) = r^{2k} = (r\xi)^{2k}$, $\sigma_{H_0^{2k-1}}(x, r\xi) = \begin{pmatrix} 0 & (r\xi)^{2k-1} \\ (r\xi)^{2k-1} & 0 \end{pmatrix}$,

and

$$\sigma_{\exp\left(\frac{-i}{\hbar}H_0T\right)}(x, r\xi) = \begin{pmatrix} \cos\left(\frac{T}{\hbar}r\xi\right) & -i\sin\left(\frac{T}{\hbar}r\xi\right) \\ -i\sin\left(\frac{T}{\hbar}r\xi\right) & \cos\left(\frac{T}{\hbar}r\xi\right) \end{pmatrix}.$$

Since mc^2 and H_0 commute, $e^{\frac{-i}{\hbar}H_mT} = e^{\frac{-i}{\hbar}mc^2T}e^{\frac{-i}{\hbar}H_0T}$ yields

$$\sigma_{\exp\left(\frac{-i}{\hbar}H_mT\right)}(x,r\xi) = e^{\frac{-i}{\hbar}mc^2T}\begin{pmatrix} \cos\left(\frac{T}{\hbar}r\xi\right) & -i\sin\left(\frac{T}{\hbar}r\xi\right) \\ -i\sin\left(\frac{T}{\hbar}r\xi\right) & \cos\left(\frac{T}{\hbar}r\xi\right) \end{pmatrix}$$

and

$$\begin{aligned}&\sigma_{\exp\left(\frac{-i}{\hbar}H_mT\right)H_m}(x,r\xi) \\ &= e^{\frac{-i}{\hbar}mc^2T}\begin{pmatrix} mc^2\cos\left(\frac{T}{\hbar}r\xi\right) - i\xi\sin\left(\frac{T}{\hbar}r\xi\right) & \xi\cos\left(\frac{T}{\hbar}r\xi\right) - imc^2\sin\left(\frac{T}{\hbar}r\xi\right) \\ \xi\cos\left(\frac{T}{\hbar}r\xi\right) - imc^2\sin\left(\frac{T}{\hbar}r\xi\right) & mc^2\cos\left(\frac{T}{\hbar}r\xi\right) - i\xi\sin\left(\frac{T}{\hbar}r\xi\right) \end{pmatrix}.\end{aligned}$$

We can gauge these using an $\mathcal{M}$-gauge, i.e., $\sigma(x,r\xi) \rightsquigarrow \sigma(x,r\xi)r^z$. Furthermore, we need to introduce a cut-off function $1_{B(0,X)} \leq \chi \leq 1_{B(0,X+1)}$ to compactify the spacial domain. Then the ground state energy of the system is given by (replacing $r\xi$ by $\xi \in \mathbb{R}$)

$$\langle H_m\rangle = \lim_{X,T\to\infty}\lim_{z\to 0}\frac{\frac{1}{2\pi}\int_{\mathbb{R}}\int_{\mathbb{R}}\chi(x)\,\mathrm{tr}\,\sigma_{\exp\left(\frac{-i}{\hbar}H_mT\right)H_m}(\xi)\,|\xi|^z\,d\xi dx}{\frac{1}{2\pi}\int_{\mathbb{R}}\int_{\mathbb{R}}\chi(x)\,\mathrm{tr}\,\sigma_{\exp\left(\frac{-i}{\hbar}H_mT\right)}(\xi)\,|\xi|^z\,d\xi dx}.$$

Replacing $\cos x$ and $\sin x$ by $\frac{e^{ix}+e^{-ix}}{2}$ and $\frac{e^{ix}-e^{-ix}}{2i}$, we obtain the known gauged Fourier Integral Operator form, and Theorem 8.7 in [14] implies that these are distributional trace integrals of Hilbert–Schmidt operators with continuous kernels. Computing the traces by hand or using the Python 2.7 implementation supplied in [13] yields the physically correct ground state energy

$$\langle H_m\rangle = E = mc^2$$

for the free massive Schwinger model.

8. Free relativistic Fermions

Moving on to four space-time dimensions, we need to add Pauli matrices to the Hamiltonian (now in natural units and using Einstein summation convention to ease notation)

$$H_m = \begin{pmatrix} m & -i\sigma_k\partial_k \\ -i\sigma_k\partial_k & m \end{pmatrix},$$

and replacing $r\xi$ by $\xi \in \mathbb{R}^3$ we obtain

$$\sigma_{e^{-iH_mT}}(x,\xi) = e^{-imT}\left(\cos(T\,\|\xi\|_{\ell_2}) - \frac{i\sin(T\,\|\xi\|_{\ell_2})}{\|\xi\|_{\ell_2}}\begin{pmatrix} 0 & \sigma_k\xi_k \\ \sigma_k\xi_k & 0 \end{pmatrix}\right)$$

and

$$\begin{aligned}&\sigma_{\exp(-iH_mT)H_m}(x,\xi) \\ &= e^{-imT}\left(\cos(T\,\|\xi\|_{\ell_2(3)})\begin{pmatrix} m & \sigma_k\xi_k \\ \sigma_k\xi_k & m \end{pmatrix} - i\,\|\xi\|_{\ell_2(3)}\sin(T\,\|\xi\|_{\ell_2(3)})\right).\end{aligned}$$

Then we can compute the ground state energy (by hand this time and having already canceled the $\mathrm{vol}_X(X)$ terms from the integration over the spatial variable x and taken the limit $X \to \mathbb{R}^3$ similar to the one-dimensional case):

$$\begin{aligned}
\langle H_m \rangle &= \lim_{T\to\infty} \lim_{z\to 0} \frac{\int_{\mathbb{R}^3} \left(4m\cos\left(T\|\xi\|_{\ell_2}\right) - 4i\|\xi\|_{\ell_2} \sin\left(T\|\xi\|_{\ell_2}\right)\right) \|\xi\|_{\ell_2}^z \, d\xi}{\int_{\mathbb{R}^3} 4\cos\left(T\|\xi\|_{\ell_2}\right) \|\xi\|_{\ell_2}^z \, d\xi} \\
&= m + \lim_{T\to\infty} \lim_{z\to 0} \frac{\int_{\mathbb{R}^3} \|\xi\|_{\ell_2(3)}^{z+1} \left(e^{iT\|\xi\|_{\ell_2(3)}} - e^{-iT\|\xi\|_{\ell_2(3)}}\right) d\xi}{\int_{\mathbb{R}^3} \|\xi\|_{\ell_2(3)}^{z} \left(e^{iT\|\xi\|_{\ell_2(3)}} + e^{-iT\|\xi\|_{\ell_2(3)}}\right) d\xi} \\
&= m + \lim_{T\to\infty} \lim_{z\to 0} \frac{\mathrm{vol}\left(\partial B_{\mathbb{R}^3}\right) \int_{\mathbb{R}_{>0}} r^{z+3} \left(e^{iTr} - e^{-iTr}\right) dr}{\mathrm{vol}\left(\partial B_{\mathbb{R}^3}\right) \int_{\mathbb{R}_{>0}} r^{z+2} \left(e^{iTr} + e^{-iTr}\right) dr} \qquad (*) \\
&= m + \lim_{T\to\infty} \lim_{z\to 0} \frac{\left(-e^{-i\frac{\pi(z+3)}{2}} - e^{-3i\frac{\pi(z+3)}{2}}\right) \Gamma(z+4) T^{-z-4}}{\left(-e^{-i\frac{\pi(z+2)}{2}} + e^{-3i\frac{\pi(z+2)}{2}}\right) \Gamma(z+3) T^{-z-3}} \\
&= m + \lim_{T\to\infty} \lim_{z\to 0} \frac{\left(-e^{-i\frac{\pi(z+3)}{2}} - e^{-i\frac{\pi(z+3)}{2}} e^{-i\pi(z+3)}\right) \Gamma(z+4) T^{-z-4}}{\left(-e^{-i\frac{\pi(z+2)}{2}} + e^{-i\frac{\pi(z+2)}{2}} e^{-i\pi(z+2)}\right) \Gamma(z+3) T^{-z-3}} \\
&= m + \lim_{T\to\infty} \lim_{z\to 0} \frac{e^{-i\frac{\pi(z+3)}{2}} \left(-1 - e^{-i\pi(z+2)} e^{-i\pi}\right) (z+3) \Gamma(z+3) T^{-z-4}}{e^{-i\frac{\pi(z+2)}{2}} \left(e^{-i\pi(z+2)} - 1\right) \Gamma(z+3) T^{-z-3}} \\
&= m + \lim_{T\to\infty} \lim_{z\to 0} \frac{e^{-i\frac{\pi(z+2)}{2}} e^{-i\frac{\pi}{2}} (z+3)}{e^{-i\frac{\pi(z+2)}{2}} T} \\
&= m + \lim_{T\to\infty} \lim_{z\to 0} \frac{-i(z+3)}{T} \\
&= m.
\end{aligned}$$

In other words, we have correctly computed $E = mc^2$ again.

Remark. This example highlights a number of properties. These will become very important in the case of the massive Schwinger model (Section 9). The "observable" m of the first summand depends on none of the variables. This leads to cancelations just like the results of the integration over X canceled. In the massive Schwinger model this will be paramount since the part of the model that is not analytically solvable will vanish in one such cancelation.

The latter summand exhibits the other important property; namely, these integrals can be written as Laplace transforms of homogeneous distributions and the time extent T of the time-torus is part of the point evaluation of the Laplace transform. Thus, the limit $T \to \infty$ depends primarily on the asymptotic expansion of the observable. In particular, this allows us to determine many terms that will vanish in the limit $T \to \infty$ simply by looking at the asymptotic expansion of the observable.

Remark. Note that (*) can be implemented directly since the Laplace transform satisfies

$$\mathcal{L}(r \mapsto r^q)(s) = \frac{\Gamma(q+1)}{s^{q+1}}$$

for $\Re(s) > 0$ and $\Re(q) > -1$, and hence for $s \in \mathbb{C} \setminus \{0\}$ and $q \in \mathbb{C} \setminus (-\mathbb{N})$ by analytic extension. In particular,

$$\int_{\mathbb{R}_{>0}} r^z e^{iTr} dr = \frac{-ie^{-i\frac{\pi z}{2}}\Gamma(z+1)}{T^{z+1}}$$

and

$$\int_{\mathbb{R}_{>0}} r^z e^{-iTr} dr = \frac{ie^{-3i\frac{\pi z}{2}}\Gamma(z+1)}{T^{z+1}}.$$

Furthermore, using stationary phase approximation (cf. Chapter 8 in [14]) on the results of Theorem 4.1, we obtain that the regularization is given solely in terms of such Laplace transforms. In other words, computing $\langle \Omega \rangle_T(z)$ reduces to the numerical problem of computing integrals over $\partial B_{\mathbb{R}^N}$ and possibly computing the limits $z \to 0$ and $T \to \infty$.

Here, the integrals over $\partial B_{\mathbb{R}^N}$ are trivial and a Python 2.7 implementation is given in [13] which, too, outputs `<H_m>=m`.

9. Gauge boson mass in the Schwinger model

Let us now return to the two-dimensional Schwinger model. With the addition of an Abelian gauge field, the model is fully interacting and has non-trivial dynamics which lead to confinement of charges and bound states. This model is no longer analytically solvable, but as we are going to consider the gauge boson mass as our observable, non-integrable parts of the system will cancel in the quotient, allowing us to compute the gauge boson mass analytically.

The Hamiltonian (in the temporal gauge and using natural units) is comprised of two parts: the fermionic Hamiltonian

$$H_F = \begin{pmatrix} m & -i\partial_1 - eA \\ -i\partial_1 - eA & m \end{pmatrix}$$

and the Hamiltonian describing the self-interaction of the gauge field

$$H_S = -\frac{1}{4}F_{\mu\nu}F^{\mu\nu} = \frac{1}{2}E^2,$$

where $F_{\mu\nu} = \partial_\mu A_\nu - \partial_\nu A_\mu$, $A = A_1$ ($A_0 = 0$ is the temporal gauge), and $E = -\partial_0 A$.

At this point, we should recall the nature of the path integral being a formal inductive limit of discretized space-time. As such, it is convenient to consider the family $((A(x), E(x)))_{x \in X}$ as canonical coordinates on a space-torus X (recall that the ζ-regularization needs a compact manifold) while time dependence of the gauge fields is implicit.

As the limit $X \to \mathbb{R}^3$ will be trivial, we will suppress it in the following computation. Then we observe

$$\sigma_{e^{-i\int_0^T H}} = e^{-imT} e^{-\frac{i}{2}TE^2} \begin{pmatrix} \cos(T\xi - e\int_0^T A) & -i\sin(T\xi - e\int_0^T A) \\ -i\sin(T\xi - e\int_0^T A) & \cos(T\xi - e\int_0^T A) \end{pmatrix}.$$

Schwinger [32] computed the observable Ω for the squared gauge boson mass in terms of the Green's function of the Abelian gauge field and we can extract the symbol of the squared gauge boson mass to be

$$\sigma_\Omega(E) = E^2 + \frac{e^2}{\pi}.$$

Hence, the gauge boson mass m_g is given by (suppressing the differential $d\xi\mathcal{D}AdEdx$, as well as gauges for ξ, x, and A)

$$\begin{aligned} m_g^2 &= \langle\Omega\rangle \\ &= \lim_{\substack{T\to\infty \\ z\to 0}} \frac{\int e^{-imT-\frac{i}{2}TE^2}\left(e^{iT\xi}e^{-ie\int_0^T A} + e^{-iT\xi}e^{ie\int_0^T A}\right)\left(E^2+\frac{e^2}{\pi}\right)|E|^z}{\int e^{-imT-\frac{i}{2}TE^2}\left(e^{iT\xi}e^{-ie\int_0^T A} + e^{-iT\xi}e^{ie\int_0^T A}\right)|E|^z} \\ &= \frac{e^2}{\pi} + \lim_{T\to\infty}\lim_{z\to 0} \frac{\int e^{-\frac{i}{2}TE^2}\left(e^{iT\xi}e^{-ie\int_0^T A} + e^{-iT\xi}e^{ie\int_0^T A}\right)|E|^{z+2}}{\int e^{-\frac{i}{2}TE^2}\left(e^{iT\xi}e^{-ie\int_0^T A} + e^{-iT\xi}e^{ie\int_0^T A}\right)|E|^z} \\ &= \frac{e^2}{\pi} + \lim_{T\to\infty}\lim_{z\to 0} \frac{\int e^{-\frac{i}{2}TE^2}|E|^{z+2}\,dE}{\int e^{-\frac{i}{2}TE^2}|E|^z\,dE} \\ &= \frac{e^2}{\pi} + \lim_{T\to\infty}\lim_{z\to 0} \underbrace{\frac{-ie^{-\frac{3i\pi}{2}\frac{z+1}{2}}\Gamma\left(\frac{z+3}{2}\right)\left(\frac{2}{T}\right)^{\frac{z+3}{2}}}{-ie^{-\frac{3i\pi}{2}\frac{z-1}{2}}\Gamma\left(\frac{z+1}{2}\right)\left(\frac{2}{T}\right)^{\frac{z+1}{2}}}}_{\propto\frac{1}{T}} \\ &= \frac{e^2}{\pi}. \end{aligned}$$

Remark. Again, we observe the cancelations and asymptotic properties as seen in Section 8. However, now they are vital to the computation since the integrals with respect to A are not analytically solvable. Yet, the structure of the observable forced these integrals in the numerator and denominator to coincide. Even more so, suppose they did not cancel, then the asymptotic behavior in E still implies that the second term vanishes in the limit $T \to \infty$.

10. Spontaneous symmetry breaking and mass – the φ^4 model

Spontaneous symmetry breaking is another very important showcase to consider. Physically, it is fundamental to the Higgs mechanism and, thus, exhibits fundamental physical behavior we want the ζ-regularized theory to reproduce. From a mathematical point of view it is interesting since it is our first example that is not of the form $\langle\Omega\rangle = \lim_{T\to\infty}\lim_{z\to 0}\frac{N(T,z)}{D(T,z)}$.

For simplicity, let us consider the simplest, non-trivial, relativistic example. Suppose we have scalar fields $\varphi = (\varphi^1, \ldots, \varphi^k)$ and a Lagrangian depending on φ. Then we obtain our ζ-regularized partition function $Z_T(\varphi)$ which is the denominator $D(T,0)$ in $\langle \Omega \rangle = \lim_{T\to\infty} \lim_{z\to 0} \frac{N(T,z)}{D(T,z)}$. We now consider $Z_T(\varphi)$ to be the partition function with respect to an effective potential (Lagrangian) $V_e^T(\varphi)$, i.e.,

$$Z_T(\varphi) = \exp\left(-i \int V_e^T(\varphi) d(t,x)\right).$$

The vacuum expectation values φ_0^j of the φ^j are constant local minimizers of V_e^T in the limit $T \to \infty$, and the corresponding field masses are the square roots of the eigenvalues of $\left(\partial_i \partial_j V_e^T(\varphi_0)\right)_{i,j\in\mathbb{N}_{\le k}}$ in the limit $T \to \infty$.

As we are looking for constant solutions φ_0^j, we can write

$$V_e^T(\varphi) = \frac{\ln Z_T(\varphi)}{-iTX},$$

where TX is the volume of our space time, and, if we assume $k = 1$ for simplicity, we want to find $\varphi_0^T \in \mathbb{R}$ such that $\partial V_e^T(\varphi_0^T) = 0$. Then $\lim_{T\to\infty} \varphi_0^T$ yields the vacuum expectation of φ and $\lim_{T\to\infty} \sqrt{\partial^2 V_e^T(\varphi_0^T)}$ the field mass.

It should be noted that this is a particularly interesting and highly non-trivial behavior of QFTs. Usually, when someone writes down a QFT, they start with a classical field theory that then will be quantized. However, if this classical field theory contains mass terms for the classical field, quantization is impossible. Thus, considering a QFT as a classical field theory necessarily implies that all field masses vanish. Hence, non-zero field masses are an entirely quantum mechanical effect and they warrant that our ζ-regularized theory is quantized.

As an example for spontaneous symmetry breaking, we would like to consider the one-dimensional φ^4 model whose Hamiltonian is given by

$$H = \int \frac{p^2}{2} - \frac{1}{2}\varphi\Delta\varphi - \frac{1}{2}\mu^2\varphi^2 + \frac{\lambda}{4!}\varphi^4 dx.$$

Then we can ζ-regularize the partition function to obtain[3]

$$Z_T(z,\varphi) = \frac{1}{2\pi} \int_{\mathbb{R}} e^{-iTX\left(\frac{p^2}{2} - \frac{\mu^2}{2}\varphi^2 + \frac{\lambda}{4!}\varphi^4\right)} |p|^z \, dp.$$

Gauge independence of $Z_T(0,\varphi)$ implies that $\ln Z_T(z,\varphi)$ exists in a neighborhood of $z = 0$ provided $Z_T(0,\varphi) \neq 0$ and a suitable branch cut of ln is chosen. As this is the case here, we may use the Python 2.7 implementation from [13] and obtain the physically correct values of $\pm\sqrt{\frac{6}{\lambda}}\mu$ for the vacuum expectation and $\sqrt{2}\mu$ for the field mass.

[3]Note that the factor $\frac{1}{2\pi}$ has to be included now. Previously, the $(2\pi)^{-\dim}$ factors used to cancel as they appeared in both numerator and denominator.

Remark. We made a priori use of the fact that we are looking for (spatially) constant φ and therefore dropped the term $\varphi\Delta\varphi$. If we were to compute an observable for instance, we would not be able to do this and the computation will be a lot more complex. In such cases, it may be advantageous to write $\varphi\Delta\varphi$ as $-\langle\nabla\varphi, \nabla\varphi\rangle$, which can be combined with the p^2 term as a space-time version of canonically conjugated momentum with respect to φ. The best choice will but depend on the specific problem at hand and even space-discretization, i.e., replacing φ by a vector $(\varphi(x_j))_{j\in\mathbb{N}_{\leq n}} \in \mathbb{R}^n$ can be an interesting option resulting in a path integral that is ζ-regularizable. In fact, this idea is fundamental to Lattice QFT which we will discuss in Section 11.

11. Application to Lattice QFT

One of the most important frameworks in computational physics is called Lattice QFT. Lattice QFTs treat the discretized path integral. In other words, we are working with the path integral before taking the formal inductive limit to continuous space time:

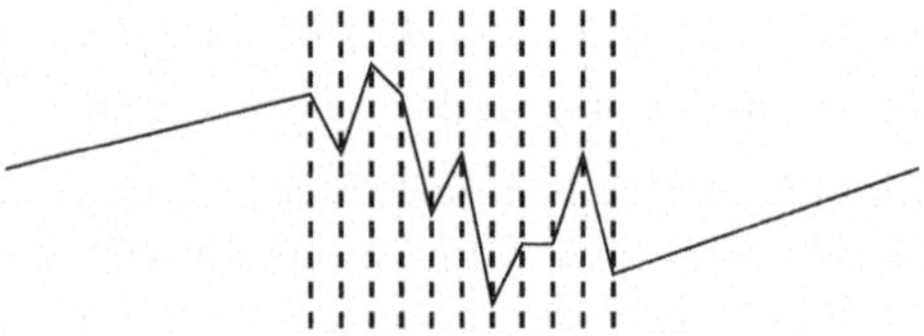

Using the discretized Lagrangian L and an observable Ω, the aim is to compute quotients

$$\langle\Omega\rangle = \frac{\mathcal{Z}(\Omega)}{\mathcal{Z}(1)} \tag{5}$$

where both numerator and denominator have the structure

$$\mathcal{Z}(\omega) = \int_{\mathbb{R}^n} e^{\frac{i}{\hbar}\int L(p)(t,x)d\mathrm{vol}(t,x)} a^\omega(p)dp, \tag{6}$$

a^ω is an amplitude depending on ω, and vol is a discrete measure depending on the discretization. However, in general, these integrals are still not well-defined as the integrands are not in $L_1(\mathbb{R}^n)$. Usually this problem is solved using a Wick rotation which turns Minkowski space-time into Euclidean space-time and results in the integrands being integrable now. More precisely, the Wick rotation replaces the time-evolution semigroup $\exp(-i\int H)$ with the "heat" semigroup $\exp(-\int H)$, which is trace-class on compact manifolds. Hence, the Wick rotated path integral is well-defined. In fact, this Wick rotation plays such an important role in QFTs that Osterwalder and Schrader [27, 28] developed axioms that a Euclidean path integral has to satisfy for it to be Wick rotated by analytic continuation to define a QFT in Minkowski space-time satisfying the Wightman axioms [34].

However, some physically interesting problems cannot be solved using Wick rotations on a lattice. Such an example would be the computation of parton distribution functions of a nucleon, i.e., the probability density of measuring a certain number of quarks of a certain type in a proton for instance. These are inaccessible on the lattice because the dynamics are localized on the light cone and using a Wick rotation $ds^2 = 0$ turns the light cone into a single point (the origin). Thus, the Wick rotation destroys the physically interesting properties. While parton distribution functions can still be recovered using very sophisticated methods [36], no such methods are known for many other high energy situations (like the early universe). Hence, being able to compute the objects in Minkowski space-time (or more precisely with a Lorentzian background) would be a great step forward. A ζ-regularized Lattice QFT might be able to fill this gap.

A "ζ-Lattice QFT" can be constructed on the background of the Hamiltonian theory we have already described; although, in many cases it is advantageous to stay in the Lagrangian picture (which is the much more common notation for QFTs) rather than explicitly computing the Hamiltonian picture. Of course, this is only possible since the discretized path integral is notationally well-posed (up to the fact that the integrals don't exist) and we can rely on the fact that we are equivalent to the Hamiltonian picture at all times.

Hence, given the Lagrangian L of a Lattice QFT, we obtain a "ζ-Lattice QFT" as follows.

(i) Choose a gauge $\mathfrak{g}$, that is, a holomorphic family $\mathfrak{g}$ such that $\forall z \in \mathbb{C}:$ $\mathfrak{g}(z) \in C^\infty(\mathbb{R}^n \setminus \{0\})$ and

$$\forall z \in \mathbb{C}\ \forall p \in \mathbb{R}^n \setminus \{0\}: \ \mathfrak{g}(z)(p) = \|p\|_{\ell_2(n)}^z\, \mathfrak{g}(z)\left(\frac{p}{\|p\|_{\ell_2(n)}}\right).$$

Then we consider the family

$$\mathcal{Z}(\omega)(z) = \int_{\mathbb{R}^n} e^{\frac{i}{\hbar}\int L(p)(t,x)\mathrm{dvol}(t,x)} a^\omega(p)\mathfrak{g}(z)(p)dp.$$

(ii) Construct a gauged Fourier Integral Operator A_ω (according to the proof of Theorem 4.1) and a holomorphic function τ_ω such that

$$\forall z \in \mathbb{C}: \ \mathcal{Z}(\omega)(z) \text{ is distributionally equivalent to } \tau_\omega(z) + \operatorname{tr} A_\omega(z).$$

Thus, we obtain a representation

$$\mathcal{Z}(\omega)(z) = \tau_\omega(z) + \sum_{\iota\in I}\int_{\partial B_{\mathbb{R}^n}}\int_{\mathbb{R}_{>0}} e^{ir\vartheta_\iota(\xi)} r^{d_\iota+z}\alpha_{\omega,\iota}(z)(\xi)drd\xi.$$

(iii) Now there are two possible cases.

(a) If ϑ_ι has no zeros on $\partial B_{\mathbb{R}^n}$, analytically continuing the Laplace transform yields the ζ-regularization of the inner integral; namely,

$$\vartheta_\iota > 0 \Rightarrow \int_{\mathbb{R}_{>0}} r^{d_\iota+z}e^{ir\vartheta_\iota(\xi)}dr = \frac{-ie^{-i\frac{\pi(d_\iota+z)}{2}}\Gamma(d_\iota+z+1)}{\vartheta_\iota(\xi)^{d_\iota+z+1}}$$

and

$$\vartheta_\iota < 0 \Rightarrow \int_{\mathbb{R}_{>0}} r^{d_\iota+z} e^{ir\vartheta_\iota(\xi)} dr = \frac{ie^{-3i\frac{\pi(d_\iota+z)}{2}}\Gamma(d_\iota+z+1)}{|\vartheta_\iota(\xi)|^{d_\iota+z+1}}.$$

(b) If ϑ_ι has zeros, the computation of these traces is more involved (see Chapters 4 and 8 of [14]).

(iv) At last, we need to take the limit $z \to 0$ (assuming $\vartheta > 0$ for notational simplicity), obtaining

$$\mathcal{Z}(\omega)(0) = \tau_\omega(0) - \sum_{\iota\in I} ie^{-i\frac{\pi d_\iota}{2}}\Gamma(d_\iota+1)\int_{\partial B_{\mathbb{R}^n}} \frac{\alpha_{\omega,\iota}(0)(\xi)}{\vartheta_\iota(\xi)^{d_\iota+1}} d\xi.$$

It should be noted that this contains the entire regularization. The remaining integrals are integrals of continuous functions over the sphere $\partial B_{\mathbb{R}^n}$. In particular, even though these still may be very difficult to compute numerically, at least we are now left with a well-posed numerical problem (as opposed to trying to compute these integrals of the non-integrable functions that we started out with).

Hence, ζ-regularization is a viable tool to study Lattice QFT without Wick rotations.

Conclusion

Having considered Feynman's path integral in the light of some properties of axiomatic QFTs, we have seen that the ill-definedness of Feynman's path integral can be expressed in terms of formal operator traces acting on operators that are not of trace-class. From this point of view, the main obstacle to understanding the path integral is the construction of a suitable trace for the occurring operators. Given the physically important methods and instances for which the path integral is well-defined (e.g., Wick rotations, lattice formulations, or trace-class operators), we proposed ζ-regularization since, for some of the special cases, it is known that ζ-regularization yields the unique trace extension (the Kontsevich–Vishik trace) we are looking for. Hence, we applied the ζ-function calculus to Feynman's path integral and obtained a new definition of ζ-regularized partition functions and ζ-regularized expectation values of observables (Definition 5.1).

In order to construct a ζ-regularization, we replaced the time-evolution semigroup $T(t) := e^{-\frac{i}{\hbar}\int_0^t H(s)ds}$ by a holomorphic family $z \mapsto \mathfrak{G}(t,z)$ satisfying $\mathfrak{G}(t,0) = T(t)$. If $\mathfrak{G}$ is suitably chosen (gauging), Feynman's path integral with time-evolution $t \mapsto \mathfrak{G}(t,z)$ is well-defined for $\Re(z) \ll 0$ and can be extended meromorphically to $\mathbb{C}$ (Theorem 4.1). Hence, the ζ-regularized path integral is obtained evaluating this meromorphic extension at $z = 0$ provided the meromorphic extension is holomorphic in a neighborhood of zero (Definition 5.1).

Since there is no particular reason to assume that this construction yields physically correct results, we considered a number of physical examples. These specifically addressed three points. First and foremost, we agree with physics (at least on these examples). Secondly, the regularization is computable and explicit in a way that allows for it to be used in computational physics since it reduces the numerical effort to evaluating integrals of continuous functions over compact manifolds (and possibly the limits $z \to 0$ and physical volume $\to \infty$). Thirdly, the ζ-regularized theory is a true extension of Feynman's path integral which allows for applications that were previously inaccessible (e.g., Lattice QFT with Lorentzian signatures).

An important question we had to address though was that of gauge dependence. Since the ζ-construction contains a large amount of choice, we need to make sure that our choices don't change the results. In many cases, we obtain gauge independence from the theory of Fourier Integral Operator ζ-functions that we utilized behind the scenes. However, there are cases (e.g., the free massive Schwinger model in Section 7) in which gauge independence is not warranted. Here, we argued from physical intuition. Since we are working with quotients of Fourier Integral Operator ζ-functions, choosing different gauges in numerator and denominator can lead to different results if the leading order Laurent coefficients in both numerator and denominator vanish. However, gauging the denominator can be interpreted as the construction of a holomorphic family of QFTs $\mathfrak{Q}$. In this interpretation, the choice of gauge for the numerator is fixed and rather than considering numerator and denominator separately, we can consider the quotient $\langle \Omega \rangle (z) = \frac{N(z)}{D(z)}$ to be a function of $\mathfrak{Q}$ rather than z. In other words, the idea is that physically reasonable results cannot be expected a priori, but should be obtained in the limit $\langle \Omega \rangle (\mathfrak{Q}(z)) \to \langle \Omega \rangle (\mathfrak{Q}(0))$; that is, to say that our QFT $\mathfrak{Q}(0)$ is consistent with the holomorphic family $\mathfrak{Q}$ of QFT it is embedded in, especially since some of these $\Re(z) \ll 0$ have path integrals that are well-defined in Feynman's sense.

References

[1] M.C. Bañuls, K. Cichy, J.I. Cirac, and K. Jansen, The mass spectrum of the Schwinger model with matrix product states, *Journal of High Energy Physics* **158** (2013).

[2] S.M. Christensen and S.A. Fulling, Trace anomalies and the Hawking effect, *Physical Review D* **15** (1977), 2088–2104.

[3] M. Creutz and B. Freedman, A statistical approach to Quantum Mechanics, *Annals of Physics* **132** (1981), 427–462.

[4] T. Degrand and C. Detar, *Lattice methods for Quantum Chromodynamics*, World Scientific, 2006.

[5] P.A.M. Dirac, The quantum theory of the emission and absorption of radiation, *Proceedings of the Royal Society A* **114** (1927), 243–265.

[6] J.J. Duistermaat, *Fourier Integral Operators*, Birkhäuser, 1996.

[7] R.P. Feynman, Space-time approach to non-relativistic Quantum Mechanics, *Reviews of Modern Physics* **20** (1948), 367–387.

[8] R.P. Feynman, A.R. Hibbs, and D.F. Styer, *Quantum Mechanics and path integrals*, Emended Edition, Dover Publications, Inc., 2005.

[9] C. Gattringer and C.B. Lang, *Quantum Chromodynamics on the lattice*, Springer, 2010.

[10] G.W. Gibbons, S.W. Hawking, and M.J. Perry, Path integrals and the indefiniteness of the gravitational action, *Nuclear Physics* **B138** (1978), 141–150.

[11] V. Guillemin, Gauged Lagrangian distributions, *Advances in Mathematics* **102** (1993), 184–201.

[12] V. Guillemin, Residue traces for certain algebras of Fourier Integral Operators, *Journal of Functional Analysis* **115** (1993), 391–417.

[13] T. Hartung, Regularizing Feynman path integrals using the generalized Kontsevich-Vishik trace, *Journal of Mathematical Physics* **58** (2017), 1–19.

[14] T. Hartung, *ζ-functions of Fourier Integral Operators*, Ph.D. thesis, King's College London, 2015.

[15] T. Hartung and S. Scott, A generalized Kontsevich-Vishik trace for Fourier Integral Operators and the Laurent expansion of ζ-functions, arXiv:1510.07324v2[math.AP], 2015.

[16] S.W. Hawking, Zeta function regularization of path integrals in curved spacetime, *Communications in Mathematical Physics* **55** (1977), 133–148.

[17] L. Hörmander, *The analysis of linear partial differential operators I–IV*, Springer, 1990.

[18] T. Johnson-Freyd, The formal path integral and quantum mechanics, *Journal of Mathematical Physics* **51** 122103 (2010), 1–31.

[19] M. Kontsevich and S. Vishik, Determinants of elliptic pseudo-differential operators, Max Planck Preprint, arXiv:hep-th/9404046 (1994).

[20] M. Kontsevich and S. Vishik, Geometry of determinants of elliptic operators, in: *Functional Analysis on the Eve of the XXI century, Vol. I*, Progress in Mathematics **131**, 173–197, Birkhäuser, 1994.

[21] N. Kumano-go, Phase space Feynman path integrals with smooth functional derivatives by time slicing approximation, *Bulletin des sciences mathematiques* **135** (2011), 936–987.

[22] N. Kumano-go and D. Fujiwara, Phase space Feynman path integrals via piecewise bicharacteristic paths and their semiclassical approximations, *Bulletin des sciences mathematiques* **132** (2008), 313–357.

[23] N. Kumano-go and A.S. Vasudeva Murthy, Phase space Feynman path integrals of higher order parabolic type with general functional as integrand, *Bulletin des sciences mathematiques* **139** (2015), 495–537.

[24] C.H. Lee, The exponential calculus of pseudodifferential operators of minimum type. I, *Proceedings of the Japan Academy* **89 (A)** (2013), 6–10.

[25] L. Maniccia, E. Schrohe, and J. Seiler, Uniqueness of the Kontsevich-Vishik trace, *Proceedings of the American Mathematical Society* **136** (2) (2008), 747–752.

[26] I. Montvay and G. Münster, *Quantum fields on a lattice*, Cambridge University Press, 1994.

[27] K. Osterwalder and R. Schrader, Axioms for Euclidean Green's functions, *Communications in Mathematical Physics* **31** (1973), 83–112.

[28] K. Osterwalder and R. Schrader, Axioms for Euclidean Green's functions II, *Communications in Mathematical Physics* **42** (1975), 281–305.

[29] S. Paycha, Zeta-regularized traces versus the Wodzicki residue as tools in quantum field theory and infinite dimensional geometry, *Proceedings of the International Conference on Stochastic Analysis and Applications*, 2001, 69–84.

[30] D.B. Ray, Reidemeister torsion and the Laplacian on lense spaces, *Advances in Mathematics* **4** (1970), 109–126.

[31] D.B. Ray and I.M. Singer, *R*-torsion and the Laplacian on Riemannian manifolds, *Advances in Mathematics* **7** (1971), 145–210.

[32] J. Schwinger, Gauge invariance and mass II, *Physical Review* **128** (1962), 2425–2429.

[33] R.T. Seeley, Complex powers of an elliptic operator, *Proceedings of Symposia in Pure Mathematics* **10** (1967), 288–307.

[34] R.F. Streater and A.S. Wightman, *PCT, spin and statistics, and all that*, Advanced Book Classics, Addison-Wesley, 1989.

[35] L.A. Takhtajan, *Quantum Mechanics for mathematicians*, American Mathematical Society, 2008.

[36] C. Wiese, *Investigating new lattice approaches to the momentum and spin structure of the nucleon*, Ph.D. thesis, Humboldt-Universität zu Berlin, 2016.

[37] K.G. Wilson, Confinement of quarks, *Physical Review D* **10** (1974), 2445.

[38] M. Wodzicki, Noncommutative residue. I. Fundamentals, in: *K-theory, arithmetic and geometry (Moscow,* 1984–1986*)*, Lecture Notes in Mathematics **1289**, Springer, 1987, 320–399.

[39] M. Wodzicki, *Spectral asymmetry and noncommutative residue*, Ph.D. thesis, Steklov Institute of Mathematics, 1984.

Tobias Hartung
Department of Mathematics
King's College London
Strand
London WC2R 2LS
United Kingdom
e-mail: tobias.hartung@kcl.ac.uk

Improving Monte Carlo integration by symmetrization

Tobias Hartung, Karl Jansen, Hernan Leövey and Julia Volmer

Abstract. The error scaling for Markov chain Monte Carlo (MCMC) techniques with N samples behaves like $1/\sqrt{N}$. This scaling makes it often very time intensive to reduce the error of calculated observables, in particular for applications in 4-dimensional lattice quantum chromodynamics as our theory of the interaction between quarks and gluons. Even more, for certain cases, where the infamous sign problem appears, MCMC methods fail to provide results with a reliable error estimate. It is therefore highly desirable to have alternative methods at hand which show an improved error scaling and have the potential to overcome the sign problem. One candidate for such an alternative integration technique we used is based on a new class of polynomially exact integration rules on $U(N)$ and $SU(N)$ which are derived from polynomially exact rules on spheres. We applied these rules successfully to a non-trivial, zero-dimensional model with a sign problem and obtained arbitrary precision results. In this article we test a possible way to apply the integration rules for spheres to the case of a one-dimensional $U(1)$ model, the topological rotor, which already leads to a problem of very high dimensionality.

Mathematics Subject Classification (2010). Primary 65C40; Secondary 28C10; Tertiary 65Z05.

Keywords. Numerical sign-problem, Markov chain Monte Carlo, Haar measure, quantum mechanical oscillator, lattice systems.

1. Introduction

High-dimensional integration problems appear in vast areas of physics and mathematics. In the field of high energy physics (HEP), which aims at exploring and understanding the fundamental interactions of elementary particles, such problems occur when the underlying models are evaluated with Feynman's path integrals [1]. Since the theories in high energy physics are

© Springer International Publishing AG, part of Springer Nature 2018

A. Böttcher et al. (eds.), *The Diversity and Beauty of Applied Operator Theory*, Operator Theory: Advances and Applications 268, https://doi.org/10.1007/978-3-319-75996-8_15

formulated in three space and one time dimensions, the number of degrees of freedom to be treated is very large leading to very high-dimensional integrals, which can easily reach $O(10^8)$ dimensions.

By formulating the involved theoretical models on a space-time crystal and using Feynman's path integral one arrives at a lattice field theory version of systems in high energy physics, see e.g. [2] for an introduction to this approach. The 4-dimensional discrete formulation within lattice field theory gives the considered theories a conceptually clean meaning, since the resulting integrals are well defined, although they are very high-dimensional.

For a number of physical questions these integrals are solvable by using importance sampling via Markov chain Monte Carlo (MCMC) methods. And, in fact, this approach has proven to be very successful: E.g., the low lying bound state spectrum of hadrons (the proton and the neutron being prominent examples) could be determined from such ab-initio calculations [3], a result that could not be computed analytically. Also, present MCMC simulations in high energy physics are reaching fully physical conditions [4, 5], requiring, however, the evaluation of even higher-dimensional integrals. In fact, the dimensionality of the involved integrals become easily of the order of hundreds of millions, so that only state of the art supercomputers with nowadays reaching hundreds of Petaflops [6] have to be employed. Still, typical projects run at the order of 1–2 years or even longer. Thus, finding methods to improve and accelerate such calculations is highly desirable.

In addition, and more severely, there are certain problems, where MCMC cannot be applied straightforward. This is the case when a highly oscillatory, complex integrand describes the problem under consideration. In this case, the interpretation of a probability density, on which MCMC methods rely, is lost and MCMC calculations do not deliver meaningful results, leading to the infamous sign problem [7]. It is important to note that this problem is not academic in high energy physics but that such situations occur when we want to understand the very early universe [8] or why there is more matter than anti-matter in the universe [9]. These so far unexplained phenomena address therefore our sheer existence and clearly need to be understood.

In this project, we have tried to solve the sign problem by employing fully symmetric and polynomially exact integration rules over integration variables from a compact group. In particular, we have considered Abelian groups $U(1)$ and also non-Abelian groups $U(N)$ and $SU(N)$ for $N > 1$, as they are used in models of high energy physics. For a simple test model, the so-called 1-dimensional quantum chromodynamics (QCD)[1], we could clearly demonstrate that this fully symmetric integration rules work at (arbitrary) machine precision level, outperforming MCMC methods by far [10]. In particular, we could demonstrate numerically that the polynomial exactness of this method also holds for the case when a phase factor appears in the integrand, i.e. when the sign problem occurs and when MCMC completely fails.[2]

[1] QCD is the theory of the strong interaction between quarks and gluons.
[2] For a related approach see refs. [11, 12].

However, for the model considered in [10] the complexity and the corresponding dimensionality was very small. And, when the integration dimension is increased, the complexity of the full symmetrization approach and the associated computational cost fast becomes prohibitively large such that realistic models cannot be treated with this method. Still, the extremely high precision we could reach in [10], owing to the polynomial exactness of the integration rule, motivated us strongly to try to apply these integration rules also in higher physical dimensions.

The idea we follow in this article is to combine the symmetrized integration rules with MCMC. We expected that we can sample sufficiently many symmetrized integration points to achieve precise results, even in cases when a phase factor appears. In order to overcome the MCMC problem for a phase factor, we included the phase factor into the physical quantity to be computed leading to a modified expectation value. The correct expectation value of the observables of interest can then be reconstructed through a reweighting procedure as described in this article.

The model we have analyzed is the so-called quantum mechanical rotor defined through integration variables from the compact, Abelian group $U(1)$ and formulated on a 1-dimensional discrete time lattice [13]. We provide a mathematical description of the model in the following section. The quantum mechanical rotor is still a very simple model with respect to realistic theories in high energy physics, but shows already some basic features of these models such as the use of compact integration variables and very similar type of interactions. We remark that we used this model already to test successfully a recursive numerical integration technique [13]. By applying a polynomially exact, full symmetrization rule to a very small system of only four discrete time points, having thus a four-dimensional integration problem, we could achieve a very high precision in the evaluation of the model with and without a phase factor. However, combining MCMC with some particular symmetrization schemes, as discussed in this article, did not lead to an improvement compared to MCMC. By analyzing in detail the way how the full symmetrization reaches the observed accurate results, we were able to find a possible explanation of this failure as we discuss at the end of this article. For an alternative to solve high-dimensional integration problems presented at the IWOTA 2017 we refer to [14]. An excellent introduction to the approach of quasi Monte Carlo methods for solving high-dimensional integrals has been given in the plenary presentation by F. Kuo at this conference [15].

This article is organized as follows. In Section 2, we give an introduction to the quantum mechanical rotor. We then proceed in Section 3 to describe the used symmetrized integration rules applied to the quantum mechanical rotor. Section 4 is devoted to numerical experiments combining MCMC and a symmetrization technique when no phase factor appears. Section 5 discusses the case when a highly oscillatory, complex integrand is added. Section 6 aims at an explanation of the failure we observe for the case of Section 5, and in Section 7 we formulate our conclusions.

2. The quantum mechanical rotor

Let us consider the coordinate $x(t) \in \mathbb{R}$ describing the trajectory of a particle in time. In classical mechanics, the path of such a particle starting at $x_0 = x(0)$ and ending at $x_1 = x(T)$ is given by the minimum of the action $S(x)$ (with respect to $x(t)$),

$$S(x) = \int_0^T L(x,t)\,\mathrm{d}t \in \mathbb{R},$$

where the Lagrangian $L(x,t)$ defines the model under consideration.

In order to analyze a particular model, or, equivalently, a Lagrangian, we will define the trajectory $x(t)$ on an Euclidean, equidistant time lattice with spacing a. Hence, we replace the continuum trajectories $x(t)$ by their discretized counterparts:

$$\begin{aligned} t \to t_i &:= i \cdot a,\ i \in \{0,1,2,\dots,d-1\}, \\ x(t) \to x_i &:= x(t_i),\ x_i \in \mathbb{R}, \end{aligned} \tag{2.1}$$

with $T = a \cdot d$, such that

$$\int_0^T \mathrm{d}t \to a \sum_{i=0}^{d-1}. \tag{2.2}$$

The continuum limit is reached by sending $a \to 0$ and $d \to \infty$, keeping $a \cdot d$ fixed. We will use cyclic boundary conditions $x_d := x_{d(\bmod d)} = x_0$ throughout this article.

In quantum mechanics, the system does not follow a single path starting at $x_0 = x(0)$ and ending at $x_T = x(T)$ with a stationary action, but the behavior of the system depends on all possible paths and the value of their actions. Following Feynman [1], the quantum mechanical system is then given by the (Euclidean) path integral

$$\int_{D^d} \mathrm{e}^{-S[x]} \mathrm{d}x, \tag{2.3}$$

where D is a domain in $\mathbb{R}$. The performed time discretization of the quantum mechanical system results in a well-defined integral (2.3). Note, however, that already for such a quantum mechanical system we may be faced with a high dimension d, which can easily reach several thousands.

Exactly the same prescription to quantize a system as described above is applied to models in high energy physics. There, in addition to discretized time, also three additional (discretized) space dimensions have to be taken into account. With these additional internal degrees of freedom it becomes clear that for such models the dimensionality after lattice discretization is very high. Typical space time lattices considered today involve mesh sizes like $64^3 \cdot 128$, with 12 internal degrees of freedom at each space-time grid point, see e.g. [4].

Coming back to the quantum mechanical model considered in this study, a physical quantity, i.e. an expectation value $\langle O[x] \rangle$ of an observable

$O[x] = O(x_0, x_2, \ldots, x_{d-1})$ can be calculated using the path integral formalism

$$\langle O[x] \rangle = \frac{\int_{D^d} O[x]\, \mathrm{e}^{-S[x]} \mathrm{d}x}{\int_{D^d} \mathrm{e}^{-S[x]} \mathrm{d}x}. \tag{2.4}$$

These observables are the objects which are of main interest in quantum physics. For example such an observable could be the mass of a proton which is the nucleus of a hydrogen atom.

The model

The particular model we are considering in this article is the topological oscillator or quantum mechanical rotor [16, 13]. It describes a particle with mass M_0 moving on a circle with radius R. Hence, its moment of inertia is given by $I = M_0 R^2$. As mentioned above, this model has already some characteristic features of gauge theories (conjugacy classes) which are of prime importance in high energy physics.

The discretized action of the model is given by

$$S[\phi] = \frac{I}{a} \sum_{i=0}^{L-1} \left(1 - \cos\left(\phi_{i+1} - \phi_i\right)\right), \tag{2.5}$$

where we take a periodic boundary condition, $\phi_L = \phi_0$, L being the number of discrete time points, i.e. the dimensionality of the problem. In the following we will, if not stated otherwise, set the lattice spacing size to $a = 1$. This means in particular that the continuum limit can be reached by taking $T = L \cdot a \to \infty$.

The action given in eq. (2.5) can be rewritten using variables from the compact group $U(1)$, $\Phi_l = e^{i\phi_l}$ as

$$S[\Phi] = \frac{I}{a} \sum_{l=1}^{L} \Re(1 - \Phi_l^* \Phi_{l+1}). \tag{2.6}$$

In this form the model can be also generalized to variables $\Phi_i \in U(N)$ or even $\Phi_i \in SU(N)$. The observable we consider in this work is the real part of the *link variable*

$$O = \mathscr{L} = \frac{1}{L} \sum_{l=1}^{L} \Phi_l^* \Phi_{l+1}, \tag{2.7}$$

and we will be interested in the expectation value $\Re\langle O \rangle$ of this observable.

3. Symmetrization

In this section we introduce the concept of symmetrized integration. We will distinguish two methods using the symmetrization idea. In the first approach, we apply a *full symmetrization* which has delivered very satisfactory results in a different 1-dimensional QCD model considered in [10]. We consider in the second method a combination of symmetrization technique and MCMC, which we term *locally full symmetrization*. The name originates from the fact that not all lattice points are fully symmetrized globally, but only local points are symmetrized, see the following discussion.

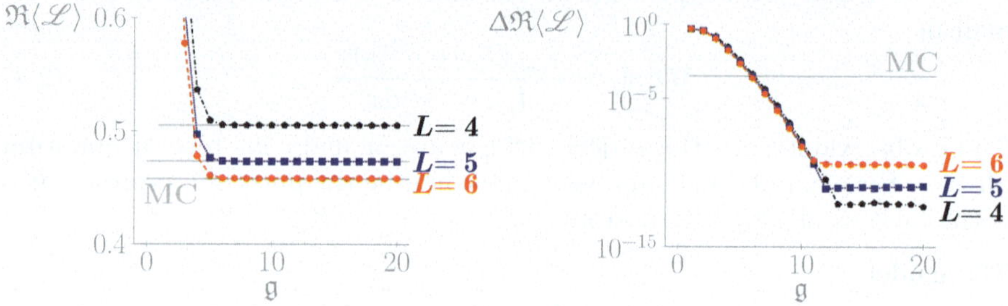

FIGURE 1. The full symmetrization results compared to the corresponding MCMC calculation results for three different lattice sizes L ($\frac{I}{a} = 1$). For $\mathfrak{g} \gtrsim 10$ we always find consistent values (left). The error of the full symmetrization method shrinks exponentially until it reaches a plateau which is orders of magnitude below the MCMC error.

As detailed below, we found that without an oscillating phase factor the full symmetrization method produces right results on machine precision for fewer than ten lattice points and different values of I in eq. (2.5). Also adding a phase factor, the full symmetrization method can reach much higher precision than MCMC when the number of symmetrization points is large enough.

The one-variable symmetrization points, as applied by us to one-dimensional QCD [10], for $\mathcal{U}(1)$ are the $\mathfrak{g}$ roots of unity

$$\{e^{(\frac{2\pi i k}{\mathfrak{g}})}, k \in \{1, \dots, \mathfrak{g}\}\} = \mathscr{G}_{\mathfrak{g}}. \tag{3.1}$$

The variables Φ_i of eq. (2.6) at all lattice points i of the quantum mechanical rotor are in $\mathcal{U}(1)$ as well, which motivates to use the symmetrization entries $G^k \in \mathscr{G}_{\mathfrak{g}}$ as sampling points for each variable. As a generalization of the one-dimensional QCD case we use a hyperrectangular rule to approximate the multi-dimensional integrals in (2.4) to

$$\langle O \rangle \approx \frac{\frac{1}{\mathfrak{g}}\sum_{k_1=1}^{\mathfrak{g}} \cdots \frac{1}{\mathfrak{g}}\sum_{k_L=1}^{\mathfrak{g}} O[G^{k_1}, \dots, G^{k_L}]\, w[G^{k_1}, \dots, G^{k_L}]}{\frac{1}{\mathfrak{g}}\sum_{k_1=1}^{\mathfrak{g}} \cdots \frac{1}{\mathfrak{g}}\sum_{k_L=1}^{\mathfrak{g}} w[G^{k_1}, \dots, G^{k_L}]}, \tag{3.2}$$

with $w[G^{k_1}, \dots, G^{k_L}] = e^{-S[G^{k_1}, \dots, G^{k_L}]}$. Note that the fully symmetrized form of (3.2) includes all possible combinations of the symmetrization sampling points in the sums and is therefore the product quadrature rule.

We applied the full symmetrization method to compute the expectation value of the link observable in eq. (2.7) following the prescription in eq. (3.2). Our results are shown in Fig. 1 as a function of the number of symmetrization points $\mathfrak{g}$. There, we also show the corresponding truncation error which we take to be the difference to the computed expectation value for $\mathfrak{g} = 30$.

Although the symmetrized integration clearly outperforms the MCMC method as can be seen in the right graph of Fig. 1, from inspecting eq. (3.2) it is obviously unfeasible to apply this method when systems with a high dimension are considered since the scaling is proportional to $L^{\mathfrak{g}}$. For high-dimensional problems, still some kind of importance sampling would therefore be required and, following this observation, we will combine in the following symmetrization with MCMC as our method of choice.

4. Applying locally full symmetrization in combination with MCMC

As we saw in the previous section, the method of full symmetrization gives much smaller error estimates than MCMC applied to the quantum mechanical rotor when the dimension, i.e. the number of time lattice points is small, $L \leq 10$. Unfortunately, the computational cost prevents us to apply this method to higher-dimensional systems. On the other hand, MCMC is usually very efficient when applied to high-dimensional integrals. It seems therefore a very natural idea to try to combine both methods. As a first step, in which we leave out a phase factor in the integrand, we used an MCMC algorithm on a symmetrized weight $w_{\text{sym}}[\mathbf{\Phi}]$ to compute the expectation value of a symmetrized observable function $O_{\text{sym}}[\mathbf{\Phi}]$. The concrete form of the symmetrized weight and observable are given further below and here we denote them in a general form as

$$[\langle O_{\text{sym}} \rangle] \equiv \frac{\int O_{\text{sym}}\, w_{\text{sym}}\, \mathrm{d}\mathbf{\Phi}}{\int w_{\text{sym}}\, \mathrm{d}\mathbf{\Phi}}. \tag{4.1}$$

The square brackets in $[\langle O_{\text{sym}} \rangle]$ express that this is not the standard expectation value with respect to the weight function $w = \mathrm{e}^{-S}$ in (2.4), but the modified *symmetrized* expectation value with respect to a *symmetrized* weight. Our expectation has been that by using this form of the observable large cancelations would occur for each configuration generated in the MCMC process and therefore that a much better accuracy could be reached, as in the case of full symmetrization discussed in Section 3.

To compute O_{sym} and w_{sym} we used the following particular form of a locally full symmetrization, which has been inspired by the work in [17]: At each lattice point i (that is for each variable Φ_i) our realization of a locally full symmetrization used a random permutation P_i to permute the symmetrization entries $G^1, \ldots, G^{\mathfrak{g}}$ to $G^{P_i(1)}, \ldots, G^{P_i(\mathfrak{g})}$. All first entries of these permuted symmetrizations at each lattice point are combined to form a symmetrized lattice $[\boldsymbol{G}^{\boldsymbol{P}(1)}] = [G^{P_1(1)}, G^{P_2(1)}, \ldots, G^{P_L(1)}]$. This works similarly for all kth entries of the permuted symmetrizations, such that $\mathfrak{g}$ symmetrized lattices are created (see Fig. 2).

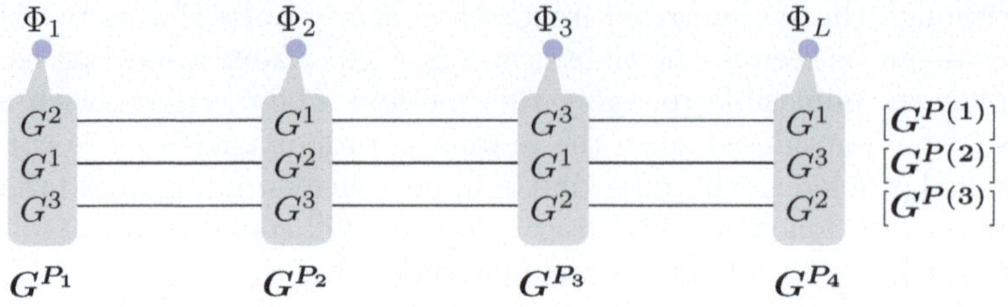

FIGURE 2. An example set in the locally full symmetrization for $L = 4$ and $\mathfrak{g} = 3$: at each lattice point i all symmetrization entries $\{G^1, G^2, G^3\}$ are permuted using a random permutation $\boldsymbol{P_i} = (P_i(1), P_i(2), P_i(3))$ to form $\boldsymbol{G^{P_i}}$. The combinations of all kth entries of these permutations, $k \in \{1, 2, 3\}$, are taken into account as the symmetrized lattices $[\boldsymbol{G^{P(k)}}] = [G^{P_1(k)}, G^{P_2(k)}, G^{P_3(k)}, G^{P_4(k)}]$.

The combination with MCMC consists in the step that at each lattice point i the symmetrization entries $G^{P_i(k)}, k \in \{1, \ldots, \mathfrak{g}\}$ are multiplied with the real dynamical degree of freedom Φ_i. The so generated symmetrized lattices are therefore of the form

$$[\boldsymbol{G^{P(k)}}\Phi] = [G^{P_1(k)}\Phi_1, G^{P_2(k)}\Phi_2, \ldots, G^{P_L(k)}\Phi_L], \qquad k \in \{1, \ldots, \mathfrak{g}\}. \quad (4.2)$$

We then perform a Metropolis test to either accept or reject these symmetrized configurations at each lattice point separately. Note that the computational effort of this combination of locally full symmetrization and MCMC scales only linearly with the number of variables used instead of exponentially in the case of full symmetrization.

We used these symmetrized lattices to compute the symmetrized observable and weight separately,

$$O_{\text{sym}} = \sum_{k=1}^{\mathfrak{g}} O[\boldsymbol{G^{P(k)}\Phi}], \quad (4.3)$$

$$w_{\text{sym}} = \sum_{k=1}^{\mathfrak{g}} w[\boldsymbol{G^{P(k)}\Phi}], \quad (4.4)$$

with the weight function $w \in \mathbb{R}$. We also developed formulae to retrieve the expectation value $\langle O \rangle$ from the symmetrized expectation value $[\langle O_{\text{sym}} \rangle]$, given that the observable function is a polynomial in the lattice variables Φ_i,

$$O = \sum_{t=1}^{L} O_t, \text{ with } O_t = \prod_{i=1}^{L} \Phi_i^{\alpha_i^t}, \quad (4.5)$$

where α_i^t is the ith component of the vectors $\alpha^t \in \mathbb{Z}^L$, one for each summand O_t. Note that the expectation value of the quantity of interest, the link, is a polynomial in the Φ_i's. Its α^t's are given by $\alpha_t^t = -1, \alpha_{t+1}^t = 1$ and all other $\alpha_k^t = 0$. The desired expectation value of O is the sum of the expectation values of the O_t's and they can be derived from the symmetrized expectation values $[\langle O_{\text{sym},t}\rangle]$. Following the appendix, we find for the true expectation value

$$\langle O_t \rangle = \frac{1}{c_t}[\langle O_{\text{sym},t}\rangle], \tag{4.6}$$

where the correction factor c_t is given by

$$c_t = 1 + \frac{1}{\mathfrak{g}} \sum_{i \neq j} \left(G^{P_t(i)}(G^{P_t(j)})^*\right)^* \left(G^{P_{t+1}(i)}(G^{P_{t+1}(j)})^*\right). \tag{4.7}$$

Note that this correction factor is independent of the variables Φ_i, which is the underlying reason to allow to relate the symmetrized expectation value to the true one.

The concrete computation of the expectation value $\langle O \rangle$ is shown in Section 4.1 and results are presented in Section 4.3.

4.1. Computing the expectation value

Combining (4.5) with (4.6) to compute the expectation value of O results in

$$\langle O \rangle = \sum_{t=1}^{L} \langle O_t \rangle = \sum_{t=1}^{L} \frac{[\langle O_{\text{sym},t}\rangle]}{c_t}. \tag{4.8}$$

In the following, one *set* is a given random permutation of the symmetrization points, as written in (4.2). For one chosen set of symmetrized lattices we used N_{MC} MCMC steps to obtain an estimate for

$$[\langle O_{\text{sym},t}\rangle] \approx \frac{1}{N_{\text{MC}}} \sum_{i=1}^{N_{\text{MC}}} O_{\text{sym},t\,i}. \tag{4.9}$$

We used N_{sets} different sets of symmetrized lattices to reduce the error on the considered observable. Then the final observable is computed by

$$\langle O \rangle_{\text{average}} = \frac{1}{N_{\text{sets}}} \sum_{j=1}^{N_{\text{sets}}} \left(\frac{1}{N_{\text{MC}}} \sum_{i=1}^{N_{\text{MC}}} \sum_{t=1}^{L} \frac{O_{\text{sym},t\,ij}}{c_t} \right). \tag{4.10}$$

In the following, this averaged observable is called $\langle O \rangle$ for simplicity. The error follows from the standard deviation of the mean observable over all sets.

The two important variables in equation (4.10), c_t and $O_{\text{sym},t}$, are now discussed further. First the distribution of the correction factor values is investigated and then the MCMC-step to compute $O_{\text{sym},t}$ is explained.

Choice of a cut value for the correction factor

Before starting test simulations to compute $O_{\text{sym},t}$ with MCMC, we needed to check that the correction factors c_t in (4.7) are not becoming too small for the sets of symmetrizations we are going to use, because there is no reason for c_t not to be very small. And, since it appears in the denominator in (4.10), it would afflict our sample by introducing unnaturally large contributions to the targeted observable if c_t becomes too small.

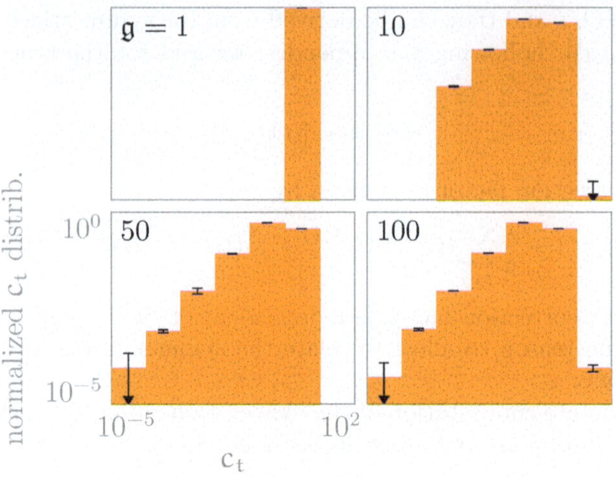

FIGURE 3. At least 90% of the correction factor values c_t, for $t \in \{1, 2, 3, 4\}$, lie in the region $c_t \in [10^{-1}, 10^1)$ for all tested $\mathfrak{g}$. For better visibility the binwidth and the distribution-axis scale exponentially.

We found that at least 90% of the computed values of the correction factor c_t, $t \in \{1, \ldots, T\}$ for $T = 4$, lie in $[10^{-1}, 10^1)$ for the tested $\mathfrak{g}$'s (Fig. 3). Most of the remaining c_t values are smaller. Here we computed 10 times the correction factors of 10000 sets to get an error estimate. To avoid the small c_t's, we applied different lower bounds $c_{\min}$ on the correction factors and rejected all sets which contained at least one c_t value smaller than $c_{\min}$. For each $c_{\min}$ we checked how many sets are rejected and how this affects the final link variable result. Of course, it is preferable to reject as few sets as possible because this cut is not physically motivated but just computationally necessary. Additionally, a larger amount of rejected sets leads to longer runtimes because new sets have to be created and checked again. And the larger amount of rejected sets could induce a systematic bias. Hence, our criterion for the final choice of the cut parameter $c_{\min}$ has been that only a very small amount of sets are to be rejected. In Fig. 4 we show the number of rejected sets as a function of $c_{\min}$.[3] As can be seen, for $c_{\min} \lesssim 10^{-2}$ the number of rejected sets is very small for all numbers of symmetrizations $\mathfrak{g}$ used. We also checked explicitly that for this value of $c_{\min} = 10^{-2}$ the result of the link expectation value is compatible with the corresponding standard-MCMC value using the cluster algorithm of [18] and the error is comparable. We therefore will use a value of $c_{\min} = 10^{-2}$ for all our further experiments.

[3]The simulations ran until 100 sets were accepted. For each data point we repeated the simulation 10 times to get an error estimate. Data points for all $\mathfrak{g} > 1$ are slightly shifted for better visibility.

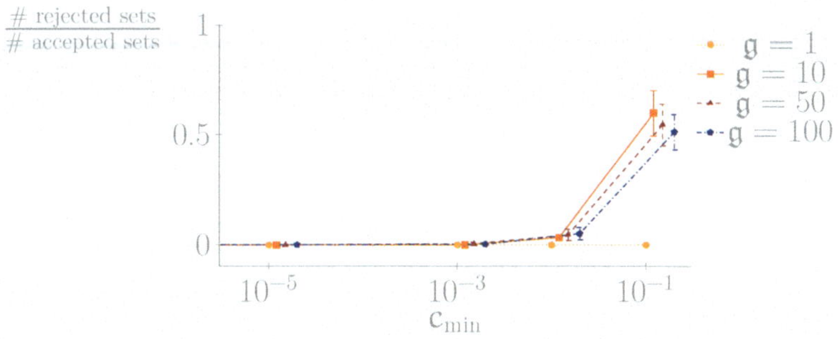

FIGURE 4. For large lower bounds $c_{\min} \gtrsim 10^{-1}$ on the correction factor c_t, given by (4.7), a significant number of sets is rejected for $\mathfrak{g} > 1$. For all smaller lower bounds tested, less than 10% of the sets are rejected.

4.2. Metropolis algorithm

As explained above, it is our goal to combine the symmetrization with the MCMC method. For this purpose, we will use a very simple MCMC method, namely the Metropolis algorithm, although the same principle can be applied to other MCMC algorithms as well. The Metropolis step in this algorithm consists of three substeps:

a. At lattice point i we choose

$$\Phi_i^{\text{new}} = \Phi_i^{\text{old}} \cdot \mathrm{e}^{i\pi \cdot \text{random}(-1,1)}, \tag{4.11}$$

with a uniform random number.

b. Then we compute the Boltzmann factor of the old and new variable,

$$B_i^{\text{old}} = \sum_{k=1}^{\mathfrak{g}} \mathrm{e}^{-S(G^{P_1(k)}\Phi_1,\ldots,G^{P_i(k)}\Phi_i^{\text{old}},\ldots,G^{P_L(k)}\Phi_L)}, \tag{4.12}$$

$$B_i^{\text{new}} = \sum_{k=1}^{\mathfrak{g}} \mathrm{e}^{-S(G^{P_1(k)}\Phi_1,\ldots,G^{P_i(k)}\Phi_i^{\text{new}},\ldots,G^{P_L(k)}\Phi_L)}. \tag{4.13}$$

c. Finally we accept Φ_i^{new} if

$$\frac{B_i^{\text{new}}}{B_i^{\text{old}}} > \text{random}(0,1)\,. \tag{4.14}$$

Step b seems very time intensive. This is especially important because the Metropolis step is repeated very often in the simulation. But the full Boltzmann factor does not need to be recomputed for every Φ_i^{new}. The action $S[\Phi]$ is a sum over local actions $S_i[\Phi_i, \Phi_{i+1}]$ and if one Φ_i changes, only two terms, S_i and S_{i-1}, are changing. Therefore the new action can be computed from the old action and two local action changes

$$S[\Phi]^{\text{new}} = S[\Phi]^{\text{old}} - (S_i^{\text{old}} + S_{i-1}^{\text{old}}) + (S_i^{\text{new}} + S_{i-1}^{\text{new}})\,. \tag{4.15}$$

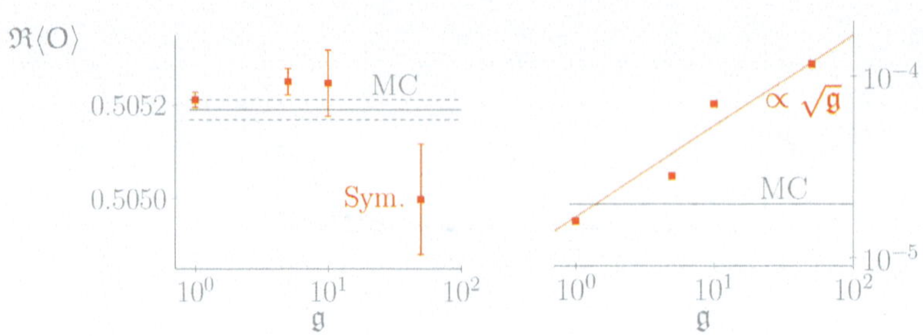

FIGURE 5. The real link variable results of the combined MCMC/symmetrization method (10 sets with 10^8 configurations each) as a function of $\mathfrak{g}$. We compare the results from this simulation to a conventional cluster algorithm simulation value (10^9 configurations) (left graph). The error grows proportional to $\sqrt{\mathfrak{g}}$, necessitating significantly larger statistics for increasing $\mathfrak{g}$ to match the accuracy of the cluster algorithm (right graph).

The same is true for the symmetrized actions $S[G^{P(k)}\Phi]^{\text{new}}$. For all symmetrization indices k we computed and saved $S[G^{P(k)}\Phi]^{\text{old}}$ once at the start. Then for each Metropolis step we computed $S[G^{P(k)}\Phi]^{\text{new}}$ with (4.15) for all k, saved them, and summed them up to get B_i^{old} and B_i^{new}. If Φ_i^{new} is accepted, all $S[G^{P(k)}\Phi]^{\text{old}}$ have to be overwritten by $S[G^{P(k)}\Phi]^{\text{new}}$.

4.3. Results

In Fig. 5 we show our results from our combined MCMC/symmetrization simulation as a function of $\mathfrak{g}$, as well as the corresponding error. Although we find a good agreement with the cluster algorithm result in general, we observe that for larger values of $\mathfrak{g}$ the expectation value fluctuates more. These results were obtained with 10^8 configurations times 10 permutation sets and compared to conventional MCMC using the cluster algorithm with 10^9 configurations. The growing of the error suggests that still significantly larger statistics are needed to reach the precision of the cluster algorithm. The error increase proportional to $\sqrt{\mathfrak{g}}$ is probably due to the fact that for larger $\mathfrak{g}$ more terms are added up to compute the link observable (4.3). Therefore the variation of the sum grows with larger $\mathfrak{g}$, and the errors get larger.

5. The topological rotor with a phase factor

Besides the desire to find methods that can compute physical observables in a more efficient way than MCMC the main motivation of this work was to try to overcome the sign problem MCMC has to face when the integrand contains a phase factor. In this section, we will address this question again for the two cases where we perform a full symmetrization and a locally full symmetrization by adding a phase factor to the integrand of the path integral of the quantum mechanical rotor.

5.1. Full symmetrization in presence of a phase factor

From our experience in the case of 1-dimensional QCD [10] we expect that also for a situation with a phase factor the full symmetrization method should outperform MCMC. We recall that the expectation value of an observable function $O[\mathbf{\Phi}]$ is generally an integral over two functions, the observable function $O[\mathbf{\Phi}]$ and the normalized weight function $\frac{w[\mathbf{\Phi}]}{\int w[\mathbf{\Phi}]\mathrm{d}\mathbf{\Phi}}$.

In case we have a phase factor, the weight function splits into two parts $w[\mathbf{\Phi}] = w_r[\mathbf{\Phi}]w_p[\mathbf{\Phi}]$, the real weight function $w_r[\mathbf{\Phi}]$ and the complex phase weight function $w_p[\mathbf{\Phi}]$. In the previous sections, we have only considered the case $w_p[\mathbf{\Phi}] = 1$, where the weight function can be interpreted as a probability distribution. The presence of a factor $w_p[\mathbf{\Phi}] \in \mathbb{C}$ (and not in $\mathbb{R}$) is responsible for the sign problem of standard MCMC methods, see e.g. [7] and references within.[4] With this weight function the expectation value has the form

$$\langle O \rangle = \frac{\int O\, w_r\, w_p\, \mathrm{d}\mathbf{\Phi}}{\int w_r\, w_p\, \mathrm{d}\mathbf{\Phi}}. \tag{5.1}$$

The particular form of the phase factor w_p we have chosen is given as

$$w_p = \prod_{l=1}^{L} (\Phi_l)^{\theta}. \tag{5.2}$$

This term emulates the situation when a complex phase factor appears and provides a suitable testbench for addressing the sign problem, in particular for the physically most relevant cases of a chemical potential or a topological term in the action. We have evaluated the expectation value $\Re\langle \mathscr{L} \rangle$ with the method of full symmetrization using (3.2). In Fig. 6 we show the result of this expectation value as a function of the parameter θ. As we can see, although the behavior of $\Re\langle \mathscr{L} \rangle$ is rather irregular in θ, with this full symmetrization approach the oscillating integral can be evaluated very well. Here the oscillations cancel each other, such that the result is very accurate.

The behavior of the truncation error, i.e. the difference of the real part of the link expectation value compared to the value at $N_{\mathrm{sym}} = 100$, is shown in Fig. 7 as a function of N_{sym} for various values of the parameter θ. We see that for larger θ the truncation error increases and reaches the value at $N_{\mathrm{sym}} = 100$ more slowly than for $\theta = 0$. Still, the absolute error is very small and outperforms the error from the cluster algorithm simulation, especially for larger θ values not shown in Fig. 7, where the MCMC error is large.

---

[4]For readability the dependence of the mentioned functions on $\mathbf{\Phi}$ in the following is only explicitly written out when needed.

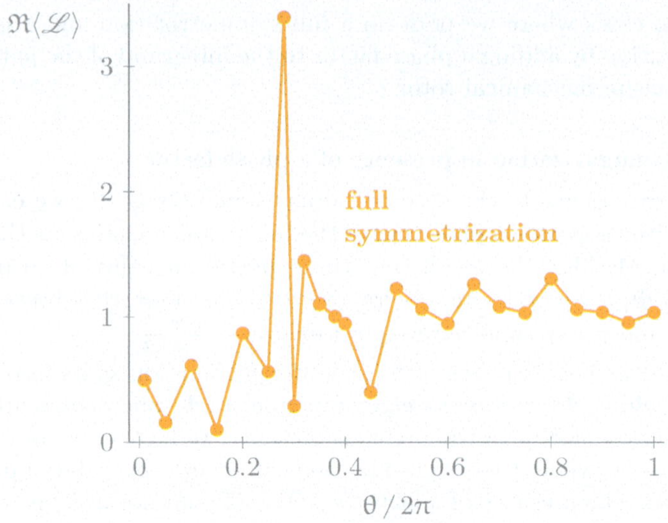

FIGURE 6. With full symmetrization ($\mathfrak{g} = 80$) the real part of the link observable is computed very accurately, such that also small fluctuations are resolved. The irregular behavior is due to the choice of the phase weight function (5.2).

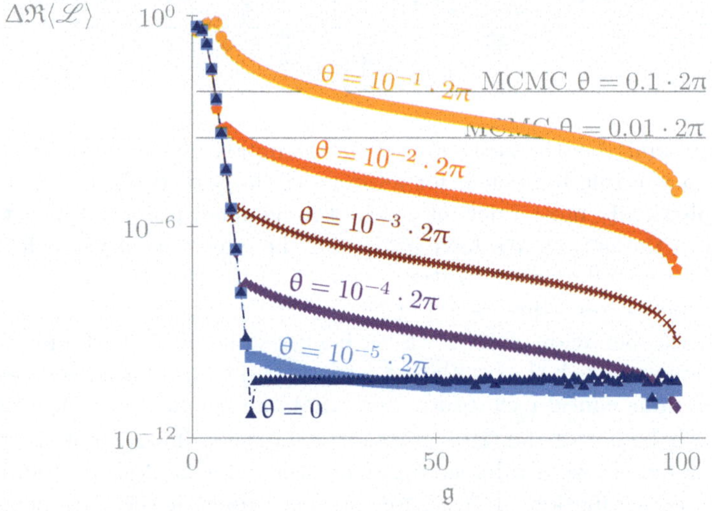

FIGURE 7. The error of the real part of the link variable changes its behavior for $\theta > 0$ and does not reach the machine precision plateau as in the case of $\theta = 0$. The grey lines are the errors resulting from the MCMC calculation with the cluster algorithm at $\theta = 0.01 \cdot 2\pi$ and $0.1 \cdot 2\pi$ for 10^6 configurations.

Given our experience from the 1-dimensional QCD model the very good behavior of the full symmetrization was, of course, expected. But, it is very reassuring that full symmetrization works in this more complicated model of the quantum mechanical rotor with a phase factor, in particular, since the integrand is non-polynomial. In addition, the accurate results for the link expectation value in Fig. 6 from the full symmetrization can serve to test the validity of the calculations where we combine MCMC with symmetrization, as we will discuss in the next section.

5.2. Results for combining MCMC with symmetrization in presence of a phase factor

With (5.1) we can evaluate the real expectation value of the link variable by taking the phase factor w_p of (5.2) as part of the observable. For a MCMC method, this means that we generate configurations of rotor variables with a positive weight $w_r \mathrm{d}\Phi / \int w_r \mathrm{d}\Phi$ and then compute the desired expectation value with the help of (5.1), a procedure known as *reweighting* [19]. Note that the here described reweighting procedure can also be applied in a straightforward way to the symmetrized observables and weights.

We first applied the reweighting procedure to a pure MCMC simulation. For a small value of $\theta = 0.01 \cdot 2\pi$, we found that the reweighting procedure works well and we obtain $\Re\langle\mathscr{L}\rangle = 0.49881(34)$. However, when we increase θ, the error increases steadily and for $\theta = 0.5{\cdot}2\pi$ we find $\Re\langle\mathscr{L}\rangle = 0.78(15)$. Thus, at this value of θ we cannot compute the link expectation value accurately anymore, which is just a manifestation of the sign problem in the MCMC approach.

Proceeding with the combined MCMC/symmetrization method, we then wanted to see whether this technique can improve this situation. Inserting (4.6) into (5.1) results in

$$\langle O \rangle_\theta = \frac{\sum_{t=1}^{L} [\langle O_{\mathrm{sym},t}\, w_p \rangle]/a_t}{[\langle w_p \rangle]/b} \tag{5.3}$$

with correction factors a_t and b. Here $[\langle \ldots \rangle]$ is the symmetrized expectation value defined through the symmetrized weight w_{sym} in (4.1). Note that here $w_{\mathrm{sym}} = \sum_{k=1}^{g} w_r[\boldsymbol{G}^{\boldsymbol{P(k)}}\boldsymbol{\Phi}]$ depends only on the real part, w_r, of w and hence is real itself (compare (4.4)). By inserting the specific forms of the link variable and the phase factor, the symmetrized expectation value of the numerator is

$$[\langle O_t\, w_p \rangle] = \left[\left\langle \prod_{k=1}^{L} \Phi_k^{\alpha_k^t} \prod_{l=1}^{L} \phi_t^{-\theta} \right\rangle\right] = \left[\left\langle \prod_{k=1}^{L} \Phi_k^{\alpha_k^t - \theta} \right\rangle\right]. \tag{5.4}$$

This is a monomial of non-integer degree. Therefore the connection between symmetrized expectation value $[\langle \ldots \rangle]$ and standard expectation value $\langle \ldots \rangle$ (derived in the appendix) is still valid, but the formula of the correction factor (4.7) has to be slightly modified for a_t and b. Comparing the observable in (5.4) with the observable (4.5) we find that we have to rescale $\alpha_k^t \to \alpha_k^t - \theta$ to compute a_t, and additionally, we need to rescale $\alpha_k^t \to -\theta$ to obtain b.

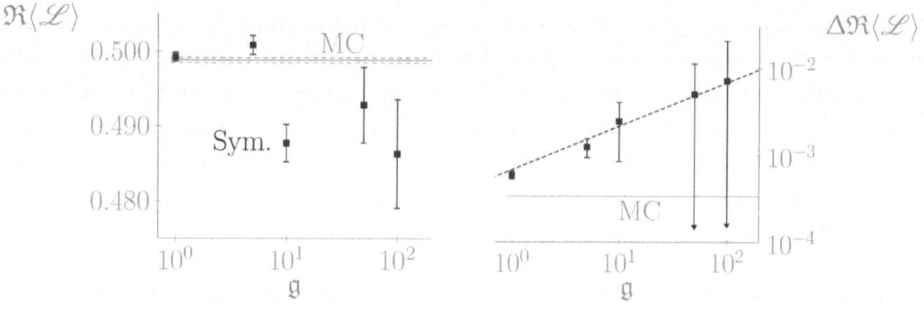

FIGURE 8. In the left graph we show the real link variable results of the combined MCMC/symmetrization method as a function of $\mathfrak{g}$. Here we use $\theta = 0.01 \cdot 2\pi$. The data from the combined method fluctuate around the MCMC value but are consistent within the error. Here, the number of symmetrization sets used is 10^4 with 100 configurations for each set. In the right graph we show the error of the combined method as function of $\mathfrak{g}$. This error grows with $\sqrt{\mathfrak{g}}$ and is clearly larger than the MCMC error for 10^6 configurations.

Therefore the correction factors for the link observable are given explicitly by

$$a_t = 1 + \frac{1}{\mathfrak{g}} \sum_{k=1}^{\mathfrak{g}} \sum_{m \neq k} \left(G^{P_t(k)}(G^{P_t(m)})^*\right)^* \left(G^{P_{t+1}(k)}(G^{P_{t+1}(m)})^*\right) \cdot \prod_{l=1}^{L} \left(G^{P_l(k)}(G^{P_l(m)})^*\right)^{-\theta} \tag{5.5}$$

$$b = 1 + \frac{1}{\mathfrak{g}} \sum_{k=1}^{\mathfrak{g}} \sum_{m \neq k} \prod_{l=1}^{L} \left(G^{P_l(k)}(G^{P_l(m)})^*\right)^{-\theta}. \tag{5.6}$$

Hence, also when adding the θ-term, the correction factor remains independent of the rotor variables Φ_i.

In Fig. 8 we show our results from the combined method as a function of $\mathfrak{g}$ and compare them with the corresponding MCMC result for 10^6 configurations. Here we choose $\theta = 0.01 \cdot 2\pi$. For the combined method, we used 10^4 symmetrization sets and 100 configurations for each set. We see that the data from the combined method are compatible with the MCMC result but have significantly larger errors. In fact, the error resulting from the combined method increases with $\sqrt{\mathfrak{g}}$, as can be seen in the right graph of Fig. 8. We have to conclude therefore that already for the here chosen small value of $\theta = 0.01 \cdot 2\pi$ the combined method cannot compete with a standard MCMC method for values of, say, $\mathfrak{g} \sim O(10)$.

When we increase the value of θ the situation becomes worse and, similarly to the case of the pure MCMC simulation, at $\theta = 0.5 \cdot 2\pi$ we cannot reliably compute the expectation value of the real link variable, see also the discussion in the next paragraph. Thus, as in the case of $\theta = 0$, the combined method is not able to beat MCMC or even solve the sign problem of MCMC.

6. A possible explanation

As we discussed above, the proposed method of locally full symmetrizations in combination with MCMC does not improve the error estimates. Especially for the case of a complex weight factor and, in particular, for larger θ-values we did not see any improvement over MCMC. In addition, we observed that the error grows with the number of symmetrizations used. In this section, we attempt to find an explanation for these findings.

In striking contrast to the combination of symmetrization and MCMC, the full symmetrization was always found to be superior to MCMC. We therefore decided to have a more detailed look at the mechanism that leads to this superior behavior of the full symmetrization. Let us recall that in the full symmetrization approach the numerator and denominator are computed separately by summing over all possible combinations of the symmetrization sampling points (3.2).

To gain an insight into how the full symmetrization method produces the very precise results we obtained, we concentrate here only on the denominator, i.e. the partition function. However, we obtained similar, corresponding results for the numerator, which includes the observable, too.

The partition function is given by the summation of all possible combinations of the symmetrization entries G^k (the denominator of (3.2)),

$$Z = \frac{1}{\mathfrak{g}} \sum_{k_1=1}^{\mathfrak{g}} \cdots \frac{1}{\mathfrak{g}} \sum_{k_L=1}^{\mathfrak{g}} w[G^{k_1}, \dots, G^{k_L}] \tag{6.1}$$

$$= \frac{1}{\mathfrak{g}^L} \sum_{k=1}^{\mathfrak{g}^L} w[\boldsymbol{G}^{\boldsymbol{k}}]. \tag{6.2}$$

The second form renames the kth possible combination of symmetrization entries $\boldsymbol{G}^{\boldsymbol{k}}$ and sums over all these $\mathfrak{g}^L$ combinations.

In the following, we will investigate the distribution of the real and imaginary summands of this complex Z separately. We define the distribution of the real summands, $\mathcal{F}_{\Re(Z)}(w)$, to be the number of summands in $\Re(Z)$ with summand entry w, divided by the number of all summands in $\Re(Z)$,

$$\mathcal{F}_{\Re(Z)}(w_i) = \frac{1}{\mathfrak{g}^L} \sum_{k=1}^{\mathfrak{g}^L} \delta\left(\Re(w[\boldsymbol{G}^{\boldsymbol{k}}]) - \Re(w_i)\right). \tag{6.3}$$

The imaginary summands' distribution, $\mathcal{F}_{\Im(Z)}(w)$, is computed accordingly.

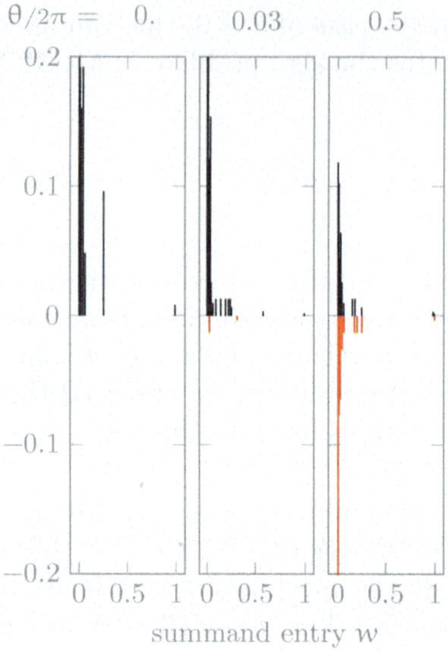

(A) $\mathfrak{g} = 5$

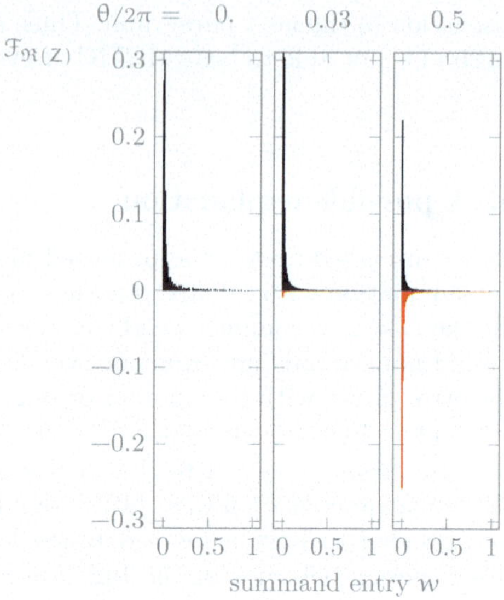

(B) $\mathfrak{g} = 30$

FIGURE 9. The distribution of summands in $\Re(Z)$, $\mathcal{F}_{\Re(Z)}$ of eq. (6.3), in the full symmetrization method as a function of the summand entry w for different θ-values. Left plot: $\mathfrak{g} = 5$, right plot: $\mathfrak{g} = 30$. By increasing θ, more and more negative summands arise in the sum to compute $\Re(Z)$. Especially the left plot for $\mathfrak{g} = 5$ shows that these negative summands approximately cancel with some of the positive summands. For $\mathfrak{g} = 30$ we find that many positive and negative summand entries between 0.5 and 1 occur in the sum of $\Re(Z)$, which cancel each other.

We show $\mathcal{F}_{\Re(Z)}$ for $\mathfrak{g} = 5$, where Z has only few different summands, in the left graph of Fig. 9 for a number of θ-values. For smaller values of θ there is a concentration of $\mathcal{F}_{\Re(Z)}$ at small summand entries, and only small fractions occur at summand entries larger than 0.5. For $\theta = 0.5$ there occur negative entries which are basically canceled by the positive entries, leading thus to a stable result. For $\mathfrak{g} = 30$ (right graph of Fig. 9) more and more negative entries occur for larger θ. Again we observe that at $\theta = 0.5$ the positive and negative summands are distributed very similarly, leading thus again to large cancelations. The distribution of the imaginary part of the summands is similar for all values of $\theta > 0$, see Fig. 10. For each $\theta > 0$ the distribution of the positive entries is similar to the distribution of negative entries, leading to large cancelations in the full symmetrization method. Of course, at $\theta = 0$ no imaginary part occurs.

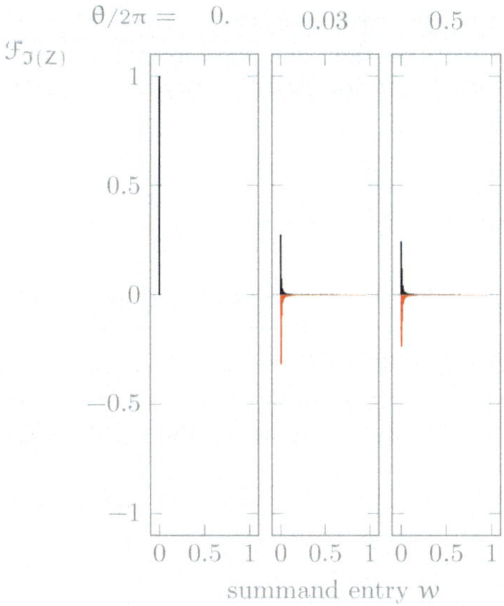

FIGURE 10. The distribution of summands in $\Im(Z)$, $\mathcal{F}_{\Im(Z)}$, i.e. the corresponding expression similar to (6.3), in the full symmetrization method as a function of the summand entry w for different θ-values. Already for a small value $\theta = 0.03$ imaginary summands occur in the sum of the $\Im(Z)$. The picture suggests that positive and negative entries occur with very similar probability, thus canceling each other; $\mathfrak{g} = 30$ is used here.

The above findings indicate that for $\theta > 0$ a lot of cancelations of summands occur in the full symmetrization method. To assure that all of these cancelations take place, all summands have to be taken into account. Although not shown here, we found a very similar behavior for the numerator in (5.1). This hints already at the reason for the problem to obtain stable results in the combined symmetrization and MCMC method (where we do not use all summands) if the sampled statistics are insufficient. Our results for the combined symmetrization and MCMC method, discussed above, suggest that the statistics required are, in fact, extremely large.

In order to better understand whether this conclusion makes sense, we computed the partition function by truncating the summands to some maximal value. The idea is to test how the value of the partition function behaves when we vary this maximal value. The real part of the so truncated partition function is the sum over all real summands smaller than some value w_{trunc},

$$\Re(Z_{\text{trunc}}(w_{\text{trunc}})) = \frac{1}{\mathfrak{g}^L} \sum_{k=1}^{\mathfrak{g}^L} \theta\left[\Re(w[\boldsymbol{G}^{\boldsymbol{k}}]) - \Re(w_{\text{trunc}})\right] . \qquad (6.4)$$

Similarly, we computed the corresponding imaginary part of the truncated partition function as well as the real and imaginary part of the truncated numerator A_{trunc} of the link observable, correspondingly. Finally, we then calculated the truncated real part of the link,

$$\mathscr{L}_{\text{trunc}} = \frac{\Re(A_{\text{trunc}}) + \Im(A_{\text{trunc}})}{\Re(Z_{\text{trunc}}) + \Im(Z_{\text{trunc}})} = \Re(\mathscr{L}_{\text{trunc}}) + \Im(\mathscr{L}_{\text{trunc}}). \tag{6.5}$$

In Fig. 11, top, we show the result for $\Re(\mathscr{L}_{\text{trunc}})$ as a function of the truncation parameter w_{trunc}. For the cases of $\theta = 0$ and $\theta = 0.01 \cdot 2\pi$, $\Re(\mathscr{L}_{\text{trunc}})$ shows a similar behavior: the smaller the value of the truncation parameter is, the more $\Re(\mathscr{L}_{\text{trunc}})$ changes. Thus, the small summands contribute more to $\Re(\mathscr{L})$, and the contribution of summands larger than 0.8 appears to be small. Therefore, $\Re(\mathscr{L})$ shows a saturation to the final value already at $w_{\text{trunc}} \approx 0.8$.

On the other hand, the imaginary part of the corresponding truncated link for $\theta = 0.01$ becomes non-zero, see Fig. 11, bottom, in contrast to the case of $\theta = 0$ where the imaginary part is zero. It appears that most contributions to $\Im(\mathscr{L}_{\text{trunc}})$ for $\theta = 0.01$ come from the small summands and that only for a truncation parameter larger about 0.8 a cancelation of the imaginary part occurs, giving finally a zero value for $\Im(\mathscr{L})$. This finding supports our explanation that a large number of summands have to be added to find a stable result and this happens to be difficult already for $\theta = 0.01$ with the combined symmetrization/MCMC method.

Let us finally discuss the case for $\theta = 0.5$. Inspecting Fig. 12, we see that the behavior of $\Re(\mathscr{L}_{\text{trunc}})$ and $\Im(\mathscr{L}_{\text{trunc}})$ is very irregular as a function of the truncation parameter. There is no hierarchy visible between small and large summands and $\mathscr{L}_{\text{trunc}}$ can assume very different values, depending on where the sum is truncated, which is also a measure for how many summands are used. Only if all summands are used, the correct result of the link observable is obtained. Thus, basically all summands are needed to obtain a sufficient cancelation of the various contributions and eventually a stable result.

How can the above observations help to understand why the combination of locally full symmetrizations and MCMC does not work? Initially, we expected that in this combined method, large oscillations would cancel and a stable result with small errors could be obtained. In particular, we anticipated that the relevant contributions would be centered around the final expectation value with a not too large spread. But Figs. 5 and 8 show that this picture does not seem to be correct and that the desired cancelations do not happen.

In order to find an explanation for this observation, we went back to the full symmetrization method. The way the final stable results are obtained here from the individual summands is very instructive to understand our observations, despite the fact that the computation of observables differ between both methods: the full symmetrization computes numerator and denominator separately, whereas the combined method uses MCMC and hence calculates an observable directly from a given probability density. Furthermore,

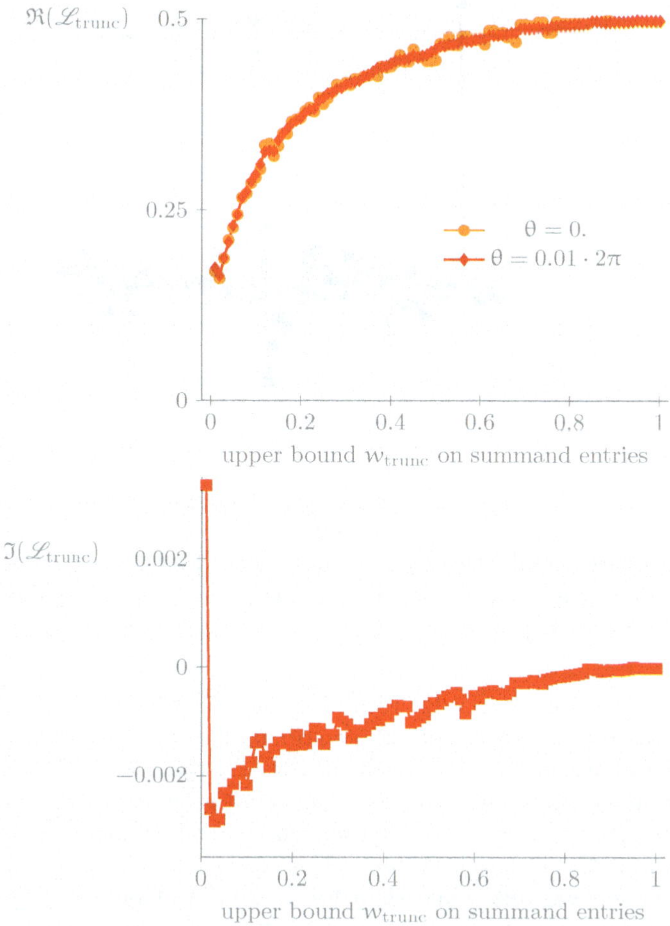

FIGURE 11. Top: for $\theta = 0$ and $\theta = 0.01$ mostly the small summand entries are relevant for computing the real part of the link variable. Here we used $\mathfrak{g} = 30$. Bottom: For the parameter $\theta = 0.01 \cdot 2\pi$ mostly the small summand entries are relevant for computing the imaginary part of the link variable. Here we used $\mathfrak{g} = 30$.

in the combined method we summed the symmetrized observable (4.3) and the symmetrized weight (4.4) over a number of combinations of symmetrization entries, but not over all possible ones, as done in the full symmetrization method. To improve the accuracy of our results, we additionally averaged the result over different combinations of symmetrization sets, which are the randomly chosen permutation sets of (4.10). Therefore, by using more permutation sets, more summands are taken into account, leading to a result which is closer to the full symmetrization result.

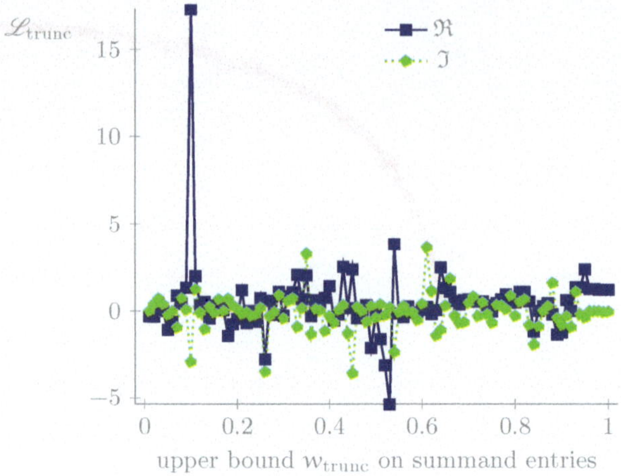

FIGURE 12. For $\theta = 0.5$ all real and imaginary summand entries contribute to the real and imaginary parts of the link observable. Only the sum of all real or imaginary entries result in the correct final real or imaginary link observable. Here we used $\mathfrak{g} = 30$.

With $\theta = 0$ this gives valid results, as seen in Fig. 5, if we use enough sets, because here it is possible to come close to the full symmetrization result. The full symmetrization result in Fig. 11, top, shows that if enough, but not too many summands are used, the result saturates to the final value. For a small value of $\theta = 0.01 \cdot 2\pi$ there are already imaginary summands contributing to the link observable, which means that we need more sets to compute the final link value than for $\theta = 0$. And indeed, Fig. 8 shows significantly larger errors than for $\theta = 0$ even with an already increased number of permutation sets used. The fluctuations are then increasing with larger values of θ since more and more summands are needed to eventually obtain stable results with small errors.

Finally, for $\theta = 0.5 \cdot 2\pi$ Fig. 12 shows that here only all summands together give the correct value. But with the combined method this situation can only be achieved by using many more permutation sets, which is inefficient to compute. Therefore we were not able to obtain results for $\theta = 0.5$ with statistically controlled errors.

7. Conclusion and summary

In this article we have investigated the quantum mechanical rotor in Feynman's (Euclidean) path integral formulation with a special emphasis on adding a phase factor. Such a setup not only easily leads to the problem of a high dimensionality, but through the presence of the phase factor, it also gives rise to the infamous sign problem appearing in MCMC methods.

Although, through the use of importance sampling, MCMC is able to address the problem of a high dimensionality, it fails when a too strongly oscillating phase factor is added. On the other hand, when we apply a polynomially exact full symmetrization method [10] to solve the corresponding path integral, for low-dimensional problems we can overcome the sign problem. We demonstrated this at the example of the quantum mechanical oscillator with a phase factor, and at the simpler model of 1-dimensional QCD [10]. However, for the latter method, the dimensionality problem strikes back and we can apply the full symmetrization method only to systems with small dimensions.

It is therefore a natural idea to try to combine the methods of MCMC and symmetrization, and in this article we tested this approach. The desired combination of both methods could be achieved by giving up on a full symmetrization and, following ideas of [17], replacing it by only local symmetrizations which were realized by sets of random permutations of the symmetrization points. Going first to a situation where the phase factor is absent, we found that the combined method gives fully compatible results with MCMC calculations. However, we could not find an improvement in the statistical accuracy of the combined method over MCMC. In contrast, when we have chosen a large number of symmetrization points, the error deteriorated and the combined method could not compete with pure MCMC.

Switching to the phase factor, again the combined method did not beat MCMC and, in particular, could not solve the sign problem. The errors of the combined method remained to be large and for the cases where the phase factor showed large oscillations, expectation values could not be computed reliably anymore.

To find a possible explanation of this negative outcome of our investigations, we analyzed the full symmetrization method in more detail. To this end, we looked which summands in the full symmetrization method are particularly important to give the final result. We found that even if the phase factor oscillates only little, that is at small θ, an imaginary part is building up, necessitating a significant number of summands to reach the final result. When the phase factor starts to oscillate strongly, we observed that basically all summands have to be taken into account to obtain the correct required expectation value. Truncating the sum leads to highly irregular behavior in the so computed expectation value with results that can be far away from the final value.

We suspect that it is this effect that made our combined method failing: in the local symmetrization approach we can simply not sample sufficiently many relevant symmetrization points to reach a stable situation. Depending on which sets of symmetrization we have chosen, the results of the individual sets can fluctuate strongly, leading to the large errors we observe. Thus, although the combined method as such is exact, it does not seem to solve the problem of dimensionality or overcome the sign problem of standard MCMC methods. Whether this shortcoming can be overcome by combining our idea with other methods for solving high-dimensional problems, such as Quasi-Monte Carlo techniques, needs further investigations in the future.

Acknowledgment

We thank Dirk Nuyens and Frances Kuo for interesting discussions during the IWOTA 2017 conference. We also thank the organizers of the IWOTA 2017 for inviting us to the minisymposium *"New approaches for high-dimensional integration in light of physics applications"* which led to very fruitful and inspiring discussions. J.V. acknowledges financial support by the Project No. JA 674/6-1, funded by the Deutsche Forschungsgemeinschaft.

Appendix

Derivation of $\langle O \rangle$. We derived an equation to retrieve the expectation value $\langle O \rangle$ from the symmetrized expectation value $[\langle O_{\text{sym}} \rangle]$, given that the observable function is a polynomial in the lattice variables Φ_i,

$$O = \sum_{t=1}^{L} O_t \quad \text{with } O_t = \prod_{i=1}^{L} \Phi_i^{\alpha_i^t}, \tag{7.1}$$

where α_i^t is the ith component of the vectors $\alpha^t \in \mathbb{Z}^L$, one for each summand O_t. We chose to compute the expectation value of the link, because it is a polynomial in the Φ_i's. Its α^t's are given by $\alpha_t^t = -1, \alpha_{t+1}^t = 1$, and all other $\alpha_k^t = 0$. The expectation value of O is the sum of the expectation values of the O_t's and they can be derived from the symmetrized expectation values $[\langle O_{\text{sym},t} \rangle]$,

$$\begin{aligned}
[\langle O_{\text{sym},t} \rangle] &= \frac{\int O_{\text{sym},t} \cdot w_{\text{sym}} \mathrm{d}\boldsymbol{\Phi}}{\int w_{\text{sym}} \mathrm{d}\boldsymbol{\Phi}} \\
&= \frac{\int \left(\sum_{k=1}^{\mathfrak{g}} O_t[\boldsymbol{G}^{\boldsymbol{P}(k)}\boldsymbol{\Phi}] \right) \left(\sum_{m=1}^{\mathfrak{g}} \exp(-S[\boldsymbol{G}^{\boldsymbol{P}(m)}\boldsymbol{\Phi}]) \right) \mathrm{d}\boldsymbol{\Phi}}{\int \left(\sum_{n=1}^{\mathfrak{g}} \exp(-S[\boldsymbol{G}^{\boldsymbol{P}(n)}\boldsymbol{\Phi}]) \right) \mathrm{d}\boldsymbol{\Phi}}.
\end{aligned}$$

Numerator and denominator can be calculated separately by using the properties of the Haar measure. For the numerator we have

$$\begin{aligned}
&\text{Numerator}([\langle O_{\text{sym},t} \rangle]) \\
&\quad = \int \sum_{k=1}^{\mathfrak{g}} O_t[\boldsymbol{G}^{\boldsymbol{P}(k)}\boldsymbol{\Phi}] \exp(-S[\boldsymbol{G}^{\boldsymbol{P}(k)}\boldsymbol{\Phi}]) \mathrm{d}\boldsymbol{\Phi} \\
&\qquad + \int \sum_{k=1}^{\mathfrak{g}} \sum_{m \neq k} O_t[\boldsymbol{G}^{\boldsymbol{P}(k)}\boldsymbol{\Phi}] \exp(-S[\boldsymbol{G}^{\boldsymbol{P}(m)}\boldsymbol{\Phi}]) \mathrm{d}\boldsymbol{\Phi} \\
&\quad = \mathfrak{g} \int O_t[\boldsymbol{\Phi}] \, \mathrm{e}^{-S[\boldsymbol{\Phi}]} \mathrm{d}\boldsymbol{\Phi} \\
&\qquad + \int \sum_{k=1}^{\mathfrak{g}} \sum_{m \neq k} O_t[\boldsymbol{G}^{\boldsymbol{P}(k)}(\boldsymbol{G}^{\boldsymbol{P}(m)})^* \boldsymbol{\Phi}] \mathrm{e}^{-S[\boldsymbol{\Phi}]} \mathrm{d}\boldsymbol{\Phi},
\end{aligned}$$

and the denominator is

$$\begin{aligned}&\text{Denominator}([\langle O_{\text{sym},t}\rangle])\\&=\mathfrak{g}\int \mathrm{e}^{-S[\boldsymbol{G}^{\boldsymbol{P}(m)}\boldsymbol{\Phi}]}\mathrm{d}\boldsymbol{\Phi}.\end{aligned}$$

Numerator and denominator combine to

$$\begin{aligned}[\langle O_{\text{sym},t}\rangle] &= \frac{\int O_t[\boldsymbol{\Phi}]\,\mathrm{e}^{-S[\boldsymbol{\Phi}]}\mathrm{d}\boldsymbol{\Phi}}{\int \mathrm{e}^{-S[\boldsymbol{\Phi}]}\mathrm{d}\boldsymbol{\Phi}}\\&\quad+\frac{1}{\mathfrak{g}}\sum_{k=1}^{\mathfrak{g}}\sum_{m\neq k}\frac{\int O_t[\boldsymbol{G}^{\boldsymbol{P}(k)}(\boldsymbol{G}^{\boldsymbol{P}(m)})^*\boldsymbol{\Phi}]\mathrm{e}^{-S[\boldsymbol{\Phi}]}\mathrm{d}\boldsymbol{\Phi}}{\int \mathrm{e}^{-S[\boldsymbol{\Phi}]}\mathrm{d}\boldsymbol{\Phi}}.\end{aligned}$$

The first term is $\langle O_t\rangle$. The second term needs the form of O_t from (7.1) to give

$$\begin{aligned}[\langle O_{\text{sym},t}\rangle] &= \langle O_t\rangle+\frac{1}{\mathfrak{g}}\sum_{k=1}^{\mathfrak{g}}\sum_{m\neq k}\frac{\int \prod_{l=1}^{L}\left(G^{P_l(k)}(G^{P_l(m)})^*\Phi_l\right)^{\alpha_l^t}\mathrm{e}^{-S[\boldsymbol{\Phi}]}\mathrm{d}\boldsymbol{\Phi}}{\int \mathrm{e}^{-S[\boldsymbol{\Phi}]}\mathrm{d}\boldsymbol{\Phi}}\\
&= \langle O_t\rangle+\frac{1}{\mathfrak{g}}\sum_{k=1}^{\mathfrak{g}}\sum_{m\neq k}\prod_{l=1}^{L}\left(G^{P_l(k)}(G^{P_l(m)})^*\right)^{\alpha_l^t}\frac{\int \prod_{p=1}^{L}\Phi_p^{\alpha_p^t}\mathrm{e}^{-S[\boldsymbol{\Phi}]}\mathrm{d}\boldsymbol{\Phi}}{\int \mathrm{e}^{-S[\boldsymbol{\Phi}]}\mathrm{d}\boldsymbol{\Phi}}\\
&= \langle O_t\rangle+\frac{1}{\mathfrak{g}}\sum_{k=1}^{\mathfrak{g}}\sum_{m\neq k}\prod_{l=1}^{L}\left(G^{P_l(k)}(G^{P_l(m)})^*\right)^{\alpha_l^t}\frac{\int O_t[\boldsymbol{\Phi}]\mathrm{d}\boldsymbol{\Phi}}{\int \mathrm{e}^{-S[\boldsymbol{\Phi}]}\mathrm{d}\boldsymbol{\Phi}}\\
&= \langle O_t\rangle\left(1+\frac{1}{\mathfrak{g}}\sum_{k=1}^{\mathfrak{g}}\sum_{m\neq k}\prod_{l=1}^{L}\left(G^{P_l(k)}(G^{P_l(m)})^*\right)^{\alpha_l^t}\right).\end{aligned}$$

If the correction factor

$$c_t\equiv 1+\frac{1}{\mathfrak{g}}\sum_{k=1}^{\mathfrak{g}}\sum_{m\neq k}\prod_{l=1}^{L}\left(G^{P_l(k)}(G^{P_l(m)})^*\right)^{\alpha_l^t}$$

is non-zero, the expectation value of Q_t can be derived from the symmetrized expectation value $[\langle Q_{\text{sym},t}\rangle]$ via

$$\langle O_t\rangle=\frac{1}{c_t}[\langle O_{\text{sym},t}\rangle].$$

This formula is true because the correction factor is independent of the lattice values Φ_i. With the proper α^t's for the link variable, the correction factor for the link is

$$c_t=1+\frac{1}{\mathfrak{g}}\sum_{i\neq j}\left(G^{P_t(i)}(G^{P_t(j)})^*\right)^*\left(G^{P_{t+1}(i)}(G^{P_{t+1}(j)})^*\right).$$

References

[1] R.P. Feynman, Space-time approach to nonrelativistic quantum mechanics, *Rev. Mod. Phys.* **20** (1948), 367–387.

[2] Christof Gattringer and Christian B. Lang, Quantum chromodynamics on the lattice, *Lect. Notes Phys.* **788** (2010), 1–343.

[3] S. Durr et al, Ab-Initio Determination of Light Hadron Masses, *Science* **322** (2008), 1224–1227.

[4] A. Abdel-Rehim et al., Nucleon and pion structure with lattice QCD simulations at physical value of the pion mass, *Phys. Rev.*, **D92(11)** (2015), 114513. [Erratum: *Phys. Rev.* **D93(3)** (2016), 039904].

[5] A. Abdel-Rehim et al., First physics results at the physical pion mass from $N_f = 2$ Wilson twisted mass fermions at maximal twist, *Phys. Rev.* **D95(9)** (2017), 094515.

[6] Haohuan Fu, Junfeng Liao, Jinzhe Yang, Lanning Wang, Zhenya Song, Xiaomeng Huang, Chao Yang, Wei Xue, Fangfang Liu, Fangli Qiao, Wei Zhao, Xunqiang Yin, Chaofeng Hou, Chenglong Zhang, Wei Ge, Jian Zhang, Yangang Wang, Chunbo Zhou, and Guangwen Yang, The sunway taihulight supercomputer: system and applications, *Science China Information Sciences* **59(7)** (2016), 072001.

[7] Matthias Troyer and Uwe-Jens Wiese, Computational complexity and fundamental limitations to fermionic quantum Monte Carlo simulations, *Phys. Rev. Lett.* **94** (2005), 170201.

[8] Christian Schmidt, Lattice QCD at finite density, *PoS* LAT2006:021 (2006).

[9] Andrew G. Cohen, D.B. Kaplan, and A.E. Nelson, Progress in electroweak baryogenesis, *Ann. Rev. Nucl. Part. Sci.* **43** (1993), 27–70.

[10] A. Ammon, T. Hartung, K. Jansen, H. Levey, and J. Volmer, Overcoming the sign problem in one-dimensional QCD by new integration rules with polynomial exactness, *Phys. Rev.* **D94(11)** (2016), 114508.

[11] Jacques Bloch, Falk Bruckmann, and Tilo Wettig, Subset method for one-dimensional QCD, *JHEP* **10** (2013), 140.

[12] Jacques Bloch and Falk Bruckmann, Positivity of center subsets for QCD, *Phys. Rev.* **D93(1)** (2016), 014508. [Addendum: *Phys. Rev.* **D93(3)** (2016), 039907].

[13] A. Ammon, A. Genz, T. Hartung, K. Jansen, H. Levey, and J. Volmer, On the efficient numerical solution of lattice systems with low-order couplings, *Comput. Phys. Commun.* **198** (2016), 71–81.

[14] Tobias Hartung, Regularizing Feynman path integrals using the generalized Kontsevich-Vishik trace, *J. Math. Phys.* **58** (2017), 123505, 1–19.

[15] F Kuo, High dimensional integration: the Quasi-Monte Carlo (QMC) way, *Plenary presentation, 28th IWOTA conference, August 14 to August 18, 2017, Technical University Chemnitz, Germany.*

[16] W. Bietenholz, U. Gerber, M. Pepe, and U.-J. Wiese, Topological Lattice Actions, *JHEP* **1012** (2010), 020.

[17] Kerstin Hesse, Frances Y. Kuo, and Ian H. Sloan, A component-by-component approach to efficient numerical integration over products of spheres, *Journal of Complexity* **23(1)** (2007), 25–51.

[18] Ulli Wolff, Collective Monte Carlo Updating for Spin Systems, *Phys. Rev. Lett.* **62** (1989), 361.

[19] A.M. Ferrenberg and R.H. Swendsen, New Monte Carlo Technique for Studying Phase Transitions, *Phys. Rev. Lett.* **61** (1988), 2635–2638.

Tobias Hartung
Department of Mathematics
King's College London
Strand
London WC2R 2LS
United Kingdom
e-mail: tobias.hartung@kcl.ac.uk

Karl Jansen
DESY
John von Neumann Institute for Computing
Platanenallee 6
15738 Zeuthen
Germany
e-mail: Karl.Jansen@desy.de

Hernan Leövey
Structured Energy Management
Axpo Trading
Parkstrasse 23
5400 Baden, Germany
e-mail: HernanEugenio.Leoevey@axpo.com

Julia Volmer
DESY
John von Neumann Institute for Computing
Platanenallee 6
15738 Zeuthen
Germany
e-mail: Julia.Volmer@desy.de

More on the density of analytic polynomials in abstract Hardy spaces

Alexei Karlovich and Eugene Shargorodsky

Abstract. Let $\{F_n\}$ be the sequence of the Fejér kernels on the unit circle $\mathbb{T}$. The first author recently proved that if X is a separable Banach function space on $\mathbb{T}$ such that the Hardy–Littlewood maximal operator M is bounded on its associate space X', then $\|f * F_n - f\|_X \to 0$ for every $f \in X$ as $n \to \infty$. This implies that the set of analytic polynomials $\mathcal{P}_A$ is dense in the abstract Hardy space $H[X]$ built upon a separable Banach function space X such that M is bounded on X'. In this note we show that there exists a separable weighted L^1 space X such that the sequence $f * F_n$ does not always converge to $f \in X$ in the norm of X. On the other hand, we prove that the set $\mathcal{P}_A$ is dense in $H[X]$ under the assumption that X is merely separable.

Mathematics Subject Classification (2010). Primary 46E30, Secondary 42A10.

Keywords. Banach function space, abstract Hardy space, analytic polynomial, Fejér kernel.

1. Preliminaries and the main results

For $0 < p \leq \infty$, let $L^p := L^p(\mathbb{T})$ be the Lebesgue space on the unit circle $\mathbb{T} := \{z \in \mathbb{C} : |z| = 1\}$ in the complex plane $\mathbb{C}$. For $f \in L^1$, let

$$\widehat{f}(n) := \frac{1}{2\pi} \int_{-\pi}^{\pi} f(e^{i\theta}) e^{-in\theta} \, d\theta, \quad n \in \mathbb{Z},$$

be the sequence of the Fourier coefficients of f. Let X be a Banach space continuously embedded in L^1. Following [13, p. 877], we will consider the abstract Hardy space $H[X]$ built upon the space X, which is defined by

$$H[X] := \{f \in X : \ \widehat{f}(n) = 0 \quad \text{for all} \quad n < 0\}.$$

This work was partially supported by the Fundação para a Ciência e a Tecnologia (Portuguese Foundation for Science and Technology) through the project UID/MAT/00297/2013 (Centro de Matemática e Aplicações).

© Springer International Publishing AG, part of Springer Nature 2018
A. Böttcher et al. (eds.), *The Diversity and Beauty of Applied Operator Theory*, Operator Theory: Advances and Applications 268, https://doi.org/10.1007/978-3-319-75996-8_16

It is clear that if $1 \le p \le \infty$, then $H[L^p]$ is the classical Hardy space H^p.

A function of the form

$$q(t) = \sum_{k=0}^{n} \alpha_k t^k, \quad t \in \mathbb{T}, \quad \alpha_0, \dots, \alpha_n \in \mathbb{C},$$

is said to be an analytic polynomial on $\mathbb{T}$. The set of all analytic polynomials is denoted by $\mathcal{P}_A$. It is well known that the set $\mathcal{P}_A$ is dense in H^p whenever $1 \le p < \infty$ (see, e.g., [3, Chap. III, Corollary 1.7(a)]). The density of the set $\mathcal{P}_A$ in the abstract Hardy spaces $H[X]$ was studied by the first author [8] for the case when X is a so-called Banach function space.

Let us recall the definition of a Banach function space. We equip $\mathbb{T}$ with the normalized Lebesgue measure $dm(t) = |dt|/(2\pi)$. Let L^0 be the space of all measurable complex-valued functions on $\mathbb{T}$. As usual, we do not distinguish functions which are equal almost everywhere (for the latter we use the standard abbreviation a.e.). Let L^0_+ be the subset of functions in L^0 whose values lie in $[0, \infty]$. The characteristic function of a measurable set $E \subset \mathbb{T}$ is denoted by χ_E.

Following [1, Chap. 1, Definition 1.1], a mapping $\rho : L^0_+ \to [0, \infty]$ is called a Banach function norm if, for all functions $f, g, f_n \in L^0_+$ with $n \in \mathbb{N}$, for all constants $a \ge 0$, and for all measurable subsets E of $\mathbb{T}$, the following properties hold:

(A1) $\rho(f) = 0 \Leftrightarrow f = 0$ a.e., $\rho(af) = a\rho(f)$, $\rho(f + g) \le \rho(f) + \rho(g)$,

(A2) $0 \le g \le f$ a.e. $\Rightarrow \rho(g) \le \rho(f)$ (the lattice property),

(A3) $0 \le f_n \uparrow f$ a.e. $\Rightarrow \rho(f_n) \uparrow \rho(f)$ (the Fatou property),

(A4) $m(E) < \infty \Rightarrow \rho(\chi_E) < \infty$,

(A5) $\displaystyle\int_E f(t)\, dm(t) \le C_E \rho(f)$

with a constant $C_E \in (0, \infty)$ that may depend on E and ρ, but is independent of f. When functions differing only on a set of measure zero are identified, the set X of all functions $f \in L^0$ for which $\rho(|f|) < \infty$ is called a Banach function space. For each $f \in X$, the norm of f is defined by $\|f\|_X := \rho(|f|)$. The set X under the natural linear space operations and under this norm becomes a Banach space (see [1, Chap. 1, Theorems 1.4 and 1.6]). If ρ is a Banach function norm, its associate norm ρ' is defined on L^0_+ by

$$\rho'(g) := \sup\left\{ \int_{\mathbb{T}} f(t)g(t)\, dm(t) \ : \ f \in L^0_+, \ \rho(f) \le 1 \right\}, \quad g \in L^0_+.$$

It is a Banach function norm itself [1, Chap. 1, Theorem 2.2]. The Banach function space X' determined by the Banach function norm ρ' is called the associate space (Köthe dual) of X. The associate space X' can be viewed as a subspace of the (Banach) dual space X^*.

Recall that L^1 is a commutative Banach algebra under the convolution multiplication defined for $f, g \in L^1$ by

$$(f * g)(e^{i\theta}) = \frac{1}{2\pi} \int_{-\pi}^{\pi} f(e^{i\theta - i\varphi}) g(e^{i\varphi})\, d\varphi, \quad e^{i\theta} \in \mathbb{T}.$$

For $n \in \mathbb{N}$, let

$$F_n(e^{i\theta}) := \sum_{k=-n}^{n} \left(1 - \frac{|k|}{n+1}\right) e^{i\theta k} = \frac{1}{n+1} \left(\frac{\sin \frac{n+1}{2}\theta}{\sin \frac{\theta}{2}} \right)^2, \quad e^{i\theta} \in \mathbb{T},$$

be the n-th Fejér kernel. For $f \in L^1$, the n-th Fejér mean of f is defined as the convolution $f * F_n$.

Given $f \in L^1$, the Hardy–Littlewood maximal function is defined by

$$(Mf)(t) := \sup_{I \ni t} \frac{1}{m(I)} \int_I |f(\tau)|\, dm(\tau), \quad t \in \mathbb{T},$$

where the supremum is taken over all arcs $I \subset \mathbb{T}$ containing $t \in \mathbb{T}$. The operator $f \mapsto Mf$ is called the Hardy–Littlewood maximal operator.

Theorem 1.1 ([8, Theorem 3.3]). *Suppose X is a separable Banach function space on $\mathbb{T}$. If the Hardy–Littlewood maximal operator is bounded on the associate space X', then for every $f \in X$,*

$$\lim_{n \to \infty} \|f * F_n - f\|_X = 0. \tag{1.1}$$

It is well known that for $f \in L^1$ one has

$$(f * F_n)(e^{i\theta}) = \sum_{k=-n}^{n} \widehat{f}(k) \left(1 - \frac{|k|}{n+1}\right) e^{i\theta k}, \quad e^{i\theta} \in \mathbb{T}$$

(see, e.g., [9, Chap. I]). This implies that if $f \in H[X] \subset H[L^1] = H^1$, then $f * F_n \in \mathcal{P}_A$. Combining this observation with Theorem 1.1, we arrive at the following.

Corollary 1.2 ([8, Theorem 1.2]). *Suppose X is a separable Banach function space on $\mathbb{T}$. If the Hardy–Littlewood maximal operator M is bounded on its associate space X', then the set of analytic polynomials $\mathcal{P}_A$ is dense in the abstract Hardy space $H[X]$ built upon the space X.*

Note that if a Banach function space X is, in addition, rearrangement-invariant, then the requirement of the boundedness of M on the space X' can be omitted in Corollary 1.2 (see [8, Theorem 1.1] or [11, Lemma 1.3(c)]). Leśnik [10] conjectured that the same fact should be true for arbitrary, not necessarily rearrangement-invariant, Banach function spaces.

In this note, we first observe that Theorem 1.1 does not hold for arbitrary separable Banach function spaces. For a function $K \in L^1$, consider the convolution operator C_K with kernel K defined by

$$C_K f = f * K, \quad f \in L^1.$$

It follows from [12, Theorem 2] that there exists a continuous function $p : \mathbb{T} \to [1, \infty)$ such that the sequence of the convolution operators C_{F_n} is not uniformly bounded in the variable Lebesgue space $L^{p(\cdot)}$ defined as the set of all $f \in L^0$ such that

$$\int_{\mathbb{T}} |f(t)|^{p(t)} dm(t) < \infty.$$

It is well known (see, e.g., [4, Propostion 2.12, Theorem 2.78, Section 2.10.3]) that if $p : \mathbb{T} \to [1, \infty)$ is continuous, then $L^{p(\cdot)}$ is a separable Banach function space equipped with the norm

$$\|f\|_{L^{p(\cdot)}} = \inf\left\{\lambda > 0 \ : \ \int_{\mathbb{T}} \left|\frac{f(t)}{\lambda}\right|^{p(t)} dm(t) \le 1\right\}.$$

Since the norms of the convolution operators C_{F_n} may not be uniformly bounded on $L^{p(\cdot)}$, the standard argument, based on the uniform boundedness principle, leads us to the following.

Theorem 1.3. *There exist a separable Banach function space X on $\mathbb{T}$ and a function $f \in X$ such that* (1.1) *is not fulfilled.*

We show that the separable Banach function space in Theorem 1.3 can be chosen as a weighted L^1 space, that is, the techniques of variable Lebesgue spaces can be omitted.

Theorem 1.4 (Main result 1). *There exist a nonnegative function $w \in L^1$ such that $w^{-1} \in L^\infty$ and a function f in the separable Banach function space*

$$X = L^1(w) = \{f \in L^0 : fw \in L^1\}$$

such that (1.1) *is not fulfilled.*

In spite of the observation made in Theorems 1.3 and 1.4, we show that the requirement of the boundedness of the Hardy–Littlewood maximal operator M on the associate space X' of a separable Banach function space X in Corollary 1.2 can be omitted. Thus, Leśnik's conjecture [10] is, indeed, true.

Theorem 1.5 (Main result 2). *If X is a separable Banach function space on $\mathbb{T}$, then the set of analytic polynomials $\mathcal{P}_A$ is dense in the abstract Hardy space $H[X]$ built upon the space X.*

The paper is organized as follows. In Section 2, we prove that a convolution operator C_K with a nonnegative symmetric kernel $K \in L^1$ is bounded on a Banach function space X if and only if it is bounded on its associate space X'. Further, we consider a special weight $w \in L^1$ such that $w^{-1} \in L^\infty$. Then $X = L^1(w)$ is a separable Banach function space with the associate space $X' = L^\infty(w^{-1})$. We show that the sequence of convolution operators $\{C_{K_n}\}$ with nonnegative bounded symmetric kernels K_n, satisfying $\|K_n\|_{L^1} = 1$ and a natural localization property, is not uniformly bounded on $X' = L^\infty(w^{-1})$, and therefore, on its associate space $X'' = X = L^1(w)$. Applying this result to the sequence of the Fejér kernels $\{F_n\}$, we prove Theorem 1.4 with the aid of the uniform boundedness principle.

In Section 3, we recall that the separability of a Banach function space X is equivalent to $X^* = X'$. Further, we collect some facts on the identification of the Hardy spaces H^p on the unit circle and the Hardy spaces $H^p(\mathbb{D})$ of analytic functions in the unit disk $\mathbb{D}$. Finally, we give a proof of Theorem 1.5 based on an application of the Hahn–Banach theorem, a corollary of the Smirnov theorem and properties of the identification of H^1 with $H^1(\mathbb{D})$.

2. Proof of the first main result

2.1. Norms of convolution operators on X and on its associate space X'

The Banach space of all bounded linear operators on a Banach space E is denoted by $\mathcal{B}(E)$.

Lemma 2.1. *Let X be a Banach function spacc on $\mathbb{T}$ and $K \in L^1$ be a nonnegative function such that $K(e^{i\theta}) = K(e^{-i\theta})$ for almost all $\theta \in [-\pi, \pi]$. Then the convolution operator C_K is bounded on the Banach function X if and only if it is bounded on its associate space X'. In that case*

$$\|C_K\|_{\mathcal{B}(X')} = \|C_K\|_{\mathcal{B}(X)}. \tag{2.1}$$

Proof. Suppose C_K is bounded on X'. Fix $f \in X \setminus \{0\}$. Since $K \geq 0$, we have $|f * K| \leq |f| * K$. According to the Lorentz–Luxemburg theorem (see, e.g., [1, Chap. 1, Theorem 2.7]), $X = X''$ with equality of the norms. Hence

$$\begin{aligned}\|f * K\|_X &\leq \| |f| * K\|_X = \| |f| * K\|_{X''} \\ &= \sup\left\{\int_{\mathbb{T}} (|f| * K)(t)|g(t)|\,dm(t) \ : \ g \in X',\ \|g\|_{X'} \leq 1\right\}.\end{aligned}$$

Then for every $\varepsilon > 0$ there exists a function $h \in X'$ such that $h \geq 0$, $\|h\|_{X'} \leq 1$, and

$$\|f * K\|_X \leq (1+\varepsilon) \int_{\mathbb{T}} (|f| * K)(t)h(t)\,dm(t). \tag{2.2}$$

Taking into account that $K(e^{i\theta}) = K(e^{-i\theta})$ for almost all $\theta \in \mathbb{R}$, by Fubini's theorem, we get

$$\int_{\mathbb{T}} (|f| * K)(t)h(t)\,dm(t) = \int_{\mathbb{T}} (h * K)(t)|f(t)|\,dm(t).$$

From this identity, Hölder's inequality for X (see, e.g., [1, Chap. 1, Theorem 2.4]), and the boundedness of C_K on X', we obtain

$$\int_{\mathbb{T}} (|f| * K)(t)h(t)\,dm(t) \leq \|f\|_X \|h * K\|_{X'} \leq \|f\|_X \|C_K\|_{\mathcal{B}(X')}. \tag{2.3}$$

It follows from (2.2)–(2.3) that

$$\|C_K\|_{\mathcal{B}(X)} = \sup_{f \in X, f \neq 0} \frac{\|f * K\|_X}{\|f\|_X} \leq (1+\varepsilon)\|C_K\|_{\mathcal{B}(X')}$$

for every $\varepsilon > 0$, which implies the boundedness of C_K on X and the inequality

$$\|C_K\|_{\mathcal{B}(X)} \leq \|C_K\|_{\mathcal{B}(X')}. \tag{2.4}$$

If C_K is bounded on X, then using the Lorentz–Luxemburg theorem and (2.4) with X' in place of X, we obtain that C_K is bounded on X' and

$$\|C_K\|_{\mathcal{B}(X')} \le \|C_K\|_{\mathcal{B}(X'')} = \|C_K\|_{\mathcal{B}(X)}. \tag{2.5}$$

Combining (2.4)–(2.5), we arrive at (2.1). □

2.2. Spaces $L^1(w)$ and $L^\infty(w^{-1})$ with a special weight w

Lemma 2.2. *Let*

$$w\left(e^{i\theta}\right) := \begin{cases} \sqrt{m}, & \frac{\pi}{2m} \le |\theta| \le \frac{\pi}{2m-1}, \quad m \in \mathbb{N}, \\ 1, & \frac{\pi}{2m+1} < |\theta| < \frac{\pi}{2m}, \quad m \in \mathbb{N}. \end{cases} \tag{2.6}$$

Then the spaces

$$L^1(w) = \{f \in L^0 : fw \in L^1\}, \quad L^\infty(w^{-1}) = \{f \in L^0 : fw^{-1} \in L^\infty\}$$

are Banach function spaces on $\mathbb{T}$ *with respect to the norms*

$$\|f\|_{L^1(w)} = \|fw\|_{L^1}, \quad \|f\|_{L^\infty(w^{-1})} = \|fw^{-1}\|_{L^\infty},$$

and $(L^1(w))' = L^\infty(w^{-1})$. *Moreover, the space* $L^1(w)$ *is separable.*

Proof. It is clear that $w^{-1} \in L^\infty$ and, since

$$\begin{aligned} \|w\|_{L^1} &= \frac{1}{2\pi}\int_{-\pi}^{\pi} w(e^{i\theta})\,d\theta \\ &= \sum_{m=1}^{\infty}\left(\frac{1}{2m} - \frac{1}{2m+1}\right) + \sum_{m=1}^{\infty}\sqrt{m}\left(\frac{1}{2m-1} - \frac{1}{2m}\right) < \infty, \end{aligned} \tag{2.7}$$

we also have $w \in L^1$. Then it follows from [7, Lemma 2.5] that $L^1(w)$ and $L^\infty(w^{-1})$ are Banach function spaces and $(L^1(w))' = L^\infty(w^{-1})$. Finally, the separability of the space $L^1(w)$ follows from [7, Proposition 2.6] and [1, Chap. 1, Corollary 5.6]. □

2.3. Norms of convolution operators are not uniformly bounded on the spaces $L^1(w)$ and $L^\infty(w^{-1})$ with the special weight w

Theorem 2.3. *Let* $\{K_n\}$ *be a sequence of bounded functions* $K_n : \mathbb{T} \to \mathbb{C}$ *such that*

$$K_n(e^{i\theta}) \ge 0, \quad K_n(e^{i\theta}) = K_n(e^{-i\theta}) \quad \text{a.e. on} \quad [-\pi, \pi], \tag{2.8}$$

$$\frac{1}{2\pi}\int_{-\pi}^{\pi} K_n\left(e^{i\theta}\right)\,d\theta = 1, \tag{2.9}$$

and

$$\lim_{n\to\infty} \sup_{\varepsilon \le |\theta| \le \pi} K_n\left(e^{i\theta}\right) = 0 \quad \text{for each} \quad \varepsilon > 0. \tag{2.10}$$

If w *is the weight given by* (2.6), *then the convolution operators* C_{K_n} *are bounded on* $L^\infty(w^{-1})$ *and on* $L^1(w)$ *for all* $n \in \mathbb{N}$, *however,*

$$\sup_{n\in\mathbb{N}} \|C_{K_n}\|_{\mathcal{B}(L^\infty(w^{-1}))} = \infty, \tag{2.11}$$

$$\sup_{n\in\mathbb{N}} \|C_{K_n}\|_{\mathcal{B}(L^1(w))} = \infty. \tag{2.12}$$

Proof. By (2.6)–(2.7), $w \in L^1$ and $w^{-1} \in L^\infty$. Therefore, for every $n \in \mathbb{N}$,

$$\begin{aligned}\|C_{K_n} f\|_{L^1(w)} &\le \frac{1}{2\pi}\left\|\int_{-\pi}^{\pi} K_n(e^{i(\cdot-\theta)})\left|f\left(e^{i\theta}\right)\right| d\theta\right\|_{L^1(w)}\\ &\le \frac{1}{2\pi}\int_{-\pi}^{\pi}\left\|K_n(e^{i(\cdot-\theta)})\right\|_{L^1(w)}\left|f\left(e^{i\theta}\right)\right| d\theta\\ &\le \frac{1}{2\pi}\|K_n\|_{L^\infty}\|w\|_{L^1}\|f\|_{L^1}\\ &= \frac{1}{2\pi}\|K_n\|_{L^\infty}\|w\|_{L^1}\|w^{-1} f w\|_{L^1}\\ &\le \frac{1}{2\pi}\|K_n\|_{L^\infty}\|w\|_{L^1}\|w^{-1}\|_{L^\infty}\|f\|_{L^1(w)}.\end{aligned}$$

Hence

$$\|C_{K_n}\|_{\mathcal{B}(L^1(w))} \le \frac{1}{2\pi}\|K_n\|_{L^\infty}\|w\|_{L^1}\|w^{-1}\|_{L^\infty}, \quad n \in \mathbb{N}.$$

It follows from (2.8) and Lemmas 2.1–2.2 that the operators C_{K_n} are bounded on $L^\infty(w^{-1})$ for all $n \in \mathbb{N}$. Moreover, (2.11) implies (2.12).

Let us prove (2.11). Consider the sequence

$$v_m\left(e^{i\theta}\right) := \begin{cases}\sqrt{m}, & \frac{\pi}{2m} \le \theta \le \frac{\pi}{2m-1},\\ 0, & \theta \in [-\pi,\pi] \setminus \left[\frac{\pi}{2m}, \frac{\pi}{2m-1}\right],\end{cases} \qquad m \in \mathbb{N}.$$

Then it follows from (2.6) that $\|v_m\|_{L^\infty(w^{-1})} = 1$ for all $m \in \mathbb{N}$.

Fix $m \in \mathbb{N}$. According to (2.9) and the localization property (2.10), there exists $n(m) \in \mathbb{N}$ such that

$$\int_{-\frac{\pi}{(2m)^2}}^{0} K_n\left(e^{i\theta}\right) d\theta = \frac{1}{2}\int_{-\frac{\pi}{(2m)^2}}^{\frac{\pi}{(2m)^2}} K_n\left(e^{i\theta}\right) d\theta \ge \frac{1}{3} \quad \text{for all} \quad n \ge n(m).$$

Since $K_n \in L^1$, for every $n \ge n(m)$, there exists $\delta_n > 0$ such that

$$\int_{-\frac{\pi}{(2m)^2}}^{-\delta_n} K_n\left(e^{i\theta}\right) d\theta \ge \frac{1}{4}.$$

Therefore, for almost all $\vartheta \in \left[\frac{\pi}{2m} - \delta_n, \frac{\pi}{2m}\right]$, one gets

$$\begin{aligned}\left(C_{K_n} v_m\right)\left(e^{i\vartheta}\right) &= \frac{\sqrt{m}}{2\pi}\int_{\frac{\pi}{2m}}^{\frac{\pi}{2m-1}} K_n\left(e^{i\vartheta-i\theta}\right) d\theta\\ &\ge \frac{\sqrt{m}}{2\pi}\int_{\frac{\pi}{2m}}^{\frac{\pi}{2m}+\frac{\pi}{(2m)^2}} K_n\left(e^{i\vartheta-i\theta}\right) d\theta\\ &= \frac{\sqrt{m}}{2\pi}\int_{\vartheta-\frac{\pi}{2m}-\frac{\pi}{(2m)^2}}^{\vartheta-\frac{\pi}{2m}} K_n\left(e^{i\eta}\right) d\eta\\ &\ge \frac{\sqrt{m}}{2\pi}\int_{-\frac{\pi}{(2m)^2}}^{-\delta_n} K_n\left(e^{i\eta}\right) d\eta \ge \frac{\sqrt{m}}{8\pi}.\end{aligned} \tag{2.13}$$

In view of (2.6), $w(e^{i\vartheta}) = 1$ for all $\vartheta \in \left(\max\left\{\frac{\pi}{2m} - \delta_n, \frac{\pi}{2m+1}\right\}, \frac{\pi}{2m}\right)$. Hence, it follows from (2.13) that

$$\|C_{K_n} v_m\|_{L^\infty(w^{-1})} \geq \frac{\sqrt{m}}{8\pi} \quad \text{for all} \quad n \geq n(m),$$

while $\|v_m\|_{L^\infty(w^{-1})} = 1$. So,

$$\|C_{K_n}\|_{\mathcal{B}(L^\infty(w^{-1}))} \geq \frac{\sqrt{m}}{8\pi} \quad \text{for all} \quad n \geq n(m).$$

Since $m \in \mathbb{N}$ is arbitrary, the latter inequality immediately implies (2.11). □

2.4. Proof of Theorem 1.4

Let $X = L^1(w)$, where w is the weight given by (2.6). By Lemma 2.2, X is a separable Banach function space. It is well known (and not difficult to check) that the sequence $\{F_n\}$ of the Fejér kernels is a sequence of bounded functions satisfying (2.8)–(2.10). By Theorem 2.3, the operators C_{F_n} are bounded on X for every $n \in \mathbb{N}$.

Assume that (1.1) is fulfilled for all $f \in X$. Then, for all $f \in X$, the sequence $\{C_{F_n} f\}$ is bounded in X. Therefore, by the uniform boundedness principle, the sequence $\{\|C_{F_n}\|_{\mathcal{B}(X)}\}$ is bounded, but this contradicts (2.12). Thus, there exists a function $f \in X$ such that (1.1) does not hold. □

3. Proof of the second main result

3.1. Separable Banach function spaces X are spaces for which X^* is isometrically isomorphic to X'

Combining [1, Chap. I, Corollaries 4.3 and 5.6] and observing that the measure dm is separable (for the definition of a separable measure, see, e.g., [1, p. 27] or [6, Chap. I, Section 6.10]), we arrive at the following.

Theorem 3.1. *Let X be a Banach function space on $\mathbb{T}$. Then X is separable if and only if its dual space X^* is isometrically isomorphic to the associate space X'.*

3.2. Hardy spaces on the unit disk

Let $\mathbb{D}$ denote the open unit disk in the complex plane $\mathbb{C}$. Recall that a function F analytic in $\mathbb{D}$ is said to belong to the Hardy space $H^p(\mathbb{D})$, $0 < p \leq \infty$, if the integral mean

$$M_p(r, F) = \left(\frac{1}{2\pi}\int_{-\pi}^{\pi} |F(re^{i\theta})|^p \, d\theta\right)^{1/p}, \quad 0 < p < \infty,$$
$$M_\infty(r, F) = \max_{-\pi \leq \theta \leq \pi} |F(re^{i\theta})|,$$

remains bounded as $r \to 1$. If $F \in H^p(\mathbb{D})$, $0 < p \leq \infty$, then the nontangential limit

$$f(e^{i\theta}) = \lim_{r \to 1-0} F(re^{i\theta})$$

exists for almost all $\theta \in [-\pi, \pi]$ (see, e.g., [5, Theorem 2.2]) and the boundary function $f = f(e^{i\theta})$ belongs to L^p.

The following lemma is an immediate consequence of the Smirnov theorem (see, e.g., [5, Theorem 2.11]).

Lemma 3.2. *If $F \in H^p(\mathbb{D})$ for some $p \in (0,1)$ and its boundary function f belongs to L^1, then $F \in H^1(\mathbb{D})$.*

Recall that if $f \in H^1$ then its analytic extension F into $\mathbb{D}$, given by the Poisson integral

$$F(re^{i\theta}) = \frac{1}{2\pi} \int_{-\pi}^{\pi} P(r, \theta - \varphi) f(e^{i\varphi})\, d\varphi, \quad 0 \le r < 1, \quad -\pi \le \theta \le \pi,$$

where

$$P(r, \theta) = \frac{1 - r^2}{1 - 2r\cos\theta + r^2}, \quad 0 \le r < 1, \quad -\pi \le \theta \le \pi,$$

is the Poisson kernel, belongs to $H^1(\mathbb{D})$ and the boundary function of F coincides with f a.e. on $\mathbb{T}$ (see, e.g., [5, Theorem 3.1]).

It is important to note that the Taylor coefficients of $F \in H^p(\mathbb{D})$ coincide with the Fourier coefficients of its boundary function $f \in L^p$. More precisely, one has the following.

Theorem 3.3 ([5, Theorem 3.4]). *Let $F(z) = \sum_{n=0}^{\infty} a_n z^n$ belong to $H^1(\mathbb{D})$ and let $\{\widehat{f}(n)\}$ be the sequence of the Fourier coefficients of its boundary function $f \in L^1$. Then $\widehat{f}(n) = a_n$ for all $n \ge 0$ and $\widehat{f}(n) = 0$ for $n < 0$.*

3.3. Proof of Theorem 1.5

Suppose $\mathcal{P}_A$ is not dense in $H[X]$. Take any function $f \in H[X]$ that does not belong to the closure of $\mathcal{P}_A$ with respect to the norm of X. Since X is separable, it follows from Theorem 3.1 that X^* is isometrically isomorphic to X'. Then, by a corollary of the Hahn–Banach theorem (see, e.g., [2, Chap. 7, Theorem 4.1]), there exists a function $g \in X' \subset L^1$ such that

$$\int_{-\pi}^{\pi} f\left(e^{i\theta}\right) g\left(e^{i\theta}\right) d\theta \ne 0 \tag{3.1}$$

and

$$\int_{-\pi}^{\pi} p\left(e^{i\theta}\right) g\left(e^{i\theta}\right) d\theta = 0 \quad \text{for all} \quad p \in \mathcal{P}_A.$$

In particular, if $p(e^{i\theta}) = e^{in\theta}$ with $n = 0, 1, 2, \dots$, then

$$\widehat{g}(-n) = 0 \quad \text{for all} \quad n = 0, 1, 2, \dots. \tag{3.2}$$

Hence $g \in H[X'] \subset H^1$. For functions $f \in H[X] \subset H^1$ and $g \in H[X'] \subset H^1$, let F and G denote their analytic extensions to the unit disk $\mathbb{D}$ by means of their Poisson integrals. Then $F, G \in H^1(\mathbb{D})$. It follows from (3.2) and Theorem 3.3 that $G(0) = 0$. Since $F, G \in H^1(\mathbb{D})$, by Hölder's inequality, $FG \in H^{1/2}(\mathbb{D})$. On the other hand, since $f \in X$ and $g \in X'$, it follows from Hölder's inequality for Banach function spaces (see [1, Chap. 1, Theorem 2.4]) that $fg \in L^1$. Then it follows from Lemma 3.2 that $FG \in H^1(\mathbb{D})$.

Since $(FG)(0) = F(0)G(0) = 0$, applying Theorem 3.3 to FG, we obtain $\widehat{fg}(0) = 0$, that is,

$$\int_{-\pi}^{\pi} f\left(e^{i\theta}\right) g\left(e^{i\theta}\right) d\theta = 0,$$

which contradicts (3.1). □

Acknowledgment

We would like to thank the referee for the useful remarks.

References

[1] C. Bennett and R. Sharpley, *Interpolation of Operators*, Academic Press, Boston, 1988.

[2] Yu.M. Berezansky, Z.G. Sheftel, and G.F. Us, *Functional Analysis, Vol. 1*, Birkhäuser, Basel, 1996.

[3] J.B. Conway, *The Theory of Subnormal Operators*, American Mathematical Society, Providence, RI, 1991.

[4] D. Cruz-Uribe and A. Fiorenza, *Variable Lebesgue Spaces*, Birkhäuser, Basel, 2013.

[5] P.L. Duren, *Theory of H^p Spaces*, Academic Press, New York and London, 1970.

[6] L.V. Kantorovich and G.P. Akilov, *Functional Analysis*, Pergamon Press, Oxford, 2nd ed., 1982.

[7] A. Karlovich, Fredholmness of singular integral operators with piecewise continuous coefficients on weighted Banach function spaces, *J. Integral Equations Appl.* **15** (2003), 263–320.

[8] A. Karlovich, Density of analytic polynomials in abstract Hardy spaces. *Comment. Math.*, to appear. Preprint is available at arXiv:1710.10078 [math.CA] (2017).

[9] Y. Katznelson, *An Introduction to Harmonic Analysis*, Dower Publications, Inc., New York, 1976.

[10] K. Leśnik, Personal communication to A. Karlovich, February 23, 2017.

[11] K. Leśnik, Toeplitz and Hankel operators between distinct Hardy spaces, arXiv:1708.00910 [math.FA] (2017).

[12] I.I. Sharapudinov, Uniform boundedness in L^p $(p = p(x))$ of some families of convolution operators, *Math. Notes* **59** (1996), 205–212.

[13] Q. Xu, Notes on interpolation of Hardy spaces, *Ann. Inst. Fourier* **42** (1992), 875–889.

Alexei Karlovich
Centro de Matemática e Aplicações,
Departamento de Matemática,
Faculdade de Ciências e Tecnologia,
Universidade Nova de Lisboa,
Quinta da Torre,
2829–516 Caparica
Portugal
e-mail: oyk@fct.unl.pt

Eugene Shargorodsky
Department of Mathematics
King's College London
Strand, London WC2R 2LS
United Kingdom
e-mail: eugene.shargorodsky@kcl.ac.uk

Pseudodifferential operators with compound non-regular symbols

Yuri I. Karlovich

Abstract. The boundedness and compactness of Fourier pseudodifferential operators with compound symbols in subclasses of $L^\infty(\mathbb{R}^2, L^1(\mathbb{R}))$ is studied on weighted Lebesgue spaces $L^p(\mathbb{R}, w)$ with $p \in (1, \infty)$ and Muckenhoupt weights $w \in A_p(\mathbb{R})$ by applying the techniques of oscillatory integrals. The boundedness and compactness conditions are also obtained for Mellin pseudodifferential operators with compound symbols in subclasses of $L^\infty(\mathbb{R}^2_+, L^1(\mathbb{R}))$, which act on the spaces $L^p(\mathbb{R}_+, d\mu)$, where $d\mu(t) = dt/t$ for $t \in \mathbb{R}_+$. The latter results allow one to reduce the smoothness of slowly oscillating Carleson curves Γ and slowly oscillating Muckenhoupt weights w in the Fredholm study of singular integral operators with shifts on weighted Lebesgue spaces $L^p(\Gamma, w)$.

Mathematics Subject Classification (2010). Primary 47G30; Secondary 45E05, 47G10.

Keywords. Fourier and Mellin pseudodifferential operators; compound symbol; oscillatory integral; singular integral operator; Lebesgue space; Muckenhoupt weight; boundedness; compactness.

1. Introduction

The paper is devoted to studying the Fourier and Mellin pseudodifferential operators with compound non-regular symbols of limiting smoothness. Pseudodifferential operators with non-regular symbols are intensively studied now (see, e.g., [5]–[7], [23], [28]–[31] and the references therein). Reducing the smoothness of symbols, which expands the fields of applications of pseudodifferential operators, is one of the main interests. Pseudodifferential operators of Mellin type with non-regular symbols are also actively studied along with Fourier pseudodifferential operators (see, e.g., [8], [22], and [23], and the references given there). Applications of pseudodifferential operators to the theory of singular integral operators with non-regular data are given,

The author was partially supported by the PFCE project (México).

© Springer International Publishing AG, part of Springer Nature 2018
A. Böttcher et al. (eds.), *The Diversity and Beauty of Applied Operator Theory*, Operator Theory: Advances and Applications 268, https://doi.org/10.1007/978-3-319-75996-8_17

e.g., in [1]–[4], [15], [19], [23]. For applications of pseudodifferential operators with non-regular symbols to singular integral operators with shifts and non-regular data see, e.g., [9], [11]–[13], [18], [21].

Let Γ be an oriented locally rectifiable simple open arc in the complex plane, and let $L^p(\Gamma, w)$ be the weighted Lebesgue space with the norm

$$\|f\|_{L^p(\Gamma,w)} := \left(\int_\Gamma |f(\tau)|^p |w(\tau)|^p |d\tau| \right)^{1/p},$$

where $p \in (1, \infty)$ and $w : \Gamma \to [0, \infty]$ is a measurable function with pre-image $w^{-1}(\{0, \infty\})$ of measure zero. As is known (see, e.g., [1]), the Cauchy singular integral operator S_Γ, given for $f \in L^1(\Gamma)$ and almost all $t \in \Gamma$ by

$$(S_\Gamma f)(t) := \lim_{\varepsilon \to 0} \frac{1}{\pi i} \int_{\Gamma \setminus \Gamma(t,\varepsilon)} \frac{f(\tau)}{\tau - t}\, d\tau, \quad \Gamma(t, \varepsilon) := \{\tau \in \Gamma : |\tau - t| < \varepsilon\}, \tag{1.1}$$

is bounded on the space $L^p(\Gamma, w)$ if and only if $w \in A_p(\Gamma)$, that is,

$$c_{p,w} := \sup_{\varepsilon>0} \sup_{t \in \Gamma} \frac{1}{\varepsilon} \left(\int_{\Gamma(t,\varepsilon)} w(\tau)^p |d\tau| \right)^{1/p} \left(\int_{\Gamma(t,\varepsilon)} w(\tau)^{-q} |d\tau| \right)^{1/q} < \infty, \tag{1.2}$$

where $1/p + 1/q = 1$. If (1.2) holds, then the Hölder inequality implies that

$$\sup_{\varepsilon>0} \sup_{t \in \Gamma} |\Gamma(t, \varepsilon)|/\varepsilon < \infty, \tag{1.3}$$

where $|\Gamma(t, \varepsilon)|$ stands for the Lebesgue (length) measure of $\Gamma(t, \varepsilon)$. Condition (1.2) is called the Muckenhoupt condition, and curves Γ satisfying (1.3) are named Carleson or Ahlfors–David curves (see [1]).

The present paper continues investigations started in [14]–[17] and [20], and deals with studying the boundedness and compactness of Fourier pseudodifferential operators with compound non-regular symbols on the weighted Lebesgue spaces $L^p(\mathbb{R}, w)$ with $p \in (1, \infty)$ and Muckenhoupt weights w belonging to $A_p(\mathbb{R})$. The results obtained are applied to studying the boundedness and compactness of Mellin pseudodifferential operators with compound non-regular symbols on the Lebesgue spaces $L^p(\mathbb{R}_+, d\mu)$ with the invariant measure $d\mu(t) = dt/t$. In contrast to the papers [15]–[17], we essentially reduce the smoothness of the considered compound symbols with respect to spatial variables. This, in turn, leads to decreasing the smoothness of curves and weights in the Fredholm theory of singular integral operators with shifts on weighted Lebesgue spaces.

The paper is organized as follows. Section 2 contains preliminaries: definitions of oscillatory integrals, the Banach algebra $V(\mathbb{R})$ of all absolutely continuous functions of bounded total variation on $\mathbb{R}$, and the Banach algebras $L^\infty(\mathbb{R}^n, V(\mathbb{R}))$ $(n = 1, 2)$ of bounded measurable $V(\mathbb{R})$-valued symbols. In Section 3 we study the boundedness and compactness of Fourier pseudodifferential operators with compound non-regular symbols on weighted Lebesgue spaces $L^p(\mathbb{R}, w)$, making use of the techniques of oscillatory integrals (see, e.g., [27], [32]). In Section 4 these results are carried over to the Mellin pseudodifferential operators with compound non-regular symbols on Lebesgue spaces $L^p(\mathbb{R}_+, d\mu)$. In Section 5 we illustrate an application of the results of Section 4 to Mellin pseudodifferential operators arising in studying singular integral operators on weighted Lebesgue spaces over slowly oscillating Carleson curves with slowly oscillating Muckenhoupt weights.

2. Preliminaries

2.1. Oscillatory integrals

Let $\mathcal{F}$ and $\mathcal{F}^{-1}$ be the Fourier transform and its inverse defined on functions $f \in L^2(\mathbb{R})$ by

$$(\mathcal{F}f)(x) = \int_{\mathbb{R}} f(t)e^{-ixt}dt, \quad (\mathcal{F}^{-1}f)(x) = \frac{1}{2\pi}\int_{\mathbb{R}} f(t)e^{ixt}dt, \quad x \in \mathbb{R}. \tag{2.1}$$

Let $C_0^\infty(\mathbb{R}^n)$ stand for the set of all infinitely differentiable complex-valued functions of compact support on $\mathbb{R}^n$, for $n \in \mathbb{N}$.

Let $\chi \in C_0^\infty(\mathbb{R} \times \mathbb{R})$ and $\chi(y,\eta) = 1$ in a neighborhood of the origin. Put $\chi_\varepsilon(y,\eta) := \chi(\varepsilon y, \varepsilon\eta)$ and consider a function $a : \mathbb{R} \times \mathbb{R} \to \mathbb{C}$. If the limit

$$\lim_{\varepsilon\to 0} \frac{1}{2\pi} \iint_{\mathbb{R}^2} \chi_\varepsilon(y,\eta)a(y,\eta)e^{-iy\eta}dyd\eta$$

exists and does not depend on the choice of the cut-off function χ, then it is called the (double) *oscillatory integral of* a and is denoted by $\mathrm{Os}\big[a(y,\eta)e^{-iy\eta}\big]$ (see, e.g., [27, Chapter 1] or [32, Vol. 1, Chapter 1]).

Clearly, if $a \in L^1(\mathbb{R}\times\mathbb{R})$, then, by the Lebesgue dominated convergence theorem (see, e.g., [24, Theorem I.11]), $\mathrm{Os}\big[a(y,\eta)e^{-iy\eta}\big]$ exists and

$$\mathrm{Os}\big[a(y,\eta)e^{-iy\eta}\big] = \frac{1}{2\pi}\iint_{\mathbb{R}^2} a(y,\eta)e^{-iy\eta}dyd\eta. \tag{2.2}$$

In what follows we will use the notation

$$\langle y\rangle^{-2} = (1+y^2)^{-1}, \quad \langle D_\eta\rangle^2 = I - \partial_\eta^2$$

and will apply the following regularization of oscillatory integrals, which is based on the relation

$$\langle y\rangle^{-2}\langle D_\eta\rangle^2 e^{-iy\eta} = e^{-iy\eta}, \tag{2.3}$$

integrating by parts, and on the Lebesgue dominated convergence theorem.

Lemma 2.1. *If the functions* $(y,\eta) \mapsto \partial_\eta^j a(y,\eta)$ *are in the space* $L^\infty(\mathbb{R}, L^1(\mathbb{R}))$ *for all* $j = 0,1,2$, *then the oscillatory integral* $\mathrm{Os}\big[a(y,\eta)e^{-iy\eta}\big]$ *exists and is given by*

$$\mathrm{Os}\big[a(y,\eta)e^{-iy\eta}\big] = \frac{1}{2\pi}\iint_{\mathbb{R}^2} \langle y\rangle^{-2}\langle D_\eta\rangle^2\{a(y,\eta)\}\, e^{-iy\eta}dyd\eta. \tag{2.4}$$

Proof. Choosing a cut-off function $\chi \in C_0^\infty(\mathbb{R}\times\mathbb{R})$ and integrating by parts making use of (2.3), we obtain

$$\begin{aligned}&\iint_{\mathbb{R}^2} \chi_\varepsilon(y,\eta)a(y,\eta)e^{-iy\eta}dyd\eta\\ &\quad= \iint_{\mathbb{R}^2} \chi_\varepsilon(y,\eta)a(y,\eta)\Big\{\langle y\rangle^{-2}\langle D_\eta\rangle^2\big\{e^{-iy\eta}\big\}\Big\}dyd\eta \qquad (2.5)\\ &\quad= \iint_{\mathbb{R}^2} \langle y\rangle^{-2}\langle D_\eta\rangle^2\big\{\chi_\varepsilon(y,\eta)a(y,\eta)\big\}\, e^{-iy\eta}dyd\eta.\end{aligned}$$

As $\partial_\eta^j\big\{\chi_\varepsilon(y,\eta)a(y,\eta)\big\} \in L^\infty\big(\mathbb{R},L^1(\mathbb{R})\big)$ for all $j=0,1,2$, and hence the function $(y,\eta)\mapsto \langle y\rangle^{-2}\langle D_\eta\rangle^2\big\{\chi_\varepsilon(y,\eta)a(y,\eta)\big\}$ belongs to $L^1(\mathbb{R}\times\mathbb{R})$, we conclude that the last double integral in (2.5) exists.

Taking into account the facts that the function

$$(y,\eta)\mapsto \langle y\rangle^{-2}\langle D_\eta\rangle^2\big\{a(y,\eta)\big\}\, e^{-iy\eta}$$

belongs to $L^1(\mathbb{R}\times\mathbb{R})$ and, therefore, the double integral on the right of (2.4) exists, we proceed to prove the equality (2.4).

By definition of oscillatory integrals and by (2.5),

$$\begin{aligned}\mathrm{Os}\big[a(y,\eta)e^{-iy\eta}\big] &= \lim_{\varepsilon\to 0}\frac{1}{2\pi}\iint_{\mathbb{R}^2} \chi_\varepsilon(y,\eta)a(y,\eta)e^{-iy\eta}dyd\eta\\ &= \lim_{\varepsilon\to 0}\frac{1}{2\pi}\iint_{\mathbb{R}^2} \langle y\rangle^{-2}\langle D_\eta\rangle^2\big\{\chi_\varepsilon(y,\eta)a(y,\eta)\big\}\, e^{-iy\eta}dyd\eta.\end{aligned} \qquad (2.6)$$

Applying the Lebesgue dominated convergence theorem and taking into account the fact that for $j=1,2$ the partial derivatives $\partial_\eta^j\chi_\varepsilon(y,\eta)$ uniformly converge on $\mathbb{R}\times\mathbb{R}$ to zero as $\varepsilon\to 0$, we infer from (2.6) that

$$\begin{aligned}\mathrm{Os}\big[a(y,\eta)e^{-iy\eta}\big] &= \lim_{\varepsilon\to 0}\frac{1}{2\pi}\iint_{\mathbb{R}^2} \chi_\varepsilon(y,\eta)\langle y\rangle^{-2}\langle D_\eta\rangle^2\big\{a(y,\eta)\big\}\, e^{-iy\eta}dyd\eta\\ &= \frac{1}{2\pi}\iint_{\mathbb{R}^2} \langle y\rangle^{-2}\langle D_\eta\rangle^2\big\{a(y,\eta)\big\}\, e^{-iy\eta}dyd\eta.\end{aligned}$$

Thus the oscillatory integral $\mathrm{Os}\big[a(y,\eta)e^{-iy\eta}\big]$ exists, equals the double integral in (2.4), and hence does not depend on the particular choice of the cut-off function χ. □

2.2. Absolutely continuous functions of bounded total variation

Let a be an absolutely continuous function on $\mathbb{R}$ of bounded total variation $V(a)$, where $V(a)$ is given by

$$\sup\left\{\sum_{k=1}^{n}\big|a(x_k)-a(x_{k-1})\big| :\ -\infty < x_0 < x_1 < \ldots < x_n < +\infty,\quad n\in\mathbb{N}\right\}.$$

Hence (see, e.g., [10, Chapter 9]), there exist the finite limits

$$a(\pm\infty) = \lim_{x\to\pm\infty} a(x),$$

and therefore a is continuous on $\overline{\mathbb{R}} := [-\infty, +\infty]$, and

$$a' \in L^1(\mathbb{R}), \quad V(a) = \int_{\mathbb{R}} |a'(y)|dy.$$

The set $V(\mathbb{R})$ of all absolutely continuous functions of bounded total variation on $\mathbb{R}$ is a Banach algebra equipped with the norm $\|a\|_V := \|a\|_{L^\infty(\mathbb{R})} + V(a)$.

2.3. Bounded measurable $V(\mathbb{R})$-valued symbols

For every $n = 1, 2$, we denote by $L^\infty(\mathbb{R}^n, V(\mathbb{R}))$ the set of all functions $a : \mathbb{R}^n \times \overline{\mathbb{R}} \to \mathbb{C}$ such that $x \mapsto a(x, \cdot)$ is a bounded measurable $V(\mathbb{R})$-valued function on $\mathbb{R}^n$. Since the Banach space $V(\mathbb{R})$ is separable, we conclude according to [26, Chapter IV, Theorem 23_2] that every measurable $V(\mathbb{R})$-valued function is a limit a.e. of a sequence of simple measurable functions $\widetilde{a}_k : \mathbb{R}^n \to V(\mathbb{R})$ having only finite sets of values $b_i \in V(\mathbb{R})$, with measurable pre-images $\widetilde{a}_k^{-1}(b_i)$ of these values. This implies that, for all $\lambda \in \overline{\mathbb{R}}$, the functions $x \mapsto a(x, \lambda)$ and the function $x \mapsto \|a(x, \cdot)\|_V$ are measurable on $\mathbb{R}^n$ as limits a.e. of suitable sequences of simple measurable functions. Therefore, $a(\cdot, \lambda) \in L^\infty(\mathbb{R}^n)$ for every $\lambda \in \overline{\mathbb{R}}$, and the function $x \mapsto \|a(x, \cdot)\|_V$, where

$$\|a(x, \cdot)\|_V := \max_{\lambda \in \overline{\mathbb{R}}} |a(x, \lambda)| + \int_{\mathbb{R}} |\partial_\lambda a(x, \lambda)| d\lambda, \tag{2.7}$$

belongs to $L^\infty(\mathbb{R}^n)$. Clearly, $L^\infty(\mathbb{R}^n, V(\mathbb{R}))$ is a Banach algebra with the norm

$$\|a\|_{L^\infty(\mathbb{R}^n, V(\mathbb{R}))} := \operatorname*{ess\,sup}_{x \in \mathbb{R}^n} \|a(x, \cdot)\|_V. \tag{2.8}$$

3. The Fourier pseudodifferential operators

Lemma 3.1. *If the function $a(x, y, \lambda)$ belongs to the algebra $L^\infty(\mathbb{R} \times \mathbb{R}, V(\mathbb{R}))$ and the function*

$$\widetilde{a}(x, y, \lambda) := a(x, y, \lambda) - a(x, x, \lambda), \tag{3.1}$$

belongs to the space $L^\infty(\mathbb{R} \times \mathbb{R}, L^1(\mathbb{R}))$, then the Fourier pseudodifferential operator A is well defined for every $u \in C_0^\infty(\mathbb{R})$ by the iterated integral

$$(Au)(x) := \frac{1}{2\pi} \int_{\mathbb{R}} d\lambda \int_{\mathbb{R}} a(x, y, \lambda) e^{i(x-y)\lambda} u(y) dy, \quad x \in \mathbb{R}. \tag{3.2}$$

Proof. We need to prove that the iterated integral on the right of (3.2) converges for every $u \in C_0^\infty(\mathbb{R})$ and almost all $x \in \mathbb{R}$.

Since the function $\widehat{a}(x, \lambda) := a(x, x, \lambda)$ belongs to $L^\infty(\mathbb{R}, V(\mathbb{R}))$, we see from [14, Theorem 3.1] that the Fourier pseudodifferential operator $\widehat{a}(x, D)$ defined for functions $u \in C_0^\infty(\mathbb{R})$ by the iterated integral

$$[\widehat{a}(x, D)u](x) := \frac{1}{2\pi} \int_{\mathbb{R}} d\lambda \int_{\mathbb{R}} \widehat{a}(x, \lambda) e^{i(x-y)\lambda} u(y) dy, \quad x \in \mathbb{R}, \tag{3.3}$$

extends to a bounded linear map on every Lebesgue space $L^p(\mathbb{R})$, $p \in (1, \infty)$, and therefore the iterated integral in (3.3) converges for almost all $x \in \mathbb{R}$. Hence, by (3.1),

$$A = \widehat{a}(x, D) + \widetilde{A}, \tag{3.4}$$

where the Fourier pseudodifferential operator $\widetilde{A}$ for every $u \in C_0^\infty(\mathbb{R})$ is defined by the iterated integral

$$(\widetilde{A}u)(x) := \frac{1}{2\pi}\int_{\mathbb{R}} d\lambda \int_{\mathbb{R}} \widetilde{a}(x, y, \lambda)e^{i(x-y)\lambda}u(y)dy, \quad x \in \mathbb{R}. \tag{3.5}$$

Since the functions $\widetilde{a}(x, y, \lambda)$ and $u(y)$ belong to the spaces $L^\infty(\mathbb{R}\times\mathbb{R}, L^1(\mathbb{R}))$ and $C_0^\infty(\mathbb{R})$, respectively, we conclude that the function

$$(y, \lambda) \mapsto \widetilde{a}(x, y, \lambda)u(y)e^{i(x-y)\lambda} \tag{3.6}$$

belongs to the space $L^1(\mathbb{R} \times \mathbb{R})$ for almost every $x \in \mathbb{R}$, which implies the convergence of the iterated integral on the right of (3.5) for almost all $x \in \mathbb{R}$ by the Fubini theorem. This and the convergence of the iterated integral in (3.3) a.e. on $\mathbb{R}$ imply the convergence of the iterated integral in (3.2) for almost all $x \in \mathbb{R}$. □

Lemma 3.2. *Under the conditions of Lemma* 3.1, *the Fourier pseudodifferential operator $\widetilde{A}$ defined for every $u \in C_0^\infty(\mathbb{R})$ by the iterated integral* (3.5) *can be represented in the form*

$$(\widetilde{A}u)(x) = \mathrm{Os}\big[\widetilde{a}(x, y, \lambda)u(x + y)e^{-iy\lambda}\big], \tag{3.7}$$

where the oscillatory integral depends on the parameter $x \in \mathbb{R}$.

Proof. It follows from (3.5) that

$$(\widetilde{A}u)(x) = \frac{1}{2\pi}\int_{\mathbb{R}} d\lambda \int_{\mathbb{R}} \widetilde{a}(x, x + y, \lambda)u(x + y)e^{-iy\lambda}dy. \tag{3.8}$$

Since the function

$$(y, \lambda) \mapsto \widetilde{a}(x, x + y, \lambda)u(x + y)e^{-iy\lambda}$$

belongs to the space $L^1(\mathbb{R} \times \mathbb{R})$ for almost every $x \in \mathbb{R}$ along with (3.6), we infer from (2.2) that

$$\begin{aligned} &\mathrm{Os}\big[\widetilde{a}(x, x + y, \lambda)u(x + y)e^{-iy\lambda}\big] \\ &\quad = \lim_{\varepsilon\to 0} \frac{1}{2\pi}\iint_{\mathbb{R}^2} \chi_\varepsilon(y, \lambda)\widetilde{a}(x, x + y, \lambda)u(x + y)e^{-iy\lambda}dyd\lambda \\ &\quad = \frac{1}{2\pi}\iint_{\mathbb{R}^2} \widetilde{a}(x, x + y, \lambda)u(x + y)\, e^{-iy\lambda}dyd\lambda \end{aligned} \tag{3.9}$$

for any cut-off function $\chi \in C_0^\infty(\mathbb{R}\times\mathbb{R})$ and for almost every $x \in \mathbb{R}$. Applying the Fubini theorem, we obtain

$$\begin{aligned} &\frac{1}{2\pi}\iint_{\mathbb{R}^2} \widetilde{a}(x, x + y, \lambda)u(x + y)\, e^{-iy\lambda}dy \\ &\qquad = \frac{1}{2\pi}\int_{\mathbb{R}} d\lambda \int_{\mathbb{R}} \widetilde{a}(x, x + y, \lambda)u(x + y)e^{-iy\lambda}dy, \end{aligned}$$

which gives (3.7) in view of (3.8) and (3.9). □

Thus, under the conditions of Lemma 3.1, the pseudodifferential operator $\widetilde{A}$ with compound symbol $\widetilde{a}(x, y, \lambda)$ can be defined via the oscillatory integral (3.7) depending on the parameter $x \in \mathbb{R}$. In this form the operator $\widetilde{A}$ can be extended to the functions $u \in L^\infty(\mathbb{R})$ whenever the functions $(x, y, \lambda) \mapsto \partial_\lambda^j \widetilde{a}(x, y, \lambda)$ belong to the space $L^\infty\big(\mathbb{R} \times \mathbb{R}, L^1(\mathbb{R})\big)$ for all $j = 0, 1, 2$.

Let $\mathcal{B}(X)$ denote the Banach algebra of all bounded linear operators acting on a Banach space X.

Lemma 3.3. *If the functions $\partial_\lambda^j \widetilde{a}(x, y, \lambda)$ belong to the space $L^\infty\big(\mathbb{R}\times\mathbb{R}, L^1(\mathbb{R})\big)$ for all $j = 0, 1, 2$, then the pseudodifferential operator $\widetilde{A}$ given by (3.7) is bounded on the space $L^\infty(\mathbb{R})$ and*

$$\big\|\widetilde{A}\big\|_{\mathcal{B}(L^\infty(\mathbb{R}))} \le \operatorname*{ess\,sup}_{x\in\mathbb{R}} \frac{1}{2\pi} \iint_{\mathbb{R}^2} \langle y\rangle^{-2} \big|\langle D_\lambda\rangle^2 \{\widetilde{a}(x, x+y, \lambda)\}\big| dy d\lambda. \quad (3.10)$$

Proof. By the hypotheses of the lemma, for every $u \in L^\infty(\mathbb{R})$, the functions

$$(x, y, \lambda) \mapsto \langle y\rangle^{-2} \partial_\lambda^j \{\widetilde{a}(x, x+y, \lambda)\} u(x+y)$$

belong to the space $L^\infty\big(\mathbb{R}, L^1(\mathbb{R} \times \mathbb{R})\big)$ for all $j = 0, 1, 2$. Indeed, taking into account the equality $\int_{\mathbb{R}} \langle y\rangle^{-2} dy = \pi$, we obtain

$$\begin{aligned}
&\operatorname*{ess\,sup}_{x\in\mathbb{R}} \iint_{\mathbb{R}^2} \langle y\rangle^{-2} \big|\partial_\lambda^j \{\widetilde{a}(x, x+y, \lambda)\} u(x+y)\big| dy d\lambda \\
&\quad\le \operatorname*{ess\,sup}_{x\in\mathbb{R}} \iint_{\mathbb{R}^2} \langle y - x\rangle^{-2} \big|\partial_\lambda^j \{\widetilde{a}(x, y, \lambda)\} u(y)\big| dy d\lambda \\
&\quad\le \operatorname*{ess\,sup}_{x\in\mathbb{R}} \int_{\mathbb{R}} \langle y - x\rangle^{-2} \big\|\partial_\lambda^j \{\widetilde{a}(x, y, \cdot)\}\big\|_{L^1(\mathbb{R})} \|u\|_{L^\infty(\mathbb{R})} dy \\
&\quad\le \pi \operatorname*{ess\,sup}_{x,y\in\mathbb{R}} \big\|\partial_\lambda^j \{\widetilde{a}(x, y, \cdot)\}\big\|_{L^1(\mathbb{R})} \|u\|_{L^\infty(\mathbb{R})} < \infty. \qquad (3.11)
\end{aligned}$$

Hence, (3.7) and Lemma 2.1 imply that

$$\begin{aligned}
(\widetilde{A}u)(x) &= \operatorname{Os}\big[\widetilde{a}(x, x+y, \lambda) u(x+y) e^{-iy\lambda}\big] \\
&= \frac{1}{2\pi} \iint_{\mathbb{R}^2} \langle y\rangle^{-2} \langle D_\lambda\rangle^2 \{\widetilde{a}(x, x+y, \lambda)\} u(x+y)\, e^{-iy\lambda} dy d\lambda, \qquad (3.12)
\end{aligned}$$

where the latter integral is a usual double integral of an $L^1(\mathbb{R}\times\mathbb{R})$ function for almost every $x \in \mathbb{R}$. Since the function $(x, y, \lambda) \mapsto \langle y\rangle^{-2} \langle D_\lambda\rangle^2 \{\widetilde{a}(x, x+y, \lambda)\}$ belongs to the space $L^\infty\big(\mathbb{R}, L^1(\mathbb{R} \times \mathbb{R})\big)$ as well (it suffices to take $u = 1$ in (3.11)), we infer from (3.12) that

$$\big\|(\widetilde{A}u)\big\|_{L^\infty(\mathbb{R})} \le \left(\operatorname*{ess\,sup}_{x\in\mathbb{R}} \frac{1}{2\pi} \iint_{\mathbb{R}^2} \langle y\rangle^{-2} \big|\langle D_\lambda\rangle^2 \{\widetilde{a}(x, x+y, \lambda)\}\big| dy d\lambda\right) \|u\|_{L^\infty(\mathbb{R})},$$

which gives (3.10) and completes the proof. □

Given $u \in C_0^\infty(\mathbb{R})$, we denote by $\widehat{u} = \mathcal{F}u$ the Fourier transform of u (see (2.1)).

Lemma 3.4. *If the functions $\partial_\lambda^j\widetilde{a}(x,y,\lambda)$ belong to the space $L^\infty(\mathbb{R}\times\mathbb{R},L^1(\mathbb{R}))$ for all $j=0,1,2,3$, and the functions $\partial_y\partial_\lambda^j\widetilde{a}(x,y,\lambda)$ and $\lambda\partial_\lambda^j\widetilde{a}(x,y,\lambda)$ belong to the space $L^\infty(\mathbb{R}\times\mathbb{R},L^1(\mathbb{R}))$ for all $j=1,2,3$, then the pseudodifferential operator $\widetilde{A}$, given by (3.7) for all $u\in C_0^\infty(\mathbb{R})$, can be represented in the form*

$$\big[\sigma_{\widetilde{A}}(x,D)u\big](x)=\frac{1}{2\pi}\int_{\mathbb{R}}d\lambda\int_{\mathbb{R}}\sigma_{\widetilde{A}}(x,\lambda)e^{i(x-y)\lambda}u(y)dy,\quad x\in\mathbb{R},\tag{3.13}$$

where the symbol $\sigma_{\widetilde{A}}(x,\lambda)$ is given by the oscillatory integral

$$\sigma_{\widetilde{A}}(x,\lambda)=\mathrm{Os}\big[\widetilde{a}(x,x+y,\lambda+\eta)e^{-iy\eta}\big],\quad (x,\lambda)\in\mathbb{R}\times\mathbb{R},\tag{3.14}$$

and belongs to the space $L^\infty(\mathbb{R},V(\mathbb{R}))$.

Proof. By Lemmas 3.2 and 3.3,

$$(\widetilde{A}u)(x)=\mathrm{Os}\big[\widetilde{a}(x,x+y,\eta)u(x+y)e^{-iy\eta}\big],$$

and the operator $\widetilde{A}$ is bounded on the space $L^\infty(\mathbb{R})$. Therefore, if we take $u(x)=e^{i\lambda x}$ in $L^\infty(\mathbb{R})$ and follow [29, Chapter 2, Theorem 3.8], we get

$$\begin{aligned}\sigma_{\widetilde{A}}(x,\lambda)&:=e^{-ix\lambda}\widetilde{A}(e^{i(\cdot)\lambda})=\mathrm{Os}\big[e^{-ix\lambda}\widetilde{a}(x,x+y,\eta)e^{i(x+y)\lambda}e^{-iy\eta}\big]\\&=\mathrm{Os}\big[\widetilde{a}(x,x+y,\eta)e^{-iy(\eta-\lambda)}\big]=\mathrm{Os}\big[\widetilde{a}(x,x+y,\lambda+\eta)e^{-iy\eta}\big],\end{aligned}\tag{3.15}$$

which gives (3.14). Substituting the first equality of (3.15) into (3.13), we infer for $u\in C_0^\infty(\mathbb{R})$ that

$$\begin{aligned}\big[\sigma_{\widetilde{A}}(x,D)u\big](x)&=\frac{1}{2\pi}\int_{\mathbb{R}}d\lambda\int_{\mathbb{R}}\sigma_{\widetilde{A}}(x,\lambda)e^{i(x-y)\lambda}u(y)dy\\&=\frac{1}{2\pi}\int_{\mathbb{R}}e^{ix\lambda}\sigma_{\widetilde{A}}(x,\lambda)\widehat{u}(\lambda)d\lambda=\frac{1}{2\pi}\int_{\mathbb{R}}\big[\widetilde{A}(e^{i(\cdot)\lambda})\big](x)\,\widehat{u}(\lambda)d\lambda\\&=\left[\widetilde{A}\left(\frac{1}{2\pi}\int_{\mathbb{R}}\widehat{u}(\lambda)e^{i(\cdot)\lambda}d\lambda\right)\right](x)=(\widetilde{A}u)(x),\end{aligned}$$

which completes the proof of (3.13).

It remains to prove that the function $(x,\lambda)\mapsto\partial_\lambda\sigma_{\widetilde{A}}(x,\lambda)$ belongs to the space $L^\infty(\mathbb{R},L^1(\mathbb{R}))$, which implies that $\sigma_{\widetilde{A}}\in L^\infty(\mathbb{R},V(\mathbb{R}))$. Since the functions $\partial_\lambda^j\widetilde{a}(x,y,\lambda)$ belong to the space $L^\infty(\mathbb{R}\times\mathbb{R},L^1(\mathbb{R}))$ for all $j=0,1,2$, it follows similarly to the proof of Lemma 3.3 that the function

$$(x,\lambda,y,\eta)\mapsto\langle y\rangle^{-2}\langle D_\eta\rangle^2\big\{\widetilde{a}(x,x+y,\lambda+\eta)\big\}e^{-iy\eta}\tag{3.16}$$

belongs to the space $L^\infty(\mathbb{R}\times\mathbb{R},L^1(\mathbb{R}\times\mathbb{R}))$. Indeed, taking into account the equality $\int_{\mathbb{R}}\langle y\rangle^{-2}dy=\pi$, we obtain

$$\begin{aligned}
&\operatorname*{ess\,sup}_{x,\lambda\in\mathbb{R}}\frac{1}{2\pi}\iint_{\mathbb{R}^2}\langle y\rangle^{-2}\big|\langle D_\eta\rangle^2\{\widetilde{a}(x,x+y,\lambda+\eta)\}\big|dyd\eta\\
&\quad\le\operatorname*{ess\,sup}_{x\in\mathbb{R}}\frac{1}{2\pi}\iint_{\mathbb{R}^2}\langle y-x\rangle^{-2}\big|\langle D_\eta\rangle^2\{\widetilde{a}(x,y,\eta)\}\big|dyd\eta\\
&\quad\le\operatorname*{ess\,sup}_{x\in\mathbb{R}}\frac{1}{2\pi}\int_{\mathbb{R}}\langle y-x\rangle^{-2}\big\|\langle D_\eta\rangle^2\{\widetilde{a}(x,y,\cdot)\}\big\|_{L^1(\mathbb{R})}dy\\
&\quad\le\frac{1}{2}\operatorname*{ess\,sup}_{x,y\in\mathbb{R}}\big\|\langle D_\eta\rangle^2\{\widetilde{a}(x,y,\cdot)\}\big\|_{L^1(\mathbb{R})}<\infty. \qquad (3.17)
\end{aligned}$$

As the function (3.16) belongs to the space $L^\infty\big(\mathbb{R}\times\mathbb{R},L^1(\mathbb{R}\times\mathbb{R})\big)$ in view of (3.17), we deduce from (3.14) and (2.2) that

$$\begin{aligned}
\sigma_{\widetilde{A}}(x,\lambda)&=\operatorname{Os}\big[\widetilde{a}(x,x+y,\lambda+\eta)e^{-iy\eta}\big]\\
&=\operatorname{Os}\big[\langle y\rangle^{-2}\langle D_\eta\rangle^2\{\widetilde{a}(x,x+y,\lambda+\eta)\}e^{-iy\eta}\big]\\
&=\frac{1}{2\pi}\iint_{\mathbb{R}^2}\langle y\rangle^{-2}\langle D_\eta\rangle^2\{\widetilde{a}(x,x+y,\lambda+\eta)\}e^{-iy\eta}dyd\eta. \qquad (3.18)
\end{aligned}$$

Hence, we conclude from (3.18) and (3.17) that

$$\begin{aligned}
\|\sigma_{\widetilde{A}}\|_{L^\infty(\mathbb{R}\times\mathbb{R})}&\le\operatorname*{ess\,sup}_{x,\lambda\in\mathbb{R}}\frac{1}{2\pi}\iint_{\mathbb{R}^2}\langle y\rangle^{-2}\big|\langle D_\eta\rangle^2\{\widetilde{a}(x,x+y,\lambda+\eta)\}\big|dyd\eta\\
&\le\frac{1}{2}\operatorname*{ess\,sup}_{x,y\in\mathbb{R}}\big\|\langle D_\eta\rangle^2\{\widetilde{a}(x,y,\cdot)\}\big\|_{L^1(\mathbb{R})}<\infty. \qquad (3.19)
\end{aligned}$$

On the other hand, under the conditions of the lemma, we infer from (3.18) for almost all $x\in\mathbb{R}$ and some $M>0$ that

$$\begin{aligned}
V[\sigma_{\widetilde{A}}(x,\cdot)]&=\int_{\mathbb{R}}\big|\partial_\lambda\sigma_{\widetilde{A}}(x,\lambda)\big|d\lambda\\
&=\frac{1}{2\pi}\int_{\mathbb{R}}d\lambda\left|\iint_{\mathbb{R}^2}\langle y\rangle^{-2}\langle D_\eta\rangle^2\partial_\eta\{\widetilde{a}(x,x+y,\lambda+\eta)\}e^{-iy\eta}dyd\eta\right|\\
&=I_1+I_2,
\end{aligned}$$

with

$$\begin{aligned}
I_1(x)&:=\frac{1}{2\pi}\int_{\mathbb{R}\setminus[-M,M]}d\lambda\left|\iint_{\mathbb{R}^2}\langle y\rangle^{-2}\langle D_\eta\rangle^2\partial_\eta\{\widetilde{a}(x,x+y,\lambda+\eta)\}e^{-iy\eta}dyd\eta\right|,\\
I_2(x)&:=\frac{1}{2\pi}\int_{-M}^{M}d\lambda\left|\iint_{\mathbb{R}^2}\langle y\rangle^{-2}\langle D_\eta\rangle^2\partial_\eta\{\widetilde{a}(x,x+y,\lambda+\eta)\}e^{-iy\eta}dyd\eta\right|. \qquad (3.20)
\end{aligned}$$

Since the functions $\partial_\lambda^j\widetilde{a}(x,y,\lambda)$ belong to the space $L^\infty\big(\mathbb{R}\times\mathbb{R},L^1(\mathbb{R})\big)$ for all $j=1,2,3$, it follows from (3.20) for almost all $x\in\mathbb{R}$ that we have the estimate

$$\begin{aligned} I_2(x) &\le \frac{1}{2\pi}\int_{-M}^{M} d\lambda \iint_{\mathbb{R}^2} \langle y\rangle^{-2}\big|\langle D_\eta\rangle^2\partial_\eta\{\widetilde{a}(x,x+y,\lambda+\eta)\}\big|dyd\eta \\ &= \frac{1}{2\pi}\sum_{j=1,3}\int_{-M}^{M} d\lambda \int_{\mathbb{R}}\langle y\rangle^{-2}dy\int_{\mathbb{R}}\big|\partial_\eta^j\{\widetilde{a}(x,x+y,\eta)\}\big|d\eta \\ &= \frac{M}{\pi}\sum_{j=1,3}\int_{\mathbb{R}}\langle y\rangle^{-2}\operatorname*{ess\,sup}_{x,y\in\mathbb{R}}\big\|\partial_\eta^j\{\widetilde{a}(x,x+y,\cdot)\}\big\|_{L^1(\mathbb{R})}dy \\ &\le M\sum_{j=1,3}\big\|\partial_\eta^j\widetilde{a}\big\|_{L^\infty(\mathbb{R}\times\mathbb{R},L^1(\mathbb{R}))}<\infty. \end{aligned} \tag{3.21}$$

On the other hand, as the functions

$$\partial_\lambda^j\widetilde{a}(x,y,\lambda),\quad \lambda\partial_\lambda^j\widetilde{a}(x,y,\lambda),\quad \partial_y\partial_\lambda^j\widetilde{a}(x,y,\lambda)$$

belong to the space $L^\infty\big(\mathbb{R}\times\mathbb{R},L^1(\mathbb{R})\big)$ for all $j=1,2,3$, from (3.20), we infer for almost all $x\in\mathbb{R}$ that

$$\begin{aligned} &I_1(x) \\ &= \int_{\mathbb{R}\setminus[-M,M]} d\lambda\left|\frac{1}{2\pi}\iint_{\mathbb{R}^2}\langle y\rangle^{-2}\langle D_\eta\rangle^2\partial_\eta\{\widetilde{a}(x,x+y,\eta)\}e^{-iy(\eta-\lambda)}dyd\eta\right| \\ &= \int_{\mathbb{R}\setminus[-M,M]} d\lambda\left|\frac{1}{2\pi}\int_{\mathbb{R}}\langle y\rangle^{-2}\mathcal{F}_{\eta\to y}\{\langle D_\eta\rangle^2\partial_\eta\{\widetilde{a}(x,x+y,\eta)\}\}e^{iy\lambda}dy\right| \\ &= \int_{\mathbb{R}\setminus[-M,M]} d\lambda\left|\frac{1}{2\pi i\lambda}\int_{\mathbb{R}}\partial_y\Big(\langle y\rangle^{-2}\mathcal{F}_{\eta\to y}\{\langle D_\eta\rangle^2\partial_\eta\{\widetilde{a}(x,x+y,\eta)\}\}\Big)e^{iy\lambda}dy\right| \\ &= \int_{\mathbb{R}\setminus[-M,M]} \frac{d\lambda}{2\pi|\lambda|}\left|\int_{\mathbb{R}}\frac{-2y}{(1+y^2)^2}\Big(\mathcal{F}_{\eta\to y}\{\langle D_\eta\rangle^2\partial_\eta\{\widetilde{a}(x,x+y,\eta)\}\}\Big)e^{iy\lambda}dy\right| \\ &+\int_{\mathbb{R}\setminus[-M,M]} \frac{d\lambda}{2\pi|\lambda|}\left|\int_{\mathbb{R}}\langle y\rangle^{-2}\Big(\mathcal{F}_{\eta\to y}\{-i\eta\langle D_\eta\rangle^2\partial_\eta\{\widetilde{a}(x,x+y,\eta)\}\}\Big)e^{iy\lambda}dy\right| \\ &+\int_{\mathbb{R}\setminus[-M,M]} \frac{d\lambda}{2\pi|\lambda|}\left|\int_{\mathbb{R}}\langle y\rangle^{-2}\Big(\mathcal{F}_{\eta\to y}\{\partial_y\langle D_\eta\rangle^2\partial_\eta\{\widetilde{a}(x,x+y,\eta)\}\}\Big)e^{iy\lambda}dy\right|. \end{aligned} \tag{3.22}$$

It is easily seen that the functions

$$\begin{aligned} f_1 &: (x,y)\mapsto \mathcal{F}_{\eta\to y}\big\{\langle D_\eta\rangle^2\partial_\eta\{\widetilde{a}(x,x+y,\eta)\}\big\}, \\ f_2 &: (x,y)\mapsto \mathcal{F}_{\eta\to y}\big\{-i\eta\langle D_\eta\rangle^2\partial_\eta\{\widetilde{a}(x,x+y,\eta)\}\big\}, \\ f_3 &: (x,y)\mapsto \mathcal{F}_{\eta\to y}\big\{\partial_y\langle D_\eta\rangle^2\partial_\eta\{\widetilde{a}(x,x+y,\eta)\}\big\}, \end{aligned} \tag{3.23}$$

being the Fourier transforms with respect to the variable η of corresponding functions in $L^1(\mathbb{R})$, belong to the space $L^\infty(\mathbb{R}\times\mathbb{R})$ and that

$$\begin{aligned} \|f_1\|_{L^\infty(\mathbb{R}\times\mathbb{R})} &\le \operatorname*{ess\,sup}_{x,y\in\mathbb{R}}\int_{\mathbb{R}}\big|\langle D_\eta\rangle^2\partial_\eta\{\widetilde{a}(x,x+y,\eta)\}\big|d\eta \\ &\le \sum_{j=1,3}\operatorname*{ess\,sup}_{x,y\in\mathbb{R}}\int_{\mathbb{R}}\big|\partial_\eta^j\{\widetilde{a}(x,y,\eta)\}\big|d\eta<\infty, \end{aligned}$$

$$\begin{aligned}\big\|f_2\big\|_{L^\infty(\mathbb{R}\times\mathbb{R})} &\le \operatorname*{ess\,sup}_{x,y\in\mathbb{R}} \int_{\mathbb{R}} \big|\eta\langle D_\eta\rangle^2\partial_\eta\{\widetilde{a}(x,x+y,\eta)\}\big|d\eta \\ &\le \sum_{j=1,3} \operatorname*{ess\,sup}_{x,y\in\mathbb{R}} \int_{\mathbb{R}} \big|\eta\partial_\eta^j\{\widetilde{a}(x,y,\eta)\}\big|d\eta < \infty,\end{aligned} \tag{3.24}$$

$$\begin{aligned}\big\|f_3\big\|_{L^\infty(\mathbb{R}\times\mathbb{R})} &\le \operatorname*{ess\,sup}_{x,y\in\mathbb{R}} \int_{\mathbb{R}} \big|\partial_y\langle D_\eta\rangle^2\partial_\eta\{\widetilde{a}(x,x+y,\eta)\}\big|d\eta \\ &\le \sum_{j=1,3} \operatorname*{ess\,sup}_{x,y\in\mathbb{R}} \int_{\mathbb{R}} \big|\partial_y\partial_\eta^j\{\widetilde{a}(x,y,\eta)\}\big|d\eta < \infty.\end{aligned}$$

Consequently, the functions

$$\begin{aligned} g_1 &: (x,y) \mapsto \frac{-2y}{(1+y^2)^2}\Big(\mathcal{F}_{\eta\to y}\big\{\langle D_\eta\rangle^2\partial_\eta\{\widetilde{a}(x,x+y,\eta)\}\big\}\Big), \\ g_2 &: (x,y) \mapsto \langle y\rangle^{-2}\Big(\mathcal{F}_{\eta\to y}\big\{-i\eta\langle D_\eta\rangle^2\partial_\eta\{\widetilde{a}(x,x+y,\eta)\}\big\}\Big), \\ g_3 &: (x,y) \mapsto \langle y\rangle^{-2}\Big(\mathcal{F}_{\eta\to y}\big\{\partial_y\langle D_\eta\rangle^2\partial_\eta\{\widetilde{a}(x,x+y,\eta)\}\big\}\Big)\end{aligned} \tag{3.25}$$

belong to the space $L^\infty\big(\mathbb{R}, L^2(\mathbb{R})\big)$ and, in view of (3.23),

$$\begin{aligned}\big\|g_1\big\|_{L^\infty(\mathbb{R},L^2(\mathbb{R}))} &= \operatorname*{ess\,sup}_{x\in\mathbb{R}}\Big(\int_{\mathbb{R}}\Big|\frac{-2y}{(1+y^2)^2}\Big|^2\big|f_1(x,y)\big|^2dy\Big)^{1/2} \\ &\le \Big(\int_{\mathbb{R}}\Big|\frac{2y}{(1+y^2)^2}\Big|^2dy\Big)^{1/2}\big\|f_1\big\|_{L^\infty(\mathbb{R}\times\mathbb{R})} < \infty, \\ \big\|g_2\big\|_{L^\infty(\mathbb{R},L^2(\mathbb{R}))} &= \operatorname*{ess\,sup}_{x\in\mathbb{R}}\Big(\int_{\mathbb{R}}\big|\langle y\rangle^{-2}\big|^2\big|f_2(x,y)\big|^2dy\Big)^{1/2} \\ &\le \Big(\int_{\mathbb{R}}\big|\langle y\rangle^{-2}\big|^2dy\Big)^{1/2}\big\|f_2\big\|_{L^\infty(\mathbb{R}\times\mathbb{R})} < \infty, \\ \big\|g_3\big\|_{L^\infty(\mathbb{R},L^2(\mathbb{R}))} &= \operatorname*{ess\,sup}_{x\in\mathbb{R}}\Big(\int_{\mathbb{R}}\big|\langle y\rangle^{-2}\big|^2\big|f_3(x,y)\big|^2dy\Big)^{1/2} \\ &\le \Big(\int_{\mathbb{R}}\big|\langle y\rangle^{-2}\big|^2dy\Big)^{1/2}\big\|f_3\big\|_{L^\infty(\mathbb{R}\times\mathbb{R})} < \infty.\end{aligned} \tag{3.26}$$

Therefore the functions

$$\begin{aligned} h_1 &: (x,\lambda) \mapsto \frac{1}{2\pi}\int_{\mathbb{R}}\frac{-2y}{(1+y^2)^2}\Big(\mathcal{F}_{\eta\to y}\big\{\langle D_\eta\rangle^2\partial_\eta\{\widetilde{a}(x,x+y,\eta)\}\big\}\Big)e^{iy\lambda}dy, \\ h_2 &: (x,\lambda) \mapsto \frac{1}{2\pi}\int_{\mathbb{R}}\langle y\rangle^{-2}\Big(\mathcal{F}_{\eta\to y}\big\{-i\eta\langle D_\eta\rangle^2\partial_\eta\{\widetilde{a}(x,x+y,\eta)\}\big\}\Big)e^{iy\lambda}dy, \\ h_3 &: (x,\lambda) \mapsto \frac{1}{2\pi}\int_{\mathbb{R}}\langle y\rangle^{-2}\Big(\mathcal{F}_{\eta\to y}\big\{\partial_y\langle D_\eta\rangle^2\partial_\eta\{\widetilde{a}(x,x+y,\eta)\}\big\}\Big)e^{iy\lambda}dy,\end{aligned} \tag{3.27}$$

being the inverse Fourier transforms of the functions (3.25) with respect to the variable y, also belong to the space $L^\infty(\mathbb{R}, L^2(\mathbb{R}))$, and

$$\|h_s\|_{L^\infty(\mathbb{R},L^2(\mathbb{R}))} = \|g_s\|_{L^\infty(\mathbb{R},L^2(\mathbb{R}))} \quad \text{for all} \quad s = 1,2,3.$$

Hence, for every $s = 1, 2, 3$, we obtain

$$\begin{aligned}
&\operatorname*{ess\,sup}_{x\in\mathbb{R}} \int_{\mathbb{R}\setminus[-M,M]} \frac{1}{|\lambda|} |h_s(x,\lambda)| d\lambda \\
&\le \left(\int_{\mathbb{R}\setminus[-M,M]} \frac{d\lambda}{|\lambda|^2} \right)^{1/2} \operatorname*{ess\,sup}_{x\in\mathbb{R}} \left(\int_{\mathbb{R}\setminus[-M,M]} |h_s(x,\lambda)|^2 d\lambda \right)^{1/2} \\
&\le (2/M)^{1/2} \|h_s\|_{L^\infty(\mathbb{R},L^2(\mathbb{R}))} = (2/M)^{1/2} \|g_s\|_{L^\infty(\mathbb{R},L^2(\mathbb{R}))} < \infty. \qquad (3.28)
\end{aligned}$$

Finally, we infer from (3.22)–(3.28) that

$$\begin{aligned}
I_1(x) &= \frac{1}{2\pi} \int_{\mathbb{R}\setminus[-M,M]} d\lambda \left| \iint_{\mathbb{R}^2} \langle y\rangle^{-2} \langle D_\eta\rangle^2 \partial_\eta \{\widetilde{a}(x, x+y, \lambda+\eta)\} e^{-iy\eta} dy d\eta \right| \\
&\le (2/M)^{1/2} \|\varphi_1\|_{L^2(\mathbb{R})} \sum_{j=1,3} \|\partial_\eta^j \widetilde{a}\|_{L^\infty(\mathbb{R}\times\mathbb{R},L^1(\mathbb{R}))} \\
&\quad + (2/M)^{1/2} \|\varphi_2\|_{L^2(\mathbb{R})} \sum_{j=1,3} \|\eta \partial_\eta^j \widetilde{a}\|_{L^\infty(\mathbb{R}\times\mathbb{R},L^1(\mathbb{R}))} \\
&\quad + (2/M)^{1/2} \|\varphi_2\|_{L^2(\mathbb{R})} \sum_{j=1,3} \|\partial_y \partial_\eta^j \widetilde{a}\|_{L^\infty(\mathbb{R}\times\mathbb{R},L^1(\mathbb{R}))} \qquad (3.29)
\end{aligned}$$

with $\varphi_1(y) := -2y(1+y^2)^{-2}$ and $\varphi_2(y) := \langle y\rangle^{-2}$. Consequently, by (3.29), $\operatorname{ess\,sup}_{x\in\mathbb{R}} I_1(x) < \infty$, which together with (3.20) and (3.21) implies that

$$\operatorname*{ess\,sup}_{x\in\mathbb{R}} V[\sigma_{\widetilde{A}}(x,\cdot)] < \infty. \qquad (3.30)$$

Combining (3.19), (3.30), (2.7)–(2.8), we see that $\sigma_{\widetilde{A}} \in L^\infty(\mathbb{R}, V(\mathbb{R}))$. □

Lemma 3.4 and the boundedness result for the Fourier pseudodifferential operators $\sigma_{\widetilde{A}}(x, D)$ with non-regular symbols $\sigma_{\widetilde{A}} \in L^\infty(\mathbb{R}, V(\mathbb{R}))$, which was obtained in [20, Theorem 4.1], immediately imply the following.

Theorem 3.5. *If the functions $\partial_\lambda^j \widetilde{a}(x, y, \lambda)$, for all $j = 0, 1, 2, 3$, belong to the space $L^\infty(\mathbb{R} \times \mathbb{R}, L^1(\mathbb{R}))$, and the functions $\lambda \partial_\lambda^j \widetilde{a}(x, y, \lambda)$ and $\partial_y \partial_\lambda^j \widetilde{a}(x, y, \lambda)$, for $j = 1, 2, 3$, belong to the space $L^\infty(\mathbb{R}\times\mathbb{R}, L^1(\mathbb{R}))$, then the Fourier pseudodifferential operator $\widetilde{A}$ with compound symbol $\widetilde{a}(x, y, \lambda)$, defined for every $u \in C_0^\infty(\mathbb{R})$ by (3.5), rewritten in the form (3.7) and represented then by the iterated integral (3.13) for all $u \in C_0^\infty(\mathbb{R})$, where the symbol $\sigma_{\widetilde{A}}(x, \lambda)$ is given by (3.14), extends to a bounded linear operator $\widetilde{A} = \sigma_{\widetilde{A}}(x, D)$ on every weighted Lebesgue space $L^p(\mathbb{R}, w)$ with $p \in (1, \infty)$ and a Muckenhoupt weight $w \in A_p(\mathbb{R})$, and*

$$\|\widetilde{A}\|_{\mathcal{B}(L^p(\mathbb{R},w))} \le C_{p,w} \|\sigma_{\widetilde{A}}\|_{L^\infty(\mathbb{R},V(\mathbb{R}))},$$

where the constant $C_{p,w} \in (0, \infty)$ depends only on p and w.

Under the conditions of Lemma 3.1, the pseudodifferential operator A given by (3.2) is represented in the form (3.4), where $\widehat{a}(x,\lambda) = a(x,x,\lambda)$ is a function belonging to the space $L^\infty(\mathbb{R}, V(\mathbb{R}))$. Combining then [20, Theorem 4.1] and Theorem 3.5, we arrive at the following assertion.

Corollary 3.6. *If the function $a(x,y,\lambda)$ belongs to the space $L^\infty(\mathbb{R}\times\mathbb{R}, V(\mathbb{R}))$ and the function $\widetilde{a}(x,y,\lambda) = a(x,y,\lambda) - a(x,x,\lambda)$ satisfies all the conditions of Theorem 3.5, then the Fourier pseudodifferential operator A, defined for every $u \in C_0^\infty(\mathbb{R})$ by the iterated integral (3.2), extends to a bounded linear operator on every weighted Lebesgue space $L^p(\mathbb{R},w)$ with $p \in (1,\infty)$ and $w \in A_p(\mathbb{R})$.*

Lemma 3.7. *If the functions $\partial_\lambda^j \widetilde{a}(x,y,\lambda)$ belong to the space $L^\infty(\mathbb{R}\times\mathbb{R}, L^1(\mathbb{R}))$ for all $j = 0,1,2,3$, the functions $\partial_y \partial_\lambda^j \widetilde{a}(x,y,\lambda)$ and $\lambda \partial_\lambda^j \widetilde{a}(x,y,\lambda)$ belong to the space $L^\infty(\mathbb{R}\times\mathbb{R}, L^1(\mathbb{R}))$ for all $j = 1,2,3$ and, for every $j = 1,2,3$,*

$$\begin{gathered}\lim_{|x|\to\infty} \partial_\lambda^j \widetilde{a}(x,x+y,\lambda) = 0, \qquad \lim_{|x|\to\infty} \lambda\partial_\lambda^j \widetilde{a}(x,x+y,\lambda) = 0, \\ \text{and} \quad \lim_{|x|\to\infty} \partial_y \partial_\lambda^j \widetilde{a}(x,x+y,\lambda) = 0\end{gathered} \tag{3.31}$$

for all $\lambda \in \overline{\mathbb{R}}$ and almost all $y \in \mathbb{R}$, then

$$\sigma_{\widetilde{A}}(x,\pm\infty) = 0 \ \text{ for almost all } \ x \in \mathbb{R}, \qquad \lim_{x\to\pm\infty} V[\sigma_{\widetilde{A}}(x,\cdot)] = 0, \tag{3.32}$$

and for every $N > 0$,

$$\lim_{M\to+\infty} \operatorname*{ess\,sup}_{|x|\le N} \Big(V_{-\infty}^{-M}[\sigma_{\widetilde{A}}(x,\cdot)] + V_M^{+\infty}[\sigma_{\widetilde{A}}(x,\cdot)]\Big) = 0. \tag{3.33}$$

Proof. It follows from (3.18) that

$$\begin{aligned}\sigma_{\widetilde{A}}(x,\lambda) &= \frac{1}{2\pi}\iint_{\mathbb{R}^2} \langle y\rangle^{-2} \langle D_\eta\rangle^2 \{\widetilde{a}(x,x+y,\eta)\} e^{-iy(\eta-\lambda)} dy d\eta \\ &= \frac{1}{2\pi}\int_{\mathbb{R}} \langle y\rangle^{-2}\Big(\mathcal{F}_{\eta\to y}\{\langle D_\eta\rangle^2\{\widetilde{a}(x,x+y,\eta)\}\}\Big) e^{iy\lambda} dy.\end{aligned} \tag{3.34}$$

Since the function $(x,y,\eta) \mapsto \langle D_\eta\rangle^2\{\widetilde{a}(x,x+y,\eta)\}$ belongs to the space $L^\infty(\mathbb{R}\times\mathbb{R}, L^1(\mathbb{R}))$, we conclude that the function

$$f : (x,y) \mapsto \mathcal{F}_{\eta\to y}\{\langle D_\eta\rangle^2\{\widetilde{a}(x,x+y,\eta)\}\}$$

belongs to the space $L^\infty(\mathbb{R}\times\mathbb{R})$, and therefore the function

$$g : (x,y) \mapsto \langle y\rangle^{-2} \mathcal{F}_{\eta\to y}\{\langle D_\eta\rangle^2\{\widetilde{a}(x,x+y,\eta)\}\}$$

belongs to the space $L^\infty(\mathbb{R}, L^1(\mathbb{R}))$. Hence the function

$$h : (x,\lambda) \mapsto \frac{1}{2\pi}\int_{\mathbb{R}} \langle y\rangle^{-2}\Big(\mathcal{F}_{\eta\to y}\{\langle D_\eta\rangle^2\{\widetilde{a}(x,x+y,\eta)\}\}\Big) e^{iy\lambda} dy,$$

being the inverse Fourier transform of the function g with respect to the variable y, belongs to the space $L^\infty(\mathbb{R}, C_0(\dot{\mathbb{R}}))$, where $C_0(\dot{\mathbb{R}})$ is the C^*-algebra of all continuous complex-valued functions on $\dot{\mathbb{R}}$ that vanish at ∞.

This in view of (3.34) gives the first equality $\sigma_{\widetilde{A}}(x,\pm\infty)=0$ in (3.32) for almost all $x\in\mathbb{R}$.

Let us prove the second equality in (3.32). Since the functions

$$(x,y,\eta)\mapsto\big|\partial_\eta^j\widetilde{a}(x,y,\eta)\big|,\quad (x,y,\eta)\mapsto\big|\eta\partial_\eta^j\widetilde{a}(x,y,\eta)\big|,\\ (x,y,\eta)\mapsto\big|\partial_y\partial_\eta^j\widetilde{a}(x,y,\eta)\big|$$

belong to the space $L^\infty\big(\mathbb{R}\times\mathbb{R},L^1(\mathbb{R})\big)$ for all $j=1,2,3$, we infer from (3.24), (3.31), and the Levi monotone convergence theorem that

$$\begin{aligned}0&\le\lim_{|x|\to\infty}\Big|\mathcal{F}_{\eta\to y}\big\{\langle D_\eta\rangle^2\partial_\eta\big\{\widetilde{a}(x,x+y,\eta)\big\}\big\}\Big|\\ &\le\lim_{|x|\to\infty}\int_{\mathbb{R}}\big|\langle D_\eta\rangle^2\partial_\eta\big\{\widetilde{a}(x,x+y,\eta)\big\}\big|d\eta\\ &=\int_{\mathbb{R}}\lim_{|x|\to\infty}\big|\langle D_\eta\rangle^2\partial_\eta\big\{\widetilde{a}(x,x+y,\eta)\big\}\big|d\eta=0.\end{aligned}$$

Applying analogous reasonings to all functions (3.23), we conclude that

$$\begin{aligned}\lim_{|x|\to\infty}\Big\{\mathcal{F}_{\eta\to y}\big\{\langle D_\eta\rangle^2\partial_\eta\big\{\widetilde{a}(x,x+y,\eta)\big\}\big\}\Big\}&=0,\\ \lim_{|x|\to\infty}\Big\{\mathcal{F}_{\eta\to y}\big\{-i\eta\langle D_\eta\rangle^2\partial_\eta\big\{\widetilde{a}(x,x+y,\eta)\big\}\big\}\Big\}&=0,\\ \lim_{|x|\to\infty}\Big\{\mathcal{F}_{\eta\to y}\big\{\partial_y\langle D_\eta\rangle^2\partial_\eta\big\{\widetilde{a}(x,x+y,\eta)\big\}\big\}\Big\}&=0\end{aligned}\tag{3.35}$$

for almost all $y\in\mathbb{R}$. Further, the functions

$$\begin{aligned}&(x,y)\mapsto\big(2y(1+y^2)^{-2}\big)\Big|\mathcal{F}_{\eta\to y}\big\{\langle D_\eta\rangle^2\partial_\eta\big\{\widetilde{a}(x,x+y,\eta)\big\}\big\}\Big|,\\ &(x,y)\mapsto\langle y\rangle^{-2}\Big|\mathcal{F}_{\eta\to y}\big\{-i\eta\langle D_\eta\rangle^2\partial_\eta\big\{\widetilde{a}(x,x+y,\eta)\big\}\big\}\Big|,\\ &(x,y)\mapsto\langle y\rangle^{-2}\Big|\mathcal{F}_{\eta\to y}\big\{\partial_y\langle D_\eta\rangle^2\partial_\eta\big\{\widetilde{a}(x,x+y,\eta)\big\}\big\}\Big|\end{aligned}\tag{3.36}$$

also belong to the space $L^\infty\big(\mathbb{R}\times\mathbb{R},L^1(\mathbb{R})\big)$, because their second factors on the right are uniformly bounded according to (3.23) and (3.24). Then, applying the Levi monotone convergence theorem to the first function in (3.36), we deduce from (3.35) that

$$\begin{aligned}0&\le\lim_{|x|\to\infty}\bigg|\frac{1}{2\pi}\int_{\mathbb{R}}\frac{-2y}{(1+y^2)^2}\Big(\mathcal{F}_{\eta\to y}\big\{\langle D_\eta\rangle^2\partial_\eta\big\{\widetilde{a}(x,x+y,\eta)\big\}\big\}\Big)e^{iy\lambda}dy\bigg|\\ &\le\lim_{|x|\to\infty}\frac{1}{2\pi}\int_{\mathbb{R}}\frac{2|y|}{(1+y^2)^2}\Big|\mathcal{F}_{\eta\to y}\big\{\langle D_\eta\rangle^2\partial_\eta\big\{\widetilde{a}(x,x+y,\eta)\big\}\big\}\Big|dy\\ &=\frac{1}{2\pi}\int_{\mathbb{R}}\frac{2|y|}{(1+y^2)^2}\lim_{|x|\to\infty}\Big|\mathcal{F}_{\eta\to y}\big\{\langle D_\eta\rangle^2\partial_\eta\big\{\widetilde{a}(x,x+y,\eta)\big\}\big\}\Big|dy=0.\end{aligned}$$

Applying the same arguments to all functions (3.36), we obtain for almost all $\lambda \in \mathbb{R}$ the equalities

$$\begin{aligned} &\lim_{|x|\to\infty} \frac{1}{2\pi} \int_{\mathbb{R}} \frac{-2y}{(1+y^2)^2} \Big(\mathcal{F}_{\eta\to y}\big\{\langle D_\eta\rangle^2 \partial_\eta \big\{\widetilde{a}(x, x+y, \eta)\big\}\big\}\Big) e^{iy\lambda} dy = 0, \\ &\lim_{|x|\to\infty} \frac{1}{2\pi} \int_{\mathbb{R}} \langle y\rangle^{-2} \Big(\mathcal{F}_{\eta\to y}\big\{-i\eta\langle D_\eta\rangle^2 \partial_\eta \big\{\widetilde{a}(x, x+y, \eta)\big\}\big\}\Big) e^{iy\lambda} dy = 0, \quad (3.37) \\ &\lim_{|x|\to\infty} \frac{1}{2\pi} \int_{\mathbb{R}} \langle y\rangle^{-2} \Big(\mathcal{F}_{\eta\to y}\big\{\partial_y\langle D_\eta\rangle^2 \partial_\eta \big\{\widetilde{a}(x, x+y, \eta)\big\}\big\}\Big) e^{iy\lambda} dy = 0. \end{aligned}$$

By the proof of Lemma 3.4, the functions $h_1(x,\lambda), h_2(x,\lambda), h_3(x,\lambda)$ given by (3.27) belong to the space $L^\infty(\mathbb{R}, L^2(\mathbb{R}))$, and therefore the functions

$$(x,\lambda) \mapsto |h_s(x,\lambda)|/|\lambda| \quad (s = 1,2,3)$$

in view of (3.28) belong to the space $L^\infty(\mathbb{R}, L^1(\mathbb{R}))$. Applying then (3.27), (3.37), and the Levi monotone convergence theorem, we infer that, abbreviating $G := \mathbb{R} \setminus [-M, M]$,

$$\begin{aligned} &\lim_{|x|\to\infty} \int_G \frac{d\lambda}{2\pi|\lambda|} \left| \int_{\mathbb{R}} \frac{-2y}{(1+y^2)^2} \Big(\mathcal{F}_{\eta\to y}\big\{\langle D_\eta\rangle^2 \partial_\eta \big\{\widetilde{a}(x, x+y, \eta)\big\}\big\}\Big) e^{iy\lambda} dy \right| \\ &\quad = \int_G \frac{d\lambda}{2\pi|\lambda|} \left| \lim_{|x|\to\infty} \int_{\mathbb{R}} \frac{-2y}{(1+y^2)^2} \Big(\mathcal{F}_{\eta\to y}\big\{\langle D_\eta\rangle^2 \partial_\eta \big\{\widetilde{a}(x, x+y, \eta)\big\}\big\}\Big) e^{iy\lambda} dy \right| \\ &\quad = 0, \\ &\lim_{|x|\to\infty} \int_G \frac{d\lambda}{2\pi|\lambda|} \left| \int_{\mathbb{R}} \langle y\rangle^{-2} \Big(\mathcal{F}_{\eta\to y}\big\{-i\eta\langle D_\eta\rangle^2 \partial_\eta \big\{\widetilde{a}(x, x+y, \eta)\big\}\big\}\Big) e^{iy\lambda} dy \right| \\ &\quad = \int_G \frac{d\lambda}{2\pi|\lambda|} \left| \lim_{|x|\to\infty} \int_{\mathbb{R}} \langle y\rangle^{-2} \Big(\mathcal{F}_{\eta\to y}\big\{-i\eta\langle D_\eta\rangle^2 \partial_\eta \big\{\widetilde{a}(x, x+y, \eta)\big\}\big\}\Big) e^{iy\lambda} dy \right| \\ &\quad = 0, \\ &\lim_{|x|\to\infty} \int_G \frac{d\lambda}{2\pi|\lambda|} \left| \int_{\mathbb{R}} \langle y\rangle^{-2} \Big(\mathcal{F}_{\eta\to y}\big\{\partial_y\langle D_\eta\rangle^2 \partial_\eta \big\{\widetilde{a}(x, x+y, \eta)\big\}\big\}\Big) e^{iy\lambda} dy \right| \\ &\quad = \int_G \frac{d\lambda}{2\pi|\lambda|} \left| \lim_{|x|\to\infty} \int_{\mathbb{R}} \langle y\rangle^{-2} \Big(\mathcal{F}_{\eta\to y}\big\{\partial_y\langle D_\eta\rangle^2 \partial_\eta \big\{\widetilde{a}(x, x+y, \eta)\big\}\big\}\Big) e^{iy\lambda} dy \right| \\ &\quad = 0. \end{aligned} \quad (3.38)$$

Combining (3.38) and (3.22), we obtain

$$\lim_{|x|\to\infty} \frac{1}{2\pi} \int_G d\lambda \left| \iint_{\mathbb{R}^2} \langle y\rangle^{-2} \langle D_\eta\rangle^2 \partial_\eta \big\{\widetilde{a}(x, x+y, \lambda+\eta)\big\} e^{-iy\eta} dy d\eta \right| = 0. \quad (3.39)$$

Similarly and even easier, one can prove from (3.21) by applying the Levi monotone convergence theorem that

$$\lim_{|x|\to\infty} \frac{1}{2\pi} \int_{-M}^{M} d\lambda \left| \iint_{\mathbb{R}^2} \langle y\rangle^{-2} \langle D_\eta\rangle^2 \partial_\eta \big\{\widetilde{a}(x, x+y, \lambda+\eta)\big\} e^{-iy\eta} dy d\eta \right| = 0. \quad (3.40)$$

Finally, combining (3.39) and (3.40), we infer from (3.20) that

$$\lim_{|x|\to\infty} V[\sigma_{\widetilde{A}}(x,\cdot)]$$
$$= \lim_{|x|\to\infty} \frac{1}{2\pi}\int_{\mathbb{R}} d\lambda \left| \iint_{\mathbb{R}^2} \langle y\rangle^{-2}\langle D_\eta\rangle^2 \partial_\eta\{\widetilde{a}(x,x+y,\lambda+\eta)\}e^{-iy\eta}dyd\eta \right| = 0,$$

which gives the second equality in (3.32).

It remains to prove (3.33). Applying (3.20) and (3.29), for every $N > 0$, we obtain

$$\operatorname*{ess\,sup}_{|x|\le N} \Big(V_{-\infty}^{-M}[\sigma_{\widetilde{A}}(x,\cdot)] + V_M^{+\infty}[\sigma_{\widetilde{A}}(x,\cdot)] \Big)$$
$$= \operatorname*{ess\,sup}_{|x|\le N} \frac{1}{2\pi}\int_G d\lambda \left| \iint_{\mathbb{R}^2} \langle y\rangle^{-2}\langle D_\eta\rangle^2 \partial_\eta\{\widetilde{a}(x,x+y,\lambda+\eta)\}e^{-iy\eta}dyd\eta \right|$$
$$\le \operatorname*{ess\,sup}_{x\in\mathbb{R}} I_1(x) \le C\big(2/M\big)^{1/2}, \tag{3.41}$$

where the constant $C \in (0,\infty)$ is independent of $M > 0$. Passing to limits as $M \to +\infty$ in (3.41), we immediately get (3.33), which completes the proof. □

Lemma 3.7 and [20, Theorem 5.4] imply the following compactness result for the Fourier pseudodifferential operators with non-regular symbols.

Theorem 3.8. *Under the conditions of Lemma 3.7, the Fourier pseudodifferential operator $\widetilde{A}$ with compound symbol $\widetilde{a}(x,y,\lambda)$, defined for every $u \in C_0^\infty(\mathbb{R})$ by (3.5) and represented in the form $\widetilde{A} = \sigma_{\widetilde{A}}(x,D)$ with a symbol $\sigma_{\widetilde{A}}(x,\lambda)$ in $L^\infty(\mathbb{R}, V(\mathbb{R}))$, is a compact operator on every weighted Lebesgue space $L^p(\mathbb{R},w)$ with $p \in (1,\infty)$ and $w \in A_p(\mathbb{R})$.*

Corollary 3.6 and Theorem 3.8 imply the following.

Corollary 3.9. *If the function $a(x,y,\lambda)$ belongs to the space $L^\infty(\mathbb{R}\times\mathbb{R}, V(\mathbb{R}))$ and the function $\widetilde{a}(x,y,\lambda) = a(x,y,\lambda) - a(x,x,\lambda)$ satisfies all the conditions of Lemma 3.7, then the pseudodifferential operator A, defined for every $u \in C_0^\infty(\mathbb{R})$ by the iterated integral (3.2) and extended to a bounded linear operator on every weighted Lebesgue space $L^p(\mathbb{R},w)$ with $p \in (1,\infty)$ and $w \in A_p(\mathbb{R})$, is represented in the form*

$$A = \widehat{a}(x,D) + K, \tag{3.42}$$

where the Fourier pseudodifferential operator $\widehat{a}(x,D)$ with symbol $\widehat{a}(x,\lambda) \in L^\infty(\mathbb{R}, V(\mathbb{R}))$ given by $\widehat{a}(x,\lambda) = a(x,x,\lambda)$ is bounded on the space $L^p(\mathbb{R},w)$ and $K = \widetilde{A}$ is a compact operator on the space $L^p(\mathbb{R},w)$.

4. The Mellin pseudodifferential operators

Let $\mathbb{R}_+ := (0,\infty)$ and let $L^\infty(\mathbb{R}_+\times\mathbb{R}_+, V(\mathbb{R}))$ stand for the set of all functions $a : \mathbb{R}_+ \times \mathbb{R}_+ \times \overline{\mathbb{R}} \to \mathbb{C}$ such that $(r,\varrho) \mapsto a(r,\varrho,\cdot)$ is a bounded measurable

$V(\mathbb{R})$-valued function on $\mathbb{R}_+ \times \mathbb{R}_+$. The set $L^\infty\big(\mathbb{R}_+ \times \mathbb{R}_+, V(\mathbb{R})\big)$ becomes a Banach algebra if we equip it with the norm

$$\|a\|_{L^\infty(\mathbb{R}_+\times\mathbb{R}_+,V(\mathbb{R}))} := \operatorname*{ess\,sup}_{(r,\varrho)\in\mathbb{R}_+\times\mathbb{R}_+} \|a(r,\varrho,\cdot)\|_V < \infty.$$

Theorem 4.1 ([15, Theorem 6.1]). *If $\widehat{b} \in L^\infty\big(\mathbb{R}_+, V(\mathbb{R})\big)$, then the Mellin pseudodifferential operator $\mathrm{Op}(\widehat{b})$, defined for functions $f \in C_0^\infty(\mathbb{R}_+)$ by the iterated integral*

$$\big[\mathrm{Op}(\widehat{b})f\big](r) = \frac{1}{2\pi}\int_{\mathbb{R}} d\lambda \int_{\mathbb{R}_+} \widehat{b}(r,\lambda)\left(\frac{r}{\varrho}\right)^{i\lambda} f(\varrho)\frac{d\varrho}{\varrho}, \quad r \in \mathbb{R}_+, \tag{4.1}$$

extends to a bounded linear operator on every Lebesgue space $L^p(\mathbb{R}_+, d\mu)$ with $p \in (1,\infty)$ and the measure $d\mu = d\varrho/\varrho$, and

$$\|\mathrm{Op}(\widehat{b})\|_{\mathcal{B}(L^p(\mathbb{R}_+,d\mu))} \le C_{p,1}\|\widehat{b}\|_{L^\infty(\mathbb{R}_+,V(\mathbb{R}))},$$

where the constant $C_{p,1}$ for $w = 1$ is taken from Theorem 3.5.

Theorem 4.1 immediately follows from [14, Theorem 3.1] by applying the transform $A \mapsto EAE^{-1}$, where E is the isometric isomorphism

$$E : L^p(\mathbb{R}_+, d\mu) \to L^p(\mathbb{R}), \quad (Ef)(x) = f(e^x), \quad x \in \mathbb{R}, \tag{4.2}$$

to the Mellin pseudodifferential operator $\mathrm{Op}(\widehat{b})$ defined by (4.1), which gives

$$E\,\mathrm{Op}(\widehat{b})\,E^{-1} = \widehat{a}(x,D), \tag{4.3}$$

where $\widehat{a}(x,\lambda) := \widehat{b}(e^x,\lambda)$, $\|\widehat{a}\|_{L^\infty(\mathbb{R},V(\mathbb{R}))} = \|\widehat{b}\|_{L^\infty(\mathbb{R}_+,V(\mathbb{R}))}$, and $\widehat{a}(x,D)$ is the Fourier pseudodifferential operator defined for functions $\varphi \in C_0^\infty(\mathbb{R})$ by

$$[\widehat{a}(x,D)\varphi](x) = \frac{1}{2\pi}\int_{\mathbb{R}} d\lambda \int_{\mathbb{R}} \widehat{a}(x,\lambda)e^{i(x-y)\lambda}\varphi(y)dy, \quad x \in \mathbb{R}.$$

Let $a(x,y,\lambda) \in L^\infty\big(\mathbb{R}\times\mathbb{R}, V(\mathbb{R})\big)$, let $\widetilde{a}(x,y,\lambda) \in L^\infty\big(\mathbb{R}\times\mathbb{R}, L^1(\mathbb{R})\big)$ be given by (3.1), and let

$$b(r,\varrho,\lambda) := a(\log r, \log\varrho, \lambda), \quad \widetilde{b}(r,\varrho,\lambda) := \widetilde{a}(\log r, \log\varrho, \lambda) \tag{4.4}$$

for all $(r,\varrho,\lambda) \in \mathbb{R}_+ \times \mathbb{R}_+ \times \mathbb{R}$. Obviously, by (4.4), the conditions

$$\partial_\lambda^j \widetilde{a}(x,y,\lambda) \in L^\infty\big(\mathbb{R}\times\mathbb{R}, L^1(\mathbb{R})\big) \ \text{ for all } \ j = 0,1,2,3;$$
$$\lambda\partial_\lambda^j \widetilde{a}(x,y,\lambda),\ \partial_y\partial_\lambda^j \widetilde{a}(x,y,\lambda) \in L^\infty\big(\mathbb{R}\times\mathbb{R}, L^1(\mathbb{R})\big) \ \text{ for all } \ j = 1,2,3$$

are equivalent, respectively, to the conditions

$$\begin{aligned} &\partial_\lambda^j \widetilde{b}(r,\varrho,\lambda) \in L^\infty\big(\mathbb{R}_+\times\mathbb{R}_+, L^1(\mathbb{R})\big) \ \ (j = 0,1,2,3);\\ &\lambda\partial_\lambda^j \widetilde{b}(r,\varrho,\lambda),\ (\varrho\partial_\varrho)\partial_\lambda^j \widetilde{b}(r,\varrho,\lambda) \in L^\infty\big(\mathbb{R}_+\times\mathbb{R}_+, L^1(\mathbb{R})\big) \ \ (j = 1,2,3). \end{aligned} \tag{4.5}$$

Similarly to Theorem 4.1, applying (4.2)–(4.3) to the Mellin pseudodifferential operators with compound non-regular symbols and using Corollary 3.6 with $w = 1$, we easily obtain the following.

Theorem 4.2. *If a function $b(r,\varrho,\lambda)$ belongs to the space $L^\infty(\mathbb{R}_+\times\mathbb{R}_+,V(\mathbb{R}))$ and the function $\widetilde{b}(r,\varrho,\lambda):=b(r,\varrho,\lambda)-b(r,r,\lambda)$ satisfies conditions* (4.5), *then the Mellin pseudodifferential operator $B=\mathrm{Op}(b)$ defined for functions $f\in C_0^\infty(\mathbb{R}_+)$ by the iterated integral*

$$\big[\mathrm{Op}(b)f\big](r)=\frac{1}{2\pi}\int_{\mathbb{R}}d\lambda\int_{\mathbb{R}_+}b(r,\varrho,\lambda)\left(\frac{r}{\varrho}\right)^{i\lambda}f(\varrho)\frac{d\varrho}{\varrho},\quad r\in\mathbb{R}_+,\tag{4.6}$$

extends to a bounded linear operator on every Lebesgue space $L^p(\mathbb{R}_+,d\mu)$ with $p\in(1,\infty)$.

Since conditions (3.31) for all $j=1,2,3$ are equivalent to the conditions

$$\begin{aligned}&\lim_{r\to s}\partial_\lambda^j\widetilde{b}(r,r\varrho,\lambda)=0\ \text{ for all }\ \lambda\in\overline{\mathbb{R}}\ \text{ and almost all }\ \varrho\in\mathbb{R}_+,\\&\lim_{r\to s}\lambda\partial_\lambda^j\widetilde{b}(r,r\varrho,\lambda)=0\ \text{ for all }\ \lambda\in\overline{\mathbb{R}}\ \text{ and almost all }\ \varrho\in\mathbb{R}_+,\\&\lim_{r\to s}(\varrho\partial_\varrho)\partial_\lambda^j\widetilde{b}(r,r\varrho,\lambda)=0\ \text{ for all }\ \lambda\in\overline{\mathbb{R}}\ \text{ and almost all }\ \varrho\in\mathbb{R}_+,\end{aligned}\tag{4.7}$$

where $j=1,2,3$ and $s\in\{0,\infty\}$, we immediately infer from Theorem 3.8 with the aid of the transform (4.3) the following compactness result for the Mellin pseudodifferential operators with compound non-regular symbols.

Theorem 4.3. *If conditions* (4.5) *and* (4.7) *are fulfilled, then the Mellin pseudodifferential operator $\widetilde{B}=\mathrm{Op}(\widetilde{b})$, defined for functions $f\in C_0^\infty(\mathbb{R}_+)$ by the iterated integral*

$$\big[\mathrm{Op}(\widetilde{b})f\big](r)=\frac{1}{2\pi}\int_{\mathbb{R}}d\lambda\int_{\mathbb{R}_+}\widetilde{b}(r,\varrho,\lambda)\left(\frac{r}{\varrho}\right)^{i\lambda}f(\varrho)\frac{d\varrho}{\varrho},\quad r\in\mathbb{R}_+,$$

and extended to a bounded linear operator on every Lebesgue space $L^p(\mathbb{R}_+,d\mu)$ with $p\in(1,\infty)$, is a compact operator on the space $L^p(\mathbb{R}_+,d\mu)$.

Note that $\widetilde{b}(r,\varrho,\lambda)\in L^\infty(\mathbb{R}_+\times\mathbb{R}_+,V(\mathbb{R}))$ since the function $\partial_\lambda\widetilde{b}(r,\varrho,\lambda)$ belongs to the space $L^\infty(\mathbb{R}_+\times\mathbb{R}_+,L^1(\mathbb{R}))$. Then, by Theorem 4.2, the operator $\widetilde{B}=\mathrm{Op}(\widetilde{b})$ is bounded along with $B=\mathrm{Op}(b)$ because

$$\widetilde{\widetilde{b}}(r,\varrho,\lambda)=\widetilde{b}(r,\varrho,\lambda)-\widetilde{b}(r,r,\lambda)=\widetilde{b}(r,\varrho,\lambda).$$

On the other hand, due to Theorem 4.1, the operator $\mathrm{Op}(\widetilde{b})=\mathrm{Op}(b)-\mathrm{Op}(\widehat{b})$ is bounded along with $B=\mathrm{Op}(b)$ and $\widehat{B}=\mathrm{Op}(\widehat{b})$, where $\widehat{b}(r,\lambda)=b(r,r,\lambda)$.

Finally, making use of transform (4.3), we immediately deduce from Corollary 3.9 the following.

Corollary 4.4. *If a function $b(r,\varrho,\lambda)$ belongs to the space $L^\infty(\mathbb{R}_+\times\mathbb{R}_+,V(\mathbb{R}))$ and the function $\widetilde{b}(r,\varrho,\lambda)=b(r,\varrho,\lambda)-b(r,r,\lambda)$ satisfies conditions* (4.5) *and* (4.7), *then the Mellin pseudodifferential operator $B=\mathrm{Op}(b)$, defined for functions $f\in C_0^\infty(\mathbb{R}_+)$ by the iterated integral* (4.6) *and extended to a bounded linear operator on every Lebesgue space $L^p(\mathbb{R}_+,d\mu)$ with $p\in(1,\infty)$, is represented similarly to* (3.42) *in the form $B=\mathrm{Op}(\widehat{b})+K$, where $\widehat{b}(r,\lambda)=b(r,r,\lambda)$ and K is a compact operator on the space $L^p(\mathbb{R}_+,d\mu)$.*

5. Applications

The results of Section 4 applied to Mellin pseudodifferential operators with compound non-regular symbols, which are related to singular integral operators on weighted Lebesgue spaces $L^p(\Gamma, w)$, where Γ are slowly oscillating Carleson curves and w are slowly oscillating Muckenhoupt weights (cf. [2] and [18]), allow one to reduce the smoothness of curves Γ and weights w, which is important for studying singular integral operators with shifts (see, e.g., [9], [11]–[13], and [21]).

Let $C_b(\mathbb{R}_+)$ and $C^1(\mathbb{R}_+)$ denote, respectively, the set of bounded continuous functions $a : \mathbb{R}_+ \to \mathbb{C}$ and the set of continuously differentiable functions $b : \mathbb{R}_+ \to \mathbb{C}$. Given $b \in C^1(\mathbb{R}_+)$, let $(rD_r)b := rb'(r)$.

Let $SO(\mathbb{R}_+)$ stand for the set of all functions $a \in C_b(\mathbb{R}_+)$ that slowly oscillate at 0 and ∞, that is (see, e.g., [25]), satisfy the condition

$$\lim_{r\to s} \max\big\{|a(x) - a(y)| : x, y \in [r, 2r]\big\} = 0 \quad \text{for} \quad s \in \{0, \infty\}. \tag{5.1}$$

We remark that (5.1) holds if and only if, for some $\lambda > 1$ (equivalently, for any $\lambda > 1$),

$$\lim_{r\to s} \max\big\{|a(r) - a(\nu r)| : \nu \in [\lambda^{-1}, \lambda]\big\} = 0 \quad \text{for} \quad s \in \{0, \infty\}. \tag{5.2}$$

Obviously, $SO(\mathbb{R}_+)$ is a unital C^*-subalgebra of $L^\infty(\mathbb{R}_+)$.

Following [2], we define slowly oscillating curves and weights, where we reduce their smoothness in comparison with [2] and [18].

An unbounded oriented simple open arc Γ with the starting point t is called a slowly oscillating curve (at t and ∞) if

$$\Gamma = \big\{\tau = t + re^{i\theta(r)} : \ r \in \mathbb{R}_+\big\} \tag{5.3}$$

where θ is a real-valued function in $C^1(\mathbb{R}_+)$, the function $(rD_r)\theta = r\theta'(r)$ belongs to $SO(\mathbb{R}_+)$, and the function $(rD_r)^2\theta$ belongs to $L^\infty(\mathbb{R}_+)$. Note that $t, \infty \notin \Gamma$ and that the function $\theta(r)$ can be unbounded as $r \to 0$ or $r \to \infty$. Obviously, Γ is a Carleson curve (see, e.g., [3]).

Let Γ be a slowly oscillating curve given by (5.3). We call a function $w : \Gamma \to (0, \infty)$ a slowly oscillating weight (at t and ∞) if

$$w\big(t + re^{i\theta(r)}\big) = e^{v(r)} \quad \text{for} \quad r \in \mathbb{R}_+, \tag{5.4}$$

where v is a real-valued function in $C^1(\mathbb{R}_+)$, the function $(rD_r)v$ belongs to $SO(\mathbb{R}_+)$, and the function $(rD_r)^2 v$ belongs to $L^\infty(\mathbb{R}_+)$. Let $1 < p < \infty$. One can show (cf. [1, Theorem 2.36]) that $w \in A_p(\Gamma)$ if and only if

$$-1/p < \liminf_{r\to s} rv'(r) \le \limsup_{r\to s} rv'(r) < 1/q \quad \text{for} \quad s \in \{0, \infty\}.$$

Let A_p^{SO} denote the set of all pairs (Γ, w) such that Γ is a slowly oscillating curve and w is a slowly oscillating weight in $A_p(\Gamma)$.

By [2] (see also [15]), studying singular integral operators on weighted Lebesgue spaces $L^p(\Gamma, w)$ over Carleson curves Γ given by (5.3) with slowly oscillating Muckenhoupt weights w given by (5.4) is reduced to studying Mellin pseudodifferential operators with compound symbols on $L^p(\mathbb{R}_+, d\mu)$.

In particular, with the operator S_Γ given by (1.1) and bounded on the space $L^p(\Gamma, w)$, one associates the compound symbol of the form

$$\sigma(r, \varrho, \lambda) := \frac{1 + i\varrho\theta'(\varrho)}{1 + im_\theta(r, \varrho)} \coth\left[\pi\frac{\lambda + i(1/p + m_v(r, \varrho))}{1 + im_\theta(r, \varrho)}\right], \tag{5.5}$$

where

$$m_v(r, \varrho) = \frac{v(r) - v(\varrho)}{\log r - \log \varrho}, \quad m_\theta(r, \varrho) = \frac{\theta(r) - \theta(\varrho)}{\log r - \log \varrho}, \tag{5.6}$$

and

$$0 < \frac{1}{p} + \inf_{r,\varrho\in\mathbb{R}_+} m_v(r, \varrho) \le \frac{1}{p} + \sup_{r,\varrho\in\mathbb{R}_+} m_v(r, \varrho) < 1. \tag{5.7}$$

Let us illustrate the results of Section 4 for Mellin pseudodifferential operators with compound and related to (5.5) symbols of the form

$$b(r, \varrho, \lambda) = \coth\left[\pi\frac{\lambda + i(1/p + m_v(r, \varrho))}{1 + im_\theta(r, \varrho)}\right]. \tag{5.8}$$

Lemma 5.1. *If the functions $r \mapsto rv'(r)$ and $r \mapsto r\theta'(r)$ belong to the algebra $SO(\mathbb{R}_+)$, then, for every $\varrho \in \mathbb{R}_+$,*

$$\lim_{r\to s}[m_v(r, r\varrho) - rv'(r)] = 0, \quad \lim_{r\to s}[m_\theta(r, r\varrho) - r\theta'(r)] = 0, \tag{5.9}$$

$$\lim_{r\to s}(\varrho\partial_\varrho)m_v(r, r\varrho) = 0, \quad \lim_{r\to s}(\varrho\partial_\varrho)m_\theta(r, r\varrho) = 0. \tag{5.10}$$

Proof. Obviously, it suffices to prove the first equalities in (5.9) and (5.10). Since the function $r \mapsto rv'(r)$ belongs to the algebra $SO(\mathbb{R}_+)$, we conclude that the function $x \mapsto e^x v'(e^x)$ belongs to the algebra $SO(\mathbb{R})$. Replacing $r = e^x$ and $\varrho = e^y$ in (5.6), we obtain

$$\begin{aligned} m_v(r, r\varrho) - rv'(r) &= \frac{v(e^{x+y}) - v(e^x)}{y} - e^x v'(e^x) \\ &= \frac{1}{y}\int_0^y \left[e^{x+t}v'(e^{x+t}) - e^x v'(e^x)\right]dt. \end{aligned} \tag{5.11}$$

As the function $x \mapsto e^x v'(e^x)$ belongs to the algebra $SO(\mathbb{R})$ and hence

$$\lim_{x\to\pm\infty}\max_{t,\tau\in[x,x+h]}\left|e^t v'(e^t) - e^\tau v'(e^\tau)\right| = 0 \quad \text{for every} \quad h > 0$$

in view of (5.1) or (5.2), we infer from (5.11) that

$$\lim_{x\to\pm\infty}\frac{1}{y}\int_0^y \left[e^{x+t}v'(e^{x+t}) - e^x v'(e^x)\right]dt = 0 \quad \text{for every} \quad y \in \mathbb{R}, \tag{5.12}$$

which gives the first equality in (5.9). Similarly, we deduce that

$$\begin{aligned} (\varrho\partial_\varrho)m_v(r, r\varrho) &= \partial_y\left(\frac{v(e^{x+y}) - v(e^x)}{y}\right) \\ &= \frac{e^{x+y}v'(e^{x+y})y - \left(v(e^{x+y}) - v(e^x)\right)}{y^2} \\ &= \int_0^1 \frac{e^{x+y}v'(e^{x+y}) - e^{x+ty}v'(e^{x+ty})}{y}\,dt, \end{aligned}$$

which implies by analogy with (5.12) that

$$\lim_{x\to\pm\infty}\int_0^1 \frac{e^{x+y}v'(e^{x+y}) - e^{x+ty}v'(e^{x+ty})}{y}\,dt = 0 \quad \text{for every} \quad y\in\mathbb{R}.$$

This gives the first equality in (5.10) and completes the proof. □

Applying Theorems 4.2 and 4.3, Corollary 4.4, and Lemma 5.1, we arrive at the following.

Theorem 5.2. *Let $p\in(1,\infty)$ and let $(\Gamma,w)\in A_p^{SO}$ be given by (5.3) and (5.4), respectively, where*

$$(rD_r)\theta,\ (rD_r)v\in SO(\mathbb{R}_+), \qquad (rD_r)^2\theta,\ (rD_r)^2 v\in L^\infty(\mathbb{R}_+),$$

and (5.7) holds. Then the Mellin pseudodifferential operator $\mathrm{Op}(b)$ *with compound symbol b given by (5.8) and (5.6) is bounded on the space $L^p(\mathbb{R}_+,d\mu)$, and* $\mathrm{Op}(b)=\mathrm{Op}(\widehat{b})+K$, *where the function $\widehat{b}(r,\lambda)\in C_b\big(\mathbb{R}_+,V(\mathbb{R})\big)$ is given by*

$$\widehat{b}(r,\lambda) := \coth\left[\pi\frac{\lambda+i(1/p+rv'(r))}{1+ir\theta'(r)}\right]$$

and K is a compact operator on the space $L^p(\mathbb{R}_+,d\mu)$.

References

[1] A. Böttcher and Yu.I. Karlovich, *Carleson Curves, Muckenhoupt Weights, and Toeplitz Operators,* Progress in Mathematics **154**, Birkhäuser, Basel, 1997.

[2] A. Böttcher, Yu.I. Karlovich, and V.S. Rabinovich, Mellin pseudodifferential operators with slowly varying symbols and singular integrals on Carleson curves with Muckenhoupt weights, *Manuscripta Math.* **95** (1998), 363–376.

[3] A. Böttcher, Yu.I. Karlovich, and V.S. Rabinovich, The method of limit operators for one-dimensional singular integrals with slowly oscillating data, *J. Operator Theory* **43** (2000), 171–198.

[4] A. Böttcher, Yu.I. Karlovich, and V.S. Rabinovich, Singular integral operators with complex conjugation from the viewpoint of pseudodifferential operators, *Operator Theory: Advances and Applications* **121**, Birkhäuser, 2001, 36–59.

[5] A.P. Calderón and R. Vaillancourt, On the boundedness of pseudodifferential operators, *J. Math. Soc. Japan* **23** (1971), 374–378.

[6] R.R. Coifman and Y. Meyer, Au delà des opérateurs pseudodifférentiels, *Astérisque* **57** (1978), 1–184.

[7] H.O. Cordes, On compactness of commutators of multiplications and convolutions, and boundedness of pseudodifferential operators, *J. Funct. Anal.* **18** (1975), 115–131.

[8] H.O. Cordes, *Elliptic Pseudo-Differential Operators – An Abstract Theory,* Lecture Notes in Math. **756**, Springer, Berlin, 1979.

[9] G. Fernández-Torres and Yu.I. Karlovich, Fredholmness of nonlocal singular integral operators with slowly oscillating data, In: *Integral Methods in Science and Engineering*, vol. 1, eds. C. Constanda et al., Birkhäuser, Cham, 2017, pp. 95–105.

[10] N.B. Haaser and J.A. Sullivan, *Real Analysis,* Dover, New York, 1991.

[11] A.Yu. Karlovich, Yu.I. Karlovich, and A.B. Lebre, On a weighted singular integral operator with shifts and slowly oscillating data, *Complex Anal. Oper. Theory* **10** (2016), 1101–1131.

[12] A.Yu. Karlovich, Yu.I. Karlovich, and A.B. Lebre, The index of weighted singular integral operators with shifts and slowly oscillating data, *J. Math. Anal. Appl.* **450** (2017), 606–630.

[13] A.Yu. Karlovich, Yu.I. Karlovich, and A.B. Lebre, Necessary Fredholm conditions for weighted singular integral operators with shifts and slowly oscillating data, *J. Integral Equ. Appl.* **29** (2017), 365–399.

[14] Yu.I. Karlovich, An algebra of pseudodifferential operators with slowly oscillating symbols, *Proc. London Math. Soc.* **92** (2006), 713–761.

[15] Yu.I. Karlovich, Pseudodifferential operators with compound slowly oscillating symbols, *Operator Theory: Advances and Applications* **171**, Birkhäuser, 2007, 189–224.

[16] Yu.I. Karlovich, Algebras of pseudodifferential operators with discontinuous symbols, *Operator Theory: Advances and Applications* **172**, Birkhäuser, 2007, 207–233.

[17] Yu.I. Karlovich, Pseudodifferential operators with compound non-regular symbols, *Math. Nachr.* **280** (2007), 1128–1144.

[18] Yu.I. Karlovich, Nonlocal singular integral operators with slowly oscillating data, *Operator Theory: Advances and Applications* **181**, Birkhäuser, 2008, 229–261.

[19] Yu.I. Karlovich, An algebra of shift-invariant singular integral operators with slowly oscillating data and its application to operators with a Carleman shift, *Operator Theory: Advances and Applications* **193**, Birkhäuser, 2009, 81–95.

[20] Yu.I. Karlovich, Boundedness and compactness of pseudodifferential operators with non-regular symbols on weighted Lebesgue spaces, *Integr. Equ. Oper. Theory* **73** (2012), 217–254.

[21] Yu.I. Karlovich, The Haseman boundary value problem with slowly oscillating coefficients and shifts, *Operator Theory: Advances and Applications* **259**, Birkhäuser, 2017, 463–500.

[22] J.E. Lewis and C. Parenti, Pseudodifferential operators of Mellin type, *Comm. Part. Diff. Equ.* **8** (1983), 477–544.

[23] V.S. Rabinovich, S. Roch, and B. Silbermann, *Limit Operators and Their Applications in Operator Theory,* Operator Theory: Advances and Applications **150**, Birkhäuser, Basel, 2004.

[24] M. Reed and B. Simon, *Methods of Modern Mathematical Physics. 1. Functional Analysis*, Academic Press, New York, 1972.

[25] D. Sarason, Toeplitz operators with piecewise quasicontinuous symbols, *Indiana Univ. Math. J.* **26** (1977), 817–838.

[26] L. Schwartz, *Analyse Mathématique*, Vol. 1, Hermann, Paris, 1967.

[27] M.A. Shubin, *Pseudodifferential Operators and Spectral Theory*, Springer, Berlin, 1987; Russian original: Nauka, Moscow, 1978.

[28] E.M. Stein, *Harmonic Analysis: Real-Variable Methods, Orthogonality, and Oscillatory Integrals,* Princeton Univ. Press, Princeton, NJ, 1993.

[29] M.E. Taylor, *Pseudodifferential Operators*, Princeton Univ. Press, Princeton, NJ, 1981.

[30] M.E. Taylor, *Pseudodifferential Operators and Nonlinear PDE,* Birkhäuser, Boston, 1991.

[31] M.E. Taylor, *Tools for PDE. Pseudodifferential Operators, Paradifferential Operators, and Layer Potentials,* American Math. Soc., Providence, RI, 2000.

[32] F. Treves, *Introduction to Pseudodifferential and Fourier Integral Operators,* Vols. 1, 2, Plenum Press, New York, 1982.

Yuri I. Karlovich
Centro de Investigación en Ciencias
Universidad Autónoma del Estado de Morelos
Av. Universidad 1001, Col. Chamilpa
C.P. 62209 Cuernavaca, Morelos
México
e-mail: `karlovich@uaem.mx`

Asymptotically sharp inequalities for polynomials involving mixed Hermite norms

Holger Langenau

Abstract. The paper concerns best constants in Markov-type inequalities between the norm of a higher derivative of a polynomial and the norm of the polynomial itself. The norm of the polynomial and its derivative is taken in L^2 on the real axis with the weight $|t|^{2\alpha}e^{-t^2}$ and $|t|^{2\beta}e^{-t^2}$, respectively. We determine the leading term of the asymptotics of the constants as the degree of the polynomial goes to infinity.

Mathematics Subject Classification (2010). Primary 15A60; Secondary 15A18, 26D10, 41A44.

Keywords. Markov inequality, Hermite weight, matrix norm.

1. Introduction and main result

A Markov-type inequality gives upper bounds on the norm of the νth derivative of a polynomial in terms of the norm of the polynomial itself, that is, it is an inequality of the form

$$\|D^\nu f_n\|_\beta \le C_n^{(\nu)}(\alpha,\beta)\|f_n\|_\alpha \text{ for all } f_n \in \mathcal{P}_n, \tag{1.1}$$

where $\mathcal{P}_n$ is the space of all algebraic polynomials with complex coefficients of degree at most n and D^ν denotes the operator that maps a polynomial to its νth derivative. In the following, the norms $\|\cdot\|_\alpha$ are the weighted Hermite norms

$$\|f\|_\alpha^2 = \int_{-\infty}^{\infty} |f(t)|^2|t|^{2\alpha}e^{-t^2}\,dt. \tag{1.2}$$

The study of Markov-type inequalities goes back to the 1880s, when Andrei Andreevich Markov proved that for a real polynomial $p(t)$ of degree n with $|p(t)| \le 1$ on $[-1,1]$ the inequality $|p'(t)| \le n^2$ holds, or equivalently,

$$\|Df\|_\infty \le n^2\|f\|_\infty \text{ for all real } f \in \mathcal{P}_n,$$

where D is the differential operator and $\|\cdot\|_\infty$ denotes the maximum norm on $[-1,1]$. He also showed that n^2 is best-possible [1, 2]. Not long after that,

© Springer International Publishing AG, part of Springer Nature 2018

A. Böttcher et al. (eds.), *The Diversity and Beauty of Applied Operator Theory*, Operator Theory: Advances and Applications 268, https://doi.org/10.1007/978-3-319-75996-8_18

his younger brother Vladimir Andreevich showed that

$$\|D^\nu f\|_\infty \le \frac{n^2(n^2-1)(n^2-2^2)\cdots(n^2-(\nu-1)^2)}{(2\nu-1)!!}\|f\|_\infty \text{ for all } f \in \mathcal{P}_n.$$

This is the best possible constant for higher derivatives in this case.

Starting in the 1940s Erhard Schmidt [14] was the first to consider such inequalities with Hilbert space norms instead of the maximum norm. He found the best constants for the first derivative

$$\lambda_n \sim \frac{2}{\pi}n, \quad \gamma_n \sim \frac{1}{\pi}n^2, \quad \eta_n = \sqrt{2n},$$

in the Laguerre, Legendre, and Hermite case, respectively. Here, the expression $a_n \sim b_n$ means that the quotient $\frac{a_n}{b_n}$ converges to 1 as n goes to infinity. In the 1960s, Lawrence F. Shampine [15, 16] studied second order derivatives for the Laguerre and Legendre norms. Peter Dörfler [7, 8] then was the first to study higher order derivatives and gave bounds for the Laguerre case. After this, the development paused for almost two decades.

Shampine's and Dörfler's approach involved the study of matrices. Since this led to certain special Toeplitz matrices, Dörfler wrote a letter to Albrecht Böttcher and together they could prove asymptotic expressions for the constants in the Laguerre and Gegenbauer cases. They made use of an old (and then already forgotten) trick used by Harold Widom in the 1960s in another context [18, 19], which we will employ later, too. Not only could they show that a limit for those constants exists, they could also treat the much more general case of weighted Laguerre and Gegenbauer norms.

In 2009, Jürgen Prestin drew the attention of Böttcher and Dörfler to the problem of using two different weights in the inequality. This concerns two different norms, and as changes of norms may improve error estimates, this case could be useful in approximation theory and numerical analysis. This is not an easy problem. However, they could give asymptotic expressions for the constants under some restrictions on the differences between the parameters involved.

Subsequently, the author lifted most of these restrictions for the Laguerre [10] and Gegenbauer [12] cases. Since one part, already studied by Böttcher and Dörfler, involved the Hilbert–Schmidt norm of integral operators, a small gap in the parameter space remains. Böttcher, Widom, and the author made some preparatory studies to close this gap in [5].

The Hermite case has not been considered so far. In the unweighted case this is a trivial question, as already noted by Schmidt [14]. However, inserting weights makes things much more complicated. We here determine the best constant in (1.1) for almost all possible parameter pairs. We denote this constant by $\eta_n^{(\nu)}(\alpha,\beta)$. Our main result is the following.

Theorem 1.1. *Let $\alpha, \beta > -1/2$ be real numbers and let ν be a positive integer. Then,*

$$\eta_n^{(\nu)}(\alpha,\beta) \sim C_\nu(\alpha,\beta) n^{(|\beta-\alpha|+\nu)/2}$$

with

$$C_\nu(\alpha,\beta) = \begin{cases} 2^{(\beta-\alpha+\nu)/2}, & \beta-\alpha \ge 0, \\ 2^{(\beta-\alpha-\nu)/2} \cdot \max\Big\{ \big\|H^{(0)}_{\nu,\alpha,\beta}\big\|_\infty,\ \big\|H^{(1)}_{\nu,\alpha,\beta}\big\|_\infty \Big\}, & \beta-\alpha < -\frac{1}{2}, \end{cases}$$

where $H^{(0)}_{\nu,\alpha,\beta}$ *and* $H^{(1)}_{\nu,\alpha,\beta}$ *are the integral operators on* $L^2(0,1)$ *defined by*

$$\begin{aligned}(H^{(0)}_{\nu,\alpha,\beta}f)(x) &= \frac{2^\nu\Gamma(\lceil \nu/2\rceil+1)}{\Gamma(\alpha-\beta+\lceil \nu/2\rceil)} \int_x^1 x^{\beta/2-1/4} y^{-\alpha/2+1/4+(\lfloor \nu/2\rfloor-\lceil \nu/2\rceil)/2} \\ &\times (y-x)^{\alpha-\beta+\lceil \nu/2\rceil-1} \sum_{\ell=0}^{\lceil \nu/2\rceil} \binom{\beta}{\ell}\binom{\beta-\alpha-\ell}{\lceil \nu/2\rceil-\ell}\left(\frac{x}{y-x}\right)^{\lceil \nu/2\rceil-\ell} f(y)\mathrm{d}y\end{aligned}$$

and

$$\begin{aligned}(H^{(1)}_{\nu,\alpha,\beta}f)(x) &= \frac{2^\nu\Gamma(\lfloor \nu/2\rfloor+1)}{\Gamma(\alpha-\beta+\lfloor \nu/2\rfloor)} \int_x^1 x^{\beta/2+1/4} y^{-\alpha/2-1/4+(\lceil \nu/2\rceil-\lfloor \nu/2\rfloor)/2} \\ &\times (y-x)^{\alpha-\beta+\lfloor \nu/2\rfloor-1} \sum_{\ell=0}^{\lfloor \nu/2\rfloor} \binom{\beta}{\ell}\binom{\beta-\alpha-\ell}{\lfloor \nu/2\rfloor-\ell}\left(\frac{x}{y-x}\right)^{\lfloor \nu/2\rfloor-\ell} f(y)\mathrm{d}y.\end{aligned}$$

2. Matrix representation of the operator

As this was done for the Laguerre and Gegenbauer norms (see, e.g., [3, 4]), in order to determine the best constant in inequality (1.1), we want to estimate the spectral norm of the matrix representation of the differential operator with respect to appropriately chosen orthonormal bases.

For the norm (1.2) the polynomials $\{\widehat{H}_0(\cdot,\alpha), \widehat{H}_1(\cdot,\alpha), \ldots\}$ form an orthonormal basis. $H_n(\cdot,\alpha)$ is the nth generalized Hermite polynomial, given by

$$H_n(t,\alpha) = 2^n\Gamma(\lfloor n/2\rfloor+1) \sum_{j=0}^{\lfloor n/2\rfloor} \binom{\alpha+\lceil n/2\rceil-1/2}{j} \frac{(-1)^j}{\Gamma(\lfloor n/2\rfloor-j+1)} t^{n-2j}.$$

With the normalizing factor

$$w_n(\alpha) = \left(2^n\sqrt{\Gamma(\lfloor n/2\rfloor+1)\Gamma(\lceil n/2\rceil+\alpha+1/2)}\right)^{-1}$$

the nth normalized polynomial is given by

$$\widehat{H}_n(t,\alpha) = w_n(\alpha)H_n(t,\alpha) = \\ \sqrt{\frac{\Gamma(\lfloor n/2\rfloor+1)}{\Gamma(\lceil n/2\rceil+\alpha+1/2)}} \sum_{j=0}^{\lfloor n/2\rfloor} \binom{\alpha+\lceil n/2\rceil-1/2}{j} \frac{(-1)^j}{\Gamma(\lfloor n/2\rfloor-j+1)} t^{n-2j}.$$

For $\alpha=0$ we get back the classical Hermite polynomials. As mentioned above, the question for the best constant then is trivial. Due to the relation

$$H_n^{(\nu)}(t,0) = 2^\nu\left(n-\nu+1\right)_\nu H_{n-\nu}(t,0),$$

the matrix representation of the operator is just a diagonal matrix. The expression $(x)_n$ stands for the rising factorial, or Pochhammer symbol,

$$(x)_n = x(x+1)\cdots(x+n-1) = \frac{\Gamma(x+n)}{\Gamma(x)}.$$

Therefore, the norm and thus the smallest constant is the maximal absolute value on the diagonal. The corresponding entry is

$$\sqrt{2^\nu \frac{\Gamma(n+1)}{\Gamma(n-\nu+1)}} \sim (2n)^{\nu/2}.$$

Hence, in that case, the constant

$$\eta_n^{(\nu)}(0,0) \sim (2n)^{\nu/2}$$

is fully identified.

Theorem 2.1. *Let $C_n = \left(c_{jk}^{(\nu)}(\alpha,\beta)\right)_{j,k=0}^{n}$ be the matrix representation of the differential operator D^ν with respect to the orthonormal bases given by the generalized Hermite polynomials with the corresponding weight. The entries $c_{jk}^{(\nu)}(\alpha,\beta)$ are given by*

$$\begin{aligned} c_{jk}^{(\nu)}(\alpha,\beta) =& \\ & 2^\nu\,\Gamma(\lfloor \nu/2\rfloor + \nu_k + 1)\sqrt{\frac{\Gamma(\lceil j/2\rceil + \beta + 1/2)}{\Gamma(\lfloor j/2\rfloor + 1)}\frac{\Gamma(\lfloor k/2\rfloor + 1)}{\Gamma(\lceil k/2\rceil + \alpha + 1/2)}} \\ & \times \binom{\lceil (j+\nu)/2\rceil - 1/2}{\lfloor \nu/2\rfloor + \nu_k}\binom{\beta - \alpha - \lfloor \nu/2\rfloor - \nu_k}{(k-j-\nu)/2} \\ & \times {}_3F_2\left(\begin{matrix} -\lfloor \nu/2\rfloor - \nu_k,\ -(k-j-\nu)/2,\ \beta + \lceil j/2\rceil + 1/2 \\ \beta - \alpha - \lfloor \nu/2\rfloor - \nu_k - (k-j-\nu)/2 + 1,\ \lceil j/2\rceil + 1/2\end{matrix}; 1\right) \end{aligned} \tag{2.1}$$

if $0 \le k - \nu - j$ is even, and they are zero otherwise. Here, $\nu_k = 1$ if k and ν are odd, and $\nu_k = 0$ if k or ν is even (i.e., $\nu_k = k\nu \mod 2$).

We note that the hypergeometric series occurring here is not defined if $\beta - \alpha - \lfloor \nu/2\rfloor - \nu_k$ is a nonnegative integer smaller than $(k-j-\nu)/2$. But then, the coefficient before this term is zero, and we define the whole term to be zero. In the other cases, the series terminates naturally before a division by zero would happen. The proof can be done by a comparison of coefficients. See [11] for details.

3. The constant for integral $\beta - \alpha \geq 0$

The matrix representation of the operator D^ν has a chessboard structure above the main diagonal. Since $\eta_{n-1}^{(\nu)}(\alpha,\beta) \le \eta_n^{(\nu)}(\alpha,\beta) \le \eta_{n+1}^{(\nu)}(\alpha,\beta)$, we assume that $N = n - \nu + 1$ is an even number. Then there is some permutation matrix U_N with

$$A_N = U_N \begin{pmatrix} E_N & 0 \\ 0 & F_N \end{pmatrix} U_N,$$

where $E_N = (e_{jk})_{j,k=0}^{N/2-1}$, $F_N = (f_{jk})_{j,k=0}^{N/2-1}$ are built from the entries

$$e_{jk} = c^{(\nu)}_{2j,2k+\nu}(\alpha,\beta), \quad f_{jk} = c^{(\nu)}_{2j+1,2k+\nu+1}(\alpha,\beta).$$

We will just investigate the matrix E_N and point out that the matrix F_N can be treated likewise. First, assume $\beta-\alpha \geq \lceil \nu/2 \rceil$ is an integer. Due to the term $\binom{\beta-\alpha-\lfloor \nu/2\rfloor-\nu_k}{(k-j-\nu)/2}$ occurring in (2.1), the matrix is banded. We will employ the same idea as in the Laguerre case in [3] and the Gegenbauer case in [4]: Consider the matrix as a sum of (shifted) diagonal matrices and make use of the fact that the norm of the sum is less than the sum of the norms of these diagonals. To derive a lower estimate, we show that some scaled version of the matrix E_N converges in the norm to a given Toeplitz operator.

Let $m = \beta-\alpha-\lceil \nu/2\rceil \in \mathbb{N}_0$. Then, after some transformation, the entries $d_j^{(\ell)}$ of the ℓth diagonal in row j, $\ell = 0,\dots,m$, are given by

$$d_j^{(\ell)} = 2^\nu \Gamma(\lceil \nu/2\rceil + 1)\sqrt{\frac{\Gamma(j+\beta+1/2)}{\Gamma(j+1)}\frac{\Gamma(j+\ell+\lfloor \nu/2\rfloor+1)}{\Gamma(j+\ell+\lceil \nu/2\rceil+\alpha+1/2)}}$$
$$\times \sum_{\tau=0}^{\min\{\lceil \nu/2\rceil,\ell\}} \binom{j+\lceil \nu/2\rceil-1/2}{\lceil \nu/2\rceil-\tau}\binom{\beta+j+\tau-1/2}{\tau}\binom{m}{\ell-\tau}.$$

Fixing ℓ, we can show that $d_j^{(\ell)}$ is increasing with j. So, the maximum along each diagonal (and with this the norm of this particular diagonal matrix) is attained for $j = N/2-1-\ell$. We get the following upper estimate for the norm of E_N:

$$\|E_N\|_\infty \leq 2^\nu\,\Gamma(\lceil \nu/2\rceil+1)\sum_{\ell=0}^{m}\sqrt{\frac{\Gamma(N/2-\ell+\beta-1/2)\Gamma(N/2+\lfloor \nu/2\rfloor)}{\Gamma(N/2-\ell)\Gamma(N/2+\lceil \nu/2\rceil+\alpha-1/2)}}$$
$$\times \sum_{\tau=0}^{\min\{\lceil \nu/2\rceil,\ell\}} \binom{N/2-\ell+\lceil \nu/2\rceil-3/2}{\lceil \nu/2\rceil-\tau}\binom{\beta+N/2-\ell+\tau-3/2}{\tau}\binom{m}{\ell-\tau}.$$

Ignoring the constant factor for the moment, as well as the terms under the square root, which we will replace with the maximum over $0 \leq \ell \leq m$, the sum reduces to

$$\sum_{\ell=0}^{m}\sum_{\tau=0}^{\min\{\lceil \nu/2\rceil,\ell\}} \binom{N/2-\ell+\lceil \nu/2\rceil-3/2}{\lceil \nu/2\rceil-\tau}\binom{\beta+N/2-\ell+\tau-3/2}{\tau}\binom{m}{\ell-\tau}$$
$$\leq \sum_{\ell=0}^{m}\sum_{\tau=0}^{\min\{\lceil \nu/2\rceil,\ell\}} \binom{N/2+\lceil \nu/2\rceil-1/2}{\lceil \nu/2\rceil-\tau}\binom{\beta+N/2+\lceil \nu/2\rceil-1/2}{\tau}\binom{m}{\ell-\tau}$$
$$= \sum_{\tau=0}^{\lceil \nu/2\rceil} \binom{N/2+\lceil \nu/2\rceil-1/2}{\lceil \nu/2\rceil-\tau}\binom{\beta+N/2+\lceil \nu/2\rceil-1/2}{\tau}\sum_{\ell=0}^{m-\tau}\binom{m}{\ell}$$

$$\leq \sum_{\tau=0}^{\lceil\nu/2\rceil} \binom{N/2+\lceil\nu/2\rceil-1/2}{\lceil\nu/2\rceil-\tau} \binom{\beta+N/2+\lceil\nu/2\rceil-1/2}{\tau} \sum_{\ell=0}^{m} \binom{m}{\ell}$$

$$= \frac{\Gamma(\beta+N+2\lceil\nu/2\rceil)}{\Gamma(\beta+N+\lceil\nu/2\rceil)\Gamma(\lceil\nu/2\rceil+1)} 2^m.$$

Together with the other terms and by applying asymptotic formulas for all of them, we get the following upper bound for $\|E_N\|_\infty$:

$$\|E_N\|_\infty \leq 2^\nu \left(\tfrac{N}{2}\right)^{(\beta-1/2)/2} \left(\tfrac{N}{2}\right)^{(-\alpha+1/2+\lfloor\nu/2\rfloor-\lceil\nu/2\rceil)/2} \cdot 2^{\beta-\alpha-\lceil\nu/2\rceil} N^{\lceil\nu/2\rceil}\big(1+O(1/N)\big) \tag{3.1}$$

$$= 2^{(\beta-\alpha+\nu)/2} N^{(\beta-\alpha+\nu)/2}\big(1+O(1/N)\big). \tag{3.2}$$

Now, let J_N denote the $(N \times N)$-matrix with ones on the counter-diagonal and zeros elsewhere. We set $B_N = J_N E_N J_N$. Then obviously, $\|E_N\|_\infty = \|E_N^*\|_\infty = \|B_N\|_\infty$. We quickly confirm that the entry at position jk of B_N equals the entry at position $N/2-1-j, N/2-1-k$ from E_N. Now, let π_N be the projection

$$\pi_N : \ell^2 \to \ell^2, \quad \{x_0, x_1, x_2, \dots\} \mapsto \{x_0, x_1, \dots, x_{N/2-1}, 0, \dots\}$$

and consider the operators $T_N = 2^{(\beta-\alpha-\nu)/2} N^{(\alpha-\beta-\nu)/2} B_N \pi_N$ on ℓ^2. We can show that these operators converge strongly to the Toeplitz operator $T^*((1+z)^{\beta-\alpha})$ on ℓ^2 that is given by the infinite Toeplitz matrix $(\varphi_{jk})_{j,k=0}^\infty$ with $\varphi_{jk} = 0$ for $k > j$ and

$$\varphi_{jk} = \binom{\beta-\alpha}{j-k} \quad \text{for } k \leq j. \tag{3.3}$$

From (3.1) we infer that $\|T_N\|_\infty \leq 2^{\beta-\alpha}(1+O(1/N))$, that is, the operators T_N are uniformly bounded. In order to prove $T_N \to T((1+z)^{\beta-\alpha})$ strongly, it is therefore enough to show that $T_N e_k$ converges to $T((1+z)^{\beta-\alpha})e_k$ for every $k \geq 0$, where $e_k \in \ell^2$ has 1 at the kth position and zeros elsewhere. As all involved operators are banded with bandwidth $m+1$ independent of N, it suffices to verify that the jk entry of T_N converges to the jk entry of the matrix $T((1+z)^{\beta-\alpha})$. But, the jk entry of T_N is zero for $k > j$ and for $k < j-m$, and if $k \leq j \leq k+m$, we can show that for $N \to \infty$ it equals φ_{jk}. For details, we again refer to [11]. Together with the Banach–Steinhaus theorem we deduce

$$\liminf_{N\to\infty} \left\|2^{(\beta-\alpha-\nu)/2} N^{(\alpha-\beta-\nu)/2} B_N\right\|_\infty \geq \left\|T\big((1+z)^{\beta-\alpha}\big)\right\|_\infty.$$

By a well-known result on the norm of Toeplitz operators (see, e.g., [6, p. 10]), the latter is

$$\left\|T\big((1+z)^{\beta-\alpha}\big)\right\|_\infty = \max_{|z|=1} |1+z|^{\beta-\alpha} = 2^{\beta-\alpha}.$$

Thus,

$$\liminf_{N\to\infty} N^{(\alpha-\beta-\nu)/2} \|E_N\|_\infty \geq 2^{(\beta-\alpha+\nu)/2}. \tag{3.4}$$

Combining (3.1) and (3.4), we conclude that

$$\|E_N\|_\infty \sim (2N)^{(\beta-\alpha+\nu)/2}.$$

As above, one can show that

$$\|F_N\|_\infty \sim (2N)^{(\beta-\alpha+\nu)/2}.$$

Since $\|C_n\|_\infty = \max\{\|E_N\|_\infty, \|F_N\|_\infty\}$, we obtain for $\beta - \alpha \geq \lceil \nu/2 \rceil$, $\beta - \alpha$ an integer, the following asymptotic behavior for $n \to \infty$:

$$\eta_n^{(\nu)}(\alpha, \beta) \sim (2n)^{(\beta-\alpha+\nu)/2}.$$

We need one more integer case. If $\beta - \alpha = 0$ the matrix is in general not banded as in the above case. However, $\alpha = \beta = 0$ has been disposed of before. In this case this matrix is a diagonal matrix and it is known that

$$\eta_n^{(\nu)}(0,0) \sim (2n)^{\nu/2}.$$

Now, assume $\alpha = \beta \neq 0$. We will show that the asymptotic expressions obtained above also hold in this case. A closer look at the matrix reveals that, although it is no diagonal matrix, it is very close to one in the sense that the entries along the diagonal are significantly bigger in their absolute values than the off-diagonal entries. Indeed, applying some estimates, the last entry on the diagonal of the matrix E_N is given by

$$e_{N/2-1,N/2-1} = 2^\nu \left(\tfrac{N}{2}\right)^{\nu/2} \left(1 + O(1/N)\right).$$

This is exactly what we want. Since

$$\|E_N e_{N/2-1}\|_2 = \sqrt{\sum_{j=0}^{N/2-1} e_{j,N/2-1}^2} \geq e_{N/2-1,N/2-1},$$

this already provides a lower bound. An upper bound is harder to show. The approach we used for the banded matrices does not work here any longer. What we do instead is to use a corollary of the Gershgorin theorem [9, p. 344]. The Gershgorin theorem provides disks in the complex plane containing the eigenvalues of a matrix. The closer such a matrix is to a diagonal matrix, the more precise the location can be given. Since we look for the singular values, we could apply the theorem to $E_N^* E_N$. Its eigenvalues all are nonnegative real numbers, so the disks are in fact intervals. However, the matrix representation of $E_N^* E_N$ is not easy to work with. The paper [13] uses the ideas of the Gershgorin theorem directly with the matrix E_N to provide intervals for the location of the singular values. Even more, it also provides a scaled version of the theorem. Since the eigenvalues do not change if we multiply a matrix by an invertible matrix from the right and its inverse from the left, we may modify the matrix entries slightly to get better bounds.

Since we are interested only in the largest singular value of the matrix E_N, the combination of Theorem 2 and Theorem 4 of [13] yields

$$\|E_N\|_\infty \leq \max_{0 \leq i \leq N/2-1} \left\{ \sum_{j=0}^{N/2-1} \frac{d_j}{d_i} |e_{ij}|, \sum_{j=0}^{N/2-1} \frac{d_j}{d_i} |e_{ji}| \right\},$$

where $d_0, \ldots, d_{N/2-1}$ are positive real numbers. We set $d_j = \left(\frac{\sqrt{j+1}}{N}\right)^{\varepsilon}$ for $0 \le j \le N/2-2$ and $d_{N/2-1} = 1$, with a small positive number $\varepsilon > 0$. With the help of these entries, we can show that the off-diagonal elements are of lower order than the ones on the diagonal. While this is not of immediate use, it is sufficient for the asymptotic results. This is done in detail in [11] and we could show

$$\eta_n^{(\nu)}(\alpha, \alpha) \sim (2n)^{\nu/2}$$

as n goes to infinity.

4. The constant for non-integral $\beta - \alpha > 0$

In this section, we make use of the results shown above for integral differences $\beta - \alpha$ to show that the asymptotic behavior

$$\eta_n^{(\nu)}(\alpha, \beta) \sim (2n)^{(\beta-\alpha+\nu)/2}$$

holds for all $\beta - \alpha \ge 0$.

First, we employ the interpolation theorem of Stein [17], or rather a corollary thereof. With the help of this theorem, we can show that if

$$\eta_n^{(\nu)}(\alpha, \beta_i) = C_n^{(\nu)}(\alpha, \beta_i), \quad i = 0, 1$$

and if

$$C_n^{(\nu)}(\alpha, (1-\theta)\beta_0 + \theta\beta_1) = \left(C_n^{(\nu)}(\alpha, \beta_0)\right)^{1-\theta} \left(C_n^{(\nu)}(\alpha, \beta_1)\right)^{\theta}$$

for all $\theta \in [0, 1]$, then

$$\eta_n^{(\nu)}(\alpha, (1-\theta)\beta_0 + \theta\beta_1) \le C_n^{(\nu)}(\alpha, (1-\theta)\beta_0 + \theta\beta_1)$$

for all $\theta \in [0, 1]$. As one easily verifies, the constant

$$\eta_n^{(\nu)}(\alpha, \beta) = (2n)^{(\beta-\alpha+\nu)/2}$$

satisfies this relation. With the appropriate choice of the parameter θ we have proven an upper bound for the constant. The same approach was used in the Laguerre and Gegenbauer cases, although we have to take special care for the gap between $\beta = \alpha$ and $\beta = \alpha + \lceil \nu/2 \rceil$. For details we refer to [11].

A little more work has to be done for the lower bound. We will concentrate on the matrix $B_N = J_N E_N J_N$, where E_N and J_N are the matrices from above. The entries b_{jk} of the matrix B_N can be written as

$$\begin{aligned} b_{jk} &= 2^{\nu}\Gamma\left(\lceil \nu/2 \rceil + 1\right) \\ &\quad \times \sqrt{\frac{\Gamma(N/2 - j + \beta - 1/2)}{\Gamma(N/2 - j)}} \sqrt{\frac{\Gamma(N/2 - k + \lfloor \nu/2 \rfloor)}{\Gamma(N/2 - k + \lceil \nu/2 \rceil + \alpha - 1/2)}} \\ &\quad \times \binom{N/2 - j + \lceil \nu/2 \rceil - 3/2}{\lceil \nu/2 \rceil} \binom{\beta - \alpha - \lceil \nu/2 \rceil}{j - k} \\ &\quad \times {}_3F_2\left(\begin{matrix} -\lceil \nu/2 \rceil,\ k - j,\ \beta + N/2 - j - 1/2 \\ \beta - \alpha - \lceil \nu/2 \rceil + k - j + 1,\ N/2 - j - 1/2 \end{matrix}; 1 \right). \end{aligned}$$

We define vectors $v^+ = (v_j^+)_{j=0}^{N/2-1}$ and $v^- = (v_j^-)_{j=0}^{N/2-1}$ for

$$\alpha + \lceil \nu/2 \rceil - \lfloor \nu/2 \rfloor \geq 1/2 \text{ and } \alpha + \lceil \nu/2 \rceil - \lfloor \nu/2 \rfloor < 1/2,$$

respectively, by

$$v_j^+ = \begin{cases} \prod_{\ell=1}^{j} \sqrt{\frac{N/2 - \ell + \lfloor \nu/2 \rfloor}{N/2 - \ell + \lceil \nu/2 \rceil + \alpha + 1/2}}, & 0 \leq j \leq \mu - 1, \\ 0, & \text{otherwise,} \end{cases}$$

$$v_j^- = \begin{cases} \prod_{\ell=j}^{j+\mu-1} \sqrt{\frac{N/2 - \ell + \lceil \nu/2 \rceil + \alpha - 3/2}{N/2 - \ell + \lceil \nu/2 \rceil - 1}}, & 0 \leq j \leq \mu - 1, \\ 0, & \text{otherwise,} \end{cases}$$

with the parameter $\mu = \mu(N) \in \mathbb{N}$ increasing with N while $\mu/N \to 0$ for N to infinity. We will only investigate $\|B_N v^+\|_2$ and note that the norm $\|B_N v^-\|_2$ can be estimated similarly. Analogously to the treatment of this norm in the Laguerre case [10], we have

$$\sum_{j=0}^{\mu-1} \rho_j \left(\sum_{k=0}^{j} \binom{\beta - \alpha}{k} \right)^2 \geq 2 \left\lfloor \frac{\mu - \lceil \beta - \alpha \rceil}{2} \right\rfloor 2^{2\beta - 2\alpha} \cdot \min_{0 \leq j \leq \mu - 1} \rho_j$$

for $\beta - \alpha \in (0, \infty)$, $\mu \in \mathbb{N}$, $\mu > \lceil \beta - \alpha \rceil$, and $\rho_j > 0, j = 0, \ldots, \mu - 1$. Consider the jth entry of $B_N v^+$,

$$\begin{aligned}(B_N v^+)_j &= 2^\nu \Gamma\big(\lceil \nu/2 \rceil + 1\big) \\ &\quad \times \sqrt{\frac{\Gamma(N/2 - j + \beta - 1/2)}{\Gamma(N/2 - j)}} \sqrt{\frac{\Gamma(N/2 + \lfloor \nu/2 \rfloor + 1)}{\Gamma(N/2 + \lceil \nu/2 \rceil + \alpha + 1/2)}} \\ &\quad \times \binom{N/2 - j + \lceil \nu/2 \rceil - 3/2}{\lceil \nu/2 \rceil} \binom{\beta - \alpha - \lceil \nu/2 \rceil}{j - k} \\ &\quad \times {}_3F_2 \left(\begin{matrix} -\lceil \nu/2 \rceil, \; k - j, \; \beta + N/2 - j - 1/2 \\ \beta - \alpha - \lceil \nu/2 \rceil + k - j + 1, \; N/2 - j - 1/2 \end{matrix} ; 1 \right). \end{aligned}$$

We are only interested in large values of N. Then, the last upper and lower arguments in the hypergeometric series are almost the same and cancel out each other. Hence, the ${}_3F_2$ transforms to ${}_2F_1$, and, by the Chu–Vandermonde identity,

$${}_2F_1 \left(\begin{matrix} -\lceil \nu/2 \rceil, \; k - j \\ \beta - \alpha - \lceil \nu/2 \rceil + k - j + 1 \end{matrix} ; 1 \right) = \frac{(\beta - \alpha - \lceil \nu/2 \rceil + 1)_{\lceil \nu/2 \rceil}}{(\beta - \alpha - \lceil \nu/2 \rceil + k - j + 1)_{\lceil \nu/2 \rceil}}.$$

Together with the coefficient $\binom{\beta - \alpha - \lceil \nu/2 \rceil}{j-k}$, this turns into $\binom{\beta - \alpha}{j-k}$.

As can easily be seen, the vectors v^+ and v^- were chosen in such a way that $\|v^+\|_2^2 \leq \mu$ and $\|v^-\|_2^2 \leq \mu$ for the values of α for which they will be applied. Therefore, we can estimate the norm of B_N together with the aforementioned relation for the hypergeometric series by

$$\begin{aligned}\|B_N\|_\infty^2 &\geq \frac{\|B_N v^+\|_2^2}{\|v^+\|_2^2} \\ &\geq 2^{2\nu} \frac{\Gamma(N/2+\lfloor \nu/2 \rfloor+1)}{\Gamma(N/2+\lceil \nu/2 \rceil+\alpha+1/2)} \sum_{j=0}^{\mu-1} \frac{\Gamma(N/2-j+\beta-1/2)}{\Gamma(N/2-j)} \\ &\qquad \times \left[\binom{N/2-j+\lceil \nu/2 \rceil-3/2}{\lceil \nu/2 \rceil}\right]^2 \left(\sum_{k=0}^{j} \binom{\beta-\alpha}{j-k}\right)^2.\end{aligned}$$

Applying the result from above, this is not smaller than

$$\begin{aligned}&2^{2\nu} \frac{\Gamma(N/2+\lfloor \nu/2 \rfloor+1)}{\Gamma(N/2+\lceil \nu/2 \rceil+\alpha+1/2)} \cdot \mu^{-1} \cdot 2 \left\lfloor \frac{\mu-\lceil \beta-\alpha \rceil}{2} \right\rfloor 2^{2\beta-2\alpha} \\ &\quad \times \min_{0\leq j\leq \mu-1} \left\{ \frac{\Gamma(N/2-j+\beta-1/2)}{\Gamma(N/2-j)} \frac{\Gamma^2(N/2-j+\lceil \nu/2 \rceil-1/2)}{\Gamma^2(N/2-j-1/2)} \right\}.\end{aligned}$$

Letting $\mu(N)$ go to infinity controlled by $N-2\mu \sim N$ and $\mu-\lceil \beta-\alpha \rceil-\lceil \nu/2 \rceil \sim \mu$, this is asymptotically equal to

$$2^{2\beta-2\alpha+2\nu} \left(\tfrac{N}{2}\right)^{\lfloor \nu/2 \rfloor-\lceil \nu/2 \rceil-\alpha+\beta+2\lceil \nu/2 \rceil} = (2N)^{\beta-\alpha+\nu}.$$

In the same way we get the result for v^- and the matrix F_N. Together with the upper estimate, we have shown

$$\eta_n^{(\nu)}(\alpha,\beta) \sim (2n)^{(\beta-\alpha+\nu)/2}$$

as n goes to infinity, for arbitrary $\beta-\alpha \geq 0$.

5. The constant for $\beta-\alpha<0$

Finally, we turn to the case where $\beta-\alpha$ is a negative real number. In contrast to the cases above, the large entries are not on the diagonal anymore, but can be found in the full upper triangle. Similar to the Laguerre and Gegenbauer cases, we again employ a result by Harold Widom [18, 19], which was independently discovered by Lawrence Shampine [15, 16].

If A_N is an $(N \times N)$-matrix, we define $k_N(x,y) = (A_N)_{\lfloor Nx \rfloor, \lfloor Ny \rfloor}$, a simple function, and the integral operator K_N on $L^2(0,1)$ that is given by

$$(K_N f)(x) = \int_0^1 k_N(x,y) f(y) \mathrm{d}y.$$

Then $\|A_N\|_\infty = N\|K_N\|_\infty$. Therefore, if we can show that the sequence of appropriately scaled operators converges in the norm to some integral operator, we have found an asymptotic expression for the matrix norm, which, in our case, is just the best constant in (1.1).

As in the Laguerre and Gegenbauer cases, we have to restrict ourselves a little bit, this time to $\beta-\alpha < -1/2$. The reason for this will be explained soon. In contrast to the aforementioned cases, the sign between two entries of the matrix E_N from above, which we will concentrate on, changes and we cannot immediately work with these. Instead, we consider the matrix $\tilde{E}_N = SE_NS$, with the diagonal matrix $S = \operatorname{diag}\big(\{(-1)^j\}_{j=0}^{N/2-1}\big)$. Because $\|SAS\| = \|A\|$ holds for any matrix A, this does not change our claim. Now let K_N be the integral operator from above, but insert $\tilde{E}_N$ as matrix (which is an $(N/2 \times N/2)$-matrix) defining the kernel. Then it can be proven that the scaled operators $(N/2)^{(\beta-\alpha-\nu)/2+1}K_N$ converge in the Hilbert–Schmidt norm (and thus in the operator norm) to the integral operator $H^{(0)}_{\nu,\alpha,\beta}$, on $L^2(0,1)$ given by

$$(H^{(0)}_{\nu,\alpha,\beta}f)(x) = \frac{2^\nu\Gamma(\lceil \nu/2\rceil + 1)}{\Gamma(\alpha-\beta+\lceil \nu/2\rceil)} \int_x^1 x^{\beta/2-1/4} y^{-\alpha/2+1/4+(\lfloor \nu/2\rfloor - \lceil \nu/2\rceil)/2} \times (y-x)^{\alpha-\beta+\lceil \nu/2\rceil-1} \sum_{\ell=0}^{\lceil \nu/2\rceil} \binom{\beta}{\ell}\binom{\beta-\alpha-\ell}{\lceil \nu/2\rceil-\ell}\left(\frac{x}{y-x}\right)^{\lceil \nu/2\rceil-\ell} f(y)\mathrm{d}y,$$

provided that $\beta-\alpha < -1/2$. That we handle the study of the convergence in the Hilbert–Schmidt norm, which then yields the convergence in the operator norm, is the reason for this restriction. If $\beta-\alpha \geq -1/2$, the operator $H^{(0)}_{\nu,\alpha,\beta}$ is no longer Hilbert–Schmidt.

For the matrix F_N we can do the same and find convergence to

$$(H^{(1)}_{\nu,\alpha,\beta}f)(x) = \frac{2^\nu\Gamma(\lfloor \nu/2\rfloor + 1)}{\Gamma(\alpha-\beta+\lfloor \nu/2\rfloor)} \int_x^1 x^{\beta/2+1/4} y^{-\alpha/2-1/4+(\lceil \nu/2\rceil - \lfloor \nu/2\rfloor)/2} \times (y-x)^{\alpha-\beta+\lfloor \nu/2\rfloor-1} \sum_{\ell=0}^{\lfloor \nu/2\rfloor} \binom{\beta}{\ell}\binom{\beta-\alpha-\ell}{\lfloor \nu/2\rfloor-\ell}\left(\frac{x}{y-x}\right)^{\lfloor \nu/2\rfloor-\ell} f(y)\mathrm{d}y.$$

Putting all of the above together, we have proven our claim. In full detail, this is done in [11].

Moreover, since it seems that the restriction to $\beta-\alpha < -1/2$ is due to the technique used, we conjecture that

$$C_\nu(\alpha,\beta) = 2^{(\beta-\alpha-\nu)/2} \cdot \max\left\{\left\|H^{(0)}_{\nu,\alpha,\beta}\right\|_\infty, \left\|H^{(1)}_{\nu,\alpha,\beta}\right\|_\infty\right\}$$

for $\beta-\alpha<0$.

Acknowledgment

I thank Albrecht Böttcher for his support and the helpful discussions.

References

[1] R.P. Boas, Jr, Inequalities for the derivatives of polynomials, *Math. Mag.* **42** (1969), 165–174.

[2] Albrecht Böttcher, Best constants for Markov type inequalities in Hilbert space norms, in: *Recent Trends in Analysis, Proceedings of the Conference in Honor of Nikolai Nikolski*, pp. 73–83, Theta, Bucharest, 2013.

[3] Albrecht Böttcher and Peter Dörfler, On the best constants in Markov-type inequalities involving Laguerre norms with different weights, *Monatsh. Math.* **161** (2010), no. 4, 357–367.

[4] Albrecht Böttcher and Peter Dörfler, On the best constants in Markov-type inequalities involving Gegenbauer norms with different weights, *Oper. Matrices* **5** (2011), no. 2, 261–272.

[5] Albrecht Böttcher, Holger Langenau, and Harold Widom, Schatten class integral operators occurring in Markov-type inequalities, *Operator Theory: Advances and Applications* **255**, Birkhäuser, 2016, 91–104.

[6] Albrecht Böttcher and Bernd Silbermann, *Introduction to large truncated Toeplitz matrices*, Universitext, Springer, New York, 1999.

[7] P. Dörfler, New inequalities of Markov type, *SIAM J. Math. Anal.* **18** (1987), no. 2, 490–494.

[8] P. Dörfler, A Markov type inequality for higher derivatives of polynomials, *Monatsh. Math.* **109** (1990), no. 2, 113–122.

[9] Roger A. Horn and Charles R. Johnson, *Matrix analysis*, Cambridge University Press, Cambridge, 1990. Corrected reprint of the 1985 original.

[10] Holger Langenau, Asymptotically sharp inequalities for polynomials involving mixed Laguerre norms, *Linear Algebra Appl.* **458** (2014), 116–127.

[11] Holger Langenau, *Best constants in Markov-type inequalities with mixed weights*, Universitätsverlag Chemnitz, Chemnitz, 2016.

[12] Holger Langenau, Asymptotically sharp inequalities for polynomials involving mixed Gegenbauer norms, *Asymptot. Anal.* **103** (2017), no. 4, 221–233.

[13] Li Qun Qi, Some simple estimates for singular values of a matrix, *Linear Algebra Appl.* **56** (1984), 105–119.

[14] Erhard Schmidt, Über die nebst ihren Ableitungen orthogonalen Polynomensysteme und das zugehörige Extremum, *Math. Ann.* **119** (1944), 165–204.

[15] Lawrence F. Shampine, Some L_2 Markoff inequalities, *J. Res. Nat. Bur. Standards Sect. B* **69B** (1965), 155–158.

[16] Lawrence F. Shampine, An inequality of E. Schmidt, *Duke Math. J.* **33** (1966), 145–150.

[17] Elias M. Stein, Interpolation of linear operators, *Trans. Amer. Math. Soc.* **83** (1956), 482–492.

[18] Harold Widom, Rapidly increasing kernels, *Proc. Amer. Math. Soc.* **14** (1963), 501–506.

[19] Harold Widom, Hankel matrices, *Trans. Amer. Math. Soc.* **121** (1966), 1–35.

Holger Langenau
Fakultät für Mathematik
Technische Universität Chemnitz
09107 Chemnitz
Germany
e-mail: holger.langenau@mathematik.tu-chemnitz.de

A two-parameter eigenvalue problem for a class of block-operator matrices

Michael Levitin and Hasen Mekki Öztürk

Abstract. We consider a symmetric block operator spectral problem with two spectral parameters. Under some reasonable restrictions, we state localisation theorems for the pair-eigenvalues and discuss relations to a class of non-self-adjoint spectral problems.

Mathematics Subject Classification (2010). Primary 15A18; Secondary 47A25.

Keywords. Multiparametric spectral problems, eigenvalues, block-operator matrices, non-self-adjoint problems.

1. Introduction

The Multiparameter Eigenvalue Problems (MEPs) are the generalisation of the one-parameter standard eigenvalue problem $M\mathbf{x} = \lambda\mathbf{x}$ and the generalised one-parameter eigenvalue problem $M\mathbf{x} = \lambda V\mathbf{x}$. MEPs can be written in the following abstract form:

$$M\mathbf{x} = \sum_{i=1}^{k} \lambda_i V_i \mathbf{x}, \tag{1.1}$$

where $\lambda_i \in \mathbb{C}$, $i = 1, 2, \ldots, k$, are spectral parameters, and M and V_i are self-adjoint linear operators in some Hilbert space $\mathcal{H}$. Then $\lambda = (\lambda_1, \ldots, \lambda_k)$ is called a *multi-parametric eigenvalue* (or *k-tuple*, or *eigentuple*) if there exists an $\mathbf{x} \in \mathcal{H} \setminus \{0\}$, called an *eigenvector*, such that (1.1) holds.

MEPs arise in numerous applications, in particular in mathematical physics when the method of separation of variables is used to solve boundary value problems for partial differential equations. In the 1960s, an abstract algebraic setting for MEPs was introduced by Atkinson [1, 2], see also [3, 5] and references therein.

In this paper, we consider a special class of two-parameter eigenvalue problems in a block-operator setting.

© Springer International Publishing AG, part of Springer Nature 2018
A. Böttcher et al. (eds.), *The Diversity and Beauty of Applied Operator Theory*, Operator Theory: Advances and Applications 268, https://doi.org/10.1007/978-3-319-75996-8_19

Let $\mathcal{H}_1$ and $\mathcal{H}_2$ be Hilbert spaces. Let in (1.1), with $k = 2$,

$$\mathbf{x} = \begin{pmatrix} \mathbf{u} \\ \mathbf{v} \end{pmatrix} \in \mathcal{H}_1 \oplus \mathcal{H}_2,$$

and

$$M = \begin{pmatrix} A & C \\ C^* & B \end{pmatrix}, \qquad V_1 = \begin{pmatrix} I & 0 \\ 0 & 0 \end{pmatrix}, \qquad V_2 = \begin{pmatrix} 0 & 0 \\ 0 & I \end{pmatrix},$$

where A, B are self-adjoint operators in the Hilbert spaces $\mathcal{H}_1$, $\mathcal{H}_2$, respectively, and C is a linear operator from $\mathcal{H}_2$ to $\mathcal{H}_1$. Hence (1.1) becomes the equation

$$M(\alpha, \beta) \begin{pmatrix} \mathbf{u} \\ \mathbf{v} \end{pmatrix} = \begin{pmatrix} A - \alpha & C \\ C^* & B - \beta \end{pmatrix} \begin{pmatrix} \mathbf{u} \\ \mathbf{v} \end{pmatrix} = \mathbf{0}. \tag{1.2}$$

In this paper, the operators A, B, and C are assumed to be bounded, with further restrictions imposed starting from Section 2. The case of unbounded operators will be considered elsewhere.

Definition 1.1. We call $(\alpha, \beta) \in \mathbb{C}^2$ a *pair-eigenvalue* of M if there exists a non-trivial solution $\left(\begin{smallmatrix} \mathbf{u} \\ \mathbf{v} \end{smallmatrix}\right) \in \mathcal{H}$ of (1.2). We denote by $\operatorname{Spec}_p(M)$ the set of all pair-eigenvalues of M. If both $\alpha, \beta \in \mathbb{R}$, then we will call (α, β) a *real pair-eigenvalue* of (1.2).

The equation (1.2) can be re-written as

$$(A - \alpha)\mathbf{u} = -C\mathbf{v}, \tag{1.3}$$

$$(B - \beta)\mathbf{v} = -C^*\mathbf{u}. \tag{1.4}$$

If $\alpha \notin \operatorname{Spec}(A)$, then (1.3) can be re-written as $\mathbf{u} = -(A - \alpha)^{-1}C\mathbf{v}$, and substituting this into (1.4) yields

$$(B - C^*(A - \alpha)^{-1}C)\mathbf{v} = \beta\mathbf{v}. \tag{1.5}$$

This also means that if $\alpha \notin \operatorname{Spec}(A)$ and $\beta(\alpha)$ is an eigenvalue of

$$B - C^*(A - \alpha)^{-1}C,$$

then $(\alpha, \beta(\alpha)) \in \operatorname{Spec}_p(M)$.

2. Basics and statements

2.1. Restrictions and notation

Suppose that $\mathcal{H}_1$ and $\mathcal{H}_2$ are finite-dimensional, and therefore we are dealing with matrices. In addition, for simplicity, take $\mathcal{H}_1 = \mathcal{H}_2 = \mathcal{H}$ and $\dim \mathcal{H} = n$. Our main results (Theorem 2.5 and its special case Theorem 2.3) are stated below.

Remark 2.1. Most of our results transfer rather seamlessly to the cases when $\mathcal{H}_1$ and $\mathcal{H}_2$ have either different finite dimensions, or are infinite-dimensional, but we exclude these from this paper for clarity.

The eigenvalues of A and B will be denoted by

$$\alpha_1 \leq \ldots \leq \alpha_n, \qquad \beta_1 \leq \ldots \leq \beta_n,$$

respectively, and their corresponding eigenvectors will be denoted by φ_j and ψ_j, $j = 1, \ldots, n$.

In stating most of our results, we restrict our attention to the case where C has rank one. Take $C = \kappa P$, where $\kappa \in \mathbb{R}$, and P is a projection onto a one-dimensional subspace $Z = \operatorname{Span}\{\mathbf{z}\}$, $\|\mathbf{z}\| = 1$. In the basis $\{\varphi_j\}$, P will have the matrix representation $(\langle \mathbf{z}, \varphi_k\rangle\langle \mathbf{z}, \varphi_j\rangle)_{k,j=1}^n$. The equation (1.2) then becomes

$$\begin{pmatrix} A-\alpha & \kappa P \\ \kappa P & B-\beta \end{pmatrix} \begin{pmatrix} \mathbf{u} \\ \mathbf{v} \end{pmatrix} = \mathbf{0}, \tag{2.1}$$

and (1.5) becomes

$$(B - \kappa^2 P (A-\alpha)^{-1} P)\mathbf{v} = \beta \mathbf{v}. \tag{2.2}$$

Thus, by (2.2), for every $\alpha \in \mathbb{C} \setminus \operatorname{Spec}(A)$, there are n complex values $\beta(\alpha)$, and the corresponding curves are continuous in α.

Let $\Phi_{X,\lambda}$ denote the eigenspace of a self-adjoint operator X corresponding to an eigenvalue λ, simple or multiple. Further denote

$$\Gamma_X := \{\lambda \in \operatorname{Spec}(X) \mid \exists \varphi \in \Phi_{X,\lambda} : \varphi \neq 0 \text{ and } \langle \mathbf{z}, \varphi\rangle = 0\},$$
$$\widetilde{\Gamma}_X := \{\lambda \in \operatorname{Spec}(X) \mid \langle \mathbf{z}, \varphi\rangle = 0 \ \forall \varphi \in \Phi_{X,\lambda}\}.$$

Note that $\widetilde{\Gamma}_X \subseteq \Gamma_X$. If λ is a simple eigenvalue of X, then $\lambda \in \Gamma_X \Leftrightarrow \lambda \in \widetilde{\Gamma}_X$. Also, Γ_X contains all the multiple eigenvalues of X.

Let $Q := I - P$ be the orthogonal projection onto $Z^\perp$. For a self-adjoint operator $X : \mathcal{H} \to \mathcal{H}$, denote

$$X_{\parallel,\parallel} = PX|_Z \ : Z \to Z, \qquad X_{\perp,\parallel} = PX|_{Z^\perp} \ : Z^\perp \to Z,$$
$$X_{\parallel,\perp} = QX|_Z \ : Z \to Z^\perp, \quad X_{\perp,\perp} = QX|_{Z^\perp} \ : Z^\perp \to Z^\perp.$$

The eigenvalues of $A_{\perp,\perp}$ and $B_{\perp,\perp}$ will be denoted by

$$\widehat{\alpha}_1 \leq \ldots \leq \widehat{\alpha}_{n-1}, \qquad \widehat{\beta}_1 \leq \ldots \leq \widehat{\beta}_{n-1},$$

respectively, and their corresponding eigenvectors will be denoted by $\widehat{\varphi}_k$ and $\widehat{\psi}_k$, $k = 1, \ldots, n-1$.

Remark 2.2. By the variational principle, the eigenvalues of A and $A_{\perp,\perp}$ interlace,

$$\alpha_k \leq \widehat{\alpha}_k \leq \alpha_{k+1}, \tag{2.3}$$

and similarly the eigenvalues of B and $B_{\perp,\perp}$ interlace,

$$\beta_k \leq \widehat{\beta}_k \leq \beta_{k+1}.$$

2.2. Statement of the simple Chess Board Theorem

Assume for the moment that $\Gamma_A = \Gamma_B = \varnothing$, which in particular implies that all the eigenvalues of A and B are simple. Denote

$$x_0 := -\infty, \qquad x_{2n} := \infty, \qquad x_{2j-1} := \alpha_j, \qquad x_{2k} := \widehat{\alpha}_k,$$

and similarly for β,

$$y_0 := -\infty, \qquad y_{2n} := \infty, \qquad y_{2j-1} := \beta_j, \qquad y_{2k} := \widehat{\beta}_k,$$

where $j = 1, \dots, n$ and $k = 1, \dots, n-1$. Then, the numbers $x_0, \dots, x_{2n}$ divide the α-line into $2n$ intervals, finite or infinite, and similarly for β. Combination of these lines divides the (α, β)-plane into rectangles, some of them semi-infinite,

$$R_{p,q} := r_p \times r_q, \quad r_p := (x_{p-1}, x_p), \quad r_q := (y_{q-1}, y_q), \quad p, q = 1, \dots, 2n,$$

see Figure 1.

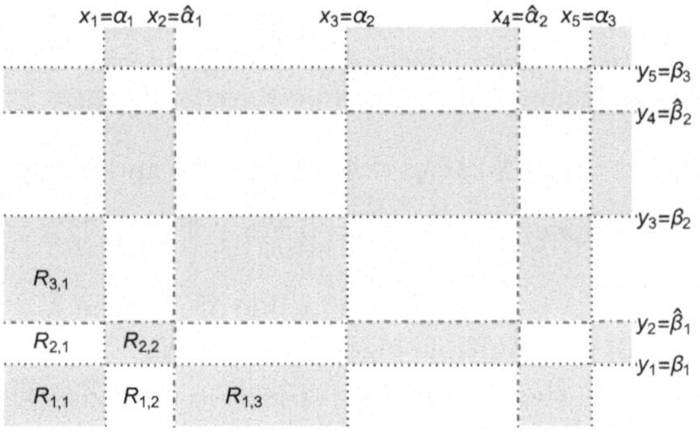

FIGURE 1. In the (α, β)-plane, black dotted and red dot-dashed lines in the vertical direction represent $\alpha = \alpha_j$ and $\alpha = \widehat{\alpha}_k$, respectively; in the horizontal direction they represent $\beta = \beta_j$ and $\beta = \widehat{\beta}_k$, respectively. Here $n = 3$, and the rectangles $R_{p,q}$ with even $p + q$ are shaded.

Theorem 2.3 (Simple Chess Board Theorem). *Let $\Gamma_A = \Gamma_B = \varnothing$. Then all the real pair-eigenvalues (α, β) of M lie on a family of curves $(\alpha, \beta(\alpha))$ with the following properties:*

(a) *each curve may pass only through rectangles $R_{p,q}$ with $p + q$ even;*
(b) *each curve may cross from rectangle to rectangle only through the corner points (x_i, y_j) with $i + j$ odd;*
(c) *each curve $\beta(\alpha)$ is continuous in α except at eigenvalues of A; and at each eigenvalue of A exactly one curve blows up in the sense that $\beta(\alpha) \to \pm\infty$ as $\alpha \to \alpha_i \pm 0$, $\alpha_i \in \mathrm{Spec}(A)$;*

(d) *each curve $\beta(\alpha)$ is monotone decreasing in α on its domain of continuity; more precisely, we have*

$$\frac{d\beta}{d\alpha} = -\kappa^2 \frac{\langle (A-\alpha)^{-2}\mathbf{z},\mathbf{z}\rangle (\langle (B-\beta)^{-1}\mathbf{z},\mathbf{z}\rangle)^2}{\langle (B-\beta)^{-2}\mathbf{z},\mathbf{z}\rangle} < 0. \tag{2.4}$$

Remark 2.4. As α and β are in fact interchangeable, Theorem 2.3 can be equivalently reformulated in terms of curves $(\alpha(\beta),\beta)$ with the only modification being that exactly one curve $\alpha(\beta)$ blows up at each eigenvalue of B in the sense that $\alpha(\beta) \to \pm\infty$ as $\beta \to \beta_j \pm 0$, $\beta_j \in \operatorname{Spec}(A)$.

2.3. Statement of the full Chess Board Theorem

In this section, we assume that either $\Gamma_A \neq \varnothing$ or $\Gamma_B \neq \varnothing$. Denote additionally, for $X : \mathcal{H} \to \mathcal{H}$,

$$\Delta_X := \{\lambda \in \operatorname{Spec}(X) \mid \lambda \in \operatorname{Spec}(X) \cap \operatorname{Spec}(X_{\perp,\perp}) \quad \text{and} \\ \dim(\Phi_{X_{\perp,\perp},\lambda}) > \dim(\Phi_{X,\lambda})\}.$$

We will state formally an analogue of Theorem 2.3 below, but we start with summarising the principle changes: first, we exclude from the dividing mesh the points of $\widetilde{\Gamma}_A \setminus \Delta_A$ and $\widetilde{\Gamma}_B \setminus \Delta_B$; and secondly, the real pair-spectrum of M will, in addition to the curves, contain the lines $(\Gamma_A \times \mathbb{R})$ and $(\mathbb{R} \times \Gamma_B)$.

More precisely, let x_i, $i = 1,\dots,s$, denote the points of

$$\left(\left(\operatorname{Spec}(A) \cup \operatorname{Spec}(A_{\perp,\perp})\right) \setminus \widetilde{\Gamma}_A\right) \cup \Delta_A$$

enumerated in increasing order without account of multiplicities, and similarly y_j, $j = 1,\dots,t$, denote the points of the analogue for B enumerated in increasing order without account of multiplicities. Set additionally $x_0 = y_0 = -\infty$, $x_{s+1} = y_{t+1} = +\infty$, and

$$R_{p,q} = (x_{p-1},x_p) \times (y_{q-1},y_q), \quad p = 1,\dots,s+1, \quad q = 1,\dots,t+1.$$

Theorem 2.5 (Full Chess Board Theorem). *All the real pair-eigenvalues (α,β) of M lie either on the straight lines $(\Gamma_A \times \mathbb{R}) \cup (\mathbb{R} \times \Gamma_B)$ or on a family of curves $(\alpha,\beta(\alpha))$ with the following properties:*

(a) *each curve may pass only through rectangles $R_{p,q}$ with $p+q$ even;*
(b) *each curve may cross from rectangle to rectangle only through the corner points (x_i,y_j) with $i+j$ odd;*
(c) *each curve $\beta(\alpha)$ is continuous in α except at eigenvalues of A not belonging to $\widetilde{\Gamma}_A$; at each such eigenvalue of A exactly one curve blows up in the sense that $\beta(\alpha) \to \pm\infty$ as $\alpha \to \alpha_i \pm 0$, $\alpha_i \in \operatorname{Spec}(A)$;*
(d) *each curve $\beta(\alpha)$ is monotone decreasing in α on its domain of continuity with* (2.4).

2.4. Limit cases

In this section, we show that when $\kappa \to 0$, the components of the real pair-eigenvalues of (2.1) approach the eigenvalues of A and B, and when $\kappa \to \infty$, they approach the eigenvalues of $A_{\perp,\perp}$ and $B_{\perp,\perp}$. For brevity, we will work under the restrictions of the simple Chess Board Theorem.

Theorem 2.6. *Suppose* $\Gamma_A = \Gamma_B = \varnothing$. *As* $\kappa \to 0$, *the real pair-eigenvalue spectrum of* M *converges to* $(\mathrm{Spec}(A) \times \mathbb{R}) \cup (\mathbb{R} \times \mathrm{Spec}(B))$, *and similarly, as* $\kappa \to \infty$, *the real pair-eigenvalue spectrum of* M *converges, but in this case to* $(\mathrm{Spec}(A_{\perp,\perp}) \times \mathbb{R}) \cup (\mathbb{R} \times \mathrm{Spec}(B_{\perp,\perp}))$.

3. Auxiliary results

The statements in this section are for a single matrix, and mostly very elementary. We shall use them later in the proof of the Chess Board Theorem. We shall frequently use the Fourier representation of the resolvent,

$$(A-\alpha)^{-1}\mathbf{f} = \sum_j \frac{\langle \mathbf{f}, \varphi_j\rangle}{\alpha_j - \alpha}\varphi_j, \qquad \alpha \notin \mathrm{Spec}(A). \tag{3.1}$$

We also set

$$R(\alpha) := \langle (A-\alpha)^{-1}\mathbf{z}, \mathbf{z}\rangle = \sum_j \frac{|\langle \mathbf{z}, \varphi_j\rangle|^2}{\alpha_j - \alpha}. \tag{3.2}$$

Lemma 3.1. *Let* $\alpha \notin \mathrm{Spec}(A)$. *Then* $R(\alpha) = 0$ *if and only if* $\alpha \in \mathrm{Spec}(A_{\perp,\perp})$ *and* $(A-\alpha)^{-1}\mathbf{z} = c\widehat{\varphi}$, *where* $\widehat{\varphi}$ *is an eigenfunction of* $A_{\perp,\perp}$ *corresponding to* α *and* $c \neq 0$.

Proof. Set $\boldsymbol{\zeta} = (A-\alpha)^{-1}\mathbf{z}$. Then

$$(A-\alpha)\boldsymbol{\zeta} = \mathbf{z} \quad \Leftrightarrow \quad \begin{pmatrix} A_{\perp,\perp} - \alpha & A_{\|,\perp} \\ A_{\perp,\|} & A_{\|,\|} - \alpha \end{pmatrix} \begin{pmatrix} \mathbf{q} \\ \mathbf{p} \end{pmatrix} = \begin{pmatrix} \mathbf{0} \\ \mathbf{z} \end{pmatrix}, \tag{3.3}$$

where $\mathbf{q} = Q\boldsymbol{\zeta}$ and $\mathbf{p} = P\boldsymbol{\zeta}$. Note that $\langle \boldsymbol{\zeta}, \mathbf{z}\rangle = 0$ iff $\mathbf{p} = P\boldsymbol{\zeta} = 0$. Substituting this into (3.3) gives us

$$\begin{cases} (A_{\perp,\perp} - \alpha)\mathbf{q} &= \mathbf{0}, \\ A_{\perp,\|}\mathbf{q} &= \mathbf{z}. \end{cases} \tag{3.4}$$

By the second equation, $\mathbf{q}$ is non-zero, and then $\alpha \in \mathrm{Spec}(A_{\perp,\perp})$ and $\mathbf{q} = c\widehat{\varphi}$, with $c \neq 0$, by the first equation. Also, we have $\mathbf{q} = Q\boldsymbol{\zeta} = (I-P)\boldsymbol{\zeta} = \boldsymbol{\zeta}$, and so $\boldsymbol{\zeta} = (A-\alpha)^{-1}\mathbf{z} = c\widehat{\varphi}$. □

Lemma 3.2. $\Gamma_A = \varnothing$ *if and only if* $\mathrm{Spec}(A) \cap \mathrm{Spec}(A_{\perp,\perp}) = \varnothing$.

Proof. If there exits an $\alpha \in \Gamma_A$, then there is an eigenfunction $\varphi \in \Phi_{A,\alpha}$ such that $\langle \mathbf{z}, \varphi\rangle = 0$, and therefore $P\varphi = 0$ and so $Q\varphi = \varphi$. Thus

$$A_{\perp,\perp}\varphi = QA\varphi = \alpha Q\varphi = \alpha\varphi,$$

and so $\alpha \in \mathrm{Spec}(A) \cap \mathrm{Spec}(A_{\perp,\perp})$.

On the other hand, let $\alpha \in \mathrm{Spec}(A) \cap \mathrm{Spec}(A_{\perp,\perp})$. Then

$$A\varphi = \alpha\varphi \quad \Rightarrow \quad \langle A\varphi, \widehat{\varphi}\rangle = \alpha\langle \varphi, \widehat{\varphi}\rangle = \alpha\langle Q\varphi, \widehat{\varphi}\rangle. \tag{3.5}$$

Also, since $\widehat{\varphi} \perp z$,

$$A\widehat{\varphi} = A(\widehat{\varphi} + 0z) = A_{\perp,\perp}\widehat{\varphi} + A_{\perp,\|}\widehat{\varphi}, \tag{3.6}$$

therefore

$$\begin{aligned}\langle A\varphi, \widehat{\varphi}\rangle = \langle \varphi, A\widehat{\varphi}\rangle &= \langle \varphi, A_{\perp,\perp}\widehat{\varphi}\rangle + \langle \varphi, A_{\perp,\|}\widehat{\varphi}\rangle \\ &= \langle Q\varphi, A_{\perp,\perp}\widehat{\varphi}\rangle + \langle P\varphi, A_{\perp,\|}\widehat{\varphi}\rangle \\ &= \alpha\langle Q\varphi, \widehat{\varphi}\rangle + \langle P\varphi, A_{\perp,\|}\widehat{\varphi}\rangle,\end{aligned}$$

which implies by (3.5) that $\langle P\varphi, A_{\perp,\|}\widehat{\varphi}\rangle = 0$. Now, if $P\varphi = 0$, then $\varphi \perp z$, and so $\alpha \in \Gamma_A$. If $A_{\perp,\|}\widehat{\varphi} = 0$, then we have from (3.6) that $A\widehat{\varphi} = \alpha\widehat{\varphi}$, and therefore $\widehat{\varphi}$ is an eigenfunction of A such that $z \perp \widehat{\varphi}$, so again $\alpha \in \Gamma_A$. □

Lemma 3.3. *If $\alpha \in \operatorname{Spec}(A) \setminus \widetilde{\Gamma}_A$, then $R(t)$ has a singularity at $t = \alpha$. The function $R(t)$ changes sign when t passes through an α_j, $j = 1, \dots, n$, or an $\widehat{\alpha}_k$, $k = 1, \dots, n-1$.*

If $\alpha \in \widetilde{\Gamma}_A$, then $(A-\alpha)^{-1}\mathbf{z}$ exists, and $R(t)$ is continuous at $t = \alpha$. It changes sign at this α if and only if additionally $\alpha \in \Delta_A$.

Proof. If $\alpha_j \in \operatorname{Spec}(A) \setminus \widetilde{\Gamma}_A$, then there exits at least one $\varphi_j \in \Phi_{A,\alpha}$ such that $\langle \mathbf{z}, \varphi_j\rangle \neq 0$, and it can be seen from (3.2) that $R(t)$ goes to $\pm\infty$ as $\alpha \to \alpha_j \mp 0$. Furthermore, since $R(t)$ has zeros at $\alpha = \widehat{\alpha}_k$ by Lemma 3.1, and also is a continuous function except at the poles $\alpha = \alpha_j$, it changes sign every time t passes through $\widehat{\alpha}_j$ as well.

The second statement follows immediately from (3.2) and the fact that $\mathbf{z} \perp \Phi_{A,\alpha}$, and the last statement can be shown by considering $A|_{\Phi^{\perp}_{A,\alpha}}$ and repeating the above argument. □

4. Proofs of the main results

We now proceed to the proof of Theorem 2.5; Theorem 2.3 then follows from Theorem 2.5 immediately as a special case.

We first derive the characteristic equation of (2.1).

Theorem 4.1. *If $\alpha \notin \operatorname{Spec}(A)$ and $\beta \notin \operatorname{Spec}(B)$, then the characteristic equation of (2.1) for $\beta(\alpha)$ is*

$$\kappa^2\langle (A-\alpha)^{-1}\mathbf{z}, \mathbf{z}\rangle\langle (B-\beta)^{-1}\mathbf{z}, \mathbf{z}\rangle = 1. \tag{4.1}$$

Proof. Re-writing the equation (2.2) as

$$(B-\beta)\mathbf{v} = \kappa^2 P(A-\alpha)^{-1}P\mathbf{v} \tag{4.2}$$

and then using the information that P is a projection, we obtain

$$(B-\beta)\mathbf{v} = \kappa^2\langle \mathbf{v}, \mathbf{z}\rangle P(A-\alpha)^{-1}\mathbf{z} = \kappa^2\langle \mathbf{v}, \mathbf{z}\rangle\langle (A-\alpha)^{-1}\mathbf{z}, \mathbf{z}\rangle\mathbf{z},$$

which implies

$$\mathbf{v} = \kappa^2\langle \mathbf{v}, \mathbf{z}\rangle\langle (A-\alpha)^{-1}\mathbf{z}, \mathbf{z}\rangle (B-\beta)^{-1}\mathbf{z}.$$

Now since the term $\kappa^2\langle \mathbf{v}, \mathbf{z}\rangle\langle (A-\alpha)^{-1}\mathbf{z}, \mathbf{z}\rangle$ is a constant, we can fix it as

$$\kappa^2\langle \mathbf{v}, \mathbf{z}\rangle\langle (A-\alpha)^{-1}\mathbf{z}, \mathbf{z}\rangle = 1 \tag{4.3}$$

by setting $\mathbf{v} := (B-\beta)^{-1}\mathbf{z}$. Substituting this into (4.3), we arrive at (4.1). □

The next lemma shows that $(\Gamma_A \times \mathbb{C}) \cup (\mathbb{C} \times \Gamma_B) \subset \mathrm{Spec}_p(M)$, strengthening in fact the claim of Theorem 2.5.

Lemma 4.2. *If $\alpha \in \Gamma_A$, then $(\alpha, \beta) \in \mathrm{Spec}_p(M)$ for all $\beta \in \mathbb{C}$. Similarly if $\beta \in \Gamma_B$, then $(\alpha, \beta) \in \mathrm{Spec}_p(M)$ for all $\alpha \in \mathbb{C}$.*

Proof. We prove the first of these statements, the second one is similar. Let $\alpha \in \Gamma_A$, and let $\varphi \in \Phi_{A,\alpha}$ such that $\langle \varphi, \mathbf{z} \rangle = 0$. An immediate check shows that

$$\begin{pmatrix} \mathbf{u} \\ \mathbf{v} \end{pmatrix} = \begin{pmatrix} \varphi \\ \mathbf{0} \end{pmatrix}$$

is a pair-eigenvector of (2.1) for a pair-eigenvalue (α, β) with an arbitrary $\beta \in \mathbb{C}$. □

In Lemma 4.2 we show what happens when $\alpha \in \Gamma_A$ or $\beta \in \Gamma_B$; our next result shows which points (α, β) may lie in $\mathrm{Spec}_p(M)$ when α is an eigenvalue of A outside of Γ_A.

Lemma 4.3. *Let $\alpha \in \mathrm{Spec}(A) \setminus \Gamma_A$ and $\beta \notin \Gamma_B$. Then $(\alpha, \beta) \in \mathrm{Spec}_p(M)$ if and only if $\beta = \widehat{\beta} \in \mathrm{Spec}(B_{\perp,\perp})$. Similarly, if $\beta \in \mathrm{Spec}(B) \setminus \Gamma_B$ and $\alpha \notin \Gamma_A$, then $(\alpha, \beta) \in \mathrm{Spec}_p(M)$ if and only if $\alpha = \widehat{\alpha} \in \mathrm{Spec}(A_{\perp,\perp})$.*

Proof. Once more, we only prove the first statement. Let $\alpha \in \mathrm{Spec}(A) \setminus \Gamma_A$. Let us re-write (1.3), (1.4) as

$$(A - \alpha)\mathbf{u} = -\kappa \langle \mathbf{v}, \mathbf{z} \rangle \mathbf{z}, \tag{4.4}$$

$$\kappa \langle \mathbf{u}, \mathbf{z} \rangle \mathbf{z} + (B - \beta)\mathbf{v} = \mathbf{0}. \tag{4.5}$$

Multiplying (4.4) by $\varphi \in \Phi_{A,\alpha}$, we get

$$\langle (A - \alpha)\mathbf{u}, \varphi \rangle = \langle \mathbf{u}, (A - \alpha)\varphi \rangle = 0 = -\kappa \langle \mathbf{v}, \mathbf{z} \rangle \langle \mathbf{z}, \varphi \rangle.$$

Since $\alpha \notin \Gamma_A$, we have $\langle \mathbf{z}, \varphi \rangle \neq 0$, and so $\langle \mathbf{v}, \mathbf{z} \rangle = 0$ (and so $P\mathbf{v} = \mathbf{0}$), and by (4.4), $\mathbf{u} = a\varphi$, where the constant a may or may not be zero.

Substituting now $\mathbf{u} = a\varphi$ into (4.5), and applying the projections Q and P to the result, we obtain

$$B_{\perp,\perp}\mathbf{v} = \beta \mathbf{v}, \tag{4.6}$$

$$B_{\perp,\parallel}\mathbf{v} = -\kappa a \langle \mathbf{z}, \varphi \rangle \mathbf{z}. \tag{4.7}$$

If $\beta \notin \mathrm{Spec}(B_{\perp,\perp})$, then by (4.6), $\mathbf{v} = \mathbf{0}$, and thus $a = 0$, and so $\mathbf{u} = \mathbf{0}$, and $(\alpha, \beta) \notin \mathrm{Spec}_p(M)$, proving the "only if" part of the statement.

If $\beta = \widehat{\beta} \in \mathrm{Spec}(B_{\perp,\perp})$, and $\widehat{\psi} \in \Phi_{B_{\perp,\perp},\widehat{\beta}}$, we choose $\mathbf{v} = b\widehat{\psi}$; we claim that a, b with $a^2 + b^2 \neq 0$ may be chosen now to satisfy (4.7). After multiplying by $\mathbf{z}$, it becomes

$$b \langle B_{\perp,\parallel}\widehat{\psi}, \mathbf{z} \rangle = -\kappa a \langle \mathbf{z}, \varphi \rangle. \tag{4.8}$$

The scalar product on the right-hand side is non-zero by our assumption $\alpha \notin \Gamma_A$. The scalar product on the left-hand side is non-zero since otherwise $\widehat{\beta} \in \mathrm{Spec}(B)$, and therefore $\beta \in \Gamma_B$ by Lemma 3.2, again contradicting our assumptions. Thus we can always choose a, b with $a^2 + b^2 \neq 0$ in order to satisfy (4.8). □

We can now prove our main result.

Proof of the full Chess Board Theorem. The eigenvalues that are inside of $(\Gamma_A \times \mathbb{R}) \cup (\mathbb{R} \times \Gamma_B)$ have been already accounted for by Lemma 4.2, so we will be working outside this set.

Recall the characteristic equation (4.1). Since it needs to be satisfied, $\langle (A-\alpha)^{-1}\mathbf{z}, \mathbf{z}\rangle$ and $\langle (B-\beta)^{-1}\mathbf{z}, \mathbf{z}\rangle$ have to have the same sign for real pair-eigenvalues. It can be seen from (3.2) that $\langle (A-\alpha)^{-1}\mathbf{z}, \mathbf{z}\rangle$ is positive when $\alpha < \alpha_1$, and by Lemma 3.3, it only changes sign every time when α passes through x_p, $p = 1, \dots, s$. Similarly, $\langle (B-\beta)^{-1}\mathbf{z}, \mathbf{z}\rangle$ is positive when $\beta < \beta_1$ and it only changes sign every time when β passes through y_q, $q = 1, \dots, t$. Thus the only allowed regions for real α and β are when $(\alpha, \beta) \in R_{p,q}$ with even $p+q$, proving, with account of Lemma 4.3, the statements (a) and (b).

Statement (c) follows immediately from (4.1) and Lemma 3.3.

To prove (d), we differentiate the characteristic equation (4.1) with respect to α, arriving at

$$\kappa^2\langle (A-\alpha)^{-2}\mathbf{z}, \mathbf{z}\rangle\langle (B-\beta)^{-1}\mathbf{z}, \mathbf{z}\rangle + \kappa^2\langle (A-\alpha)^{-1}\mathbf{z}, \mathbf{z}\rangle\langle (B-\beta)^{-2}\mathbf{z}, \mathbf{z}\rangle\frac{d\beta}{d\alpha} = 0,$$

so that

$$\frac{d\beta}{d\alpha} = -\frac{\langle (A-\alpha)^{-2}\mathbf{z}, \mathbf{z}\rangle\langle (B-\beta)^{-1}\mathbf{z}, \mathbf{z}\rangle}{\langle (A-\alpha)^{-1}\mathbf{z}, \mathbf{z}\rangle\langle (B-\beta)^{-2}\mathbf{z}, \mathbf{z}\rangle},$$

and re-arranging with account of (4.1), we can re-write β' as in (2.4).

To see $\beta' < 0$, we observe from (4.1) that $\langle (A-\alpha)^{-1}\mathbf{z}, \mathbf{z}\rangle \neq 0$ and $\langle (B-\beta)^{-1}\mathbf{z}, \mathbf{z}\rangle \neq 0$. Also,

$$\langle (A-\alpha)^{-2}\mathbf{z}, \mathbf{z}\rangle = \langle (A-\alpha)^{-1}\mathbf{z}, (A-\alpha)^{-1}\mathbf{z}\rangle = \|(A-\alpha)^{-1}\mathbf{z}\|,$$

which is always positive by (3.1), and similarly $\langle (B-\beta)^{-2}\mathbf{z}, \mathbf{z}\rangle > 0$, and therefore $d\beta/d\alpha < 0$. □

Proof of Theorem 2.6. By the characteristic equation (4.1), we have that either $\langle (A-\alpha_\kappa)^{-1}\mathbf{z}, \mathbf{z}\rangle \to \infty$ or $\langle (B-\beta_\kappa)^{-1}\mathbf{z}, \mathbf{z}\rangle \to \infty$ as $\kappa \to 0$, and the first statement follows by Lemma 3.3 and standard perturbation techniques. Similarly, if $\kappa \to \infty$, then either $\langle (A-\alpha_\kappa)^{-1}\mathbf{z}, \mathbf{z}\rangle \to 0$ or $\langle (B-\beta_\kappa)^{-1}\mathbf{z}, \mathbf{z}\rangle \to 0$, and the result follows from Lemma 3.1. □

5. Examples

5.1. Motivation and Example 1

The main motivation of this paper comes from the particular non-self-adjoint problem which was considered in [4], with corresponding change of notations. Consider the $n \times n$ matrices

$$A_1 = \begin{pmatrix} 0 & 1 & & \\ 1 & 0 & \ddots & \\ & \ddots & \ddots & 1 \\ & & 1 & 0 \end{pmatrix}, \qquad P_1 = \begin{pmatrix} 0 & & & \\ & \ddots & & \\ & & 0 & \\ & & & 1 \end{pmatrix}.$$

We set $A = B = A_1$ and $C = \kappa P_1$ (i.e. $\mathbf{z} = (0, \dots, 0, 1)^T$). The eigenvalues of A_1 are given by

$$\alpha_j = 2\cos\left(\frac{\pi j}{n+1}\right), \quad j = 1, \dots, n,$$

and the eigenvalues of $(A_1)_{\perp,\perp}$ are given by the same formula with n replaced by $n-1$.

In fact, [4] studied the spectrum of a non-self-adjoint problem

$$\begin{pmatrix} A_1 + \gamma & P_1 \\ -P_1 & -A_1 - \gamma \end{pmatrix} \begin{pmatrix} \mathbf{u} \\ \mathbf{v} \end{pmatrix} = \lambda \begin{pmatrix} \mathbf{u} \\ \mathbf{v} \end{pmatrix}, \tag{5.1}$$

where λ is a spectral parameter and $\gamma \in \mathbb{R}$ is fixed; the problem (5.1) relates to (2.1) by setting $\kappa = 1$ and

$$\alpha = \lambda - \gamma, \qquad \beta = -\lambda - \gamma. \tag{5.2}$$

We shall return to the comparison of the two problems and especially to non-real λ in Section 6.

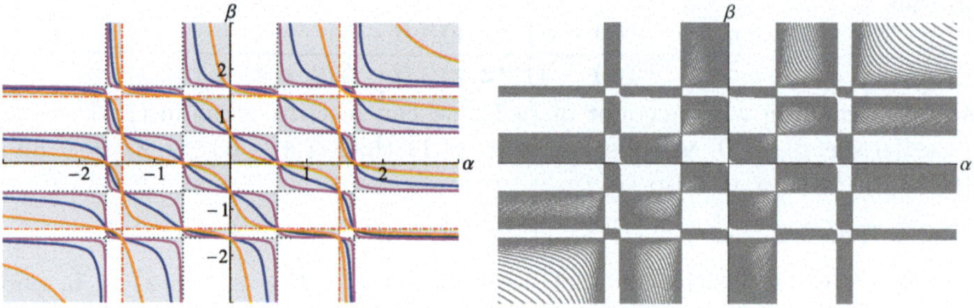

FIGURE 2. $A = B = A_1$ for $n = 4$. Left: $\mathrm{Spec}_p(M)$ with $\kappa = 0.4$ (magenta curves), $\kappa = 1$ (blue curves) and $\kappa = 2$ (orange curves). Right: the superimposition of $\mathrm{Spec}_p(M)$ for the values of κ from 0.001 to 10 with the step-size of 0.1.

Note that $\Gamma_{A_1} = \varnothing$ and the general spectral picture in the (α, β)-plane including the rectangular mesh can be seen in Figures 2 and 5. We see that the results of the simple Chess Board Theorem hold.

5.2. Example 2

This example illustrates the case when $\Gamma = \varnothing$. We denote by $\mathrm{diag}(\lambda_1, \dots, \lambda_n)$ a diagonal matrix composed of the entries $\lambda_1, \dots, \lambda_n$. Let $A_2 = \mathrm{diag}(-1, 1)$ and $B_1 = \mathrm{diag}(1, 3)$. Set $\mathbf{z} = \left(\frac{1}{\sqrt{5}}, \frac{2}{\sqrt{5}}\right)^T$. Then $\Gamma_{A_2} = \Gamma_{B_1} = \varnothing$. We also have $\mathrm{Spec}((A_2)_{\perp,\perp}) = \left\{-\frac{3}{5}\right\}$ and $\mathrm{Spec}((B_1)_{\perp,\perp}) = \left\{\frac{7}{5}\right\}$. The spectral picture can be seen in the left of Fig. 3, and we see that the simple Chess Board Theorem (Theorem 2.3) holds.

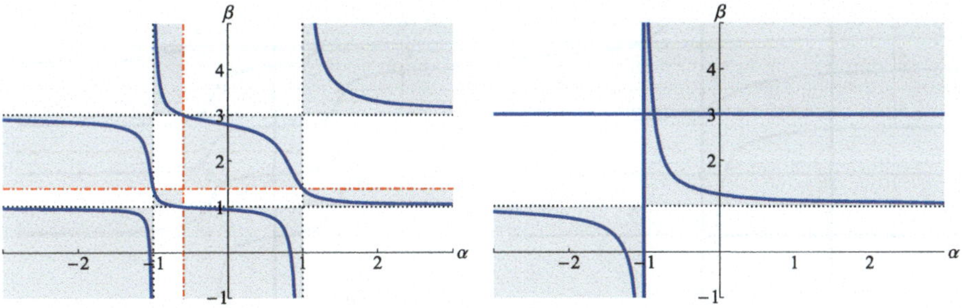

FIGURE 3. $\mathrm{Spec}_p(M)$ for two cases. Left: with $A = A_2$, $B = B_1$, and $\kappa = 2/3$. Right: with $A = A_3$, $B = B_1$, and $\kappa = 1/2$.

5.3. Example 3

This example illustrates two cases; the case when $\Gamma \neq \varnothing$ and $\widetilde{\Gamma} = \varnothing$, and also the case when $\Gamma = \widetilde{\Gamma} \neq \varnothing$. Consider $A_3 = \mathrm{diag}(-1,-1)$ and B_1. Set $\mathbf{z} = (1,0)^T$. Then $\Gamma_{A_3} = \{-1\}$, $\widetilde{\Gamma}_{A_3} = \varnothing$, and $\Gamma_{B_1} = \widetilde{\Gamma}_{B_1} = \{3\}$. Also $\mathrm{Spec}((A_3)_{\perp,\perp}) = \{-1\}$, $\mathrm{Spec}((B_1)_{\perp,\perp}) = \{3\}$. The spectral picture is shown in the right of Fig. 3. We see that $\mathrm{Spec}_p(M)$ has an additional vertical straight line at $\alpha = -1$, and there is also a blow-up at $\alpha = -1$. This line is included in the mesh since $\mathbf{z}$ is orthogonal to one eigenvector but $\mathbf{z} \not\perp \Phi_{A_3,-1}$. On the other hand, there is a horizontal straight line passing through $\beta = 3$ which is not included in the mesh since B_1 has simple eigenvalues and $\mathbf{z} \perp \Phi_{B_1,3}$.

5.4. Example 4

This example illustrates the case when $\Gamma, \widetilde{\Gamma} \neq \varnothing$ and $\Gamma \neq \widetilde{\Gamma}$. Take

$$A_4 = \mathrm{diag}(1,1,3,3), \qquad B_2 = \begin{pmatrix} -2 & 1 & 0 & 0 \\ 1 & -1 & 0 & 0 \\ 0 & 0 & 2 & 1 \\ 0 & 0 & 1 & 3 \end{pmatrix}. \tag{5.3}$$

Set $\mathbf{z} = (0,0,0,1)^T$. Then we have $\mathrm{Spec}(B_2) = \{-(3 \pm \sqrt{5})/2, (5 \pm \sqrt{5})/2\}$ and $\Gamma_{A_4} = \{1,3\}$, $\widetilde{\Gamma}_{A_4} = \{1\}$, $\Gamma_{B_2} = \widetilde{\Gamma}_{B_2} = \{-\frac{1}{2}(3 \pm \sqrt{5})\}$. We also obtain $\mathrm{Spec}((A_4)_{\perp,\perp}) = \{1,3\}$, where the eigenvalue 1 has multiplicity two, and $\mathrm{Spec}((B_2)_{\perp,\perp}) = \{(-3 \pm \sqrt{5})/2, 2\}$. The spectral picture is shown in the left of Fig. 4. As expected, there are two additional vertical straight lines: at $\alpha = 1$, where is no blow-up and the line is not included in the mesh since $\mathbf{z} \perp \Phi_{A_4,1}$; and at $\alpha = 3$, where is a blow-up and the line is included in the mesh since $\mathbf{z} \not\perp \Phi_{A_4,3}$. On the other hand, there are two additional horizontal lines at $\beta = -(3 \pm \sqrt{5})/2$ which are not included in the mesh as $\mathbf{z}$ is orthogonal to the corresponding eigenspaces.

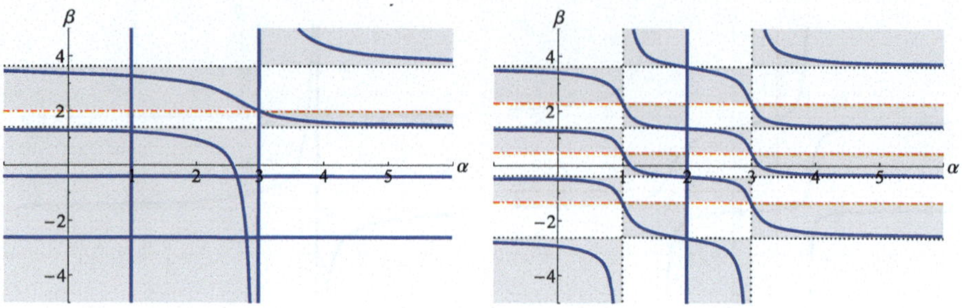

FIGURE 4. Left: $\mathrm{Spec}_p(M)$ with $A = A_4$, $B = B_2$, and $\kappa = 1$. Right: $\mathrm{Spec}_p(M)$ with $A = A_5$, $B = B_2$, and $\kappa = 1$.

5.5. Example 5

This example illustrates the case when $\Delta \neq \varnothing$. Consider $A_5 = \mathrm{diag}(1,2,2,3)$, and B_2. Set $\mathbf{z} = \left(\frac{1}{\sqrt{2}}, 0, 0, \frac{1}{\sqrt{2}}\right)^T$. Then $\Gamma_{A_5} = \widetilde{\Gamma}_{A_5} = \{2\}$ and $\Gamma_{B_2} = \widetilde{\Gamma}_{B_2} = \varnothing$. Also $\mathrm{Spec}((A_5)_{\perp,\perp}) = \{2\}$, where the eigenvalue 2 has multiplicity three, and $\mathrm{Spec}((B_2)_{\perp,\perp}) = \{(1 \pm \sqrt{13})/2, 1/2\}$. The spectral picture is shown on the right side of Fig. 4. Since $\mathbf{z} \perp \Phi_{A_5,2}$, there is no blow-up at $\alpha = 2$. Nevertheless, this line is also included in the mesh as $\Delta_{A_5} = \{2\}$, that is, $\dim(\Phi_{(A_5)_{\perp,\perp},2}) > \dim(\Phi_{A_5,2})$.

6. Relation to a non-self-adjoint problem

We now return to the example studied in [4]. Generally speaking, there are n complex $\beta(\alpha) \in \mathbb{C}$ for every $\alpha \in \mathrm{Spec}(A)\setminus\mathbb{C}$. We therefore limit our attention to pair-eigenvalues subject to the additional restriction

$$\mathrm{Im}(\alpha + \beta) = 0, \tag{6.1}$$

which is equivalent to introducing the additional restriction $\gamma \in \mathbb{R}$, see (5.2).

A general spectral picture of this non-self-adjoint problem in the (α, β)-plane is illustrated in Figure 5. Light blue curves depict the real parts of non-real pair-eigenvalues $\mathrm{Re}\ \beta(\mathrm{Re}\ \alpha)$ such that (6.1) holds, which keeps all $(\alpha, \beta) \in \mathbb{R}^2$ in the picture (shown in dark blue) and also some non-real pair-eigenvalues. It is easily verified that the spectra are symmetric with respect to $(\alpha, \beta) \leftrightarrow (\beta, \alpha)$ and $(\alpha, \beta) \leftrightarrow (-\alpha, -\beta)$.

The real and non-real eigenvalue curves $\lambda(\gamma)$ may collide, with two possible types of collisions: those when two real eigenvalues collide and produce a complex conjugate pair, called Type-A, and those when a pair of complex conjugate eigenvalues collide and become real, called Type-B, see Fig. 6 for equivalents in the (α, β)-plane.

Lemma 6.1. *The collisions happen at the points where* $\frac{d\beta}{d\alpha} = -1$.

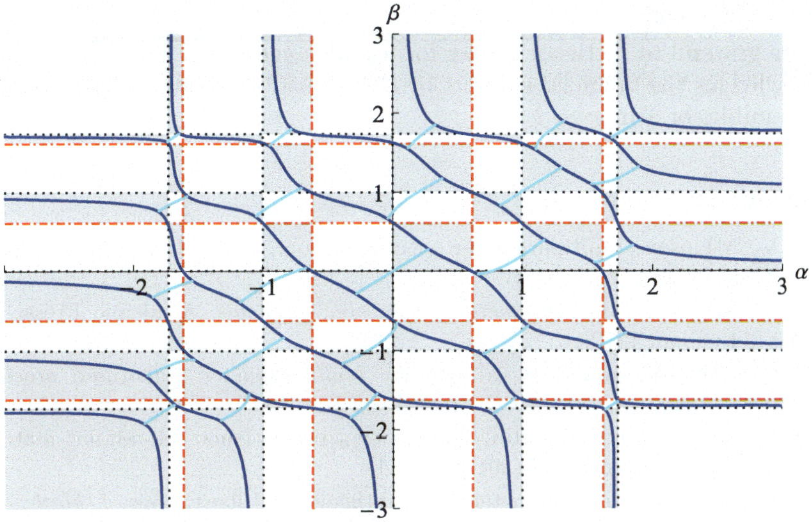

FIGURE 5. $\mathrm{Spec}_p(M)$ with $A = B = A_1$ and $C = P_1$ for $n = 5$.

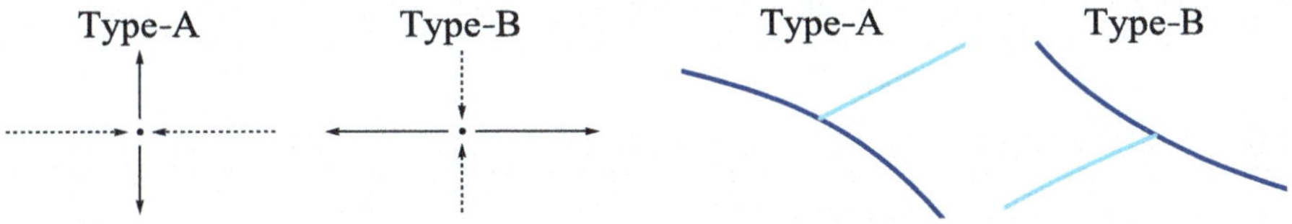

FIGURE 6. Left: the collisions in the $(\mathrm{Re}(\lambda), \mathrm{Im}(\lambda))$-plane. Right: the collisions in the (α, β)-plane.

Proof. Consider

$$\begin{pmatrix} A - \lambda + \gamma & C \\ C^* & B + \lambda + \gamma \end{pmatrix} \begin{pmatrix} \mathbf{u} \\ \mathbf{v} \end{pmatrix} = \mathbf{0}.$$

Considering the curves $\gamma(\lambda)$ instead of $\lambda(\gamma)$ and differentiating with respect to λ we arrive at

$$\begin{pmatrix} -1 + \gamma' & \\ & 1 + \gamma' \end{pmatrix} \begin{pmatrix} \mathbf{u} \\ \mathbf{v} \end{pmatrix} = - \begin{pmatrix} A - \lambda + \gamma & C \\ C^* & B + \lambda + \gamma \end{pmatrix} \begin{pmatrix} \mathbf{u}' \\ \mathbf{v}' \end{pmatrix}, \tag{6.2}$$

which is solvable if and only if the right hand side of (6.2) is perpendicular to $\binom{\mathbf{u}}{\mathbf{v}}$. Therefore multiplying by $\binom{\mathbf{u}}{\mathbf{v}}$ we obtain

$$\begin{aligned} (-1 + \gamma')\|\mathbf{u}\|^2 + (1 + \gamma')\|\mathbf{v}\|^2 = 0 \quad &\Leftrightarrow \quad \gamma'(\|\mathbf{u}\|^2 + \|\mathbf{v}\|^2) = \|\mathbf{u}\|^2 - \|\mathbf{v}\|^2 \\ &\Leftrightarrow \quad \gamma' = 0, \end{aligned}$$

and since at critical points $\frac{d\gamma}{d\lambda} = 0$, using (5.2) we obtain

$$\frac{d\gamma}{d\lambda} = \frac{\frac{d\gamma}{d\alpha}}{\frac{d\lambda}{d\alpha}} = \frac{-\frac{1}{2} - \frac{1}{2}\beta'}{\frac{1}{2} - \frac{1}{2}\beta'} = 0 \quad \Leftrightarrow \quad \beta' = -1. \qquad \square$$

Acknowledgments

We are grateful to E. Brian Davies for useful suggestions. The second author acknowledges the financial support by the Ministry of National Education of the Republic of Turkey.

References

[1] F.V. Atkinson, Multiparameter spectral theory, *Bull. Amer. Math. Soc.* **74** (1968), no. 1, 1–27.

[2] F.V. Atkinson, *Multiparameters eigenvalue problems,* Academic Press, New York–London, 1972.

[3] F.V. Atkinson and A.B. Mingarelli, *Multiparameter eigenvalue problems: Sturm–Liouville theory,* CRC Press, Boca Raton, FL, 2011.

[4] E.B. Davies and M. Levitin, Spectra of a class of non-self-adjoint matrices, *Linear Algebra Appl.* **448** (2014), 55–84.

[5] B.D. Sleeman, Multiparameter spectral theory in Hilbert space, *J. Math. Anal. Appl.* **65** (1978), no. 3, 511–530.

Michael Levitin and Hasen Mekki Öztürk
Department of Mathematics and Statistics
University of Reading
Whiteknights, PO Box 220
Reading RG6 6AX
UK
e-mail: M.Levitin@reading.ac.uk
H.M.Ozturk@pgr.reading.ac.uk

Finite sections of the Fibonacci Hamiltonian

Marko Lindner and Hagen Söding

Abstract. We study finite but growing principal square submatrices A_n of the one- or two-sided infinite Fibonacci Hamiltonian A. Our results show that such a sequence (A_n), no matter how the points of truncation are chosen, is always stable – implying that A_n is invertible for sufficiently large n and $A_n^{-1} \to A^{-1}$ pointwise.

Mathematics Subject Classification (2010). Primary 65J10, Secondary 47A35, 47B36.

Keywords. Finite section method, Fibonacci Hamiltonian, Jacobi operator, limit operators.

1. Introduction

The 1D Schrödinger operator $-\Delta + b\cdot$ with a bounded potential $b \in L^\infty(\mathbb{R})$ can be discretized, via finite differences on a uniform grid on $\mathbb{R}$, by the second order difference operator

$$(Ax)_n = x_{n-1} + v_n x_n + x_{n+1}, \quad n \in \mathbb{Z}, \tag{1}$$

acting on a sequence space like $\ell^p(\mathbb{Z})$. The discrete potential $v = (v_n) \in \ell^\infty(\mathbb{Z})$ corresponds to evaluations of the potential b on the grid (subtracted by a two that comes from the discretization of the Laplace operator). A is commonly referred to as a discrete 1D Schrödinger operator.

A particularly beautiful example, the so-called *Fibonacci Hamiltonian*, arises when the discrete potential v is given by the formula

$$v_n = \chi_{[1-\alpha,1)}(n\alpha \mod 1), \quad n \in \mathbb{Z}, \tag{2}$$

where $\alpha = \frac{\sqrt{5}-1}{2}$ is the golden ratio and $\chi_{\mathcal{I}}$ is the characteristic function of an interval $\mathcal{I}$.

The sequence v from (2) is not periodic (as α is irrational); it displays a so-called quasiperiodic pattern. Here are its values $v_1, \ldots, v_{55}$ and

© Springer International Publishing AG, part of Springer Nature 2018
A. Böttcher et al. (eds.), *The Diversity and Beauty of Applied Operator Theory*, Operator Theory: Advances and Applications 268, https://doi.org/10.1007/978-3-319-75996-8_20

three attempts to identify basic building blocks of the sequence (one with normal/bold face, one with separation by minus signs and one with under/overlines):

$$\underline{\mathbf{10}110\text{-}\mathbf{10}110\text{-}110}\text{-}\underline{\mathbf{10}110\text{-}\mathbf{10}110\text{-}110}\overline{\text{-}\mathbf{10}110\text{-}110}\text{-}\underline{\mathbf{10}110\text{-}\mathbf{10}110\text{-}110}\overline{\text{-}\mathbf{10}110\text{-}101}.$$

The global pattern of these building blocks (on each scale) is the same as the pattern formed by 1 and 0 on the finest scale. The Fibonacci potential shows self-similarity on many levels.

The Fibonacci Hamiltonian is the standard model in 1D for physical properties of so-called quasicrystals and is therefore heavily studied in mathematical physics. Most of the research deals with the spectrum of A, which is a Cantor set of measure zero without any eigenvalues (purely singular continuous spectrum).

Our focus is different. The operator (1) acts via multiplication with a two-sided infinite tridiagonal matrix $(a_{ij})_{i,j\in\mathbb{Z}}$. The main diagonal carries the sequence v, the first sub- and super-diagonal are constant to one, and the rest is zero. We study the applicability of the so-called finite section method to that infinite matrix.

The *finite section method* (*FSM*) looks at finite submatrices

$$A_n = (a_{ij})_{i,j=l_n}^{r_n}, \quad n \in \mathbb{N}$$

of an infinite matrix $A = (a_{ij})_{i,j\in\mathbb{Z}}$ with integer cut-off points

$$l_n \to -\infty \quad \text{and} \quad r_n \to +\infty$$

and asks whether

a) the matrices A_n are invertible for all sufficiently large n and
b) their inverses (after embedding them into an infinite matrix again) converge pointwise in $\ell^p(\mathbb{Z})$ to the inverse of A.

Assuming invertibility of A on $\ell^p(\mathbb{Z})$, property **b)** is equivalent to the uniform boundedness of the inverses A_n^{-1}.

As a consequence, one can solve an infinite system $Ax = b$ approximately by solving large but finite systems $A_n x_n = b_n$. For one-sided infinite matrices $A = (a_{ij})_{i,j\in\mathbb{N}}$, all of the above remains true but l_n should be fixed at 1.

Our result. For the Fibonacci Hamiltonian A from (1) with potential (2) as well as for its one-sided infinite submatrix $A_{\mathbb{N}} := (a_{ij})_{i,j\in\mathbb{N}}$, we first show that both operators are invertible on every space $\ell^p(\mathbb{Z})$, resp. $\ell^p(\mathbb{N})$, with $p \in [1,\infty]$ before proving that the FSM with arbitrary cut-off points is applicable for A as well as for $A_{\mathbb{N}}$.

Historic remarks. Quasicrystals are materials that show features of periodicity (so-called Bragg peaks in diffraction experiments) and aperiodicity (symmetries that rule out periodicity) at the same time – so-called quasiperiodicity. They have first been observed by D. Shechtman in 1982 in his laboratory [29] but are meanwhile also found to be occurring in nature. Physicists and mathematicians quickly developed an interest in this topic. In particular the spectrum of the corresponding Hamiltonian is of huge interest for the understanding of electrical properties of the material [1, 13, 30]. The most famous quasiperiodic ensemble in 2D is the Penrose tiling [4]. The understanding of the corresponding spectrum seems however currently out of reach, so that one resorts to 1D ensembles, the most common of which is the Fibonacci sequence (2). A detailed history and the current state of the art on the extensively studied spectral analysis of the Fibonacci Hamiltonian can be found in [9].

The idea of the FSM is so natural that it is difficult to give a historical starting point. First rigorous treatments are from Baxter [2] and Gohberg & Feldman [12] on Wiener–Hopf and convolution operators in dimension $N = 1$ in the early 1960's. For convolution equations in higher dimensions $N \geq 2$, the FSM goes back to Kozak & Simonenko [14, 15], and for general band-dominated operators with scalar [23] and operator-valued [24, 25] coefficients, most results arc duc to Rabinovich, Roch & Silbermann. For the state of the art see, e.g., [3, 27, 20, 28].

2. The finite section method

As usual, for an index set $\mathbb{I} \subset \mathbb{Z}$, let $\ell^p(\mathbb{I})$ denote the set of all complex sequences $(x_k)_{k\in\mathbb{I}}$ with $\sum_{k\in\mathbb{I}} |x_k|^p < \infty$ for $p \in [1, \infty)$, and $\ell^\infty(\mathbb{I})$ be the set of all bounded complex sequences over $\mathbb{I}$.

Let $A = (a_{ij})_{i,j\in\mathbb{Z}}$ be a band matrix (i.e., a matrix with only finitely many nonzero diagonals) with uniformly bounded complex entries. Then A acts, via matrix-vector multiplication, as a bounded linear operator on all spaces $\ell^p(\mathbb{Z})$ with $p \in [1, \infty]$. Denote that operator again by A.

For integer cut-off points $l_1, l_2, \dots$ and $r_1, r_2, \dots$ with

$$l_n \to -\infty \quad \text{and} \quad r_n \to +\infty,$$

we look at the finite submatrices

$$A_n = (a_{ij})_{i,j=l_n}^{r_n}, \quad n \in \mathbb{N}, \tag{3}$$

of A and call the sequence $(A_n)_{n\in\mathbb{N}}$ *stable* if there exists an $n_0 \in \mathbb{N}$ such that A_n is invertible for all $n \geq n_0$ and $\sup_{n\geq n_0} \|A_n^{-1}\| < \infty$.

Invertibility of A and stability of (A_n) together are sufficient and necessary for the *applicability* of (A_n), that is, for the pointwise convergence (i.e., column-wise convergence of the matrices) $A_n^{-1} \to A^{-1}$, when A_n^{-1} is extended to an infinite matrix again. This approximation of A^{-1} can be used for solving equations $Ax = b$ approximately via the solutions of growing finite systems.

We see that it is crucial to know about the stability of (A_n). This stability is closely connected to a family of one-sided infinite matrices that are associated to A and to the cut-off sequences (l_n) and (r_n). Those associated one-sided infinite matrices are partial limits of the upper left and the lower right corner of the finite matrix A_n as $n \to \infty$. Precisely, the associated matrices are the entrywise limits

$$(a_{i+l'_n, j+l'_n})_{i,j=0}^{\infty} \to B_+ \quad \text{and} \quad (a_{i+r'_n, j+r'_n})_{i,j=-\infty}^{0} \to C_- \quad \text{as } n \to \infty \tag{4}$$

of one-sided infinite submatrices of A, where $(l'_n)_{n=1}^\infty$ and $(r'_n)_{n=1}^\infty$ are subsequences of $(l_n)_{n=1}^\infty$ and $(r_n)_{n=1}^\infty$, respectively, such that the limits (4) exist. The boundedness of the diagonals of A ensures (by Bolzano–Weierstrass and a Cantor diagonal argument) the existence of such subsequences and the corresponding limits (4). Here is the result.

Lemma 2.1. [*Lemma* 1.2 *of* [7]] *For a band matrix* $A = (a_{ij})_{i,j\in\mathbb{Z}}$ *and two cut-off sequences* $(l_n)_{n=1}^\infty$ *and* $(r_n)_{n=1}^\infty$ *in* $\mathbb{Z}$ *with* $l_n \to -\infty$ *and* $r_n \to +\infty$, *the following are equivalent:*

(i) *the FSM* $(A_n)_{n=1}^\infty$ *with* A_n *from* (3) *is applicable to* A,
(ii) *the FSM* $(A_n)_{n=1}^\infty$ *with* A_n *from* (3) *is stable,*
(iii) A *and the limits* B_+ *and* C_- *from* (4) *are invertible for all subsequences* (l'_n) *of* (l_n) *and* (r'_n) *of* (r_n).

So by the choice of the cut-off sequences (l_n) and (r_n), one can control the selection of associated matrices B_+ and C_- and hence control the stability and applicability of the FSM.

The construction of the B_+ and C_- brings us to the notion of a limit operator [23, 26, 19].

Definition 2.2. Let $\mathbb{I}$ be either $\mathbb{Z}$ or $\mathbb{N}$. For a bounded one- or two-sided infinite band matrix $A = (a_{ij})_{i,j\in\mathbb{I}}$ and a sequence $h_1, h_2, \ldots$ in $\mathbb{I}$ with $|h_n| \to \infty$ we say that $B = (b_{ij})_{i,j\in\mathbb{Z}}$ is a *limit operator* of A if, for all $i, j \in \mathbb{Z}$,

$$a_{i+h_n,j+h_n} \to b_{ij} \quad \text{as} \quad n \to \infty. \tag{5}$$

We write A_h instead of B.

Note that limit operators are always given by a two-sided infinite matrix, no matter if the matrix A to start with is one- or two-sided infinite.

So in this language, our associated matrices B_+ and C_- from (4) are one-sided truncations of limit operators of A: each $B_+ = (b_{ij})_{i,j=0}^\infty$ is a submatrix of a limit operator $B = (b_{ij})_{i,j\in\mathbb{Z}}$ of A w.r.t. a subsequence h of (l_n), and each $C_- = (c_{ij})_{i,j=-\infty}^0$ is a submatrix of a limit operator $C = (c_{ij})_{i,j\in\mathbb{Z}}$ of A w.r.t. a subsequence h of (r_n). To be able to rephrase Lemma 2.1 in that language, we introduce the following notations.

Definition 2.3. **a)** For a bounded one- or two-sided infinite band matrix $A = (a_{ij})_{i,j\in\mathbb{I}}$ with $\mathbb{I} \in \{\mathbb{Z}, \mathbb{N}\}$ and a sequence $g_1, g_2, \ldots$ in $\mathbb{I}$ with $|g_n| \to \infty$ we write $\mathrm{Lim}_g(A)$ for the set of all limit operators A_h with respect to a subsequence h of g, and we write $\mathrm{Lim}(A)$ for the set of all limit operators of A. Moreover, put

$$\mathrm{Lim}_+(A) := \mathrm{Lim}_{(1,2,3,\ldots)}(A) \quad \text{and} \quad \mathrm{Lim}_-(A) := \mathrm{Lim}_{(-1,-2,-3,\ldots)}(A).$$

b) For a two-sided infinite matrix $A = (a_{ij})_{i,j\in\mathbb{Z}}$, write

$$A_\pm := (a_{ij})_{i,j\in\mathbb{Z}_\pm},$$

respectively, where $\mathbb{Z}_- := \{\ldots, -2, -1, 0\}$ and $\mathbb{Z}_+ := \{0, 1, 2, \ldots\}$.

Note that A_+ and A_- overlap in a_{00}. Here is the announced reformulation of Lemma 2.1.

Corollary 2.4. *For a bounded band matrix* $A = (a_{ij})_{i,j\in\mathbb{Z}}$ *and two cut-off sequences* $l = (l_n)_{n=1}^{\infty}$ *and* $r = (r_n)_{n=1}^{\infty}$ *in* $\mathbb{Z}$ *with* $l_n \to -\infty$ *and* $r_n \to +\infty$, *the following are equivalent:*

(i) *the FSM* $(A_n)_{n=1}^{\infty}$ *with* A_n *from* (3) *is applicable to* A,
(ii) *the FSM* $(A_n)_{n=1}^{\infty}$ *with* A_n *from* (3) *is stable,*
(iii) A *and all operators* B_+ *and* C_- *with* $B \in \operatorname{Lim}_l(A)$ *and* $C \in \operatorname{Lim}_r(A)$ *are invertible.*

If stability holds for $l = (-1, -2, -3, \dots)$ and $r = (1, 2, 3, \dots)$ then it holds for arbitrary cut-off sequences (l_n) and (r_n).

Corollary 2.5. *For a bounded band matrix* $A = (a_{ij})_{i,j\in\mathbb{Z}}$, *the following are equivalent:*

(i) *the FSM* $(A_n)_{n=1}^{\infty}$ *with* A_n *from* (3) *is applicable for arbitrary cut-offs* (l_n) *and* (r_n),
(ii) *the FSM* $(A_n)_{n=1}^{\infty}$ *with* A_n *from* (3) *is stable for arbitrary cut-offs* (l_n) *and* (r_n),
(iii) A *and all operators* B_+ *and* C_- *with* $B \in \operatorname{Lim}_-(A)$ *and* $C \in \operatorname{Lim}_+(A)$ *are invertible.*

The one-sided infinite case, in which $A = (a_{ij})_{i,j\in\mathbb{N}}$, only requires minimal changes to what was written above: The sequence (l_n) is then constant at 1 and therefore the operators B_+ do not appear in (iii) of Lemma 2.1 and Corollary 2.4.

Limit operators are not only good for detecting stability[1] of the FSM. Their primary purpose is to characterize the coset $A + \mathcal{K}(X)$ of A modulo the ideal of all compact operators $\mathcal{K}(X)$, where we abbreviate $\ell^p(\mathbb{I}) =: X$.

Recall that a bounded linear operator A on X, we write $A \in \mathcal{L}(X)$, is a *Fredholm operator* if its coset $A + \mathcal{K}(X)$ is invertible in the so-called Calkin algebra $\mathcal{L}(X)/\mathcal{K}(X)$, which holds iff the nullspace of A has finite dimension and the range of A has finite codimension in X.

Lemma 2.6. *For a bounded band matrix* $A = (a_{ij})_{i,j\in\mathbb{I}}$ *with* $\mathbb{I} \in \{\mathbb{Z}, \mathbb{N}\}$ *and* $X = \ell^p(\mathbb{I})$ *with any* $p \in [1, \infty]$, *the following are equivalent:*

(i) A *is a Fredholm operator on* X,
(ii) *all limit operators of* A *are invertible on* $\ell^p(\mathbb{Z})$ [23, 21],
(iii) *all limit operators of* A *are injective on* $\ell^\infty(\mathbb{Z})$ [5, 6].

3. The Fibonacci word

Recall the infinite sequence $v = (v_n)_{n\in\mathbb{Z}}$ of zeros and ones from (2). In this section we interpret v as an infinite word over the alphabet $\Sigma = \{0, 1\}$. Let us recall some basic notions on words. For a detailed discussion, including on the Fibonacci word, see e.g. [22].

[1] They only come into play here because the stability of the sequence (A_n) is equivalent to the operator $D := \operatorname{Diag}(A_1, A_2, \dots)$ being a Fredholm operator. Then Lemma 2.6 below is applied to D.

3.1. Some words on words

An *alphabet* is a nonempty set Σ. A finite vector $w = (w_1, \dots, w_n) \in \Sigma^n$ is called a *word* of *length* n over Σ. We write $|w| = n$ for its length. Sequences $(w_1, w_2, \dots)$ and $(\dots, w_{-2}, w_{-1})$ are *one-sided infinite words* over Σ and $(\dots, w_{-2}, w_{-1}, w_0, w_1, w_2, \dots)$ is a *two-sided infinite word* over Σ when $w_i \in \Sigma$ for all i. The word of length zero is denoted by ε and is called *the empty word.*

Let $\Sigma^* := \cup_{n=0}^{\infty} \Sigma^n$ denote the set of all finite words over Σ. Moreover, for an infinite index set $\mathbb{I} \in \{\mathbb{Z}, \mathbb{N}, -\mathbb{N}, \mathbb{Z}_+, \mathbb{Z}_-\}$, let $\Sigma^{\mathbb{I}}$ denote the set of all infinite words $(w_n)_{n\in\mathbb{I}}$ over Σ.

The word $(w_1, w_2, \dots, w_n)$ is often simply written as $w_1 w_2 \dots w_n$. Similarly for infinite words. For two words $u = u_1 \dots u_m$ and $v = v_1 \dots v_n$, the word $u_1 \dots u_m v_1 \dots v_n$ is denoted by $u \circ v$ or just uv. This operation, called concatenation, is associative on Σ^*, with ε as the neutral element of Σ^*. Concatenation is also defined between two oppositely directed one-sided infinite words (at their finite endpoints) and between finite and one-sided infinite words in the natural way.

A word w is called a *subword* (or *factor*) of a word u if u can be written as xwy with (possibly empty) words x and y. We write $w \prec u$ if w is a subword of u. Then $\varepsilon \prec u$ holds for all words u. The reversed word of $u = u_1 \dots u_m$ and $w = w_1 w_2 \dots$ is $u^R := u_m \dots u_1$ and $w^R := \dots w_2 w_1$, respectively.

3.2. Finite Fibonacci words: substitution, recursion, and limit

Let $\Sigma = \{0, 1\}$ and $\varphi : \Sigma^* \to \Sigma^*$ be the homomorphism (w.r.t. concatenation $\circ$) with $\varphi : 0 \mapsto 1$ and $\varphi : 1 \mapsto 10$. Then put $f_1 := 1$, $f_2 := \varphi(f_1)$, $f_3 := \varphi(f_2)$, etc. In particular, we get

$$\begin{array}{lclcl} f_1 & = & \varphi(0) & = & 1, \\ f_2 & = & \varphi(1) & = & 10, \\ f_3 & = & \varphi(10) = \varphi(1)\varphi(0) & = & 101, \\ f_4 & = & \varphi(101) = \varphi(10)\varphi(1) & = & 10110, \\ f_5 & = & \varphi(10110) = \varphi(101)\varphi(10) & = & 10110101, \\ f_6 & = & \varphi(10110101) = \varphi(10110)\varphi(101) & = & 1011010110110, \\ & \vdots & & & \end{array}$$

This leads to the list of *finite Fibonacci words* $f_1, f_2, \dots$. It is easy to see (by induction) that

$$f_{n+1} = f_n f_{n-1} \tag{6}$$

holds for $n \geq 2$, so that the length of f_n is the n-th Fibonacci number; let us denote it by F_n.

The pointwise limit of this sequence (f_n) is the one-sided infinite Fibonacci word $v_+ = (v_n)_{n\in\mathbb{N}}$ with each v_n from (2). More precisely, equip Σ with the discrete topology, $\Sigma^{\mathbb{N}}$ with the product topology and extend each f_n (by anything) to the right to a word in $\Sigma^{\mathbb{N}}$; then (f_n) converges, by (6), and the limit is $v_+ = 1011010110110101101011011010110110\dots$.

3.3. The rotation formula and symmetry

The above mechanisms define the positive half $v_+ = v_1v_2\dots$ of the two-sided infinite Fibonacci word $v = (v_n)_{n\in\mathbb{Z}}$. The missing entries $\dots, v_{-2}, v_{-1}, v_0$ can, of course, be computed from the "rotation formula" (2) but they can also be expressed in terms of v_+: for $n \in \mathbb{Z}$, put

$$t_n := n\alpha \mod 1 \in [0,1), \quad \text{where } \alpha = \tfrac{\sqrt{5}-1}{2} \text{ is the golden ratio,}$$

so that $v_n = \chi_{[1-\alpha,1)}(t_n)$ by (2). For arithmetics modulo 1 it is of course useful to think of the interval $[0,1)$ as a circle with $0 \cong 1$.

Because $t_{-1} = 1-\alpha$ and $t_0 = 0 \cong 1$ exactly mark the two endpoints of the interval $[1-\alpha, 1)$ and the sequences $(t_n)_{n\le -1}$ and $(t_n)_{n\ge 0}$ evolve from there, equispaced in opposite directions along our circle, one observes the symmetry $v_{-2} = v_1$, $v_{-3} = v_2$, $\dots$, in short:

$$v = v_+^R 10 v_+, \tag{7}$$

where the 10 in the middle refers to $v_{-1}v_0$.

Note that, by the irrationality of α, all t_n are pairwise distinct. So the asymmetry that is caused by the different brackets of the interval $[1-\alpha, 1)$ only shows for $n = -1$ and $n = 0$, where t_n exactly hits the two interval endpoints. For $n \in \mathbb{Z} \setminus \{-1, 0\}$, one has $v_n = v_{-1-n}$.

3.4. Subwords of length n

Another intriguing feature of the Fibonacci word is its small number of subwords.

Let $\Sigma = \{0,1\}$. A random word $u \in \Sigma^{\mathbb{Z}}$ would, almost surely, contain every one of the 2^n words $w \in \Sigma^n$ as a subword, for every $n \in \mathbb{N}$. For the Fibonacci word $v \in \Sigma^{\mathbb{Z}}$, the situation is very different:

length	subwords of v of that length	count
1	0, 1	2
2	01, 10, 11	3
3	010, 011, 101, 110	4
4	0101, 0110, 1010, 1011, 1101	5
$\vdots$		$\vdots$
n	$\dots$	$n+1$

(8)

The number, say $\mathrm{sub}_v(n)$, of subwords of v of any length $n \in \mathbb{N}$ is exactly $n+1$.

For general words $u \in \Sigma^{\mathbb{Z}}$, it is easy to see that sub_u grows monotonically, and if $\mathrm{sub}_u(n) = \mathrm{sub}_u(n+1)$ for some n then $\mathrm{sub}_u(m)$ will remain at that value, say p, for all $m \ge n$. The latter says that v is p-periodic (up to a finite perturbation).

So for an aperiodic word u, the function sub_u grows strictly monotonically (by at least 1 for each n), starting from $\mathrm{sub}_u(1) = |\Sigma| = 2$. So the subword count function with minimal growth (among the unbounded functions) is given by $\mathrm{sub}_u(n) = n+1$. This is exactly what is observed for the Fibonacci word $u = v$.

4. Finite sections of the Fibonacci Hamiltonian

Let $v = (v_n)_{n\in\mathbb{Z}}$ be the Fibonacci sequence (2) and let

$$A := S_{-1} + M_v + S_1 \;:\; \ell^p(\mathbb{Z}) \to \ell^p(\mathbb{Z}) \tag{9}$$

be the Fibonacci Hamiltonian (1), where

$$S_k : \ell^p(\mathbb{Z}) \to \ell^p(\mathbb{Z}), \quad (S_k x)_{n+k} = x_n, \quad n \in \mathbb{Z}$$

denotes the shift by $k \in \mathbb{Z}$ components and

$$M_b : \ell^p(\mathbb{Z}) \to \ell^p(\mathbb{Z}), \quad (M_b x)_n = b_n x_n, \quad n \in \mathbb{N}$$

denotes the operator of pointwise multiplication by $b = (b_n)_{n\in\mathbb{Z}} \in \ell^\infty(\mathbb{Z})$.

We identify A with its two-sided infinite matrix $(a_{ij})_{i,j\in\mathbb{Z}}$ with $a_{nn} = v_n$ and $a_{n,n\pm1} = 1$ for all $n \in \mathbb{Z}$ and zeros everywhere else. Corollary 2.5 connects the FSM of A with the limit operators of A. So we need to get a hand on these limit operators.

4.1. Limit operators of the Fibonacci Hamiltonian

Let $h = (h_1, h_2, \ldots)$ be a sequence in $\mathbb{Z}$ with $h_k \to \pm\infty$, so that the limit operator A_h of the Fibonacci Hamiltonian A from (9) exists. Then

$$A_h \;=\; (S_{-1})_h + (M_v)_h + (S_1)_h \;=:\; S_{-1} + M_{v_h} + S_1$$

with a new potential

$$v_h \;:=\; \lim_{k\to\infty} S_{-h_k} v,$$

where the limit is taken w.r.t. pointwise convergence on $\Sigma^{\mathbb{Z}}$ for $\Sigma = \{0, 1\}$.

The set $\mathcal{F}$ of all such potentials v_h is translation invariant (translations of limit operators of A are limit operators of translations of A) and closed under pointwise convergence. $\mathcal{F}$ is the so-called *Fibonacci subshift*. By our definition of $\mathcal{F}$,

$$\mathrm{Lim}(A) = \{S_{-1} + M_{v_h} + S_1 \;:\; v_h \in \mathcal{F}\}.$$

The set $\mathcal{F}$ is explicitly known (see, e.g., Theorem 2.14 in [8] and the appendix of [11]):

$$\mathcal{F} = \{v^\theta, w^\theta : \theta \in [0, 1)\}, \tag{10}$$

where

$$v_n^\theta := \chi_{[1-\alpha,1)}(\theta + n\alpha \bmod 1), \quad w_n^\theta := \chi_{(1-\alpha,1]}(\theta + n\alpha \bmod 1), \quad n \in \mathbb{Z}.$$

In particular,

$$A \in \mathrm{Lim}(A), \tag{11}$$

since $v = v^0 \in \mathcal{F}$. In fact, we do not need this explicit description (10) of $\mathcal{F}$. The following lemma is sufficient (and much more handy) for us. It expresses the well-known minimality of the Fibonacci subshift.

Lemma 4.1. *Every $v_h \in \mathcal{F}$ has the same list* (8) *of subwords as v. So for every $w \in \Sigma^*$ and every $v_h \in \mathcal{F}$ it holds that*

$$w \prec v \quad\Longleftrightarrow\quad w \prec v_h.$$

Proof. Take arbitrary $w \in \Sigma^*$ and $v_h \in \mathcal{F}$. So there exists a sequence $h = (h_1, h_2, \dots)$ in $\mathbb{Z}$ with $h_k \to \pm\infty$ and $v_h = \lim_{k\to\infty} S_{-h_k} v$, pointwise.

$\boxed{\Leftarrow}$ If $w \prec v_h$ then $w \prec S_{-h_k} v$ for large k (strict topology on Σ), so that $w \prec v$.

$\boxed{\Rightarrow}$ Now let $w \prec v$. W.l.o.g. assume $w \prec v_+$. Choose $n \in \mathbb{N}$ so that w appears in the first F_n letters of v_+, i.e., $w \prec f_n \prec f_{n+1}$ (recall the notations from §3.2).

By (6), we have $f_{n+2} = f_{n+1} f_n$ and $f_{n+3} = f_{n+2} f_{n+1} = f_{n+1} f_n f_{n+1}$. By induction, every f_m with $m \ge n$, and hence v_+, is composed of f_n and f_{n+1}. Since w appears as a subword in f_n and f_{n+1}, it appears infinitely often in v_+, where two appearances of w are at most $|f_{n+1}| = F_{n+1}$ letters away from each other. So every translate $S_{-h_k} v$ of v contains w in an F_{n+1}-neighborhood of zero. Hence, every limit $v_h \in \mathcal{F}$ contains w (in an F_{n+1}-neighborhood of zero). □

4.2. Main results

Now we are ready to state and prove our two main results.

Theorem 4.2. *The FSM of the two-sided infinite Fibonacci Hamiltonian* (9) *is stable for any choice of cut-off points and in every space $\ell^p(\mathbb{Z})$ with $p \in [1, \infty]$.*

The compression $A_{\mathbb{N}}$ of A from (9) to $\ell^p(\mathbb{N})$ is called *one-sided infinite Fibonacci Hamiltonian.* Its matrix $(a_{ij})_{i,j\in\mathbb{N}}$ is the submatrix of A consisting of all rows and columns with $i, j \in \mathbb{N}$.

Theorem 4.3. *The FSM of the one-sided infinite Fibonacci Hamiltonian $A_{\mathbb{N}}$ is stable for any choice of cut-off points and in every space $\ell^p(\mathbb{N})$ with $p \in [1, \infty]$.*

The rest of this paper is devoted to the proof of these two theorems. The main ingredient, besides Corollary 2.5 and Lemma 2.6, is the following lemma.

Lemma 4.4. *For the Fibonacci Hamiltonian A from* (9)*, the following statements hold:*

a) *All $B \in \mathrm{Lim}(A)$ are injective on $\ell^\infty(\mathbb{Z})$.*
b) *For all $B \in \mathrm{Lim}_+(A)$, the compression B_- is injective on $\ell^\infty(\mathbb{Z}_-)$.*
c) *For all $B \in \mathrm{Lim}_-(A)$, the compression B_+ is injective on $\ell^\infty(\mathbb{Z}_+)$.*

Here we use the notations $B_\pm$ and $\mathbb{Z}_\pm$ from Definition 2.3 b). We now show how this lemma implies Theorems 4.2 and 4.3 before we come to its proof (in Section 4.3).

Proof of Theorem 4.2. Let $p \in [1, \infty]$. By Corollary 2.5, we have to show that

1) A is invertible on $\ell^p(\mathbb{Z})$,
2) for all $B \in \mathrm{Lim}_+(A)$, the compression B_- is invertible on $\ell^p(\mathbb{Z}_-)$, and
3) for all $B \in \mathrm{Lim}_-(A)$, the compression B_+ is invertible on $\ell^p(\mathbb{Z}_+)$.

It is sufficient to study the case $p = 2$ as A and all B_+ and B_- are band matrices, and so their invertibility is independent of $p \in [1, \infty]$ (see e.g. [16, §5.2.7]).

Property **a)** of Lemma 4.4 implies the invertibility of all $B \in \mathrm{Lim}(A)$, by Lemma 2.6. Since $A \in \mathrm{Lim}(A)$, by (11), also $B = A$ is invertible. So **1)** is shown.

To show **2)**, take an arbitrary $B \in \mathrm{Lim}_+(A)$ and look at B_- as an operator on $\ell^2(\mathbb{Z}_-)$. Since B_- is injective on $\ell^\infty(\mathbb{Z}_-)$, by Lemma 4.4 **b)**, it is also injective on the subset $\ell^2(\mathbb{Z}_-)$ of $\ell^\infty(\mathbb{Z}_-)$. Its adjoint is also injective on $\ell^2(\mathbb{Z}_-)$ since B_- is self-adjoint (by $A = A^*$). So it remains to show that the range of B_- is closed: From **1)** it follows that A is Fredholm. By Lemma 2.6, B is invertible, hence Fredholm. By Lemma 2.6 again, all operators in $\mathrm{Lim}(B) \supset \mathrm{Lim}(B_-)$ are invertible, whence also B_- is Fredholm (by Lemma 2.6 again) and hence has a closed range.

3) follows from Lemma 4.4 **c)** and Lemma 2.6 in the very same way. □

Proof of Theorem 4.3. This time we have to show that

4) $A_\mathbb{N}$ is invertible on $\ell^p(\mathbb{Z})$,

5) for all $B \in \mathrm{Lim}_+(A_\mathbb{N})$, the compression B_- is invertible on $\ell^p(\mathbb{Z}_-)$.

Statement **4)** follows from **3)** because $A_\mathbb{N} = B_+$ for $B = S_{-1}AS_1 \in \mathrm{Lim}_-(A)$. Statement **5)** follows from **2)** because $\mathrm{Lim}_+(A_\mathbb{N}) = \mathrm{Lim}_+(A)$, by the construction of $A_\mathbb{N}$. □

Let us point out that the presence of (iii) in Lemma 2.6 is vital here. With only (i) and (ii) at hand, we would be stuck in a vicious circle. The study of the invertibility of A can be reduced to the following, presumably easier, problems: injectivity of A, injectivity of A^*, Fredholmness of A. The latter again splits into many, presumably easier, problems: invertibility of all limit operators B of A, by Lemma 2.6 (ii). But now A is one of those operators B, by (11), which brings us back to the original problem! So it is good to have – and use – Lemma 2.6 (iii) instead of (ii) here.

Now all that remains to be done is the proof of Lemma 4.4.

4.3. Proof of Lemma 4.4

First notice that one can restrict consideration to real sequences in both the one- and two-sided infinite case. Since B (and the compressions B_+ and B_-) correspond to real matrices, it holds

$$Bx = 0 \iff 0 = \mathrm{Re}(Bx) = B(\mathrm{Re}(x)) \text{ and } 0 = \mathrm{Im}(Bx) = B(\mathrm{Im}(x))$$

with $\mathrm{Re}(\cdot)$ and $\mathrm{Im}(\cdot)$ denoting the real and imaginary part of a sequence. So the injectivity of B on the space of real bounded sequences implies the injectivity on the space $\ell^\infty(\mathbb{I})$ of complex bounded sequences. One is left with proving $Bx = 0 \Rightarrow x = 0$ for all bounded real sequences. The idea is most transparent in the one-sided infinite case. So let us start with the proof of **c)**.

To show that an operator B_+ is injective on $\ell^\infty(\mathbb{Z}_+)$, derive the entries $x_1, x_2, \ldots$ of a solution $x = (x_n)_{n\in\mathbb{Z}_+}$ of the homogeneous equation $B_+x = 0$, starting from a nonzero initial entry x_0, and prove that some entry x_n will eventually exceed (in modulus) any previously given bound $r > 0$. Because, for every $r > 0$, this computation will only take finitely many steps $x_1, \ldots, x_n$, it is enough to know about finite subwords of the potential of B_+. (Our proof does not use the explicit formula (10).)

Identify B_+ with its matrix $(b_{ij})_{i,j\in\mathbb{Z}_+}$. Because of the tridiagonal structure, the value of x_0 is sufficient to calculate the whole solution vector x. More precisely, $x_1 = -b_{00}\, x_0$ and $x_{n+1} = -b_{nn}\, x_n - x_{n-1}$ for $n \in \mathbb{N}$. As usual, rewrite this recurrence with transfer matrices:

$$T_{b_{nn}} \begin{pmatrix} x_{n-1} \\ x_n \end{pmatrix} = \begin{pmatrix} x_n \\ x_{n+1} \end{pmatrix}, \text{ where } T_{b_{nn}} = \begin{pmatrix} 0 & 1 \\ -1 & -b_{nn} \end{pmatrix} \text{ with } b_{nn} \in \{0,1\}.$$

W.l.o.g. we can assume $x_0 = 1$. Here is an example computation for a certain diagonal (b_{nn}):

n	0	1	2	3	4	5	6	7	8	9	10	11	12	13	14
b_{nn}	1	0	1	1	0	1	0	1	1	0	1	1	0	1	0
x_n	1	−1	−1	2	−1	−2	3	2	−5	3	5	−8	3	8	−11

In this example, with a bit of optimism, we seem to observe that

- the diagonal (b_{nn}) of B is composed of blocks "101" and "01", and
- the entries x_n at the beginning of each block grow unboundedly in modulus.

We will prove that this is always the case. The following lemma is a special case of a partition of general Sturmian words as introduced in [17] (also see [10, 18]).

Lemma 4.5. *The diagonal* $b := (b_{nn})_{n\in\mathbb{Z}_+}$ *of* B_+ *with* $B \in \mathrm{Lim}_-(A)$ *is of the form*

$$b = p\, w_1\, w_2\, w_3 \dots \quad \text{with } p \in \{\varepsilon, 1\} \text{ and } w_i \in \{101, 01\} \text{ for all } i \in \mathbb{N}.$$

Proof. By Lemma 4.1 and (8), b contains neither 00 nor 111 as a subword. So 0 is always followed by 1, and 1 is always followed by 101 or 01. □

So we are particularly interested in the patterns "101" and "01" and their corresponding transfer matrices

$$T_{101} := T_1 T_0 T_1 = \begin{pmatrix} 0 & -1 \\ 1 & 2 \end{pmatrix} \quad \text{and} \quad T_{01} := T_1 T_0 = \begin{pmatrix} -1 & 0 \\ 1 & -1 \end{pmatrix}. \tag{12}$$

Let us say that a vector $\binom{y_1}{y_2} \in \mathbb{R}^2$ has property **C** if $y_1 \cdot y_2 < 0$ and $|y_1| < |y_2|$.

Lemma 4.6. *Both* T_{101} *and* T_{01} *preserve property* **C**. *More precisely, if* $\binom{y_1}{y_2}$ *in* $\mathbb{R}^2$ *has property* **C** *then* $\binom{z_1}{z_2} := T_w \binom{y_1}{y_2}$ *with* $w \in \{101, 01\}$ *has properties*

A) $|z_2| > |y_2|$ *with* $|z_2| - |y_2| \ge \min\{|y_1 + y_2|, |y_1|\} > 0$,
B) $|z_1 + z_2| \ge |y_1 + y_2|$ *and* $|z_1| \ge |y_1|$, *and*
C) $z_1 \cdot z_2 < 0$ *and* $|z_1| < |z_2|$.

Proof. This is a straightforward computation using

$$T_{101} \begin{pmatrix} y_1 \\ y_2 \end{pmatrix} = \begin{pmatrix} -y_2 \\ y_2 + (y_1 + y_2) \end{pmatrix} \quad \text{and} \quad T_{01} \begin{pmatrix} y_1 \\ y_2 \end{pmatrix} = \begin{pmatrix} -y_1 \\ y_1 - y_2 \end{pmatrix}. \quad □$$

Property **A** shows a growth (in modulus) of the second vector component after applying T_{101} or T_{01}. By property **B**, the amount of growth is non-decreasing when applying T_{101} or T_{01} again. The fact that property **C** is preserved keeps the argument working for the next application of T_{101} or T_{01}, leading to unbounded growth of the second vector component, by induction.

So all what we need is one first occurrence of property **C** for a vector $\binom{x_n}{x_{n+1}}$ in our computation of a sequence $x = (x_0, x_1, x_2, \dots)$ that solves $B_+ x = 0$.

We start with the case $p = \varepsilon$. Besides $x_0 = 1$ (see above), we put $x_{-1} := 0$ to start our recurrence and account for the non-existence of column number -1 in the matrix B_+. Depending on which of T_{101} and T_{01} from (12) we apply to $\binom{x_{-1}}{x_0} = \binom{0}{1}$, we get

$$T_{01}\begin{pmatrix}0\\1\end{pmatrix} = \begin{pmatrix}0\\-1\end{pmatrix} \quad \text{or} \quad T_{101}\begin{pmatrix}0\\1\end{pmatrix} = \begin{pmatrix}-1\\2\end{pmatrix}.$$

So repeated application of T_{01} leads to $\pm\binom{0}{1}$ but after the first application of T_{101}, which will eventually happen since b is not periodic, one gets $\binom{x_n}{x_{n+1}} = \pm\binom{-1}{2}$ for some $n \in \mathbb{N}$. This vector has property **C**. From our arguments above it follows that the sequence x is unbounded.

If the prefix p of b is 1, it follows from $x_0 = 1$ that $x_1 = -1$, so that our recurrence starts with $\binom{x_0}{x_1} = \binom{1}{-1}$. The application of the transfer matrices from (12) yields

$$T_{01}\begin{pmatrix}1\\-1\end{pmatrix} = \begin{pmatrix}-1\\2\end{pmatrix} \quad \text{and} \quad T_{101}\begin{pmatrix}1\\-1\end{pmatrix} = \begin{pmatrix}1\\-1\end{pmatrix}.$$

So repeated application of T_{101} leads to $\binom{1}{-1}$ but after the first application of T_{01}, which will eventually happen since b is not periodic, one gets $\binom{x_n}{x_{n+1}} = \binom{-1}{2}$ for some $n \in \mathbb{N}$. This vector has property **C**. From our arguments above it follows that the sequence x is unbounded.

For both possibilities of the prefix $b \in \{\varepsilon, 1\}$ and all possibilities of the following blocks $w_i \in \{101, 01\}$ (recall Lemma 4.5), we have shown that all nontrivial solutions x of the homogenous system $B_+ x = 0$ are unbounded, so that every B_+ with $B \in \mathrm{Lim}_-(A)$ is injective on $\ell^\infty(\mathbb{Z}_+)$. Thus, **c)** is proved.

To see **b)**, consider the three flip operators

$$J_\leftarrow : \ell^\infty(\mathbb{Z}_+) \to \ell^\infty(\mathbb{Z}_-), \; J_\rightarrow : \ell^\infty(\mathbb{Z}_-) \to \ell^\infty(\mathbb{Z}_+), \; \text{and} \; J_\leftrightarrow : \ell(\mathbb{Z}) \to \ell(\mathbb{Z}),$$

all three acting by the rule $x \mapsto y$ with $y_n = x_{-n}$ for n in $\mathbb{Z}_-$, $\mathbb{Z}_+$ and $\mathbb{Z}$, respectively.

The formula $v = v_+^R 10 v_+$ from (7) implies that $\mathrm{Lim}_+(A)$ exactly consists of the reflections $C = J_\leftrightarrow B J_\leftrightarrow$ of operators $B \in \mathrm{Lim}_-(A)$, so that

$$\{C_- \;:\; C \in \mathrm{Lim}_+(A)\} \;=\; \{(\underbrace{J_\leftrightarrow B J_\leftrightarrow}_{C})_+ = J_\leftarrow B_+ J_\rightarrow \;:\; B \in \mathrm{Lim}_-(A)\}.$$

So, clearly, **b)** follows from **c)**.

Finally, we are going to prove **a)**. To do this, we will use a two-sided version of the proof of **c)**.

Let $B \in \mathrm{Lim}(A)$ and again let $b \in \{0,1\}^{\mathbb{Z}}$ be the diagonal of B. Let x be a nontrivial solution of the homogeneous equation $Bx = 0$. We will prove that the sequence x grows unboundedly in at least one direction, left or right. The growth to the right is studied as in the proof of **c)** above – growth to the left by symmetric arguments. Here is the analogue of Lemma 4.5.

Lemma 4.7. *The diagonal $b := (b_{nn})_{n\in\mathbb{Z}}$ of $B \in \mathrm{Lim}(A)$ is of the form*

$$b = \cdots w_{-4}\, w_{-3}\, \underbrace{101}_{w_{-2}}\,\underbrace{101}_{w_{-1}}\,\underbrace{01}_{w_0}\,\underbrace{101}_{w_1}\,\underbrace{101}_{w_2}\, w_3\, w_4 \cdots \tag{13}$$

with $w_{-i} \in \{101, 10\}$ and $w_i \in \{101, 01\}$ for all $i \in \mathbb{N}$.

Proof. The word 101 101 01 101 101 is contained in the Fibonacci word v (as $v_{-6}\cdots v_7$) and therefore, by Lemma 4.1, also in b. By Lemma 4.1 and (8), b contains neither 00 nor 111 as a subword. So, as argued in Lemma 4.5, 0 is always followed by 1, and 1 is always followed by 101 or 01. Moreover, 0 is always preceded by 1, and 1 is always preceded by 101 or 10. □

So besides T_{101} and T_{01}, we now also have to look at the transfer matrix $T_{10} := T_0T_1$. Note that, when we study the asymptotics of x towards $-\infty$ (going backward in "time"), we will have to look at inverses of the transfer matrices.

Therefore, let us say that a vector $\binom{y_1}{y_2} \in \mathbb{R}^2$ has property **F** if $y_1 \cdot y_2 < 0$ and $|y_1| > |y_2|$. Here is the "leftward" analogue of Lemma 4.6.

Lemma 4.8. *Both T_{101}^{-1} and T_{10}^{-1} preserve property* **F**. *More precisely, if $\binom{y_1}{y_2}$ in $\mathbb{R}^2$ has property* **F** *then $\binom{z_1}{z_2} := T_w^{-1}\binom{y_1}{y_2}$ with $w \in \{101, 10\}$ has properties*

D) $|z_1| > |y_1|$ *with* $|z_1| - |y_1| \geq \min\{|y_1 + y_2|, |y_2|\} > 0$,
E) $|z_1 + z_2| \geq |y_1 + y_2|$ *and* $|z_2| \geq |y_2|$, *and*
F) $z_1 \cdot z_2 < 0$ *and* $|z_1| > |z_2|$.

Proof. This is again a straightforward computation using

$$T_{101}^{-1}\begin{pmatrix} y_1 \\ y_2 \end{pmatrix} = \begin{pmatrix} y_1 + y_2 + y_1 \\ -y_1 \end{pmatrix} \quad \text{and} \quad T_{10}^{-1}\begin{pmatrix} y_1 \\ y_2 \end{pmatrix} = \begin{pmatrix} -y_1 + y_2 \\ -y_2 \end{pmatrix}. \qquad \square$$

As before, property **D** states a growth (in modulus) of the first component. Property **E** ensures that the amount of this growth is non-decreasing in further applications of T_w^{-1} with $w \in \{101, 10\}$, and the fact that property **F** is preserved makes sure that the same argument keeps working for further applications of T_w^{-1}, leading to unbounded growth.

So again, we just need a first occurrence of property **F** for a vector $\binom{x_n}{x_{n+1}}$ with $n < 0$ or a first occurrence of property **C** for a vector $\binom{x_n}{x_{n+1}}$ with $n \geq 0$ in our computation of a sequence $x = (\ldots, x_{-2}, x_{-1}, x_0, x_1, x_2, \ldots)$ that solves $Bx = 0$. Then x will be unbounded.

This time, one entry, say x_0, does not determine the whole sequence x, but two entries do. Let the two entries of x that are associated to the entries 0 and 1 of w_0 in (13) be equal to α and β, respectively, with arbitrary $\alpha, \beta \in \mathbb{R}$. W.l.o.g label them as x_0 and x_1.

Using the adjacent entries of $w_0 = 01$ in b, see (13), the corresponding entries $x_{-4}, \dots, x_5$ in x turn out to be as follows:

x_{-4}	x_{-3}	x_{-2}	x_{-1}	x_0	x_1	x_2	x_3	x_4	x_5
$\alpha - 2\beta$	β	$-\alpha + \beta$	$-\beta$	α	β	$-\alpha - \beta$	α	$\alpha + \beta$	$-2\alpha - \beta$

With respect to α and β, we have to distinguish the following cases:

1. If $\alpha = \beta = 0$ then $x = 0$ follows.
2. If $\alpha = 0$ and $\beta \neq 0$ then $\binom{x_{-4}}{x_{-3}} = \binom{-2\beta}{\beta}$ has property **F**.
3. If $\alpha \neq 0$ and $\beta = 0$ then $\binom{x_4}{x_5} = \binom{\alpha}{-2\alpha}$ has property **C**.
4. If $\alpha \neq 0$ and $\beta \neq 0$ we have to look at two more cases:
   (a) If $\alpha \cdot \beta > 0$ then $\binom{x_1}{x_2} = \binom{\beta}{-\alpha-\beta}$ has property **C**.
   (b) If $\alpha \cdot \beta < 0$ then $\binom{x_{-4}}{x_{-3}} = \binom{\alpha-2\beta}{\beta}$ has property **F**.

This completes the study of all cases. Each nontrivial solution of the homogenous equation $Bx = 0$ is unbounded, thus B is injective on $\ell^\infty(\mathbb{Z})$. This completes the proof of **a)** and thus of Lemma 4.4.

Acknowledgements. The first author thanks Albrecht Böttcher, Daniel Potts, Peter Stollmann and David Wenzel for organizing a wonderful IWOTA conference 2017 in Chemnitz with many inspiring talks and countless other pleasant moments.

References

[1] M. Baake and U. Grimm, *Aperiodic Order. Vol. 1. A Mathematical Invitation*, Encyclopedia of Mathematics and its Applications **149**, Cambridge University Press, Cambridge, 2013.

[2] G. Baxter, A norm inequality for a 'finite-section' Wiener-Hopf equation, *Illinois J. Math.* **7** (1963), 97–103.

[3] A. Böttcher and S.M. Grudsky, *Spectral Properties of Banded Toeplitz Matrices*, SIAM, Philadelphia, 2005.

[4] N.G. de Bruijn, Algebraic theory of Penrose's non-periodic tilings of the plane. I, *Nederl. Akad. Wetensch. Indag. Math.* **43** (1981), 39–52.

[5] S.N. Chandler-Wilde and M. Lindner, Sufficiency of Favard's condition for a class of band-dominated operators on the axis, *J. Funct. Anal.* **254** (2008), 1146–1159.

[6] S.N. Chandler-Wilde and M. Lindner, *Limit Operators, Collective Compactness, and the Spectral Theory of Infinite Matrices*, Memoirs of the AMS **210**, no. 989, Amer. Math. Soc., Providence, 2011.

[7] S.N. Chandler-Wilde and M. Lindner, Coburn's lemma and the finite section method for random Jacobi operators, *J. Funct. Anal.* **270** (2016), 802–841.

[8] D. Damanik, Strictly ergodic subshifts and associated operators, *Proceedings of Symposia in Pure Math. (Festschrift for Barry Simon's 60th birthday)*, 505–538, Amer. Math. Soc., Prividence, 2007.

[9] D. Damanik, M. Embree and A. Gorodetski, Spectral properties of Schrödinger operators arising in the study of quasicrystals, *Mathematics of Aperiodic Order*, Progress in Mathematics **309**, Birkhäuser, 2015, 307–370.

[10] D. Damanik and D. Lenz, Uniform spectral properties of one-dimensional quasicrystals. I. Absence of eigenvalues, *Comm. Math. Phys.* **207** (1999), 687–696.

[11] D. Damanik and D. Lenz, Half-line eigenfunction estimates and purely singular continuous spectrum of zero Lebesgue measure, *Forum Math.* **16** (2004), 109–128.

[12] I. Gohberg and I.A. Feldman, *Convolution Equations and Projection Methods for their Solution*, Transl. of Math. Monographs, **41**, AMS, Providence, R.I., 1974 [Russian original: Nauka, Moscow, 1971].

[13] J. Kellendonk, D. Lenz and J. Savinien (eds.), *Mathematics of Aperiodic Order*, Progress in Mathematics **309**, Birkhäuser, Basel, 2015.

[14] A.V. Kozak, A local principle in the theory of projection methods (Russian), *Dokl. Akad. Nauk SSSR* **212** (1973), 1287–1289; English transl. *Soviet Math. Dokl.* **14** (1974), 1580–1583.

[15] A.V. Kozak and I.V. Simonenko, Projection methods for solving multidimensional discrete convolution equations (Russian), *Sib. Mat. Zh.* **21** (1980), 119–127.

[16] V.G. Kurbatov, *Functional Differential Operators and Equations*, Kluwer Academic Publishers, Dordrecht, Boston, London, 1999.

[17] D. Lenz, *Aperiodische Ordnung und gleichmäßige spektrale Eigenschaften von Quasikristallen*, PhD thesis, Frankfurt, Logos Verlag, Berlin, 2000.

[18] D. Lenz, Hierarchical structures in Sturmian dynamical systems, *Theoret. Comput. Sci.* **303** (2003), 463–490.

[19] M. Lindner, *Infinite Matrices and their Finite Sections: An Introduction to the Limit Operator Method*, Frontiers in Mathematics, Birkhäuser, Basel, 2006.

[20] M. Lindner, *Fredholm Theory and Stable Approximation of Band Operators and Generalisations*, Habilitation thesis, TU Chemnitz, 2009.

[21] M. Lindner and M. Seidel, An affirmative answer to a core issue on limit operators, *J. Funct. Anal.* **267** (2014), 901–917.

[22] M. Lothaire, *Algebraic Combinatorics on Words*, Cambridge Univ. Press, Cambridge, 2002.

[23] V.S. Rabinovich, S. Roch and B. Silbermann, Fredholm theory and finite section method for band-dominated operators, *Integral Equations Operator Theory* **30** (1998), 452–495.

[24] V.S. Rabinovich, S. Roch and B. Silbermann, Band-dominated operators with operator-valued coefficients, their Fredholm properties and finite sections, *Integral Equations Operator Theory* **40** (2001), no. 3, 342–381.

[25] V.S. Rabinovich, S. Roch and B. Silbermann, Algebras of approximation sequences: Finite sections of band-dominated operators, *Acta Appl. Math.* **65** (2001), 315–332.

[26] V.S. Rabinovich, S. Roch and B. Silbermann, *Limit Operators and Their Applications in Operator Theory*, Birkhäuser, Basel, Boston, Berlin, 2004.

[27] S. Roch, *Finite Sections of Band-dominated Operators*, Memoirs of the AMS **191**, no. 895, Amer. Math. Soc., Providence, 2008.

[28] M. Seidel, *On Some Banach Algebra Tools in Operator Theory*, PhD thesis, TU Chemnitz, 2012.

[29] P.J. Steinhardt, Quasicrystals: a brief history of the impossible, *Rend. Fis. Acc. Lincei* **24** (2013), 85–91.

[30] P. Stollmann, *Caught by Disorder. Bound states in random media,* Progress in Mathematical Physics **20**, Birkhäuser, Boston, 2001.

Marko Lindner
Hamburg University of Technology (TUHH)
Institute of Mathematics
21073 Hamburg
Germany
e-mail: `lindner@tuhh.de`

Hagen Söding
Student of Technomathematics
at Hamburg University and TUHH
e-mail: `hagen.soeding@studium.uni-hamburg.de`

Spectral asymptotics for Toeplitz operators and an application to banded matrices

Alexander Pushnitski

Abstract. We consider a class of compact Toeplitz operators on the Bergman space on the unit disc. The symbols of the operators in our class are assumed to have a sufficiently regular power-like behaviour near the boundary of the disc. We compute the asymptotics of the singular values of Toeplitz operators in this class. We use this result to obtain the asymptotics of the singular values for a class of compact banded matrices.

Mathematics Subject Classification (2010). Primary 47B32; Secondary 47B36.

Keywords. Toeplitz operators, Bergman space, banded matrix, spectral asymptotics.

1. Introduction and main results

1.1. Introduction

Let $\mathbb{D}$ be the unit disc in the complex plane, and let $L^2(\mathbb{D})$ be the Hilbert space of all square integrable functions with respect to the normalised Lebesgue area measure on $\mathbb{D}$. Next, let $B^2(\mathbb{D})$ be the Bergman space, i.e., the closure of the linear span of the functions $\{z^n\}_{n=0}^{\infty}$ in $L^2(\mathbb{D})$. Denote by $P : L^2(\mathbb{D}) \to B^2(\mathbb{D})$ the Bergman projection, i.e., the orthogonal projection in L^2 onto B^2. For a *symbol* $\varphi \in L^\infty(\mathbb{D})$, the Toeplitz operator $T(\varphi)$ in $B^2(\mathbb{D})$ is defined by

$$T(\varphi)f = P(\varphi \cdot f), \quad f \in B^2(\mathbb{D}).$$

It is well known that if $\varphi(z) \to 0$ as $|z| \to 1$, then $T(\varphi)$ is compact. Moreover, roughly speaking, the rate of convergence of $\varphi(z) \to 0$ as $|z| \to 1$ determines the rate of convergence of the sequence of singular values $s_n(T(\varphi)) \to 0$ as $n \to \infty$. For radial symbols $\varphi(z) = \varphi(|z|)$ it is very easy to make this statement precise. Indeed, in this case the Toeplitz operator $T(\varphi)$ is diagonal in the standard orthonormal basis $\{\sqrt{n+1}z^n\}_{n=0}^{\infty}$ of $B^2(\mathbb{D})$, and so

© Springer International Publishing AG, part of Springer Nature 2018
A. Böttcher et al. (eds.), *The Diversity and Beauty of Applied Operator Theory*, Operator Theory: Advances and Applications 268, https://doi.org/10.1007/978-3-319-75996-8_21

the sequence of the singular values of $T(\varphi)$ is given by

$$s_n(T(\varphi)) = (n+1)\left|(T(\varphi)z^n, z^n)_{L^2(\mathbb{D})}\right| = 2(n+1)\left|\int_0^1 r^{2n}\varphi(r)rdr\right|.$$

Specialising to the power behaviour $\varphi(r) = (1-r)^\gamma$, $\gamma > 0$, by an elementary calculation one obtains

$$s_n(T(\varphi)) = 2^{-\gamma}\Gamma(\gamma+1)n^{-\gamma} + O(n^{-1-\gamma}), \quad n \to \infty. \tag{1.1}$$

The purpose of this paper is

(*i*) to present a method that allows one to extend this calculation to symbols φ that have some sufficiently regular angular dependence;

(*ii*) to give an application to the spectral analysis of banded matrices.

1.2. Spectral asymptotics for Toeplitz operators

Theorem 1.1. *Let $\varphi \in L^\infty(\mathbb{D})$, and let $\gamma > 0$. Assume that for some continuous function φ_∞ on the unit circle one has*

$$\sup_{0<\theta\leq 2\pi} |(1-r)^{-\gamma}\varphi(re^{i\theta}) - \varphi_\infty(e^{i\theta})| \to 0, \quad r \to 1.$$

Then the singular values of $T(\varphi)$ satisfy

$$s_n(T(\varphi)) = C_\gamma(\varphi_\infty)n^{-\gamma} + o(n^{-\gamma}), \quad n \to \infty,$$

where

$$C_\gamma(\varphi_\infty) = 2^{-\gamma}\Gamma(\gamma+1)\left(\int_0^{2\pi} |\varphi_\infty(e^{i\theta})|^{1/\gamma}\frac{d\theta}{2\pi}\right)^\gamma.$$

As will be clear from the proof, the requirement of the continuity of φ_∞ can be considerably relaxed. For example, any Riemann integrable φ_∞ is admissible.

If φ is real-valued, then it is evident that $T(\varphi)$ is self-adjoint. In this case, let us denote by $\{\lambda_n^+(T(\varphi))\}_{n=0}^\infty$ the sequence of positive eigenvalues of $T(\varphi)$ and set $a^+ = \max\{a, 0\}$ for $a \in \mathbb{R}$. We have an analogous result for eigenvalues.

Theorem 1.2. *Assume the hypothesis of Theorem* 1.1 *and let φ be real-valued. Then the positive eigenvalues of $T(\varphi)$ satisfy*

$$\lambda_n^+(T(\varphi)) = C_\gamma^+(\varphi_\infty)n^{-\gamma} + o(n^{-\gamma}), \quad n \to \infty, \tag{1.2}$$

where

$$C_\gamma^+(\varphi_\infty) = 2^{-\gamma}\Gamma(\gamma+1)\left(\int_0^{2\pi} \varphi_\infty^+(e^{i\theta})^{1/\gamma}\frac{d\theta}{2\pi}\right)^\gamma.$$

Of course, a similar result holds true for the negative eigenvalues.

By using conformal mapping, it is possible to extend Theorems 1.1 and 1.2 to other domains. However, we prefer to focus on the simplest case, as our aim is to emphasise the method rather than the result.

A result similar to Theorem 1.2 is known in the context of Toeplitz operators on the Fock space. In this case, the symbol $\varphi(z)$ depends on $z \in \mathbb{C}$, and the rate of convergence of the eigenvalues of the corresponding Toeplitz operator to zero depends on the rate of convergence of $\varphi(z) \to 0$ as $|z| \to \infty$.

In [6], symbols that behave as $|z|^{-\gamma}\varphi_\infty(e^{i\theta})$ when $|z|\to\infty$ are considered and a spectral asymptotics of the type (1.2) is proved. The proof is achieved through a reduction to a pseudodifferential operator in $L^2(\mathbb{R})$.

Another closely related recent result is [3], where the authors consider harmonic Toeplitz operators in a bounded domain in $\mathbb{R}^d$ with symbols that have a power decay near the boundary. They obtain asymptotics of eigenvalues very similar to (1.2). The method of proof of [3] is quite different from the one in this paper and consists in a reduction to a pseudodifferential operator on the boundary. The same method can probably be applied to give an alternative proof of Theorems 1.1 and 1.2.

1.3. Application: spectral asymptotics for banded matrices

Let A be the operator on ℓ^2 corresponding to an infinite matrix $\{a_{j,k}\}_{j,k=0}^\infty$ of the following form. Our first assumption is that the operator A is *banded*, i.e., for some $M\in\mathbb{N}$, we have

$$a_{j,j+m}=0 \quad \text{if } |m|>M.$$

Our second assumption is that for each m with $|m|\le M$, the sequence of entries $a_{j,j+m}$ has a power asymptotics as $j\to\infty$. More precisely, we fix an exponent $\gamma>0$ and complex numbers b_m, $m=-M,\dots,M$, and assume that

$$a_{j,j+m}=b_m j^{-\gamma}+o(j^{-\gamma}),\quad j\to\infty,\quad |m|\le M.$$

It is easy to see that under these assumptions the matrix A is compact. It is also not difficult to see that $s_n(A)=O(n^{-\gamma})$. The theorem below gives the asymptotic behaviour of the singular values.

Theorem 1.3. *Under the above assumptions, the singular values of A satisfy*

$$s_n(A)=\left(\int_{-\pi}^{\pi}|b(e^{i\theta})|^{1/\gamma}\frac{d\theta}{2\pi}\right)^{\gamma}n^{-\gamma}+o(n^{-\gamma}),\quad n\to\infty,$$

where

$$b(e^{i\theta})=\sum_{k=-M}^{M}b_k e^{ik\theta},\quad -\pi<\theta\le\pi. \tag{1.3}$$

If A is self-adjoint, then b is necessarily real-valued. In this case one has a similar asymptotic formula for the positive eigenvalues of A.

Theorem 1.4. *Assume the hypothesis of Theorem 1.3 and suppose that A is self-adjoint. Then the positive eigenvalues of A satisfy*

$$\lambda_n^+(A)=\left(\int_{-\pi}^{\pi}b^+(e^{i\theta})^{1/\gamma}\frac{d\theta}{2\pi}\right)^{\gamma}n^{-\gamma}+o(n^{-\gamma}),\quad n\to\infty,$$

where b is given by (1.3) *and* $b^+=\max\{b,0\}$.

Again, a similar result holds for the negative eigenvalues of A.

Example 1.5. *Lömmel polynomials* (see e.g. [7, Section 3] and references therein) are orthogonal polynomials associated with the Jacobi matrix (i.e., the tridiagonal matrix) which in our notation corresponds to $M = 1$, $a_{j,j} = 0$, and

$$a_{j,j+1} = a_{j+1,j} = \frac{1}{2\sqrt{(j+\nu+1)(j+\nu)}}, \quad j \geq 0$$

(here $\nu > 0$ is a parameter). Since the entries on the main diagonal are identically zero, the spectrum of this Jacobi matrix is symmetric with respect to reflection around zero. Further, it is known that its eigenvalues are given by $1/j_{k,\nu-1}$, $k \in \mathbb{Z}$, where $j_{k,\nu-1}$ are the zeros of the Bessel function $J_{\nu-1}$. These zeros have the asymptotic behaviour $j_{k,\nu-1} \sim \pi k$ as $|k| \to \infty$, and so the positive eigenvalues of our Jacobi matrix satisfy

$$\lambda_n^+(A) = \frac{1}{\pi n} + o\Big(\frac{1}{n}\Big), \quad n \to \infty.$$

This agrees with Theorem 1.4, which gives $b(e^{i\theta}) = \cos\theta$, $\gamma = 1$, and

$$\int_{-\pi}^{\pi} b^+(e^{i\theta})\frac{d\theta}{2\pi} = \int_{-\pi/2}^{\pi/2} \cos\theta \frac{d\theta}{2\pi} = \frac{1}{\pi}.$$

Example 1.6. *Tricomi–Carlitz polynomials* (see [7, Section 3]) are orthogonal polynomials associated with the Jacobi matrix with $M = 1$, $a_{j,j} = 0$, and

$$a_{j,j+1} = a_{j+1,j} = \frac{\sqrt{j+1}}{\sqrt{(j+\alpha+1)(j+\alpha)}}, \quad j \geq 0$$

(α is a real parameter). Again, the spectrum of this Jacobi matrix is symmetric with respect to reflection around zero, and it is known that its eigenvalues are given by $\pm 1/\sqrt{n+\alpha}$, $n \geq 0$. This agrees with Theorem 1.4, which gives $b(e^{i\theta}) = 2\cos\theta$, $\gamma = 1/2$, and

$$\left(\int_{-\pi}^{\pi} b^+(e^{i\theta})^2 \frac{d\theta}{2\pi}\right)^{1/2} = \int_{-\pi/2}^{\pi/2} 4(\cos\theta)^2 \frac{d\theta}{2\pi} = 1.$$

1.4. Key ideas of the proof and the structure of the paper

We focus on the case of the singular value asymptotics, i.e., on Theorems 1.1 and 1.3; the eigenvalue asymptotics is considered in a similar fashion. It will be convenient to work with the singular value counting function:

$$n(s;T) = \#\{n : s_n(T) > s\}, \quad s > 0. \tag{1.4}$$

We recall that for a compact operator T, the relations

$$s_n(T) = \frac{C}{n^\gamma} + o(n^{-\gamma}),\ n \to \infty, \ \text{and}\ n(s;T) = \frac{C^{1/\gamma}}{s^{1/\gamma}} + o(s^{-1/\gamma}),\ s \to 0, \tag{1.5}$$

are equivalent.

The main issue is to prove Theorem 1.1 for symbols of the form

$$\varphi(re^{i\theta}) = (1-r)^\gamma \varphi_\infty(e^{i\theta}).$$

For such symbols, we shall write $T_\gamma(\varphi_\infty)$ instead of $T(\varphi)$. By a limiting argument, the problem reduces to replacing the continuous function φ_∞ by step functions that are constant on each arc

$$\delta_\ell = \left\{e^{i\theta} : \tfrac{2\pi\ell}{L} < \theta < \tfrac{2\pi(\ell+1)}{L}\right\},$$

where $L \in \mathbb{N}$ is fixed. Let us denote by $\mathbb{1}_{\delta_\ell}$ the characteristic function of δ_ℓ and write

$$\varphi_\infty = \sum_{\ell=1}^{L} c_\ell \mathbb{1}_{\delta_\ell} \tag{1.6}$$

with some coefficients c_ℓ. Then

$$T_\gamma(\varphi_\infty) = \sum_{\ell=1}^{L} c_\ell T_\gamma(\mathbb{1}_{\delta_\ell}). \tag{1.7}$$

Our main observation is that the terms in the sum (1.7) are *asymptotically orthogonal* in the sense that the products $T_\gamma(\mathbb{1}_{\delta_\ell})T_\gamma(\mathbb{1}_{\delta_m})$, $\ell \neq m$, satisfy certain Schatten class properties. An operator theoretic lemma from [2] (Theorem 2.2) then shows that the leading term coefficient in the spectral asymptotics is *additive* for the sum (1.7). This means that the relation

$$n(s; T_\gamma(\varphi_\infty)) = \sum_{\ell=1}^{L} n(s; c_\ell T_\gamma(\mathbb{1}_{\delta_\ell})) + o(s^{-1/\gamma}), \quad s \to 0,$$

holds true. Using this fact, it is not difficult to derive the required spectral asymptotics for piecewise constant functions φ_∞ of the form (1.6) from the same asymptotics for characteristic functions φ_∞.

We note that this construction is almost purely operator theoretic and uses very little specific features of the problem, apart from rotational symmetry and asymptotic orthogonality. Thus, it can be used in other contexts, for example for Toeplitz operators associated with multi-dimensional domains with spherical symmetry. This method of proof, based on asymptotic orthogonality, was developed in [4, 5] for a different purpose.

Let us explain the key idea of the proof of Theorem 1.3. Consider a symbol of the form

$$\varphi(re^{i\theta}) = (1-r)^\gamma e^{im\theta},$$

where $m \in \mathbb{Z}$. Then the matrix of $T(\varphi)$ in the standard orthonormal basis $\{\sqrt{k+1}z^k\}_{k=0}^\infty$ of the Bergman space is

$$a_{j,k} = \sqrt{j+1}\sqrt{k+1}(\varphi z^j, z^k)_{L^2(\mathbb{D})}.$$

It is easy to see that $a_{j,k} \neq 0$ only if $k = j+m$ and

$$a_{j,j+m} = 2^{-\gamma}\Gamma(\gamma+1)j^{-\gamma} + o(j^{-\gamma})$$

as $j \to \infty$. This corresponds to a matrix A as in Theorem 1.3 with only one non-zero diagonal. Taking a finite linear combination of such matrices gives the case of a general banded matrix. Thus, the proof of Theorem 1.3 reduces to Theorem 1.1.

In Section 2 we recall some background facts concerning singular value and eigenvalue asymptotics for compact operators and state the result from [2] on asymptotically orthogonal operators. In Section 3 we prove Theorems 1.1 and 1.2 and in Section 4 we prove Theorems 1.3 and 1.4.

Acknowledgements

The author is grateful to P. Honoré and M. Reguera for useful discussions at the early stage of this work, to the organisers of IWOTA 2017 for the encouragement, and to G. Raikov and L. Golinskii for valuable remarks on the text of the paper.

2. Operator theoretic tools

Here we collect some general operator theoretic tools related to the singular value and eigenvalue estimates and asymptotics for compact operators.

2.1. Definitions

For a compact operator T in a Hilbert space, we denote by $\{s_n(T)\}_{n=0}^{\infty}$ the non-increasing sequence of singular values of T, enumerated with multiplicities taken into account. Recall that $s_n(T)$ is defined as the n'th eigenvalue of the positive semi-definite self-adjoint operator $|T| = \sqrt{T^*T}$. We will work with the singular value counting function (1.4). For $p > 0$, the standard Schatten class $\mathbf{S}_p$ is defined by the condition

$$\sum_{n=0}^{\infty} s_n(T)^p < \infty.$$

In terms of the counting function $n(s;T)$ this can be equivalently rewritten as

$$\int_0^{\infty} n(s;T)s^{p-1}ds < \infty.$$

The weak Schatten class $\mathbf{S}_{p,\infty}$ is defined by the condition

$$n(s;T) = O(s^{-p}), \quad s \to 0.$$

The quantity

$$\|T\|_{\mathbf{S}_{p,\infty}} = \sup_{s>0} sn(s;T)^{1/p}$$

is a quasi-norm on $\mathbf{S}_{p,\infty}$, and we will be considering $\mathbf{S}_{p,\infty}$ with respect to the topology given by this quasi-norm. The subclass $\mathbf{S}_{p,\infty}^0 \subset \mathbf{S}_{p,\infty}$ is defined by the condition

$$n(s;T) = o(s^{-p}), \quad s \to 0;$$

it can be characterised as the closure of all operators of finite rank in $\mathbf{S}_{p,\infty}$.

In order to work with the singular value asymptotics, it will be convenient to use the following functionals in $\mathbf{S}_{p,\infty}$:

$$\Delta_p(T) := \limsup_{s\to 0} s^p n(s;T), \quad \delta_p(T) := \liminf_{s\to 0} s^p n(s;T). \tag{2.1}$$

In applications, one usually has $\Delta_p(T) = \delta_p(T)$, but it is technically convenient to treat the upper and lower limits separately. The functionals Δ_p and δ_p are continuous in $\mathbf{S}_{p,\infty}$.

We will denote $\mathbf{S}_0 = \cap_{p>0}\mathbf{S}_p$. In other words, $\mathbf{S}_0$ consists of compact operators T such that for all $C > 0$, one has

$$s_n(T) = O(n^{-C}), \quad n \to \infty.$$

2.2. Additive and multiplicative estimates

Below we recall some estimates for singular values of sums and products of compact operators.

The following fundamental result is known as the Ky Fan lemma; see e.g. [1].

Lemma 2.1. *Let $A \in \mathbf{S}_{p,\infty}$ and $B \in \mathbf{S}^0_{p,\infty}$ for some $p > 0$. Then*

$$\Delta_p(A + B) = \Delta_p(A), \quad \delta_p(A + B) = \delta_p(A).$$

In Section 3 we will also need more advanced information about the quantities Δ_p and δ_p. One has the following additive esimates, see e.g. [1, formulas (11.6.12), (11.6.14), (11.6.15)]:

$$\Delta_p(A_1 + A_2)^{1/(p+1)} \leq \Delta_p(A_1)^{1/(p+1)} + \Delta_p(A_2)^{1/(p+1)}, \quad (2.2)$$

$$|\Delta_p(A_1)^{1/(p+1)} - \Delta_p(A_2)^{1/(p+1)}| \leq (\Delta_p(A_1 - A_2))^{1/(p+1)}, \quad (2.3)$$

$$|\delta_p(A_1)^{1/(p+1)} - \delta_p(A_2)^{1/(p+1)}| \leq (\Delta_p(A_1 - A_2))^{1/(p+1)}. \quad (2.4)$$

We will also need some multiplicative estimates. One has (see [1, (11.1.19), (11.1.12)])

$$n(s_1 s_2; A_1 A_2) \leq n(s_1; A_1) + n(s_2; A_2), \quad s_1 > 0, s_2 > 0, \quad (2.5)$$

$$n(s; A_1 A_2) \leq n(s; \|A_1\| A_2), \quad s > 0. \quad (2.6)$$

From (2.5) it is not difficult to obtain (see [1, (11.6.18)]) the bound

$$\Delta_{p/2}(A_1 A_2) \leq 2\Delta_p(A_1)\Delta_p(A_2). \quad (2.7)$$

2.3. Asymptotically orthogonal operators

The theorem below is the key operator theoretic ingredient of our construction. It has first appeared (under slightly more restrictive assumptions) in [2, Theorem 3]. Here we follow the presentation of [4, Theorem 2.2].

Theorem 2.2. [2, 4] *Let $p > 0$. Assume that $A_1, \dots, A_L \in \mathbf{S}_\infty$ and*

$$A_\ell^* A_j \in \mathbf{S}^0_{p/2,\infty}, \quad A_\ell A_j^* \in \mathbf{S}^0_{p/2,\infty} \quad \text{for all } \ell \neq j. \quad (2.8)$$

Then for $A = A_1 + \dots + A_L$, we have

$$\Delta_p(A) = \limsup_{s\to 0} s^p \sum_{\ell=1}^{L} n(s, A_\ell),$$

$$\delta_p(A) = \liminf_{s\to 0} s^p \sum_{\ell=1}^{L} n(s, A_\ell).$$

Proof. Put

$$\mathcal{H}^L = \underbrace{\mathcal{H} \oplus \cdots \oplus \mathcal{H}}_{L \text{ terms}}$$

and let $A_0 = \operatorname{diag}\{A_1, \dots, A_L\}$ in $\mathcal{H}^L$, i.e.,

$$A_0(f_1, \dots, f_L) = (A_1 f_1, \dots, A_L f_L).$$

Since

$$A_0^* A_0 = \operatorname{diag}\{A_1^* A_1, \dots, A_L^* A_L\},$$

we see that

$$n(s; A_0) = \sum_{\ell=1}^{L} n(s; A_\ell).$$

Thus, we need to prove the relations

$$\Delta_p(A) = \Delta_p(A_0), \quad \delta_p(A) = \delta_p(A_0).$$

We will focus on the functionals Δ_p; the functionals δ_p are considered in the same way.

Next, let $J : \mathcal{H}^L \to \mathcal{H}$ be the operator given by

$$J(f_1, \dots, f_L) = f_1 + \cdots + f_L \quad \text{so that} \quad J^* f = (f, \dots, f).$$

Then

$$J A_0(f_1, \dots, f_L) = A_1 f_1 + \cdots + A_L f_L$$

and

$$(J A_0)^* f = (A_1^* f, \dots, A_L^* f).$$

It follows that

$$(J A_0)(J A_0)^* f = (A_1 A_1^* + \cdots + A_L A_L^*) f \tag{2.9}$$

and the operator $(JA_0)^*(JA_0)$ is the "matrix" in $\mathcal{H}^L$ given by

$$(JA_0)^*(JA_0) = \begin{pmatrix} A_1^* A_1 & A_1^* A_2 & \cdots & A_1^* A_L \\ A_2^* A_1 & A_2^* A_2 & \cdots & A_2^* A_L \\ \vdots & \vdots & \ddots & \vdots \\ A_L^* A_1 & A_L^* A_2 & \cdots & A_L^* A_L \end{pmatrix}.$$

By our assumption (2.8), we have

$$(JA_0)^*(JA_0) - A_0^* A_0 \in \mathbf{S}^0_{p/2,\infty}. \tag{2.10}$$

Indeed, the "matrix" of the operator in (2.10) has zeros on the diagonal, and its off-diagonal entries are given by $A_\ell^* A_j$, $\ell \neq j$. Now Lemma 2.1 implies that

$$\Delta_{p/2}((JA_0)^*(JA_0)) = \Delta_{p/2}(A_0^* A_0)$$

or

$$\Delta_{p/2}((JA_0)(JA_0)^*) = \Delta_{p/2}(A_0^* A_0), \tag{2.11}$$

because for any compact operator T the non-zero singular values of T^*T and TT^* coincide.

Further, since $AA^* = \sum_{\ell,j=1}^{L} A_\ell A_j^*$, it follows from (2.9) and the second assumption (2.8) that

$$AA^* - (JA_0)(JA_0)^* = \sum_{j\neq\ell} A_\ell A_j^* \in \mathbf{S}^0_{p/2,\infty}.$$

Using Lemma 2.1 again, from here we obtain

$$\Delta_p(A) = \Delta_{p/2}(AA^*) = \Delta_{p/2}((JA_0)(JA_0)^*).$$

Combining the last equality with (2.11), we see that

$$\Delta_p(A) = \Delta_{p/2}(A_0^* A_0) = \Delta_p(A_0).$$

The same reasoning also proves $\delta_p(A) = \delta_p(A_0)$. □

Corollary 2.3. *Under the hypothesis of the theorem above, assume in addition that*

$$n(s; A_1) = n(s; A_2) = \cdots = n(s; A_L), \quad s > 0.$$

Then

$$\Delta_p(A) = L\Delta_p(A_1), \quad \delta_p(A) = L\delta_p(A_1).$$

Finally, we shall briefly discuss the corresponding result for the asymptotics of the positive eigenvalues for compact self-adjoint operators. We let $n_+(s; T)$ stand for the positive eigenvalue counting function for a compact self-adjoint operator T:

$$n_+(s; T) = \#\{n : \lambda_n^+(T) > s\}, \quad s > 0.$$

Similarly to the quantities (2.1), we set

$$\Delta_p^+(T) := \limsup_{s\to 0} s^p n_+(s; T), \quad \delta_p^+(T) := \liminf_{s\to 0} s^p n_+(s; T).$$

Then we have the following theorem, which is proven in [5, Theorem 2.3].

Theorem 2.4. *Let $p > 0$. Assume that $A_1, \dots, A_L \in \mathbf{S}_\infty$ are self-adjoint operators such that*

$$A_\ell A_j \in \mathbf{S}^0_{p/2,\infty}, \quad \textit{for all } \ell \neq j.$$

Then for $A = A_1 + \cdots + A_L$, we have

$$\Delta_p^+(A) = \limsup_{s\to 0} s^p \sum_{\ell=1}^{L} n_+(s, A_\ell),$$

$$\delta_p^+(A) = \liminf_{s\to 0} s^p \sum_{\ell=1}^{L} n_+(s, A_\ell).$$

In particular, if

$$n_+(s; A_1) = n_+(s; A_2) = \cdots = n_+(s; A_L), \quad s > 0,$$

then

$$\Delta_p^+(A) = L\Delta_p^+(A_1), \quad \delta_p^+(A) = L\delta_p^+(A_1).$$

3. Proof of Theorems 1.1 and 1.2

3.1. Preliminary remarks

By the equivalence (1.5), the statement of Theorem 1.1 can be equivalently rewritten in terms of the singular value counting function as

$$\lim_{s\to 0} s^{1/\gamma} n(s; T(\varphi)) = \frac{1}{2}\Gamma(\gamma+1)^{1/\gamma} \int_0^{2\pi} |\varphi_\infty(e^{i\theta})|^{1/\gamma} \frac{d\theta}{2\pi}.$$

Throughout the proof, we use the shorthand notation $\varkappa_\gamma$ for the coefficient appearing on the right-hand side:

$$\varkappa_\gamma := \frac{1}{2}\Gamma(\gamma+1)^{1/\gamma}.$$

Using this notation and the functionals Δ_p and δ_p defined in (2.1), one can rewrite the statement of Theorem 1.1 as

$$\Delta_{1/\gamma}(T(\varphi)) = \delta_{1/\gamma}(T(\varphi)) = \varkappa_\gamma \int_0^{2\pi} |\varphi_\infty(e^{i\theta})|^{1/\gamma} \frac{d\theta}{2\pi}.$$

As in Section 1, for a symbol φ of the form

$$\varphi(re^{i\theta}) = (1-r)^\gamma g(e^{i\theta}), \quad g \in L^\infty(\mathbb{T}),$$

we will write $T_\gamma(g)$ instead of $T(\varphi)$. The case of a radially symmetric φ corresponds to the choice $g = 1$. In this case, the asymptotics of the singular values is given by (1.1). In terms of the asymptotic functionals Δ_p, δ_p this can be rewritten as

$$\Delta_{1/\gamma}(T_\gamma(1)) = \delta_{1/\gamma}(T_\gamma(1)) = \varkappa_\gamma.$$

Finally, we need some notation: for a symbol φ we denote by $M(\varphi)$ the operator of multiplication by $\varphi(z)$ in $L^2(\mathbb{D})$. Then the Toeplitz operator $T(\varphi)$ can be written as

$$T(\varphi) = PM(\varphi)P^* \quad \text{in } B^2(\mathbb{D}), \tag{3.1}$$

where the orthogonal projection P is understood to act from $L^2(\mathbb{D})$ to $B^2(\mathbb{D})$, and P^* acts from $B^2(\mathbb{D})$ to $L^2(\mathbb{D})$.

3.2. Asymptotic orthogonality

The main analytic ingredient of our construction is the following lemma.

Lemma 3.1. *Let $g_1, g_2 \in L^\infty(\mathbb{T})$ be such that the distance between the supports of g_1 and g_2 on $\mathbb{T}$ is positive. Then*

$$T_\gamma(g_1)T_\gamma(g_2)^* \in \mathbf{S}_0.$$

Proof. For $j = 1, 2$, denote

$$\varphi_j(re^{i\theta}) = (1-r)^\gamma g_j(e^{i\theta}),$$
$$\psi_j(re^{i\theta}) = (1-r)^\gamma \mathbb{1}_{[1/2,1]}(r) g_j(e^{i\theta}),$$

where $\mathbb{1}_{[1/2,1]}$ is the characteristic function of the interval $[1/2, 1]$. Since the difference $\varphi_j - \psi_j$ is bounded and supported in the disc $|z| \le 1/2$, it is easy to conclude that

$$T(\varphi_j - \psi_j) \in \mathbf{S}_0, \quad j = 1, 2.$$

Thus, it suffices to prove the inclusion

$$T(\psi_1)T(\psi_2)^* \in \mathbf{S}_0.$$

We have

$$T(\psi_1)T(\psi_2)^* = PM(\psi_1)P^*PM(\overline{\psi_2})P^*,$$

and so it suffices to prove the inclusion

$$M(\psi_1)P^*PM(\overline{\psi_2}) \in \mathbf{S}_0.$$

Further, let $\omega_1, \omega_2 \in C^\infty(\overline{\mathbb{D}})$ be such that the distance between the supports $\operatorname{supp}\omega_1$, $\operatorname{supp}\omega_2$ is positive and

$$\omega_1\psi_1 = \psi_1, \quad \omega_2\psi_2 = \psi_2.$$

Such functions exist by our assumption on the supports of g_1, g_2. We have

$$M(\psi_1)P^*PM(\overline{\psi_2}) = M(\psi_1)M(\omega_1)P^*PM(\overline{\omega_2})M(\overline{\psi_2}).$$

So it suffices to prove that $M(\omega_1)P^*PM(\overline{\omega_2}) \in \mathbf{S}_0$. Clearly, P^*P is the orthogonal projection in $L^2(\mathbb{D})$ whose integral kernel is the Bergman kernel. Using the explicit formula for the Bergman kernel, we see that $M(\omega_1)P^*PM(\overline{\omega_2})$ is the integral operator in $L^2(\mathbb{D})$ with the kernel

$$\frac{\omega_1(z)\overline{\omega_2(\zeta)}}{(1-z\overline{\zeta})^2}, \quad z,\zeta \in \mathbb{D}.$$

Since ω_1 and ω_2 have disjoint supports, we see that this kernel is C^∞-smooth. It is a well known fact that integral operators with C^∞ kernels on compact domains belong to $\mathbf{S}_0$ (it can be proven, for example, by approximating the integral kernel by polynomials). Thus, the operator $M(\omega_1)P^*PM(\overline{\omega_2})$ is in the class $\mathbf{S}_0$. □

We would like to have an analogous statement where g_1 and g_2 are characteristic functions of disjoint (but possibly "touching") open intervals. We will obtain it from Lemma 3.1 by an approximation argument. To this end, in the next subsection we develop some rather crude estimates.

3.3. Auxiliary estimates

Lemma 3.2. *If $|g| \le g_0$, where g_0 is a constant, then*

$$\Delta_{1/\gamma}(T_\gamma(g)) \le 2\kappa_\gamma |g_0|^{1/\gamma}.$$

Proof. Let us write our symbol φ as

$$\varphi = \varphi_0^{1/2}\varphi_1\varphi_0^{1/2}, \quad \text{where} \quad \varphi_0(z) = |g_0|(1-|z|)^\gamma \quad \text{and} \quad |\varphi_1(z)| \le 1.$$

Then by (3.1) we have

$$n(s; T_\gamma(g)) = n(s; T(\varphi)) = n(s; GM(\varphi_1)G^*), \quad G = PM(\varphi_0^{1/2})$$

and

$$n(s; |g_0|T_\gamma(1)) = n(s; T(\varphi_0)) = n(s; PM(\varphi_0)P^*) = n(s; GG^*).$$

Applying the estimates (2.5) and (2.6), we obtain

$$\begin{aligned} n(s; GM(\varphi_1)G^*) &\le n(\sqrt{s}; G) + n(\sqrt{s}; M(\varphi_1)G^*) \\ &\le n(\sqrt{s}; G) + n(\sqrt{s}; G^*) \\ &= 2n(\sqrt{s}; G) = 2n(s; GG^*) = 2n(s; T(\varphi_0)). \end{aligned}$$

Multiplying by $s^{1/\gamma}$ and taking lim sup yields

$$\Delta_{1/\gamma}(T_\gamma(g)) \le 2\Delta_{1/\gamma}(|g_0|T_\gamma(1)) = 2|g_0|^{1/\gamma}\kappa_\gamma,$$

as required. □

Lemma 3.3. *Let $\delta \subset \mathbb{T}$ be an arc with arclength $|\delta| < 2\pi$. Then*

$$\Delta_{1/\gamma}(T_\gamma(\mathbb{1}_\delta)) \le \kappa_\gamma|\delta|.$$

Proof. Let $L \in \mathbb{N}$ be such that $2\pi/(L+1) \le |\delta| < 2\pi/L$. For $\ell = 1, \dots, L$, let δ_ℓ be the arc δ rotated by the angle $2\pi\ell/L$:

$$\delta_\ell = e^{2\pi i\ell/L}\delta_\ell. \tag{3.2}$$

In particular, $\delta_L = \delta$. Then the arcs $\delta_1, \dots, \delta_L$ are disjoint and so

$$g := \sum_{\ell=1}^{L} \mathbb{1}_{\delta_\ell} \le 1.$$

By Lemma 3.2, it follows that

$$\Delta_{1/\gamma}(T_\gamma(g)) \le 2\kappa_\gamma.$$

Further, it is easy to see that the operators $T_\gamma(\mathbb{1}_{\delta_\ell})$ are unitarily equivalent to each other by rotation. Thus,

$$n(s; T_\gamma(\mathbb{1}_{\delta_\ell})) = n(s; T_\gamma(\mathbb{1}_\delta)), \quad s > 0,$$

for all ℓ. Finally, we have

$$T_\gamma(\mathbb{1}_{\delta_\ell})T_\gamma(\mathbb{1}_{\delta_j}) \in \mathbf{S}_0, \quad \ell \ne j,$$

by Lemma 3.1. Thus, we can apply Corollary 2.3 to $A_\ell = T_\gamma(\mathbb{1}_{\delta_\ell})$ and to $A = T_\gamma(g)$. This yields

$$\Delta_{1/\gamma}(T_\gamma(\mathbb{1}_\delta)) = \Delta_{1/\gamma}(T_\gamma(g))/L \le 2\kappa_\gamma/L \le 2\pi\kappa_\gamma/(L+1) \le \kappa_\gamma|\delta|,$$

as claimed. □

Lemma 3.4. *Let δ and δ' be two arcs in $\mathbb{T}$ such that the symmetric difference $\delta \triangle \delta'$ has total length $< \varepsilon$. Then*

$$\Delta_{1/\gamma}(T_\gamma(\mathbb{1}_\delta)) - T_\gamma(\mathbb{1}_{\delta'})) \le 2^{1+1/\gamma}\kappa_\gamma\varepsilon.$$

Proof. Let $\delta \triangle \delta' = \delta_1 \cup \delta_2$, where δ_1, δ_2 are intervals with $|\delta_1| < \varepsilon$, $|\delta_2| < \varepsilon$. Then

$$T_\gamma(\mathbb{1}_\delta) - T_\gamma(\mathbb{1}_{\delta'}) = \pm T_\gamma(\mathbb{1}_{\delta_1}) \pm T_\gamma(\mathbb{1}_{\delta_2}),$$

where the signs depend on the relative location of δ, δ'. Using the estimate (2.2), we get

$$\Delta_{1/\gamma}(T_\gamma(\mathbb{1}_\delta) - T_\gamma(\mathbb{1}_{\delta'}))^{\gamma/(\gamma+1)} \\ \leq \Delta_{1/\gamma}(T_\gamma(\mathbb{1}_{\delta_1}))^{\gamma/(\gamma+1)} + \Delta_{1/\gamma}(T_\gamma(\mathbb{1}_{\delta_2}))^{\gamma/(\gamma+1)},$$

and so, applying Lemma 3.3, we get

$$\Delta_{1/\gamma}(T_\gamma(\mathbb{1}_\delta) - T_\gamma(\mathbb{1}_{\delta'})) \leq \kappa_\gamma(|\delta_1|^{\gamma/(\gamma+1)} + |\delta_2|^{\gamma/(\gamma+1)})^{1+1/\gamma} \leq \varepsilon\kappa_\gamma 2^{1+1/\gamma},$$

as required. □

Now we can prove a refined version of Lemma 3.1, where the supports of g_1, g_2 are allowed to "touch".

Lemma 3.5. *Let δ and δ' be disjoint open arcs in $\mathbb{T}$: $\delta \cap \delta' = \varnothing$. Then*

$$T_\gamma(\mathbb{1}_\delta)T_\gamma(\mathbb{1}_{\delta'}) \in \mathbf{S}^0_{1/2\gamma,\infty}.$$

Proof. Let us "shrink" δ a little: for $\varepsilon > 0$, let δ_ε be an arc such that the distance between δ_ε and δ' is positive and the symmetric difference $\delta_\varepsilon \triangle \delta$ has a total length $< \varepsilon$. By Lemma 3.1, we have

$$T_\gamma(\mathbb{1}_{\delta_\varepsilon})T_\gamma(\mathbb{1}_{\delta'}) \in \mathbf{S}_0 \subset \mathbf{S}^0_{1/2\gamma,\infty}.$$

By Lemma 2.1, it follows that

$$\Delta_{1/2\gamma}(T_\gamma(\mathbb{1}_\delta)T_\gamma(\mathbb{1}_{\delta'})) = \Delta_{1/2\gamma}\big((T_\gamma(\mathbb{1}_\delta) - T_\gamma(\mathbb{1}_{\delta_\varepsilon}))T_\gamma(\mathbb{1}_{\delta'})\big).$$

Applying the estimate (2.7), we get

$$\Delta_{1/2\gamma}\big((T_\gamma(\mathbb{1}_\delta) - T_\gamma(\mathbb{1}_{\delta_\varepsilon}))T_\gamma(\mathbb{1}_{\delta'})\big) \leq \Delta_{1/\gamma}(T_\gamma(\mathbb{1}_\delta) - T_\gamma(\mathbb{1}_{\delta_\varepsilon}))\Delta_{1/\gamma}(T_\gamma(\mathbb{1}_{\delta'})).$$

By Lemma 3.4, we get

$$\Delta_{1/2\gamma}(T_\gamma(\mathbb{1}_\delta)T_\gamma(\mathbb{1}_{\delta'})) \leq C_\gamma\varepsilon.$$

Since ε can be chosen arbitrarily small, we get

$$\Delta_{1/2\gamma}(T_\gamma(\mathbb{1}_\delta)T_\gamma(\mathbb{1}_{\delta'})) = 0,$$

which is exactly what is required. □

3.4. Step functions g

Lemma 3.6. *Let δ be an arc with $|\delta| = 2\pi/L$, $L \in \mathbb{N}$. Then*

$$\Delta_{1/\gamma}(T_\gamma(\mathbb{1}_\delta)) = \delta_{1/\gamma}(T_\gamma(\mathbb{1}_\delta)) = \kappa_\gamma/L.$$

Proof. Let δ_ℓ be as in (3.2). Then $\delta_j \cap \delta_\ell = \varnothing$ for $j \neq \ell$ and

$$1 = \sum_{\ell=1}^{L} \mathbb{1}_{\delta_\ell} \quad \text{a.e. on } \mathbb{T}.$$

Thus,

$$T_\gamma(1) = \sum_{\ell=1}^{L} T_\gamma(\mathbb{1}_{\delta_\ell}),$$

and by Lemma 3.5

$$T_\gamma(\mathbb{1}_{\delta_\ell})T_\gamma(\mathbb{1}_{\delta_j}) \in \mathbf{S}^0_{1/2\gamma,\infty}.$$

Consequently, we can apply Corollary 2.3, which yields

$$\Delta_{1/\gamma}(T_\gamma(\mathbb{1}_\delta)) = \Delta_{1/\gamma}(T_\gamma(1))/L,$$

and similarly for the lower limits $\delta_{1/\gamma}$. □

Lemma 3.7. *Let $\delta \subset \mathbb{T}$ be an arc of length $|\delta| = 2\pi/L$, $L \in \mathbb{N}$, and let δ_ℓ be as in* (3.2). *Let*

$$g = \sum_{\ell=1}^{L} c_\ell \mathbb{1}_{\delta_\ell} \tag{3.3}$$

for some coefficients $c_1, \dots, c_\ell \in \mathbb{C}$. Then

$$\Delta_{1/\gamma}(T_\gamma(g)) = \delta_{1/\gamma}(T_\gamma(g)) = \kappa_\gamma \int_0^{2\pi} |g(e^{i\theta})|^{1/\gamma} \frac{d\theta}{2\pi}. \tag{3.4}$$

Proof. We have

$$T_\gamma(g) = \sum_{\ell=1}^{L} c_\ell T_\gamma(\mathbb{1}_{\delta_\ell}),$$

and

$$T_\gamma(\mathbb{1}_{\delta_\ell}) T_\gamma(\mathbb{1}_{\delta_j}) \in \mathbf{S}^0_{1/2\gamma,\infty}, \quad j \neq \ell.$$

By Theorem 2.2, we get

$$\begin{aligned}\Delta_{1/\gamma}(T_\gamma(g)) &\leq \sum_{\ell=1}^{L} \Delta_{1/\gamma}(c_\ell T_\gamma(\mathbb{1}_{\delta_\ell})) = \sum_{\ell=1}^{L} |c_\ell|^{1/\gamma} \Delta_{1/\gamma}(T_\gamma(\mathbb{1}_{\delta_\ell}))\\ &= \frac{1}{L}\sum_{\ell=1}^{L} |c_\ell|^{1/\gamma} \Delta_{1/\gamma}(T_\gamma(1)) = \kappa_\gamma \int_0^{2\pi} |g(e^{i\theta})|^{1/\gamma} \frac{d\theta}{2\pi}\end{aligned}$$

and similarly

$$\delta_{1/\gamma}(T_\gamma(g)) \geq \frac{1}{L}\sum_{\ell=1}^{L} |c_\ell|^{1/\gamma} \delta_{1/\gamma}(T_\gamma(1)) = \kappa_\gamma \int_0^{2\pi} |g(e^{i\theta})|^{1/\gamma} \frac{d\theta}{2\pi}. \qquad \square$$

3.5. Concluding the proof

Lemma 3.8. *Let $g \in C(\mathbb{T})$. Then formula* (3.4) *holds true.*

Proof. For any $\varepsilon > 0$, there exists a step function g_ε of the form (3.3) such that $\|g - g_\varepsilon\|_\infty \leq \varepsilon$. By Lemma 3.7, the identity

$$\Delta_{1/\gamma}(T_\gamma(g_\varepsilon)) = \delta_{1/\gamma}(T_\gamma(g_\varepsilon)) = \kappa_\gamma \int_{-\pi}^{\pi} |g_\varepsilon(e^{i\theta})|^{1/\gamma} \frac{d\theta}{2\pi} \tag{3.5}$$

holds true for all $\varepsilon > 0$; our task is to pass to the limit as $\varepsilon \to 0$. It is obvious that one can pass to the limit on the right-hand side of (3.5). As for the left-hand side, note first that by Lemma 3.2 we have

$$\Delta_{1/\gamma}(T_\gamma(g - g_\varepsilon)) \leq 2\kappa_\gamma \varepsilon^{1/\gamma}.$$

Applying the estimate (2.3), we then get

$$\begin{aligned}&|\Delta_{1/\gamma}(T_\gamma(g))^{\gamma/(\gamma+1)} - \Delta_{1/\gamma}(T_\gamma(g_\varepsilon))^{\gamma/(\gamma+1)}|\\ &\leq \Delta_{1/\gamma}(T_\gamma(g - g_\varepsilon))^{\gamma/(\gamma+1)} \leq C_\gamma \varepsilon^{1/(\gamma+1)}.\end{aligned}$$

It follows that

$$\lim_{\varepsilon\to 0}\Delta_{1/\gamma}(T_\gamma(g_\varepsilon))=\Delta_{1/\gamma}(T_\gamma(g)).$$

Similarly, using (2.4) instead of (2.3), we obtain

$$\lim_{\varepsilon\to 0}\delta_{1/\gamma}(T_\gamma(g_\varepsilon))=\delta_{1/\gamma}(T_\gamma(g)).$$

Now we can pass to the limit $\varepsilon\to 0$ in (3.5), which gives the desired result. □

Proof of Theorem 1.1. Write $\varphi=\varphi_0+\varphi_1$, where

$$\varphi_0(re^{i\theta})=(1-r)^\gamma g(e^{i\theta}),$$

and

$$\varphi_1(z)=o((1-|z|)^\gamma),\quad |z|\to 1.$$

By the previous step, we have that $T(\varphi_0)$ satisfies the required asymptotics. It remains to prove that $T(\varphi_1)\in\mathbf{S}^0_{1/\gamma,\infty}$. In order to do this, for any $\varepsilon>0$ write $\varphi_1=\psi_\varepsilon+\widetilde{\psi}_\varepsilon$, where $\widetilde{\psi}_\varepsilon$ is supported inside the smaller disc $|z|<a$, $a<1$, and ψ_ε satisfies the estimate

$$|\psi_\varepsilon(z)|\le\varepsilon(1-|z|)^\gamma,\quad |z|<1.$$

It is easy to see that $T(\widetilde{\psi}_\varepsilon)\in\mathbf{S}_0$. On the other hand, by Lemma 3.2, we have

$$\Delta_{1/\gamma}(T(\psi_\varepsilon))\le 2\kappa_\gamma\varepsilon^{1/\gamma}.$$

By Lemma 2.1, we get

$$\Delta_{1/\gamma}(T(\varphi_1))=\Delta_{1/\gamma}(T(\psi_\varepsilon))\le 2\kappa_\gamma\varepsilon^{1/\gamma}.$$

Since ε is arbitrary, we get the equality $\Delta_{1/\gamma}(T(\varphi_1))=0$, which means $T(\varphi_1)\in\mathbf{S}^0_{1/\gamma,\infty}$. □

The proof of Theorem 1.2 repeats the above proof verbatim. The only differences are that (i) instead of using Corollary 2.3, we use Theorem 2.4; (ii) instead of working with the functionals Δ_p, δ_p, we work with Δ_p^+, δ_p^+; (iii) all the symbols appearing in the proof are real-valued.

4. Proof of Theorems 1.3 and 1.4

Let $\varphi(e^{i\theta})=(1-|z|)^\gamma b(e^{i\theta})$ with b as in (1.3) and consider the corresponding Toeplitz operator $T(\varphi)$ in $B^2(\mathbb{D})$. Let $T=\{t_{n,m}\}_{n,m=0}^\infty$ be the matrix of $T(\varphi)$ in the orthonormal basis $\{\sqrt{k+1}z^k\}_{k=0}^\infty$:

$$t_{j,k}=\sqrt{j+1}\sqrt{k+1}(T(\varphi)z^j,z^k).$$

We have

$$t_{j,j+m}=0\quad\text{if } |m|>M.$$

Further, for $|m|\le M$ we have

$$\begin{aligned}t_{j,j+m}&=b_m\sqrt{j+1}\sqrt{j+m+1}((1-|z|)^\gamma e^{im\theta}z^j,z^{j+m})\\&=2^{-\gamma}\Gamma(\gamma+1)b_m j^{-\gamma}+o(j^{-\gamma})\qquad\text{as } j\to\infty.\end{aligned}$$

This calculation shows that $2^\gamma(\Gamma(\gamma+1))^{-1}T = A + A'$, where A' is a banded matrix with $a'_{j,j+m} = o(j^{-\gamma})$ as $j \to \infty$ for all $|m| \leq M$. Considering A' as a sum of $2M+1$ matrices, each of which has non-zero entries only on the "off-diagonal" $k = j+m$, it is easy to see that $A' \in \mathbf{S}^0_{1/\gamma,\infty}$. Thus, by Lemma 2.1,

$$\Delta_{1/\gamma}(A) = \Delta_{1/\gamma}(2^\gamma(\Gamma(\gamma+1))^{-1}T) = 2(\Gamma(\gamma+1))^{-1/\gamma}\Delta_{1/\gamma}(T).$$

Finally, by Theorem 1.1,

$$\Delta_{1/\gamma}(A) = 2(\Gamma(\gamma+1))^{-1/\gamma}\Delta_{1/\gamma}(T) = \int_{-\pi}^{\pi} |b(e^{i\theta})|^{1/\gamma}\frac{d\theta}{2\pi}.$$

The same calculation applies to $\delta_{1/\gamma}(A)$. This completes the proof of Theorem 1.3.

To prove Theorem 1.4, one repeats the above arguments for the functionals $\Delta^+_{1/\gamma}$, $\delta^+_{1/\gamma}$ instead of $\Delta_{1/\gamma}$, $\delta_{1/\gamma}$, and uses Theorem 1.4 instead of Theorem 1.3.

References

[1] M.Sh. Birman and M.Z. Solomyak, *Spectral theory of self-adjoint operators in Hilbert space,* Reidel, Dordrecht, 1987.

[2] M.Sh. Birman and M. Z. Solomyak, Compact operators with power asymptotic behavior of the singular numbers, *J. Sov. Math.* **27** (1984), 2442–2447.

[3] V. Bruneau and G. Raikov, Spectral properties of harmonic Toeplitz operators and applications to the perturbed Krein Laplacian, preprint, arXiv:1609.08229.

[4] A. Pushnitski and D. Yafaev, Localization principle for compact Hankel operators, *J. Funct. Anal.* **270** (2016), 3591–3621.

[5] A. Pushnitski and D. Yafaev, Spectral asymptotics for compact self-adjoint Hankel operators, *J. Operator Theory* **74** (2015), no. 2, 417–455.

[6] G.D. Raikov, Eigenvalue asymptotics for the Schrödinger operator, *Commun. PDE* **15** (1990), no. 3, 407–434.

[7] W. Van Assche, Compact Jacobi matrices: from Stieltjes to Krein and $M(a, b)$, 100 ans après Th.-J. Stieltjes, *Ann. Fac. Sci. Toulouse Math.* **6** (1996), Special issue, 195–215.

Alexander Pushnitski
Department of Mathematics
King's College London
Strand, London WC2R 2LS
United Kingdom
e-mail: `alexander.pushnitski@kcl.ac.uk`

Beyond fractality: piecewise fractal and quasifractal algebras

Steffen Roch

Abstract. Fractality is a property of C^*-algebras of approximation sequences with several useful consequences: for example, if (A_n) is a sequence in a fractal algebra, then the pseudospectra of the A_n converge in the Hausdorff metric. The fractality of a *separable* algebra of approximation sequences can always be forced by a suitable restriction. This observation leads to the question to describe the possible fractal restrictions of a given algebra. In this connection we define two classes of algebras beyond the class of fractal algebras (piecewise fractal and quasifractal algebras), give examples for algebras with these properties, and present some first results on the structure of quasifractal algebras (being continuous fields over the set of their fractal restrictions).

Mathematics Subject Classification (2010). Primary 47N40; Secondary 65J10, 46L99.

Keywords. Finite sections discretization, block Toeplitz operators, fractal restriction, continuous fields.

1. Introduction

Fractality is a special property of algebras of approximation sequences which typically arise as follows.

Let H be a Hilbert space and $\mathcal{P} = (P_n)_{n\geq 1}$ a filtration on H, i.e., a sequence of orthogonal projections of finite rank that converges strongly to the identity operator on H. Let $\mathcal{F}^{\mathcal{P}}$ denote the set of all bounded sequences $(A_n)_{n\geq 1}$ of operators $A_n \in L(\operatorname{im} P_n)$ and $\mathcal{G}^{\mathcal{P}}$ the set of all sequences $(A_n) \in \mathcal{F}^{\mathcal{P}}$ with $\|A_n\| \to 0$. Provided with the operations

$$(A_n) + (B_n) := (A_n + B_n), \quad (A_n)(B_n) := (A_n B_n), \quad (A_n)^* := (A_n^*) \quad (1.1)$$

and the norm $\|(A_n)\| := \sup \|A_n P_n\|$, $\mathcal{F}^{\mathcal{P}}$ becomes a unital C^*-algebra and $\mathcal{G}^{\mathcal{P}}$ a closed ideal of $\mathcal{F}^{\mathcal{P}}$. The importance of the quotient algebra $\mathcal{F}^{\mathcal{P}}/\mathcal{G}^{\mathcal{P}}$ in numerical analysis stems from the fact that a coset $(A_n) + \mathcal{G}^{\mathcal{P}}$ is invertible in $\mathcal{F}^{\mathcal{P}}/\mathcal{G}^{\mathcal{P}}$ if and only if the A_n are invertible for all sufficiently large n and

© Springer International Publishing AG, part of Springer Nature 2018
A. Böttcher et al. (eds.), *The Diversity and Beauty of Applied Operator Theory*, Operator Theory: Advances and Applications 268, https://doi.org/10.1007/978-3-319-75996-8_22

if the norms of the inverses are uniformly bounded, which is equivalent to saying that (A_n) is a stable sequence.

With every non-empty subset A of $L(H)$, we associate the smallest C^*-subalgebra $\mathcal{S}^{\mathcal{P}}(\mathsf{A})$ of $\mathcal{F}^{\mathcal{P}}$ that contains all sequences $(P_nAP_n)_{n\geq 1}$ with $A \in \mathsf{A}$. Algebras of this form are the prototypes of algebras of approximation sequences mentioned above.

To make this concrete, consider the algebra $\mathcal{S}(\mathsf{T}(C))$ of the finite sections discretization (FSD) for Toeplitz operators with continuous generating function. Here, H is the Hilbert space $l^2(\mathbb{Z}^+)$, P_n is the projection on H sending $(x_0, x_1, \ldots)$ to $(x_0, \ldots, x_{n-1}, 0, 0, \ldots)$ (we agree to omit the superscript $\mathcal{P}$ when the filtration is specified in this way), and A is the C^*-algebra $\mathsf{T}(C)$ generated by all Toeplitz operators $T(a)$ with a a continuous function on the complex unit circle $\mathbb{T}$. Recall that $T(a)$ is given by the matrix representation $(a_{i-j})_{i,j\geq 0}$ with respect to the standard basis of $l^2(\mathbb{Z}^+)$, where

$$a_k := \frac{1}{2\pi} \int_0^{2\pi} a(e^{is})e^{-iks}\, ds, \quad k \in \mathbb{Z}, \tag{1.2}$$

denotes the kth Fourier coefficient of a. It is well known that the algebra $\mathsf{T}(C)$ has a nice description, as follows.

Theorem 1.1. $\mathsf{T}(C) = \{T(a) + K : a \in C(\mathbb{T}) \text{ and } K \in K(l^2(\mathbb{Z}^+))\}$.

Here, $K(l^2(\mathbb{Z}^+))$ is the ideal of the compact operators on $l^2(\mathbb{Z}^+)$.

Similarly, the sequences in the algebra $\mathcal{S}(\mathsf{T}(C))$ are completely characterized in the following theorem by Böttcher and Silbermann [2] (see also [3], [4, Section 1.4.2], and the pioneering paper [10]). Therein R_n stands for the operator $(x_0, x_1, \ldots) \mapsto (x_{n-1}, \ldots, x_0, 0, 0, \ldots)$ on $l^2(\mathbb{Z}^+)$. It is not hard to see that for each sequence $\mathbf{A} = (A_n) \in \mathcal{S}(\mathsf{T}(C))$, the strong limits $W(\mathbf{A}) :=$ s-lim A_nP_n and $\widetilde{W}(\mathbf{A}) :=$ s-lim $R_nA_nR_nP_n$ exist and that W and $\widetilde{W}$ are unital *-homomorphisms from $\mathcal{S}(\mathsf{T}(C))$ to $L(l^2(\mathbb{Z}^+))$ (actually, to $\mathsf{T}(C)$).

Theorem 1.2. (a) *The algebra $\mathcal{S}(\mathsf{T}(C))$ consists of all sequences $(A_n)_{n\geq 1}$ of the form*

$$(A_n) = (P_nT(a)P_n + P_nKP_n + R_nLR_n + G_n) \tag{1.3}$$

where $a \in C(\mathbb{T})$, K and L are compact operators on $l^2(\mathbb{Z}^+)$, and $(G_n) \in \mathcal{G}$. The representation of a sequence $(A_n) \in \mathcal{S}(\mathsf{T}(C))$ in this form is unique.

(b) *For every sequence $\mathbf{A} \in \mathcal{S}(\mathsf{T}(C))$, the coset $\mathbf{A} + \mathcal{G}$ is invertible in the quotient algebra $\mathcal{S}(\mathsf{T}(C))/\mathcal{G}$ (equivalently, $\mathbf{A} + \mathcal{G}$ is invertible in $\mathcal{F}/\mathcal{G}$ or, again equivalently, $\mathbf{A}$ is stable) if and only if the operators $W(\mathbf{A})$ and $\widetilde{W}(\mathbf{A})$ are invertible.*

The algebra $\mathcal{S}(\mathsf{T}(C))$ of the FSD of the Toeplitz operators gives a first example of a fractal algebra. The idea behind the notion of a fractal algebra comes from a remarkable property of the algebra $\mathcal{S}(\mathsf{T}(C))/\mathcal{G}$: the structure of this algebra is determined by the two representations W and $\widetilde{W}$. These representations are defined by certain strong limits, hence, the operators

$W(\mathbf{A})$ and $\widetilde{W}(\mathbf{A})$ can be determined from each subsequence of the sequence $\mathbf{A} \in \mathcal{S}(\mathsf{T}(C))$. This observation implies that whenever a *subsequence* of a sequence $\mathbf{A} \in \mathcal{S}(\mathsf{T}(C))$ is stable, then the operators $W(\mathbf{A})$ and $\widetilde{W}(\mathbf{A})$ are already invertible and, hence, the *full* sequence $\mathbf{A}$ is stable by Theorem 1.2.

One can state this observation in a slightly different way: every sequence in $\mathcal{S}(\mathsf{T}(C))$ can be rediscovered from each of its (infinite) subsequences up to a sequence tending to zero in the norm. In that sense, the essential information on a sequence in $\mathcal{S}(\mathsf{T}(C))$ is stored in each of its subsequences. Subalgebras of $\mathcal{F}$ with this property were called *fractal* in [9] (see also [6]) in order to emphasize this self-similarity aspect. We will recall some basic properties of fractal algebras that will be needed in what follows and start with the official definition of a fractal algebra. We will state this definition in the slightly more general context where $\mathcal{C} = (\mathcal{C}_n)_{n\in\mathbb{N}}$ is a sequence of unital C^*-algebras and $\mathcal{F}^{\mathcal{C}}$ is the set of all bounded sequences (A_n) with $A_n \in \mathcal{C}_n$. With the operations as in (1.1) and with the supremum norm, $\mathcal{F}^{\mathcal{C}}$ becomes a unital C^*-algebra and the set $\mathcal{G}^{\mathcal{C}}$ of all sequences in $\mathcal{F}^{\mathcal{C}}$ tending to zero in the norm forms a closed ideal of $\mathcal{F}^{\mathcal{C}}$. Again, we will often simply write $\mathcal{F}$ and $\mathcal{G}$ in place of $\mathcal{F}^{\mathcal{C}}$ and $\mathcal{G}^{\mathcal{C}}$. Note in that connection that

$$\|(A_n) + \mathcal{G}\|_{\mathcal{F}/\mathcal{G}} = \limsup_{n\to\infty} \|A_n\|_{\mathcal{C}_n} \tag{1.4}$$

for every sequence $(A_n) \in \mathcal{F}$.

The sequences in $\mathcal{G}$ are often called *zero sequences.* Thus, $(G_n) \in \mathcal{F}$ is a zero sequence if $\lim_{n\to\infty} \|G_n\| = 0$. We call a sequence $(G_n) \in \mathcal{F}$ a *partial zero sequence* if $\liminf_{n\to\infty} \|G_n\| = 0$. The perhaps simplest way to define fractal algebras is the following (which is equivalent to the original definition in [9]).

Definition 1.3. A C^*-subalgebra $\mathcal{A}$ of $\mathcal{F}$ is called *fractal* if every partial zero sequence in $\mathcal{A}$ is a zero sequence.

The fractality of the algebra $\mathcal{S}(\mathsf{T}(C))$ can be seen as follows. Suppose $\mathbf{A} := (P_nT(a)P_n + P_nKP_n + R_nLR_n + G_n)$ is a partial zero sequence in $\mathcal{S}(\mathsf{T}(C))$. Then, necessarily, $W(\mathbf{A}) = T(a)+K = 0$ and $\widetilde{W}(\mathbf{A}) = T(\tilde{a})+L = 0$ with $\tilde{a}(t) := a(t^{-1})$. Hence, $\mathbf{A} \in \mathcal{G}$.

Here are some facts which illustrate the importance of the notion of fractality.

(F1) For a sequence (A_n) in a fractal subalgebra $\mathcal{A}$ of $\mathcal{F}$, the sets of the singular values (the pseudospectra, the numerical ranges) of the A_n converge with respect to the Hausdorff metric to the set of the singular values (the pseudospectrum, the numerical range) of the coset $(A_n)+\mathcal{G}$ (see [4], Chapter 3).

(F2) The ideal of the compact sequences in a fractal algebra has a nice structure: it is a dual subalgebra of $\mathcal{A}/\mathcal{G}$ as shown in [7] (see the part before Corollary 4.10 for the definition of a compact sequence and the result).

(F3) If (A_n) is a sequence in a fractal algebra, then $\lim_{n\to\infty} \|A_n\|$ exists (compare this fact with (1.4), which holds for an arbitrary sequence in the algebra $\mathcal{F}$).

Property (F3) is crucial for the present paper. It follows easily from the definition of a fractal algebra and, conversely, the existence of $\lim_{n\to\infty} \|A_n\|$ for every sequence (A_n) in $\mathcal{A}$ implies that $\mathcal{A}$ is fractal.

It is certainly not true that every subalgebra of $\mathcal{F}$ is fractal ($\mathcal{F}$ itself is not fractal), but it is a remarkable consequence of (F3) that *every separable C^*-subalgebra of $\mathcal{F}$ has a fractal restriction.* To state this precisely, we need some more notation. Let $\eta : \mathbb{N} \to \mathbb{N}$ be a strictly increasing sequence. By $\mathcal{F}_\eta$ we denote the set of all subsequences $(A_{\eta(n)})$ of sequences (A_n) in $\mathcal{F}$. One can make $\mathcal{F}_\eta$ to a C^*-algebra in a natural way. The mapping $R_\eta : \mathcal{F} \to \mathcal{F}_\eta$, $(A_n) \mapsto (A_{\eta(n)})$ is called the *restriction* of $\mathcal{F}$ onto $\mathcal{F}_\eta$. For every subset $\mathcal{S}$ of $\mathcal{F}$, we abbreviate $R_\eta \mathcal{S}$ by $\mathcal{S}_\eta$. It is easy to see that $\mathcal{G}_\eta$ coincides with the ideal of the sequences in $\mathcal{F}_\eta$ which tend to zero in the norm. Since the strictly increasing sequences $\eta : \mathbb{N} \to \mathbb{N}$ are in one-to-one correspondence to the infinite subsets $\mathbb{M} := \eta(\mathbb{N})$ of $\mathbb{N}$, we will also use the notation $\mathcal{A}|_{\mathbb{M}}$ in place of $R_\eta \mathcal{A} = \mathcal{A}_\eta$. With these notations, we can formulate the following result of [6] (a shorter proof is in [8]).

Theorem 1.4 (Fractal restriction theorem). *If $\mathcal{A}$ is a separable C^*-subalgebra of $\mathcal{F}$, then there is a strictly increasing sequence $\eta : \mathbb{N} \to \mathbb{N}$ such that the restricted algebra $\mathcal{A}_\eta$ is fractal.*

One cannot expect that Theorem 1.4 holds for arbitrary C^*-subalgebras of $\mathcal{F}$; for example it is certainly not true for the algebra $\mathcal{F}$. On the other hand, non-separable fractal algebras exist: the algebra of the FSD for Toeplitz operators with *piecewise continuous* generating function can serve as an example.

The goal of this paper is to present some first steps into the world beyond fractal algebras. Repeated use of the fractal restriction theorem will lead us to the fractal exhaustion theorem, which then will give rise to single out two classes of non-fractal algebras, the piecewise fractal and the quasifractal algebras. For both classes, we present typical examples and study some properties. For piecewise fractal algebras, this will be quite simple: they are just constituted by a finite number of fractal algebras, and (F1) – (F3) hold for each of the finite restrictions separately. For quasifractal algebras, it is our first goal to get an overview of the possible fractal restrictions. In particular, we will define a topology on the set of all (equivalence classes of) fractal restrictions which makes this set to a compact Hausdorff space. Then we show that every quasifractal algebra can be considered as a continuous field of C^*-algebras over this space.

2. Fractal exhaustion of C^*-subalgebras of $\mathcal{F}$

The restriction process in Theorem 1.4 can be iterated to yield a complete decomposition of a separable subalgebra of $\mathcal{F}$ into fractal restrictions.

Theorem 2.1 (Fractal exhaustion theorem). *Let $\mathcal{A}$ be a separable C^*-subalgebra of $\mathcal{F}$. Then there exist a (finite or infinite) number of infinite subsets $\mathbb{M}_1, \mathbb{M}_2, \ldots$ of $\mathbb{N}$ with*

$$\mathbb{M}_i \cap \mathbb{M}_j = \emptyset \quad \textit{for } i \neq j \quad \textit{and} \quad \cup_i \mathbb{M}_i = \mathbb{N}$$

such that every restriction $\mathcal{A}|_{\mathbb{M}_i}$ is a fractal subalgebra of $\mathcal{F}|_{\mathbb{M}_i}$.

Proof. With Theorem 1.4, we find an infinite subset $\mathbb{M}_1$ of $\mathbb{N}$ such that $\mathcal{A}|_{\mathbb{M}_1}$ is fractal. Without loss of generality we may assume that $1 \in \mathbb{M}_1$ (otherwise we include 1 into $\mathbb{M}_1$). If $\mathbb{N} \setminus \mathbb{M}_1$ is a finite set, we include these finitely many points into $\mathbb{M}_1$. The algebra $\mathcal{A}|_{\mathbb{M}_1}$ is still fractal, and we are done.

If $\mathbb{N} \setminus \mathbb{M}_1$ is an infinite set, we apply Theorem 1.4 to the restriction $\mathcal{A}|_{\mathbb{N} \setminus \mathbb{M}_1}$ and get an infinite subset $\mathbb{M}_2$ of $\mathbb{N} \setminus \mathbb{M}_1$ such that $\mathcal{A}|_{\mathbb{M}_2}$ is fractal. Without loss we may assume that the smallest number in $\mathbb{N} \setminus \mathbb{M}_1$ belongs to $\mathbb{M}_2$. If now $\mathbb{N} \setminus (\mathbb{M}_1 \cup \mathbb{M}_2)$ is finite, we include these finitely many points into $\mathbb{M}_2$ and are done.

If $\mathbb{N} \setminus (\mathbb{M}_1 \cup \mathbb{M}_2)$ is infinite, we proceed in this way and obtain a finite (in case one of the sets $\mathbb{N} \setminus (\mathbb{M}_1 \cup \ldots \cup \mathbb{M}_k)$ is finite) or infinite sequence $\mathcal{A}|_{\mathbb{M}_1}, \mathcal{A}|_{\mathbb{M}_2}, \ldots$ of fractal restrictions of $\mathcal{A}$. It follows from our construction that the $\mathbb{M}_i$ are pairwise disjoint, and the inclusion of the smallest number of $\mathbb{N} \setminus (\mathbb{M}_1 \cup \ldots \cup \mathbb{M}_k)$ into $\mathbb{M}_{k+1}$ guaranties that $k \in \mathbb{M}_1 \cup \ldots \cup \mathbb{M}_k$, which gives the exhausting property. □

If the number of restrictions in Theorem 2.1 is infinite, then the relation between the algebra $\mathcal{A}$ and its restrictions may be quite loose. For example, there could be a sequence $\mathbf{A}$ in $\mathcal{A}$ such that every restriction $\mathbf{A}|_{\mathbb{M}_k}$ tends to zero in the norm, but $\mathbf{A}$ does not belong to $\mathcal{G}$ (consider a sequence the restriction of which to $\mathbb{M}_k$ is $(P_1, 0, \ldots)$ for every k). This cannot happen if the number of restrictions is finite, which leads to the following definition.

Definition 2.2. A C^*-subalgebra $\mathcal{A}$ of $\mathcal{F}$ is called *piecewise fractal* if there are finitely many infinite subsets $\mathbb{M}_1, \ldots, \mathbb{M}_k$ of $\mathbb{N}$ with

$$\mathbb{M}_i \cap \mathbb{M}_j = \emptyset \quad \text{for } i \neq j \quad \text{and} \quad \cup_{i=1}^k \mathbb{M}_i = \mathbb{N} \tag{2.1}$$

such that every restriction $\mathcal{A}|_{\mathbb{M}_i}$ is a fractal subalgebra of $\mathcal{F}|_{\mathbb{M}_i}$.

A typical example of a piecewise fractal algebra (in fact, a close relative of the algebra of the FSD for Toeplitz operators) will be examined in the following section. It is clear that, in piecewise fractal algebras, properties (F1) – (F3) hold separately on each of the finitely many fractal restrictions.

It turns out that several important properties of a sequence $\mathbf{A}$ in $\mathcal{F}$ can be expressed in terms of the family of *all* fractal restrictions of $\mathbf{A}$. To explain this observation, we introduce a class of subalgebras of $\mathcal{F}$ which is still small enough to own a useful fractality property, but which is also large enough to cover all separable subalgebras of $\mathcal{F}$.

Definition 2.3. A C^*-subalgebra $\mathcal{A}$ of $\mathcal{F}$ is called *quasifractal* if every restriction of $\mathcal{A}$ has a fractal restriction.

Lemma 2.4. (a) *Piecewise fractal C^*-subalgebras of $\mathcal{F}$ are quasifractal.*
(b) *Separable C^*-subalgebras of $\mathcal{F}$ are quasifractal.*

Proof. Let $\mathcal{A}$ be piecewise fractal and let $\mathbb{M}$ be an infinite subset of $\mathbb{N}$. Then there is an infinite subset $\mathbb{M}_i$ of $\mathbb{N}$ as in (2.1) such that the intersection $\mathbb{M} \cap \mathbb{M}_i =: \mathbb{K}$ is infinite. Then $\mathbb{K}$ defines a fractal restriction of $\mathcal{A}|_{\mathbb{M}}$, which proves (a). Assertion (b) is a direct consequence of the fractal restriction theorem. □

Proposition 2.5. *Let $\mathcal{A}$ be a quasifractal C^*-subalgebra of $\mathcal{F}$. Then a sequence $\mathbf{A} \in \mathcal{A}$ is a zero sequence (is stable) if and only if every fractal restriction of $\mathbf{A}$ goes to zero (is stable, respectively).*

Proof. If $\mathbf{A}$ is a zero sequence, then every restriction of $\mathbf{A}$ goes to zero as well. If $\mathbf{A} = (A_n)$ is not in $\mathcal{G}$, there are a restriction η and a positive constant C such that $\|A_{\eta(n)}\| \geq C$ for all $n \in \mathbb{N}$. Due to the quasifractality of $\mathcal{A}$, there is a fractal restriction μ of η. The restricted sequence $(A_{\mu(n)})$ does not tend to zero. The argument for stability is similar. □

In particular, this result holds when $\mathbf{A}$ is a sequence in $\mathcal{F}$ and $\mathcal{A}$ is the smallest C^*-subalgebra of $\mathcal{F}$ which contains $\mathbf{A}$. Since $\mathcal{A}$ is separable, it is quasifractal. Note also that the fractal exhaustion theorem (Theorem 2.1) holds for general quasifractal algebras (in place of separable algebras) as well.

3. The FSD for block Toeplitz operators

We are now going to extend the results cited in the introduction to the FSD for Toeplitz operators with matrix-valued generating functions, which will provide us with an archetypal example of a piecewise fractal algebra.

3.1. Block Toeplitz operators

Throughout this section, N denotes a fixed positive integer. For a C^*-subalgebra B of $L^\infty(\mathbb{T})$, we write $B^{N\times N}$ for the C^*-algebra of all $N \times N$-matrices with entries in B. The elements of $B^{N\times N}$ are considered as functions on $\mathbb{T}$ with values in $\mathbb{C}^{N\times N}$.

Let $a \in L^\infty(\mathbb{T})^{N\times N}$. The kth Fourier coefficient a_k of a is given as in (1.2). We define the Toeplitz operator $T(a)$ and the Hankel operator $H(a)$ with generating function a via their matrix representations $(a_{i-j})_{i,j\geq 0}$ and $(a_{i+j+1})_{i,j\geq 0}$ with respect to the standard basis of $l^2(\mathbb{Z}^+)$ in verbatim the same way as for $N = 1$, having in mind that in the present setting, the a_k are $N \times N$-matrices. To emphasize the latter fact, $T(a)$ and $H(a)$ are usually referred to as *block Toeplitz* and *block Hankel operators*. For B as above, we write $\mathsf{T}(B^{N\times N})$ for the smallest closed subalgebra of $L(l^2(\mathbb{Z}^+))$ which contains all Toeplitz operators $T(a)$ with $a \in B^{N\times N}$.

Every Toeplitz operator $T(a)$ generated by a (scalar-valued) function $a \in L^\infty(\mathbb{T})$ can also be viewed as an $N \times N$-block Toeplitz operator generated by a certain function $a_{\langle N\rangle} \in L^\infty(\mathbb{T})^{N\times N}$. In particular, if a is a trigonometric polynomial a, then $a_{\langle N\rangle}$ has only finitely many non-vanishing Fourier coefficients and is, hence, a function in $C^{N\times N} := C(\mathbb{T})^{N\times N}$. Since the trigonometric polynomials are dense in $C(\mathbb{T})$ we obtain the following.

Proposition 3.1. $\mathsf{T}(B) \subseteq \mathsf{T}(B^{N\times N})$ *for* $B = C(\mathbb{T}), L^\infty(\mathbb{T})$.

This inclusion holds for other function classes as well, e.g., for $B = PC$, the algebra of the piecewise continuous functions. We will not need these results in the present paper.

The analogue of Theorem 1.1 reads as follows.

Theorem 3.2. $\mathsf{T}(C^{N\times N}) = \{T(a) + K : a \in C(\mathbb{T})^{N\times N}$ *and* $K \in K(l^2(\mathbb{Z}^+))\}$.

Proof. As in the case $N = 1$ one can show that the right-hand side is a C^*-subalgebra of $L(l^2(\mathbb{Z}^+))$. Since this algebra contains all Toeplitz operators $T(a)$ with $a \in C(\mathbb{T})^{N\times N}$, the inclusion $\subseteq$ follows. For the reverse inclusion, we have to show $K(l^2(\mathbb{Z}^+)) \subseteq \mathsf{T}(C^{N\times N})$. This follows from $K(l^2(\mathbb{Z}^+)) \subseteq \mathsf{T}(C)$ by Theorem 1.1 and $\mathsf{T}(C) \subseteq \mathsf{T}(C^{N\times N})$ by Proposition 3.1. □

3.2. An adapted FSD for block Toeplitz operators

Let the filtration $\mathcal{P} = (P_n)$ and the reflection operators R_n on $l^2(\mathbb{Z}^+)$ be as in Theorem 1.2, and let $a \in C(\mathbb{T})^{N\times N}$. In contrast to the case $N = 1$, where every finite section $P_nT(a)P_n$ is a finite Toeplitz matrix again, the block Toeplitz structure of the $P_nT(a)P_n$ gets lost when $N > 1$ and n is not divisible by N. It is therefore only natural to consider the *adapted* or restricted sequence $(P_{nN}T(a)P_{nN})_{n\ge1}$ instead of the full sequence $(P_nT(a)P_n)_{n\ge1}$ of all finite sections of $T(a)$. Accordingly, we set $\mathcal{P}_N := (P_{nN})_{n\ge1}$ and write $\mathcal{S}_{N\mathbb{N}}(\mathsf{T}(C^{N\times N}))$ for the smallest closed subalgebra of $\mathcal{F}^{\mathcal{P}_N}$ which contains all sequences $(P_{nN}T(a)P_{nN})_{n\ge1}$ with $a \in C^{N\times N}$. The algebra $\mathcal{S}(\mathsf{T}(C^{N\times N}))$ of the *full* FSD for block Toeplitz operators, which is generated by the sequences $(P_nT(a)P_n)_{n\ge1}$, will be the subject of the following section.

A common basis both for the adapted and the full FSD is provided by the following lemma.

Lemma 3.3. *Let* $0 \le i < N$. *The strong limits*

$$W(\mathbf{A}) := \operatorname*{s-lim}_{n\to\infty} A_nP_n, \quad \widetilde{W}_i(\mathbf{A}) := \operatorname*{s-lim}_{n\to\infty} R_{nN+i}A_{nN+i}R_{nN+i}$$

exist for every sequence $\mathbf{A} = (A_n) \in \mathcal{S}(\mathsf{T}(C^{N\times N}))$. *In particular, if* $\mathbf{A}$ *is* $(P_nT(a)P_n)$ *with* $a \in C(\mathbb{T})^{N\times N}$, *then* $W(\mathbf{A}) = T(a)$ *and* $\widetilde{W}_i(\mathbf{A}) = T(\widetilde{a}_i)$ *with*

$$\widetilde{a}_i(t) := \begin{cases} R_N\, a(t^{-1})\, R_N & \text{if } i = 0, \\ \begin{pmatrix} R_i & 0 \\ 0 & tR_{N-i} \end{pmatrix} a(t^{-1}) \begin{pmatrix} R_i & 0 \\ 0 & t^{-1}R_{N-i} \end{pmatrix} & \text{if } i > 0. \end{cases} \tag{3.1}$$

The operators R_k in (3.1) are understood as $k \times k$ matrices.

Proof. The existence of the strong limits is either evident or follows from (3.1), which on its hand rests on the equality

$$R_{nN+i}T(a)R_{nN+i} = P_{nN+i}T(\widetilde{a}_i)P_{nN+i}, \tag{3.2}$$

holding for general $a \in L^\infty(\mathbb{T})^{N\times N}$. Note that it is *clear* that (3.2) holds with a *certain* function $\widetilde{a_i}$. The concrete form of these functions, as shown in (3.1), follows by straightforward, but somewhat tedious, calculations showing that the kth Fourier coefficient of $\widetilde{a_i}$ coincides with the kth Fourier coefficient of the function on the right-hand side of the equality (3.1). □

Theorem 3.4. (a) *The algebra $\mathcal{S}_{N\mathbb{N}}(\mathsf{T}(C^{N\times N}))$ of the adapted FSD coincides with the set of all sequences*

$$(P_{nN}T(a)P_{nN} + P_{nN}KP_{nN} + R_{nN}LR_{nN} + G_{nN})_{n\geq 1} \tag{3.3}$$

where $a \in C(\mathbb{T})^{N\times N}$, $K,\, L \in K(l^2(\mathbb{Z}^+))$, and $(G_n) \in \mathcal{G}^{\mathcal{P}}$.

(b) *The sequence (3.3) is stable if and only if the operators $T(a) + K$ and $T(\widetilde{a_0}) + L$ are invertible.*

Proof. Let $\mathcal{S}$ denote the set of all sequences (3.3). Proceeding as in the proof of Theorem 1.2 (a) and using Lemma 3.3, which we need here for $i = 0$ only, we obtain that $\mathcal{S}$ is a C^*-subalgebra of $\mathcal{F}$ and that W and $\widetilde{W_0}$ are *-homomorphisms on $\mathcal{S}$. Since $\mathcal{S}$ contains all sequences $(P_{nN}T(a)P_{nN})$ with $a \in C(\mathbb{T})^{N\times N}$, we conclude that $\mathcal{S}_{N\mathbb{N}}(\mathsf{T}(C^{N\times N})) \subseteq \mathcal{S}$.

For the reverse inclusion we have to show that all sequences

$$(P_{nN}KP_{nN} + R_{nN}LR_{nN} + G_{nN})_{n\geq 1}$$

with $K,\, L \in K(l^2(\mathbb{Z}^+))$ and $(G_n) \in \mathcal{G}$ belong to the algebra $\mathcal{S}_{N\mathbb{N}}(\mathsf{T}(C^{N\times N}))$. From Theorem 1.2 (a) and Lemma 3.1 we know that

$$(P_nKP_n + R_nLR_n + G_n)_{n\geq 1} \in \mathcal{S}(\mathsf{T}(C)) \subseteq \mathcal{S}(\mathsf{T}(C^{N\times N})),$$

hence the restriction of that sequence to $N\mathbb{N}$ belongs to $\mathcal{S}_{N\mathbb{N}}(\mathsf{T}(C^{N\times N}))$. This settles the proof of (a). Assertion (b) follows as in the proof of Theorem 1.2 (b). □

As a by-product we obtain that the algebra $\mathcal{S}_{N\mathbb{N}}(\mathsf{T}(C^{N\times N}))$ of the adapted FSD can also be characterized as the smallest closed subalgebra of $\mathcal{F}^{\mathcal{P}_N}$ which contains all sequences $(P_{nN}AP_{nN})_{n\geq 1}$ with $A \in \mathsf{T}(C^{N\times N})$.

3.3. The full FSD for block Toeplitz operators

Now we turn our attention to the algebra $\mathcal{S}(\mathsf{T}(C^{N\times N}))$ of the full FSD for block Toeplitz operators. In analogy with Theorems 1.2 (b) and 3.4, we will derive a complete description of that algebra. For that goal, we define the remainder function $\kappa : \mathbb{N} \to \{0, 1, \ldots, N-1\}$ such that N divides $n - \kappa(n)$.

Theorem 3.5. (a) *The algebra $\mathcal{S}(\mathsf{T}(C^{N\times N}))$ of the full FSD coincides with the set of all sequences*

$$(P_nT(a)P_n + P_nKP_n + R_nL_{\kappa(n)}R_n + G_n)_{n\geq 1} \tag{3.4}$$

where $a \in C(\mathbb{T})^{N\times N}$, $K,\, L_0,\, L_1,\, \ldots,\, L_{N-1} \in K(l^2(\mathbb{Z}^+))$, and $(G_n) \in \mathcal{G}^{\mathcal{P}}$.

(b) *The sequence (3.4) is stable if and only if the operators $T(a) + K$ and $T(\widetilde{a_i}) + L_i$ are invertible for every $0 \leq i < N$.*

Proof. (a) Let again $\mathcal{S}$ denote the set of all sequences (3.4). The inclusion $\mathcal{S}(\mathsf{T}(C^{N\times N})) \subseteq \mathcal{S}$ follows as in the proof of Theorem 3.4, using now Lemma 3.3 in its general form. The more interesting part of the proof is the reverse inclusion $\mathcal{S} \subseteq \mathcal{S}(\mathsf{T}(C^{N\times N}))$.

The sequences $(P_n T(a) P_n)$ with $a \in C(\mathbb{T})^{N\times N}$ belong to $\mathcal{S}(\mathsf{T}(C^{N\times N}))$ by definition. From Proposition 3.1 we infer that $\mathcal{S}(\mathsf{T}(C)) \subseteq \mathcal{S}(\mathsf{T}(C^{N\times N}))$; hence, the sequences $(P_n K P_n)$ with $K \in K(l^2(\mathbb{Z}^+))$ and the sequences in $\mathcal{G}^{\mathcal{P}}$ belong to $\mathcal{S}(\mathsf{T}(C^{N\times N}))$ by Theorem 1.2 (a). It remains to show that the sequences

$$(0, \ldots, 0, R_j L R_j, 0, \ldots, 0, R_{j+N} L R_{j+N}, 0, \ldots)$$

(starting with a block of $j-1$ zeros; all subsequent blocks of zeros have length $N-1$) belong to $\mathcal{S}(\mathsf{T}(C^{N\times N}))$ for every $1 \le j \le N$ and $L \in K(l^2(\mathbb{Z}^+))$. Since the algebra $\mathcal{S}(\mathsf{T}(C^{N\times N}))$ is closed, it is sufficient to show that all sequences

$$(0, \ldots, 0, R_j P_k L P_k R_j, 0, \ldots, 0, R_{j+N} P_k L P_k R_{j+N}, 0, \ldots)$$

with $k \in \mathbb{N}$ belong to $\mathcal{S}(\mathsf{T}(C^{N\times N}))$. This sequence is the product of the sequence $(R_n P_k L P_k R_n)_{n\ge1}$, which is in $\mathcal{S}(\mathsf{T}(C))$ by Theorem 1.2 (a) and hence also in $\mathcal{S}(\mathsf{T}(C^{N\times N}))$, with the sequence

$$(0, \ldots, 0, R_j P_k R_j, 0, \ldots, 0, R_{j+N} P_k R_{j+N}, 0, \ldots). \tag{3.5}$$

So it remains to show that these sequences are in $\mathcal{S}(\mathsf{T}(C^{N\times N}))$ for every $k \in \mathbb{N}$. This task can be further reduced to showing that the sequence

$$(0, \ldots, 0, R_j P_1 R_j, 0, \ldots, 0, R_{j+N} P_1 R_{j+N}, 0, \ldots) \tag{3.6}$$

is in $\mathcal{S}(\mathsf{T}(C^{N\times N}))$ for every $1 \le j \le N$. Indeed, with the shift operators $V_{\pm1}$ defined on $L^2(\mathbb{Z}^+)$ by

$$V_1 : (x_0, x_1, \ldots) \mapsto (0, x_0, x_1, \ldots), \quad V_{-1} : (x_0, x_1, \ldots) \mapsto (x_1, x_2, \ldots),$$

we have $(P_n V_{\pm1} P_n)_{n\ge1} \in \mathcal{S}(\mathsf{T}(C))$ and

$$P_n V_{-1} P_n \cdot R_n P_1 R_n \cdot P_n V_1 P_n = R_n (P_2 - P_1) R_n.$$

Thus, if the sequence (3.6) is in $\mathcal{S}(\mathsf{T}(C^{N\times N}))$, then the sequence

$$(0, \ldots, 0, R_j(P_2 - P_1)R_j, 0, \ldots, 0, R_{j+N}(P_2 - P_1)R_{j+N}, 0, \ldots),$$

obtained by multiplying (3.6) by $(P_n V_{-1} P_n)$ from the left and by $(P_n V_1 P_n)$ from the right, is in $\mathcal{S}(\mathsf{T}(C^{N\times N}))$; hence, the sequence (3.5) is in $\mathcal{S}(\mathsf{T}(C^{N\times N}))$ when $k = 2$. Repeating this argument we get the assertion for general k.

So we are left with verifying that (3.6) is in $\mathcal{S}(\mathsf{T}(C^{N\times N}))$. Now for $1 \le j \le N$, let $B_j = (b_{kl})_{k,l=1}^N$ and $D_j = (d_{kl})_{k,l=1}^N$ be the $N \times N$-matrices with $b_{j1} = d_{jj} = 1$ and with all other entries being zero, and set

$$A_j := \begin{pmatrix} 0 & B_j & 0 & 0 & \\ 0 & 0 & B_j & 0 & \\ 0 & 0 & 0 & B_j & \\ & & & & \ddots \end{pmatrix}, \quad C_j := \begin{pmatrix} C_j & 0 & 0 & \\ 0 & C_j & 0 & \\ 0 & 0 & C_j & \\ & & & \ddots \end{pmatrix}.$$

These are block Toeplitz operators with polynomial generating function, hence the sequences $(P_nA_jP_n)$ and $(P_nC_jP_n)$ belong to $\mathcal{S}(\mathsf{T}(C^{N\times N}))$. A straightforward computation gives

$$P_nC_jP_n - P_nA_jP_nA_j^*P_n$$
$$= \begin{cases} \operatorname{diag}(0,\ldots,0) & \text{if } 1 \le n < j, \\ \operatorname{diag}(0,\ldots,0,1,0,\ldots,0) & \text{if } j \le n \le N, \\ \operatorname{diag}(0,\ldots,0) & \text{if } N+1 \le n < N+j, \\ \operatorname{diag}(0,\ldots,0,1,0,\ldots,0) & \text{if } N+j \le n \le 2N, \\ \operatorname{diag}(0,\ldots,0) & \text{if } 2N+1 \le n < 2N+j, \\ \operatorname{diag}(0,\ldots,0,1,0,\ldots,0) & \text{if } 2N+j \le n \le 3N \end{cases}$$

and so on, with the ones standing at the jth, $(N+j)$th and $(2N+j)$th position in lines 2, 4 and 6, respectively. For $j = N$ we conclude that the sequence $(E_n^N)_{n=1}^\infty$ with

$$E_n^N := \begin{cases} \operatorname{diag}(0,\ldots,0,1) & \text{if } n = kN, \\ \operatorname{diag}(0,\ldots,0) & \text{else} \end{cases}$$

belongs to $\mathcal{S}(\mathsf{T}(C^{N\times N}))$. Similarly, for $j = N-1$ and $j = N-2$, the sequences (E_n^j) with

$$E_n^{N-1} := \begin{cases} \operatorname{diag}(0,\ldots,0,1) & \text{if } n = kN-1, \\ \operatorname{diag}(0,\ldots,0,1,0) & \text{if } n = kN, \\ \operatorname{diag}(0,\ldots,0) & \text{else} \end{cases}$$

and

$$E_n^{N-2} := \begin{cases} \operatorname{diag}(0,\ldots,0,1) & \text{if } n = kN-2, \\ \operatorname{diag}(0,\ldots,0,1,0) & \text{if } n = kN-1, \\ \operatorname{diag}(0,\ldots,0,1,0,0) & \text{if } n = kN, \\ \operatorname{diag}(0,\ldots,0) & \text{else} \end{cases}$$

are elements of $\mathcal{S}(\mathsf{T}(C^{N\times N}))$. Employing a shift argument as before we conclude that with (E_n^N) also the sequence $(E_n^{N,1})_{n=1}^\infty$ with

$$E_n^{N,1} := \begin{cases} \operatorname{diag}(0,\ldots,0,1,0) & \text{if } n = kN, \\ \operatorname{diag}(0,\ldots,0) & \text{else} \end{cases}$$

and hence the sequence $(F_n^{N-1}) := (E_n^{N-1}) - (E_n^{N,1})$ with

$$F_n^{N-1} := \begin{cases} \operatorname{diag}(0,\ldots,0,1) & \text{if } n = kN-1, \\ \operatorname{diag}(0,\ldots,0) & \text{else} \end{cases}$$

belongs to $\mathcal{S}(\mathsf{T}(C^{N\times N}))$. Similarly, with (E_n^N) and (F_n^{N-1}), also the shifted sequences $(E_n^{N,2})$ and $(F_n^{N-1,1})$ with

$$E_n^{N,2} := \begin{cases} \operatorname{diag}(0,\ldots,0,1,0,0) & \text{if } n = kN, \\ \operatorname{diag}(0,\ldots,0) & \text{else} \end{cases}$$

and

$$F_n^{N-1,1} := \begin{cases} \operatorname{diag}(0,\ldots,0,1,0) & \text{if } n = kN-1, \\ \operatorname{diag}(0,\ldots,0) & \text{else} \end{cases}$$

belong to $\mathcal{S}(\mathsf{T}(C^{N\times N}))$. Then also the sequence

$$(F_n^{N-2}) := (E_n^{N-2}) - (E_n^{N,2}) - (F_n^{N-1,1})$$

with

$$F_n^{N-2} := \begin{cases} \operatorname{diag}(0, \ldots, 0, 1) & \text{if } n = kN - 2, \\ \operatorname{diag}(0, \ldots, 0) & \text{else} \end{cases}$$

lies in $\mathcal{S}(\mathsf{T}(C^{N\times N}))$. So we have found that the sequences $(F_n^N) := (E_n^N)$, (F_n^{N-1}), and (F_n^{N-2}), i.e., the sequences (3.6) with $j = N$, $j = N-1$, and $j = N-2$, are in $\mathcal{S}(\mathsf{T}(C^{N\times N}))$. Continuing in this way, we get the assertion for general j. This finishes the proof of assertion (a); assertion (b) follows again as in the proof of Theorem 1.2 (b). □

Corollary 3.6. *The algebra $\mathcal{S}(\mathsf{T}(C^{N\times N}))/\mathcal{G}$ is *-isomorphic to the C^*-algebra of all $(N+1)$-tuples $(W(\mathbf{A}), \widetilde{W}_0(\mathbf{A}), \ldots, \widetilde{W}_{N-1}(\mathbf{A}))$ with $\mathbf{A} \in \mathcal{S}(\mathsf{T}(C^{N\times N}))$.*

4. Quasifractal algebras

4.1. An example

We start with a concrete example of a quasifractal algebra which we will obtain by a discretization of continuous functions of Toeplitz operators. Let $X = [0, 1]$ (or another compact metric, hence separable, space) and $(\xi_n)_{n\geq 1}$ a dense sequence in X. Let $\mathcal{S}(X, \mathsf{T}(C))$ stand for the smallest C^*-subalgebra of $\mathcal{F}$ which contains all sequences $(P_n A(\xi_n) P_n)$ where $A : X \to \mathsf{T}(C)$ is a continuous function. If we apply this discretization to a constant function $A : X \to \mathsf{T}(C)$, we just get the usual FSD for A. In this sense, we have $\mathcal{S}(\mathsf{T}(C)) \subseteq \mathcal{S}(X, \mathsf{T}(C))$.

Theorem 4.1. *The algebra $\mathcal{S}(X, \mathsf{T}(C))$ is quasifractal.*

Proof. Consider an arbitrary restriction of $\mathcal{S}(X, \mathsf{T}(C))$ given by a strictly increasing sequence η. By compactness, the sequence $(\xi_{\eta(n)})$ has a convergent subsequence $(\xi_{\mu(n)})$ with limit $\mu^* \in X$. Let $A : X \to \mathsf{T}(C)$ be continuous. Then $\|A(\mu(n)) - A(\mu^*)\| \to 0$. Hence, the sequence $(P_{\mu(n)} A(\mu(n)) P_{\mu(n)})$ differs from the sequence $(P_{\mu(n)} A(\mu^*) P_{\mu(n)}) \in \mathcal{S}(\mathsf{T}(C))_\mu$ by a zero sequence. This shows that $\mathcal{S}(X, \mathsf{T}(C))_\mu = \mathcal{S}(\mathsf{T}(C))_\mu$. Since $\mathcal{S}(\mathsf{T}(C))$ is fractal, this implies the fractality of the restriction $\mathcal{S}(X, \mathsf{T}(C))_\mu$. Since η was arbitrary, the algebra $\mathcal{S}(X, \mathsf{T}(C))$ is quasifractal. □

4.2. The fractal variety of an algebra

Let $\mathcal{C}$ be a sequence of unital C^*-algebras and $\mathcal{A}$ be a C^*-subalgebra of the algebra $\mathcal{F}^{\mathcal{C}}$. By $\operatorname{fr}\mathcal{A}$ we denote the set of all infinite subsets $\mathbb{M}$ of $\mathbb{N}$ such that the restriction $\mathcal{A}|_{\mathbb{M}}$ is fractal. We say that $\mathbb{M}_1, \mathbb{M}_2 \in \operatorname{fr}\mathcal{A}$ are *equivalent* if $\mathbb{M}_1 \cup \mathbb{M}_2 \in \operatorname{fr}\mathcal{A}$. This relation is reflexive and symmetric. The following lemma implies that it is also transitive and, hence, an equivalence relation.

Lemma 4.2. *If $\mathbb{M}_1, \mathbb{M}_2 \in \operatorname{fr}\mathcal{A}$ and $\mathbb{M}_1 \cap \mathbb{M}_2$ is infinite, then $\mathbb{M}_1 \cup \mathbb{M}_2 \in \operatorname{fr}\mathcal{A}$.*

Proof. Let $\mathbb{M}$ be an infinite subset of $\mathbb{M}_1 \cup \mathbb{M}_2$, and let $\mathbf{A} \in \mathcal{A}$ be a sequence for which $\mathbf{A}|_\mathbb{M}$ is a zero sequence. We show that then $\mathbf{A}|_{\mathbb{M}_1 \cup \mathbb{M}_2}$ is a zero sequence, whence the fractality of $\mathcal{A}|_{\mathbb{M}_1 \cup \mathbb{M}_2}$ by definition.

One of the sets $\mathbb{M} \cap \mathbb{M}_1$, $\mathbb{M} \cap \mathbb{M}_2$ is infinite; say $\mathbb{M} \cap \mathbb{M}_1$. Then $\mathbf{A}|_{\mathbb{M} \cap \mathbb{M}_1}$ is a zero subsequence of $\mathbf{A}|_{\mathbb{M}_1} \in \mathcal{A}|_{\mathbb{M}_1}$. Since $\mathcal{A}|_{\mathbb{M}_1}$ is fractal, $\mathbf{A}|_{\mathbb{M}_1}$ is a zero sequence. But then $\mathbf{A}|_{\mathbb{M}_1 \cap \mathbb{M}_2}$ is a zero subsequence of $\mathbf{A}|_{\mathbb{M}_2} \in \mathcal{A}|_{\mathbb{M}_2}$. Since $\mathcal{A}|_{\mathbb{M}_2}$ is fractal, we conclude that $\mathbf{A}|_{\mathbb{M}_2}$ is a zero sequence. Thus, $\mathbf{A}|_{\mathbb{M}_1 \cup \mathbb{M}_2}$ is a zero sequence. □

We write $\mathbb{M}_1 \sim \mathbb{M}_2$ if $\mathbb{M}_1, \mathbb{M}_2 \in \operatorname{fr} \mathcal{A}$ are equivalent, denote the set of all equivalence classes of the relation $\sim$ by $(\operatorname{fr} \mathcal{A})^\sim$, and call $(\operatorname{fr} \mathcal{A})^\sim$ the *fractal variety* of $\mathcal{A}$. If $\mathcal{A}$ is fractal, then $(\operatorname{fr} \mathcal{A})^\sim$ is a singleton, consisting of the equivalence class of $\mathbb{N}$.

Our goal is to define a topology on $(\operatorname{fr} \mathcal{A})^\sim$ which makes $(\operatorname{fr} \mathcal{A})^\sim$ to a compact Hausdorff space. For $\mathcal{A}$ as above, let $\mathcal{L}(\mathcal{A})$ denote the smallest closed complex subalgebra of $l^\infty := l^\infty(\mathbb{N})$ which contains all sequences $(\|A_n\|)$ where (A_n) is a sequence in $\mathcal{A}$. Clearly, $\mathcal{L}(\mathcal{A})$ is a commutative C^*-algebra, and $\mathcal{L}(\mathcal{A})$ is unital if $\mathcal{A}$ is unital.

For a C^*-subalgebra $\mathcal{L}$ of l^∞, we let $\operatorname{cr} \mathcal{L}$ stand for the set of all infinite subsets $\mathbb{M}$ of $\mathbb{N}$ such that all sequences in the restriction $\mathcal{L}|_\mathbb{M}$ converge. The algebra $\mathcal{L}$ is called *quasiconvergent* if every infinite subset of $\mathbb{N}$ has an infinite subset in $\operatorname{cr} \mathcal{L}$.

Proposition 4.3. *If $\mathcal{A}$ is a C^*-subalgebra of $\mathcal{F}$, then* $\operatorname{fr} \mathcal{A} = \operatorname{cr} \mathcal{L}(\mathcal{A})$.

Proof. If $\mathbb{M} \in \operatorname{fr} \mathcal{A}$, then the sequence $(\|A_n\|)_{n \in \mathbb{M}}$ converges for every sequence $(A_n) \in \mathcal{A}$ by Fact (F3) in the introduction; hence, $\mathbb{M} \in \operatorname{cr} \mathcal{L}(\mathcal{A})$. Conversely, let $\mathbb{M} \in \operatorname{cr} \mathcal{L}(\mathcal{A})$, and let $(A_n)_{n \in \mathbb{M}}$ be a partial zero sequence in $\mathcal{A}|_\mathbb{M}$. Then the sequence $(\|A_n\|)$ is in $\mathcal{L}(\mathcal{A})$; hence the sequence $(\|A_n\|)_{n \in \mathbb{M}}$ converges. The limit of this sequence is necessarily equal to 0; hence $(A_n)_{n \in \mathbb{M}}$ is a zero sequence, and $\mathcal{A}|_\mathbb{M}$ is fractal by definition. □

Corollary 4.4. *A C^*-subalgebra $\mathcal{A}$ of $\mathcal{F}$ is quasifractal if and only if the associated C^*-subalgebra $\mathcal{L}(\mathcal{A})$ of l^∞ is quasiconvergent.*

4.3. Quasiconvergent algebras

Let c and c_0 denote the algebras of the convergent sequences and of the zero sequences on $\mathbb{N}$, respectively. The restrictions of l^∞, c, and c_0 to an infinite subset $\mathbb{M}$ of $\mathbb{N}$ can be identified with $l^\infty(\mathbb{M})$, $c(\mathbb{M})$, and $c_0(\mathbb{M})$.

Let $\mathcal{L}$ be a C^*-subalgebra of l^∞ and $\mathbb{M}$ be an infinite subset of $\mathbb{N}$. The restriction $\mathcal{L}|_\mathbb{M}$ is called *non-degenerated* if $\mathcal{L}|_\mathbb{M}$ is not contained in $c_0(\mathbb{M})$. The algebra $\mathcal{L}$ is called *non-degenerated* if no restriction of $\mathcal{L}$ to an infinite subset of $\mathbb{N}$ is degenerated. Every unital algebra $\mathcal{L}$ is non-degenerated.

For every $\mathbb{M} \in \operatorname{cr} \mathcal{L}$, the mapping

$$\varphi_\mathbb{M} : \mathcal{L} \to \mathbb{C}, \quad a \mapsto \lim(a|_\mathbb{M}) \tag{4.1}$$

is a continuous linear functional on $\mathcal{L}$ which is a character if $\mathbb{M}$ is non-degenerated. Since $\mathcal{L} \cap c_0$ is in the kernel of the mapping (4.1), the quotient mapping

$$\varphi_{\mathbb{M}} : \mathcal{L}/(\mathcal{L} \cap c_0) \to \mathbb{C}, \quad a + (\mathcal{L} \cap c_0) \mapsto \lim(a|_{\mathbb{M}}) \tag{4.2}$$

is well defined. This mapping is a character of $\mathcal{L}/(\mathcal{L} \cap c_0)$ if $\mathbb{M}$ is non-degenerated.

Proposition 4.5. *Let $\mathcal{L}$ be a unital and quasiconvergent C^*-subalgebra of l^∞. Then the set $\{\varphi_{\mathbb{M}} : \mathbb{M} \in \operatorname{cr} \mathcal{L}\}$ is strictly spectral for $\mathcal{L}/(\mathcal{L} \cap c_0)$, i.e., if $b \in \mathcal{L}/(\mathcal{L} \cap c_0)$ and $\varphi_{\mathbb{M}}(b)$ is invertible for all $\mathbb{M} \in \operatorname{cr} \mathcal{L}$, then b is invertible.*

Proof. Suppose that $a + (\mathcal{L} \cap c_0)$ is not invertible in $\mathcal{L}/(\mathcal{L} \cap c_0)$. Then $a + c_0$ is not invertible in $\mathcal{L}/c_0$; hence, a is a partial zero sequence. Let $\mathbb{M}'$ be an infinite subset of $\mathbb{N}$ such that $a|_{\mathbb{M}'} \to 0$. Since $\mathcal{L}$ is quasiconvergent, there is an infinite subset $\mathbb{M}$ of $\mathbb{M}'$ which belongs to $\operatorname{cr} \mathcal{L}$. The character associated with $\mathbb{M}$ satisfies $\varphi_{\mathbb{M}}(a) = 0$. Conversely, if $a \in \mathcal{L}$ and $\varphi_{\mathbb{M}}(a) \neq 0$ for all $\mathbb{M} \in \operatorname{cr} \mathcal{L}$, then $a + (\mathcal{L} \cap c_0)$ is invertible in $\mathcal{L}/(\mathcal{L} \cap c_0)$. This is the strict spectral property. □

To conclude that $\{\varphi_{\mathbb{M}} : \mathbb{M} \in \operatorname{cr} \mathcal{L}\}$ is all of the maximal ideal space $\operatorname{Max}(\mathcal{L}/(\mathcal{L} \cap c_0))$ we need a further property of $\mathcal{L}$: separability.

Proposition 4.6. *Let $\mathcal{L}$ be a unital, separable, and quasiconvergent C^*-subalgebra of l^∞. Then $\{\varphi_{\mathbb{M}} : \mathbb{M} \in \operatorname{cr} \mathcal{L}\} = \operatorname{Max}(\mathcal{L}/(\mathcal{L} \cap c_0))$.*

Proof. The assertion is a consequence of an observation by Nistor and Prudhon: since $\mathcal{L}/(\mathcal{L} \cap c_0)$ is separable, every strictly spectral family for $\mathcal{L}/(\mathcal{L} \cap c_0)$ is exhaustive (see [5] for the terminology and a proof). A short direct proof of the proposition goes as follows. Let φ be a character of $\mathcal{L}/(\mathcal{L} \cap c_0)$. We extend φ to a character on $\mathcal{L}$ by $\varphi : a \mapsto \varphi(a + (\mathcal{L} \cap c_0))$. Since $\mathcal{L}$ is separable, the kernel of φ is separable. Let $(j_n)_{n \in \mathbb{N}}$ be a sequence which is dense in $\ker \varphi$. Then the element

$$j := \sum_{j=1}^{\infty} \frac{1}{2^n} \frac{j_n^* j_n}{\|j_n\|^2}$$

belongs to $\ker \varphi$, implying that $j + c_0$ is not invertible in $\mathcal{L}/c_0$. By Proposition 4.5, there is a set $\mathbb{M} \in \operatorname{cr} \mathcal{L}$ such that $\varphi_{\mathbb{M}}(j) = 0$. Since characters are positive, we conclude that $\varphi_{\mathbb{M}}(j_n^* j_n) = 0$, hence $\varphi_{\mathbb{M}}(j_n) = 0$ for all $n \in \mathbb{N}$. The continuity of $\varphi_{\mathbb{M}}$ and the density of (j_n) in $\ker \varphi$ imply that $\varphi_{\mathbb{M}}$ vanishes on $\ker \varphi$. Thus, the characters φ and $\varphi_{\mathbb{M}}$ coincide. □

To make the equality established in the previous proposition to a bijection between (cosets of) $\operatorname{cr} \mathcal{L}$ and $\operatorname{Max}(\mathcal{L}/(\mathcal{L} \cap c_0))$, we need to understand which sets $\mathbb{M} \in \operatorname{cr} \mathcal{L}$ generate the same character $\varphi_{\mathbb{M}}$. Proceeding similarly as in the previous section, we call $\mathbb{M}_1, \mathbb{M}_2 \in \operatorname{cr} \mathcal{L}$ *equivalent* if $\mathbb{M}_1 \cup \mathbb{M}_2 \in \operatorname{cr} \mathcal{L}$. The so-defined relation $\sim$ is an equivalence relation, and $\mathbb{M}_1 \sim \mathbb{M}_2$ if and only if $\varphi_{\mathbb{M}_1} = \varphi_{\mathbb{M}_2}$. We denote the equivalence class of $\mathbb{M} \in \operatorname{cr} \mathcal{L}$ by $\mathbb{M}^\sim$ and write $(\operatorname{cr} \mathcal{L})^\sim$ for the set of all equivalence classes. Then, by construction, the mapping

$$(\operatorname{cr} \mathcal{L})^\sim \to \{\varphi_{\mathbb{M}} : \mathbb{M} \in \operatorname{cr} \mathcal{L}\}, \quad \mathbb{M}^\sim \mapsto \varphi_{\mathbb{M}}$$

is a (well defined) bijection. Combining this observation with the result of Proposition 4.6 we obtain the following.

Corollary 4.7. *Let $\mathcal{L}$ be a unital, separable, and quasiconvergent C^*-subalgebra of l^∞. Then* $\mathbb{M}^\sim \mapsto \varphi_\mathbb{M}$ *is a bijection from* $(\operatorname{cr}\mathcal{L})^\sim$ *onto* $\operatorname{Max}(\mathcal{L}/(\mathcal{L}\cap c_0))$.

4.4. Quasifractal algebras as continuous fields

Recall from Proposition 4.3 that $\operatorname{fr}\mathcal{A} = \operatorname{cr}\mathcal{L}(\mathcal{A})$ for every C^*-subalgebra $\mathcal{A}$ of $\mathcal{F}$. If $\mathcal{A}$ is quasifractal, then $\mathcal{L}(\mathcal{A})$ is quasiconvergent by Corollary 4.4, and the relations $\sim$ on $\operatorname{fr}\mathcal{A}$ and $\operatorname{cr}\mathcal{L}(\mathcal{A})$ are compatible in the sense that $(\operatorname{fr}\mathcal{A})^\sim = (\operatorname{cr}\mathcal{L}(\mathcal{A}))^\sim$. Thus, if $\mathcal{A}$ is unital and quasifractal and $\mathcal{L}(\mathcal{A})$ is separable, then there is a (well defined) bijection

$$(\operatorname{fr}\mathcal{A})^\sim \to \operatorname{Max}(\mathcal{L}(\mathcal{A})/(\mathcal{L}(\mathcal{A})\cap c_0)), \quad \mathbb{M}^\sim \mapsto \varphi_\mathbb{M}. \tag{4.3}$$

This bijection transfers the Gelfand topology of $\operatorname{Max}(\mathcal{L}(\mathcal{A})/(\mathcal{L}(\mathcal{A})\cap c_0))$ onto $(\operatorname{fr}\mathcal{A})^\sim$, thus making the latter to a compact Hausdorff space.

We claim that the algebra $\mathcal{A}/(\mathcal{A}\cap\mathcal{G})$ is *-isomorphic to a continuous field of C^*-algebras over the base space $(\operatorname{fr}\mathcal{A})^\sim$ in the following sense[1].

Definition 4.8. Let X be a compact Hausdorff space and let $\mathcal{B}$ be the direct product of a family $\{\mathcal{B}_x\}_{x\in X}$ of C^*-algebras, labeled by X. A *continuous field of C^*-algebras over X* is a C^*-subalgebra $\mathcal{C}$ of $\mathcal{B}$ with the following properties:

(a) $\mathcal{C}$ is maximal, i.e., $\mathcal{B}_x = \{c(x) : c \in \mathcal{C}\}$ for every $x \in X$,
(b) the function $X \to \mathbb{C}$, $x \mapsto \|c(x)\|$ is continuous for every $c \in \mathcal{C}$.

The algebras $\mathcal{B}_x$ are called the fibers of $\mathcal{A}$, and X is the base space.

Set $X = (\operatorname{fr}\mathcal{A})^\sim$, for $\mathbb{M} \in \operatorname{fr}\mathcal{A}$ define $\mathcal{B}_\mathbb{M}$ as $\mathcal{A}|_\mathbb{M}/(\mathcal{A}|_\mathbb{M}\cap\mathcal{G}|_\mathbb{M})$ (note that these algebras depend on the equivalence class $\mathbb{M}^\sim$ of $\mathbb{M}$ only), and let $\mathcal{B}$ be the direct product of the family $\{\mathcal{B}_\mathbb{M}\}_{\mathbb{M}\in\operatorname{fr}\mathcal{A}}$. Every sequence $\mathbf{A} \in \mathcal{A}$ determines a function in $\mathcal{B}$ via

$$\mathbb{M} \mapsto \mathbf{A}|_\mathbb{M} + (\mathcal{A}|_\mathbb{M}\cap\mathcal{G}|_\mathbb{M}) \in \mathcal{A}|_\mathbb{M}/(\mathcal{A}|_\mathbb{M}\cap\mathcal{G}|_\mathbb{M}). \tag{4.4}$$

Let $\mathcal{C}$ be the set of all functions (4.4) with $\mathbf{A} \in \mathcal{A}$.

Theorem 4.9. *Let $\mathcal{A}$ be a unital and quasifractal C^*-subalgebra of $\mathcal{F}$ for which $\mathcal{L}(\mathcal{A})$ is separable. Then*

(a) *$\mathcal{C}$ is a continuous field of C^*-algebras over* $(\operatorname{fr}\mathcal{A})^\sim$,
(b) *the mapping which sends $\mathbf{A} + (\mathcal{A}\cap\mathcal{G})$ to the function* (4.4) *is a *-isomorphism from $\mathcal{A}/(\mathcal{A}\cap\mathcal{G})$ onto $\mathcal{C}$.*

Proof. (a) Evidently, $\mathcal{C}$ is maximal. Let $\mathbf{A} = (A_n) \in \mathcal{A}$. Then

$$\|\mathbf{A}|_\mathbb{M} + (\mathcal{A}|_\mathbb{M}\cap\mathcal{G}|_\mathbb{M})\| = \lim_{n\in\mathbb{M}} \|A_n\| = \varphi_\mathbb{M}(\mathbf{a} + (\mathcal{L}(\mathcal{A})\cap c_0));$$

where $\mathbf{a} := (\|A_n\|) \in \mathcal{L}$. Since $\mathcal{M} \mapsto \varphi_\mathbb{M}(\mathbf{a} + (\mathcal{L}(\mathcal{A})\cap c_0))$ is a continuous function, it follows that condition (b) of Definition 4.8 is also satisfied.

[1]Note that one usually adds a third condition to the definition of a continuous field, namely that $\mathcal{C}$ is a $C(X)$-algebra.

(b) It is evident that this mapping is a surjective *-homomorphism. If $\mathbf{A} \in \mathcal{A}$ and $\mathbf{A}|_{\mathbb{M}} \in \mathcal{G}|_{\mathbb{M}}$ for every $\mathbb{M} \in \text{fr}\,\mathcal{A}$, then $\mathbf{A} \in \mathcal{G}$ by Proposition 2.5. Thus, the mapping in assertion (b) is also injective. □

To state our last result, we need some more notation. For $\mathcal{F}$ as in the setting of the FSD for Toeplitz operators, let $\mathcal{K}$ denote the smallest closed ideal of $\mathcal{F}$ which contains all sequences (K_n) with $\sup \text{rank}\, K_n < \infty$. The sequences in $\mathcal{K}$ are called *compact*. Further, a C^*-algebra is called *elementary* if it is *-isomorphic to an algebra $K(H)$, the compact operators on a certain Hilbert space H, and a C^*-algebra is called *dual* if it is *-isomorphic to a direct sum of elementary algebras. See [1] for more on dual algebras.

For example, the compact sequences in $\mathcal{S}(\mathsf{T}(C))$ are just the sequences

$$(A_n) = (P_n K P_n + R_n L R_n + G_n)$$

where K and L are compact operators on $l^2(\mathbb{Z}^+)$ and $(G_n) \in \mathcal{G}$, and the algebra $(\mathcal{S}(\mathsf{T}(C)) \cap \mathcal{K})/\mathcal{G}$ is isomorphic to the algebra of all pairs (K, L), hence, to the direct sum of two copies of $K(l^2(\mathbb{Z}^+))$.

A basic observation in [7] states that $(\mathcal{A} \cap \mathcal{K})/\mathcal{G}$ is a dual algebra whenever $\mathcal{A}$ is fractal. Combining this observation with Theorem 4.9 we obtain the following.

Corollary 4.10. *Let $\mathcal{A}$ be a unital and quasifractal C^*-subalgebra of $\mathcal{F}$ for which $\mathcal{L}(\mathcal{A})$ is separable. Then $\mathcal{A}/(\mathcal{A} \cap \mathcal{K})$ is *-isomorphic to a continuous field of dual algebras over* $(\text{fr}\,\mathcal{A})^\sim$.

References

[1] M.C.F. Berglund, Ideal C^*-algebras, *Duke Math. J.* **40** (1973), 241–257.

[2] A. Böttcher and B. Silbermann, The finite section method for Toeplitz operators on the quarter-plane with piecewise continuous symbols, *Math. Nachr.* **110** (1983), 279–291.

[3] A. Böttcher and B. Silbermann, *Introduction to Large Truncated Toeplitz Matrices*, Springer, New York 1999.

[4] R. Hagen, S. Roch, and B. Silbermann, *C^*-Algebras and Numerical Analysis*, Marcel Dekker, Inc., New York, 2001.

[5] V. Nistor and N. Prudhon, Exhausting families of representations and spectra of pseudodifferential operators, http://front.math.ucdavis.edu/1411.7921.

[6] S. Roch, Algebras of approximation sequences: Fractality, in: *Problems and Methods in Mathematical Physics*, Oper. Theory: Adv. Appl. **121**, Birkhäuser, Basel, 2001, 471–497.

[7] S. Roch, Algebras of approximation sequences: Fredholm theory in fractal algebras, *Studia Math.* **150** (2002), no. 1, 53–77.

[8] S. Roch, Extension-restriction theorems for algebras of approximation sequences, In *Proc. WOAT* 2016, Lisbon (to appear).

[9] S. Roch and B. Silbermann, C^*-algebra techniques in numerical analysis, *J. Oper. Theory* **35** (1996), no. 2, 241–280.

[10] B. Silbermann, Lokale Theorie des Reduktionsverfahrens für Toeplitzoperatoren, *Math. Nachr.* **104** (1981), 137 – 146.

Steffen Roch
Technische Universität Darmstadt
Fachbereich Mathematik
Schlossgartenstrasse 7
64289 Darmstadt
Germany
e-mail: `roch@mathematik.tu-darmstadt.de`

Unbounded operators on Hilbert C^*-modules and C^*-algebras

Konrad Schmüdgen

Abstract. Hilbert C^*-modules are generalizations of Hilbert spaces equipped with scalar products taking values in C^*-algebras. The failure of the projection theorem leads to new difficulties for the operator theory on Hilbert C^*-modules compared to the Hilbert space setting. In this paper we discuss two classes of unbounded operators (regular operators, graph regular operators) on Hilbert C^*-modules and C^*-algebras.

Mathematics Subject Classification (2010). Primary 46L08, Secondary 47C15.

Keywords. Hilbert C^*-module, unbounded operator, regular operator, affiliated operator.

1. Why operators on Hilbert C^*-modules?

Unbounded operators on Hilbert C^*-modules play an important role in several fields of mathematics and mathematical physics. They appear as *unbounded Fredholm modules or Kasparov modules* in K-theory of C^*-algebras and as *Dirac operators* in noncommutative geometry. In the C^*-approach to quantum field theory (Haag–Kastler axioms) *observables* (for instance, the field operators) can be considered as operators on the local C^*-algebras.

The main driving force for developing a theory of "well-behaved" (that is, regular) unbounded operators on C^*-modules was the theory of noncompact quantum groups. In the C^*-approach pioneered by S.L. Woronowicz, the *coordinate functions* act as (in general, unbounded) operators on the corresponding C^*-algebra. We briefly explain this for the quantum $(ax+b)$-group. Let q be a complex number of modulus one. The coordinate algebra of the quantum $(ax+b)$-group is the unital $*$-algebra X with two Hermitian generators a, b satisfying the relation

$$ab = qba. \tag{1.1}$$

In addition, this algebra is equipped with the structure of a Hopf $*$-algebra, but we will not carry out this here. Suppose that $q^2 \neq 1$. In this case, all

© Springer International Publishing AG, part of Springer Nature 2018
A. Böttcher et al. (eds.), *The Diversity and Beauty of Applied Operator Theory*, Operator Theory: Advances and Applications 268, https://doi.org/10.1007/978-3-319-75996-8_23

self-adjoint operators a and b satisfying (1.1) are unbounded and a class of "good" representations of (1.1) has to be selected. Then a C^*-algebra A is constructed which carries the Hopf $*$-algebra structure such that these self-adjoint operators act as regular operators on A. Since a large part of the algebraic quantum group theory is encoded in the coordinate algebra X, a close relationship between the operators a, b and the C^*-algebra A is required. Details can be found in [23]. For noncompact quantum groups unbounded operators on C^*-algebras are crucial and advanced results of their theory are really needed.

Apart from the class of regular operators, unbounded operator theory on C^*-algebras, more generally on Hilbert C^*-modules, is still at the very beginning and it has not (yet) obtained much attention among operator theorists. In this article I will give a short introduction into this subject addressed also to non-experts. I will put the main emphasis on operator-theoretic aspects, not on C^*-algebras! I will avoid technicalities as far as possible and point out the difference to ordinary Hilbert space theory. Among others, I will discuss a new class of unbounded operators (graph regular operators) invented in a recent joint paper [6] of the author with R. Gebhardt, see also [7] for more details. In my opinion, graph regular operators are an important class of operators beyond regular operators. An elaboration of this concept (for instance, its relations to representation theory) and the search for further important and useful classes of operators should be the next steps towards an operator theory on Hilbert C^*-modules.

Regular (unbounded) operators on Hilbert C^*-modules were introduced by S. Baaj [1] and in a different setting (as affiliated operators on C^*-algebras) by S.L. Woronowicz [21]. Fundamental results on affiliated operators on C^*-algebras were obtained in [22]. Regular operators are treated in Chapters 9 and 10 of C. Lance's book [13]. A slightly larger class of densely defined operators are the semiregular operators invented by A. Pal [16]. Operators on Hilbert C^*-modules have been studied in [2], [8], [11], [12], [16], [17], [4], [9], [6], [15], and others. A comprehensive overview on the literature concerning Hilbert C^*-modules and related topics was compiled by M. Frank [5].

Proofs of results and facts stated with no explicit reference can be found in [13] for regular operators and in [6] for graph regular operators.

Throughout this paper, A denotes a (not necessarily unital) C^*-algebra and E, F, G are Hilbert C^*-modules over A.

The author would like to thank R. Gebhardt for useful discussions on the subject of this paper.

2. Hilbert C^*-modules

Roughly speaking, a Hilbert C^*-module over A is a right A-module which is equipped with an A-valued scalar product and complete in the corresponding norm. The precise definition is the following.

Definition 2.1. A *pre-Hilbert C^*-module* E over the C^*-algebra A is a complex vector space E which is also right A-module, together with a sesquilinear map $\langle \cdot, \cdot \rangle_E : E \times E \to A$, such that for $x, y, z \in E$ and $a \in A$:

$$\langle x, ya \rangle_E = \langle x, y \rangle_E \, a,$$
$$\langle x, y \rangle_E = (\langle y, x \rangle_E)^*,$$
$$\langle x, x \rangle_E \geq 0 \text{ and } \langle x, x \rangle_E = 0 \implies x = 0.$$

If $(E, \|.\|_E)$ is complete, where $\|.\|_E$ is the norm

$$\|x\|_E := \|\langle x, x \rangle_E\|_A^{1/2}, \; x \in E,$$

then E is called a *Hilbert C^*-module* over A, briefly a *Hilbert A-module.*

If no confusion can arise we write $\langle \cdot, \cdot \rangle$ instead of $\langle \cdot, \cdot \rangle_E$. In the case $A = \mathbb{C}$, Hilbert A-modules are just "ordinary" complex Hilbert spaces.

Example 1. *First standard example: $E = A$.*

Clearly, $E := A$ is a Hilbert A-module with multiplication as right action and A-valued scalar product

$$\langle a, b \rangle := a^* b, \; a, b \in E.$$

In this case, $\|a\|_E = \|a\|_A$ for $a \in A$.

Example 2. *Second standard example: $E = l_2(A)$.*

$$l_2(A) := \left\{ (x_n)_{n=1}^{\infty} : x_n \in A, \sum_{n=1}^{\infty} x_n^* x_n \text{ converges in } A \right\}$$

is a Hilbert A-module with pointwise operations and

$$\langle (x_n), (y_n) \rangle := \sum_{n=1}^{\infty} x_n^* y_n, \; (x_n), (y_n) \in l_2(A).$$

If the C^*-algebra A is infinite-dimensional, then $l_2(A)$ is different from

$$H_1 := \left\{ (x_n)_{n=1}^{\infty} : \sum_{n=1}^{\infty} \|x_n\|^2 < \infty \right\} \subset l_2(A).$$

Definition 2.2. The *orthogonal complement* of a subset M of E is

$$M^{\perp} := \{ x \in E : \langle x, y \rangle = 0 \text{ for } y \in M \}.$$

Obviously, $M^{\perp}$ is a closed submodule of E and hence itself a Hilbert A-module.

Definition 2.3. A submodule G of E is called

- *orthogonally closed* if $G = G^{\perp\perp}$,
- *orthogonally complemented* if $G \oplus G^{\perp} = E$,
- *essential* if $G^{\perp} = \{0\}$.

Each orthogonally closed submodule is closed, but the converse is not true.

Example 3. *A proper closed submodule which is essential.*

Let $E = A = C([0,1])$. Then $G = \{f \in E : f(0) = 0\}$ is a closed submodule of E such that $G \neq E$ and $G^\perp = \{0\}$. In particular, $G \oplus G^\perp \neq E$.

The preceding example shows that a projection theorem does not hold for Hilbert C^*-modules! From the technical side, this failure is the main reason for most of the difficulties and pathologies in operator theory on Hilbert C^*-modules.

3. Operators on Hilbert C^*-modules

Definition 3.1. An *operator* $t : E \to F$ is a $\mathbb{C}$-linear A-linear map of E into F defined on a right submodule $\mathcal{D}(t)$ of E, that is,

$$t(\lambda x) = \lambda t(x) \quad \text{and} \quad t(xa) = t(x)a \quad \text{for} \quad \lambda \in \mathbb{C},\ x \in \mathcal{D}(t),\ a \in A.$$

Note that the A-linearity $t(xa) = t(x)a$ is a very strong requirement.

Suppose that $t : E \to F$ is an *essentially defined* operator, that is, $\mathcal{D}(t)^\perp = \{0\}$. Set

$$\mathcal{D}(t^*) := \{y \in F | \exists\ z \in E : \langle tx, y\rangle_F = \langle x, z\rangle_E \text{ for } x \in \mathcal{D}(t)\}.$$

Since $\mathcal{D}(t)^\perp = \{0\}$, z is uniquely determined by y. Define $t^*y := z$. It is easily verified that $t^* : F \to E$ is an operator, called the *adjoint* of t, and

$$\langle tx, y\rangle_F = \langle x, t^*y\rangle_E \quad \text{for} \quad x \in \mathcal{D}(t),\ y \in \mathcal{D}(t^*).$$

Definition 3.2. An essentially defined operator $t : E \to E$ is called *symmetric* if $t \subseteq t^*$ and *self-adjoint* if $t = t^*$.

These definitions are very similar to the corresponding definitions of operators on Hilbert spaces. However, in contrast to ordinary Hilbert space theory there are two crucial differences:

- Self-adjoint operators are not necessarily densely defined!
- Self-adjoint operators are not necessarily "good" operators!

(For instance, $(t + \mathrm{i})E$ and $(t^2 + 1)E$ are not dense in general!)

In order to remedy these drawbacks, regularity conditions are needed, see Definition 4.1 below.

Definition 3.3. An operator $t : E \to F$ is called *orthogonally closed* if its *graph* $\mathcal{G}(t) := \{(x, tx) : x \in \mathcal{D}(t)\}$ satisfies $\mathcal{G}(t)^{\perp\perp} = \mathcal{G}(t)$.

An orthogonally closed operator is closed, the converse does not hold.

It seems that the notion of an "orthogonally closed operator" is more important than that of a "closed operator", because the former is better adapted to the A-valued scalar products.

4. Regular operators

The following definition introduces the first fundamental notion of this paper.

Definition 4.1. A closed operator $t : E \to F$ is called *regular* if

$$\mathcal{D}(t) \text{ is dense in } E,\ \mathcal{D}(t^*) \text{ is dense in } F, \text{ and } (1 + t^*t)E \text{ is dense in } E.$$

The set of regular operators of E into F is denoted by $\operatorname{Reg}(E, F)$.

Regular operators have been invented by S. Baaj [1]. In the special case $E = F = A$ (see Example 2) the regular operators $t : E \to F$ are precisely the *affiliated operators* of A in the sense of S.L. Woronowicz [21].

The next theorem contains basic results on regular operators.

Theorem 4.2. *Suppose that $t : E \to F$ is a closed operator such that t and t^* are densely defined. Then the following are equivalent:*

(i) $t \in \operatorname{Reg}(E, F)$.
(ii) *$\mathcal{G}(t)$ is orthogonally complemented: $\mathcal{G}(t) \oplus \mathcal{G}(t)^\perp = E \oplus F$.*
(iii) $(1 + t^*t)E = E$.

If $t \in \operatorname{Reg}(E, F)$, then $t^* \in \operatorname{Reg}(F, E)$, $t = t^{**}$, and $(1 + tt^*)F$ is dense in F.

As noted in [16], $t^* \in \operatorname{Reg}(F, E)$ does not imply that $t \in \operatorname{Reg}(E, F)$. (The corresponding implication of [13, Corollary 9.6] does not hold.) However, if t is densely defined and orthogonally closed, then $t = t^{**}$ and hence, by the last statement of Theorem 4.2, $t^* \in \operatorname{Reg}(F, E)$ implies $t \in \operatorname{Reg}(E, F)$.

The following theorem (proved in [18]) gives a regularity criterion by means of resolvents. For this we assume that the C^*-algebra A is realized on a Hilbert space $\mathcal{H}$ and that $E = A$. The multiplier algebra of A is (isomorphic to) the C^*-algebra

$$M(A) = \{a \in \mathbf{B}(\mathcal{H}) : aA \subseteq A,\ Aa \subseteq A\}. \tag{4.1}$$

Theorem 4.3. *Let T be a closed operator on $\mathcal{H}$ with non-empty resolvent set $\rho(T)$ and let $\lambda \in \rho(T)$. Then T is a regular operator on $E = A$ if and only if $(T - \lambda I)^{-1} \in M(A)$ and the ranges $(T - \lambda I)^{-1}A$ and $(T^* - \overline{\lambda} I)^{-1}A$ are dense in A.*

Since Theorem 4.3 is formulated in terms of resolvents, it is better adapted to standard operator theory. In general, the criteria in Theorems 4.2 and 4.3 are difficult to verify. In Theorem 4.3 the denseness of the ranges $(T - \lambda I)^{-1}A$ and $(T^* - \overline{\lambda} I)^{-1}A$ in A is crucial. It can be shown that these ranges are dense in A (with respect to the norm of A!) if for each (irreducible) representation of A their images are dense in the image of A in the corresponding Hilbert space norm. The latter denseness condition is much easier to deal with.

In the next example A is the C^*-algebra $\mathcal{K}(\mathcal{H})$ of compact operators on a Hilbert space $\mathcal{H}$.

Example 4. $A = \mathcal{K}(\mathcal{H})$.

Obviously, $M(A) = \mathbf{B}(\mathcal{H})$. Then each densely defined closed operator T on $\mathcal{H}$ acts as a regular operator on $E = A = \mathcal{K}(\mathcal{H})$.

That is, operator theory on an ordinary Hilbert space $\mathcal{H}$ is just the theory of regular operators on the C^*-algebra of compact operators on $\mathcal{H}$.

In the seminal paper [22], K. Napiórkowski and S.L. Woronowicz have developed the basics of an operator theory for *regular* operators on $E = A$. It includes the following results (the last one was proved in [6]):

- polar decomposition;
- functional calculus of normal operators;
- self-adjoint extension theory of symmetric operators via Cayley transform;
- existence of self-adjoint extensions of positive symmetric operators;
- Nelson's integrability theorem of Lie algebra representations;
- Stone's theorem;
- Kato–Rellich theorem for relatively bounded symmetric operators.

5. Graph regular operators

The second main concept of this paper is the following notion. It was introduced and studied in [6].

Definition 5.1. An orthogonally closed operator $t : E \to F$ is *graph regular* if

$$\mathcal{D}(t)^\perp = \{0\},\ (1 + t^*t)E \text{ is dense in } E, \text{ and } (1 + tt^*)F \text{ is dense in } F.$$

Let $\mathrm{Reg}_{\mathrm{gr}}(E, F)$ denote the set of graph regular operators of E into F.

It is not difficult to show that if $t : E \to F$ is orthogonally closed and $\mathcal{D}(t)^\perp = \{0\}$, then also $\mathcal{D}(t^*)^\perp = \{0\}$. However, in contrast to regular operators, both denseness conditions in Definition 5.1 are needed.

The following theorems are the counter-part of Theorem 4.2 for graph regular operators.

Theorem 5.2. *Let $t : E \to F$ be an operator such that $\mathcal{D}(t)^\perp = \{0\}$ and $\mathcal{D}(t^*)^\perp = \{0\}$. Then the following conditions are equivalent:*

(i) $t \in \mathrm{Reg}_{\mathrm{gr}}(E, F)$.
(ii) *$\mathcal{G}(t)$ is orthogonally complemented: $\mathcal{G}(t) \oplus \mathcal{G}(t)^\perp = E \oplus F$.*
(iii) *$(1 + t^*t)E = E$ and $(1 + tt^*)F = F$.*

Theorem 5.3. *Let $t : E \to F$ be an orthogonally closed operator such that $\mathcal{D}(t)^\perp = \{0\}$. Then $t \in \mathrm{Reg}_{\mathrm{gr}}(E, F)$ if and only if $t^* \in \mathrm{Reg}_{\mathrm{gr}}(F, E)$.*

Comparing conditions (ii) in Theorems 4.2 and 5.2 yields the following.

Corollary 5.4. *Each regular operator is graph regular.*

There is no such nice theory as for regular operators and many pathologies can occur (see e.g. the multiplication operator t_m in Example 6 below).

As indicated above, there exist densely defined self-adjoint operators on Hilbert C^*-modules which are not regular. The first example of this kind was constructed by M. Hilsum [8]; this phenomenon was further elaborated in [9]. (The corresponding example in [13, pp. 103–104] is not correct.)

6. Some Examples

Because of the A-linearity condition in Definition 3.1, it is not completely obvious how to get examples of unbounded graph regular or regular operators. Each densely defined closed operator on a Hilbert space is a quotient of two bounded operators. Hence the first guess might be to look for a similar result on Hilbert modules.

Definition 6.1. An *adjointable operator* is an operator of

$$\mathcal{L}(E,F) := \{t : E \to F \text{ with } \mathcal{D}(t) = E,\ \mathcal{D}(t^*) = F\}.$$

Adjointable operators are always bounded, but a bounded operator defined on E is not necessarily adjointable.

Example 5. *Quotients "ba^{-1}" for $a \in \mathcal{L}(G,E)$, $b \in \mathcal{L}(G,F)$.*

Let $a \in \mathcal{L}(G,E)$, $b \in \mathcal{L}(G,F)$. Suppose $\ker(a) \subseteq \ker(b)$, $\ker(a^*) = \{0\}$. Define

$$\mathcal{D}(t) = aG, \quad t(ax) = bx,\ x \in G.$$

If t is closed, then $t : E \to F$ is graph regular and $t^* = (a^*)^{-1}b^*$. In particular, if $a \in \mathcal{L}(F,E)$ and $\ker(a) = \ker(a^*) = \{0\}$, then $a^{-1} : E \to F$ is graph regular.

Important examples of regular operators are provided by the next theorem; it follows from [22, Theorem 2.1].

Theorem 6.2. *Let G be a Lie group. Then each element of the Lie algebra of G acts as a regular operator on the C^*-algebra $E = C^*(G)$.*

Let X be a locally compact topological Hausdorff space and E the C^*-algebra $A = C_0(X)$ of continuous functions on X vanishing at infinity. For any function $m : X \to \mathbb{C}$ the multiplication operator t_m on E is defined by

$$\mathcal{D}(t_m) := \{f \in C_0(X) : m \cdot f \in C_0(X)\},\ t_m f := m \cdot f,\ f \in \mathcal{D}(t_m).$$

The following fact is proved in [21].

Proposition 6.3. *For each continuous function $m \in C(X)$ the operator t_m is regular on E. Each regular operator on E is of the form t_m with $m \in C(X)$.*

In sharp contrast, multiplication operators with discontinuous functions can be graph regular on E. We illustrate this with three examples on $X = \mathbb{R}$.

Example 6.

$$m(x) := \begin{cases} x^{-1}\exp(\mathrm{i}x^{-1}), & x \neq 0, \\ 0, & x = 0. \end{cases}$$

The operator t_m is graph regular on E. Moreover, $t_m^* t_m = t_m t_m^*$, that is, t_m is normal, but $\mathcal{D}(t_m) \neq \mathcal{D}(t_m^*)$.

Example 7. Both operators t_{m_0} and t_{m_1} are graph regular on E.

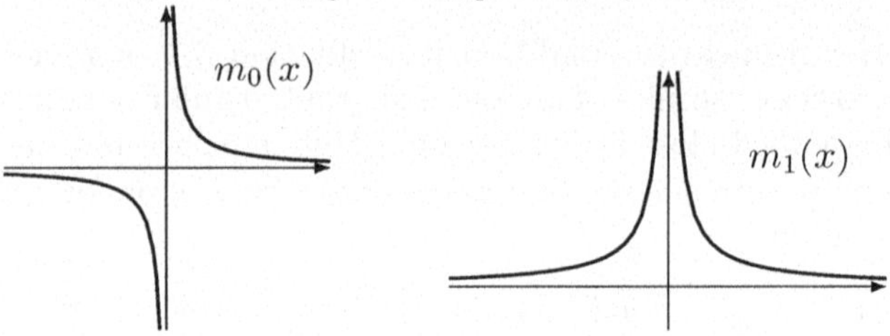

Example 8. The operator t_{m_2} is not graph regular on E.

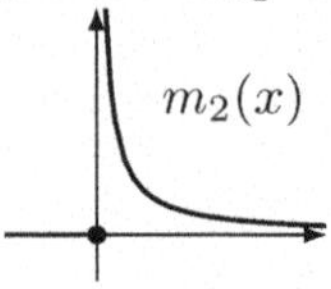

7. The bounded transform for Hilbert space operators

The bounded transform for Hilbert space operators was invented by W.F. Kaufman [10], see [20, Section 7.3] for proofs of the following facts.

Definition 7.1. Let T be a densely defined closed operator on a Hilbert space $\mathcal{H}$. The *bounded transform* of T is the operator

$$Z_T := T(I + T^*T)^{-1/2}.$$

The operator Z_T is a contraction defined on the whole Hilbert space $\mathcal{H}$ such that

$$\ker(I - (Z_T)^* Z_T) = \ker(I + T^*T)^{-1} = \{0\}.$$

The operator T can be recovered from Z_T by

$$T = Z_T\big(I - (Z_T)^* Z_T\big)^{-1/2}.$$

Conversely, if Z is a contraction such that $\ker(I - Z^*Z) = \{0\}$, then

$$T := Z(I - Z^*Z)^{-1/2}$$

is a densely defined closed operator T such that $Z = Z_T$.

Further, the mapping $T \to Z_T$ preserves adjoints, that is, $Z_{T^*} = (Z_T)^*$, and normality, that is, T is normal if and only if so is Z_T. In particular, Z_T is self-adjoint if and only if T is self-adjoint.

8. Graph regular operators and bounded transform

Now we develop the bounded transform for operators on Hilbert C^*-modules and use it to characterize graph regular operators and regular operators.

Definition 8.1. $\mathcal{Z}(E,F) := \{z \in \mathcal{L}(E,F) : \|z\| \le 1,\ \ker(I - z^*z) = \{0\}\}$,
$\mathcal{Z}_r(E,F) := \{z \in \mathcal{L}(E,F) : \|z\| \le 1,\ (I - z^*z)E \text{ is dense in } E\}$.

It is easily seen that $\mathcal{Z}_r(E,F)$ is a (in general proper) subset of $\mathcal{Z}(E,F)$. For $z \in \mathcal{Z}(E,F)$, we set

$$t_z := z(I - z^*z)^{-1/2}.$$

The operator $I-z^*z$ belongs to the C^*-algebra $\mathcal{L}(E,E)$. Since $z \in \mathcal{Z}(E,F)$, it is nonnegative and has trivial kernel, so the inverse $(I - z^*z)^{-1/2}$ of its square root $(I - z^*z)^{1/2}$ is a well-defined operator on the Hilbert C^*-module E with domain $(I - z^*z)^{1/2}E$.

Conversely, for $t \in \mathrm{Reg}_{\mathrm{gr}}(E,F)$ we define

$$E_t := \overline{\mathcal{D}(t^*t)},\ F_{t^*} := \overline{\mathcal{D}(tt^*)},\ z_t := t(I + t^*t)^{-1/2} \restriction_{E_t}.$$

The operator z_t is called the *bounded transform* of t. It can be shown that $\mathcal{D}(t^*t)$ is essential in E and that $\mathcal{D}(tt^*)$ is essential in F.

The following theorems are the main results concerning the bounded transform.

Theorem 8.2. *The map* $z \mapsto t_z$ *is a bijection of* $\mathcal{Z}_r(E,F)$ *onto* $\mathrm{Reg}(E,F)$. *In particular,* $z = t_z(I + t_z^*t_z)^{-1/2}$ *for* $z \in \mathcal{Z}_r(E,F)$.

Theorem 8.3. *The map* $z \mapsto t_z$ *is an injection of* $\mathcal{Z}(E,F)$ *into* $\mathrm{Reg}_{\mathrm{gr}}(E,F)$. *If* $t \in \mathrm{Reg}_{\mathrm{gr}}(E,F)$, *then* $t_{z_t} : E_t \to F_{t^*}$ *is a regular operator, called the* regular part *of the graph regular operator* t.

For the next theorem we assume that $E = A$. Then $\mathcal{L}(E,E)$ is the multiplier algebra $M(A)$. One may think of A as being realized on a Hilbert space $\mathcal{H}$; then $M(A)$ is given by (4.1). Recall from C^*-algebra theory that each $*$-representation of A extends uniquely to a $*$-representation of $M(A)$. Therefore, if $t \in \mathrm{Reg}(E)$, then $z_t \in M(A)$, so $\pi(z_t)$ is well-defined. The following basic result was proved in [21].

Theorem 8.4. *Let* $E = A$ *and* $t \in \mathrm{Reg}(E)$. *For each* $*$*-representation* π *of* A *there is a unique densely defined closed operator* $\pi(t)$ *on the Hilbert space* $\mathcal{H}(\pi)$ *such that* $Z_{\pi(t)} = \pi(z_t)$, *that is, the bounded transform of* $\pi(t)$ *is equal to* $\pi(z_t)$. *Moreover,*

$$\pi(t)(\pi((I - (z_t)^*z_t)^{1/2}a)\varphi) = \pi(z_ta)\varphi,\ a \in A,\ \varphi \in \mathcal{H}(\pi). \tag{8.1}$$

Theorem 8.4 says that a regular operator can be "mapped" to a densely defined closed operator (defined by (8.1)) in *each* representation of A. In contrast, graph regular operators can be "transported" to densely defined Hilbert space operators only in *certain* representations. For instance, for the graph regular operator t_m in Example 6 the point evaluation at 0 cannot be defined.

9. Further examples

9.1. Lie algebra of the Heisenberg group

Let H be the 3-dimensional Heisenberg group, that is, H is the Lie group of matrices

$$\begin{pmatrix} 1 & a & c \\ 0 & 1 & b \\ 0 & 0 & 1 \end{pmatrix}, \quad \text{where } a, b, c \in \mathbb{R}.$$

The Lie algebra of H has a basis $\{X, Y, Z\}$ with the commutation relations

$$[X, Y] = Z, \ [X, Z] = [Y, Z] = 0.$$

The family of irreducible unitary representations U of H consists of a series U_λ, $\lambda \in \mathbb{R}^\times$, of infinite-dimensional representations acting on $L^2(\mathbb{R})$ and of a series U_a, $a \in \mathbb{R}^2$, of one-dimensional representations. The corresponding actions of the Lie algebra generators are given by the formulas

$$\lambda \neq 0: \ dU_\lambda(X) = -\mathrm{i}\lambda x, \ \ dU_\lambda(Y) = \frac{d}{dx}, \ \ dU_\lambda(Z) = \mathrm{i}\lambda \cdot I,$$
$$\lambda = 0: \ dU_a(X) = \mathrm{i}a_1, \ \ dU_a(Y) = \mathrm{i}a_2, \ \ dU_a(Z) = 0, \ \ a = (a_1, a_2) \in \mathbb{R}^2.$$

The C^*-algebra $C^*(H)$ of H is completely described in [14, Theorem 2.16]. It consists of operator fields $F = (F(\lambda); \lambda \in \mathbb{R})$ such that $F(\lambda)$, $\lambda \neq 0$, is a compact operator on $L^2(\mathbb{R})$, $F(0) \in C_0(\mathbb{R}^2)$, and

$$\lim_{\lambda \to 0} \|F(\lambda) - \nu_\lambda(F(0))\| = 0, \tag{9.1}$$

where ν_λ is a linear map of $C_0(\mathbb{R}^2)$ into $L^2(\mathbb{R})$. Then we have the following.

Proposition 9.1. $(\mathrm{i}Z)^{-1}$ *is a graph regular self-adjoint operator on* $E = C^*(H)$.

Note that the operator $(\mathrm{i}Z)^{-1}$ is not regular, because $dU_a(\mathrm{i}Z) = 0$ for $a \in \mathbb{R}^2$ and hence $(\mathrm{i}Z)^{-1}$ is not densely defined.

9.2. Unbounded Toeplitz operators

For $\phi \in L^\infty(\mathbb{T})$ let T_ϕ denote the corresponding Toeplitz operator on $H^2(\mathbb{T})$. The C^*-algebra generated by the unilateral shift $S = T_z$ is the Toeplitz algebra:

$$\mathcal{T} := \{T_\phi : \phi \in C(\mathbb{T})\} \dot{+} \mathcal{K}(H^2(\mathbb{T})).$$

Let $p, q \in \mathbb{C}[z]$ be relatively prime polynomials such that $q \neq 0$ in $\mathbb{D}$. The Toeplitz operator $T_{p/q}$ is defined by

$$\mathcal{D}(T_{p/q}) := \{f \in H^2(\mathbb{T}) : (p/q)f \in H^2(\mathbb{T})\}, \ T_{p/q}f := (p/q)f, \ f \in \mathcal{D}(T_{p/q}).$$

Proposition 9.2. *If q has a zero on $\mathbb{T}$, then the operator $T_{p/q}$ on $E = \mathcal{T}$ is graph regular, but not regular.*

For instance, the operator $(S - I)^{-1}$ is graph regular, but not regular.

9.3. A fraction algebra related to the canonical commutation relations

Let $Q = x$ and $P = -\mathrm{i}\frac{d}{dx}$ be the position and momentum operators, respectively, on $L^2(\mathbb{R})$. Then the bounded operators

$$a := (Q - \mathrm{i}I)^{-1} \quad \text{and} \quad b := (P - \mathrm{i}I)^{-1}. \tag{9.2}$$

satisfy the commutation relations

$$a - a^* = 2\mathrm{i}a^*a = 2\mathrm{i}aa^*, \quad b - b^* = 2\mathrm{i}b^*b = 2\mathrm{i}bb^*, \tag{9.3}$$

$$ab - ba = -\mathrm{i}ab^2a = -\mathrm{i}ba^2b, \quad ab^* - b^*a = -\mathrm{i}a(b^*)^2a = -\mathrm{i}b^*a^2b^*. \tag{9.4}$$

Let A denote the universal unital C^*-algebra with generators a and b and defining relations (9.3) and (9.4). This algebra appeared in [19] and independently in the paper [3] of D. Buchholz and H. Grundling. In [19] it was used as a tool for proving a noncommutative Positivstellensatz for the Weyl algebra.

The set of irreducible representations of A consists of a single infinite-dimensional representation on $L^2(\mathbb{R})$ given by the operators (9.2) and of series of one-dimensional representations given by the points of the circles

$$K_1 := \{(a, 0) \in \mathbb{C}^2 : a - \overline{a} = 2\mathrm{i}|a|^2\}, \quad K_2 := \{(0, b) \in \mathbb{C}^2 : b - \overline{b} = 2\mathrm{i}|b|^2\}.$$

Suggested by equation (9.2) we define operators q and p on $E = A$ by

$$q := \mathrm{i}I + a^{-1}, \ \mathcal{D}(q) := aA, \quad \text{and} \quad p := \mathrm{i} + b^{-1}, \ \mathcal{D}(p) := bA.$$

Proposition 9.3. *q and p are graph regular self-adjoint operators on $E := A$.*

Looking at the one-dimensional representations it follows that the domains aA and bA are not dense in A. Hence q and p are not regular.

The C^*-algebra A contains the compacts $\mathcal{K}(L^2(\mathbb{R}))$ as an essential ideal and the restrictions of q and p are regular operators for $\mathcal{K}(L^2(\mathbb{R}))$. In fact, these restrictions are the regular parts of q and p.

10. Why graph regular operators?

In complex function theory isolated singularities of holomorphic functions such as

$$f_1(z) = z^{-1}, \quad f_2(z) = \exp(z^{-1})$$

are studied by the behavior of the functions in a neighborhood of the singularity.

Many C^*-algebras consist of operator fields $z \to a(z) \in A_z$, where A_z is a C^*-algebra on a Hilbert space $\mathcal{H}_z$. For such a C^*-algebra let us consider an operator field $z \to t(z)$ such that $t(z), z \neq z_0$, is a *regular operator* for the C^*-algebra A_z. In general, $t(z_0)$ is not defined. Often graph regular operators are of this form; for instance, the multiplication operator t_{m_0} with $m_0(x) = x^{-1}$ for $A = C_0(\mathbb{R})$ in Example 7, the operator $(\mathrm{i}Z)^{-1}$ for the C^*-algebra $C^*(H)$ in Example 9.1, and the operator $(S-I)^{-1}$ for the Toeplitz algebra in Example 9.2. Because of these examples, it is hoped that *graph regular operators become useful tools for the study of operator fields with isolated singularities.*

References

[1] S. Baaj, *Multiplicateurs non bórnes*, Thesis, Universite Pierre et Marie Curie, Paris, 1981.

[2] S. Baaj and P. Julg, Théorie bivariante de Kasparov et opérateurs non bornés dans les C*-modules hilbertiens, *C.R. Acad. Sci. Paris Ser. I Math.* **296** (1983), 875–878.

[3] D. Buchholz and H. Grundling, The resolvent algebras: a new approach to canonical quantum systems, *J. Funct. Anal.* **254** (2010), 2725–2779.

[4] M. Frank and K. Sharifi, Generalized inverses and polar decomposition of unbounded regular operators on Hilbert C^*-modules, *J. Operator Theory* **64** (2010), 377–386.

[5] M. Frank, *Hilbert C^*-modules and related subjects – a guided reference overview*, HTWK Leipzig, 89 pages, last update 31.3.2017.

[6] R. Gebhardt and K. Schmüdgen, Unbounded operators on Hilbert C^*-modules, *Intern. J. Math.* **26** (2015), 197–255.

[7] R. Gebhardt, *Unbounded operators on Hilbert C^*-modules: graph regular operators*, Thesis, University of Leipzig, 2016.

[8] M. Hilsum, Fonctorialité en K-théorie bivariante pour les varietés lipschitziennes, K-*Theory* **3** (1987), 401–440.

[9] J. Kaad and M. Lesch, A local global principle for regular operators in Hilbert C^*-modules, *J. Funct. Anal.* **262** (2012), 4540–4569.

[10] W.F. Kaufman, Representing a closed operator as a quotient of continuous operators, *Proc. Amer. Math. Soc.* **72** (1978), 531–534.

[11] D. Kucerovsky, The KK-product of unbounded modules, K-*Theory* **11** (1997), 17–34.

[12] D. Kucerovsky, Functional calculus and representations of $C_0(X)$ on a Hilbert module, *Quart. J. Math.* **53** (2002), 467–477.

[13] E.C. Lance, *Hilbert C^*-modules*, Cambridge Univ. Press, 1995.

[14] J. Ludwig and L. Turowska, The C^*-algebras of the Heisenberg group and of thread-like Lie groups, *Math. Z.* **268** (2011), 897–930.

[15] R. Meyer, Representations by unbounded operators, C^*-hulls, local-global principle, and induction, Preprint, University of Göttingen, 2016.

[16] A. Pal, Regular operators on Hilbert C^*-modules, *J. Operator Theory* **42** (1999), 331–350.

[17] F. Pierrot, Opérateurs réguliers dans les C^*-modules et structure des C^*-algèbres de groups de Lie semisimples complexes simplement connexes, *J. Lie Theory* **16** (2006), 651–689.

[18] K. Schmüdgen, Unbounded operators affiliated with C^*-algebras, Preprint, University of Leipzig, 2005.

[19] K. Schmüdgen, Algebras of fractions and strict Positivstellensätze for $*$-algebras, *J. reine angew. Math.* **647** (2010), 57–88.

[20] K. Schmüdgen, *Unbounded self-adjoint operators on Hilbert space*, Graduate Texts, Springer-Verlag, Dordrecht, 2012.

[21] S.L. Woronowicz, Unbounded elements affiliated with C^*-algebras and non-compact quantum groups, *Commun. Math. Phys.* **136** (1991), 399–432.

[22] S.L. Woronowicz and K. Napiórkowski, Operator theory in the C^*-algebra framework, *Reports Math. Phys.* **31** (1992), 353–371.

[23] S.L. Woronowicz and S. Zakrewski, Quantum $ax + b$ group, *Rev. Math. Phys.* **14** (2002), 797–828.

Konrad Schmüdgen
Universität Leipzig
Mathematisches Institut
Augustusplatz 10/11
04109 Leipzig
Germany
e-mail: schmuedgen@math.uni-leipzig.de

A characterization of positive normal functionals on the full operator algebra

Zoltán Sebestyén, Zsigmond Tarcsay and Tamás Titkos

Abstract. Using the recent theory of Krein–von Neumann extensions for positive functionals we present several simple criteria to decide whether a given positive functional on the full operator algebra $B(H)$ is normal. We also characterize those functionals defined on the left ideal of finite rank operators that have a normal extension.

Mathematics Subject Classification (2010). Primary 46K10, Secondary 46A22.

Keywords. Krein–von Neumann extension, normal functionals, trace.

The aim of this short note is to present a theoretical application of the generalized Krein–von Neumann extension, namely to offer a characterization of positive normal functionals on the full operator algebra. To begin with, let us fix our notations. Given a complex Hilbert space H, denote by $B(H)$ the full operator algebra, i.e., the C^*-algebra of continuous linear operators on H. The symbols $B_F(H), B_1(H), B_2(H)$ are referring to the ideals of continuous finite rank operators, trace class operators, and Hilbert–Schmidt operators, respectively. Recall that $B_2(H)$ is a complete Hilbert algebra with respect to the inner product

$$(X \,|\, Y)_2 = \mathrm{Tr}(Y^*X) = \sum_{e\in\mathcal{E}} (Xe \,|\, Ye), \qquad X, Y \in B_2(H).$$

Here Tr refers to the the trace functional and $\mathcal{E}$ is an arbitrary orthonormal basis in H. Recall also that $B_1(H)$ is a Banach *-algebra under the norm $\|X\|_1 := \mathrm{Tr}(|X|)$, and that $B_F(H)$ is dense in both $B_1(H)$ and $B_2(H)$, with respect to the norms $\|\cdot\|_1$ and $\|\cdot\|_2$, respectively. It is also known that $X \in B_1(H)$ holds if and only if X is the product of two elements of $B_2(H)$. For the proofs and further basic properties of Hilbert–Schmidt and trace class operators we refer the reader to [1, 2, 6].

Zsigmond Tarcsay was supported by the Hungarian Ministry of Human Capacities, NTP-NFTÖ-17. Corresponding author: Tamás Titkos.

© Springer International Publishing AG, part of Springer Nature 2018

A. Böttcher et al. (eds.), *The Diversity and Beauty of Applied Operator Theory*, Operator Theory: Advances and Applications 268, https://doi.org/10.1007/978-3-319-75996-8_24

Before dealing with normal functionals, we recall the notion of a representable functional. Let $\mathscr{B}$ be a complex *-algebra. A linear functional $g : \mathscr{B} \to \mathbb{C}$ is called *representable* if there exist a complex Hilbert space H_g with inner product $(\cdot \mid \cdot)_g$, a *-representation $\pi_g : \mathscr{B} \to B(H_g)$, and a cyclic vector $\zeta_g \in H_g$ such that

$$g(b) = (\pi_g(b)\zeta_g \mid \zeta_g)_g$$

holds for all $b \in \mathscr{B}$. We refer the reader to [5] for more details about representable functionals.

Let $\mathscr{A}$ be a von Neumann algebra, that is, a strongly closed *-subalgebra of $B(H)$ containing the identity. A continuous linear functional $f : \mathscr{A} \to \mathbb{C}$ is called normal if it is continuous in the ultraweak topology, that is, if f belongs to the predual of $\mathscr{A}$. It is well known that the predual of $B(H)$ is $B_1(H)$, hence every normal functional can be represented by a trace class operator. We will use this property as the definition.

Definition. A linear functional $f : B(H) \to \mathbb{C}$ is called a normal functional if there exists a trace class operator F such that

$$f(X) := \operatorname{Tr}(XF) = \operatorname{Tr}(FX), \quad X \in B(H).$$

We remark that such a functional is always continuous due to the inequality

$$|\operatorname{Tr}(XF)| \leq \|F\|_1 \cdot \|X\|.$$

Our main tool is a canonical extension theorem for linear functionals which is analogous to the well-known operator extension theorem named after the pioneers of the 20th century operator theory, M.G. Krein [3] and J. von Neumann [4]. For the details see Section 5 in [8], especially Theorem 5.6 and the subsequent comments.

Next, we recall the cited theorem.

A Krein–von Neumann type extension. *Let $\mathscr{I}$ be a left ideal of a complex Banach *-algebra $\mathscr{A}$, and consider a linear functional $\varphi : \mathscr{I} \to \mathbb{C}$. The following statements are equivalent:*

(a) *There is a representable positive functional $\varphi^\bullet : \mathscr{A} \to \mathbb{C}$ extending φ which is minimal in the sense that*

$$\varphi^\bullet(x^*x) \leq \widetilde{\varphi}(x^*x) \ \textit{holds for all } x \in \mathscr{A}$$

whenever $\widetilde{\varphi} : \mathscr{A} \to \mathbb{C}$ is a representable extension of φ.

(b) *There is a constant $C \geq 0$ such that $|\varphi(a)|^2 \leq C \cdot \varphi(a^*a)$ for all $a \in \mathscr{I}$.*

We remark that the construction used in the proof of the above theorem is closely related to the one developed in [7] for Hilbert space operators. The main advantage of that construction is that we can compute the values of the smallest extension $\varphi^\bullet$ on positive elements, namely

$$\varphi^\bullet(x^*x) = \sup\left\{|\varphi(x^*a)|^2 \,\middle|\, a \in \mathscr{I},\ \varphi(a^*a) \leq 1\right\} \text{ for all } x \in \mathscr{A}. \qquad (*)$$

The minimal extension $\varphi^\bullet$ is called the *Krein–von Neumann extension of* φ.

The characterization we are going to prove is as follows.

Main Theorem. *For a given positive functional $f : B(H) \to \mathbb{C}$ the following statements are equivalent:*

(i) *f is normal.*
(ii) *There exists a normal positive functional g such that $f \leq g$.*
(iii) *$f \leq g$ holds for every positive functional g that agrees with f on $B_F(H)$.*
(iv) *For every $X \in B(H)$ we have*

$$f(X^*X) = \sup\{|f(X^*A)|^2 \mid A \in B_F(H), f(A^*A) \leq 1\}. \qquad (**)$$

(v) $f(I) \leq \sup\{|f(A)|^2 \mid A \in B_F(H), f(A^*A) \leq 1\}$.

Proof. The proof is divided into three claims, which might be interesting on their own right. Before doing that, we make some observations. For a given trace class operator S let us denote by f_S the normal functional defined by

$$f_S(X) := \operatorname{Tr}(XS), \quad X \in B(H).$$

The map $S \mapsto f_S$ is order preserving between positive trace class operators and normal positive functionals. Indeed, if $S \geq 0$, then

$$f_S(A^*A) = \operatorname{Tr}(A^*AS) = \|AS^{1/2}\|_2^2 \geq 0.$$

Conversely, if f_S is a positive functional and $P_{\langle h\rangle}$ denotes the orthogonal projection onto the subspace spanned by $h \in H$, we obtain $S \geq 0$ by

$$(Sh \mid h) = \operatorname{Tr}(P_{\langle h\rangle}S) = f_S(P_{\langle h\rangle}^* P_{\langle h\rangle}) \geq 0 \text{ for all } h \in H.$$

Our first two claims will prove that (i) and (iv) are equivalent.

<u>*Claim 1*</u>. Let f be a normal positive functional and set $\varphi := f|_{B_F(H)}$. Then f is the smallest positive extension of φ, i.e., $\varphi^\bullet = f$.

Proof of Claim 1. Since $f \geq 0$ is normal, there is a positive $S \in B_1(H)$ such that $f = f_S$. By assumption φ has a positive extension (namely f itself is one), thus there exists also the Krein–von Neumann extension denoted by $\varphi^\bullet$. As $f_S - \varphi^\bullet$ is a positive functional due to the minimality of $\varphi^\bullet$, its norm is attained at the identity I. Therefore it is enough to show that

$$\varphi^\bullet(I) \geq f_S(I) = \operatorname{Tr}(S).$$

We know from $(*)$ that

$$\varphi^\bullet(X^*X) = \sup\{|\varphi(X^*A)|^2 \mid A \in B_F(H), \varphi(A^*A) \leq 1\}$$

for any $X \in B(H)$. Choosing $A = \operatorname{Tr}(S)^{-1/2}P$ for a projection P with finite rank, we see that $\varphi(A^*A) = \operatorname{Tr}(S)^{-1}\operatorname{Tr}(PS) \leq 1$, whence

$$\varphi^\bullet(I) \geq |\varphi(A)|^2 = \frac{\operatorname{Tr}(PS)^2}{\operatorname{Tr}(S)}.$$

Taking the supremum over P on the right-hand side we obtain $\varphi^\bullet(I) \geq \operatorname{Tr}(S)$, which proves the claim.

<u>*Claim 2*</u>. The smallest positive extension of φ, i.e. $(f|_{B_F(H)})^\bullet$ is normal.

Proof of Claim 2. First observe that the restriction of f to $B_2(H)$ defines a continuous linear functional on $B_2(H)$ with respect to the norm $\|\cdot\|_2$. Due to the Riesz representation theorem, there exists a unique representing operator $S \in B_2(H)$ such that

$$f(A) = (A \,|\, S)_2 = \operatorname{Tr}(S^* A) \text{ for all } A \in B_2(H). \qquad (***)$$

We are going to show that $S \in B_1(H)$. Indeed, let $\mathcal{E}$ be an orthonormal basis in H and let $\mathcal{F}$ be any non-empty finite subset of $\mathcal{E}$. Denoting by $P_{\mathcal{F}}$ the orthogonal projection onto the subspace spanned by $\mathcal{F}$ we get

$$\sum_{e \in \mathcal{F}} (Se \,|\, e) = (P_{\mathcal{F}} \,|\, S)_2 = f(P_{\mathcal{F}}) \leq f(I).$$

Taking the supremum over $\mathcal{F}$ we obtain that S is in the trace class. By Claim 1, the smallest positive extension $\varphi^\bullet$ of φ equals f_S, which is normal. This proves Claim 2.

Now, we are going to prove (ii)⇒(i).

Claim 3. If there exists a normal positive functional g such that $f \leq g$ holds, then f is normal as well.

Proof of Claim 3. Let g be a normal positive functional dominating f, and let T be a trace class operator such that $g = f_T$. According to Claim 2 it is enough to prove that $f = \varphi^\bullet$. Since $h := f - \varphi^\bullet$ is positive, this will follow by showing that $h(I) = 0$. We see from $(***)$ that $h(A) = 0$ for every finite rank operator A. Consequently, as $h \leq f \leq f_T$, it follows that

$$h(I) = h(I - P) \leq f_T(I - P) = \operatorname{Tr}(T) - \operatorname{Tr}(TP)$$

for every finite rank projection P. Taking the infimum over P we obtain $h(I) = 0$, and therefore Claim 3 is established.

To complete the proof we mention all the missing trivial implications. Taking $g := f$, we see that (i) implies (ii). As $(**)$ means that $\varphi^\bullet = f$, the equivalence of (iii) and (iv) follows from the minimality of the Krein–von Neumann extension. Replacing X with I in $(**)$ we obtain that (iv) implies (v). Conversely, (v) implies (iv) as $\varphi^\bullet \leq f$ and $f - \varphi^\bullet$ attains its norm at the identity I. □

Finally, we remark that the above proof contains a characterization of the property of having a normal extension for a functional defined on $B_F(H)$.

Corollary. *Let $\varphi : B_F(H) \to \mathbb{C}$ be a linear functional. The following statements are equivalent to the existence of a normal positive extension:*

(a) *There is a $C \geq 0$ such that $|\varphi(A)|^2 \leq C \cdot \varphi(A^* A)$ for all $A \in B_F(H)$.*
(b) *There is a positive functional f such that $f|_{B_F(H)} = \varphi$.*
(c) *There is a positive operator $F \in B_1(H)$ such that $\varphi(A) = \operatorname{Tr}(FA)$ for all $A \in B_F(H)$.*

References

[1] I.C. Gohberg and M.G. Krein, *Introduction to the Theory of Linear Nonselfadjoint Operators*, Translations of Mathematical Monographs **18**, Amer. Math. Soc., Providence, RI, 1969.

[2] R.V. Kadison and J.R. Ringrose, *Fundamentals of the theory of operator algebras I.*, Academic Press, New York, 1983.

[3] M.G. Krein, The theory of self-adjoint extensions of semi-bounded Hermitian transformations and its applications, I–II, *Mat. Sbornik* **20** (1947), 431–495, *Mat. Sbornik* **21** (1947), 365–404 (Russian).

[4] J. von Neumann, Allgemeine Eigenwerttheorie Hermitescher Funktionaloperatoren, *Math. Ann.* **102** (1930), 49–131.

[5] T.W. Palmer, *Banach Algebras and the General Theory of *-Algebras II*, Cambridge University Press, Cambridge, 2001.

[6] R. Schatten, *Norm Ideals of Completely Continuous Operators*, Ergebnisse der Mathematik und ihrer Grenzgebiete **27**, Springer, Berlin, 1960.

[7] Z. Sebestyén, Operator extensions on Hilbert space, *Acta Sci. Math. (Szeged)* **57** (1993), 233–248.

[8] Z. Sebestyén, Zs. Szűcs, and Zs. Tarcsay, Extensions of positive operators and functionals, *Linear Algebra Appl.* **472** (2015), 54–80.

Zoltán Sebestyén and Zsigmond Tarcsay
Department of Applied Analysis
Eötvös Loránd University
Pázmány Péter sétány 1/c
Budapest H-1117
Hungary
e-mail: sebesty@cs.elte.hu
tarcsay@cs.elte.hu

Tamás Titkos
Alfréd Rényi Institute of Mathematics
Hungarian Academy of Sciences
Reáltanoda utca 13–15
Budapest H-1053
Hungary
and
BBS University of Applied Sciences
Alkotmány u. 9
Budapest H-1054
Hungary
e-mail: titkos.tamas@renyi.mta.hu

The linearised Korteweg–de Vries equation on general metric graphs

Christian Seifert

Abstract. We consider the linearised Korteweg–de Vries equation, sometimes called Airy equation, on general metric graphs with edge lengths bounded away from zero. We show that properties of the induced dynamics can be obtained by studying boundary operators in the corresponding boundary space induced by the vertices of the graph. In particular, we characterise unitary dynamics and contractive dynamics. We demonstrate our results on various special graphs, including those recently treated in the literature.

Mathematics Subject Classification (2010). Primary 35Q53; Secondary 47B25, 81Q35.

Keywords. metric graphs, linearised KdV-equation, generators of C_0-semigroups.

1. Introduction

The Korteweg–de Vries equation [4, 3]

$$\frac{\partial u}{\partial t} = \frac{3}{2}\sqrt{\frac{g}{\ell}}\left(\frac{\sigma}{3}\frac{\partial^3 u}{\partial x^3} + \frac{2a}{3}\frac{\partial u}{\partial x} + \frac{1}{2}\frac{\partial u^2}{\partial x}\right)$$

models shallow water waves in channels, where u describes the elevation of the water w.r.t. the average water depth and g, ℓ , σ, and a are constants. Due to the last term on the right-hand side it becomes non-linear. Assuming only small elevations u (i.e. u is close to zero) and/or long waves (i.e. $\frac{\partial u}{\partial x}$ is close to zero), the linear approximation neglecting the non-linearity (which is the linearisation around the stationary solution $u = 0$) yields an equation of the form

$$\frac{\partial u}{\partial t} = \alpha\frac{\partial^3 u}{\partial x^3} + \beta\frac{\partial u}{\partial x} \tag{1}$$

with appropriate constants α and β. In this paper we are going to study the evolution equation on general networks, i.e. metric graphs, from a functional analytic point of view.

© Springer International Publishing AG, part of Springer Nature 2018

A. Böttcher et al. (eds.), *The Diversity and Beauty of Applied Operator Theory*, Operator Theory: Advances and Applications 268, https://doi.org/10.1007/978-3-319-75996-8_25

Evolution equations (or, more generally, differential operators) on metric graphs have been intensively studied during the last two decades. However, the focus was put on Schrödinger type operators and the corresponding heat and Schrödinger evolution equations, see [1] and references therein. Recently, also KdV-type equations on star graphs have gained interest, see [2, 9, 7, 8, 5]. Such star graphs give rise to model either singular interactions at one point, i.e. interface conditions, but can also be interpreted as models for junctions of channels. The drawback of star graphs is that one only has exactly one vertex (i.e. junction), and the channels are modelled by halflines.

In this paper, we consider the linearised KdV-equation (1) on general metric graphs, i.e., we model a whole network, including channels of finite length. We will describe the evolution equation in a functional analytic setup, namely in the framework of strongly continuous semigroups. Thus, we are left to study the spacial operator describing the right-hand side of (1). The aim is to obtain a "reasonable" dynamics in an L_2-setting, meaning either unitary C_0-semigroups or contractive C_0-semigroups (which resembles the fact that the spacial derivatives appear only in odd order). In order to do that we will employ the framework of boundary systems developed in [6].

In Section 2 we introduce the metric graph and the operator setup for the spacial derivatives. Section 3 summarises notions from Krein space theory which we will need to describe the right boundary conditions at the vertices. We then focus on unitary and contractive dynamics in Section 4. In Section 5 we specialise our framework (which does not take the concrete graph structure into account) to the graph structure setting. We end this paper by briefly listing some examples in Section 6, where we refer to the corresponding literature, but also explaining new examples which have not been dealt with before.

2. Notation and model

Let $\Gamma = (V, E, a, b, \gamma_{\mathrm{l}}, \gamma_{\mathrm{r}})$ be a metric graph, i.e., V is the set of vertices of Γ, and E is the set of edges of Γ. Moreover, $a, b\colon E \to [-\infty, \infty]$ are such that $a_e < b_e$ for all $e \in E$, and each edge $e \in E$ is identified with the interval $(a_e, b_e) \subseteq \mathbb{R}$. Let $E_{\mathrm{l}} := \{e \in E;\ a_e > -\infty\}$ and $E_{\mathrm{r}} := \{e \in E;\ b_e < \infty\}$ be the sets of edges with finite starting and termination point, respectively, and let $\gamma_{\mathrm{l}}\colon E_{\mathrm{l}} \to V$, $\gamma_{\mathrm{r}}\colon E_{\mathrm{r}} \to V$ assign to each $e \in E_{\mathrm{l}}$ or $e \in E_{\mathrm{r}}$ the starting vertex $\gamma_{\mathrm{l}}(e)$ and the termination vertex $\gamma_{\mathrm{r}}(e)$, respectively. Note that we do not assume V or E to be finite or countable.

We assume to have a positive lower bound on the edge lengths, that is,

$$\ell := \inf_{e \in E} (b_e - a_e) > 0. \tag{2}$$

For $k \in \{0, 1, 2\}$ we define the mappings $\mathrm{tr}_{\mathrm{l}}^k\colon \bigoplus_{e\in E} W_2^{k+1}(a_e, b_e) \to \ell_2(E_{\mathrm{l}})$ and $\mathrm{tr}_{\mathrm{r}}^k\colon \bigoplus_{e\in E} W_2^{k+1}(a_e, b_e) \to \ell_2(E_{\mathrm{r}})$ by

$$(\mathrm{tr}_{\mathrm{l}}^k u)(e) := u_e^{(k)}(a_e+) \ \ (e \in E_{\mathrm{l}}), \quad (\mathrm{tr}_{\mathrm{r}}^k u)(e) := u_e^{(k)}(b_e-) \ \ (e \in E_{\mathrm{r}}).$$

Furthermore, define the trace maps $\mathrm{Tr}_\mathrm{l}\colon \bigoplus_{e\in E} W_2^3(a_e,b_e) \to \ell_2(E_\mathrm{l})^3$ and $\mathrm{Tr}_\mathrm{r}\colon \bigoplus_{e\in E} W_2^3(a_e,b_e) \to \ell_2(E_\mathrm{r})^3$ by

$$\mathrm{Tr}_\mathrm{l}\, u := (\mathrm{tr}_\mathrm{l}^0\, u, \mathrm{tr}_\mathrm{l}^1\, u, \mathrm{tr}_\mathrm{l}^2\, u), \quad \mathrm{Tr}_\mathrm{r}\, u := (\mathrm{tr}_\mathrm{r}^0\, u, \mathrm{tr}_\mathrm{r}^1\, u, \mathrm{tr}_\mathrm{r}^2\, u).$$

Let $\mathcal{H}_\Gamma := \bigoplus_{e\in E} L_2(a_e,b_e)$ be the Hilbert space we are going to consider. Let $(\alpha_e)_{e\in E}$ in $(0,\infty)$ be bounded and bounded away from zero, let $(b_e)_{e\in E}$ be bounded, and also abbreviate $\alpha_\mathrm{l} := (\alpha_e)_{e\in E_\mathrm{l}}$, $\alpha_\mathrm{r} := (\alpha_e)_{e\in E_\mathrm{r}}$, $\beta_\mathrm{l} := (\beta_e)_{e\in E_\mathrm{l}}$, and $\beta_\mathrm{r} := (\beta_e)_{e\in E_\mathrm{r}}$. By the same symbol we will denote the corresponding multiplication operators in $\ell_2(E_\mathrm{l})$ and $\ell_2(E_\mathrm{r})$, respectively.

Remark 2.1. One could choose $\alpha_e \in \mathbb{R}\setminus\{0\}$. However, for edges $e\in E$ with $\alpha_e < 0$ we can just change the orientation of the edge (by setting $\tilde{\alpha}_e := -\alpha_e$, $\tilde{\beta}_e := -\beta_e$, and $\tilde{u} := u(-\cdot)$). Thus, w.l.o.g. we may (and will) assume $\alpha_e > 0$ for all $e \in E$. Moreover, by scaling the variables appropriately, it would suffice to deal with the case $\beta_e = 0$ for all $e\in E$. In order to do this uniformly for all edges, one only needs boundedness of (β_e) and $\left(\frac{1}{\alpha_e}\right)$. However, we will keep the β_e's possibly non-zero.

Definition. We define the *minimal operator* A_0 in $\mathcal{H}_\Gamma$ by

$$D(A_0) := \bigoplus_{e\in E} C_\mathrm{c}^\infty(a_e,b_e),$$
$$(A_0 u)_e := \alpha_e \partial^3 u_e + \beta_e \partial u_e \quad (e\in E, u\in D(A_0)).$$

And we define the *maximal operator* $\hat{A}$ in $\mathcal{H}_\Gamma$ by

$$D(\hat{A}) := \bigoplus_{e\in E} W_2^3(a_e,b_e),$$
$$(\hat{A} u)_e := \alpha_e \partial^3 u_e + \beta_e \partial u_e \quad (e\in E, u\in D(\hat{A})).$$

Applying integration by parts we obtain the following lemma.

Lemma 2.2. *We have* $-A_0^* = \hat{A}$.

Define $F\colon G(\hat{A}) \to \ell_2(E_\mathrm{r})^3 \oplus \ell_2(E_\mathrm{l})^3$ by

$$F(u, \hat{A}u) := (\mathrm{Tr}_\mathrm{r}\, u, \mathrm{Tr}_\mathrm{l}\, u).$$

Lemma 2.3. *F is linear and surjective.*

Proof. Linearity of F is clear. In order to show that F is surjective first note that there exists $\varphi \in C_\mathrm{c}^\infty[0,\ell)$, where ℓ is as in (2), such that φ equals 1 in a neighbourhood of 0 and $\varphi(x) = 0$ for $x \geqslant \frac{\ell}{2}$. For $(t^0,t^1,t^2)\in\mathbb{C}^3$ let $u(x) := (t^0 + t^1 x + \frac{1}{2}t^2x^2)\varphi(x)$ for $x\in(0,\ell)$. Then there exists a $c \geqslant 0$ (independent of (t^0,t^1,t^2)) such that

$$\|u\|_{L_2(0,\ell)}^2 + \|u'\|_{L_2(0,\ell)}^2 + \|u'''\|_{L_2(0,\ell)}^2 \leqslant c\big\|(t^0,t^1,t^2)\big\|_2^2.$$

Since (α_e) and (β_e) are bounded, as a consequence, F is surjective. □

For $u, v \in D(\hat{A})$ we obtain by integration by parts

$$\left(u \middle| \hat{A}v\right) + \left(\hat{A}u \middle| v\right) = -\left(\begin{pmatrix} -\beta_\mathrm{r} & 0 & -\alpha_\mathrm{r} \\ 0 & \alpha_\mathrm{r} & 0 \\ -\alpha_\mathrm{r} & 0 & 0 \end{pmatrix} \mathrm{Tr}_\mathrm{r}\, u \middle| \mathrm{Tr}_\mathrm{r}\, v\right) + \left(\begin{pmatrix} -\beta_\mathrm{l} & 0 & -\alpha_\mathrm{l} \\ 0 & \alpha_\mathrm{l} & 0 \\ -\alpha_\mathrm{l} & 0 & 0 \end{pmatrix} \mathrm{Tr}_\mathrm{l}\, u \middle| \mathrm{Tr}_\mathrm{l}\, v\right). \tag{3}$$

Let us define $\Omega\colon G(\hat{A}) \times G(\hat{A}) \to \mathbb{C}$ by

$$\Omega\big((u, \hat{A}u), (v, \hat{A}v)\big) := \left(\begin{pmatrix} 0 & 1 \\ 1 & 0 \end{pmatrix} \begin{pmatrix} u \\ \hat{A}u \end{pmatrix} \middle| \begin{pmatrix} v \\ \hat{A}v \end{pmatrix}\right),$$

and $\omega\colon \big(\ell_2(E_\mathrm{r})^3 \oplus \ell_2(E_\mathrm{l})^3\big) \times \big(\ell_2(E_\mathrm{r})^3 \oplus \ell_2(E_\mathrm{l})^3\big) \to \mathbb{C}$ by

$$\omega\big((x, y), (u, v)\big) := \left(\begin{pmatrix} -B_\mathrm{r} & 0 \\ 0 & B_\mathrm{l} \end{pmatrix} \begin{pmatrix} x \\ y \end{pmatrix} \middle| \begin{pmatrix} u \\ v \end{pmatrix}\right),$$

abbreviating

$$B_\mathrm{l} := \begin{pmatrix} -\beta_\mathrm{l} & 0 & -\alpha_\mathrm{l} \\ 0 & \alpha_\mathrm{l} & 0 \\ -\alpha_\mathrm{l} & 0 & 0 \end{pmatrix}, \qquad B_\mathrm{r} := \begin{pmatrix} -\beta_\mathrm{r} & 0 & -\alpha_\mathrm{r} \\ 0 & \alpha_\mathrm{r} & 0 \\ -\alpha_\mathrm{r} & 0 & 0 \end{pmatrix}.$$

By rewriting (3) we obtain

$$\Omega\big((u, \hat{A}u), (v, \hat{A}v)\big) = \omega\big(F(u, \hat{A}u), F(v, \hat{A}v)\big) \tag{4}$$

for all $(u, \hat{A}u), (v, \hat{A}v) \in G(\hat{A})$.

Let L be a densely defined linear operator from $\ell_2(E_\mathrm{r})^3$ to $\ell_2(E_\mathrm{l})^3$. Then $A_0 \subseteq A_L \subseteq \hat{A} = -A_0^*$ is defined by

$$D(A_L) := \big\{u \in D(\hat{A});\ \mathrm{Tr}_\mathrm{r}\, u \in D(L),\ L\,\mathrm{Tr}_\mathrm{r}\, u = \mathrm{Tr}_\mathrm{l}\, u\big\},$$

that is, $G(A_L) = F^{-1}(G(L))$.

3. Operators in Krein spaces

Remark 3.1. Let $\mathcal{K}$ be a vector space and $\langle\cdot\,|\,\cdot\rangle$ an (indefinite) inner product on $\mathcal{K}$ such that $(\mathcal{K}, \langle\cdot\,|\,\cdot\rangle)$ is a Krein space. Then $(\cdot\,|\,\cdot) := \langle J\cdot\,|\,\cdot\rangle$, where J is the fundamental symmetry of $\mathcal{K}$, defines an inner product on $\mathcal{K}$ such that $(\mathcal{K}, (\cdot\,|\,\cdot))$ is a Hilbert space. Notions such as closedness or continuity are then defined by the underlying Hilbert space structure.

Definition. Let $\mathcal{K}_-$, $\mathcal{K}_+$ be Krein spaces and $\omega\colon (\mathcal{K}_- \oplus \mathcal{K}_+) \times (\mathcal{K}_- \oplus \mathcal{K}_+) \to \mathbb{C}$ be sesquilinear.

(a) Let $X \subseteq \mathcal{K}_- \oplus \mathcal{K}_+$ be a subspace. Then X is called *ω-self-orthogonal* if $X = X^{\perp_\omega}$, where

$$X^{\perp_\omega} := \big\{(x, y) \in \mathcal{K}_- \oplus \mathcal{K}_+;\ \omega\big((x, y), (u, v)\big) = 0 \text{ for all } (u, v) \in X\big\}.$$

(b) Let L be a densely defined linear operator from $\mathcal{K}_-$ to $\mathcal{K}_+$. Then its $(\mathcal{K}_-, \mathcal{K}_+)$-*adjoint* $L^\sharp$ is defined by

$$D(L^\sharp) := \{y \in \mathcal{K}_+;\ \exists z \in \mathcal{K}_- : \langle Lx \,|\, y\rangle_+ = \langle x \,|\, z\rangle_- \ \text{ for all } x \in D(L)\},$$
$$L^\sharp y := z.$$

Clearly, $L^\sharp$ is then a linear operator from $\mathcal{K}_+$ to $\mathcal{K}_-$.

(c) Let L be a linear operator from $\mathcal{K}_-$ to $\mathcal{K}_+$. Then L is called a $(\mathcal{K}_-, \mathcal{K}_+)$-*contraction* if

$$\langle Lx \,|\, Lx\rangle_+ \leqslant \langle x \,|\, x\rangle_- \quad \text{for all } x \in D(L).$$

(d) Let L be a linear operator from $\mathcal{K}_-$ to $\mathcal{K}_+$. Then L is called $(\mathcal{K}_-, \mathcal{K}_+)$-*unitary* if $D(L)$ and $R(L)$ are dense, L is injective, and finally $L^\sharp = L^{-1}$.

If $\mathcal{K}_-$, $\mathcal{K}_+$ are Hilbert spaces, then obviously $(\mathcal{K}_-, \mathcal{K}_+)$-adjoint (-contraction, -unitary) operators are the usual objects of Hilbert space operator theory.

Remark 3.2. Let $\mathcal{K}_\pm$ be Krein spaces. Let L be $(\mathcal{K}_-, \mathcal{K}_+)$-unitary. Then

$$\langle Lx \,|\, Ly\rangle_+ = \langle x \,|\, y\rangle_- \quad \text{for all } x, y \in D(L). \tag{5}$$

However, L may not be bounded.

Note that $\ell_2(E_\mathrm{l})^3$ equipped with $\langle \cdot \,|\, \cdot\rangle_\mathrm{l} : \ell_2(E_\mathrm{l})^3 \times \ell_2(E_\mathrm{l})^3 \to \mathbb{C}$,

$$\langle (x^0, x^1, x^2) \,|\, (y^0, y^1, y^2)\rangle_\mathrm{l} := \left(B_\mathrm{l}(x^0, x^1, x^2) \,|\, (y^0, y^1, y^2)\right)_{\ell_2(E_\mathrm{l})^3}$$

yields a Krein space $\mathcal{K}_\mathrm{l} := (\ell_2(E_\mathrm{l})^3, \langle \cdot \,|\, \cdot\rangle_\mathrm{l})$. Analogously, $\ell_2(E_\mathrm{r})^3$ equipped with $\langle \cdot \,|\, \cdot\rangle_\mathrm{r} : \ell_2(E_\mathrm{r})^3 \times \ell_2(E_\mathrm{r})^3 \to \mathbb{C}$,

$$\langle (x^0, x^1, x^2) \,|\, (y^0, y^1, y^2)\rangle_\mathrm{r} := \left(B_\mathrm{r}(x^0, x^1, x^2) \,|\, (y^0, y^1, y^2)\right)_{\ell_2(E_\mathrm{r})^3}$$

yields a Krein space $\mathcal{K}_\mathrm{r} := (\ell_2(E_\mathrm{r})^3, \langle \cdot \,|\, \cdot\rangle_\mathrm{r})$.

4. Dynamics

Let us study different types of dynamics for the equation.

Generating unitary groups

We are now interested in generators of unitary groups. By Stone's theorem this is equivalent to looking for skew-self-adjoint realisations of A_L.

Theorem 4.1. *Let L be a linear operator from $\ell_2(E_\mathrm{r})^3$ to $\ell_2(E_\mathrm{l})^3$ such that $D(L)$ and $R(L)$ are dense. Then A_L is skew-self-adjoint if and only if L is $(\mathcal{K}_\mathrm{r}, \mathcal{K}_\mathrm{l})$-unitary.*

Proof. By [6, Corollary 2.3 and Example 2.7 (b)], see also [5, Theorem 3.7], we have to show that $G(A_L)$ is Ω-self-orthogonal if and only if $G(L)$ is ω-self-orthogonal. But this is an easy consequence of (4) and the definition of the operator A_L. □

Generating contraction semigroups

Instead of unitary dynamics let us characterise generators of contraction semigroups.

Theorem 4.2. *Let L be a densely defined closed linear operator from $\ell_2(E_{\mathrm{r}})^3$ to $\ell_2(E_{\mathrm{l}})^3$. Then A_L is the generator of a semigroup of contractions if and only if L is $(\mathcal{K}_{\mathrm{r}}, \mathcal{K}_{\mathrm{l}})$-contractive and $L^\sharp$ is $(\mathcal{K}_{\mathrm{l}}, \mathcal{K}_{\mathrm{r}})$-contractive.*

Proof. Note that since L is closed, also A_L is closed. First, it is easy to see that A_L^* is given by

$$\begin{aligned} D(A_L^*) &= \{u \in D(A_0^*);\ \mathrm{Tr}_{\mathrm{l}}\, u \in D(L^\sharp),\ L^\sharp\, \mathrm{Tr}_{\mathrm{l}}\, u = \mathrm{Tr}_{\mathrm{r}}\, u\}, \\ A_L^* u &= A_0^* u. \end{aligned}$$

For $u \in D(A_L)$ we compute

$$\begin{aligned} 2\,\mathrm{Re}\,(A_L u \,|\, u) &= \Omega\big((u, A_L u), (u, A_L u)\big) \\ &= \omega\big(F(u, A_L u), F(u, A_L u)\big) \\ &= \omega\big((\mathrm{Tr}_{\mathrm{r}}\, u, L\, \mathrm{Tr}_{\mathrm{r}}\, u), (\mathrm{Tr}_{\mathrm{r}}\, u, L\, \mathrm{Tr}_{\mathrm{r}}\, u)\big) \\ &= -\left\langle \mathrm{Tr}_{\mathrm{r}}\, u \,|\, \mathrm{Tr}_{\mathrm{r}}\, u\right\rangle_{\mathrm{r}} + \left\langle L\, \mathrm{Tr}_{\mathrm{r}}\, u \,|\, L\, \mathrm{Tr}_{\mathrm{r}}\, u\right\rangle_{\mathrm{l}}. \end{aligned}$$

Hence, A_L is dissipative, i.e. $\mathrm{Re}\,(A_L u \,|\, u) \leqslant 0$ for all $u \in D(A_L)$, if and only if L is $(\mathcal{K}_{\mathrm{r}}, \mathcal{K}_{\mathrm{l}})$-contractive. Similarly, A_L^* is dissipative if and only if $L^\sharp$ is $(\mathcal{K}_{\mathrm{l}}, \mathcal{K}_{\mathrm{r}})$-contractive. Thus, the Lumer–Phillips theorem in Hilbert spaces yields the assertion. □

5. Local boundary conditions

So far, we did not take into account the graph structure. Now, we ask for boundary conditions at each vertex $v \in V$ separately. For $v \in V$ let

$$E_{\mathrm{l},v} := \{e \in E_{\mathrm{l}};\ \gamma_{\mathrm{l}}(e) = v\}, \quad E_{\mathrm{r},v} := \{e \in E_{\mathrm{r}};\ \gamma_{\mathrm{r}}(e) = v\}.$$

Then $\ell_2(E_{\mathrm{l},v})$ equipped with $\langle \cdot \,|\, \cdot \rangle_{\mathrm{l},v} := \langle \cdot \,|\, \cdot \rangle_{\mathrm{l}}\,|_{\ell_2(E_{\mathrm{l},v})^3 \times \ell_2(E_{\mathrm{l},v})^3}$ and $\ell_2(E_{\mathrm{r},v})$ equipped with $\langle \cdot \,|\, \cdot \rangle_{\mathrm{r},v} := \langle \cdot \,|\, \cdot \rangle_{\mathrm{r}}\,|_{\ell_2(E_{\mathrm{r},v})^3 \times \ell_2(E_{\mathrm{r},v})^3}$ yield Krein spaces $\mathcal{K}_{\mathrm{l},v}$ and $\mathcal{K}_{\mathrm{r},v}$, respectively, such that

$$\langle \cdot \,|\, \cdot \rangle_{\mathrm{l}} = \sum_{v \in V} \langle \cdot \,|\, \cdot \rangle_{\mathrm{l},v}$$

and analogously for $\langle \cdot \,|\, \cdot \rangle_{\mathrm{r}}$.

For $v \in V$ let L_v be a densely defined linear operator from $\ell_2(E_{\mathrm{r},v})^3$ to $\ell_2(E_{\mathrm{l},v})^3$, and define $A_0 \subseteq A_{(L_v)_{v \in V}} \subseteq \hat{A} = -A_0^*$ by

$$\begin{aligned} D(A_L) := \big\{u \in D(\hat{A});\ \forall v \in V : (\mathrm{Tr}_{\mathrm{r}}\, u)|_{E_{\mathrm{r},v}} \in D(L_v), \\ L_v\big((\mathrm{Tr}_{\mathrm{r}}\, u)|_{E_{\mathrm{r},v}}\big) = \mathrm{Tr}_{\mathrm{l}}\, u|_{E_{\mathrm{l},v}}\big\}. \end{aligned}$$

For the case of unitary dynamics we obtain the following corollary.

Corollary 5.1. *For $v \in V$ let L_v be a densely defined linear operator from $\ell_2(E_{\mathrm{r},v})^3$ to $\ell_2(E_{\mathrm{l},v})^3$. Then $A_{(L_v)_{v\in V}}$ is the generator of a unitary group if and only if L_v is $(\mathcal{K}_{\mathrm{r},v}, \mathcal{K}_{\mathrm{l},v})$-unitary for all $v \in V$.*

Proof. We show that (L_v) is $(\mathcal{K}_\mathrm{r}, \mathcal{K}_\mathrm{l})$-unitary if and only if L_v is $(\mathcal{K}_{\mathrm{r},v}, \mathcal{K}_{\mathrm{l},v})$-unitary for all $v \in V$. Then the result follows from Theorem 4.1. Note that (L_v) acts as a block-diagonal operator with block L_v from $\ell_2(E_{\mathrm{r},v})^3$ to $\ell_2(E_{\mathrm{l},v})^3$ for all $v \in V$. Hence, clearly, (L_v) is densely defined with dense range if and only if L_v is densely defined with dense range for all $v \in V$. Moreover, (L_v) is injective if and only if L_v is injective for all $v \in V$. Since $(L_v)^\sharp = (L_v^\sharp)$ we also obtain $(L_v)^\sharp = (L_v)^{-1}$ if and only if $L_v^\sharp = L_v^{-1}$ for all $v \in V$. □

Analogously, in the case of contractive dynamics we have the following result.

Corollary 5.2. *For $v \in V$ let L_v be a densely defined linear operator from $\ell_2(E_{\mathrm{r},v})^3$ to $\ell_2(E_{\mathrm{l},v})^3$ such that $L := (L_v)_{v\in V}$ is closed. Then $A_{(L_v)_{v\in V}}$ is the generator of a semigroup of contractions if and only if L_v is $(\mathcal{K}_{\mathrm{r},v}, \mathcal{K}_{\mathrm{l},v})$-contractive and $L_v^\sharp$ is $(\mathcal{K}_{\mathrm{l},v}, \mathcal{K}_{\mathrm{r},v})$-contractive for all $v \in V$.*

Proof. We show that (L_v) is $(\mathcal{K}_\mathrm{r}, \mathcal{K}_\mathrm{l})$-contractive and $(L_v)^\sharp$ is $(\mathcal{K}_\mathrm{l}, \mathcal{K}_\mathrm{r})$-contractive if and only if L_v is $(\mathcal{K}_{\mathrm{r},v}, \mathcal{K}_{\mathrm{l},v})$-contractive and $L_v^\sharp$ is $(\mathcal{K}_{\mathrm{l},v}, \mathcal{K}_{\mathrm{r},v})$-contractive for all $v \in V$. Then the result follows from Theorem 4.2. Again, (L_v) acts as a block-diagonal operator with block L_v from $\ell_2(E_{\mathrm{r},v})^3$ to $\ell_2(E_{\mathrm{l},v})^3$ for all $v \in V$. Hence, (L_v) is $(\mathcal{K}_\mathrm{r}, \mathcal{K}_\mathrm{l})$-contractive if and only if L_v is $(\mathcal{K}_{\mathrm{r},v}, \mathcal{K}_{\mathrm{l},v})$-contractive for all $v \in V$. Since $(L_v)^\sharp = (L_v^\sharp)$, we have that $(L_v)^\sharp$ is $(\mathcal{K}_\mathrm{l}, \mathcal{K}_\mathrm{r})$-contractive if and only if $L_v^\sharp$ is $(\mathcal{K}_{\mathrm{l},v}, \mathcal{K}_{\mathrm{r},v})$-contractive for all $v \in V$. □

6. Examples

In this section we will specialise to particular examples of graphs. For those special cases already treated in the literature we just explain the setup. We ask the reader to go to the corresponding references for more details in these cases.

Two semi-infinite edges

Let us consider two semi-infinite edges attached to one vertex v, and let the two edges correspond to the intervals $(-\infty, 0]$ and $[0, \infty)$:

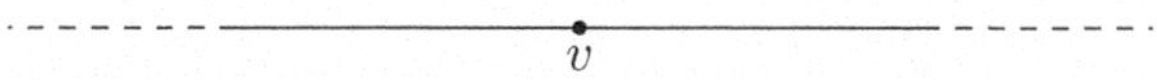

If the coefficients (α_e) and (β_e) are constant, we can interpret the equation as the linearised KdV-equation on the real line with a generalised point interaction at 0 (which corresponds to the vertex v). This situation was considered in [2].

As a particular example, if $\alpha_e = 1$ and $\beta_e = 0$ for all e, then

$$L := \begin{pmatrix} 1 & 0 & 0 \\ \sqrt{2} & 1 & 0 \\ 1 & \sqrt{2} & 1 \end{pmatrix}$$

yields a unitary dynamics, since $L\colon \mathcal{K}_{\mathrm{r}} \to \mathcal{K}_{\mathrm{l}}$ is $(\mathcal{K}_{\mathrm{r}}, \mathcal{K}_{\mathrm{l}})$-unitary, i.e., L is bijective and

$$\langle Lx \mid Ly \rangle_{\mathrm{l}} = \langle x \mid y \rangle_{\mathrm{r}} \quad (x, y \in \mathcal{K}_{\mathrm{r}}).$$

An example yielding contractive (but not unitary) dynamics in the case $\alpha_e = 1$ and $\beta_e = 0$ for all e is given by

$$L := \begin{pmatrix} 2 & 0 & 1 \\ 0 & 2 & 0 \\ 1 & 0 & 2 \end{pmatrix}.$$

Star graphs

The special case of star graphs was considered in [7, 8, 9, 5]. Here, we have one vertex v, each edge is described by a semi-infinite interval, say $(-\infty, 0]$ or $[0, \infty)$, and each edge is adjacent to v with its endpoint corresponding to the value 0 for the interval:

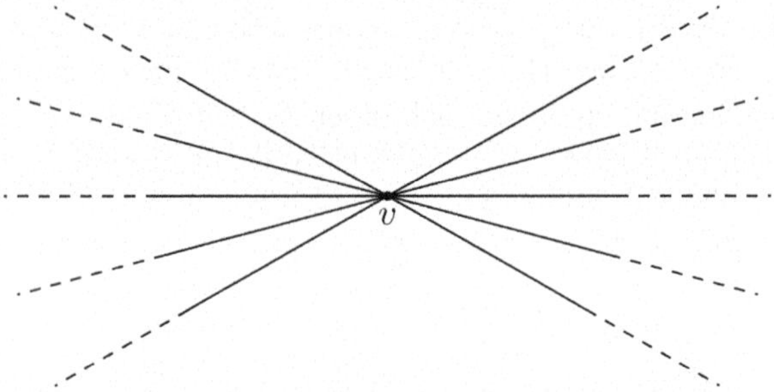

Star graphs generalise graphs with two semi-infinite edges to more than two edges. For those graphs, there is a necessary condition on the graph for existence of unitary dynamics, namely $|E_{\mathrm{l}}| = |E_{\mathrm{r}}|$, where $|\cdot|$ denotes the cardinality for sets.

A Loop

Let Γ be a loop, i.e., $|V| = |E| = 1$, and the edge corresponds to the interval $[0, 1]$, and both endpoints of the edge are attached to the vertex v:

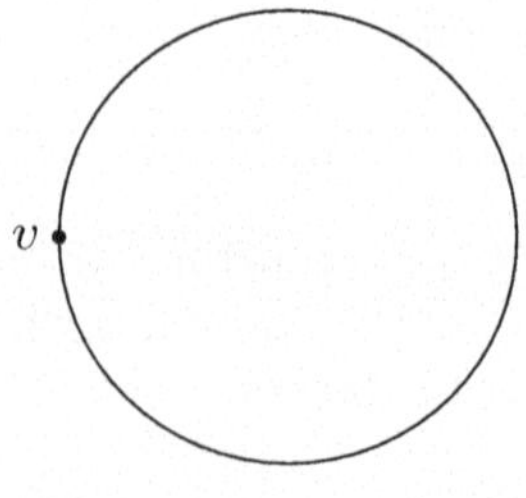

Here, we can model generalised periodic boundary conditions. Indeed, consider L to be represented by the identity matrix in the usual basis. Then this results in periodic boundary conditions

$$u^{(k)}(0+) = u^{(k)}(1-) \quad (k \in \{0,1,2\}).$$

Moreover, L becomes Krein space unitary in this case, so the dynamics is unitary.

Graphs with more than one vertex

We will now consider graphs with more than one vertex. By the results of Section 5 we can study the dynamics for the linearised KdV-equation on a general graph by studying the 'local' behaviour at each vertex. Since locally around each vertex a graph looks like a (local part of a) star graph, this brings us back to studying boundary couplings for star graphs.

As an example, let us consider the following graph, where the edge lengths are all equal to 1 for simplicity:

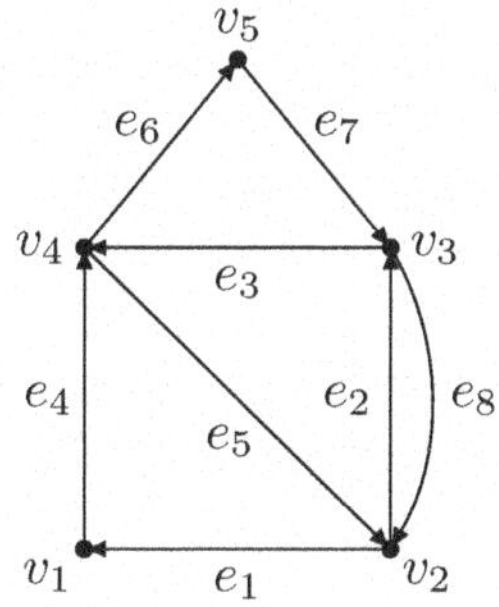

Since $|E_{\mathrm{l},v_k}| = |E_{\mathrm{r},v_k}|$ for $k \in \{1,\ldots,5\}$, by defining suitable boundary coupling operators L_{v_k} $(k \in \{1,\ldots,5\})$, we can obtain unitary or contractive dynamics. Let us assume that $\alpha_{e_k} = 1$ and $\beta_{e_k} = 0$ for all $k \in \{1,\ldots,8\}$. Then, as an example for unitary dynamics we can choose

$$L_{v_1} = L_{v_5} := \begin{pmatrix} 1 & 0 & 0 \\ \sqrt{2} & 1 & 0 \\ 1 & \sqrt{2} & 1 \end{pmatrix}$$

and

$$L_{v_2} = L_{v_3} = L_{v_4} := \begin{pmatrix} 1 & 0 & 0 & 0 & 0 & 0 \\ 0 & 1 & 0 & 0 & 0 & 0 \\ \sqrt{2} & 0 & 1 & 0 & 0 & 0 \\ 0 & 0 & 0 & 1 & 0 & 0 \\ 1 & 0 & \sqrt{2} & 0 & 1 & 0 \\ 0 & 0 & 0 & 0 & 0 & 1 \end{pmatrix}.$$

Acknowledgment

The author thanks Delio Mugnolo and Diego Noja for many useful discussions on the linearised KdV-equation.

References

[1] G. Berkolaiko and P. Kuchment, *Introduction to Quantum Graphs*, Mathematical Surveys and Monographs **186**, Amer. Math. Soc., Providemce, 2013.

[2] B. Deconinck, N.E. Sheils, and D.A. Smith, The linear KdV equation with an interface, *Comm. Math. Phys.* **347** (2016), no. 2, 489–509.

[3] D. Lannes, *The Water Waves Problem: Mathematical Analysis and Asymptotics*, Mathematical Surveys and Monographs **188**, Amer. Math. Soc., Providence, 2013.

[4] D.J. Korteweg and G. de Vries, On the change of form of long waves advancing in a rectangular canal, and on a new type of long stationary waves, *The London, Edinburgh, and Dublin Phil. Mag. and Journal of Sci.* **39** (1895), 422–443.

[5] D. Mugnolo, D. Noja, and C. Seifert, Airy-type evolutuion equations of star graphs, submitted. arXiv-Preprint 1608.01461.

[6] C. Schubert, C. Seifert, J. Voigt, and M. Waurick, Boundary systems and (skew-)self-adjoint operators on infinite metric graphs. *Math. Nachr.* **288** (2015), 1776–1785.

[7] Z.A. Sobirov, M.I. Akhmedov, and H. Uecker, Cauchy problem for the linearized KdV equation on general metric star graph. *Nanosystems* **6** (2015), 198–204.

[8] Z.A. Sobirov, M.I. Akhmedov, and H. Uecker, Exact solution of the Cauchy problem for the linearized KdV equation on metric star graph. *Uzbek. Math. J.* **3** (2015), 143–154.

[9] Z.A. Sobirov, M.I. Akhmedov, O.V. Karpova, and B. Jabbarova, Linearized KdV equation on a metric graph. *Nanosystems* **6** (2015), 757–761.

Christian Seifert
Technische Universität Hamburg
Insitut für Mathematik
Am Schwarzenberg-Campus 3 (E)
21073 Hamburg
Germany
e-mail: `christian.seifert@tuhh.de`

Bounded multiplicative Toeplitz operators on sequence spaces

Nicola Thorn

Abstract. In this paper, we study the linear mapping which sends the sequence $x = (x_n)_{n\in\mathbb{N}}$ to $y = (y_n)_{n\in\mathbb{N}}$ where $y_n = \sum_{k=1}^{\infty} f(n/k)x_k$ for $f : \mathbb{Q}^+ \to \mathbb{C}$. This operator is the multiplicative analogue of the classical Toeplitz operator, and as such we denote the mapping by $\mathscr{M}_f$. We show that for $1 \leq p \leq q \leq \infty$, if $f \in \ell^r(\mathbb{Q}^+)$, then $\mathscr{M}_f : \ell^p \to \ell^q$ is bounded where $\frac{1}{r} = 1 - \frac{1}{p} + \frac{1}{q}$. Moreover, for the cases when $p = 1$ with any q, $p = q$, and $q = \infty$ with any p, we find that the operator norm is given by $\|\mathscr{M}_f\|_{p,q} = \|f\|_{r,\mathbb{Q}^+}$ when $f \geq 0$. Finding a necessary condition and the operator norm for the remaining cases highlights an interesting connection between the operator norm of $\mathscr{M}_f$ and elements in ℓ^p that have a multiplicative structure, when considering $f : \mathbb{N} \to \mathbb{C}$. We also provide an argument suggesting that $f \in \ell^r$ may not be a necessary condition for boundedness when $1 < p < q < \infty$.

Mathematics Subject Classification (2010). Primary 47B37; Secondary 47B35, 11N99.

Keywords. Bounded multiplicative Toeplitz operators, multiplicative sequences, sequence spaces.

1. Introduction

In this paper, we study the multiplicative Toeplitz operator, denoted by $\mathscr{M}_f$, which sends a sequence $(x_n)_{n\in\mathbb{N}}$ to $(y_n)_{n\in\mathbb{N}}$ where

$$y_n = \sum_{k=1}^{\infty} f\left(\frac{n}{k}\right) x_k, \tag{1.1}$$

and f is a function defined from the positive rationals, $\mathbb{Q}^+$, to $\mathbb{C}$. We can think of $\mathscr{M}_f$ as being given by the infinite matrix A_f whose entries are $a_{i,j} = f(i/j)$ for $i, j \in \mathbb{N}$:

© Springer International Publishing AG, part of Springer Nature 2018
A. Böttcher et al. (eds.), *The Diversity and Beauty of Applied Operator Theory*, Operator Theory: Advances and Applications 268, https://doi.org/10.1007/978-3-319-75996-8_26

$$A_f = \begin{pmatrix} f(1) & f(1/2) & f(1/3) & f(1/4) & \cdots \\ f(2) & f(1) & f(2/3) & f(1/2) & \cdots \\ f(3) & f(3/2) & f(1) & f(3/4) & \cdots \\ f(4) & f(2) & f(4/3) & f(1) & \cdots \\ f(5) & f(5/2) & f(5/3) & f(5/4) & \cdots \\ \vdots & \vdots & \vdots & \vdots & \ddots \end{pmatrix}.$$

Characterised by matrices with constants on skewed diagonals, these mappings are the "multiplicative" analogue of the vastly studied classical Toeplitz operators on sequence spaces. The topic of multiplicative analogues of Toeplitz operators, discussed in [1], [2] and [3] for example, has grown in recent years, with the study of other multiplicative constructions; for example, [4] and [5] investigate the multiplicative Hankel matrix, otherwise known as Helson matrices.

Toeplitz operators, $\mathscr{T}_\phi$, are most often studied via the function ϕ, which is referred to as the symbol. In a similar manner, we shall be considering $\mathscr{M}_f$ in terms of the function f and asking for which f do certain properties hold[1]. By taking f supported only on $\mathbb{N}$, we have $y_n = \sum_{d|n} f\left(\frac{n}{d}\right) x_d = (f * x)(n)$, where $*$ is the Dirichlet convolution [6]. In this case, A_f becomes a lower triangular matrix given by

$$A_f = \begin{pmatrix} f(1) & 0 & 0 & 0 & \cdots \\ f(2) & f(1) & 0 & 0 & \cdots \\ f(3) & 0 & f(1) & 0 & \cdots \\ f(4) & f(2) & 0 & f(1) & \cdots \\ f(5) & 0 & 0 & 0 & \cdots \\ \vdots & \vdots & \vdots & \vdots & \ddots \end{pmatrix}.$$

We shall denote the mapping induced by this matrix by $\mathscr{D}_f$.

Interesting connections to analytic number theory and many open questions have fuelled recent research. For example, in [1] the author illustrates a connection between these operators and the Riemann zeta function. Namely, by choosing f to be supported on $\mathbb{N}$ where $f(n) = \frac{1}{n^\alpha}$ (denoted by $\mathscr{D}_\alpha$), we have that $\mathscr{D}_\alpha : \ell^2 \to \ell^2$ is bounded iff $\alpha > 1$, in which case $\|\mathscr{D}_\alpha\|_{2,2} = \zeta(\alpha)$. Thus when $\alpha \leq 1$, then $\mathscr{D}_\alpha$ is unbounded. By restricting the range of the mapping when $\alpha \in \left(\frac{1}{2}, 1\right]$ and considering

$$Y_\alpha(N) = \sup_{\|x\|_2=1} \left(\sum_{n=1}^{N} |y_n|^2 \right)^{\frac{1}{2}},$$

it can be shown that $Y_\alpha(N)$ is a lower bound for the maximal order of the Riemann zeta function. Specifically, for $\alpha \in \left(\frac{1}{2}, 1\right)$,

$$Z_\alpha(T) = \max_{t\in[0,T]} |\zeta(\alpha + it)| \geq Y_\alpha\left(T^{2/3(\alpha-1/2)-\epsilon}\right),$$

[1]The symbol of $\mathscr{M}_f$ would be given by $F(t) = \sum_{q\in\mathbb{Q}} f(q)q^{it}$ where $t \in \mathbb{R}$.

for sufficiently large T. Moreover, an estimate for $Y_\alpha(N)$ leads to

$$\log Z_\alpha(T) \gg \frac{(\log T)^{1-\alpha}}{\log\log T},$$

a known estimate for the maximal order of ζ. There have since been some improvements upon this estimate, and new estimates for the case when $\alpha = \frac{1}{2}$ have been found, which interestingly utilise a similar method [7], [8]. For other literature on the connections to the Riemann zeta function see also [9], [2].

The authors of [3] also highlight an application of analytic number theory to these operators, by using the properties of smooth numbers to ascertain $\|\mathscr{D}_f x\|_{p,p} = \|f\|_1$ when f is expressible in terms of completely multiplicative and non-negative functions (see the preliminaries for the definitions).

One can also consider the matrix properties of these mappings. For example, [10] considers the determinants of multiplicative Toeplitz matrices. By taking an $N \times N$ truncation, denoted by $A_f(N)$, the author is able to show that if f is multiplicative, then the determinant of $A_f(N)$ can be given as a product over the primes up to N, of determinants of Toeplitz matrices.

In Section 2, we generalise results on the boundedness of $\mathscr{D}_f$ contained in [1] and [3], giving a partial criterion for $\mathscr{M}_f$ to be bounded as a mapping from $\ell^p \to \ell^q$. In an attempt to find a full criterion, we present a relationship between the sets of multiplicative sequences and the operator norm $\|\mathscr{D}_f\|_{p,q}$ in Section 3. By considering $\mathscr{D}_f$ acting upon these subsets, we are able to give a further boundedness result which, due to this connection, indicates that the extension of the partial criterion may not hold. As such, we speculate whether the result can be generalised to $\mathscr{M}_f$ acting on ℓ^p spaces, which is then followed by a discussion on the existence of a possible counterexample to this generalisation. We end the paper with a summary of the open problems that arise within this paper, and also some unanswered questions which are concerned with other operator properties of multiplicative Toeplitz operators such as the spectral points of $\mathscr{M}_f$.

Preliminaries and notation

Sequences and arithmetic functions. We use the terms "sequences" (real or complex valued) and "functions" interchangeably, as we can write any arithmetical function $f(n)$ as a sequence indexed by the natural numbers $f = (f_n)_{n\in\mathbb{N}}$.

Multiplicative functions. First, we say that f (not identically zero) is multiplicative if $f(nm) = f(n)f(m)$ for every $n, m \in \mathbb{N}$ such that $(n, m) = 1$. Secondly, we say f is completely multiplicative if this holds for all $n, m \in \mathbb{N}$. Finally, if $g(n) = cf(n)$ where f is multiplicative, we call g constant multiplicative.

Euler products. If f is multiplicative such that $\sum_{n\in\mathbb{N}} |f(n)| < \infty$, then

$$\sum_{n=1}^{\infty} f(n) = \prod_{t\in\mathbb{P}} \sum_{k=1}^{\infty} f(t^k),$$

where $\mathbb{P}$ is the set of prime numbers. If f is completely multiplicative, we can write

$$\sum_{n=1}^{\infty} f(n) = \prod_{t\in\mathbb{P}} \frac{1}{1-f(t)}.$$

GCD and LCM. We use (n,m) and $[n,m]$ to denote the greatest common divisor and least common multiple of n and m in $\mathbb{N}$, respectively. We let $d(n)$ stand for the number of divisors of n, including 1 and n itself.

O-notation. We say that f is of the order of g and write $f = O(g)$ if, for some constant, $|f(n)| \leq C\,|g(n)|$ as $n \to \infty$. We also write $f \ll g$ to mean $f = O(g)$.

Sequence spaces. For $p \in [1,\infty]$, let ℓ^p denote the usual space of sequences $x = (x_n)_{n\in\mathbb{N}}$ for which the norm $\|x\|_p := (\sum_{n=1}^{\infty} |x_n|^p)^{1/p}$ converges or $\|x\|_\infty = \sup_{n\in\mathbb{N}} |x_n|$ exists (if $p \in [1,\infty)$ or $p = \infty$ respectively). Analogously, we define $\ell^p(\mathbb{Q}^+)$ to be the space of sequences $x = (x_s)_{s\in\mathbb{Q}^+}$ for which $\|x\|_{p,\mathbb{Q}^+} = (\sum_{s\in\mathbb{Q}^+} |x_s|^p)^{1/p}$ converges or $\|x\|_{\infty,\mathbb{Q}^+} = \sup_{s\in\mathbb{Q}^+} |x_s|$ exists. For the case when $p = 2$, we also have that $\langle x, y\rangle = \sum_{n\in\mathbb{N}} x_n \overline{y_n}$.

Operator norm. Given a bounded linear operator L, we use the usual notation $\|L\|_{p,q}$ to denote the operator norm of $L : \ell^p \to \ell^q$ which is given by $\|L\|_{p,q} = \sup_{\|x\|_p=1} \|Lx\|_q$.

2. Partial criterion for boundedness

The following results extend theorems contained in [1] and [3].

Theorem 2.1. *For $1 \leq p \leq q \leq \infty$, define $r \in [1,\infty]$ by*

$$\frac{1}{r} = 1 - \frac{1}{p} + \frac{1}{q}$$

where $\frac{1}{\infty} = 0$. If $f \in \ell^r(\mathbb{Q}^+)$, then $\mathscr{M}_f : \ell^p \to \ell^q$ is bounded. More precisely, we have

$$\|\mathscr{M}_f x\|_q \leq \|x\|_p \|f\|_{r,\mathbb{Q}^+}.$$

Theorem 2.1 gives a partial criterion for boundedness between ℓ^p and ℓ^q; partial in the sense that $f \in \ell^r(\mathbb{Q}^+)$ is a sufficient condition. It is natural to ask whether this is also a necessary condition, i.e., does $\mathscr{M}_f : \ell^p \to \ell^q$ bounded imply that $f \in \ell^r$? Moreover, can we find the operator norm, $\|\mathscr{M}_f\|_{p,q}$? For f positive, both of these questions can be answered by Theorem 2.2 for the cases where $p = q$, $p = 1$ with any q, and $q = \infty$ with any p. We refer to these as the "edge" cases.

Theorem 2.2. *Let us define r as in Theorem 2.1. For $p = q$, $p = 1$ (any q), $q = \infty$ (any p) with $f \in \ell^r(\mathbb{Q}^+)$ positive, we have*

$$\|\mathscr{M}_f\|_{p,q} = \|f\|_{r,\mathbb{Q}^+}.$$

Proof of Theorem 2.1. Let y_n be given by (1.1). The proof proceeds by considering the cases separately.

- $1 \le p \le q < \infty$

By Hölder's inequality,

$$|y_n| \le \sum_{k=1}^{\infty} \left| f\left(\frac{n}{k}\right) x_k \right| = \sum_{k=1}^{\infty} \left| f\left(\frac{n}{k}\right) \right|^{r\left(1-\frac{1}{p}\right)} \left| f\left(\frac{n}{k}\right) \right|^{\frac{r}{q}} |x_k|^{\frac{p}{q}} |x_k|^{1-\frac{p}{q}}$$

$$\le \left(\sum_{k=1}^{\infty} \left| f\left(\frac{n}{k}\right) \right|^{r} \right)^{\left(1-\frac{1}{p}\right)} \left(\sum_{k=1}^{\infty} |x_k|^p \right)^{\frac{1}{p}-\frac{1}{q}} \left(\sum_{k=1}^{\infty} \left| f\left(\frac{n}{k}\right) \right|^{r} |x_k|^p \right)^{\frac{1}{q}}$$

$$\le \|f\|_{r,\mathbb{Q}^+}^{r\left(1-\frac{1}{p}\right)} \|x\|_p^{1-\frac{p}{q}} \left(\sum_{k=1}^{\infty} \left| f\left(\frac{n}{k}\right) \right|^{r} |x_k|^p \right)^{\frac{1}{q}}.$$

Hence,

$$\sum_{n=1}^{\infty} |y_n|^q \le \|f\|_{r,\mathbb{Q}^+}^{rq\left(1-\frac{1}{p}\right)} \|x\|_p^{q-p} \sum_{n=1}^{\infty}\sum_{k=1}^{\infty} \left| f\left(\frac{n}{k}\right) \right|^{r} |x_k|^p .$$

Considering only the summation on the RHS above,

$$\sum_{n=1}^{\infty}\sum_{k=1}^{\infty} \left| f\left(\frac{n}{k}\right) \right|^{r} |x_k|^p \le \sum_{s\in\mathbb{Q}^+} |f(s)|^r \sum_{k=1}^{\infty} |x_k|^p = \|f\|_{r,\mathbb{Q}^+}^{r} \|x\|_p^p.$$

Therefore,

$$\|\mathscr{M}_f x\|_q^q = \sum_{n=1}^{\infty} |y_n|^q \le \|f\|_{r,\mathbb{Q}^+}^{qr\left(1-\frac{1}{p}\right)+r} \|x\|_p^{q-p+p} = \|f\|_{r,\mathbb{Q}^+}^{q} \|x\|_p^q.$$

- $p = 1$ and $q = \infty$ (so $r = \infty$)

By the triangle inequality,

$$|y_n| \le \sum_{k=1}^{\infty} \left| f\left(\frac{n}{d}\right) x_k \right| \le \|f\|_{\infty,\mathbb{Q}^+} \sum_{k=1}^{\infty} |x_k| \le \|f\|_{\infty,\mathbb{Q}^+} \|x\|_1.$$

Hence, $\|\mathscr{M}_f x\|_\infty \le \|f\|_{\infty,\mathbb{Q}^+} \|x\|_1$.

- $q = \infty$ with $1 < p < \infty$ (so $r = \frac{p}{p-1}$)

By Hölder's inequality, we have

$$|y_n| \le \sum_{k=1}^{\infty} \left| f\left(\frac{n}{k}\right) x_k \right| \le \left(\sum_{k=1}^{\infty} \left| f\left(\frac{n}{k}\right) \right|^{r} \right)^{\frac{1}{r}} \left(\sum_{k=1}^{\infty} |x_k|^p \right)^{\frac{1}{p}} \le \|f\|_{r,\mathbb{Q}^+} \|x\|_p.$$

Thus, $\|\mathscr{M}_f x\|_\infty \le \|f\|_{r,\mathbb{Q}^+} \|x\|_p$.

- $p = q = \infty$ (so $r = 1$)

We now have $|y_n| \le \|x\|_\infty \sum_{k=1}^{\infty} \left| f\left(\frac{n}{k}\right) \right| \le \|x\|_\infty \|f\|_{1,\mathbb{Q}^+}$, which gives the desired inequality $\|\mathscr{M}_f x\|_\infty \le \|x\|_\infty \|f\|_{1,\mathbb{Q}^+}$. □

Proof of Theorem 2.2. We consider each edge case separately.

1. We first embark on the case when $p = 1$ with any q.

- Let $q \in [1, \infty)$, so that $r = q$.

Fix $c \in \mathbb{N}$, and let $x_n = 1$ if $n = c$ and 0 otherwise. Then $\|x\|_1 = 1$, and so

$$|y_n|^q = \left|\sum_{k=1}^{\infty} f\left(\frac{n}{k}\right) x_k\right|^q = \left|f\left(\frac{n}{c}\right)\right|^q.$$

Therefore,

$$\begin{aligned}\|\mathscr{M}_f x\|_q^q &= \sum_{n=1}^{\infty} |y_n|^q = \sum_{n=1}^{\infty} \left|f\left(\frac{n}{c}\right)\right|^q = \sum_{d|c} \sum_{\substack{n=1\\(n,c)=d}}^{\infty} \left|f\left(\frac{n}{c}\right)\right|^q \\ &= \sum_{d|c} \sum_{\substack{m=1\\(m,\frac{c}{d})=1}}^{\infty} \left|f\left(\frac{md}{c}\right)\right|^q \qquad \text{by writing } n = md, \\ &= \sum_{d|c} \sum_{\substack{m=1\\(m,d)=1}}^{\infty} \left|f\left(\frac{m}{d}\right)\right|^q \qquad \text{by writing } \frac{c}{d} \mapsto d. \end{aligned} \tag{2.1}$$

Note that we can write

$$\|f\|_{q,\mathbb{Q}^+}^q = \sum_{s\in\mathbb{Q}^+} |f(s)|^q = \sum_{v=1}^{\infty} \sum_{\substack{u=1\\(u,v)=1}}^{\infty} \left|f\left(\frac{u}{v}\right)\right|^q. \tag{2.2}$$

By computing the difference between (2.2) and (2.1), we shall show that $\|\mathscr{M}_f x\|_q$ can be made arbitrarily close to $\|f\|_{q,\mathbb{Q}^+}$. We have

$$\sum_{v=1}^{\infty} \sum_{\substack{u=1\\(u,v)=1}}^{\infty} \left|f\left(\frac{u}{v}\right)\right|^q - \sum_{d|c} \sum_{\substack{n=1\\(n,c)=d}}^{\infty} \left|f\left(\frac{n}{c}\right)\right|^q = \sum_{\substack{u,v\in\mathbb{N}\\(u,v)=1\\v\nmid c}} \left|f\left(\frac{u}{v}\right)\right|^q.$$

Now, choose $c = (2 \cdot 3 \cdot 5 \cdots T)^k$ where $k \in \mathbb{N}$ and T is prime. Then $v \nmid c$ implies $v > T$ for k large enough. Therefore, for every $\epsilon > 0$, we can choose T such that

$$\|f\|_{q,\mathbb{Q}^+}^q - \|\mathscr{M}_f x\|_q^q = \sum_{\substack{u,v\in\mathbb{N}\\(u,v)=1\\v\nmid c}} \left|f\left(\frac{u}{v}\right)\right|^q < \epsilon.$$

Hence, $\|\mathscr{M}_f\|_{1,q} = \|f\|_{q,\mathbb{Q}^+}$ as required.

- Let $q = \infty$, so $r = q = \infty$.

Fix $c \in \mathbb{N}$. Like before, choose $x_n = 1$ if $n = c$ and 0 otherwise. Again $\|x\|_1 = 1$. Now,

$$\|\mathscr{M}_f x\|_\infty = \sup_{n\in\mathbb{N}} |y_n| = \sup_{n\in\mathbb{N}} \left|f\left(\frac{n}{c}\right)\right|.$$

Note that there exist $u, v \in \mathbb{N}$ with $(u, v) = 1$ such that $\|f\|_{\infty,\mathbb{Q}^+} - \epsilon < \left|f\left(\frac{u}{v}\right)\right|$. Simply choose $n = u$ and $c = v$. Then

$$\|f\|_{\infty,\mathbb{Q}^+} - \|\mathscr{M}_f x\|_\infty < \epsilon.$$

2. Now consider the edge case where $p = q$.

- Let $1 < p = q < \infty$, so $r = 1$.

Fix $c \in \mathbb{N}$. Choose $x_n = 1/d(c)^{\frac{1}{q}}$ if $n \mid c$ and 0 otherwise. Hence, we have $\|x\|_q^q = \frac{1}{d(c)} \sum_{d|c} 1 = 1$. By Hölder's inequality,

$$\sum_{n=1}^{\infty} x_n^{q-1} y_n \leq \left(\sum_{n=1}^{\infty} |x_n|^q\right)^{1-\frac{1}{q}} \left(\sum_{n-1}^{\infty} y_n^q\right)^{\frac{1}{q}} = \left(\sum_{n-1}^{\infty} y_n^q\right)^{\frac{1}{q}} = \|\mathscr{M}_f x\|_q.$$

Consequently, it suffices to show that $\sum_{n=1}^{\infty} x_n^{q-1} y_n$ can be made arbitrarily close to $\|f\|_{1,\mathbb{Q}^+}$. We have

$$\begin{aligned}\sum_{n=1}^{\infty} x_n^{q-1} y_n &= \frac{1}{d(c)^{\frac{q-1}{q}}} \sum_{n|c} y_n = \frac{1}{d(c)^{\frac{q-1}{q}}} \sum_{n|c} \sum_{k|c} f\left(\frac{n}{k}\right) x_k \\ &= \frac{1}{d(c)} \sum_{n,k|c} f\left(\frac{n}{k}\right).\end{aligned}$$

We now follow the argument given in [2] (page 87). For $s = \frac{u}{v} \in \mathbb{Q}^+$,

$$\frac{1}{d(c)} \sum_{n,k|c} f\left(\frac{n}{k}\right) = \frac{1}{d(c)} \sum_{s \in \mathbb{Q}^+} f(s) \sum_{\substack{n,k|c \\ s=\frac{n}{k}}} 1 = \frac{1}{d(c)} \sum_{u,v \in \mathbb{N}} f\left(\frac{u}{v}\right) \sum_{\substack{n,k|c \\ nv=uk}} 1,$$

where we used that $\frac{n}{k} = \frac{u}{v}$ if and only if $nv = uk$. Since $(u, v) = 1$ we have $u \mid n$ and $v \mid k$, and for any contribution to the summation on the RHS, we must have $u, v \mid c$, that is, $uv \mid c$. Assume therefore, that $uv \mid c$. By writing $n = lu$ and $k = lv$ for some $l \in \mathbb{N}$, we get

$$\begin{aligned}\frac{1}{d(c)} \sum_{uv|c} f\left(\frac{u}{v}\right) \sum_{\substack{n,k|c \\ nv=uk}} 1 &= \frac{1}{d(c)} \sum_{uv|c} f\left(\frac{u}{v}\right) \sum_{lu,lv|c} 1 \\ &= \frac{1}{d(c)} \sum_{uv|c} f\left(\frac{u}{v}\right) \sum_{l|\frac{c}{uv}} 1 = \sum_{uv|c} f\left(\frac{u}{v}\right) \frac{d\left(c/uv\right)}{d(c)}.\end{aligned}$$

Now, by choosing c appropriately, we can show that $\frac{d(c/uv)}{d(c)}$ can be made close to 1 for all u, v less than some large constant. Fix $T \in \mathbb{P}$ and choose c to be

$$c = \prod_{\substack{t \leq T \\ t \in \mathbb{P}}} t^{\alpha_t} \quad \text{with} \quad \alpha_t = \left[\frac{\log T}{\log t}\right].$$

If $uv \mid c$, then $uv = \prod_{t \le T} t^{\beta_t}$ where $\beta_t \in [0, \alpha_t]$, and hence

$$\frac{d(c/uv)}{d(c)} = \prod_{t \le T} \left(\frac{\alpha_t - \beta_t + 1}{\alpha_t + 1} \right) = \prod_{t \le T} \left(1 - \frac{\beta_t}{\alpha_t + 1} \right).$$

If we take $uv \le \sqrt{\log T}$, then $t^{\beta_t} \le \sqrt{\log T}$ for every prime divisor t of uv. Therefore, $\beta_t \le \frac{\log \log T}{2 \log t}$ and $\beta_t = 0$ if $t > \sqrt{\log T}$. It follows that

$$\begin{aligned} \frac{d(c/uv)}{d(c)} &= \prod_{t \le \sqrt{\log T}} \left(1 - \frac{\beta_t}{\alpha_t + 1} \right) \ge \prod_{t \le \sqrt{\log T}} \left(1 - \frac{\log \log T}{2 \log T} \right) \\ &= \left(1 - \frac{\log \log T}{2 \log T} \right)^{\pi(\sqrt{\log T})}, \end{aligned}$$

where $\pi(x)$ is the prime counting function up to x. As $\pi(x) \ll \frac{x}{\log x}$, we have for sufficiently large T,

$$\frac{d(c/uv)}{d(c)} = \left(1 - \frac{\log \log T}{2 \log T} \right)^{\pi(\sqrt{\log T})} \ge 1 - \frac{C}{\sqrt{\log T}},$$

for some constant C. Therefore,

$$\begin{aligned} \sum_{uv \mid c} f\left(\frac{u}{v}\right) \frac{d(c/uv)}{d(c)} &> \sum_{uv \le \sqrt{\log T}} f(s) \left(1 - \frac{C}{\sqrt{\log T}} \right) - \sum_{uv > \sqrt{\log T}} f(q) \\ &\ge \sum_{s \in \mathbb{Q}^+} f(s) - \frac{C_1}{\sqrt{\log T}} - 2 \sum_{uv > \sqrt{\log T}} f(s), \end{aligned}$$

as $f \in \ell^1(\mathbb{Q}^+)$. By choosing T to be arbitrarily large, for every $\epsilon > 0$, we have

$$\|f\|_{1,\mathbb{Q}^+} - \|\mathscr{M}_f x\|_q \le \|f\|_{1,\mathbb{Q}^+} - \sum_{n=1}^{\infty} x_n^{q-1} y_n < \epsilon.$$

- We now consider the case where $p = q = \infty$, and so $r = 1$.

Let $x_n = 1$ for all $n \in \mathbb{N}$, so that $\|x\|_\infty = 1$. Moreover, for a fixed $c \in \mathbb{N}$, we have

$$|y_c| = \sum_{k=1}^{\infty} f\left(\frac{c}{k}\right) x_k = \sum_{k=1}^{\infty} f\left(\frac{c}{k}\right).$$

Again, by applying the same methods already shown, we conclude that y_c can be arbitrarily close to $\|f\|_{1,\mathbb{Q}^+}$. Hence, $\|\mathscr{M}_f\|_{\infty,\infty} = \|f\|_{1,\mathbb{Q}^+}$.

3. Finally, we consider the case when $q = \infty$ with any p.

We have already dealt with the case when $p = 1$ and $p = \infty$. So let $p \in (1, \infty)$, giving $r = \frac{p}{p-1}$. Fix $c \in \mathbb{N}$, and let

$$x_n = f\left(\frac{c}{n}\right)^{\frac{r}{p}} F_c^{-\frac{1}{p}}, \quad \text{where } F_c = \sum_{n=1}^{\infty} f\left(\frac{c}{n}\right)^r \text{ exists as } f \in \ell^r(\mathbb{Q}^+).$$

With this choice,

$$\|x\|_p = \frac{1}{F_c}\sum_{n=1}^{\infty} f\left(\frac{c}{n}\right)^r = \frac{F_c}{F_c} = 1.$$

Now consider just the term y_c,

$$y_c = F_c^{-\frac{1}{p}} \sum_{k=1}^{\infty} f\left(\frac{c}{k}\right) f\left(\frac{c}{k}\right)^{\frac{r}{p}} = F_c^{-\frac{1}{p}} \sum_{k=1}^{\infty} f\left(\frac{c}{k}\right)^r,$$

as $1 + \frac{r}{p} = \frac{p-1+1}{p-1} = r$. Therefore,

$$y_c = F_c^{1-\frac{1}{p}} = F_c^{\frac{1}{r}} = \left(\sum_{k=1}^{\infty} f\left(\frac{c}{k}\right)^r\right)^{\frac{1}{r}}.$$

We can apply the same argument as before to show that for every $\epsilon > 0$, we can choose $c = (2 \cdot 3 \cdot 5 \cdots T)^k$ where T is prime such that y_c can be made arbitrarily close to $\|f\|_{r,\mathbb{Q}^+}$. Hence, $\|\mathscr{M}_f\|_{p,\infty} = \|f\|_{r,\mathbb{Q}^+}$. □

Remark 2.3. In [2], the author showed that if f is any, not necessarily strictly positive, sequence in $\ell^1(\mathbb{Q}^+)$, then $\mathscr{M}_f : \ell^2 \to \ell^2$ is bounded and the operator norm is given by

$$\|\mathscr{M}_f\|_{2,2} = \sup_{t\in\mathbb{R}} \left| \sum_{q\in\mathbb{Q}^+} f(q) q^{it} \right|.$$

By assuming f positive, the supremum of the above is attained when $t = 0$, and as such $\|\mathscr{M}_f\|_{2,2} = \|f\|_{1,\mathbb{Q}^+}$ as given in Theorem 2.2. The differing operator norm when f is not positive, is echoed in the work of [3], where an example is given showing that $\|\mathscr{D}_f\|_{p,p} \neq \|f\|_1$. Determining $\|\mathscr{M}_f\|_{p,q}$ for any f and general p, q remains an open question, but is not, however, the focus of this paper.

3. Connection with multiplicative sequences

Generalising Theorem 2.2 to find a necessary condition and the operator norm for all other p and q (which we will refer to as the interior cases) is challenging and is the focus of the proceeding discussions.

We start by taking f supported on $\mathbb{N}$, i.e., $\mathscr{M}_f = \mathscr{D}_f$. To understand the behaviour of the operator norm in the interior cases, we can consider where $\|\mathscr{D}_f x\|$ attains its supremum value in the edge cases. First, setting $c = 1$ in case 1 of the proof of Theorem 2.2 yields the supremum of $\|\mathscr{D}_f x\|_q$. This gives $x_n = 1$ if $n = 1$ and 0 otherwise, and as such x is completely multiplicative. Secondly, for $1 < p = q < \infty$ in case 2, we choose $x_n = \frac{1}{d(c)^{1/p}}$ whenever $n \mid c$ and 0 otherwise, which is a constant multiplicative sequence. Moreover, for $p = q = \infty$, the completely multiplicative sequence $x_n = 1$ (for all $n \in \mathbb{N}$) attains the operator norm. Finally, in case 3, for f multiplicative, x is again constant multiplicative.

It follows, for the edge cases, that $\mathscr{D}_f$ is "largest" when acting on a sequence $x \in \ell^p$ that has multiplicative structure. Why this is the case is unclear and leads to a surprising connection between the operator norm of $\mathscr{D}_f$ and the set of multiplicative elements in ℓ^p, which we denote by $\mathcal{M}^p$. Moreover, we shall denote the set of completely multiplicative sequences in ℓ^p by $\mathcal{M}_c^p$. It is interesting to ask therefore how $\mathscr{D}_f$ acts on these sets for $1 < p < q < \infty$, as from this connection, we would expect $\mathscr{D}_f : \ell^p \to \ell^q$ to attain its supreme value here. Thus, we shall investigate the boundedness of $\mathscr{D}_f : \mathcal{M}_c^p \to \ell^q$ for $1 < p < q < \infty$, with the aim of giving some insight into $\|\mathscr{D}_f\|_{p,q}$, its norm.[2]

From Theorem 2.1, it follows that $\mathscr{D}_f : \mathcal{M}_c^p \to \ell^q$ is bounded if $f \in \ell^r$. We wish to know whether this is also a necessary condition. In Theorem 3.1, we show that for f completely multiplicative, the requirement that f be $\mathcal{M}_c^r$ is not a necessary condition for $\mathscr{D}_f : \mathcal{M}_c^p \to \mathcal{M}^2$ to be bounded[3] when $p \in (1, 2)$ and $q = 2$. One can speculate therefore that $f \in \ell^r$ is not a necessary condition when considering $\mathscr{D}_f : \ell^p \to \ell^2$.

Theorem 3.1. *Let $1 < p < 2$. If $f \in \mathcal{M}_c^2$, the mapping $\mathscr{D}_f : \mathcal{M}_c^p \to \mathcal{M}^2$ is bounded.*

To highlight the difference between this criterion and that shown in the previous section, we consider the following example. Let $f(n) = \frac{1}{n^\alpha}$ for $\alpha > \frac{1}{2}$ and $p = \frac{3}{2}$, giving $\frac{1}{r} = 1 - \frac{2}{3} + \frac{1}{2} = \frac{5}{6}$. Theorem 2.1 states that if $\alpha > \frac{5}{6}$, then $\mathscr{D}_f : \mathcal{M}_c^{3/2} \to \ell^2$ is bounded. In contrast, Theorem 3.1 shows that only $\alpha > \frac{1}{2}$ is required for boundedness. For the proof of Theorem 3.1, we will require the following lemma, which will be proved below.

Lemma 3.2. *Let $f, g, h, j \in \mathcal{M}_c^2$. Then,*

$$\langle f * g, h * j \rangle = \frac{\langle g, j \rangle \langle f, h \rangle \langle f, j \rangle \langle g, h \rangle}{\langle fg, hj \rangle}. \tag{3.1}$$

Proof of Theorem 3.1. By taking $h = f$ and $g = j = x$ in (3.1), we have

$$\|\mathscr{D}_f x\|_2 = \frac{\|f\|_2 \|x\|_2 \, |\langle f, x \rangle|}{\|fx\|_2} \leq \|f\|_2 \|x\|_2 \, |\langle f, x \rangle| ,$$

as f and x are completely multiplicative, and as such we have $x_1 = 1$ and $f(1) = 1$, giving

$$\|fx\|_2 = \sum_{n=1}^{\infty} |f(n) x_n|^2 \geq 1.$$

[2] $\mathcal{M}_c^p$ and $\mathcal{M}^p$ are subsets, not subspaces of ℓ^p. For example, they are not closed under addition. Given X, Y which are subsets of some Banach space, we say $L : X \to Y$ is bounded iff $\|Lx\| \leq C\|x\|$ for all $x \in X$.

[3] The convolutions of two multiplicative sequences is also multiplicative, so we can consider $y \in \mathcal{M}^2$.

Now,

$$\frac{\|\mathscr{D}_f x\|_2}{\|x\|_p} \leq \frac{\|f\|_2 \|x\|_2 \, |\langle f, x\rangle|}{\|x\|_p} = \|f\|_2 \prod_{t\in\mathbb{P}} \frac{\left(1-|x_t|^p\right)^{\frac{1}{p}}}{\left(1-|x_t|^2\right)^{\frac{1}{2}}\left(1-|x_t f(t)|\right)}, \tag{3.2}$$

where we made use of Euler products. Therefore, it remains to show that the product over primes is bounded independently of x_t. As $0 \leq |x_t| < 1$, we can say that

$$|x_t|^2 < |x_t|^p \implies \frac{1}{1-|x_t|^2} < \frac{1}{1-|x_t|^p}.$$

Hence, the product of (3.2) is at most

$$\prod_{t\in\mathbb{P}} \frac{\left(1-|x_t|^p\right)^{\frac{1}{p}}}{\left(1-|x_t|^p\right)^{\frac{1}{2}}\left(1-|x_t f(t)|\right)} = \prod_{t\in\mathbb{P}} \frac{\left(1-|x_t|^p\right)^{\frac{2-p}{2p}}}{\left(1-|x_t f(t)|\right)}.$$

By taking logarithms, we arrive at the equality

$$\log\left(\prod_{t\in\mathbb{P}} \frac{\left(1-|x_t|^p\right)^{\frac{2-p}{2p}}}{\left(1-|x_t f(t)|\right)}\right) = \sum_{t\in\mathbb{P}} \left(\log\frac{1}{1-|x_t f(t)|} - \frac{2-p}{2p}\log\frac{1}{1-|x_t|^p}\right).$$

Note in general for $a > 0$, we have $a \leq \log\left(\frac{1}{1-a}\right) = a + O\left(a^2\right)$. Hence,

$$\sum_{t\in\mathbb{P}} \log\left(\frac{1}{1-|x_t|^p}\right) \geq \sum_{t\in\mathbb{P}} |x_t|^p,$$

and moreover,

$$\sum_{t\in\mathbb{P}} \log\left(\frac{1}{1-|x_t f(t)|}\right) = \sum_{t\in\mathbb{P}} |x_t f(t)| + O(1),$$

where the $O(1)$ term is independent of the sequence x_t. Therefore, we obtain

$$\begin{aligned}\sum_{t\in\mathbb{P}} &\left(\log\frac{1}{1-|x_t f(t)|} - \frac{2-p}{2p}\log\frac{1}{1-|x_t|^p}\right) \\ &< \sum_{t\in\mathbb{P}}\left(|x_t f(t)| - \frac{2-p}{2p}|x_t|^p\right) + O(1).\end{aligned}$$

Now, we consider the case when the terms of the above series are positive. In other words, with $\beta = \frac{1}{p-1}$,

$$|x_t f(t)| \geq \frac{2-p}{2p}|x_t|^p \iff \left(\frac{2p}{2-p}|f(t)|\right)^{\beta} \geq |x_t|.$$

Hence, by only summing over the t which yield positive terms, we have

$$\sum_{t\in\mathbb{P}}\left(|x_tf(t)|-\frac{2-p}{2p}|x_t|^p\right)<\sum_{\substack{t \text{ s.t}\\ |x_t|\le(\frac{2p}{2-p}|f(t)|)^\beta}}\left(|x_tf(t)|-\frac{2-p}{2p}|x_t|^p\right)$$

$$\le\sum_{\substack{t \text{ s.t}\\ |x_t|\le(\frac{2p}{2-p}|f(t)|)^\beta}}|x_tf(t)|\le\left(\frac{2p}{2-p}\right)^\beta\sum_{t\in\mathbb{P}}|f(t)|^\beta|f(t)|.$$

As $\beta+1=\frac{p}{p-1}>2$, we see that

$$\sum_{t\in\mathbb{P}}|f(t)|^{\beta+1}\le\sum_{t\in\mathbb{P}}|f(t)|^2<\infty,$$

since $f\in\mathcal{M}_c^2$. Hence, the product in (3.2) is bounded, which implies that the mapping $\mathscr{D}_f:\mathcal{M}_c^p\to\mathcal{M}^2$ is bounded. □

Proof of Lemma 3.2. We start by computing the LHS of (3.1):

$$\langle f*g,h*j\rangle=\sum_{n\ge1}(f*g)(n)\overline{(h*j)(n)}=\sum_{n\ge1}\sum_{c,d|n}f(c)g\Big(\frac{n}{c}\Big)\overline{h(d)j\Big(\frac{n}{d}\Big)}$$

$$=\sum_{c,d\ge1}\sum_{\substack{n\ge1\\ c,d|n}}f(c)g\Big(\frac{n}{c}\Big)\overline{h(d)j\Big(\frac{n}{d}\Big)}=\sum_{c,d\ge1}\sum_{\substack{n\ge1\\ [c,d]|n}}f(c)g\left(\frac{n}{c}\right)\overline{h(d)j\Big(\frac{n}{d}\Big)},$$

since $c,d|n\Leftrightarrow[c,d]|n$. Now, as $[c,d]|n\Leftrightarrow n=[c,d]m$, the above is given by

$$\sum_{c,d\ge1}\sum_{m\ge1}f(c)g\left(\frac{m[c,d]}{c}\right)\overline{h(d)j\left(\frac{m[c,d]}{d}\right)}$$

$$=\sum_{m\ge1}g(m)\overline{j(m)}\sum_{c,d\ge1}f(c)g\left(\frac{[c,d]}{c}\right)\overline{h(d)j\left(\frac{[c,d]}{d}\right)}$$

$$=\langle g,j\rangle\sum_{c,d\ge1}f(c)g\left(\frac{d}{(c,d)}\right)\overline{h(d)j\left(\frac{c}{(c,d)}\right)}\qquad\text{as } [c,d](c,d)=cd,$$

$$=\langle g,j\rangle\sum_{k\ge1}\sum_{\substack{c,d\ge1\\ (c,d)=k}}f(c)g\left(\frac{d}{(c,d)}\right)\overline{h(d)j\left(\frac{c}{(c,d)}\right)}.$$

If $(c,d)=k$, then $c=c'k$, $d=d'k$ where $(c',d')=1$. Therefore,

$$\langle f*g,h*j\rangle=\langle g,j\rangle\sum_{k\ge1}\sum_{\substack{c',d'\ge1\\ (c',d')=1}}f(c'k)g(d')\overline{h(d'k)j(c')},$$

which is equal to

$$\langle g, j\rangle \sum_{k\geq 1} f(k)\overline{h(k)} \sum_{\substack{c',d'\geq 1\\(c',d')=1}} f(c')g(d')\overline{h(d')j(c')}$$
$$= \langle g, j\rangle\langle f, h\rangle \sum_{\substack{c',d'\geq 1\\(c',d')=1}} f(c')g(d')\overline{h(d')j(c')}. \tag{3.3}$$

We now compute the RHS of (3.1). We have

$$\langle f, j\rangle\langle g, h\rangle = \sum_{c,d\geq 1} f(c)\overline{j(c)}g(d)\overline{h(d)} = \sum_{k\geq 1}\sum_{\substack{c,d\geq 1\\(c,d)=k}} f(c)\overline{j(c)}g(d)\overline{h(d)}$$
$$= \sum_{k\geq 1}\sum_{\substack{c',d'\geq 1\\(c',d')=1}} f(c'k)\overline{j(c'k)}g(d'k)\overline{h(d'k)}$$
$$= \sum_{k\geq 1} f(k)\overline{j(k)}g(k)\overline{h(k)} \sum_{\substack{c',d'\geq 1\\(c',d')=1}} f(c')\overline{j(c')}g(d')\overline{h(d')}$$
$$= \langle fg, hj\rangle \sum_{\substack{c',d'\geq 1\\(c',d')=1}} f(c')\overline{j(c')}g(d')\overline{h(d')}. \tag{3.4}$$

Hence, by comparing (3.3) with (3.4) we obtain (3.1). □

Naturally one can ask if Theorem 3.1 generalises to ℓ^p. In other words: *is $f \in \ell^2$ a sufficient condition for $\mathscr{D}_f : \ell^p \to \ell^2$ to be bounded for every p in $(1, 2)$?* Furthermore, Theorem 3.1 raises some interesting points of speculation regarding this question. It would perhaps be surprising if Theorem 3.1 could not be generalised to $\mathscr{M}_f$ on ℓ^p as we know that in the edge cases, the operator norm is "largest" when acting on multiplicative sequences. Why this would not also be true for the interior cases is unclear. In contrast, we know from Theorem 2.2 that when $p = 2$, for boundedness, $f \in \ell^1$ is needed. If a generalisation is possible, there would be a jump in the required value of r. That is, by considering $p = 2 - \epsilon$ for any $\epsilon > 0$, then $f \in \ell^2$ is all that is required. Why the jump between $f \in \ell^1$ to $f \in \ell^2$ would occur is also unclear. Finding a generalisation of Theorem 3.1 has not been possible, and leads to an investigation of a possible counterexample to the question raised above.

A possible counterexample

We wish to know, given $f \in \ell^2$, does there exist $x \in \ell^p$, for $p \in (1, 2)$, such that $\mathscr{D}_f x \notin \ell^2$? For simplicity, we choose $f(n) = \frac{1}{n^\alpha}$ with $\alpha > \frac{1}{2}$.

Proposition 3.3. *Let $p \in (1, 2)$, $q = 2$, and $\alpha > \frac{1}{2}$. If $(x_n) \in \ell^p$ is a sequence such that $x_n \ll 1/d(n)^{\frac{1}{2-p}}$, then $\mathscr{D}_\alpha x \in \ell^2$.*

Proof. By the Cauchy–Schwarz inequality, we have

$$y_n^2 = \left(\sum_{d|n} \frac{x_{n/d}}{d^\alpha}\right)^2 \leq \sum_{d|n} 1 \sum_{d|n} \frac{x_{n/d}^2}{d^{2\alpha}} = d(n) \sum_{d|n} \frac{x_{n/d}^2}{d^{2\alpha}}.$$

So,

$$\begin{aligned}\|\mathscr{D}_\alpha x\|_2^2 &\leq \sum_{n=1}^{\infty} d(n) \sum_{d|n} \frac{x_{n/d}^2}{d^{2\alpha}} = \sum_{d=1}^{\infty}\sum_{m=1}^{\infty} d(md)\frac{x_m^2}{d^{2\alpha}} \quad \text{by writing } dm = n, \\ &\leq \sum_{d=1}^{\infty} \frac{d(d)}{d^{2\alpha}} \sum_{m=1}^{\infty} d(m) x_m^2,\end{aligned}$$

as $d(mn) \leq d(m)d(n)$. As $\alpha > \frac{1}{2}$, the first series on the RHS is convergent (and given by $\zeta(2\alpha)^2$). Hence,

$$\|\mathscr{D}_\alpha x\|_2^2 \ll \sum_{m=1}^{\infty} d(m) x_m^2.$$

This is convergent if $x_m^2 d(m) \ll x_m^p$ (as $x \in \ell^p$). By rearranging, this is equivalent to $x_m \ll 1/d(m)^{\frac{1}{2-p}}$ as required. □

From Proposition 3.3, we can conclude that any counterexample, say $x = (x_n)$, must satisfy $x_n > 1/d(n)^{\frac{1}{2-p}}$ for infinitely many $n \in \mathbb{N}$. As such we define

$$S = \left\{n \in \mathbb{N} : x_n > 1/d(n)^{\frac{1}{2-p}}\right\},$$

and we may assume that the support of x is contained within the set S, i.e., $x_n = 0$ if $n \notin S$. However, some care must be taken in choosing S (if an example is possible), as

$$\sum_{n \in S} \frac{1}{d(n)^{\frac{p}{2-p}}} \leq \sum_{n \in S} x_n^p < \infty \tag{3.5}$$

must be satisfied due to $x \in \ell^p$. First, S must be a "sparse" set; consider the function which counts number of $n \in S$ below a c given, $S(c) = \sum_{\substack{n \leq c \\ n \in S}} 1$. Then

$$S(c) = \sum_{\substack{n \leq c \\ n \in S}} \frac{x_n^p}{x_n^p} \ll c^\epsilon \sum_{\substack{n \leq c \\ n \in S}} x_n^p \ll c^\epsilon \quad \text{for all } \epsilon > 0,$$

as $1/x_n^p \leq d(n)^{\frac{p}{2-p}} \ll n^\epsilon \leq c^\epsilon$ for all $\epsilon > 0$. For example, choosing $S = \mathbb{N}$ fails. Secondly, S must contain n with large numbers of divisors, otherwise $1/d(n)^{\frac{p}{2-p}} \not\to 0$ as $n \to \infty$, and so (3.5) will not be satisfied (S can not be a subset of $\mathbb{P}$, for example). However, the following example indicates the difficulty of choosing S to yield $\mathscr{D}_\alpha$ unbounded: define $S = \left\{2^k : k \in \mathbb{N}\right\}$.

We see that (3.5) is satisfied because

$$\sum_{n\in S}\frac{1}{d(n)^{\frac{p}{2-p}}}=\sum_{k=1}^{\infty}\frac{1}{(k+1)^{\frac{p}{2-p}}}<\infty \quad \text{as} \quad \frac{p}{2-p}>1 \text{ for } p\in(1,2).$$

Now,

$$y_n=\sum_{2^k|n}\frac{2^{k\alpha}}{n^{\alpha}}x_d.$$

Write $n=2^l m$ where m is odd. Then

$$\begin{aligned}
(y_{2^l m})^2 &= \left(\sum_{k-0}^{l}\frac{x_{2^k}}{(2^{l-k}m)^{\alpha}}\right)^2=\frac{1}{m^{2\alpha}}\left(\sum_{k-0}^{l}\frac{x_{2^k}}{2^{(l-k)\alpha}}\right)^2\\
&=\frac{1}{m^{2\alpha}}\left(\sum_{k=0}^{l}\frac{x_{2^{l-k}}}{2^{k\alpha}}\right)^2 \qquad \text{by writing } k\mapsto l-k,\\
&=\frac{1}{m^{2\alpha}}\left(\sum_{k=0}^{l}\frac{x_{2^{l-k}}}{2^{k(\alpha-\delta)}}\frac{1}{2^{k\delta}}\right)^2\le\frac{1}{m^{2\alpha}}\sum_{k=0}^{l}\left(\frac{x_{2^{l-k}}}{2^{k(\alpha-\delta)}}\right)^2\sum_{k=0}^{l}\frac{1}{2^{2k\delta}}\\
&\ll\frac{1}{m^{2\alpha}}\sum_{k=0}^{l}\left(\frac{x_{2^{l-k}}}{2^{k(\alpha-\delta)}}\right)^2.
\end{aligned}$$

We now sum over all l and m,

$$\begin{aligned}
\sum_{l=1}^{\infty}\sum_{\substack{m\in\mathbb{N}\\ m\text{ odd}}}(y_{2^l m})^2 &\ll \sum_{l=1}^{\infty}\sum_{\substack{m\in\mathbb{N}\\ m\text{ odd}}}\frac{1}{m^{2\alpha}}\sum_{k=0}^{l}\left(\frac{x_{2^{l-k}}}{2^{k(\alpha-\delta)}}\right)^2\\
&\le\zeta(2\alpha)\sum_{l=0}^{\infty}\sum_{k=0}^{l}\left(\frac{x_{2^{l-k}}}{2^{k(\alpha-\delta)}}\right)^2\\
&\ll\sum_{k=0}^{\infty}\sum_{l=0}^{\infty}\left(\frac{x_{2^l}}{2^{k(\alpha-\delta)}}\right)^2=\sum_{k=0}^{\infty}\frac{1}{2^{2k(\alpha-\delta)}}\sum_{l=0}^{\infty}x_{2^l}^2,
\end{aligned}$$

which is finite as $x\in\ell^p$. The following Proposition suggests some further structure of S.

Proposition 3.4. *Let $\alpha>\frac{1}{2}$ and $\beta=\frac{p}{(2-p)(2\alpha-1)}$. Let $y=\gamma+\mu$ where $\gamma=(\gamma_n)$ and $\mu=(\mu_n)$ are given by*

$$\gamma_n=\sum_{\substack{d|n\\ \frac{n}{d}\in S\\ d\ge d(n)^{\beta}}}\frac{x_{n/d}}{d^{\alpha}} \quad \text{and} \quad \mu_n=\sum_{\substack{d|n\\ \frac{n}{d}\in S\\ d<d(n)^{\beta}}}\frac{x_{n/d}}{d^{\alpha}}.$$

Then $\gamma\in\ell^2$.

Proof. By the Cauchy–Schwarz inequality,

$$\gamma_n^2 = \Bigg(\sum_{\substack{d|n \\ \frac{n}{d}\in S \\ d\geq d(n)^\beta}} \frac{x_{n/d}}{d^\alpha} \Bigg)^2 = \Bigg(\sum_{\substack{d|n \\ d\in S \\ d\leq \frac{n}{d(n)^\beta}}} x_d \left(\frac{d}{n}\right)^\alpha \Bigg)^2$$

$$\leq \sum_{\substack{d|n \\ d\in S \\ d\leq \frac{n}{d(n)^\beta}}} x_d^2 \sum_{\substack{d|n \\ d\in S \\ d\leq \frac{n}{d(n)^\beta}}} \left(\frac{d}{n}\right)^{2\alpha} \ll \sum_{\substack{d|n \\ d\in S \\ d\leq \frac{n}{d(n)^\beta}}} \left(\frac{d}{n}\right)^{2\alpha},$$

as $x \in \ell^2$. Therefore,

$$\sum_{n=1}^{\infty} \gamma_n^2 \ll \sum_{n=1}^{\infty} \sum_{\substack{d|n \\ d\in S \\ d\leq \frac{n}{d(n)^\beta}}} \left(\frac{d}{n}\right)^{2\alpha} \leq \sum_{d\in S} \sum_{\substack{m\geq 1 \\ d(dm)^\beta < m}} \frac{1}{m^{2\alpha}}$$

$$\leq \sum_{d\in S} \sum_{d(d)^\beta < m} \frac{1}{m^{2\alpha}} \ll \sum_{d\in S} \frac{1}{d(d)^{\beta(2\alpha-1)}},$$

as, for $s > 1$,

$$\sum_{n>m} \frac{1}{n^s} \ll m^{1-s}$$

(see [6], page 55). By assumption, we have

$$\sum_{n\in S} \frac{1}{d(n)^{\beta(2\alpha-1)}} = \sum_{n\in S} \frac{1}{d(n)^{\frac{p}{2-p}}} \leq \sum_{n\in S} x_n^p < \infty,$$

as required. □

From Proposition 3.4, we can see that any counterexample must yield $\mu \notin \ell^2$. This suggests that S must contain $n \in \mathbb{N}$ such that n has a large number of small divisors so that $d < d(n)^\beta$ is satisfied often and in turn ensuring that many divisors contribute to the summation. The investigation of finding a suitable support set S has not yet yielded $\mu \notin \ell^2$, and this gives little indication of a successful counterexample. The lack of existence of either a generalisation of Theorem 3.1 or a counterexample demonstrates perhaps the challenging nature of this problem and leaves further open questions regarding the boundedness of multiplicative Toeplitz operators.

Open questions

We conclude this paper by summarising the open problems that have risen from our discussion:

- Is $f \in \ell^r$ a necessary condition for $\mathscr{D}_f : \ell^p \to \ell^q$ to be bounded for any p and q?
- Can we generalise Theorem 3.1 from multiplicative subsets to the mapping $\mathscr{D}_f : \ell^p \to \ell^2$? Or can we find a counterexample to this?

Finally, we give some further open questions regarding multiplicative Toeplitz operators which we have not discussed in this paper:

- What is the operator norm when f can take negative values? Does it mimic that given in [2]?
- Can we compute the spectrum of $\mathscr{M}_f$? Does $\mathscr{M}_f$ have any eigenvalues and, if so, what are they?
- For which f is $\mathscr{M}_f$ Fredholm, and can we describe the essential spectrum of $\mathscr{M}_f$?

References

[1] T.W. Hilberdink, An arithmetical mapping and applications to Ω-results for the Riemann zeta function, *Acta Arithmetica* **139** (2009), 341–367.

[2] T.W. Hilberdink, Multiplicative Toeplitz matrices and the Riemann zeta function, in: *Four Faces of Number Theory,* EMS Series of Lectures in Mathematics, European Mathematical Society (EMS), Zürich, 2015, 77–121.

[3] P.K. Codeca and M. Nair, Smooth numbers and the norms of arithmetic Dirichlet convolutions, *Journal of Mathematical Analysis and Applications* **347** (2008), 400–406.

[4] O.F. Brevig, K.-M. Perfekt, K. Seip, A.G. Siskakis, and D. Vukotić, The multiplicative Hilbert matrix, *Advances in Mathematics* **302** (2016), 410–432.

[5] K.-M. Perfekt and A. Pushnitski, On Helson matrices: moment problems, non-negativity, boundedness, and finite rank, arXiv:1611.03772v2 [math.FA], to appear in *Proceedings of the London Mathematical Society.*

[6] T.M. Apostol, *Introduction to Analytic Number Theory,* Springer, New York, Heidelberg, 1976.

[7] C. Aistleitner, Lower bounds for the maximum of the Riemann zeta function along vertical lines, *Mathematische Annalen* **365** (2016), 473–496.

[8] A. Bondarenko and K. Seip, Large greatest common divisor sums and extreme values of the Riemann zeta function, *Duke Mathematical Journal* **166** (2017), 1685–1701.

[9] T.W. Hilberdink, Quasi-norm of an arithmetical convolution operator and the order of the Riemann Zeta function, *Functiones et Approximatio* **49** (2013), 201–220.

[10] T.W. Hilberdink, Determinants of multiplicative Toeplitz matrices, *Acta Arithmetica* **125** (2006), 265–284.

Nicola Thorn
Department of Mathematics
University of Reading
Whiteknights
PO Box 22,
Reading RG66AX
UK
e-mail: `n.j.b.thorn@pgr.reading.ac.uk`

On higher index differential-algebraic equations in infinite dimensions

Sascha Trostorff and Marcus Waurick

Abstract. We consider initial value problems for differential-algebraic equations in a possibly infinite-dimensional Hilbert space. Assuming a growth condition for the associated operator pencil, we prove existence and uniqueness of solutions for arbitrary initial values in a distributional sense. Moreover, we construct a nested sequence of subspaces for initial values in order to obtain classical solutions.

Mathematics Subject Classification (2010). Primary: 34A09, Secondary: 34A12, 34A30, 34G10.

Keywords. Differential-algebraic equations, higher index, infinite-dimensional state space, consistent initial values, distributional solutions.

1. Introduction and main results

In this short note, we consider two solution concepts of differential-algebraic equations (DAEs) in infinite dimensions. For this, let E and A be bounded linear operators in some possibly infinite-dimensional Hilbert space H.

We consider the implicit initial value problem

$$\begin{cases} Eu'(t) + Au(t) = 0, & t > 0, \\ u(0+) = u_0 \end{cases} \tag{$*$}$$

for some given $u_0 \in H$. In order to talk about a well-defined problem in $(*)$, we assume that the pair (E, A) is *regular*, that is,

$$\exists \nu \in \mathbb{R} \colon \mathbb{C}_{\mathrm{Re}>\nu} \subseteq \rho(E, A),$$
$$\exists C \geq 0, k \in \mathbb{N}\, \forall s \in \mathbb{C}_{\mathrm{Re}>\nu} \colon \| (sE + A)^{-1} \| \leq C|s|^k,$$

where

$$\rho(E, A) := \{s \in \mathbb{C};\, (sE + A)^{-1} \in L(H)\}.$$

We note that these two conditions are our replacements for regularity in finite dimensions. Indeed, for H finite-dimensional, (E, A) is called regular,

© Springer International Publishing AG, part of Springer Nature 2018
A. Böttcher et al. (eds.), *The Diversity and Beauty of Applied Operator Theory*, Operator Theory: Advances and Applications 268, https://doi.org/10.1007/978-3-319-75996-8_27

if $\det(sE+A)\neq 0$ for some $s\in\mathbb{C}$. Thus, $s\mapsto\det(sE+A)$ is a polynomial of degree at most $\dim H$, which is not identically zero. The growth condition is a consequence of the Weierstrass or Jordan normal form theorem valid for finite spatial dimensions, see e.g. [1, 2, 4]. The smallest possible $k\in\mathbb{N}$ occurring in the resolvent estimate is called the *index of* (E,A):

$$\operatorname{ind}(E,A):=\min\{k\in\mathbb{N};\,\exists C\geq 0\,\forall s\in\mathbb{C}_{\mathrm{Re}>\nu}:\|(sE+A)^{-1}\|\leq C|s|^k\}.$$

We shall also define a sequence of (initial value) spaces associated with (E,A):

$$\mathrm{IV}_0:=H\text{ and }\mathrm{IV}_{k+1}:=\{x\in H;\,Ax\in E[\mathrm{IV}_k]\}\quad(k\in\mathbb{N}).$$

A first observation is the following.

Proposition 1.1. *Let $k=\operatorname{ind}(E,A)$ and assume that $E[\mathrm{IV}_k]\subseteq H$ is closed. Then $\mathrm{IV}_{k+1}=\mathrm{IV}_{k+2}$.*

Since the sequence of spaces $(\mathrm{IV}_k)_k$ is decreasing (see Lemma 3.1), Proposition 1.1 leads to the following question.

Problem 1.2. Assume that $E[\mathrm{IV}_j]\subseteq H$ is closed for each $j\in\mathbb{N}$. Do we then have

$$\min\{k\in\mathbb{N};\mathrm{IV}_{k+1}=\mathrm{IV}_{k+2}\}=\operatorname{ind}(E,A)?$$

With the spaces $(\mathrm{IV}_k)_k$ at hand, we can present the main theorem of this article.

Theorem 1.3. *Assume that $E[\mathrm{IV}_{\operatorname{ind}(E,A)}]\subseteq H$ is closed, $u_0\in\mathrm{IV}_{\operatorname{ind}(E,A)+1}$. Then there exists a unique continuously differentiable function $u\colon\mathbb{R}_{>0}\to H$ with $u(0+)=u_0$ such that*

$$Eu'(t)+Au(t)=0\quad(t>0).$$

With Proposition 1.1 and Theorem 1.3, it is possible to derive the following consequence.

Corollary 1.4. *Assume that $E[\mathrm{IV}_j]\subseteq H$ is closed for each $j\in\mathbb{N}$, $u_0\in H$. Then there exists a continuously differentiable function $u\colon\mathbb{R}_{>0}\to H$ with $u(0+)=u_0$ and*

$$Eu'(t)+Au(t)=0\quad(t>0),$$

if, and only if, $u_0\in\mathrm{IV}_{\operatorname{ind}(E,A)+1}$.

Corollary 1.4 suggests that the answer to Problem 1.2 is in the affirmative for H being finite-dimensional.

Also in our main result, there is room for improvement: In applications, it is easier to show that $R(E)\subseteq H$ is closed as the IV-spaces are not straightforward to compute. Thus, we ask whether the latter theorem can be improved in the following way.

Problem 1.5. Does $R(E)\subseteq H$ closed imply that $E[\mathrm{IV}_{\operatorname{ind}(E,A)}]\subseteq H$ is closed or even the closedness of $E[\mathrm{IV}_j]\subseteq H$ for all $j\in\mathbb{N}$?

We shall briefly comment on the organization of this article. In the next section, we introduce the time-derivative operator in a suitably weighted vector-valued L_2-space. This has been used intensively in the framework of so-called 'evolutionary equations', see [6]. With this notion, it is possible to obtain a *distributional* solution of $(*)$ such that the differential algebraic equation holds in an integrated sense, where the number of integrations needed corresponds to the index of the DAE. We conclude this article with the proofs of Proposition 1.1, Theorem 1.3, and Corollary 1.4. We emphasize that we do not employ any Weierstrass or Jordan normal theory in the proofs of our main results. We address the case of unbounded A to future research. The case of index 0 is discussed in [8], where also exponential stability and dichotomies are studied.

2. The time derivative and weak solutions of DAEs

Throughout this section, we assume that H is a Hilbert space and that $E, A \in L(H)$ with (E, A) regular. We start out with the definition of the space of (equivalence classes of) vector-valued L_2 functions: Let $\nu \in \mathbb{R}$. Then we set

$$L_{2,\nu}(\mathbb{R}; H) := \left\{ f : \mathbb{R} \to H\,;\, f \text{ measurable}, \int_{\mathbb{R}} |f(t)|_H^2 \exp(-2\nu t)\,\mathrm{d}t < \infty \right\},$$

see also [6, 3, 5]. Note that $L_{2,0}(\mathbb{R}; H) = L_2(\mathbb{R}; H)$. We define $H_\nu^1(\mathbb{R}; H)$ to be the (H-valued) Sobolev space of $L_{2,\nu}(\mathbb{R}; H)$-functions with weak derivative representable as $L_{2,\nu}(\mathbb{R}; H)$-function. With this, we can define the derivative operator

$$\partial_{0,\nu} \colon H_\nu^1(\mathbb{R}; H) \subseteq L_{2,\nu}(\mathbb{R}; H) \to L_{2,\nu}(\mathbb{R}; H),\ \phi \mapsto \phi'.$$

In the next theorem we recall some properties of the operator just defined. For this, the *Fourier–Laplace transformation* $\mathcal{L}_\nu \colon L_{2,\nu}(\mathbb{R}; H) \to L_2(\mathbb{R}; H)$ is introduced as being the unitary extension of

$$\mathcal{L}_\nu \phi(t) := \frac{1}{\sqrt{2\pi}} \int_{\mathbb{R}} \phi(s) \mathrm{e}^{-(\mathrm{i}t+\nu)s} \mathrm{d}s \quad (\phi \in C_c(\mathbb{R}; H),\, t \in \mathbb{R}),$$

where $C_c(\mathbb{R}; H)$ denotes the space of compactly supported, continuous H-valued functions defined on $\mathbb{R}$. Moreover, let

$$\begin{aligned}\mathrm{m} \colon \{f \in L_2(\mathbb{R}; H); (t \mapsto tf(t)) \in L_2(\mathbb{R}; H)\} \subseteq L_2(\mathbb{R}; H) &\to L_2(\mathbb{R}; H),\\ f &\mapsto (t \mapsto tf(t))\end{aligned}$$

be the multiplication by the argument operator with maximal domain.

Theorem 2.1 ([3, Corollary 2.5]). *Let* $\nu \in \mathbb{R}$. *Then*

$$\partial_{0,\nu} = \mathcal{L}_\nu^*(\mathrm{i}\,\mathrm{m} + \nu)\mathcal{L}_\nu.$$

Remark 2.2. A direct consequence of Theorem 2.1 is the continuous invertibility of $\partial_{0,\nu}$ if $\nu \neq 0$.

Corollary 2.3. *Let $\nu > 0$ be such that $\rho(E, A) \supseteq \mathbb{C}_{\mathrm{Re}>\nu}$ and also such that $\|(sE + A)^{-1}\| \leq C|s|^{\mathrm{ind}(E,A)}$ for some $C \geq 0$ and all $s \in \mathbb{C}_{\mathrm{Re}>\nu}$. Then*

$$\partial_{0,\nu}^{-k}\left(\partial_{0,\nu}E + A\right)^{-1} \in L(L_{2,\nu}(\mathbb{R}; H)),$$

where $k = \mathrm{ind}(E, A)$. Moreover, $\partial_{0,\nu}^{-k}\left(\partial_{0,\nu}E + A\right)^{-1}$ is causal, i.e., for each $f \in L_{2,\nu}(\mathbb{R}; H)$ with $\mathrm{spt}\, f \subseteq \mathbb{R}_{\geq a}$ for some $a \in \mathbb{R}$ it follows that

$$\mathrm{spt}\, \partial_{0,\nu}^{-k}\left(\partial_{0,\nu}E + A\right)^{-1} f \subseteq \mathbb{R}_{\geq a}.$$

Proof. By Theorem 2.1 and the unitarity of $\mathcal{L}_\nu$, we obtain that the first claim is equivalent to

$$(\mathrm{i\,m} + \nu)^{-k}\left((\mathrm{i\,m} + \nu)\, E + A\right)^{-1} \in L(L_2(\mathbb{R}; H)),$$

which, in turn, would be implied by the fact that the function

$$t \mapsto (\mathrm{i}t + \nu)^{-k}\left((\mathrm{i}t + \nu)\, E + A\right)^{-1}$$

belongs to the space $L^\infty(\mathbb{R}; L(H))$. This is, however, true by regularity of (E, A). We now show the causality. As the operator $\partial_{0,\nu}^{-k}\left(\partial_{0,\nu}E + A\right)^{-1}$ commutes with translation in time, it suffices to prove the claim for $a = 0$. So let $f \in L_{2,\nu}(\mathbb{R}; H)$ with $\mathrm{spt}\, f \subseteq \mathbb{R}_{\geq 0}$. By a Paley–Wiener type result (see e.g. [7, 19.2 Theorem]), the latter is equivalent to

$$\left(\mathbb{C}_{\mathrm{Re}>\nu} \ni z \mapsto \left(\mathcal{L}_{\mathrm{Re}\, z} f\right)(\mathrm{Im}\, z)\right) \in \mathcal{H}^2(\mathbb{C}_{\mathrm{Re}>\nu}; H),$$

where $\mathcal{H}^2(\mathbb{C}_{\mathrm{Re}>\nu}; H)$ denotes the Hardy space of H-valued functions on the half-plane $\mathbb{C}_{\mathrm{Re}>\nu}$. As

$$\left(\mathcal{L}_{\mathrm{Re}\, z}\partial_{0,\nu}^{-k}\left(\partial_{0,\nu}E + A\right)^{-1} f\right)(\mathrm{Im}\, z) = z^{-k}\left(zE + A\right)^{-1}\left(\mathcal{L}_{\mathrm{Re}\, z} f\right)(\mathrm{Im}\, z)$$

for each $z \in \mathbb{C}_{\mathrm{Re}>\nu}$, we infer that also

$$\left(\mathbb{C}_{\mathrm{Re}>\nu} \ni z \mapsto \left(\mathcal{L}_{\mathrm{Re}\, z}\partial_{0,\nu}^{-k}\left(\partial_{0,\nu}E + A\right)^{-1} f\right)(\mathrm{Im}\, z)\right) \in \mathcal{H}^2(\mathbb{C}_{\mathrm{Re}>\nu}; H),$$

due to the boundedness and analyticity of

$$\left(\mathbb{C}_{\mathrm{Re}>\nu} \ni z \mapsto z^{-k}\left(zE + A\right)^{-1} \in L(H)\right).$$

This proves the claim. □

Corollary 2.3 states a particular boundedness property for the solution operator associated with (∗). This can be made more precise by introducing a scale of extrapolation spaces associated with $\partial_{0,\nu}$.

Definition 2.4. Let $k \in \mathbb{N}$, $\nu > 0$. Then we define $H_\nu^k(\mathbb{R}; H) := D(\partial_{0,\nu}^k)$ endowed with the scalar product $\langle \phi, \psi \rangle_k := \langle \partial_{0,\nu}^k \phi, \partial_{0,\nu}^k \psi \rangle_0$. Quite similarly, we define $H_\nu^{-k}(\mathbb{R}; H)$ as the completion of $L_{2,\nu}(\mathbb{R}; H)$ with respect to the inner product $\langle \phi, \psi \rangle_{-k} := \langle \partial_{0,\nu}^{-k} \phi, \partial_{0,\nu}^{-k} \psi \rangle_0$.

We observe that the spaces $(H_\nu^k(\mathbb{R}; H))_{k\in\mathbb{Z}}$ are nested in the sense that $j_{k\to\ell} \colon H_\nu^k(\mathbb{R}; H) \hookrightarrow H_\nu^\ell(\mathbb{R}; H), x \mapsto x$, whenever $k \geq \ell$.

Remark 2.5. The operator $\partial_{0,\nu}^{\ell}$ can be considered as a densely defined isometry from $H_\nu^k(\mathbb{R};H)$ to $H_\nu^{k-\ell}(\mathbb{R};H)$ with dense range for all $k\in\mathbb{Z}$. The closure of this densely defined isometry will be given the same name. In this way, we can state the boundedness property of the solution operator in Corollary 2.3 equivalently as follows:

$$(\partial_{0,\nu}E+A)^{-1}\in L\left(L_{2,\nu}(\mathbb{R};H),H_\nu^{-k}(\mathbb{R};H)\right).$$

More generally, as $(\partial_{0,\nu}E+A)^{-1}$ and $\partial_{0,\nu}^{-1}$ commute, we obtain

$$(\partial_{0,\nu}E+A)^{-1}\in L\left(H_\nu^{j}(\mathbb{R};H),H_\nu^{j-k}(\mathbb{R};H)\right)$$

for each $j\in\mathbb{Z}$.

Note that by the Sobolev embedding theorem (see e.g. [3, Lemma 5.2]) the δ-distribution of point evaluation at 0 is an element of $H_\nu^{-1}(\mathbb{R};H)$; in fact it is the derivative of $\chi_{\mathbb{R}_{\geq 0}}\in L_{2,\nu}(\mathbb{R};H)=H_\nu^0(\mathbb{R};H)$. With these preparations at hand, we consider the following implementation of the initial value problem stated in (∗): Let $u_0\in H$. Find $u\in H_\nu^{-k}(\mathbb{R};H)$ such that

$$(\partial_{0,\nu}E+A)\,u=\delta\cdot Eu_0. \tag{2.1}$$

Theorem 2.6. *Let (E,A) be regular. Then for all $u_0\in H$ there exists a unique $u\in H_\nu^{-k}(\mathbb{R};H)$ such that (2.1) holds. Moreover, we have*

$$u=\chi_{\mathbb{R}_{\geq 0}}u_0-(\partial_{0,\nu}E+A)^{-1}\chi_{\mathbb{R}_{\geq 0}}Au_0$$

and

$$\operatorname{spt}\partial_{0,\nu}^{-k}u\subseteq\mathbb{R}_{\geq 0}.$$

Proof. Note that the unique solution is given by

$$u=(\partial_{0,\nu}E+A)^{-1}\delta\cdot Eu_0\in H_\nu^{-k-1}(\mathbb{R};H).$$

Hence,

$$\begin{aligned}u-\chi_{\mathbb{R}_{\geq 0}}u_0&=(\partial_{0,\nu}E+A)^{-1}\left(\delta\cdot Eu_0-(\partial_{0,\nu}E+A)\,\chi_{\mathbb{R}_{\geq 0}}u_0\right)\\&=-(\partial_{0,\nu}E+A)^{-1}\chi_{\mathbb{R}_{\geq 0}}Au_0,\end{aligned}$$

which shows the desired formula. Since $\chi_{\mathbb{R}_{\geq 0}}u_0\in L_{2,\nu}(\mathbb{R};H)\hookrightarrow H_\nu^{-k}(\mathbb{R};H)$ and $(\partial_{0,\nu}E+A)^{-1}\chi_{\mathbb{R}_{\geq 0}}Au_0\in H_\nu^{-k}(\mathbb{R};H)$ by Corollary 2.3 we obtain the asserted regularity for u. The support statement follows from the causality statement in Corollary 2.3. □

In the concluding section, we will discuss the spaces IV_k in connection to (E,A) and will prove the main results of this paper mentioned in the introduction.

3. Proofs of the main results and initial value spaces

Again, we assume that H is a Hilbert space, and that $E, A \in L(H)$ with (E, A) regular.

At first, we turn to the proof of Proposition 1.1. For this, we note some elementary consequences of the definition of IV_k and of regularity.

Lemma 3.1. (a) *For all $k \in \mathbb{N}$, we have $\mathrm{IV}_k \supseteq \mathrm{IV}_{k+1}$.*

(b) *Let $s \in \mathbb{C} \cap \rho(E, A)$. Then*

$$E(sE + A)^{-1}A = A(sE + A)^{-1}E.$$

(c) *Let $k \in \mathbb{N}$, $x \in \mathrm{IV}_k$. Then for all $s \in \mathbb{C} \cap \rho(E, A)$ we have*

$$(sE + A)^{-1}Ex \in \mathrm{IV}_{k+1}.$$

(d) *Let $s \in \mathbb{C} \cap \rho(E, A) \setminus \{0\}$. Then*

$$(sE + A)^{-1}E = \frac{1}{s} - \frac{1}{s}(sE + A)^{-1}A.$$

(e) *Let $k \in \mathbb{N}$, $x \in \mathrm{IV}_k$. Then for all $s \in \mathbb{C} \cap \rho(E, A) \setminus \{0\}$ we have*

$$(sE + A)^{-1}Ex = \frac{1}{s}x + \sum_{\ell=1}^{k} \frac{1}{s^{\ell+1}} x_\ell + \frac{1}{s^{k+1}}(sE + A)^{-1}Aw.$$

for some $w \in H$, $x_1, \ldots, x_k \in H$.

Proof. The proof of (a) is an induction argument. The claim is trivial for $k = 0$. For the inductive step, we see that the assertion follows using the induction hypothesis by

$$\mathrm{IV}_{k+1} = A^{-1}[E[\mathrm{IV}_k]] \supseteq A^{-1}[E[\mathrm{IV}_{k+1}]] = \mathrm{IV}_{k+2}.$$

Next, we prove (b). We compute

$$\begin{aligned} E(sE + A)^{-1}A &= E(sE + A)^{-1}(sE + A - sE) \\ &= E - E(sE + A)^{-1}sE \\ &= E - (sE + A - A)\,(sE + A)^{-1}E \\ &= A(sE + A)^{-1}E. \end{aligned}$$

We prove (c) by induction on k. For $k = 0$, we let $x \in \mathrm{IV}_0 = H$ and put $y := (sE + A)^{-1} Ex$. Then, by (b), we get that

$$Ay = A\,(sE + A)^{-1}\,Ex = E\,(sE + A)^{-1}\,Ax \in R(E) = E[\mathrm{IV}_0].$$

Hence, $y \in \mathrm{IV}_1$. For the inductive step, we assume that the assertion holds for some $k \in \mathbb{N}$. Let $x \in \mathrm{IV}_{k+1}$. We need to show that $y := (sE + A)^{-1} Ex$ is in IV_{k+2}. For this, note that there exists a $w \in \mathrm{IV}_k$ such that $Ax = Ew$.

In particular, by the induction hypothesis, we have $(sE+A)^{-1}Ew \in \mathrm{IV}_{k+1}$. Then we compute, using (b) again,

$$\begin{aligned} Ay &= A(sE+A)^{-1}Ex \\ &= E(sE+A)^{-1}Ax \\ &= E(sE+A)^{-1}Ew \in E[\mathrm{IV}_{k+1}]. \end{aligned}$$

Hence, $y \in \mathrm{IV}_{k+2}$ and (c) is proved.
For (d), it suffices to observe

$$\begin{aligned} (sE+A)^{-1}E &= \frac{1}{s}(sE+A)^{-1}sE \\ &= \frac{1}{s}(sE+A)^{-1}(sE+A-A) \\ &= \frac{1}{s} - \frac{1}{s}(sE+A)^{-1}A. \end{aligned}$$

In order to prove part (e), we proceed by induction on $k \in \mathbb{N}$. The case $k=0$ has been dealt with in part (d) by choosing $w=-x$. For the inductive step, we let $x \in \mathrm{IV}_{k+1}$. By definition of IV_{k+1}, we find $y \in \mathrm{IV}_k$ such that $Ax = Ey$. By the induction hypothesis, we find $w \in H$ and $x_1, \ldots, x_k \in H$ such that

$$(sE+A)^{-1}Ey = \frac{1}{s}y + \sum_{\ell=1}^{k} \frac{1}{s^{\ell+1}}x_\ell + \frac{1}{s^{k+1}}(sE+A)^{-1}Aw.$$

With this we compute using (d)

$$\begin{aligned} (sE+A)^{-1}Ex &= \frac{1}{s}x - \frac{1}{s}(sE+A)^{-1}Ax \\ &= \frac{1}{s}x - \frac{1}{s}(sE+A)^{-1}Ey \\ &= \frac{1}{s}x - \frac{1}{s}\left(\frac{1}{s}y + \sum_{\ell=1}^{k}\frac{1}{s^{\ell+1}}x_\ell + \frac{1}{s^{k+1}}(sE+A)^{-1}Aw\right) \\ &= \frac{1}{s}x + \sum_{\ell=1}^{k+1}\frac{1}{s^{\ell+1}}\widetilde{x}_\ell + \frac{1}{s^{k+2}}(sE+A)^{-1}A\widetilde{w}, \end{aligned}$$

with $\widetilde{x}_1 = -y, \widetilde{x}_\ell = -x_{\ell-1}$ for $\ell \geq 2$ and $\widetilde{w} = -w$. □

With Lemma 3.1 (a), we obtain the following reformulation of Proposition 1.1.

Proposition 3.2. *Assume that* $E[\mathrm{IV}_{\mathrm{ind}(E,A)}] \subseteq H$ *is closed. Then*

$$\mathrm{IV}_{\mathrm{ind}(E,A)+1} \subseteq \mathrm{IV}_{\mathrm{ind}(E,A)+2}.$$

Proof. Note that the closedness of $E[\mathrm{IV}_{\mathrm{ind}(E,A)}]$ implies the same for the space $\mathrm{IV}_{\mathrm{ind}(E,A)+1}$ since A is continuous. We set $k := \mathrm{ind}(E,A)$. Now let $x \in \mathrm{IV}_{k+1}$. Then we need to find $y \in \mathrm{IV}_{k+1}$ with $Ax = Ey$. By definition there exists an $x_0 \in \mathrm{IV}_k$ with the property $Ax = Ex_0$. For $n \in \mathbb{N}$ large enough we define $y_n := n(nE+A)^{-1}Ex_0$. Since $x_0 \in \mathrm{IV}_k$, we deduce with the help of Lemma 3.1 (c) that $y_n \in \mathrm{IV}_{k+1}$. Moreover, by Lemma 3.1 (e), $(y_n)_n$ is bounded. Choosing a suitable subsequence, for which we use the same name, we may assume that $(y_n)_n$ is weakly convergent to some $y \in H$. The closedness of IV_{k+1} implies $y \in \mathrm{IV}_{k+1}$. Then, using Lemma 3.1 (e), we find $w \in H$ and $x_1, \ldots, x_{k+1} \in H$ such that

$$(nE + A)^{-1} Ex_0 = \sum_{\ell=0}^{k} \frac{1}{n^{\ell+1}} x_\ell + \frac{1}{n^{k+1}} (nE + A)^{-1} Aw.$$

Hence, we obtain

$$\begin{aligned} Ey &= \operatorname*{w-lim}_{n\to\infty} Ey_n \\ &= \operatorname*{w-lim}_{n\to\infty} E (nE + A)^{-1} nEx_0 \\ &= \operatorname*{w-lim}_{n\to\infty} nE (nE + A)^{-1} Ax \\ &= \operatorname*{w-lim}_{n\to\infty} (nE + A - A) (nE + A)^{-1} Ax \\ &= Ax - \operatorname*{w-lim}_{n\to\infty} A (nE + A)^{-1} Ex_0 \\ &= Ax - A \operatorname*{w-lim}_{n\to\infty} \left(\sum_{\ell=0}^{k} \frac{1}{n^{\ell+1}} x_\ell + \frac{1}{n^{k+1}} (nE + A)^{-1} Aw \right) = Ax, \end{aligned}$$

which yields the assertion. □

With an idea similar to the one in the proof of Proposition 1.1 (Proposition 3.2), it is possible to show that $E\colon \mathrm{IV}_{k+1} \to E[\mathrm{IV}_k]$ is an isomorphism if $k = \operatorname{ind}(E, A)$ and $E[\mathrm{IV}_k] \subseteq H$ is closed. We will need this result also in the proof of our main theorem.

Theorem 3.3. *Let (E, A) be regular and assume that $E[\mathrm{IV}_k] \subseteq H$ is closed, $k = \operatorname{ind}(E, A)$. Then*

$$E\colon \mathrm{IV}_{k+1} \to E[\mathrm{IV}_k], x \mapsto Ex$$

is a Banach space isomorphism.

Proof. Note that by the closed graph theorem, it suffices to show that the operator under consideration is one-to-one and onto. So, for proving injectivity, we let $x \in \mathrm{IV}_{k+1}$ such that $Ex = 0$. By definition, there exist $y \in \mathrm{IV}_k$ such that $Ey = Ax = Ax + nEx$ for all $n \in \mathbb{N}$. Hence, for $n \in \mathbb{N}$ large enough, we have $x = (nE + A)^{-1} Ey$. Thus, from $y \in \mathrm{IV}_k$ we deduce with the help of Lemma 3.1 (e) that there exist $w, x_1, \ldots .x_k \in H$ such that

$$x = (nE + A)^{-1} Ey = \frac{1}{n} y + \sum_{\ell=1}^{k} \frac{1}{n^{\ell+1}} x_\ell + \frac{1}{n^{k+1}} (nE + A)^{-1} Aw \to 0$$

as $n \to \infty$, which shows that $x = 0$.

Next, let $y \in E[\mathrm{IV}_k]$. For $n \in \mathbb{N}$ large enough, we put

$$w_n := (nE + A)^{-1} ny.$$

By Lemma 3.1 (c), we obtain that $w_n \in \mathrm{IV}_{k+1}$. Let $x \in \mathrm{IV}_k$ with $Ex = y$. Then, using Lemma 3.1 (e), we find $w, x_1, \ldots, x_k \in H$ such that

$$\begin{aligned} w_n &= (nE+A)^{-1}ny \\ &= (nE+A)^{-1}nEx \\ &= x + \sum_{\ell=1}^{k} \frac{1}{n^\ell} x_\ell + \frac{1}{n^k}(nE+A)^{-1}Aw, \end{aligned}$$

proving the boundedness of $(w_n)_n$. Without loss of generality, we may assume that $(w_n)_n$ weakly converges to $z \in \mathrm{IV}_{k+1} = A^{-1}[E[\mathrm{IV}_k]]$. Hence,

$$\begin{aligned} Ez &= \operatorname*{w-lim}_{n\to\infty} Ew_n \\ &= \operatorname*{w-lim}_{n\to\infty} \frac{1}{n}(nE+A)\, w_n \\ &= \operatorname*{w-lim}_{n\to\infty} \frac{1}{n}(nE+A)\,(nE+A)^{-1}ny \\ &= y. \end{aligned}$$

□

Next, we come to the proof of our main result Theorem 1.3, which we restate here for convenience.

Theorem 3.4. *Assume that $E[\mathrm{IV}_{\mathrm{ind}(E,A)}] \subseteq H$ is closed, $u_0 \in \mathrm{IV}_{\mathrm{ind}(E,A)+1}$. Then (2.1) has a unique continuously differentiable solution $u : \mathbb{R}_{>0} \to H$, satisfying $u(0+) = u_0$ and*

$$Eu'(t) + Au(t) = 0 \quad (t > 0). \tag{3.1}$$

Moreover, the solution coincides with the solution given in Theorem 2.6.

Proof. Let $u_0 \in \mathrm{IV}_{\mathrm{ind}(E,A)+1}$. We denote $\widetilde{E}\colon \mathrm{IV}_{k+1} \to E[\mathrm{IV}_k], x \mapsto Ex$, where $k = \mathrm{ind}(E,A)$. By Theorem 3.3, we have that $\widetilde{E}$ is an isomorphism. For $t > 0$, we define

$$u(t) := \exp\bigl(-t\widetilde{E}^{-1}A\bigr)u_0.$$

Then $u(0+) = u_0$. Moreover, $u(t)$ is well-defined. Indeed, if $u_0 \in \mathrm{IV}_{k+1}$, then $Au_0 \in E[\mathrm{IV}_k]$. Hence, $\widetilde{E}^{-1}Au_0 \in \mathrm{IV}_{k+1}$ is well-defined. Since $E[\mathrm{IV}_k]$ is closed and A is continuous, we infer that IV_{k+1} is a Hilbert space. Thus, we deduce that $u\colon \mathbb{R}_{>0} \to \mathrm{IV}_{k+1}$ is continuously differentiable. In particular, we obtain

$$\mathrm{IV}_{k+1} \ni u'(t) = -\widetilde{E}^{-1}Au(t).$$

If we apply $\widetilde{E}$ to both sides of the equality, we obtain (3.1). If $u : \mathbb{R}_{>0} \to H$ is a continuously differentiable solution of (3.1) with $u(0+) = u_0$, we infer that $u \in L_{2,\nu}(\mathbb{R}; H)$ for some $\nu > 0$ large enough, where we extend u to $\mathbb{R}_{<0}$ by zero. Hence,

$$\partial_{0,\nu}Eu + Au = E\partial_{0,\nu}u + Au = Eu' + Au + \delta \cdot Eu(0+) = \delta \cdot Eu_0,$$

where we have used that u is differentiable on $\mathbb{R}_{<0} \cup \mathbb{R}_{>0}$ and jumps at 0. Thus, u is the solution given in Theorem 2.6, from which we also derive the uniqueness. □

We conclude with a comment on the proof of Corollary 1.4.

Remark 3.5. We note that the condition $u_0 \in \mathrm{IV}_{\mathrm{ind}(E,A)+1}$ arises naturally if we assume that IV_j is closed for each $j \in \mathbb{N}$. Indeed, if $u : \mathbb{R}_{>0} \to H$ is a continuously differentiable solution of (3.1), we infer that

$$Au(t) = -Eu'(t) \quad (t > 0),$$

and thus $u(t) \in \mathrm{IV}_1$ for $t > 0$. Since IV_1 is closed, we derive $u'(t) \in \mathrm{IV}_1$ and hence, inductively $u(t) \in \bigcap_{j\in\mathbb{N}} \mathrm{IV}_j$ for each $t > 0$. Since $\bigcap_{j\in\mathbb{N}} \mathrm{IV}_j$ is equal to $\mathrm{IV}_{\mathrm{ind}(E,A)+1}$ by Proposition 3.2, we get

$$u_0 = u(0+) \in \mathrm{IV}_{\mathrm{ind}(E,A)+1}.$$

References

[1] T. Berger, A. Ilchmann, and S. Trenn, The quasi-Weierstraß form for regular matrix pencils, *Linear Algebra Appl.* **436** (2012), no. 10, 4052–4069.

[2] L. Dai, *Singular Control Systems*, Springer, New York, 1989.

[3] A. Kalauch, R. Picard, S. Siegmund, S. Trostorff, and M. Waurick, A Hilbert space perspective on ordinary differential equations with memory term, *Journal of Dynamics and Differential Equations* **26** (2014), no. 2, 369–399.

[4] P. Kunkel and V. Mehrmann, *Differential-Algebraic Equations. Analysis and Numerical Solution,* European Mathematical Society Publishing House, Zürich, 2006.

[5] R. Picard, *Hilbert Space Approach to Some Classical Transforms,* Pitman Research Notes in Mathematics Series **196**, Longman Scientific & Technical, Harlow; copublished in the U.S. with John Wiley & Sons, Inc., New York, 1989.

[6] R. Picard, A structural observation for linear material laws in classical mathematical physics. *Mathematical Methods in the Applied Sciences* **32** (2009), 1768–1803.

[7] W. Rudin, *Real and Complex Analysis*, Mathematics series, McGraw-Hill, 1987.

[8] S. Trostorff and M. Waurick, On differential-algebraic equations in infinite dimensions, Technical report, TU Dresden, University of Strathclyde, 2017.

Sascha Trostorff
Insitut für Analysis
Fakultät Mathematik
Technische Universität Dresden
Germany
e-mail: `sascha.trostorff@tu-dresden.de`

Marcus Waurick
Department of Mathematics and Statistics
University of Strathclyde
Glasgow
United Kingdom
e-mail: `marcus.waurick@strath.ac.uk`

Characterizations of centrality by local convexity of certain functions on C^*-algebras

Dániel Virosztek

Abstract. We provide a quite large function class which is useful to distinguish central and non-central elements of a C^*-algebra in the following sense: for each element f of this function class, a self-adjoint element a of a C^*-algebra is central if and only if the function f is locally convex at a.

Mathematics Subject Classification (2010). Primary: 46L05.

Keywords. C^*-algebra, centrality, convexity.

1. Introduction

1.1. Motivation

Connections between algebraic properties of C^*-algebras and some essential properties of functions defined on them by functional calculus have been investigated widely.

The first results concern the relation between the commutativity of a C^*-algebra and the monotonicity (with respect to the order induced by positivity) of certain functions defined on the positive cone of it. It was shown by Ogasawara in 1955 that a C^*-algebra is commutative if and only if the map $a \mapsto a^2$ is monotonic increasing on its positive cone [9]. Later on, Pedersen provided a generalization of Ogasawara's result for any power function $a \mapsto a^p$ with $p > 1$ [10]. More recently, Wu proved that the exponential function is also useful to distinguish commutative and non-commutative C^*-algebras in the above sense [13], and in 2003, Ji and Tomiyama described the class of all functions that can be used to decide whether a C^*-algebra is commutative or not [5].

The author was partially supported by the Hungarian National Research, Development and Innovation Office – NKFIH (grant no. K124152).

© Springer International Publishing AG, part of Springer Nature 2018
A. Böttcher et al. (eds.), *The Diversity and Beauty of Applied Operator Theory*, Operator Theory: Advances and Applications 268, https://doi.org/10.1007/978-3-319-75996-8_28

Some "local" results were also obtained in this topic. First, Molnár showed that a self-adjoint element a of a C^*-algebra is central if and only if the exponential function is *locally* monotone at a [8]. Later on, we managed to provide a quite large class of functions (containing all the power functions with exponent greater than 1 and also the exponential function) which has the property that each element of this function class can distinguish central and non-central elements via local monotonicity [12].

Investigating the connections between the commutativity of a C^*-algebra (or locally, the centrality of an element) and the global (or local) convexity property of some functions is of particular interest, as well.

In 2010, Silvestrov, Osaka and Tomiyama showed that a C^*-algebra $\mathcal{A}$ is commutative if and only if there exists a convex function f defined on the positive axis which is not convex of order 2 (that is, it is not convex on the C^*-algebra of the 2×2 matrices) but convex on $\mathcal{A}$ [11, Thm. 4.].

Motivated by the above mentioned result in [11], the main aim of this paper is to provide a large class of functions which have the property that they are locally convex only at central elements, that is, they characterize central elements by local convexity.

1.2. Basic notions, notation

Throughout this paper, C^*-algebras are always assumed to be unital. The spectrum of an element a of the C^*-algebra $\mathcal{A}$ is denoted by $\sigma(a)$. The symbol $\mathcal{A}_s$ stands for the set of all self-adjoint elements of $\mathcal{A}$. A self-adjoint element of a C^*-algebra is called *positive* if its spectrum is contained in $[0, \infty)$. The order induced by positivity on the self-adjoint elements is defined as follows: $a \leq b$ if $b - a$ is positive. In the sequel, the symbol $\mathcal{H}$ stands for a complex Hilbert space and $\mathcal{B}(\mathcal{H})$ denotes the algebra of all bounded linear operators on $\mathcal{H}$. The inner product on a Hilbert space is denoted by $\langle \cdot, \cdot \rangle$ and the induced norm is denoted by $\|\cdot\|$. If u and v are elements of a Hilbert space, the symbol $u \otimes v$ stands for the linear map $z \mapsto \langle z, v \rangle u$.

2. The main theorem

In this section we provide the main result of this paper. In order to do so, we first need a definition.

Definition 1 (Local convexity). *Let $\mathcal{A}$ be a C^*-algebra and let f be a continuous function defined on some open interval $I \subset \mathbb{R}$. Let $a \in \mathcal{A}_s$ with $\sigma(a) \subset I$. We say that f is* locally convex *at the point a if for every $b \in \mathcal{A}_s$ such that $\sigma(a + b) \cup \sigma(a - b) \subset I$ we have*

$$f(a) \leq \frac{1}{2} \left(f(a + b) + f(a - b) \right).$$

Remark 2. Note that in fact the above definition is the definition of the *mid-point convexity*. However, in this paper every function is assumed to be continuous, so there is no difference between mid-point convexity and convexity.

Now we are in the position to present the main result of the paper.

Theorem 3. *Let $I \subset \mathbb{R}$ be an open interval and let f be a convex function in $C^2(I)$ such that the second derivative f'' is strictly concave on I. Let $\mathcal{A}$ be a C^*-algebra and let $a \in \mathcal{A}_s$ be such that $\sigma(a) \subset I$. Then the followings are equivalent:*

(1) *The element a is central, that is, $ab = ba$ for every $b \in \mathcal{A}$.*
(2) *The function f is locally convex at a.*

Example. On the interval $I = (0, \infty)$ the functions $f(x) = x^p$ $(2 < p < 3)$ satisfy the conditions given in Theorem 3. That is, these functions are useful to distinguish central and non-central elements via local convexity.

3. Proof of the main theorem

This section is devoted to the proof of Theorem 3. We believe that some of the main ideas of the proof can be better understood if we provide the proof first only for the special case of the C^*-algebra of all 2×2 matrices and then turn to the proof of the general case.

3.1. The case of the algebra of 2×2 matrices

Let $I \subset \mathbb{R}$ be an open interval and f be a function defined on I that satisfies the conditions given in Theorem 3. Let $\mathcal{A}$ be the C^*-algebra of all 2×2 complex matrices (which is denoted by $M_2(\mathbb{C})$). Let $A \in M_2(\mathbb{C})$ be a self-adjoint matrix with $\sigma(A) \subset I$.

The proof of the direction (1)$\Longrightarrow$(2) is clear. If A is central, that is, $A = \lambda I_2$ (where I_2 denotes the identity element of $M_2(\mathbb{C})$) for some $\lambda \in I$, then $f(A) \leq \frac{1}{2}\left(f(A+B) + f(A-B)\right)$ holds for every self-adjoint $B \in M_2(\mathbb{C})$ (such that $\sigma(A+B) \cup \sigma(A-B) \subset I$) because of the convexity of f as a scalar function.

The interesting part is the direction (2)$\Longrightarrow$(1). We will prove it by contraposition, that is, we show that if A is not central, then f is not locally convex at the point A. So assume that the self-adjoint matrix A is not central, which means that it has two different eigenvalues, say, x and y in I.

Let us use the formula for the (higher order) Fréchet derivatives of matrix valued functions defined by the functional calculus given by Hiai and Petz [4, Thm. 3.33]. This formula is essentially based on the prior works of Daleckii and Krein [2], Bhatia [1], and Hiai [3].

This formula gives us that if $A = \left[\begin{smallmatrix} x & 0 \\ 0 & y \end{smallmatrix}\right]$ and $B = \left[\begin{smallmatrix} 1 & 1 \\ 1 & 1 \end{smallmatrix}\right]$, then the second order Fréchet derivative of the function f (defined by the functional calculus) at the point A with arguments (B, B) is

$$\partial^2 f(A)(B,B) = 2\begin{bmatrix} f^{[2]}[x,x,x] + f^{[2]}[x,x,y] & f^{[2]}[x,x,y] + f^{[2]}[x,y,y] \\ f^{[2]}[x,x,y] + f^{[2]}[x,y,y] & f^{[2]}[x,y,y] + f^{[2]}[y,y,y] \end{bmatrix},$$

where $f^{[2]}[\cdot,\cdot,\cdot]$ denotes the *second divided difference* with respect to f. (For the Fréchet derivatives, we use the notation of Hiai and Petz [4].)

It is well-known that

$$\begin{aligned}\partial^2 f(A)(B,B) &= \frac{\mathrm{d}^2}{\mathrm{d}t^2} f(A+tB)_{|t=0} \\ &= \lim_{t\to 0}\frac{1}{t^2}\left(f(A+tB)-2f(A)+f(A-tB)\right). \end{aligned} \tag{1}$$

Now we show that $\partial^2 f(A)(B,B)$ is not positive semidefinite. Indeed, assume $w = [1 \quad -1]^\top$, then

$$\begin{aligned}&\langle \partial^2 f(A)(B,B)w,w\rangle \\ &\quad = 2\Big(f^{[2]}[x,x,x] + f^{[2]}[x,x,y] - f^{[2]}[x,x,y] - f^{[2]}[x,y,y] \\ &\qquad -f^{[2]}[x,x,y] - f^{[2]}[x,y,y] + f^{[2]}[x,y,y] + f^{[2]}[y,y,y]\Big) \\ &\quad = 2\left(f^{[2]}[x,x,x] - f^{[2]}[x,x,y] - f^{[2]}[x,y,y] + f^{[2]}[y,y,y]\right), \end{aligned} \tag{2}$$

where $\langle\cdot,\cdot\rangle$ denotes the inner product on $\mathbb{C}^2$. Using the basic properties of the divided differences (which can be found e.g. in Section 3.4 of the book [4]) one can compute that the above expression (2) is equal to

$$\begin{aligned}&f''(x) - 2\frac{f'(x) - \frac{f(x)-f(y)}{x-y}}{x-y} - 2\frac{\frac{f(x)-f(y)}{x-y} - f'(y)}{x-y} + f''(y) \\ &\qquad = f''(x) - 2\frac{f'(x)-f'(y)}{x-y} + f''(y). \end{aligned} \tag{3}$$

And the expression (3) is negative by the strict concavity of the function f'' as one can see for example by the following calculation:

$$\begin{aligned}&2\left(\frac{1}{2}\left(f''(x)+f''(y)\right) - \frac{f'(x)-f'(y)}{x-y}\right) \\ &\qquad = 2\left(\int_0^1 tf''(x) + (1-t)f''(y)\mathrm{d}t - \int_0^1 f''\left(tx+(1-t)y\right)\mathrm{d}t\right) \\ &\qquad = 2\int_0^1 tf''(x)+(1-t)f''(y) - f''\left(tx+(1-t)y\right)\mathrm{d}t. \end{aligned} \tag{4}$$

The integrand in (4) is continuous in t and is negative for every $0 < t < 1$ because $x \neq y$ and f'' is strictly concave, hence the above integral (4) is negative. So we deduced that $\langle \partial^2 f(A)(B,B)w,w\rangle < 0$. (It is fair to remark that the above computation is essentially a possible proof of the well-known Hermite–Hadamard inequality.)

So, by (1), we have

$$\left\langle \lim_{t\to 0}\frac{1}{t^2}\left(f(A+tB)-2f(A)+f(A-tB)\right)w,w\right\rangle < 0.$$

This means that

$$\lim_{t\to 0}\frac{1}{t^2}\langle\left(f(A+tB)-2f(A)+f(A-tB)\right)w,w\rangle < 0,$$

so there exists some $t_0 > 0$ such that

$$\big\langle \big(f(A+t_0B) - 2f(A) + f(A-t_0B)\big)\, w, w\big\rangle < 0. \tag{5}$$

(For further use, let us denote the negative number in (5) by $-\delta$.) So, we obtained that $f(A+t_0B) - 2f(A) + f(A-t_0B)$ is not positive semidefinite, i.e.,

$$0 \not\leq f(A+t_0B) - 2f(A) + f(A-t_0B),$$

in other words,

$$f(A) \not\leq \frac{1}{2}\big(f(A+t_0B) + f(A-t_0B)\big).$$

This means that f is not locally convex at the point A. The proof is done.

3.2. The general case

The proof of Theorem 3 in the case of a general C^*-algebra is heavily based on our arguments given in [12]. For the convenience of the reader, we repeat some of the arguments of [12] here in this subsection instead of referring to [12] all the time.

Also in this general case, the proof of the direction (1)$\Longrightarrow$(2) is easy. As f is continuous and convex as a function of one real variable, the map $a \mapsto f(a)$ is also convex on any set of commuting self-adjoint elements of a C^*-algebra (provided that the expression $f(a)$ makes sense). So, centrality automatically implies local convexity.

To prove the direction (2)$\Longrightarrow$(1), we use contraposition again. Assume that $a \in \mathcal{A}_s$, $\sigma(a) \subset I$ and a is not central, that is, $aa' - a'a \neq 0$ for some $a' \in \mathcal{A}$. Then, by [7, 10.2.4. Corollary], there exists an irreducible representation $\pi : \mathcal{A} \to \mathcal{B}(\mathcal{H})$ such that $\pi(aa' - a'a) \neq 0$, i.e., $\pi(a)\pi(a') \neq \pi(a')\pi(a)$. Let us fix this irreducible representation π. So, $\pi(a)$ is a non-central self-adjoint (and hence normal) element of $\mathcal{B}(\mathcal{H})$ with $\sigma(\pi(a)) \subset I$ (as a representation does not increase the spectrum). By the non-centrality, $\sigma(\pi(a))$ has at least two elements, and by the normality, every element of $\sigma(\pi(a))$ is an approximate eigenvalue [6, 3.2.13. Lemma]. Let x and y be two different elements of $\sigma(\pi(a))$, and let $\{u_n\}_{n\in\mathbb{N}} \subset \mathcal{H}$ and $\{v_n\}_{n\in\mathbb{N}} \subset \mathcal{H}$ satisfy

$$\lim_{n\to\infty} (\pi(a)u_n - xu_n) = 0, \ \lim_{n\to\infty} (\pi(a)v_n - yv_n) = 0,$$
$$\langle u_m, v_n\rangle = 0 \ \text{ for all } m, n \in \mathbb{N}.$$

(As $x \neq y$, the approximate eigenvectors can be chosen to be orthogonal.) Set $\mathcal{K}_n := \operatorname{span}\{u_n, v_n\}$ and let E_n be the orthoprojection onto the closed subspace $\mathcal{K}_n^\perp \subset \mathcal{H}$. Let

$$\psi_n(a) := xu_n \otimes u_n + yv_n \otimes v_n + E_n\pi(a)E_n.$$

We intend to show that

$$\lim_{n\to\infty} \psi_n(a) = \pi(a)$$

in the operator norm topology. Let h be an arbitrary non-zero element of $\mathcal{H}$ and consider the orthogonal decompositions $h = h_1^{(n)} + h_2^{(n)}$, where $h_1^{(n)} \in \mathcal{K}_n$ and $h_2^{(n)} \in \mathcal{K}_n^\perp$ for any $n \in \mathbb{N}$. Let us further introduce the two symbols

$\varepsilon_{u,n} := \pi(a)u_n - xu_n$ and $\varepsilon_{v,n} := \pi(a)v_n - yv_n$ and recall that, in the standard topology of the Hilbert space $\mathcal{H}$, $\lim_{n\to\infty} \varepsilon_{u,n} = 0$ and $\lim_{n\to\infty} \varepsilon_{v,n} = 0$. Now,

$$\frac{1}{\|h\|} \|(\pi(a) - \psi_n(a))\, h\| \\ \leq \frac{1}{\|h\|} \left\|(\pi(a) - \psi_n(a))\, h_1^{(n)}\right\| + \frac{1}{\|h\|} \left\|(\pi(a) - \psi_n(a))\, h_2^{(n)}\right\|.$$

Both the first and the second term of the right-hand side of the above inequality are bounded by the term $\|\varepsilon_{u,n}\| + \|\varepsilon_{v,n}\|$ because

$$\frac{1}{\|h\|} \left\|(\pi(a) - \psi_n(a))\, h_1^{(n)}\right\| = \frac{1}{\|h\|} \|(\pi(a) - \psi_n(a))\,(\alpha_n u_n + \beta_n v_n)\|$$
$$= \frac{1}{\|h\|} \|\alpha_n x u_n + \alpha_n \varepsilon_{u,n} - x\alpha_n u_n + \beta_n y v_n + \beta_n \varepsilon_{v,n} - y\beta_n v_n\|$$
$$\leq \frac{|\alpha_n|}{\|h\|} \|\varepsilon_{u,n}\| + \frac{|\beta_n|}{\|h\|} \|\varepsilon_{v,n}\| \leq \|\varepsilon_{u,n}\| + \|\varepsilon_{v,n}\|$$

as the sequences $\{|\alpha_n|\}$ and $\{|\beta_n|\}$ are obviously bounded by $\|h\|$, and

$$\frac{1}{\|h\|} \left\|(\pi(a) - \psi_n(a))\, h_2^{(n)}\right\| = \frac{1}{\|h\|} \left\|(I_{\mathcal{H}} - E_n)\, \pi(a) h_2^{(n)}\right\|$$
$$= \frac{1}{\|h\|} \left\|(u_n \otimes u_n + v_n \otimes v_n)\, \pi(a) h_2^{(n)}\right\|$$
$$= \frac{1}{\|h\|} \left\|\langle \pi(a) h_2^{(n)}, u_n\rangle u_n + \langle \pi(a) h_2^{(n)}, v_n\rangle v_n\right\|$$
$$= \frac{1}{\|h\|} \left\|\langle h_2^{(n)}, \pi(a) u_n\rangle u_n + \langle h_2^{(n)}, \pi(a) v_n\rangle v_n\right\|$$
$$\leq \frac{1}{\|h\|} \left|\langle h_2^{(n)}, x u_n + \varepsilon_{u,n}\rangle\right| + \frac{1}{\|h\|} \left|\langle h_2^{(n)}, y v_n + \varepsilon_{v,n}\rangle\right|$$
$$= \frac{1}{\|h\|} \left|\langle h_2^{(n)}, \varepsilon_{u,n}\rangle\right| + \frac{1}{\|h\|} \left|\langle h_2^{(n)}, \varepsilon_{v,n}\rangle\right|$$
$$\leq \frac{\|h_2^{(n)}\|}{\|h\|} \|\varepsilon_{u,n}\| + \frac{\|h_2^{(n)}\|}{\|h\|} \|\varepsilon_{v,n}\| \leq \|\varepsilon_{u,n}\| + \|\varepsilon_{v,n}\|.$$

We used that a is self-adjoint and that hence so is $\pi(a)$. So, we found that

$$\sup\left\{ \frac{1}{\|h\|} \|(\pi(a) - \psi_n(a))\, h\| \,\middle|\, h \in \mathcal{H} \setminus \{0\} \right\} \leq 2\,(\|\varepsilon_{u,n}\| + \|\varepsilon_{v,n}\|) \to 0,$$

which means that $\psi_n(a)$ tends to $\pi(a)$ in the operator norm topology.

Let us use the notation $B_n := (u_n + v_n) \otimes (u_n + v_n)$ and $w_n := u_n - v_n$. By the result of Subsection 3.1 (the proof for the case of $\mathcal{A} = M_2(\mathbb{C})$) we have

$$\langle (f(\psi_n(a) + t_0 B_n) - 2f(\psi_n(a)) + f(\psi_n(a) - t_0 B_n))\, w_n, w_n \rangle \\ = -\delta < 0, \qquad (6)$$

where t_0 is the same as in (5) and $-\delta$ is the left-hand side of (5), for any $n \in \mathbb{N}$. That is, the left-hand side of (6) is independent of n.

The operator B_n is a self-adjoint element of $\mathcal{B}(\mathcal{H})$ and $\mathcal{K}_n$ is a finite-dimensional subspace of $\mathcal{H}$, hence by Kadison's transitivity theorem [7, 10.2.1. Theorem], there exists a self-adjoint $b_n \in \mathcal{A}$ such that

$$\pi(b_n)_{|\mathcal{K}_n} = B_{n|\mathcal{K}_n}.$$

So, we can rewrite (6) as

$$\begin{aligned}&\langle (f(\psi_n(a) + t_0\pi(b_n)) - 2f(\psi_n(a)) + f(\psi_n(a) - t_0\pi(b_n)))\, w_n, w_n\rangle \\ &\quad = -\delta < 0,\end{aligned} \tag{7}$$

A standard continuity argument which is based on the fact that $\psi_n(a)$ tends to $\pi(a)$ in the operator norm topology shows that

$$\lim_{n\to\infty} \|f(\psi_n(a)) - f(\pi(a))\| = 0. \tag{8}$$

Moreover, by Kadison's transitivity theorem, the sequence $\pi(b_n)$ is bounded (for details, the reader should consult the proof of [6, 5.4.3. Theorem]), and hence

$$\lim_{n\to\infty} \|f(\psi_n(a) \pm t_0\pi(b_n)) - f(\pi(a) \pm t_0\pi(b_n))\| = 0 \tag{9}$$

also holds. By (8) and (9), for any $\delta > 0$ one can find $n_0 \in \mathbb{N}$ such that for $n > n_0$ we have

$$\|f(\psi_n(a)) - f(\pi(a))\| < \frac{1}{16}\delta$$

and

$$\|f(\psi_n(a) \pm t_0\pi(b_n)) - f(\pi(a \pm t_0 b_n))\| < \frac{1}{16}\delta.$$

Therefore, by (7), for $n > n_0$, the inequality

$$\langle (f(\pi(a + t_0 b_n)) - 2f(\pi(a)) + f(\pi(a - t_0 b_n)))\, w_n, w_n\rangle < -\delta/2 < 0,$$

holds. In other words,

$$f(\pi(a)) \not\leq \frac{1}{2}(f(\pi(a + t_0 b_n)) + f(\pi(a - t_0 b_n))),$$

or equivalently (as functional calculus commutes with every representation of a C^*-algebra),

$$\pi(f(a)) \not\leq \pi\left(\frac{1}{2}(f(a + t_0 b_n) + f(a - t_0 b_n))\right).$$

Any representation of a C^*-algebra preserves the semidefinite order, hence this means that

$$f(a) \not\leq \frac{1}{2}(f(a + t_0 b_n) + f(a - t_0 b_n)),$$

which means that f is not locally convex at a. The proof is done.

Acknowledgement

The author is grateful to Lajos Molnár for proposing the problem discussed in this paper and for great conversations about this topic and about earlier versions of this paper. The author is grateful to Albrecht Böttcher for suggestions that helped to improve the presentation of this paper.

References

[1] R. Bhatia, *Matrix Analysis,* Springer, New York, 1996.

[2] Ju.L. Daleckii and S.G. Krein, Integration and differentiation of functions of Hermitian operators and applications to the theory of perturbations, *Amer. Math. Soc. Transl., Ser.* 2 **47** (1965), 1–30.

[3] F. Hiai, Matrix analysis: matrix monotone functions, matrix means, and majorization, *Interdisciplinary Information Sciences* **16** (2010), 139–248.

[4] F. Hiai and D. Petz, *Introduction to Matrix Analysis and Applications,* Hindustan Book Agency, New Delhi and Springer, Heidelberg, 2014.

[5] G. Ji and J. Tomiyama, On characterizations of commutativity of C^*-algebras, *Proc. Amer. Math. Soc.* **131** (2003), 3845–3849.

[6] R.V. Kadison and J.R. Ringrose, *Fundamentals of the Theory of Operator Algebras,* Vol. I, Academic Press, Orlando, 1983.

[7] R.V. Kadison and J.R. Ringrose, *Fundamentals of the Theory of Operator Algebras,* Vol. II, Academic Press, Orlando, 1986.

[8] L. Molnár, A characterization of central elements in C^*-algebras, *Bull. Austral. Math. Soc.* **95** (2017), 138–143.

[9] T. Ogasawara, A theorem on operator algebras, *J. Sci. Hiroshima Univ. Ser. A* **18** (1955), 307–309.

[10] G.K. Pedersen, *C^*-Algebras and Their Automorphism Groups,* London Mathematical Society Monographs **14**, Academic Press, Inc., London–New York, 1979.

[11] S. Silvestrov, H. Osaka and J. Tomiyama, Operator convex functions over C^*-algebras, *Proc. Eston. Acad. Sci.* **59** (2010), 48–52.

[12] D. Virosztek, Connections between centrality and local monotonicity of certain functions on C^*-algebras, *J. Math. Anal. Appl.* **453** (2017), 221–226.

[13] W. Wu, An order characterization of commutativity for C^*-algebras, *Proc. Amer. Math. Soc.* **129** (2001), 983–987.

Dániel Virosztek
Functional Analysis Research Group, Bolyai Institute
University of Szeged
H-6720 Szeged, Aradi vértanúk tere 1.
Hungary
e-mail: `virosz89@gmail.com`

Double-scaling limits of Toeplitz determinants and Fisher–Hartwig singularities

Jani A. Virtanen

Abstract. Double-scaling limits of Toeplitz determinants $D_n(f_t)$ generated by a set of functions $f_t \in L^1$ are discussed as both $n \to \infty$ and $t \to 0$ simultaneously, which is currently of great importance in mathematics and in physics. The main focus is on the cases where the number of Fisher–Hartwig singularities changes as $t \to 0$. All the results on double-scaling limits are discussed in the context of applications in random matrix theory and in mathematical physics.

Mathematics Subject Classification (2010). Primary 47B35; Secondary 15B05, 82B20.

Keywords. Toeplitz determinants, Szegő asymptotics, Fisher–Hartwig singularities, double-scaling limits.

1. Introduction

For a sequence (f_k) of complex numbers, we define the (infinite) *Toeplitz matrix* $T(f)$ and the $n \times n$ Toeplitz matrix $T_n(f)$ by setting

$$T(f) = (f_{j-k})_{j,k\geq 0} \tag{1}$$

and

$$T_n(f) = (f_{j-k})_{0\leq j,k\leq n-1} \tag{2}$$

where $n \in \mathbb{N}$. Given a function f in L^1 on the unit circle $\mathbb{T}$, we can define a Toeplitz matrix via the Fourier coefficients $f_k = \frac{1}{2\pi}\int f(e^{i\theta})e^{-i\theta k}d\theta$ of f as in (1) or (2). The function f is referred to as the *symbol* of the corresponding Toeplitz matrix. It is well known that an infinite Toeplitz matrix generates a bounded linear operator on ℓ^2 if and only if the symbol is in $L^\infty(\mathbb{T})$. For more information on Toeplitz operators and their spectral properties, we refer to [5].

This work was supported by Engineering and Physical Sciences Research Council (EPSRC) grant EP/M024784/1.

© Springer International Publishing AG, part of Springer Nature 2018
A. Böttcher et al. (eds.), *The Diversity and Beauty of Applied Operator Theory*, Operator Theory: Advances and Applications 268, https://doi.org/10.1007/978-3-319-75996-8_29

The (spectral) properties of these matrices have been studied since the early 1900s, and their importance in mathematics and in a variety of applications continues to increase. Finite Toeplitz matrices play an important role in (numerical) linear algebra while the infinite ones have contributed to considerable advances and new concepts in operator theory and functional analysis. The juxtaposition of these two cases that we are interested in is the asymptotic study of Toeplitz matrices, and, in particular, the large n limits of Toeplitz determinants $D_n(f)$ and their recent applications in mathematical physics.

In what follows we consider two types of symbols. We start with Szegő symbols, which are sufficiently smooth symbols with no winding or zeros. By contrast, Fisher–Hartwig symbols may possess zeros, discontinuities, (integrable) singularities, and nonzero winding numbers.

The study of the asymptotics of Toeplitz determinants with such symbols is a feast of beautiful mathematics with remarkable applications. In the following two sections we briefly discuss the relevant results, symbol classes and selected references as much as is needed for double-scaling limits in Section 4.

2. Szegő symbols

A function $f \in L^\infty$ is called *Szegő* if it has no zeros and

$$\sum_{n \in \mathbb{Z}} |f_n| < \infty \tag{3}$$

and

$$\sum_{n \in \mathbb{Z}} (|n| + 1)|f_n|^2 < \infty \tag{4}$$

and $\operatorname{wind} f = 0$, where $\operatorname{wind} f$ is the *winding number* of f defined by

$$\operatorname{wind} f = \frac{1}{2\pi i} \int_{f(\mathbb{T})} z^{-1} dz. \tag{5}$$

The set of all functions satisfying (3) is called the *Wiener algebra* and denoted by W. Observe that each function in W is continuous and that each Szegő symbol f has a logarithm $\log f \in W$ with $\sum_{n \in \mathbb{Z}} (|n| + 1)|(\log f)_n|^2 < \infty$. We will also need the fact that functions analytic in some annulus of $\mathbb{T}$ are in W and satisfy (4).

In the rest of this section, we recall and formulate the two Szegő limit theorems in a form suitable for our purposes. For an excellent introduction to this topic, see [4]. For a thorough treatment, see Chapter 10 of [5], which includes block Toeplitz matrices, and [19] for results on Toeplitz matrices whose symbols are measures. In [1] the asymptotics are computed using the Riemann–Hilbert method—the main tool to deal with double-scaling limits of Toeplitz determinants in Section 4.

More than a hundred years ago, at the age of nineteen, Szegő proved the following result.

Theorem 1 (First Szegő Limit Theorem). *If $f > 0$ is continuous on $\mathbb{T}$, then*

$$\lim_{n\to\infty} \frac{1}{n} \log D_n(f) = (\log f)_0 = \frac{1}{2\pi} \int_0^{2\pi} \log f(e^{i\theta}) d\theta, \tag{6}$$

equivalently,

$$D_n(f) = \exp\Big(n(\log f)_0 + o(n)\Big) \tag{7}$$

or

$$\lim_{n\to\infty} \frac{\log \lambda_1^{(n)} + \cdots + \log \lambda_n^{(n)}}{n} = (\log f)_0, \tag{8}$$

where $\lambda_k^{(n)} > 0$ are the eigenvalues of $T_n(f)$.

This is one of the few results that we mention without direct application as its motivation. However, it would become clear a few decades later that one needs to know more about the error term $o(n)$ in (7) and indeed the motivation was no other than the exact computation of the 2D Ising model, which is covered in great detail in [23]. In 1925 Ising proved that the one-dimensional Ising model with nearest-neighbor forces exhibits no phase transition; see [23, Chapter III].

The difference between the one-dimensional and two-dimensional Ising models is enormous. In one dimension both the free energy and the 2-spin correlation function can be exactly computed in closed form in the presence of a magnetic field h. By contrast, in two dimensions we need to rely on approximations. Indeed, Onsager and Kaufman managed to express the 2-spin correlation function of the 2D Ising model as a Toeplitz determinant, and worked on their asymptotics. The beautiful story surrounding these developments is best told in [11].

More than three decades after his first theorem, Szegő became aware of Onsager and Kaufman's work, and in 1952 he proved his strong limit theorem (for positive symbols in $C^{1+\epsilon}$ with $\epsilon > 0$) which finally provided the error term. Over the years, the result was generalized to complex-valued matrix symbols in Krein algebras (see Section 10.4 of [4]) and to the case in which symbols are measures (see Chapter 6 of [19]). We formulate the result for Szegő symbols. For the proofs, see Section 5.1 of [4], and, in the case of analytic symbols, Section 5.7 of [1] or [8], which employ Riemann–Hilbert methods.

Theorem 2 (Strong Szegő Limit Theorem). *If f is Szegő, then*

$$D_n(f) = \exp\Big(n(\log f)_0 + \sum_{k=1}^{\infty} k(\log f)_k(\log f)_{-k} + o(1)\Big). \tag{9}$$

3. Fisher–Hartwig symbols

In 1968, motivated by applications to statistical mechanics, M. Fisher and R. Hartwig [13, 17] singled out a class of symbols of the form

$$f(z) = e^{V(z)} z^{\beta} |z - z_1|^{2\alpha} g_{z_1,\beta}(z) z_1^{-\beta}, \tag{10}$$

where V is analytic in an annulus of $\mathbb{T}$, $\operatorname{Re}\alpha > -\frac{1}{2}$, $\beta \in \mathbb{C}$, $z_1 = e^{i\theta_1}$, and

$$g_{z_1,\beta}(z) = \begin{cases} e^{i\pi\beta} & \text{if } 0 \le \arg z < \theta_1 \\ e^{-i\pi\beta} & \text{if } \theta_1 \le \arg z < 2\pi. \end{cases} \tag{11}$$

The condition on α ensures that $f \in L^1$. If $\beta \neq 0$, $g_{z_1,\beta}$ has a jump at z_1. Suppose now that $\beta = 0$. If $\operatorname{Re}\alpha > 0$, f has a zero at z_1, and if $\operatorname{Re}\alpha < 0$, f has a pole at z_1. Finally, if $\operatorname{Re}\alpha = 0$ and $\operatorname{Im}\alpha \neq 0$, then f has a discontinuity of oscillating type. Representing Fisher–Hartwig singularities this way is more natural for Riemann–Hilbert analysis of the problem. For an alternative (equivalent) definition of these symbols, more suitable for an operator-theoretic approach, see Section 5.7 of [5].

More generally, we say that $f \in L^1$ is *Fisher–Hartwig* if

$$f(z) = e^{V(z)} z^{\sum_{k=0}^{n} \beta_k} \prod_{k=0}^{m} |z - z_k|^{2\alpha_k} g_{z_k,\beta_k}(z) z_k^{-\beta_k}, \tag{12}$$

where $m \geq 0$, $\operatorname{Re}\alpha_k > -\frac{1}{2}$, $\beta_k \in \mathbb{C}$, $z_k = e^{i\theta_k}$ with $0 = \theta_0 < \ldots < \theta_n < 2\pi$ and g_{z_k,β_k} is defined in (11).

It took more than 40 years and considerable effort of many mathematicians before the asymptotic study of Toeplitz determinants with Fisher–Hartwig singularities was finally completed in full generality; see [10]. For a comprehensive account of this exciting story, see [3, 5, 11, 12]. For the purpose of this work, we consider only two major cases, which involves the concept of the seminorm

$$\|\beta\| = \max_{j,k} |\operatorname{Re}\beta_j - \operatorname{Re}\beta_k|, \tag{13}$$

where $1 \le j, k \le n$ if $\alpha_0 = \beta_0 = 0$, and $0 \le j, k \le n$ otherwise, while $\|\beta\| = 0$ if $n = 0$. The final result (in a sense) when $\|\beta\| < 1$ was obtained in [12].

Theorem 3 (Fisher–Hartwig conjecture). *Let f be a Fisher–Hartwig symbol with $V \in C^\infty$, $\|\beta\| < 1$ and $\alpha_k \pm \beta_k \notin \mathbb{Z}_-$ for $k = 0, \ldots, m$. Then*

$$D_n(f) = \exp\left\{ nV_0 + \sum_{k=1}^{\infty} kV_k V_{-k} \right\} \prod_{j=0}^{m} b_+(z_j)^{-(\alpha_j - \beta_j)} b_-(z_j)^{-(\alpha_j + \beta_j)} \tag{14}$$

$$\times\, n^{\sum_{j=0}^{m}(\alpha_j^2 - \beta_j^2)} \prod_{0 \le j < k \le m} |z_j - z_k|^{2(\beta_j\beta_k - \alpha_j\alpha_k)} \left(\frac{z_k}{z_j e^{i\pi}} \right)^{\alpha_j\beta_k - \alpha_k\beta_j}$$

$$\times \prod_{j=0}^{m} \frac{G(1 + \alpha_j + \beta_j) G(1 + \alpha_j - \beta_j)}{G(1 + 2\alpha_j)} (1 + o(1)),$$

where G is Barnes' G-function and the product over $j < k$ is set to 1 if $m = 0$.

For important notes on this theorem, such as its validity for less smooth functions V, see the remarks to Theorem 1.1 of [10], and Notes and Comments 10.66–10.80 of [5], which gives a detailed history of the subject. We will not discuss any of these notes here as the formulation of the theorem above is all we need. What we wish to emphasize is that the approach in [10] is based on the Riemann–Hilbert method while earlier the conjecture had been verified in several cases using operator-theoretic methods and (weighted) Hardy space techniques.

We next consider the case $\|\beta\| \geq 1$. In 1991 Basor and Tracy [2] observed that the conjecture fails if $f(e^{i\theta}) = 1$ for $0 < \theta < \pi$ and $f(e^{i\theta}) = -1$ for $\pi < \theta < 2\pi$, in which case $\|\beta\| = 1$. They conjectured the following result, which is proved in [10].

Theorem 4 (Basor–Tracy conjecture). *Let f be a Fisher–Hartwig symbol and suppose that $\mathcal{M}$ is nondegenerate (see Remark 5). Then*

$$D_n(f) = \sum \left(\prod_{k=0}^{m} z_k^{n_k} \right)^n \mathcal{R}(f(z; n_0, \ldots, n_m))(1 + o(1)),$$

where the sum is over all Fisher–Hartwig representations in $\mathcal{M}$ and each $\mathcal{R}(f(z; n_0, \ldots, n_m))$ is obtained from the Fisher–Hartwig conjecture.

Remark 5. Let f be a Fisher–Hartwig symbol with parameters α_k and β_k; see (12). A Fisher–Hartwig symbol $f(z; n_0, \ldots, n_m)$ with parameters α_k and $\hat{\beta}_k$ is said to be a *FH-representation* of f if

$$\hat{\beta}_k = \begin{cases} \beta_k + n_k & \text{if } \beta_k \neq 0 \text{ or } \alpha_k \neq 0, \\ \beta_k & \text{otherwise} \end{cases} \tag{15}$$

for some $n_k \in \mathbb{Z}$ with $n_0 + \ldots + n_m = 0$. Then

$$f(z) = \prod_{k=0}^{m} z_k^{n_k} \times f(z; n_0, \ldots, n_m). \tag{16}$$

The set of FH-representations of f for which $\sum_{k=0}^{m} (\operatorname{Re} \beta_k + n_k)^2$ is minimal is denoted by $\mathcal{M}$. Finally, we say that a FH-representation is *degenerate* if $\alpha_k + (\beta_k + n_k) \in \mathbb{Z}_-$ or $\alpha_k - (\beta_k + n_k) \in \mathbb{Z}_-$ for some k. We call $\mathcal{M}$ *nondegenerate* if it contains no degenerate FH-representation. For further details on FH-representations, see Lemma 1.12 and the preceding comments in [10].

4. Double-scaling limits

Let $f_t \in L^1$ for $t \geq 0$. By a *double-scaling limit* of Toeplitz determinants $D_n(f_t)$ we mean an asymptotic expansion of $D_n(f_t)$ valid uniformly for $0 \leq t < t_0$ for some t_0, as $n \to \infty$. In general, double-scaling limits (of not necessarily Toeplitz determinants) are of great interest in mathematics and in physics, and it can be argued, as in [9], that double-scaling limits may explain universal behavior that transcends the problem at hand. Another typical application is a system of size n that has an external parameter t with a critical value t_c, and one needs to understand the transition $t \to t_c$. We have already seen how this is featured in the study of the 2D Ising model, and RMT provides yet another source of such situations, where the matrix distribution may depend on some parameter t. Most problems of this kind remain open [9].

It appears that the first study of double-scaling limits of Toeplitz determinants was done in [25] in the context of the 2D Ising model, and the phase transition was described in terms of a solution to the Painlevé III equation. In particular, with $f_t(z) = (z - e^t)^{-1/2}(z - e^{-t})^{1/2} z^{-1/2} z^{i\pi/2}$ the double-scaling limits of $D_n(f_t)$ were computed under the assumption that $x = n(e^{2t} - 1)$ remains fixed. This was later relaxed by Tracy et al. allowing for $x \to \infty$ and $x \to 0$ and obtaining the right asymptotics up to a multiplicative constant, whose exact value was computed also for small x by Tracy [20] in 1991. Under additional boundary conditions on the Ising model their study is a special case of Theorem 6 of [6] below, which actually provides uniform asymptotics in x.

If f is Fisher–Hartwig, we denote the number of its singularities by $\#f = m$; see (12). Given distinct nonnegative integers j and j_0, we consider the problem of finding double-scaling limits of $D_n(f_t)$ when $\#f_t = j$ for $t > 0$ and $\#f_0 = j_0$. We concentrate on two model cases that result in the change in the number of singularities as $t \to 0$. The term

$$h_t^+(z; w) = (z - e^t e^{i\theta_0})^{\alpha+\beta}(z - e^{-t}e^{i\theta_0})^{-\alpha+\beta} z^{-\alpha+\beta} e^{-i\pi(\alpha+\beta)} \tag{17}$$

produces an additional singularity at $w = e^{i\theta_0} \in \mathbb{T}$ as $t \to 0$, while, for moving singularities $z_1 = e^{i(\theta_0+t)}$ and $z_2 = e^{i(\theta_0-t)}$, the term

$$h_t^-(z; w) = z^{\beta_1+\beta_2} \prod_{k=1}^{2} |z - z_k|^{2\alpha_k} g_{z_k,\beta_k} z_k^{-\beta_k} \tag{18}$$

reduces the number of singularities by one as $t \to 0$. Another way of describing the actions of $h_t^{\pm}(z; w)$ is that there is an emergence of a singularity at $e^{i\theta_0}$ in (17) as $t \to 0$ while in (18) the two singularities coalesce at $e^{i\theta_0}$ as $t \to 0$.

4.1. Transition from Szegő to one Fisher–Hartwig singularity

Let $w = 1$ and consider the symbols

$$f_t(z) = e^{V(z)} h_t^+(z; w) = e^{V(z)}(z - e^t)^{\alpha+\beta}(z - e^{-t})^{-\alpha+\beta} z^{-\alpha+\beta} e^{-i\pi(\alpha+\beta)}, \tag{19}$$

where V is analytic in a neighborhood of $\mathbb{T}$ and $\alpha \pm \beta \notin \mathbb{Z}_-$. When $t > 0$, f_t is Szegő and Theorem 2 provides the asymptotics of $D_n(f_t)$, while, for $t = 0$, the symbol is Fisher–Hartwig and we may apply Theorem 3. The following result provides asymptotics of $D_n(f_t)$ as $n \to \infty$ uniformly for all t sufficiently small.

Theorem 6 ([6]). *Let $\alpha > -\frac{1}{2}$ and $\beta \in i\mathbb{R}$. Then there is a $t_0 > 0$ such that uniformly for $0 < t < t_0$,*

$$\begin{aligned}\log D_n(f_t) &= nV_0 + (\alpha+\beta)nt \\ &\quad + \sum_{k=1}^{\infty} k\left(V_k - (\alpha+\beta)\frac{e^{-tk}}{k}\right)\left(V_{-k} - (\alpha-\beta)\frac{e^{-tk}}{k}\right) \\ &\quad + \log G_{\alpha+\beta,\alpha-\beta} + \Omega(2nt) + o(1),\end{aligned}$$

where $G_{\alpha+\beta,\alpha-\beta} = \frac{G(1+\alpha+\beta)G(1+\alpha-\beta)}{G(1+2\alpha)}$, $\Omega(2nt) = \int_0^{2nt}(\sigma(x) - \alpha^2 + \beta^2)x^{-1}dx + (\alpha^2 - \beta^2)\log 2nt$ and σ is described by a special Painlevé V transcendent.

For further (technical) details and generalizations, such as complex-valued α, see Theorem 1.4 of [6]. The motivation for Theorem 6 comes from the 2D Ising model and in particular it describes the transition of the 2-spin correlation function for large n from temperature $T < T_c$ to $T = T_c$, where T_c is the critical temperature of when the spontaneous magnetization occurs. For further details, see Section 1.1 of [6]. Another application of Theorem 6 is the emptiness formation probability in a Heisenberg spin chain, which we discuss in more detail in Section 4.3.

4.2. Transition from two Fisher–Hartwig singularities to one

Again let $w = 1$ and consider the symbols $f_t(z) = e^{V(z)} h_t^-(z; w)$, which have two F-H singularities located at e^{it} and $e^{i(2\pi - t)}$ when $t > 0$ which merge into one at 1 when $t \to 0$. The double-scaling limits of $D_n(f_t)$ were computed in [7], and can be used to prove conjectures of Dyson (on the largest occupation number in the ground state of a one-dimensional Bose gas) and of Fyodorov and Keating on the second moment of powers of the characteristic polynomials of random matrices [15]). Unlike in Theorem 6, the study needs to be split into two cases depending on whether $\|\beta\| < 1$.

4.3. Transition from one Fisher–Hartwig singularities to two

Consider a symbol f_t that has one fixed F-H singularity at $\mathbb{T} \setminus \{1\}$ with parameters α_1, β_1 and one emerging singularity at 1 with parameters α_0, β_0 as in (19). For a given $t \geq 0$, we may compute the asymptotics of $D_n(f_t)$ using the F-H conjecture. It is easy to see that these asymptotics are not uniform for t. Double-scaling limits of these determinants were recently computed in [18]. As in Section 4.2, one needs to consider the two cases $\|\beta\| < 1$ and $\|\beta\| = 1$. In the former case, with $\alpha_0 \in \mathbb{R}$, $\alpha_1 \in \mathbb{C}$, $\alpha_0, \operatorname{Re} \alpha_1 > -1/2$, $\beta_0 \in i\mathbb{R}$, $\beta_1 \in \mathbb{C}$, uniformly for sufficiently small t, we get as $n \to \infty$,

$$\begin{aligned} D_n(t) &= \exp\{nV_0 + nt(\alpha_0 + \beta_0)\} \exp\left\{\sum_{k=1}^{\infty} k\left[(\log a_t)_k\right]\left[(\log a_t)_{-k}\right]\right\} \\ &\times \exp\{-(\alpha_1 - \beta_1)\log a_{t,+}(z_1)\} \exp\{(\alpha_1 + \beta_1)\log a_{t,-}(z_1)\} \\ &\times n^{(\alpha_1^2 - \beta_1^2)} G_{\alpha_0+\beta_0, \alpha_0-\beta_0} G_{\alpha_1+\beta_1, \alpha_1-\beta_1} \exp(\Omega(2nt))(1 + o(1)), \quad (20) \end{aligned}$$

where $G_{\alpha_j+\beta_j, \alpha_j-\beta_j}$ and Ω are defined as in Theorem 6.

We finish this section with an application of these double-scaling limits to a problem in quantum spin chains. In [14], the asymptotic behavior of the emptiness formation probability (EFP) for 1D anisotropic XY spin-1/2 chain in a transverse magnetic field is studied. In a sense, the EFP can be viewed as the probability of formation of a ferromagnetic string of length n in the antiferromagnetic ground state of the chain. By expressing the EFP as a Toeplitz determinant with symbol σ, one can compute its asymptotics as the number of spins goes to infinity using the SSLT and the F-H conjecture. As explained in [14], the values of the magnetic field h correspond to three regions $\Sigma_\pm$ and Σ_0, which are separated by critical regions $\Omega_\pm$ and Ω_0. The asymptotic behavior of the EFP in Σ_- is governed by Szegő symbols, in Σ_0 by σ with one F-H singularity, and Σ_+ by σ with two F-H singularities. With Theorem 6, one can describe the transition from Σ_- to Σ_0, and using the results of this section, we get a description of the transition from Σ_0 to Σ_+.

5. Open problems

The approach in [6, 7, 18] is based on the use of Riemann–Hilbert problems, associated orthogonal polynomials, Painlevé equations, special functions, such as hypergeometric functions, and overall requires heavy machinery and computations, which may seem unnecessarily long for those familiar with the operator-theoretic approach to the asymptotics of Toeplitz determinants. It would be interesting to try to compute double-scaling limits (or verify the Basor–Tracy conjecture) without the use of the Riemann–Hilbert method. One idea involves expressing the Toeplitz determinant as a Fredholm determinant using the GCBO identity [16], after which one is faced with a problem of connecting to Painlevé V (as in [6]), which depends on the structure of the kernel (see [21, 24]). These ideas combined with the deformation theory [22] may provide double-scaling limits at least in some (special) cases.

Another interesting problem is the computation of double-scaling limits for Toeplitz determinants with a mix of merging, emerging and fixed types of F-H singularities; that is, combining the model cases in [6, 7, 18]. Besides intrinsic interest, these types of results could be used to make the arguments in [15] rigorous, and complete the program started in [7].

Acknowledgments. The author wishes to thank the American Institute of Mathematics (AIM) and the organizers of the AIM workshop on Fisher–Hartwig symbols for providing an excellent scientific environment for this work and for many opportunities for stimulating discussions.

References

[1] J. Baik, P. Deift, and T. Suidan, *Combinatorics and random matrix theory*, Amer. Math. Soc., Providence, RI, 2016.

[2] E. Basor and C.A. Tracy, The Fisher-Hartwig conjecture and generalizations. *Phys. A* **177** (1991), 167–173.

[3] A. Böttcher, The Onsager formula, the Fisher-Hartwig conjecture, and their influence on research into Toeplitz operators. *J. Statist. Physics* (*Lars Onsager Festschrift*) **78** (1995), 575–585.

[4] A. Böttcher and B. Silbermann, *Introduction to large truncated Toeplitz matrices*, Springer, New York, 1999.

[5] A. Böttcher and B. Silbermann, *Analysis of Toeplitz operators*, 2nd ed., Springer-Verlag, Heidelberg, 2006.

[6] T. Claeys. A. Its and I. Krasovsky, Emergence of a singularity for Toeplitz determinants and Painlevé V, *Duke Math. J.* **160** (2011), 207–262.

[7] T. Claeys and I. Krasovsky, Toeplitz determinants with merging singularities, *Duke Math. J.* **164** (2015), 2897–2987.

[8] P. Deift, Integrable operators, In: *Differential operators and spectral theory*, 69–84, Amer. Math. Soc. Transl. Ser. 2 **189**, Amer. Math. Soc., Providence, RI, 1999.

[9] P. Deift, Some open problems in random matrix theory and the theory of integrable systems. II, *SIGMA Symmetry Integrability Geom. Methods Appl.* **13** (2017), paper no. 016, 23.

[10] P. Deift, A. Its and I. Krasovsky, Asymptotics of Toeplitz, Hankel, and Toeplitz+Hankel determinants with Fisher-Hartwig singularities, *Ann. of Math. (2)* **174** (2011), 1243–1299.

[11] P. Deift, A. Its and I. Krasovsky, Toeplitz matrices and Toeplitz determinants under the impetus of the Ising model: some history and some recent results, *Comm. Pure Appl. Math.* **66** (2013), 1360–1438.

[12] T. Ehrhardt, A status report on the asymptotic behavior of Toeplitz determinants with Fisher-Hartwig singularities, In *Recent advances in operator theory (Groningen, 1998)*, Oper. Theory Adv. Appl. **124**, Birkhäuser, Basel, 2001, 217–241.

[13] M.E. Fisher and R.E. Hartwig, Toeplitz determinants: some applications, theorems, and conjectures, In *Advances in Chemical Physics*, John Wiley & Sons, Inc., 2007, 333–353.

[14] F. Franchini and A.G. Abanov, Asymptotics of Toeplitz determinants and the emptiness formation probability for the XY spin chain, *J. Phys. A* **38** (2005), 5069–5095.

[15] Y.V. Fyodorov and J.P. Keating, Freezing transitions and extreme values: random matrix theory, and disordered landscapes, *Philos. Trans. R. Soc. Lond. Ser. A Math. Phys. Eng. Sci.* **372** (2007), 20120503, 32.

[16] J. Gravner, C.A. Tracy and H. Widom, Limit Theorems for Height Fluctuations in a Class of Discrete Space and Time Growth Models, *J. Statist. Phys.* **102** (2001), 1085–1132.

[17] R.E. Hartwig and M.E. Fisher, Asymptotic behavior of Toeplitz matrices and determinants, *Arch. Rational Mech. Anal.* **32** (1969), 190–225.

[18] K. Kozlowska and J.A. Virtanen, Transition asymptotics of Toeplitz determinants and emergence of Fisher-Hartwig representations, preprint.

[19] B. Simon, *Orthogonal polynomials on the unit circle. Part 1*, Amer. Math. Soc., Providence, RI, 2005.

[20] C.A. Tracy, Asymptotics of a τ-function arising in the two-dimensional Ising model, *Comm. Math. Phys.* **142** (1991), 297–311.

[21] C.A. Tracy and H. Widom, Fredholm determinants, differential equations and matrix models, *Comm. Math. Phys.* **163** (1994), 33–72.

[22] C.A. Tracy and H. Widom, Asymptotics in ASEP with step initial condition, *Comm. Math. Phys.* **290** (2009), 129–154.

[23] B.M. McCoy and T.T. Wu, *The two-dimensional Ising model*, Harvard University Press, Cambridge, MA, 1973.

[24] H. Widom, On the solution of a Painlevé III equation, *Math. Phys. Anal. Geom.* **3** (2000), 375–384.

[25] T.T. Wu, B.M. McCoy, C.A. Tracy, and E. Barouch, Spin-spin correlation functions for the two-dimensional Ising model: Exact theory in the scaling region, *Phys. Rev. B* **13** (1976), 316–374.

Jani A. Virtanen
Department of Mathematics
University of Reading
Whiteknights
Reading RG6 6AX
England
e-mail: j.a.virtanen@reading.ac.uk

The manufacturer's authorised representative in the EU is Springer
Nature Customer Service Centre GmbH, Europaplatz 3, 69115 Heidelberg,
Germany. If you have any concerns regarding our products, please
contact ProductSafety@springernature.com

Printed and bound by CPI Group (UK) Ltd, Croydon, CR0 4YY
30/09/2026
02225434-0017